Excitations in Two-Dimensional and Three-Dimensional Quantum Fluids

NATO ASI Series

Advanced Science Institutes Series

A series presenting the results of activities sponsored by the NATO Science Committee, which aims at the dissemination of advanced scientific and technological knowledge, with a view to strengthening links between scientific communities.

The series is published by an international board of publishers in conjunction with the NATO Scientific Affairs Division

A	**Life Sciences**	Plenum Publishing Corporation
B	**Physics**	New York and London
C	**Mathematical and Physical Sciences**	Kluwer Academic Publishers
D	**Behavioral and Social Sciences**	Dordrecht, Boston, and London
E	**Applied Sciences**	
F	**Computer and Systems Sciences**	Springer-Verlag
G	**Ecological Sciences**	Berlin, Heidelberg, New York, London,
H	**Cell Biology**	Paris, Tokyo, Hong Kong, and Barcelona
I	**Global Environmental Change**	

Recent Volumes in this Series

Volume 252—Laser Systems for Photobiology and Photomedicine
edited by A. N. Chester, S. Martellucci, and A. M. Scheggi

Volume 253—Condensed Systems of Low Dimensionality
edited by J. L. Beeby

Volume 254—Quantum Coherence in Mesoscopic Systems
edited by B. Kramer

Volume 255—Vacuum Structure in Intense Fields
edited by H. M. Fried and Berndt Muller

Volume 256—Information Dynamics
edited by Harald Atmanspacher and Herbert Scheingraber

Volume 257—Excitations in Two-Dimensional and Three-Dimensional Quantum Fluids
edited by A. F. G. Wyatt and H. J. Lauter

Volume 258—Large-Scale Molecular Systems: Quantum and Stochastic Aspects—Beyond the Simple Molecular Picture
edited by Werner Gans, Alexander Blumen, and Anton Amann

Series B: Physics

Excitations in Two-Dimensional and Three-Dimensional Quantum Fluids

Edited by

A. F. G. Wyatt

University of Exeter
Exeter, United Kingdom

and

H. J. Lauter

Institute Laue Langevin
Grenoble, France

Plenum Press
New York and London
Published in cooperation with NATO Scientific Affairs Division

Proceedings of a NATO Advanced Research Workshop on
Excitations in Two-Dimensional and Three-Dimensional Quantum Fluids,
held August 10-15, 1990,
in Exeter, United Kingdom

Library of Congress Cataloging-in-Publication Data

NATO Advanced Research Workshop on Excitations in Two-Dimensional and Three-Dimensional Quantum Fluids (1990 : Exeter, England)
Excitations in two-dimensional and three-dimensional quantum fluids / edited by A.F.G. Wyatt and H.J. Lauter.
p. cm. -- (NATO ASI series. Series B, Physics ; vol. 257)
"Proceedings of a NATO Advanced Research Workshop on Excitations in Two-Dimensional and Three-Dimensional Quantum Fluids, held August 10-15, 1990, in Exeter, United Kingdom"--T.p. verso
"Published in cooperation with NATO Scientific Affairs Division."
Includes bibliographical references and index.
ISBN-13: 978-1-4684-5939-5 e-ISBN-13: 978-1-4684-5937-1
DOI: 10.1007/978-1-4684-5937-1

1. Quantum liquids--Congresses. 2. Liquid helium--Congresses. 3. Superfluidity--Congresses. 4. Collisional excitation--Congresses. I. Wyatt, A. F. G. (Adrian Frederick George), 1938- . II. Lauter, H. J. III. North Atlantic Treaty Organization. Scientific Affairs Division. IV. Title. V. Series: NATO ASI series. Series B, Physics ; v. 257.
QC174.4.A1N38 1990
530.4'2--dc20 91-15448
CIP

Softcover reprint of the hardcover 1st edition 1991
A Division of Plenum Publishing Corporation
233 Spring Street, New York, N.Y. 10013

PREFACE

The study of quantum fluids in three dimensions has been an important area for many years as it embraces Bose-Einstein condensation, superfluidity and macroscopic quantisation. These are fundamental aspects of physics which can be studied in liquid 4He. In contrast, quantum fluids in two dimension is more recent and less developed. Nevertheless it has shown many interesting phenomena including a rich variety of phases and the Kosterlitz-Thouless transition. Intermediate between these dimensions are the restricted geometries of micro porous materials into which He may be introduced. The main quantum materials considered are ^4He, ^3He, D_2, H_2, H and electrons on the surface of ^4He. The superfluid phases of 3He were excluded, except for superfluid film flow, as ^3He involves a separate set of problems.

These proceedings arise from a lively Advanced Research Workshop on Excitations in Two-Dimensional and Three-Dimensional Quantum Fluids held in Exeter 10-15 August 1990. Fifty scientists took part and each provided a written contribution. Perhaps it is a testimony to the discussions that several papers were revised by the authors after the meeting. The order of the chapters is the same as the presentations at the workshop. This arrangement starts with ^4He in three dimensions which establishes a base from which the two dimensional properties can be viewed. At the end of each section there is a report on the discussion session. These are interesting and useful chapters as they clarify points made in the papers and define the boundary of current understanding. These discussions and the last chapter indicate where some of the unsolved problems lie.

This volume brings together different theoretical models of ^4He and clearly shows their contrasting approaches. The excitation picture of ^4He is well established although there is much to be learned about the details of phonons and rotons. It is suggested that phonons should be considered as zero sound modes and rotons as single particle excitations in order to explain the temperature dependence of neutron scattering.

Neutron scattering is a major tool for studying the excitations in ^4He and in ^3He, and the present state of the art is described in some detail. Deep inelastic neutron scattering shows that the Bose condensate fraction is ~ 10% at T = 0 K. However most properties of superfluid ^4He do not depend on its exact value but only require that there be macroscopic occupation of the k = 0 state. It is clear that the necessary conditions for superfluid behaviour are not yet defined.

Roton dynamics is discussed in several chapters. This is topical as it is now possible to create beams of ballistic rotons and perform scattering experiments. There are considerations of rotons interacting with other rotons, vortex lines and with quantum kinks which occur at the phase boundary of solid and liquid ^4He.

Dilute mixtures of ^{3}He in ^{4}He is an interesting system and its bulk properties are reviewed. The thermal conductivity of ultradilute mixtures is anomolous and a new theory is proposed.

Recent theories of 2D films of both ^{4}He and ^{3}He on substrates are discussed. These theories are in good agreement with NMR studies. The striking result is that well defined 'layers' of ^{3}He build up as ^{3}He is added and they appear to be weakly-coupled 2D Fermi systems.

The substrate for 2D layers has a profound effect on the ordered phases of light atoms. The impressive results from a variety of techniques such as neutron scattering, heat capacity and NMR are reviewed. NMR is used to explore the magnetic properties of 2D layers of ^{3}He. At low temperatures $T \sim 1$mK, 3He on graphite is a 2D ferromagnet. Thin layers of ^{4}He on substrates show that there are excitations associated with both surfaces. At the substrate-liquid interface there are localised excitations possibly related to the Kapitza resistance and at the free surface there are ripplons.

There is a vigorous discussion on whether He crystals are rough or smooth as $T \rightarrow 0$K. It is argued theoretically that the presence of thermally activated dislocations mean that surfaces are rough at all non-zero temperatures. However it was countered that the experimental evidence showed that smooth surfaces do exist in practice. The question now is whether there is thermal equilibrium when the smooth surfaces are seen.

The last section is on electrons and ions at the surface of liquid He. These are probes of the surface as well as being interesting systems in their own right and their behaviour is reviewed. Some aspects are well understood such as mobility and Wigner crystallisation. However other properties such as tunnelling rates present a mystery.

Many people contributed to the success of the workshop and to the production of the proceedings. The advisory committee were D O Edwards, P V E McClintock, W G Stirling, W F Vinen. The good humour and vitality shown by the delegates will always be remembered. Crossmead Conference Centre provided the right environment and excellent food. Special thanks go to Valerie Barnes for the secretarial work and to Anne Wyatt for the social programme. Magnex Scientific generously sponsored the welcome party.

The financial support from the North Atlantic Treaty Organisation, the Commission for the European Communities and the Institute Laue - Langevin for the workshop is gratefully acknowledged.

A F G Wyatt
H J Lauter

December 1990

CONTENTS

EXCITATIONS IN LIQUID ^{4}He

Henry R. Glyde

Department of Physics
University of Alberta
Edmonton, Alberta, Canada T6G 2J1

ABSTRACT

In this review we sketch the new understanding of collective and single particle excitations in liquid ^{4}He revealed by recent neutron scattering and theoretical studies. The aim is to set the stage for the talks by Stirling, Griffin, Sokol, Svensson and Campbell. Recent experiments have focussed chiefly on the temperature dependence of the dynamic structure factor, $S(Q,\omega)$, and on scattering at high Q to explore Final State (FS) interactions and to determine the condensate fraction, $n_0(T)$. Measurements of the temperature dependence of $S(Q,\omega)$ show that the low Q, phonon region and the higher Q, maxon and roton regions are quite different. At low Q, $S(Q,\omega)$ is concentrated in a single peak largely independent of T. At $Q = 1.13\,\mathring{A}^{-1}$ $S(Q,\omega)$ has a sharp peak and a broad component at low T. The sharp peak disappears from $S(Q,\omega)$ at $T = T_\lambda$ leaving only the broad component in the normal phase. An interpretation is sketched which suggests that the excitations are zero sound at low Q. At higher Q the zero sound mode broadens and the sharp peak in $S(Q,\omega)$ is interpreted as a single particle excitation for $Q \geq 1.1\,\mathring{A}^{-1}$. High Q data suggests $n_0 = 10 \pm 2\,\%$ at $T = 0\,K$ and that FS effects are small by $Q \approx 20\,\mathring{A}^{-1}$.

1. INTRODUCTION

In this talk we review recent developments in neutron scattering studies of liquid ^{4}He. The aim is to introduce the following talks by Stirling, Griffin, Sokol and Svensson and the later talk by Campbell. Sokol will present new data on neutron scattering from liquid ^{4}He at high momentum transfer.[1–3] This data has been used to determine the momentum distribution and condensate fraction,[1,2] $n_0(T)$, in superfluid ^{4}He and to explore final state corrections to the Impulse Approximation.[3,4] The remainder of the talks will be devoted to discussion of the temperature dependence of the dynamic structure factor,[5–12] $S(Q,\omega)$. Of particular interest is the interpretation of the observed phonon and roton excitations.[12,13]

Several reviews provide a broad picture of excitations in liquid ^{4}He. The work up to 1972, both theory and experiment, is comprehensively reviewed by Woods and Cowley.[13] More recent reviews by Price,[14] Glyde,[15] Svensson and Sears,[16] Glyde and Svensson[17] and Svensson[18] bring this up to date.

To make clear what is observed in a neutron scattering measurement, we introduce $S(Q,\omega)$ in section 2. The observed form of $S(Q,\omega)$ at low temperature is displayed in section 3. Recent data on the temperature dependence of $S(Q,\omega)$ will be sketched in section 4 to set the stage for a full presentation by Stirling. A recent interpretation of this data based on the dielectric formulation will be introduced in section 5. Griffin will discuss the dielectric formulation and the new interpretation specifically. In section 6 we present a model to describe the coupling between single particle and density excitations in $S(Q,\omega)$. Svensson will summarize the current picture and propose directions for future study.

2. NEUTRON SCATTERING FROM LIQUID ^{4}He

Neutrons interact with the ^{4}He nuclei. The potential, $V(r)$, seen by a neutron is a sum of Fermi contact potentials at each ^{4}He nucleus,

$$V(r) = \frac{2\pi\hbar^2}{M}\sum_{\ell} b_\ell \delta(\mathbf{r}-\mathbf{r}_\ell) = \frac{2\pi\hbar^2 b}{M}\sum_{\ell}\delta(\mathbf{r}-\mathbf{r}_\ell) \tag{1}$$

where $b_\ell = b$ is the ^{4}He–neutron scattering length, the same for each atom. The density operator in the fluid is

$$\rho(\mathbf{r}) = \sum_{\ell}\delta(\mathbf{r}-\mathbf{r}_\ell) \tag{2}$$

with Fourier transform

$$\rho(Q) = \int d\mathbf{r}\, e^{-i\mathbf{Q}\cdot\mathbf{r}}\rho(\mathbf{r}) = \sum_{\ell} e^{-i\mathbf{Q}\cdot\mathbf{r}_\ell}. \tag{3}$$

Thus, $V(r)$ is proportional to $\rho(r)$ and neutrons probe the number density in ^{4}He. Specifically,[19] for momentum $(\hbar Q)$ and energy $(\hbar\omega)$ transferred from the neutron to the fluid the cross-section due to $V(r)$ is proportional to the dynamic structure factor

$$S(Q,\omega) = \frac{1}{2\pi}\int dt\, e^{i\omega t} S(Q,t)$$

where

$$S(Q,t) = \frac{1}{N}\langle \rho(Q,t)\rho^{+}(Q,0)\rangle. \tag{4}$$

Thus in a neutron scattering experiment we observe excitations and time correlations in the Fourier component of the number density, $\rho(Q,t)$.

The second quantized form of $\rho(Q,t)$ is $\rho(Q) = \sum_k a_k^{+} a_{\mathbf{k}+\mathbf{Q}}$ where a_k^{+} is a single particle creation operator. In superfluid ^{4}He where $n_0 \neq 0$, we show in section 5 that $S(Q,\omega)$ has a component proportional to the single particle correlation function $\langle a_Q(t)\, a_Q^{+}(0)\rangle$.

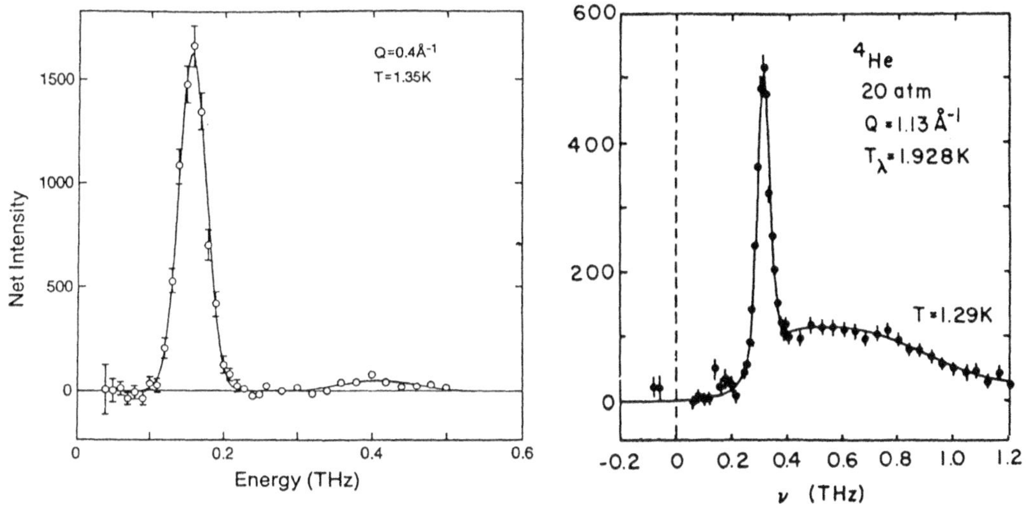

Figure 1. Observed scattering intensity from superfluid ^{4}He at low T (from Refs. 11, 12).

3. SUPERFLUID ^{4}He AT LOW TEMPERATURE

Neutrons scattering from liquid ^{4}He create (or absorb) excitations in the fluid density, as seen from (4). In Fig. 1 we show the scattering intensity at fixed wave vector transfer $Q = 0.4$ and $1.13\,\mathring{A}^{-1}$ as a function energy transfer, $\hbar\omega$. The intensity contains a sharp peak and a broad component, especially at higher Q. Miller et al.[20] proposed a two component form for $S(Q,\omega)$, $S(Q,\omega) = S_1(Q,\omega) + S_2(Q,\omega)$ with $S_1(Q,\omega)$ describing the sharp peak and $S_2(Q,\omega)$ the broad component. The position of the sharp peak in $S_1(Q,\omega)$ is defined as the energy, ω_Q, of the characteristic excitation. In Fig. 2 we show the now famous dispersion curve for ω_Q; the phonon $(Q \leq 0.8\,\mathring{A}^{-1})$-maxon $(Q \approx 1.1\,\mathring{A}^{-1})$-roton $(Q \approx 2.0\,\mathring{A}^{-1})$ curve.

Current neutron data on superfluid ^{4}He at low T is consistent with the broad picture painted by Cowley and Woods in 1971. Recent data has provided much improved precision. In Fig. 3 we show the broad map of intensity for $0 \leq Q \leq 4\,\mathring{A}^{-1}$ given by Cowley and Woods.[21] This shows the phonon-roton curve (labeled PHONON in Fig. 3) and the intensity at higher ω. It also shows that for $Q \geq 3\,\mathring{A}^{-1}$, the scattering intensity is centred near the free, single atom recoil energy $\hbar\omega_R = (\hbar Q)^2/2M$. The width of $S(Q,\omega)$ for $Q \geq 3\,\mathring{A}^{-1}$ is largely determined by the Doppler broadening set by the zero point kinetic energy. Martel et al.[22] and Stirling et al.[23] have shown that the peak position and width of $S(Q,\omega)$ oscillates with Q in the region $3 \leq Q \leq 8\,\mathring{A}^{-1}$. These oscillations can be attributed to oscillations with Q in the scattering amplitude between the interacting helium atoms in the fluid.[22,24] Scattering at high Q $(Q \geq 10\,\mathring{A}^{-1})$ has now been used to determine the condensate fraction,[4,24] $n_0(T)$. At these Q values the neutron scatters from a single atom. For $Q \geq 20\,\mathring{A}^{-1}$, $S(Q,\omega)$ approaches the Impulse Approximation,

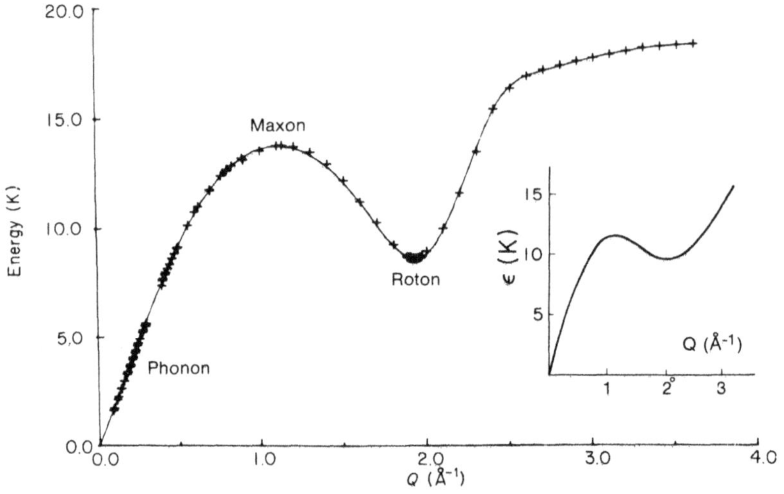

Figure 2. Phonon-Roton dispersion curve in superfluid ^{4}He at SVP and $T \leq 1.2\,K$ (from Donnelly et al. Ref 40). Inset from Landau, Ref. 27.

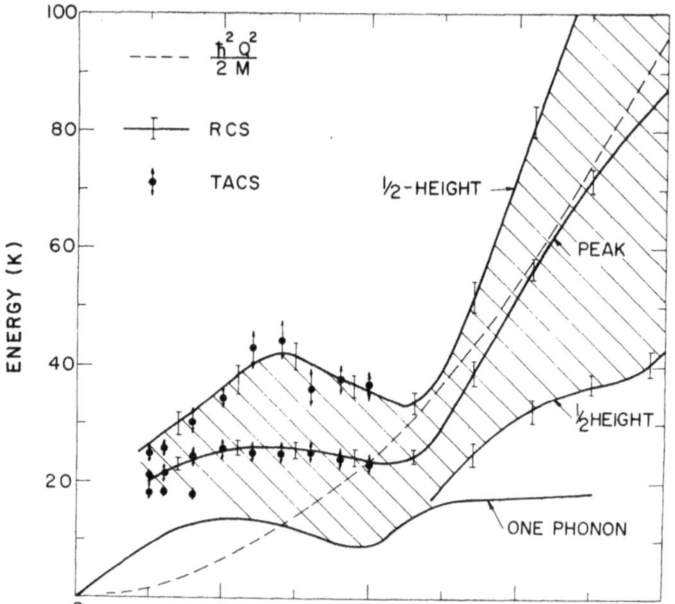

Figure 3. Phonon-Roton (PHONON) curve and the peak (PEAK) position and half heights of the broad component at higher ω for superfluid ^{4}He at SVP and $T = 1.1\,K$ (from Cowley and Woods Ref. 21).

$$S_{IA}(Q,\omega) = \int d\mathbf{p}n(\mathbf{p})\delta\left(\omega - \omega_R - \frac{\mathbf{p}\cdot\mathbf{Q}}{m}\right)$$

Beginning with a given $n(\mathbf{p})$, Sokol and collaborators[1-3] have folded $S_{IA}(Q,\omega)$ with functions to represent Final State and Instrument broadening and compared the result with their experiments at $Q = 23\,\text{Å}^{-1}$. This comparison finds the data most consistent with $n_0(T=0) = 10 \pm 2\,\%$ for reasonable choice of $n(\mathbf{p})$. This and Final State broadening[4] will be the subject of Sokol's talk.

In Fig. 4 we show the scattering intensity in the broad component at $Q = 1.13\,\text{Å}^{-1}$ observed by Stirling.[26] This shows that the intensity at SVP dips to nearly zero immediately above the sharp peak and that there are subsidiary peaks in the broad component. As pressure increases, the broad component becomes smoother.

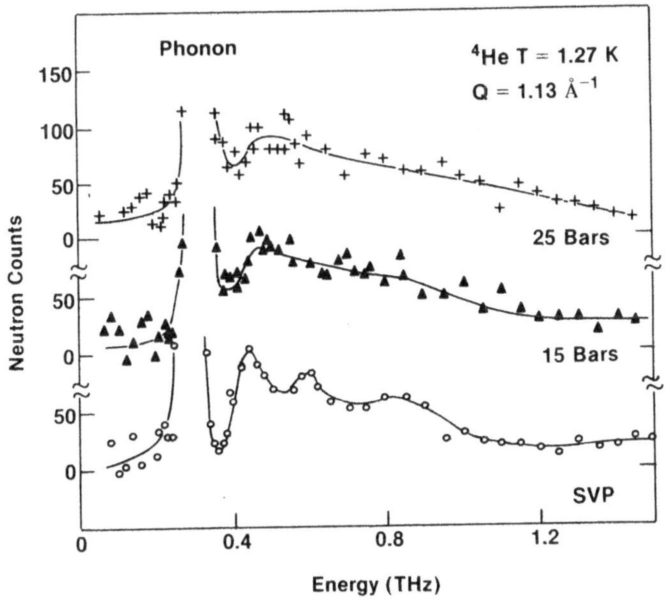

Figure 4. Scattering intensity from ^{4}He (Stirling, Ref. 26).

Landau[27] proposed that only collective excitations involving many atoms can propagate in liquid ^{4}He, since the ^{4}He atoms are strongly interacting. These are the "elementary excitations" of ^{4}He. The P–R curve is the dispersion curve for these excitations, the same at all Q. The inset in Fig. 2 is the curve proposed by Landau in 1947. Feynman[28] also proposed that the characteristic excitations are density excitations. The excitated state $|\psi_{ex}\rangle$ containing a single excitation was obtained by operating on the ground state $|\psi_0\rangle$ with the density operator $\rho(Q)$ in (3),

$$|\psi_{ex}\rangle = \rho_Q|\psi_0\rangle. \tag{5}$$

In this way, Feynman obtained a P–R curve but one which lies considerably above that in Fig. 2 in the roton region. The Feynman technique based on wave functions has evolved into the method of Correlated Basis Functions (CBF). Using CBF's, an accurate P–R curve, notably in the roton region can be obtained.[30] This will be the subject of Campbell's talk with comments by Griffin. Similar good agreement with the observed P–R curve at $T \leq 1\,K$ has been obtained by Aldrich and Pines[29] using their polarization potential theory.

Bogoliubov[31] analysed single particle excitations (SP) defined by $\langle a_Q^+(t)a_Q(0)\rangle$ in a weakly interacting Bose gas. In a gas, almost all atoms are in the condensate, $n_0 \sim 1$. At low Q, he showed that the SP excitations had a linear, sound like dispersion curve, $\omega_Q = cQ$. For a strongly interacting Bose liquid, Gavoret and Nozières[31] showed that the SP excitations, described the SP Green $G_1 = -i\langle Ta_Q(t)a_Q^+(0)\rangle$, and the density function, $\chi = -i\langle T\rho(Q,t)\rho^+(Q,0)\rangle$, had the same pole at $Q \to 0$. That is, both G and χ showed linear dispersion, $\omega_Q = cQ$ with the same "sound" velocity at $Q \to 0$. This point is discussed further by Griffin.

4. TEMPERATURE DEPENDENCE OF $S(Q,\omega)$

In a key paper, Woods and Svensson[5] presented detailed measurements of the temperature dependence of $S(Q,\omega)$ for $1.1 \leq Q \leq 2.1\,\text{Å}^{-1}$. These showed that the intensity in the sharp component of $S(Q,\omega)$ decreased with increasing temperature. The sharp peak in $S_1(Q,\omega)$, apparently disappeared entirely from $S(Q,\omega)$ at $T = T_\lambda$. This effect is displayed in Fig. 5 for the maxon ($Q = 1.13\,\text{Å}^{-1}$) and the roton ($Q = 1.923\,\text{Å}^{-1}$) in liquid ^{4}He under pressure. Above T_λ, in the normal phase, only the broad component remains. As seen at the roton Q in Fig. 5 a broad largely temperature independent $S(Q,\omega)$ is observed in the normal phase. Most significantly, these experiments showed that the sharp peak in $S(Q,\omega)$, from which the maxon and roton energies, ω_Q, are obtained, is unique to the superfluid and does not exist in the normal phase.

Svensson[18] has also shown that more consistent single excitation properties are obtained if the single excitation function $S_1(Q,\omega)$ is fitted to the sharp peak only. For example, the roton lifetime at higher T agrees better with the independent low T values observed by Mezei,[33] to be discussed in his talk here. They are also more consistent with theory[34] based on thermal roton scattering. In this event, the single excitation identified by the sharp peak is disappearing from $S(Q,\omega)$ at $T = T_\lambda$.

In Fig. 6 we show the scattering intensity at $Q = 0.4\,\text{Å}^{-1}$ as a function of T observed by Stirling and Glyde.[12] At low T, the intensity is confined essentially in a single peak, as in Fig. 1. As T increases, the peak broadens but the intensity remains within the peak. This continues to be the case in normal ^{4}He for $T > T_\lambda$. There is no qualitative change in $S(Q,\omega)$ as we go from the superfluid to the normal phase. The temperature dependence at low Q and higher Q ($Q \geq 1\,\text{Å}^{-1}$) is therefore quite different.

On the basis of this data, Stirling and Glyde[12] and Glyde and Griffin[35] have proposed two regions. At low Q, liquid ^{4}He supports collective zero sound in the density. As proposed by Landau[27] and Pines[36] this collective mode follows because the He atoms are strongly interacting. The mode is independent of statistics and therefore of Bose condensation. It exists in both normal and superfluid ^{4}He. At higher Q ($Q \geq 1.2\,\text{Å}^{-1}$), in the second region, liquid ^{4}He does not support collective zero sound. For example, the broad component at the maxon Q in Fig. 5

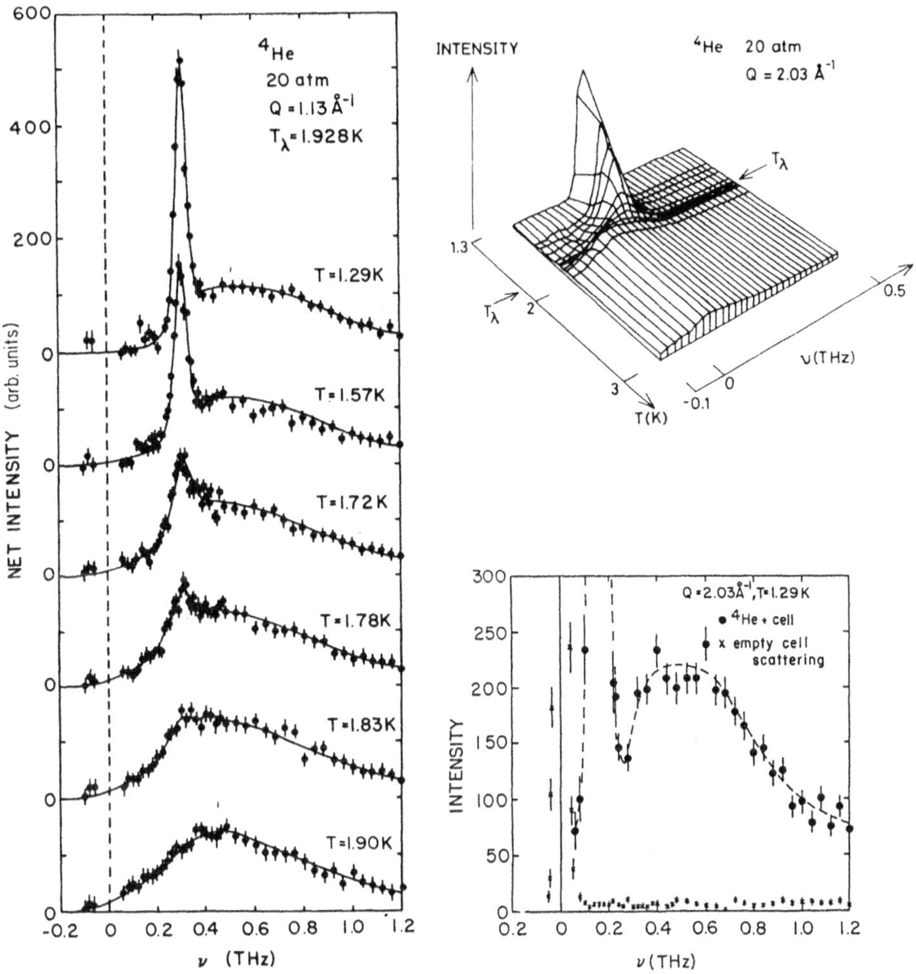

Figure 5. Temperature dependence of the scattering intensity at the maxon ($Q = 1.13\,\mathring{A}^{-1}$) and roton ($Q = 2.03\,\mathring{A}^{-1}$) and $p = 20$ bar. The broad component at the roton Q is shown, lower right. (from Ref. 11).

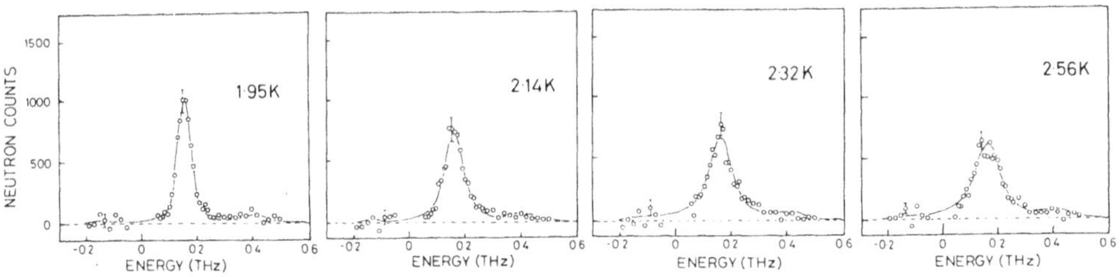

Figure 6. Temperature dependence of the scattering intensity at $Q = 0.4\,\mathring{A}^{-1}$ and SVP (from Ref. 12).

is interpreted[12] as predominantly a heavily broadened zero sound mode. The sharp peak for $Q \geq 1.2\,\mathring{A}^{-1}$ is a single particle excitation. The single particle (SP) excitation is an excitation of the single particle Green function, G. The G is a component of $S(Q,\omega)$ only in the superfluid phase and has weight proportional to $n_0(T)$. Thus the single particle peak intensity decreases with increasing T and disappears from $S(Q,\omega)$ entirely for $T \geq T_\lambda$. This interpretation is based on the dielectric formulation of Bose liquids which is outlined in the next section (see also Griffin in these proceedings).

5. DIELECTRIC FORMULATION

We sketch here the dielectric formulation of Bose fluids.[32,38,39] The aim is to illustrate how $S(Q,\omega)$ can separate into two parts in the superfluid phase where $n_0(T) \neq 0$. One part is proportional to the single particle Green function, G, and the other to the dynamic susceptibility of the atoms lying above the condensate, χ_R. If the coupling between the two terms is weak, then the single particle excitation may be seen in $S(Q,\omega)$ when $n_0(T) \neq 0$.

We begin with the total density dynamic susceptibility,

$$\chi(Q,t) = -\frac{i}{N}\langle \rho(Q,t)\rho^+(Q,0)\rangle, \tag{5}$$

related to $S(Q,\omega)$ by $S(Q,\omega) = -(n\pi)^{-1}[N(\omega)+1]\chi''(Q,\omega)$. In second quantization, the density operator is

$$\begin{aligned} \rho(Q) &= \sum_k a_k^+ a_{k+Q} \\ &= a_0^+ a_Q + a_{-Q}^+ a_0 + \sum_k{}' a_k^+ a_{k+Q}, \end{aligned} \tag{6}$$

where terms involving $k = 0$ are written separately. When there is a macroscopic number N_0 of particles in the condensate $(k = 0)$, the operators a_0 and a_0^+ may be replaced by a number, $a_0^+ = a_0 = \sqrt{N_0}$ (Bose broken symmetry). Then

$$\rho(Q) = \sqrt{N_0}A_Q + \rho'(Q) \tag{7}$$

where $A_Q = (a_Q + a_{-Q}^+)$, and $\rho'(Q) = \sum_k' a_k a_{k+Q}$ is the usual density operator involving atoms above the condensate. Substitution of (7) into (5) leads to

$$\begin{aligned} i\chi(Q,t) = n_0\langle TA_Q(t)A_Q^+(0)\rangle + \sqrt{n_0}\left[\langle TA_Q(t)\rho^{+'}(Q,0)\rangle + h.c.\right] \\ + \frac{1}{N}\langle T\rho'(Q,t)\rho^{+'}(Q,0)\rangle \end{aligned} \tag{8}$$

The first term, proportional to $G = -i\langle TA_Q(t)A_Q^+(0)\rangle$, is the Bogoliubuv term, the second and third are interference terms and the last term is the "regular" susceptibility involving atoms above the condensate, $\chi' = -i\langle T\rho'(Q,t)\rho^{+'}(Q,0)\rangle$. It can be shown[38,39] that the interference terms are also proportional to G so that (8) gives,

$$\chi(Q,\omega) = \Lambda(Q,\omega)G(Q,\omega)\Lambda(Q,\omega) + \chi_R(Q,\omega). \tag{9}$$

Thus $\chi(Q,\omega)$ contains a term proportional to $G(Q,\omega)$ with weight $\Lambda^2(Q,\omega) \propto n_0(T)$. Above T_λ, in the normal phase, the $\Lambda G \Lambda$ term vanishes and $\chi \to \chi_R$.

The sharp peak in Fig. 5 in the maxon and roton regions is interpreted as a single particle excitation in G of (9). It has weight in $S(Q,\omega)$ proportional to $n_0(T)$ and disappears from $S(Q,\omega)$ at T_λ. The broad component in Fig. 5 at the maxon is interpreted as a broadened zero sound mode in χ_R. This broad component is approximately temperature independent and persists above T_λ in the normal phase. At the roton Q, the χ_R is very broad and concentrated at low ω.

The data at $Q = 0.4\,\mathring{A}^{-1}$ shown in Fig. 6 is interpreted as scattering predominantly from a zero sound mode. In this case χ_R has a pole so that the mode exists in the normal phase above T_λ. While the mode broadens with increasing T, the scattering intensity remains chiefly within a single zero sound (ZS) peak.

When χ_R has a pole, it can be shown that Λ, G and χ_R all share the same poles. In this case the two terms in (9) are coupled and the separation in (9) is not very useful. Recall that Gavoret and Nozières showed that χ and G share the same pole at $Q \to 0$ and $T = 0\,K$. When the modes are coupled, it is better to analyse the total χ. We sketch this analysis and mode coupling in section 6. The maxon is the most interesting region because the SP and ZS modes, though coupled, are separated (e.g. as in Fig. 5).

6. MODE COUPLING

To illustrate the coupling between the single particle (SP) and zero sound (ZS) modes we begin with the full dynamic susceptibility, χ. We introduce[38,39] an irreducible $\widetilde{\chi}(Q,\omega)$ defined by

$$\chi = \frac{\widetilde{\chi}}{1 - V(Q)\widetilde{\chi}} = \frac{\widetilde{\chi}}{\epsilon} \tag{10}$$

where $V(Q)$ is the Fourier transform of the interatomic potential and $\epsilon = 1 - V(Q)\widetilde{\chi}$ is the dielectric function. Diagrammatic analysis shows that $\widetilde{\chi}$ also separates in the form (9), $\widetilde{\chi} = \widetilde{\Lambda}\widetilde{G}\widetilde{\Lambda} + \widetilde{\chi}_R$ where $\widetilde{\Lambda}$, $\widetilde{G}$ and $\widetilde{\chi}_R$ are the corresponding irreducible parts. The imaginary part of χ in (10) gives

$$S(Q,\omega) \propto -\chi''(Q,\omega) = \frac{1}{V(Q)}\left(\frac{1}{\epsilon}\right)'' \tag{11}$$

where

$$\epsilon = 1 - V\widetilde{\Lambda}\widetilde{G}\widetilde{\Lambda} - V\widetilde{\chi}_R. \tag{12}$$

To introduce coupled modes,[35] we model G as $G = 2\bar{\omega}_{SP}/(\omega^2 - \bar{\omega}_{SP}^2)$. That is, G has a sharp response at the "uncoupled" SP frequency $\bar{\omega}_{SP}$. We model χ_R by $V\chi_R = VnQ^2/m(\omega^2 + i2\bar{\omega}_0\Gamma_0) = \bar{\omega}_0/(\omega^2 + 2i\omega\Gamma_0)$. This is the high frequency limit of χ_R with "uncoupled" ZS mode frequency $\bar{\omega}_0$ and width $\overline{\Gamma}_0$. The ϵ in (12) is then

$$\epsilon = 1 - \frac{\Delta}{\omega^2 - \bar{\omega}_{SP}^2} - \frac{\bar{\omega}^2}{\omega^2 + i2\omega\overline{\Gamma}_0} \tag{13}$$

where $\Delta \equiv 2\omega_1 V(Q)\widetilde{\Lambda}^2 \propto n_0(T)$. From (11) and (13) we obtain[35] $-V(Q)\chi''(Q,\omega)$ as

$$\frac{\bar{\omega}_0^2(2\omega\overline{\Gamma}_0)(\omega^2 - \bar{\omega}_{SP}^2)^2}{[(\omega^2 - \omega_0^2)(\omega^2 - \omega_{SP}^2)]^2 + [2\omega\overline{\Gamma}_0(\omega^2 - \tilde{\omega}_{SP}^2)]^2} \tag{14}$$

where

$$D = (\omega^2 - \omega_0^2)(\omega^2 - \omega_{SP}^2)$$
$$= (\omega^2 - \bar{\omega}_0^2)(\omega^2 - \bar{\omega}_{SP}^2) - \Delta'\omega^2 \tag{15}$$

and

$$\tilde{\omega}_{SP}^2 = \bar{\omega}_{SP}^2 + \Delta. \tag{16}$$

The response function (14) has several key features which describes the data. Firstly, $\bar{\omega}_0$ and $\bar{\omega}_{SP}$ are the original uncoupled ZS and SP mode frequencies. These are coupled in the term D via $\Delta' = \Delta \propto n_0(T)$, which acts as a coupling constant in the superfluid phase where $n_0 \neq 0$. (We have distinguished Δ' from Δ in (15) since a more general treatment gives $\Delta' = \Delta + 4\Gamma_{SP}\Gamma_0$ where Γ_{SP} is the SP mode width.) The ω_0 and ω_{SP} are the shifted ZS and SP frequencies; $\bar{\omega}_0 \to \omega_0$, $\bar{\omega}_{SP} \to \omega_{SP}$. If $\Delta' > 0$, the coupling separates the modes, $|\omega_0 - \omega_{SP}| > |\bar{\omega}_0 - \bar{\omega}_{SP}|$, as in mode hybridization. If $\Delta' < 0$, the modes are clamped together $|\omega_0 - \omega_{SP}| < |\bar{\omega}_0 - \bar{\omega}_{SP}|$. The $\Delta'(Q)$ is a function of Q (and ω) and may change sign as Q increases. At the maxon $\Delta'(Q)$ is probably positive.

Secondly, the numerator in (14) vanishes at $\tilde{\omega}_{SP}$. In Figs. 4 and 5 the observed intensity dips sharply just above the main peak. If this near zero in the data can be associated with the zero in (14), then its position identifies $\tilde{\omega}_{SP}$.

Thirdly, $\chi''(Q,\omega)$ in (14) will peak where D vanishes and $2\omega_0\Gamma_0(\omega^2 - \tilde{\omega}_{SP}^2)$ is also small. D vanishes at $\omega = \omega_0$ and $\omega = \omega_{SP}$. If $\tilde{\omega}_{SP} \approx \omega_0$, χ'' peaks at the ZS energy. If $\tilde{\omega}_{SP} = \omega_{SP}$, χ'' peaks at the SP energy. This shows how intensity can be shifted between the SP and ZS peaks (if separated) depending upon $\tilde{\omega}_{SP}^2 = \bar{\omega}_{SP}^2 + \Delta$.

Finally, as T increases $n_0(T)$ decreases and $\Delta(Q,T)$ decreases. At $T = T_\lambda$, $\Delta \to 0$, $\omega_{SP} = \tilde{\omega}_{SP} = \bar{\omega}_{SP}$ and χ'' reduces to

$$-V\chi'' = \frac{\bar{\omega}_0^2(2\omega\bar{\Gamma}_0)}{(\omega^2 - \bar{\omega}_0^2)^2 + (2\omega\bar{\Gamma}_0)^2}$$

in the normal phase. This is a simple response function peaked near $\bar{\omega}_0$ with HWHM $\bar{\Gamma}_0$ as seen for $T \geq T_\lambda$ in Fig. 5.

To illustrate the properties of (14), we plot $S(Q,\omega)$ obtained from (11) and (14) in Fig. 7. The plotted $S(Q,\omega)$ includes the thermal factor $[N(\omega) + 1]$ and $S(Q,\omega)$ is convoluted with a Gaussian of full width $W = 0.0156\,THz$ to simulate a finite instrument resolution width. The parameters in (14) are chosen to simulate the maxon at 20 bars ($Q = 1.13\,\mathring{A}^{-1}$) as a function of temperature (see Fig. 5). The zero sound mode frequency and half width were chosen from the $T = 1.90\,K$ ($\approx T_\lambda$) data of Fig. 5 as $\bar{\omega}_0 = \omega_0 = 0.5\,THz$ and $\bar{\Gamma}_0 = 0.25\,THz$. The single particle frequency was fixed at $\omega_{SP} = 0.3\,THz$. The ω_0 and ω_{SP} were held constant, independent of T, and $\bar{\omega}_{SP}$ and $\bar{\omega}_0$ were determined from (15). We chose $\Delta = -\Delta'$ leaving single parameter Δ' free in (14) and (15) to be adjusted as a function of T. The resulting $S(Q,\omega)$ for three values of Δ' ($\Delta' \propto n_0(T)$) are shown in Fig. 7. The temperature used in $[N(\omega) + 1]$ is also shown.

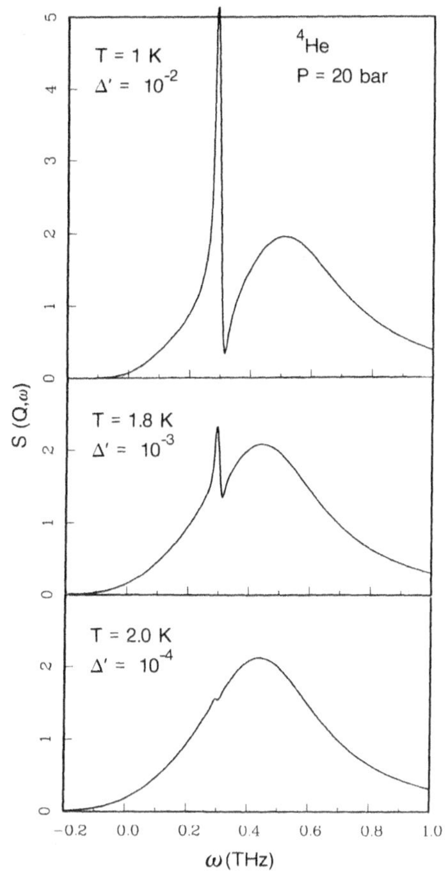

Figure 7. $S(Q,\omega)$ calculated from (11) and (14) for the maxon $Q = 1.13\,\mathring{A}^{-1}$ at $p = 20$ bars with $\omega_{SP} = 0.3THz$, $\omega_0 = 0.5\ THz$, $\overline{\Gamma}_0 = 0.25\ THz$, $\Delta = -\Delta'$ and selected values of Δ' (in THz).

Clearly, for $\Delta' = 0.01\ THz$, intended to simulate low T, $S(Q,\omega)$ in Fig. 7 has a sharp peak at the single particle frequency ω_{SP} and a broad component at higher ω. As Δ' ($\propto n_0(T)$) is reduced, the intensity in the sharp peak is reduced. For $\Delta' = 0.0001\ THz$, intended to simulate $T \approx T_\lambda$, there remains only a tiny blip at $\omega = \omega_{SP}$.

Thus the simple model in (14) simulates the broad features of the temperature dependence of $S(Q,\omega)$ observed at the maxon Q as shown in Fig. 5. The sharp peak is a single particle excitation and the broad peak is a broadened zero sound mode. Multi quasiparticle excitations need to be added to this model and other improvements are in progress.

ACKNOWLEDGEMENTS

Close collaboration with Allan Griffin and William Stirling is warmly acknowledged. This work was supported in part by NSERC of Canada, the office of Basic Energy Sciences, USDOE under contract No. DE–FG02– 84ER45082 and a NATO collaboration grant.

REFERENCES

1. T. R. Sosnick, W. M. Snow, R. N. Silver and P. E. Sokol, Phys. Rev. B (in press) (1990).
2. T. R. Sosnick, W. M. Snow and P. E. Sokol, Phys. Rev. B **41**, 11185 (1990).
3. T. R. Sosnick, W. M. Snow, P. E. Sokol and R. N. Silver, Europhys. Lett. **9**, 707 (1989).
4. See *Momentum Distributions,* edited by R. Silver and P. E. Sokol (Plenum, N. Y., 1985).
5. A. D. B. Woods and E. C. Svensson, Phys. Rev. Lett. **41**, 974 (1978).
6. J. A. Tarvin and L. Passell, Phys. Rev. B **19**, 1458 (1979).
7. K. S. Pedersen and K. Carneiro, Phys. Rev. B **22**, 191 (1980).
8. E. Talbot and A. Griffin, Phys. Rev. B **29**, 253, 3952 (1984).
9. A. Griffin, Can. J. Phys. **65**, 1368 (1987).
10. V. P. Mineev, Sov. Phys. JETP Lett. **32**, 489 (1980).
11. E. F. Talbot, H. R. Glyde, W. G. Stirling, and E. C. Svensson, Phys. Rev. B **38**, 11229 (1988).
12. W. G. Stirling and H. R. Glyde, Phys. Rev. B **41**, 4224 (1990), H. R. Glyde and W. G. Stirling in *PHONONS* 89 edited by S. Hunklinger, W. Ludwig and G. Weiss (World Scientific, Hong Kong, 1990).
13. A. D. B. Woods and R. A. Cowley, Rep. Prog. Phys. **36**, 1135 (1973).
14. D. L. Price, in *The Physics of Liquid and Solid Helium*, Part II, edited by K. H. Bennemann and J. B. Ketterson (Wiley, New York 1978) p. 675.
15. H. R. Glyde, in *Condensed Matter Research Using Neutrons*, edited by S. W. Lovesey and R. Scherm (Plenum, New York 1984) p. 95.
16. E. C. Svensson and V. F. Sears, in *Frontiers of Neutron Scattering* edited by R. J. Birgeneau, D. E. Moncton, and A. Zeillinger, p. 126, (North-Holland, Amsterdam, 1986). [Reprinted from Physica **137B**, 126 (1986)].
17. H. R. Glyde and E. C. Svensson, in *Neutron Scattering*, edited by D. L. Price and K. Sköld, Methods of Exp. Phys., Vol. 23, Part B, (Academic Press, New York 1987) p. 303.
18. E. C. Svensson, in *Elementary Excitations in Quantum Fluids,* edited by K. Ohbayashi and M. Watabe, Springer Series in Solid State Physics, Vol. 79 (Springer-Verlag, Heidelberg, 1989).
19. S. W. Lovesey, *Theory of Neutron Scattering from Condensed Matter*, Vol. 1 (OUP, Oxford, 1984).
20. A. Miller, D. Pines and P. Nozières, Phys. Rev. **127**, 1452 (1962).
21. R. A. Cowley and A. D. B. Woods, Can. J. Phys. **49**, 177 (1971).
22. P. Martel, E. C. Svensson, A. D. B. Woods, V. F. Sears and R. A. Cowley, J. Low Temp. Phys. **23**, 285 (1976).
23. W. G. Stirling, E. F. Talbot, B. Tanatar and H. R. Glyde, J. Low Temp. Phys. **73**, 33 (1988).
24. B. Tanatar, E. F. Talbot and H. R. Glyde, Phys. Rev. **36**, 8376 (1987).
25. E. C. Svensson in *75th Jubilee Conference on ^{4}He*, edited by J. G. M. Armitage (World Scientific, Singapore 1983) p. 10.
26. W. G. Stirling, in *Proc. 2nd Int. Conf. on Phonon Physics*, edited by J. Kollár, N. Kroó, N. Menyhárd and T. Siklós, p. 829, (World Scientific, Singapore, 1985).
27. L. D. Landau, J. Phys. U.S.S.R. **11**, 91 (1947).
28. R. P. Feynman, Phys. Rev. **94**, 262 (1954).
29. D. Pines, Can. J. Phys. **65**, 1357 (1987).

30. E. Manousakis and V. R. Pandharipande, Phys. Rev. B**33**, 150 (1986).
31. N. N. Bogoliubov, J. Phys. U.S.S.R. **11**, 23 (1947).
32. J. Gavoret and P. Nozières, Ann. Phys. N.Y. **28**, 349 (1964).
33. F. Mezei, Phys. Rev. Lett. **44**, 1601 (1980).
34. K. Bedell, D. Pines and A. Zawadowski, Phys. Rev. **B29**, 102 (1984).
35. H. R. Glyde and A. Griffin, Phys. Rev. Lett. (submitted).
36. D. Pines, in *Quantum Fluids* edited by D. F. Brewer, p. 257, (North-Holland, Amsterdam, 1966).
37. N. Hugenholtz and D. Pines, Phys. Rev. **116**, 489 (1959).
38. A. Griffin and T. H. Cheung, Phys. Rev. A **7**, 2086 (1973).
39. P. Szépfalusy and I. Kondor, Ann. Phys. (N.Y.) **82**, 1 (1974).
40. R. J. Donnelly, J. A. Donnelly and R. N. Hills, J. Low Temp. Phys. **44**, 471 (1981).

EXCITATIONS IN SUPERFLUID ^{4}He AND THE DIELECTRIC FORMALISM

Allan Griffin

Department of Physics
University of Toronto
Toronto, Ontario, Canada M5S 1A7

1. Introduction and Preamble

Despite great efforts over many years, a truly microscopic understanding of the phonon-maxon-roton spectrum in superfluid ^{4}He has continued to be elusive. High resolution neutron scattering studies[1,2,3] of the lineshape exhibited by $S(\vec{Q},\omega)$ have given vivid evidence of the completely different behaviour of the low Q phonon spectrum and the high Q maxon-roton spectrum as the temperature increases towards T_λ. Recently, Glyde and the author have realized that recent experimental results gave the needed clues which were missing before. We have proposed a unified theory[4] of the excitations in superfluid ^{4}He which is firmly anchored in the existence of a Bose broken symmetry (see Sokol, these proceedings). We believe that a suitably parameterized version of this microscopic formulation will provide a natural starting point for detailed comparisons with the excitations observed in high resolution neutron scattering data. For further discussion of its experimental implications, we refer to the papers by Glyde and Svensson in these proceedings. The main goals of this article are to discuss the basic elements of the new picture, to review its historical development and relation to earlier ideas, and finally, to sketch how it might be brought into contact with the very successful phenomenological approach based on variational many body wavefunctions.

The Glyde-Griffin (GG) scenario is built on the field-theoretic analysis of a Bose-condensed fluid associated with pioneering work of Bogoliubov[5], Beliaev[6], Hugenholtz and Pines[7], Gavoret and Nozières[8], and Hohenberg and Martin[9]. The key step in understanding the unique structure imposed on the dynamic structure factor $S(\vec{Q},\omega)$ and the single-particle Beliaev Green's function $G_{\alpha\beta}(\vec{Q},\omega)$ by a Bose broken symmetry was made by Gavoret and Nozières[8]. This was formally developed by Ma and Woo[10] in what is now referred to as the "dielectric formalism" (the name has its origin in the initial application to a charged Bose gas). As we discuss in Section 2, this is based on separating out the contributions to perturbation diagrams which are unique to a Bose-condensed fluid. It was later emphasized by Kondor and Szépfalusy[11] and, independently, Griffin and Cheung[12] that this formalism gave a natural way of understanding the special hybridized nature of the single-particle excitations and collective density fluctuations (zero sound) in superfluid ^{4}He, as well as how these modes changed as one passed from the superfluid to the normal phase above T_λ.

Excitations in Two-Dimensional and Three-Dimensional Quantum Fluids
Edited by A.G.F. Wyatt and H.J. Lauter, Plenum Press, New York, 1991

The main accomplishment of the dielectric formalism is that it exposes the inevitable structure of various correlation functions which is induced by a Bose-condensate. In the formal theory, the correlation functions are expressed in terms of irreducible, proper contributions, these "regular" functions being the building blocks (see Section 2). In the near future, not much progress can be expected in the difficult evaluation of these building blocks from microscopic theory for superfluid ^{4}He at finite temperature, although some of the limiting behaviour (high and low frequencies, long wavelengths) is known rigorously[8,13,14]. A more useful approach will probably lie in developing simple model parameterizations for these regular functions, and seeing how well such model calculations can describe the observed lineshapes in neutron scattering data over a wide range of Q, ω and T. GG have already introduced the simplest possible approximation which brings out the rich structure which $S(\vec{Q},\omega)$ can exhibit because of the condensate-induced coupling of the zero sound (ZS) phonon modes with the single-particle (SP) maxon-roton excitations (see also Glyde, these proceedings).

The GG scenario ^{4}He is a natural culmination of the field-theoretic analysis of Bose liquids over the last 30 years[15]. Experimentally, it has its genesis in the pioneering work of Woods and Svensson[1]. It brings together ideas that had been suggested earlier[15–19], except they had never been put together in the right combination nor their precise implications fully realized. In Section 3, we discuss some of this history and try to explain why it took so long to get it right! In Section 4, we briefly discuss some future research which should develop out of the GG picture.

2. Dielectric Formalism

In the presence of a Bose broken symmetry, the Fourier component of the density fluctuation operator splits naturally into two components,

$$\hat{\rho}_Q = \sqrt{N_0}(\hat{a}_Q + \hat{a}^+_{-Q}) + \sum_p{}' \hat{a}^+_p \hat{a}_{p+Q} \equiv \sqrt{N_0}\hat{A}_Q + \tilde{\rho}_Q \ , \tag{1}$$

where $\sqrt{N_0} = \langle \hat{a}_0 \rangle$ is the Bose order parameter. The first term in (1) corresponds to density fluctuations due to atoms going in/out of the condensate reservoir. $\tilde{\rho}_Q$ is the density fluctuation component involving excited (non-condensate) atoms and is the only term present above T_λ. Using (1), the dynamic structure factor

$$S(\vec{Q},\omega) = \frac{1}{2\pi n}\int_{-\infty}^{\infty} dt e^{i\omega t}\langle \hat{\rho}_Q(t)\hat{\rho}_{-Q}(0)\rangle \tag{2}$$

naturally splits into three components, as first emphasized in refs. 7 and 16. Working in terms of the time-ordered density response function $\chi_{\rho\rho}(\vec{Q},t) \equiv -i\langle T\hat{\rho}_Q(t)\hat{\rho}_{-Q}\rangle$, we have

$$\chi_{\rho\rho}(\vec{Q},\omega) = n_0 \sum_{\alpha,\beta} G_{\alpha\beta}(\vec{Q},\omega) + \sqrt{n_0}\chi_{int}(\vec{Q},\omega) + \chi_{\tilde{\rho}\tilde{\rho}}(\vec{Q},\omega) \ , \tag{3}$$

where $G_{\alpha\beta}(\vec{Q},\omega)$ is the matrix single-particle Green's function introduced by Beliaev[6] (α and β are the components of a 2×2 matrix). The interference term describes the mixed response functions

$$\chi_{int}(\vec{Q},t) = -i[\langle T\hat{A}_Q(t)\tilde{\rho}_{-Q}\rangle + \langle T\tilde{\rho}_Q(t)\hat{A}_{-Q}\rangle] \ . \tag{4}$$

In a weakly interacting dilute Bose gas (WIDBG) at $T = 0$, one need only keep the first term in (3) since almost none of the atoms are outside the condensate. This is the case studied in the standard textbooks. In a WIDBG at finite temperatures ($T \sim T_\lambda$) and in superfluid ^{4}He, the condensate fraction n_0/n is quite small and it is then crucial to keep the second and third terms in (3). The dielectric formalism[10–12] is a systematic procedure for doing this.

We first present a new mean field derivation of the key results of the dielectric formalism, which is much simpler than the more formal diagrammatic analysis given in refs. 10–13. We consider the linear response to an *external* potential which couples into the variable $\tilde{\rho}_Q$ and a broken symmetry *external* potential which couples into the variable $\hat{A}_Q$,

$$\hat{V}_{\text{ex}} = \int \frac{d\vec{Q}}{(2\pi)^3} \delta\phi^0(\vec{Q},\omega)\tilde{\rho}_Q + \int \frac{d\vec{Q}}{(2\pi)^3} \delta\eta^0(\vec{Q},\omega)\hat{A}_Q \ . \tag{5}$$

Thus we have

$$\begin{aligned} \delta\tilde{\rho}(\vec{Q},\omega) &= \chi_{\tilde{\rho}\tilde{\rho}}(\vec{Q},\omega)\delta\phi^0(\vec{Q},\omega) + \chi_{\tilde{\rho}A}(\vec{Q},\omega)\delta\eta^0(\vec{Q},\omega) \\ &\simeq \chi^0_{\tilde{\rho}\tilde{\rho}}(\vec{Q},\omega)\left[\delta\phi^0(Q,\omega) + V(\vec{Q})\delta\tilde{\rho}(Q,\omega) + \sqrt{n_0}V(\vec{Q})\delta A(\vec{Q},\omega)\right] \end{aligned} \tag{6}$$

$$\begin{aligned} \delta A(Q,\omega) &= \chi_{AA}(\vec{Q},\omega)\delta\eta^0(\vec{Q},\omega) + \chi_{A\tilde{\rho}}(\vec{Q},\omega)\delta\phi^0(\vec{Q},\omega) \\ &\simeq \chi^0_{AA}(\vec{Q},\omega)\left[\delta\eta^0(\vec{Q},\omega) + \sqrt{n_0}V(\vec{Q})\delta\tilde{\rho}(\vec{Q},\omega) + n_0V(\vec{Q})\delta A(\vec{Q},\omega)\right] . \end{aligned} \tag{7}$$

In (6) and (7), the second lines are written in terms of the two self-consistent-fields which arise from a mean field approximation to the various terms of the interaction Hamiltonian

$$\hat{V} = \frac{1}{2}\sum_Q V(Q)\left[n_0\hat{A}_Q\hat{A}_{-Q} + \sqrt{n_0}(\hat{A}_Q\tilde{\rho}_{-Q} + \tilde{\rho}_Q\hat{A}_{-Q}) + \tilde{\rho}_Q\tilde{\rho}_{-Q}\right] \ . \tag{8}$$

The "screened" response functions relative to these effective fields are denoted by χ^0. In the simplest approximation, these are for a non-interacting Bose gas, in which case $\chi^0_{\tilde{\rho}A} = \chi^0_{A\tilde{\rho}} = 0$ [which has been assumed in the second lines of (6) and (7)] and

$$n_0\chi^0_{AA}(\vec{Q},\omega) = \frac{n_0Q^2/m}{\omega^2 - (Q^2/2m)^2} \ . \tag{9}$$

One can easily solve the coupled equations (6) and (7) to obtain

$$\chi_{\tilde{\rho}\tilde{\rho}}(\vec{Q},\omega) = \frac{\chi^0_{\tilde{\rho}\tilde{\rho}}}{1 - \frac{V(Q)\chi^0_{\tilde{\rho}\tilde{\rho}}}{1 - n_0V(Q)\chi^0_{AA}}} \quad ; \quad \chi_{AA}(\vec{Q},\omega) = \frac{\chi^0_{AA}}{1 - \frac{n_0V(Q)\chi^0_{AA}}{1 - V(Q)\chi^0_{\tilde{\rho}\tilde{\rho}}}} \tag{10a}$$

$$\chi_{\tilde{\rho}A}(\vec{Q},\omega) = \chi_{A\tilde{\rho}}(\vec{Q},\omega) = \sqrt{n_0}V(Q)\chi^0_{AA}\chi^0_{\tilde{\rho}\tilde{\rho}}/\epsilon(\vec{Q},\omega) \ . \tag{10b}$$

We note that $\chi_{AA} = \sum_{\alpha,\beta} G_{\alpha\beta}$ and thus is directly related to the single-particle Green's functions. As expected, all three response functions are seen to share the *same* common denominator

$$\epsilon(\vec{Q},\omega) = 1 - V(\vec{Q})[n_0\chi^0_{AA} + \chi^0_{\tilde{\rho}\tilde{\rho}}] \ , \tag{11}$$

and hence they have the same excitation spectrum. Adding up the results in (10), the total density response function in (3) is given by

$$\chi_{\rho\rho}(\vec{Q},\omega)=\frac{n_0\chi^0_{AA}+\chi^0_{\tilde{\rho}\tilde{\rho}}}{\epsilon(\vec{Q},\omega)}\ , \tag{12}$$

which is recognized as a RPA-type approximation. In this language, the Bogoliubov approximation to (10)–(12) corresponds to simply setting $\chi^0_{\tilde{\rho}\tilde{\rho}}=0$. This simple mean field derivation can be easily generalized to include backflow contributions (such as is done in polarization potential theory[20]) to all three response functions in (10). As will be discussed elsewhere, one can show that $V(Q)$ in (11) can be related to $f^s_Q+(\omega^2/Q^2)f^v_Q$ involving the soft potentials of PP theory[20].

Many-body diagrammatic analysis[10–14] allows one to generalize the results in (10)–(12) in a formally rigorous manner. It is customary to introduce irreducible diagrams as those which cannot be split into two by cutting a single interaction line. Thus one has

$$\chi_{\rho\rho}(\vec{Q},\omega)=\frac{\bar{\chi}_{\rho\rho}(\vec{Q},\omega)}{1-V(Q)\bar{\chi}_{\rho\rho}(\vec{Q},\omega)}\equiv\frac{\bar{\chi}_{\rho\rho}}{\epsilon(\vec{Q},\omega)}\ , \tag{13}$$

where $\bar{\chi}_{\rho\rho}$ is the irreducible part of the full density response function. In a Bose-condensed fluid, new features arise because of the mixed correlation functions such as (4), which couple the single-particle and density fluctuations. These are proportional to Bose symmetry breaking functions $\bar{\Lambda}_\alpha(\vec{Q},\omega)$, which vanish if $n_0=0$ but lead to a hybridization of the spectrum of $G_{\alpha\beta}$ and $\chi_{\rho\rho}$ below T_λ.

We simply write down the final general expressions. One finds that the single-particle Beliaev self-energies split into irreducible ($\bar{\Sigma}_{\alpha\beta}$) and reducible parts,

$$\Sigma_{\alpha\beta}(\vec{Q},\omega)=\bar{\Sigma}_{\alpha\beta}+\frac{\bar{\Lambda}_\alpha V(Q)\bar{\Lambda}_\beta}{\epsilon^R(Q,\omega)}\ , \tag{14}$$

where

$$\epsilon^R(\vec{Q},\omega)\equiv 1-V(Q)\bar{\chi}^R_{\rho\rho}(\vec{Q},\omega)\ . \tag{15}$$

Here $\bar{\chi}^R_{\rho\rho}$ is the proper part of the irreducible density response function $\bar{\chi}_{\rho\rho}$, i.e., the contribution which cannot be split into two parts by cutting an internal single-particle line. It is convenient to introduce an irreducible Beliaev single-particle Green's function $\bar{G}_{\alpha\beta}$, which only contains the irreducible self-energies $\bar{\Sigma}_{\alpha\beta}$. Then one can show that

$$\bar{\chi}_{\rho\rho}(\vec{Q},\omega)=\sum_{\alpha,\beta}\bar{\Lambda}_\alpha\bar{G}_{\alpha\beta}(\vec{Q},\omega)\bar{\Lambda}_\beta+\bar{\chi}^R_{\rho\rho}(\vec{Q},\omega)\ . \tag{16}$$

In a loose way, we can think of these two terms as the condensate and normal parts, respectively.

Examination of (13)–(16) immediately shows how the Bose vertex function $\bar{\Lambda}_\alpha$ ties together the structure of $G_{\alpha\beta}$ and $\chi_{\rho\rho}$. In particular, (14) shows how it allows the non-condensate fluctuations (given by the zero of ϵ^R) to give rise to singularities in the self-energy. Denoting $\bar{G}_{\alpha\beta}=\bar{N}_{\alpha\beta}/\bar{D}$, one can show that the spectrum of both $\chi_{\rho\rho}$ and $G_{\alpha\beta}$ have the common denominator

$$C(\vec{Q},\omega)\equiv\bar{D}(Q,\omega)\epsilon^R(Q,\omega)-V(Q)\sum_{\alpha\beta}\bar{\Lambda}_\alpha\bar{N}_{\alpha\beta}\bar{\Lambda}_\beta\ . \tag{17}$$

This shows how the vertex function $\bar{\Lambda}_\alpha(\vec{Q},\omega)$ hybridizes the single-particle (SP) excitations given by the zeros of $\bar{D}(Q,\omega)$ and the collective zero sound (ZS) excitations given by the zeros of $\epsilon^R(\vec{Q},\omega)$ in (15). All these features are nicely illustrated in the simple mean field expressions (10)–(12). In that case, $\bar{G}_{\alpha\beta}$ and $\bar{\chi}^R_{\rho\rho}$ are those of a free Bose gas and $\bar{\Lambda}_\alpha(\vec{Q},\omega) = \sqrt{n_0(T)}$.

In analyzing $S(\vec{Q},\omega)$ data, the most useful expression for $\chi_{\rho\rho}$ is that given by (13) in conjunction with (16). One only needs $\epsilon(\vec{Q},\omega)$, since one can easily verify that

$$S(\vec{Q},\omega) = \frac{1}{\pi n}[N(\omega)+1]\frac{1}{V(Q)}Im\left[\frac{1}{\epsilon(Q,\omega)}\right] \tag{18}$$

where $N(\omega)$ is the Bose function. It is clear that in order to obtain the correct hybridized spectrum, one has to treat both terms in (16) carefully. The simplest model approximation is[4,19]

$$\epsilon(\vec{Q},\omega) = 1 - \frac{2V(Q)\bar{\Lambda}^2(Q)\bar{\omega}_{sp}}{\omega(\omega+i\Gamma_{sp})-\bar{\omega}^2_{sp}} - \frac{\bar{\omega}_0^2}{\omega(\omega+i\Gamma_0)} \ . \tag{19}$$

This describes a damped SP excitation $\bar{\omega}_{sp}$ coupled to a damped zero sound ZS mode $\bar{\omega}_0$ via the symmetry breaking Bose vertex function $\bar{\Lambda}$. Somewhat surprisingly, this simple kind of parameterization appears to be able to reproduce much of the detailed lineshape behaviour observed in recent high resolution neutron scattering data. For further discussion, see Glyde (these proceedings) and Glyde and Griffin[21]. In retrospect, one now sees that a two-component parameterization must be done on $\bar{\chi}_{\rho\rho}$ as given by (16), and *not* on the full $\chi_{\rho\rho}$ in (13) as originally advocated by Woods and Svensson[1]. In this sense, (13) is more useful than the alternative but equivalent version[10–12]

$$\bar{\chi}_{\rho\rho}(\vec{Q},\omega) = \sum_{\alpha,\beta}\frac{\bar{\Lambda}_\alpha}{\epsilon^R}G_{\alpha\beta}(\vec{Q},\omega)\frac{\bar{\Lambda}_\beta}{\epsilon^R} + \frac{\bar{\chi}^R_{\rho\rho}(\vec{Q},\omega)}{\epsilon^R(\vec{Q},\omega)} \ , \tag{20}$$

although (20) emphasizes the close relationship between $\chi_{\rho\rho}$ and $G_{\alpha\beta}$ when $\bar{\Lambda}_\alpha \neq 0$.

3. Some Historical Remarks: Why did it take so long?

The dielectric many-body formalism sketched in Section 2 is valid at all Q, ω and T. In particular, it is also valid in the low frequency, hydrodynamic region[14], where first and second sound are the hybridized branches discussed above. However, the regular functions ($\bar{\chi}^R_{\rho\rho}$, $\bar{\Sigma}_{\alpha\beta}$ and $\bar{\Lambda}_\alpha$) which are the building blocks of the theory are, in general, very difficult to evaluate in liquid ^{4}He at finite temperatures. One important limiting case was obtained in the long wavelength limit at $T = 0$ by Gavoret and Nozières[8]. Their results showed that both $\chi_{\rho\rho}$ and $G_{\alpha\beta}$ were dominated by the same phonon pole, which had the macroscopic sound velocity. However, generalization of this result to finite temperatures has remained obscure.

In lieu of microscopic calculations at large Q and at finite T, one must be guided by experimental studies. In the pioneering work of Woods[22] at $Q \simeq 0.4$Å^{-1}, the phonon excitation in $S(\vec{Q},\omega)$ clearly persisted to well above T_λ, which showed its existence as a pole of $S(\vec{Q},\omega)$ had nothing to do with Bose condensation. In contrast, similar experiments at higher Q by Woods and Svensson[1] showed that the intensity of the maxon-roton peak appeared to vanish as one passed from the superfluid to the normal phase, even though peak position only changed slightly. In early attempts by the author to understand this quite different temperature-dependent behaviour at low and high Q, the guiding assumption was that the phonon-maxon-roton quasiparticle described the *same* excitation branch. This assumption has been taken as a dogma since Landau's (tentative) suggestion[23] and was further anchored after the introduction of the Feynman-Bijl variational wavefunction which appeared to describe the essential physics at all wavevectors. Thus, the author's initial interpretation[24,18] of the Woods-Svensson high Q data in terms of a renormalized atomic-like SP excitation was in conflict with the clear fact that the low Q data must be interpreted as a collective ZS excitation[15] (as first emphasized by Pines[17]).

The resolution of this conflict only occurred very recently, stimulated by the new experimental studies[2,3] carried out at the I.L.L. in Grenoble. Stirling and Glyde[3] argued that the finite temperature scattering data strongly suggested that the phonon and maxon-roton parts of the dispersion curve described two *different* excitation branches. It was then realized by GG[4] that this dual nature was implicit all along in the dielectric formalism. In particular, the (hitherto mysterious) smooth crossover between the phonon and maxon-roton branches in the region $\sim 0.8-1.0\text{Å}^{-1}$ could be understood as a natural result of quantum mechanical condensate-induced hybridization.

In our analysis, the zero sound density fluctuation dominates the spectrum of $\chi_{\rho\rho}$ at low Q, both below and above T_λ, as observed in references 3 and 22. The characteristics of this collective mode are determined by the strong interactions and short range correlations which arise in a liquid,[17,20] and only weakly on the quantum statistics involved or the temperature (thermal effects). Because of the Bose broken symmetry which sets in below T_λ, this phonon mode has finite weight in the single particle Green's function $G_{\alpha\beta}$ and increasingly dominates its *low* energy spectrum as we go to low temperatures. Zero sound modes in liquid ^{3}He and classical liquids are not elementary excitations. Bose statistics does *not* play a crucial role in the existence of a zero sound mode but does in its role as an elementary excitation in the superfluid phase. This is consistent with the well known discussion of possible excited states in a Bose liquid by Feynman[25]. Feynman's analysis showed how Bose statistics were crucial in ensuring that there were no *other* low energy, long wavelength excitations besides a collective density fluctuation.

The maxon-roton peak in $S(\vec{Q},\omega)$ has its origin as a pole in the single-particle atomic spectrum, which probably exists above T_λ without much change in the dispersion relation[18]. These single-particle states have high energy (>8K) and exist at relatively large wavevectors ($\gtrsim 0.8\text{Å}^{-1}$). Thus it is physically reasonable that they would not be modified very much when the liquid goes into the superfluid phase. In contrast, available $S(\vec{Q},\omega)$ data imply that there are *no* well-defined intrinsic *low* energy, long wavelength single-particle excitations in liquid ^{4}He. The idea that a roton is physically quite different from a phonon has a long history, although this difference is obscured when one uses a single variational wavefunction valid at both small and large wavevectors. Feynman[25] argued that the roton corresponds to a high energy excited state involving the motion of a single ^{4}He atom in the potential well of its neighbors. Miller, Pines and Nozières[16] also viewed the roton as an atomic-like excitation, with backflow. Our present analysis simply extends these ideas to higher temperatures, to T_λ and above.

4. Future Prospects

If the new GG picture of the excitations in superfluid ^{4}He fulfills its promise, it should give a much needed impetus to future research. There is much further work needed on the adequacy of various model parameterizations for the regular functions. At present, future progress seems quite promising (see Glyde, these proceedings). Moreover there is a clear need for additional high resolution neutron scattering studies which will probe the interesting cross-over expected in the intermediate region $\sim 0.8\text{Å}^{-1}$. While finite temperature studies[1−3] have been very important in understanding the basic physics, probably future work should concentrate at low temperatures in order to sharpen up the effects due to predicted hybridization.

More generally, we think the new GG picture of the coupled nature of the phonon and maxon-roton excitations in pure superfluid ^{4}He should have important implications for the study of superfluid ^{3}He–^{4}He mixtures. In the traditional treatment of mixtures[26], the ^{4}He quasiparticle branch is taken as a given and the emphasis is put on how it hybridizes with the ^{3}He particle-hole excitations. Our interpretation of the maxon-roton as a single-particle excitation would seem to suggest a new approach. The dielectric formalism has been worked out for mixtures[27] but has not been applied in any detailed way. Here also, simple parameterizations of the various proper, irreducible (regular) functions would seem to be the best initial approach.

Another topic of interest is to generalize the work of ref. 4 to include hybridization of the single-particle excitations with the two-quasiparticle continuum. As with the coupling to the zero sound (particle-hole) modes, this also arises from the Bose broken symmetry[28]. This would allow a systematic study of the well-defined structure visible in the $S(\vec{Q},\omega)$ data associated with the maxon-roton and roton-roton multiparticle continuum[29] in addition to the zero sound contributions. We conjecture that below the crossover at $Q \sim 0.8\text{Å}^{-1}$, the high energy maxon SP excitation hybridizes with the 2-quasi-particle continuum. This would be the precise analogue of the well-known behaviour of the roton at $Q \gtrsim 2.5\text{Å}^{-1}$ first discussed by Pitaevskii[28].

Finally, one of the important remaining theoretical problems lies in understanding how the physical picture implied by GG can be related to some appropriate variational many body wavefunction. We recall that the original Feynman-Bijl wavefunction for an excited state is a pure density fluctuation

$$|FB\rangle = \hat{\rho}_Q^+|\Phi_0\rangle \ . \tag{21}$$

In contrast, at larger Q, the Feynman-Cohen excited state ansatz is given by $|FC\rangle = F_Q^+|\Phi_0\rangle$, where (see Campbell, these proceedings)

$$F_Q^+ = \hat{\rho}_Q^+ + \sum_q A(Q,q)\hat{\rho}_q^+\hat{\rho}_{Q-q}^+ \ . \tag{22}$$

Clearly $|FC\rangle$ is not a pure density fluctuation but contains an admixture with something else, whose physical significance can only be related to backflow at small Q. We speculate that the second term in (22) may have a significant contribution from the coupling into single-particle excitations induced by the condensate. Further studies concerning this are needed.

For further discussion of some of the implications of the GG picture, see also reference 30.

Acknowledgements

I would like to thank many colleagues who have shared my interest in the mysteries of superfluid ^{4}He. In particular, I acknowledge endless discussions for over a decade with Eric Svensson and more recently with Henry Glyde. This work was supported by the Natural Sciences and Engineering Research Council (NSERC) of Canada.

REFERENCES

1. A.D.B. Woods and E.C. Svensson, Phys. Rev. Letters **41**, 974 (1978).
2. E.F. Talbot, H.R. Glyde, W.G. Stirling and E.C. Svensson, Phys. Rev. **B38**, 11 229 (1988).
3. W.G. Stirling and H.R. Glyde, Phys. Rev. **B41**, 4224 (1990).
4. H.R. Glyde and A. Griffin, Phys. Rev. Letters, **65**, September 17 issue (1990).
5. N.N. Bogoliubov, J. Phys.(U.S.S.R.) **11**, 23 (1947).
6. S.T. Beliaev, Sov. Phys. JETP **7**, 289 (1958) [Zh. Eksp. Teor. Fiz. **34**, 417 (1958)].
7. N. Hugenholtz and D. Pines, Phys. Rev. **116**, 489 (1959).
8. J. Gavoret and P. Nozières, Ann. Phys. (N.Y.) **28**, 349 (1964).
9. P.C. Hohenberg and P.C. Martin, Ann. Phys. (N.Y.) **34**, 291 (1965).
10. S.K. Ma and C.W. Woo, Phys. Rev. **159**, 165 (1967).
11. P. Szépfalusy and I. Kondor, Ann. Phys. (N.Y.) **82**, 1 (1974).
12. A. Griffin and T.H. Cheung, Phys. Rev. **A7**, 2086 (1973).
13. V. Wong and H. Gould, Ann. Phys. (N.Y.) **83**, 252 (1974).
14. E. Talbot and A. Griffin, Ann. Phys. (N.Y.) **151**, 71 (1983).
15. For a review, see A. Griffin, Can. Journ. Phys. **65**, 1368 (1987).
16. A. Miller, D. Pines and P. Nozières, Phys. Rev. **127**, 1452 (1962).
17. D. Pines, in **Quantum Fluids**, ed by D.F. Brewer (John Wiley, N.Y., 1966), p. 257.
18. A. Griffin and E. Talbot, Phys. Rev. **B24**, 5075 (1981).
19. A. Griffin and S. Payne, Journ. Low Temp. Phys. **64**, 155 (1986).
20. For a review, see D. Pines, Can. Journ. Phys. **65**, 1357 (1987).
21. H.R. Glyde and A. Griffin, in preparation.

22. A.D.B. Woods, Phys. Rev. Letters **14**, 355 (1965).

23. L.D. Landau, J. Phys. (U.S.S.R.) **11**, 91 (1947).

24. A. Griffin, Phys. Rev. **B**19, 5946 (1979).

25. R.P. Feynman, Phys. Rev. **91**, 1301 (1953); **94**, 262 (1954).

26. See, for example, W. Hsu, D. Pines and C.H. Aldrich, Phys. Rev. **B32**, 7179 (1985).

27. E. Talbot and A. Griffin, Journ. Low Temp. Phys. **56**, 141 (1984).

28. A. Zawadowski, J. Ruvalds and J. Solana, Phys. Rev. **A5**, 399 (1972).

29. W.G. Stirling, in **Proc. 2nd International Conf. on Phonon Physics**, ed. by J. Kollar, N. Kroo, N. Menyhard, T. Siklos (World Scientific, Singapore, 1985), p. 829.

30. A. Griffin and E.C. Svensson, LT-19 Proceedings (Sussex), in Physica **B165 + 166**, 487 (1990).

RECENT HIGH-RESOLUTION NEUTRON SCATTERING STUDIES OF LIQUID ^{4}He

W.G. Stirling

Department of Physics, School of Physical Science and Engineering
University of Keele, Keele, Staffs., ST5 5BG, U.K.
and SERC, Daresbury Laboratory, Warrington, WA4 4AD, U.K.

1. INTRODUCTION

Neutron scattering has provided much detailed quantitative information on the nature of the elementary excitations of liquid helium and on the interactions between those excitations. It could reasonably be argued that the most significant features of the excitation spectrum of superfluid ^{4}He were known by about 1975, following the classic investigations of Cowley and Woods[1,2] at Chalk River, and of the Brookhaven group[3,4]. The phonon-maxon-roton dispersion curve shown in Fig. 1 had been studied as a function of both temperature and pressure (or equivalently, of density), while the existence of the multiphonon continuum at higher energies had been noted and its relative strength investigated. Nevertheless, neutron scattering studies have continued up to the present day, particularly at Chalk River Nuclear Laboratories and at the Institut Laue-Langevin (I.L.L.). What then was the aim of these second-generation experiments? In their excellent and comprehensive review of neutron scattering studies of solid and liquid helium, published in 1987, Glyde and Svensson[5] identified several important questions which remained unanswered in the early 1970s. These ranged from uncertainties as to the form of the dispersion relation at small wavevectors to the origin of the oscillations in the peak position and width of the scattering function $S(Q,\omega)$ at large wavevector Q and frequency ω. In addition, the relationship, if any, of the unique excitation spectrum to the existence of superfluidity below T_λ remained a question of profound importance.

In this article we summarise the relevant investigations carried out at I.L.L., Grenoble. The aim of this paper is to present the neutron results, leaving detailed theoretical interpretation to the authors of accompanying chapters. The article is laid out as follows. Section 2 presents a brief account of the experimental conditions employed for these measurements followed in Section 3 by a study of the pressure-dependence of the dispersion relation in the small Q (phonon) region[6]. A precise determination of the roton parameters[7] (roton gap, wavevector, and effective mass) is described in Section

Excitations in Two-Dimensional and Three-Dimensional Quantum Fluids
Edited by A.G.F. Wyatt and H.J. Lauter, Plenum Press, New York, 1991

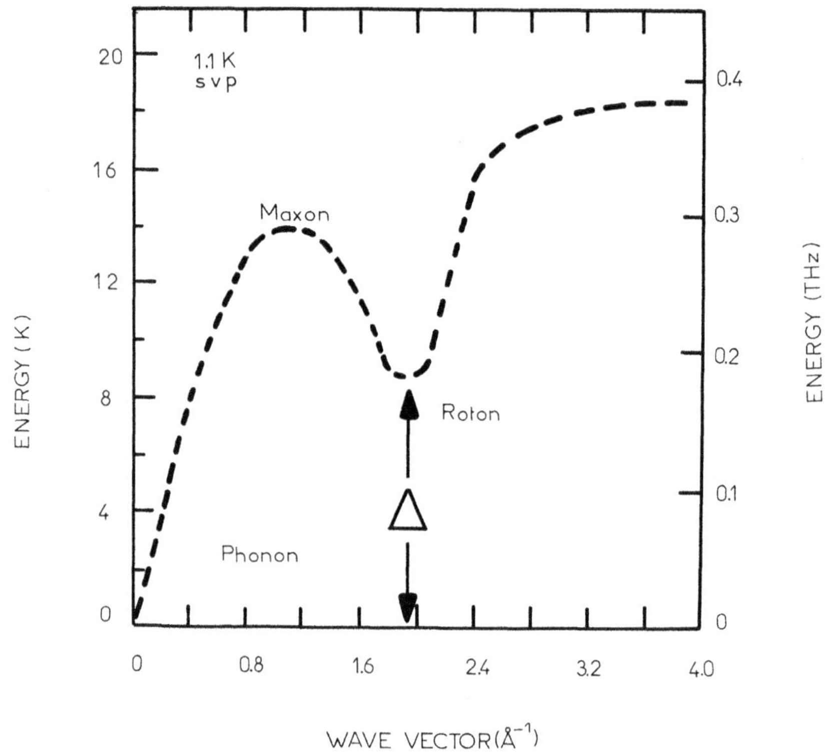

Fig. 1. Dispersion relation of superfluid ^{4}He at saturated vapour pressure.

4. Section 5 deals with the structure of the multiphonon continuum[7,8]. The temperature-dependence of S(Q,ω), below and above T_λ, is discussed in Section 6[9-11] where important new results are presented. Finally, Section 7 contains concluding remarks. A discussion of excitation lifetimes is left to the accompanying article by Mezei and his collaborators.

2. EXPERIMENTAL DETAILS

The development of specialised high-flux research reactors has led to a considerable improvement and refinement of the techniques of neutron inelastic scattering. For our purposes the most significant advances have come from the use of "cold-sources" which effectively shift the Maxwellian velocity spectrum of moderated neutrons towards longer wavelengths, or smaller energies, very well suited for studies on liquid ^{4}He. In this way, the instrumental resolution can be improved over that of conventional "thermal" instruments. Higher flux and better resolution are the two principal factors which have led to recent improvements over previous investigations on ^{4}He.

Most of the experiments discussed here were carried out using the IN12 triple-axis crystal spectrometer (TACS). This instrument is situated on a curved neutron guide tube from the (vertical) cold-source in the I.L.L. high flux reactor. The instrument has two advantages over conventional reactor instruments. The cold source provides an enhanced flux of neutrons in the energy range 1 to 10 meV (1 meV = 11.6 K = 0.242 THz = 8.07 cm^{-1}) and, further, the curvature of the guide tube suppresses troublesome higher-order neutrons (λ/2, λ/3, . . .) which might otherwise be scattered by the monochromator. Cooled beryllium filters are employed to further reduce the flux of any unwanted high energy neutrons. The instrument uses a large (12 x 10 cm^2) pyrolytic graphite monochromator which can be vertically curved to focus on the sample, where appropriate, and employs high-transmission Soller-slit collimators with gadolinium-coated stretched mylar blades.

For the phonon measurements[6] of Section 3, the instrument was operated with a fixed incident neutron energy of 43.8K, giving a measured (incoherent scattering) instrumental resolution width of 0.76K (FWHM). Appropriate energies and collimations were used for the other types of measurement described here. For example, to study the scattering function at temperatures in the superfluid and normal phases[10] [Section 6] measurements at Q = 0.4$Å^{-1}$ were made with a fixed final energy of 54.1K and corresponding resolution of 1.73K. In this case the use of the fixed <u>final</u> energy configuration was appropriate as the measured intensity is then directly proportional to S(Q,ω); for a fixed <u>incident</u> energy, a calculated correction, dependent on the analyser energy, must be made to the energy spectra.

A large body of new results[11] is introduced in Section 6. These experiments used the time-of-flight (TOF) technique, which allows the simultaneous determination of energy spectra at many different angles (and hence wavevectors). This obvious advantage is somewhat counterbalanced by the fact that energy distributions are measured at constant scattering angle, not constant Q, and that instrumental resolution corrections are more difficult than in the case of the triple-axis spectrometer. A TOF spectrometer normally has a much larger solid angle at the detector, leading to higher count rates. Generally, however, the two instruments complement each other.

The IN6 TOF spectrometer was used for some of the experiments[11] described in Section 6. This instrument, on a curved neutron guide from the I.L.L. cold-source, has three focussing pyrolytic graphite monochromators which provide a very intense monochromatic beam at the sample position. For the experiments described here, the neutron energy was 44.8K, chosen so that with detectors at angles up to 115^{0} , wavevectors up to the roton minimum could be examined. The instrumental resolution was approximately 1.4K.

For all the measurements described here, the high-purity ^4He sample was condensed into an aluminium cell of 3 cm diameter, with cadmium discs spaced 1 cm apart vertically to minimise multiple scattering[12] . The effective sample volume was approximately 40 cm^3. The sample cell was mounted in an I.L.L. ("Orange") helium-flow cryostat; the sample temperature was measured with a calibrated carbon resistor. The sample cell scattering was determined separately and subtracted from the observed energy spectra to give the net helium scattering.

3. PHONON DISPERSION UNDER PRESSURE

A detailed knowledge of the dispersion curve is an important prerequisite for an understanding of the properties of liquid ^4He (thermodynamics, interactions between elementary excitations, . . .). In particular, the important role in the kinematics of phonon decay, played by the precise form in the phonon region ($Q \lesssim 1.0\,\mathring{A}^{-1}$) has often been emphasised[13]. For 3-phonon decay processes to be kinematically permitted, the phonon phase velocity ω/Q must <u>exceed</u> the sound velocity C_0 for wavevectors less than some critical value Q_c where the dispersion becomes "negative" or normal. Neutron measurements[1,14,15] were consistent with this "anomalous dispersion" but the first clear demonstration[6] was made using the IN12 spectrometer. These results, at saturated vapour pressure, are presented in Fig. 2a as a conventional excitation energy-wavevector dispersion curve, and in Fig. 2b as phonon phase velocities.

This high-resolution study has now been extended to consider the effects of pressure. Heat pulse experiments[16] have shown, indirectly, that as the pressure is increased and the sound velocity increases the region of anomalous dispersion shrinks with the dispersion becoming normal at about 20 bars. The results of Fig. 3a, for applied pressures of 5, 15 and 20 bar, are in good agreement with these of Svensson et al[14] and confirm, at least qualitatively, the predictions of the polarisation potential calculations of Aldrich et al[17], presented in Fig. 3b. The major uncertainty in these new data lies in the absolute value of the pressure and hence of the relevant sound velocity $C_0(P)$.

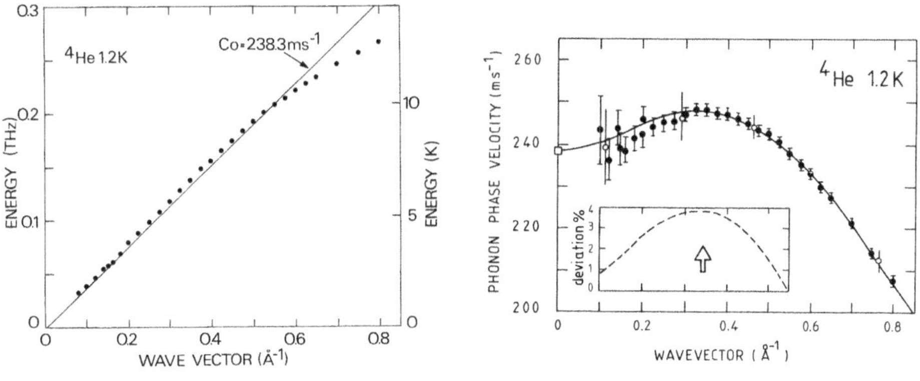

(a) **(b)**

Fig. 2. (a) Phonon dispersion curve of He at 1.2K compared with the sound velocity C_o[18].

(b) Phonon phase velocity[6] of ^{4}He at 1.2K compared with a fit to expression (1) of Maris[13] (full line). The insert shows the deviation (anomalous dispersion) from normal linear dispersion.

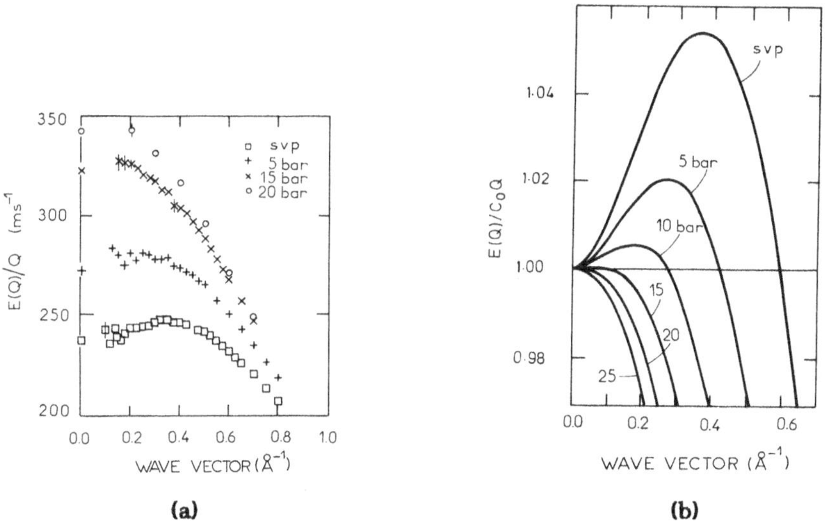

(a) **(b)**

Fig.3. (a) Pressure-dependence of the phonon phase velocity of ^{4}He at 1.2K.

(b) Calculated phonon phase velocity as a function of pressure showing the disappearance of the anomalous dispersion (after Aldrich et al[17])

Many functional forms have been proposed to describe the phonon region. Maris[13] used data from several techniques, including existing neutron scattering results, to estimate the values of the parameters in his polynomial expression

$$\hbar\omega(Q) = C_o Q \left[1 + \gamma Q^2 \quad \frac{1 - Q^2/Q_A{}^2}{1 + Q^2/Q_B{}^2}\right] \tag{1}$$

This function has been used to fit the data of Fig. 3a and the results are compared with the s.v.p. parameters of Maris[13] in Table 1 of the appendix. The sound velocities C_o were fixed at the values obtained by Maynard[18] ; clearly different sets of parameters are obtained if this restriction is lifted.

4. ROTON PARAMETERS

A precise knowledge of the roton gap energy Δ, the roton wavevector Q_R, and the roton effective mass m^* is important as, above about 1K, the thermodynamic properties of superfluid ^{4}He are dominated by these excitations. In addition, by comparing with the energy of the peak in the two-roton Raman spectrum[19] , the binding energy of a roton pair may be determined from an accurate neutron measurement of Δ. Woods et al[20] made such a measurement some years ago at T = 0.75K and s.v.p. using the IN3 TACS at I.L.L.

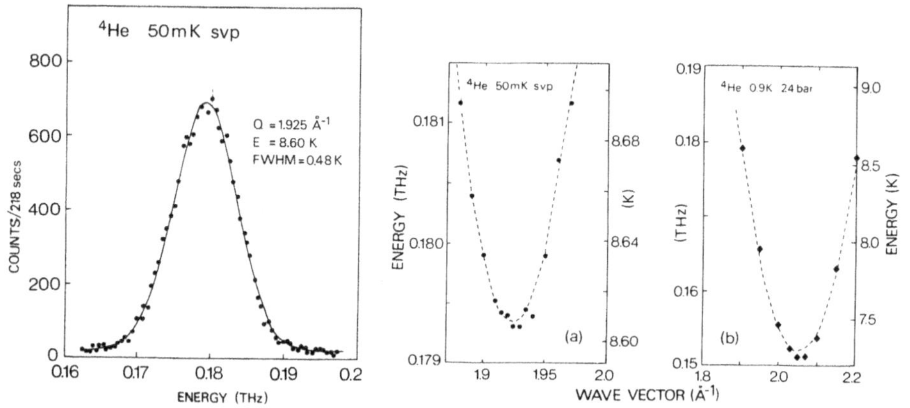

(a) (b)

Fig. 4. (a) High resolution neutron group at the roton wavevector, Q = 1.925Å^{-1}, at T = 50mK. The line is a Gaussian fit.
(b) Roton energies[7] at s.v.p. and T = 50mK, and at 24 bar and T = 0.9K. The lines are fits of a quadratic dispersion curve, Equation (2).

This work has been extended to provide "zero-temperature" roton parameters (actually at T = 50 mK) and at elevated pressures. Long wavelength neutrons were used to give the best possible instrumental resolution with good counting statistics, and particular care was taken in calibration of the instrument to minimise uncertainties in the measured frequencies. A typical low-temperature neutron group is presented in Fig. 4(a). Peak positions, extracted by fitting to Gaussian functions, were then fitted to the parabolic dispersion relation

$$\hbar\omega(Q) = \Delta + \frac{\hbar^2}{2m^*}(Q - Q_R)^2 \qquad (2)$$

Examples of the results of this procedure are shown in Fig. 4(b) for T = 50mK and s.v.p., and for T = 0.9K and p = 24 bar. The parameters obtained in this way are compared in Table 2 of the appendix with previous values[3,20] and with those tabulated by Donnelly et al[21] .

5. STRUCTURE OF THE MULTIPHONON CONTINUUM

Cowley and Woods[1] recognised that, in addition to the sharp phonon-roton excitations, there was scattering at higher energies. This "second-branch" or "multiphonon continuum" has continued to be of interest as it contains information on the interactions between the elementary excitations and, in principle, the effective roton-roton coupling can be deduced from consideration of the complete $S(Q,\omega)$, both one-phonon and multiphonon parts. A detailed investigation of the nature and form of the continuum scattering has been reported by Svensson et al[22] . An important result of this work was the realisation that the continuum was not simply a featureless "bump", but that it had a complicated shape with a tail extending to high frequencies.

The higher instrumental resolution possible at the I.L.L. has allowed us to observe clear structure in the multiphonon continuum which can be correlated with pairs of excitations. Two experiments have been performed. In the first[7], structured continuum scattering was observed for wavevectors up to $1.5Å^{-1}$. As shown in Fig. 5(a), at the maxon wavevector, a sharp peak at around 0.45 THz (21.6K) is observed at s.v.p.; this feature occurs at a somewhat lower energy at $Q = 1.5Å^{-1}$ (Fig. 5(b)). These peaks correspond with the energy expected for a roton-maxon pair. This agrees well with the calculations of Manousakis and Pandharipande[23] who show that the two-phonon density of states is strongly peaked at about 23K; this peak arises from one roton plus one maxon. As the pressure is increased, in general the experimental continuum scattering becomes smoother, although a sharp feature persists at $Q = 1.5Å^{-1}$ (Fig. 5(b)).

A more recent experiment[8] , with improved experimental conditions, has confirmed these observations. As shown in Fig. 6, at $Q = 1.13Å^{-1}$ there are two maxima close to the sum of the roton and maxon energies (R + M) and twice the maxon energy (2M). At higher Q, the 2 peaks are at about 2R and twice the maxon energy (2M), while at the highest wavevectors examined the spectrum becomes essentially a single wide peak. In general, there is good qualitative agreement with theoretical predictions[23].

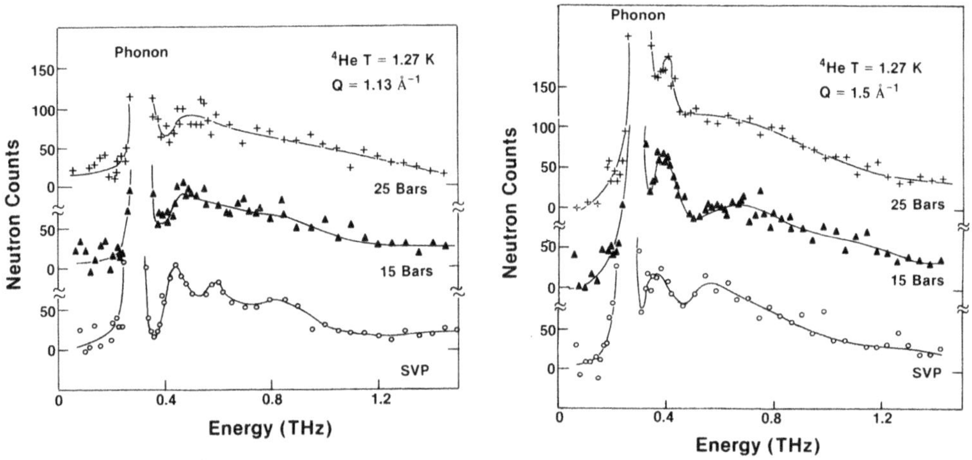

(a) (b)

Fig.5. (a) Structure of the multiphonon continuum at the maxon wavevector $Q = 1.13Å^{-1}$ as a function of pressure[7]. The lines are guides to the eye.

(b) As 5(a) at $Q = 1.5Å^{-1}$.

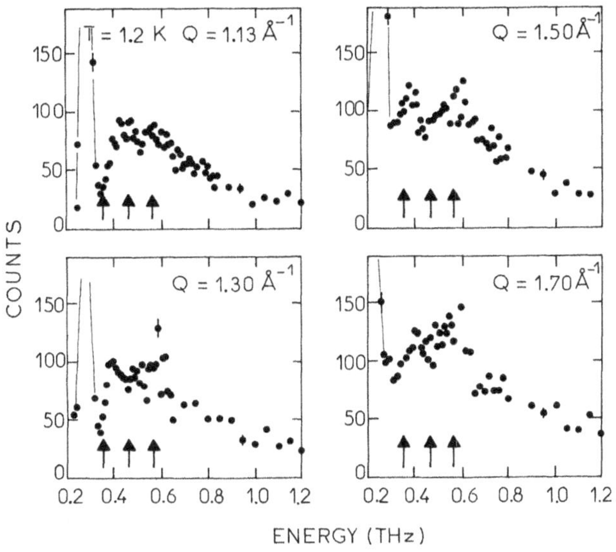

Fig. 6. Q-dependence of the multiphonon continuum scattering[8]. The arrows indicate twice the roton frequency (2R), the sum of the roton plus maxon frequencies (R + M), and twice the maxon frequency (2M).

6. TEMPERATURE DEPENDENCE OF S(Q, ω)

Probably the most exciting problem remaining in the microscopic theory of liquid ^{4}He is the relationship between superfluidity and the excitation spectrum. Are the sharp phonon-roton excitations unique to the superfluid phase (and thus associated with the Bose condensate), or do they arise simply because ^{4}He is a very cold liquid? The most important previous attempt to address this outstanding question is due to Woods and Svensson[24] (WS). Their results indicated that S(Q, ω) changes qualitatively on passing from the superfluid to the normal phase. They compared the data with a model in which S(Q, ω) is decomposed into two components, a superfluid part weighted by the superfluid density $\rho_S(T)/\rho$, and a normal fluid part weighted by the normal fluid density $\rho_N(T)/\rho$. Thus the sharp one-phonon part disappears at the lambda point. This model provides a remarkably good representation of the experimental scattering function, but there are systematic differences between theory and experiment[9,10]. It also appears that there is little a priori theoretical justification for the suggested decomposition.

To investigate further the behaviour of S(Q, ω) near T_λ, two recent experiments[9,10] have been performed using the IN12 TACS. High precision measurements have been made of the temperature dependence of S(Q, ω) at s.v.p.[10] in the phonon (Q = 0.4Å^{-1}) and roton (Q = 1.925Å^{-1}) regions. Fig. 7 presents an example of the raw data at Q = 0.4Å^{-1}. As expected, the sharp (resolution-limited) one-phonon peak is accompanied by a weak, broad multiphonon distribution. As the temperature is raised, the phonon peak broadens with very little shift in energy. Above T_λ (2.17K), the peak continues to broaden, but remains a well-defined (inelastic) excitation, as seen from Fig. 8(a). So there is little sign of any discontinuity in S(Q, ω) at or near T_λ at this wavevector.

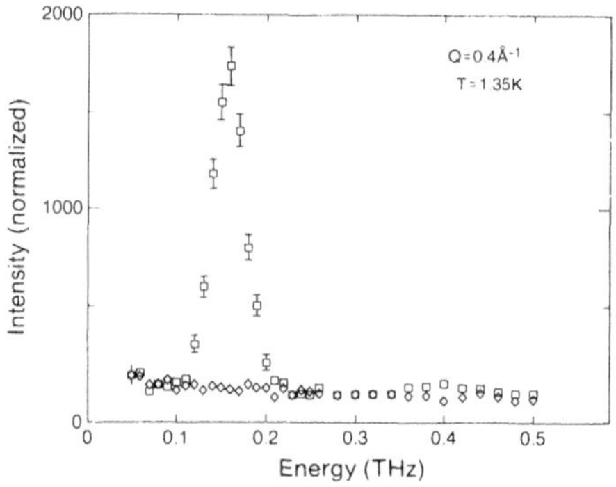

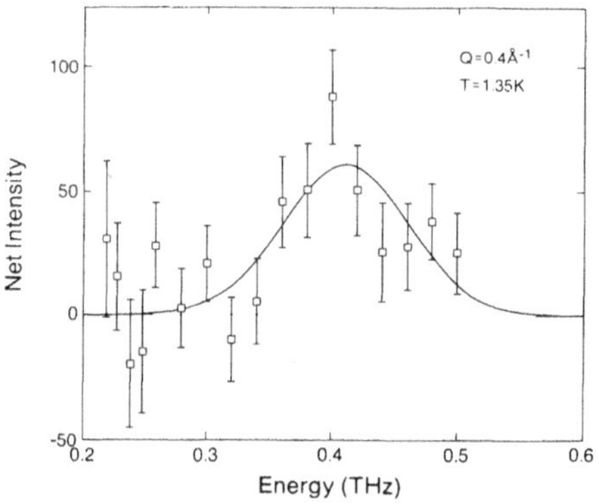

Fig. 7. (upper) High resolution energy spectrum[10] at $Q = 0.4Å^{-1}$; diamonds empty-cell; squares, cell plus liquid ^{4}He at s.v.p.
(lower) Multiphonon part of the spectrum. The line is a fitted Gaussian function.

The roton results are strikingly different. The very sharp one-roton peak at 0.18 THz (8.6K) broadens very rapidly as observed in previous studies[25] with a particularly rapid increase close to T_λ. There is, however, a marked change in the shape of the scattering function $S(Q,\omega)$ just below T_λ, as shown in Fig. 8(b). Clearly, the spectra for T = 2.09K and 2.17K are qualitatively different. The

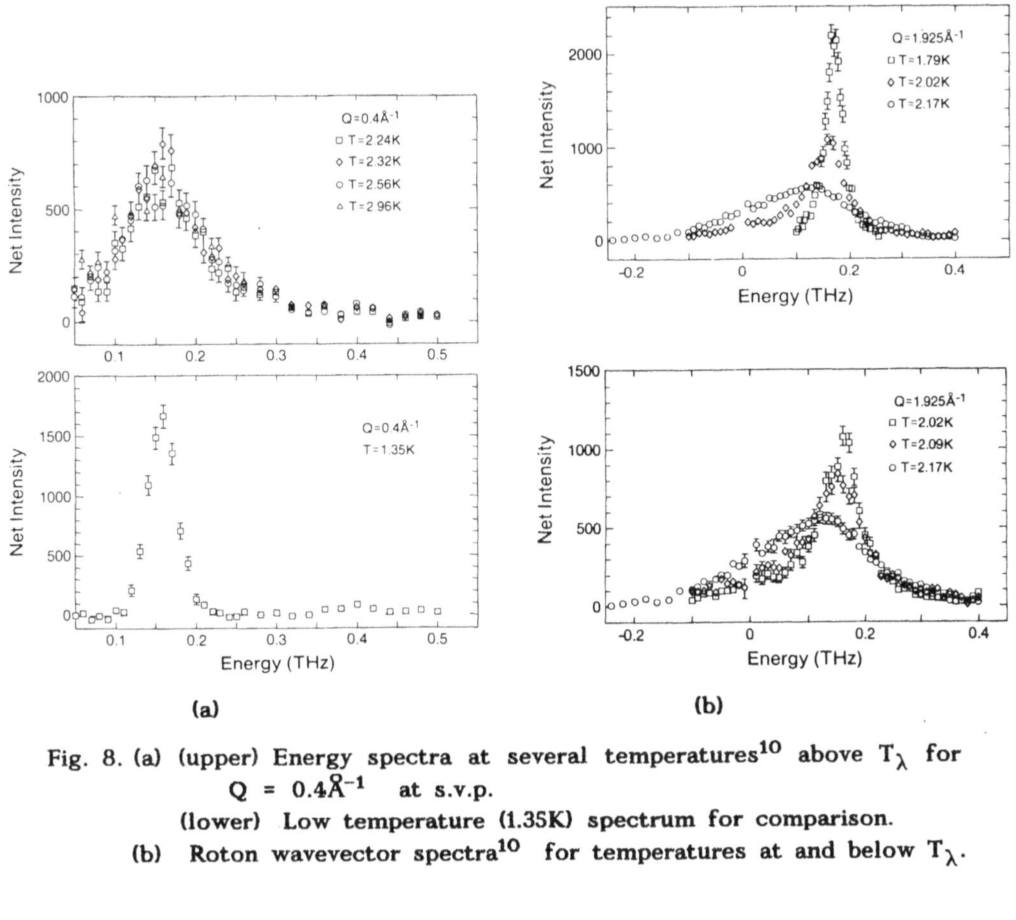

(a) (b)

Fig. 8. (a) (upper) Energy spectra at several temperatures[10] above T_λ for $Q = 0.4 Å^{-1}$ at s.v.p.
(lower) Low temperature (1.35K) spectrum for comparison.
(b) Roton wavevector spectra[10] for temperatures at and below T_λ.

inelastic spectra for the superfluid phase are much more sharply peaked than those above T_λ. Thus it appears that a sharp feature of $S(Q,\omega)$ disappears at T_λ. In the normal phase, the spectrum continues to broaden with a significant decrease in the excitation frequency. So, in contrast to the situation at the phonon wavevector, the shape of the inelastic scattering at the roton minimum is quite different in the two phases.

To obtain the temperature variation of the excitation energies and widths (inverse lifetimes), the scattering function was expressed as the sum of a one-phonon part and a multiphonon part

$$S(Q,\omega) = S_1(Q,\omega) + S_M(Q,\omega) \tag{3}$$

The $S_M(Q,\omega)$ at $Q = 0.4Å^{-1}$ was modelled as a simple Gaussian function, fitted to the lowest temperature data (Fig. 7), with the appropriate Bose factor. The weak multiphonon spectrum at the roton wavevector was modelled as a small temperature-independent "background"; the final fitted parameters were largely independent of the value of this component. As in the work of Talbot et al[9], the one-phonon scattering function $S_1(Q,\omega)$ was represented by a pair of Lorentzian functions (neutron energy loss and gain) corresponding to excitation creation and annihilation. Thus

$$S_1(Q,\omega) = \frac{1}{2\pi}\left[n_B(\omega) + 1\right] Z(Q,T)\; A_1(Q,\omega) \tag{4}$$

where $n_B(\omega)$ is the Bose factor at temperature T and frequency ω and $Z(Q,T)$ is the one-phonon intensity. The one-phonon response function $A_1(Q,\omega)$ is then given by

$$A_1(Q,\omega)\,/\,2 =$$

$$\left[\frac{\Gamma(Q,T)}{\{\omega - \omega(Q,T)\}^2 + \Gamma(Q,T)^2} - \frac{\Gamma(Q,T)}{\{\omega + \omega(Q,T)\}^2 + \Gamma(Q,T)^2}\right] \tag{5}$$

where $\Gamma(Q,T)$ is the half width at half maximum (HWHM). The function of equations (3) to (5) was convoluted with a Gaussian function, obtained by fitting to the lowest temperature spectra, to allow for the effects of instrumental resolution. However we note that, in the light of subsequent theoretical developments[28], the application of this simple model to larger wavevectors such as the roton is probably not particularly meaningful at temperatures above about 1.8K.

Values of the one-phonon frequency $\omega(Q,T)$, half width $\Gamma(Q,T)$ and intensity $Z(Q,T)$ are plotted in Fig. 9 for $Q = 0.4Å^{-1}$. Although the phonon width increases rapidly from about 1.8K, there is little sign of any "jump" at T_λ ; the phonon frequency and intensity are likewise little affected by the transition. However there is a much greater increase in linewidth at the roton wavevector. At low temperature both phonon and roton lines are extremely narrow ($<$ 0.01THz = 0.5K ; HWHM) whereas near T_λ the roton linewidth is some 3 times wider than that at $Q = 0.4Å^{-1}$.

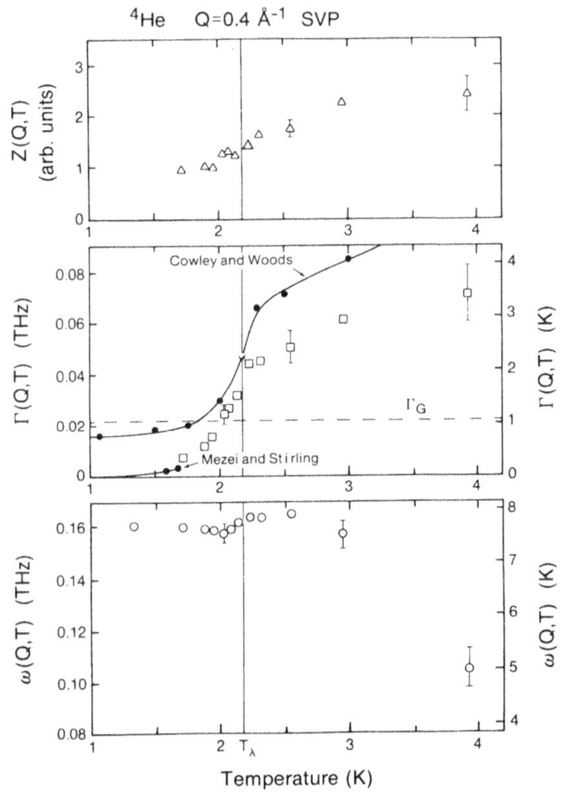

Fig. 9. Energy, halfwidth and intensity parameters[10] for Q = 0.4Å^{-1} s.v.p. data. Previous values of Cowley and Woods[1] (without allowance for instrumental resolution) and of Mezei and Stirling[27] are shown for comparison. The dashed line (Γ_G) gives the resolution half-width.

So, in summary, at s.v.p. the phonon peak broadens but remains well-defined above T_λ and the transition does not affect $S(Q,\omega)$ in any obvious way. On the other hand, the roton spectrum changes significantly near T_λ with the apparent disappearance of the sharp (superfluid) component.

As well as these s.v.p. experiments, Talbot et al[9] have studied the temperature variation of $S(Q,\omega)$ at an applied pressure of 20 bar. Under pressure, the excitation frequencies are shifted and the values of $\rho_S(T)$ and $\rho_N(T)$ are changed from their s.v.p. values, providing a valuable check of the Woods - Svensson[24] hypothesis. Figs. 10 and 11 present inelastic spectra at the maxon (1.13 Å^{-1}) and roton (2.03Å^{-1} at 20 bar) wavevectors. It is evident that the maxon spectrum has both a sharp and a broad component, the former of which disappears at, or just below, T_λ. In the normal phase (Fig. 11) the shape of the energy distribution is essentially unchanged up to 4K. Similar conclusions may be drawn from the roton spectra. The sharp roton peak decreases rapidly in intensity as the temperature approaches T_λ (Fig. 10). A small sharp component, visible in Fig. 11 at 1.90K, is absent at 1.93K (T_λ = 1.928K) and above. Figure 12 presents a two-dimensional representation of the scattering function $S(Q = 2.03\text{Å}^{-1}, \omega, T)$; the sharp roton peak broadens and weakens towards T_λ where there is a sudden change of form to a largely temperature-independent normal fluid component.

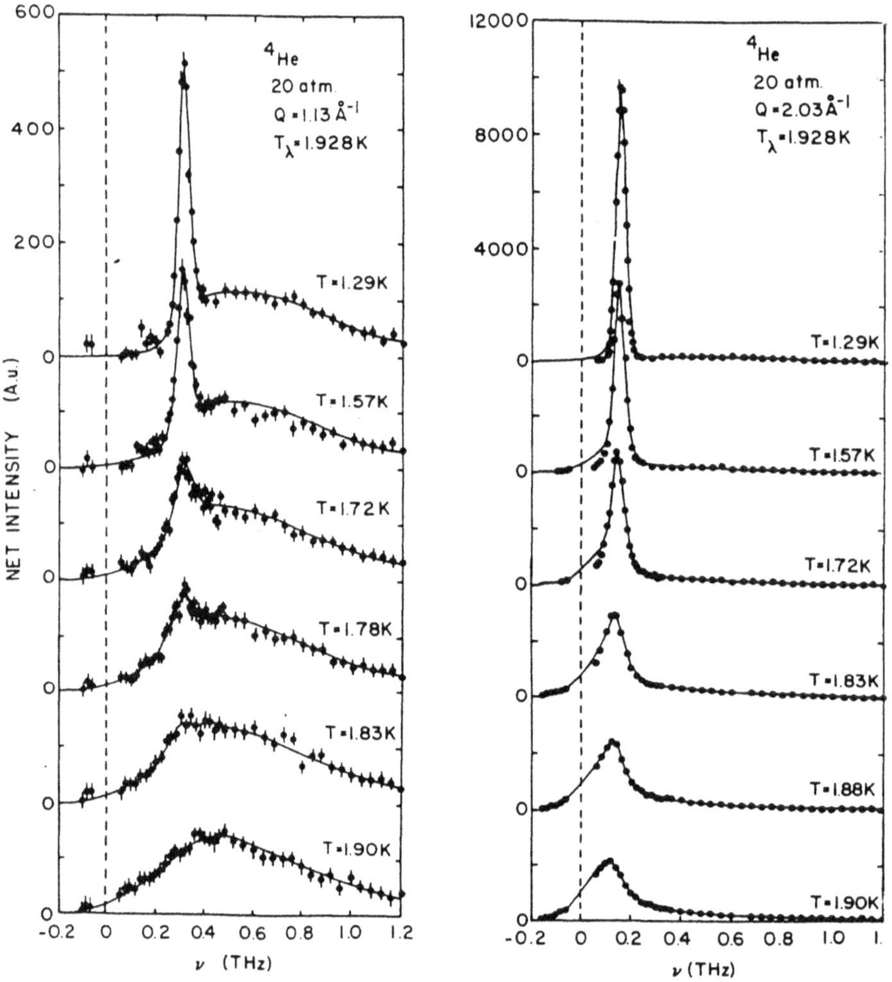

Fig. 10 Energy spectra[9] below T_λ at 20 atm for the maxon ($Q = 1.13\text{Å}^{-1}$) and roton (2.03Å^{-1}) wavevectors.

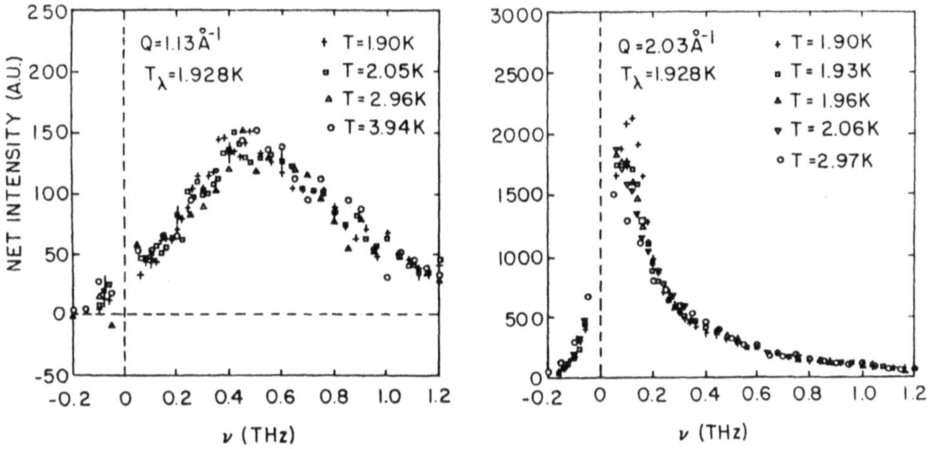

Fig. 11. **As Fig.10 for temperatures close-to and above T_{λ}.**

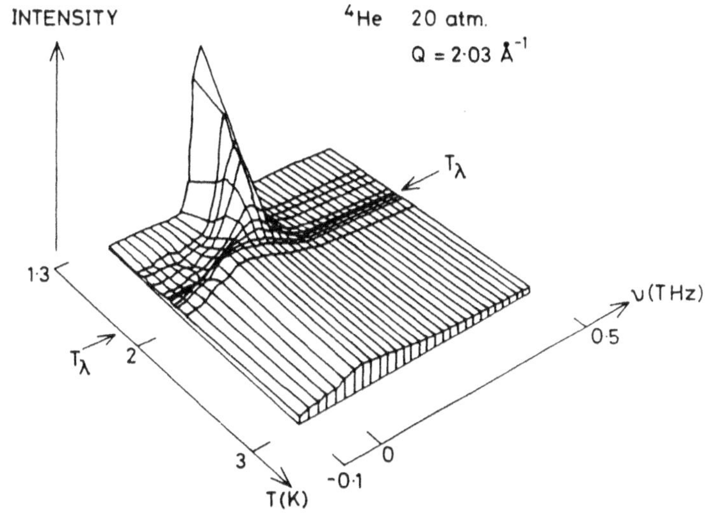

Fig. 12 Three dimensional representation of the scattering function[9] $S(Q = 2.03Å^{-1}, \omega, T)$ at the roton wavevector (P = 20 bar).

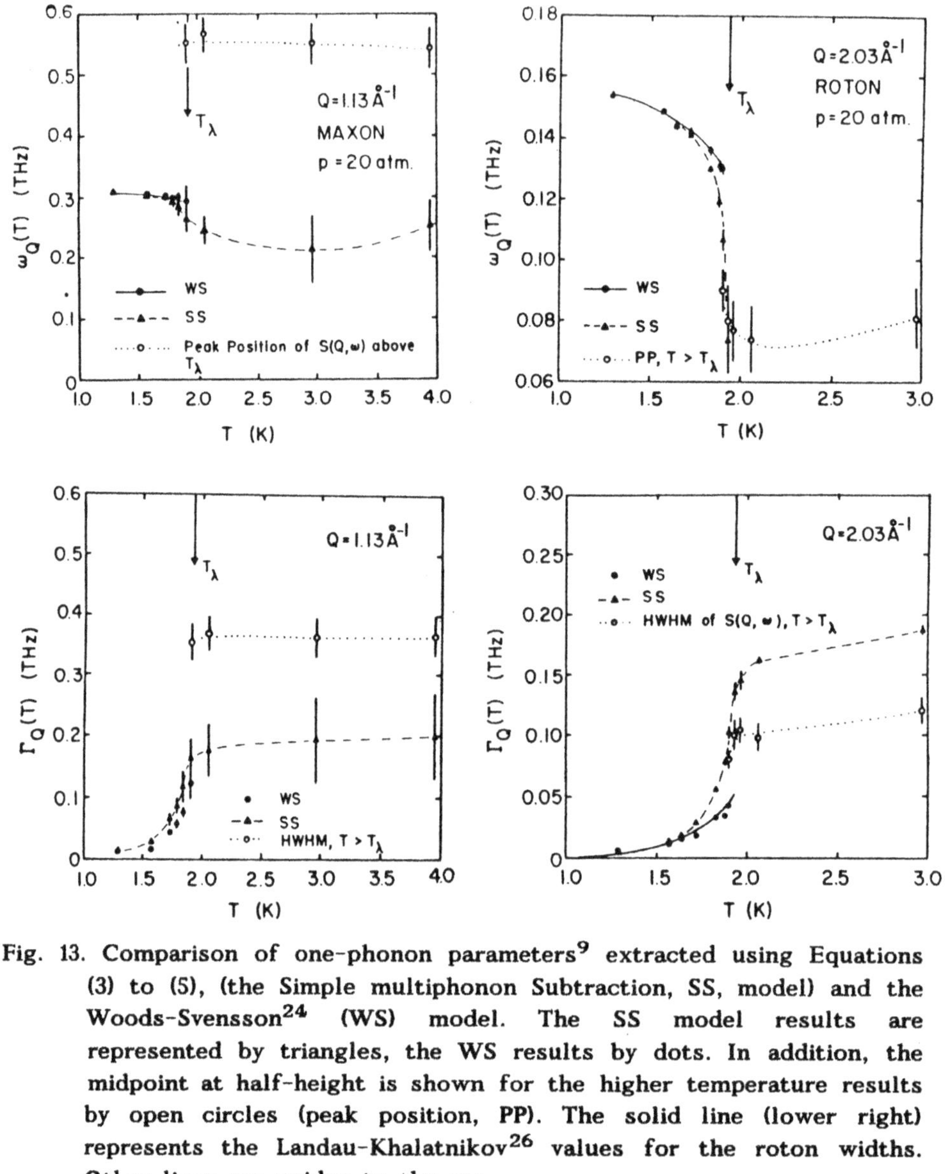

Fig. 13. Comparison of one-phonon parameters[9] extracted using Equations (3) to (5), (the Simple multiphonon Subtraction, SS, model) and the Woods-Svensson[24] (WS) model. The SS model results are represented by triangles, the WS results by dots. In addition, the midpoint at half-height is shown for the higher temperature results by open circles (peak position, PP). The solid line (lower right) represents the Landau-Khalatnikov[26] values for the roton widths. Other lines are guides to the eye.

These data at 20 bar were analysed as described above (Equations 3-5) and also following the model of Woods and Svensson[24]. For the maxon, both methods yield very similar frequencies below T_λ (Fig. 13), largely independent of temperature, with the WS one-phonon component vanishing above the transition, by definition. In contrast, the roton frequency decreases significantly below T_λ, leaving a very low peak frequency (PP) above. The linewidths at both wavevectors increase through the transition and are less strongly dependent on temperature in the normal phase. Despite differences in the parameters obtained using the two models for $S(Q,\omega)$, there is one clear conclusion which can be drawn from the experimental results of Talbot et al[9] : there is a definite change in the one-phonon contribution to the scattering function at or near T_λ for the maxon and roton wavevectors. This is in agreement with the s.v.p. roton results of Stirling and Glyde[10], but in marked contrast to what occurs at the phonon wavevector where there is little obvious effect on passing through the super-to-normal fluid transition.

7. PRECISION DETERMINATION OF $S(Q,\omega)$ AS A FUNCTION OF T AND P

To answer requests for accurate tabulated scattering function data[21] at a series of temperatures and pressures, such as exist for related thermodynamic properties[18], an ambitious programme of measurement[11] has been initiated at the I.L.L., using the TOF spectrometer IN6. The ultimate aim is to provide scattering function information of high statistical accuracy with precision in energy and wavevector of order 0.1% and 0.5% respectively. As a first step in this procedure, extensive data have been collected on $S(Q,\omega)$ at s.v.p., under experimental conditions outlined in Section 2. Since data is collected with varying excitation energy <u>and</u> wavevector, an interpolation procedure is necessary to obtain the scattering function at constant <u>Q</u>. This new data represents an important contribution to the questions discussed in the previous section as wavevectors from 0.4 to 1.8Å^{-1} have been examined simultaneously and at several temperatures. In advance of publication, some preliminary results are presented here.

Three wavevectors will be discussed here ; Q = 0.6 in the phonon region, Q = 1.0 near the maxon and Q = 1.8Å^{-1} near the roton minimum. Figures 14(a) and 15(a) (on an expanded scale) compare the spectra for T = 1.24K, 1.50K, 2.07K and 2.26K. For Q = 0.6, at the two lowest temperatures the single phonon and multiphonon peaks are clearly separated; the two components become essentially indistinguishable at the highest temperatures. As in the TACS experiments described above[10], $S(Q,\omega)$ broadens smoothly with little obvious discontinuity on passing T_λ. The lower temperature data at Q = 1.0 clearly display structure in the multiphonon continuum as discussed in Section 5. At this Q value the spectra for T = 2.07K and 2.26K look very similar except in the region of the phonon peak where significant intensity has disappeared above T_λ. The much more intense one-phonon peak dominates the spectrum at Q = 1.8. After initially broadening and decreasing in height there is a major change in the shape of the spectrum between 2.07K and 2.26K.

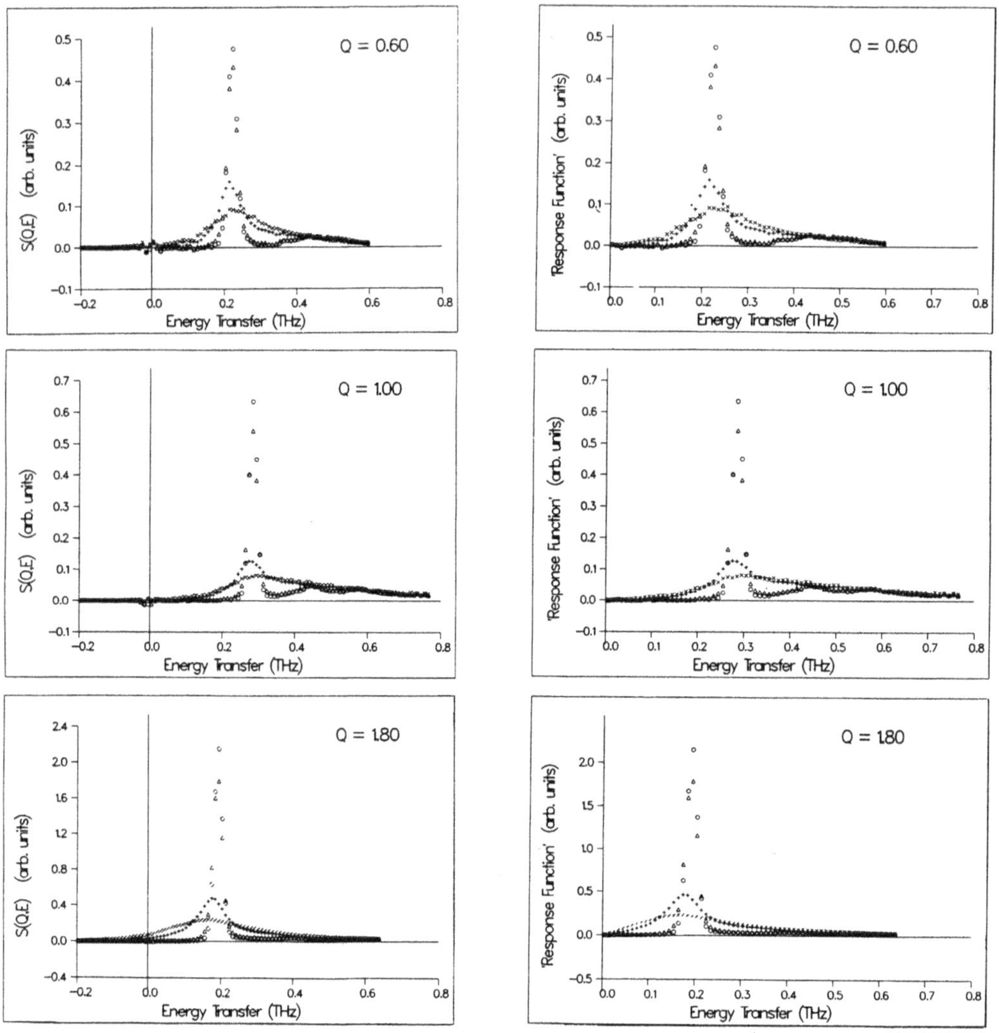

(a) **(b)**

Fig. 14 (a) **Time-of-flight data for ^{4}He at s.v.p. (preliminary data of Andersen et al[11]) for Q = 0.6, 1.0, and 1.8 $\AA^{-1}$ at several temperatures.**

T = 1.24K (o), T = 1.50K (Δ), T = 2.07K (+), T = 2.26K (X).

(b) **The one-phonon response function $A_1(Q,\omega)$ as defined in Equations 3 and 4 .**

T = 1.24K (o), T = 1.50K (Δ), T = 2.07K (+), T = 2.26K (X).

Only the positive energy part is displayed.

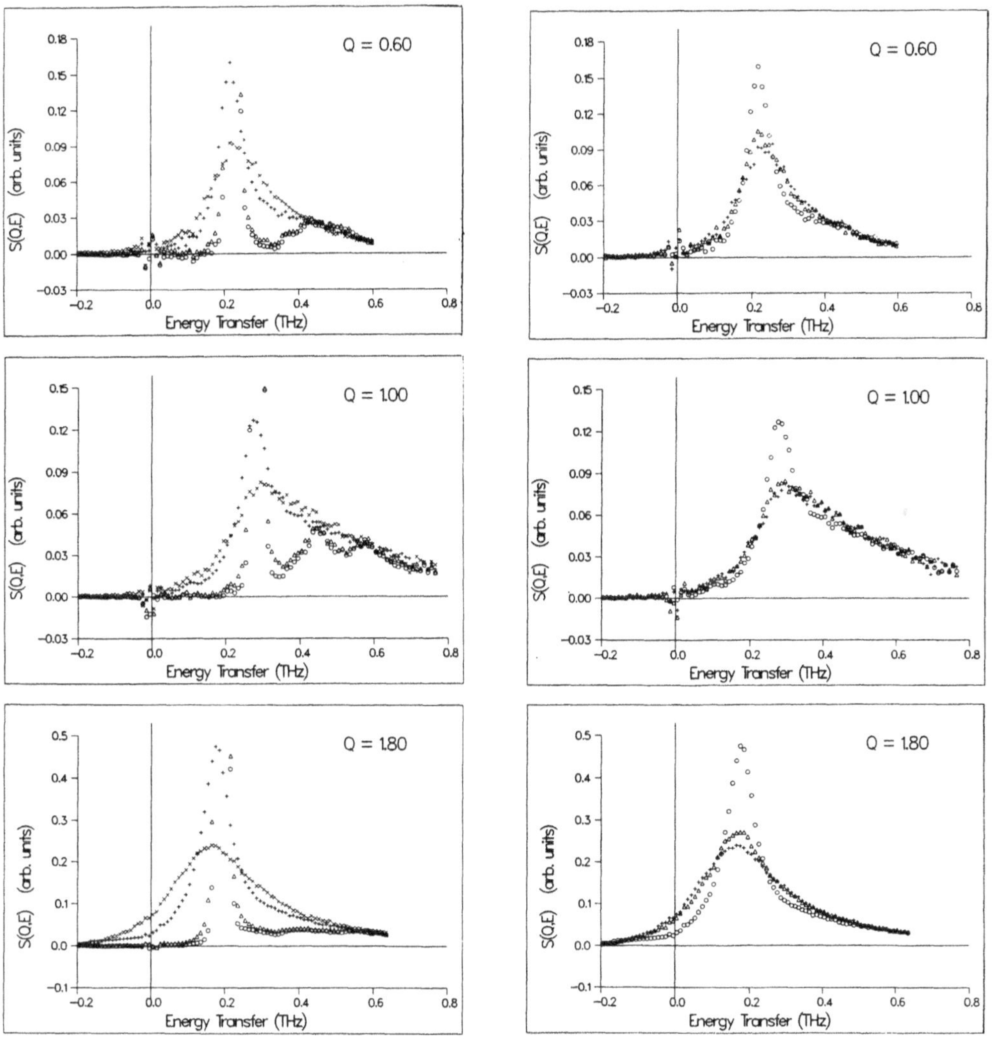

(a) (b)

Fig. 15.(a) As Fig. 14 (a) on an expanded scale
T = 1.24K (o), T = 1.50K (Δ), T = 2.07K (+), T = 2.26K (X).
(b) As Fig. 14(a) for temperatures near T_λ.
T = 2.07K (o), T = 2.17K (Δ), T = 2.26K (+).

Changes in the energy spectra close to T_λ are emphasised in Fig. 15(b) where the data for T = 2.07K, 2.17K ($T = T_\lambda$), and 2.26K are presented. A common feature at all 3 wavevectors is the larger change between T_λ - 0.1K and T_λ than between T_λ and T_λ + 0.1K. This change is qualitatively greater at Q = 1.0 and 1.8 than at the phonon wavevector of 0.6 Å^{-1}.

It is also instructive to consider the changes occuring in the response function $A_1(Q,\omega)$ of Equation 5, rather than in the scattering function $S(Q,\omega)$: these are related by the Bose (temperature) factor (Equation 4). When due allowance is made for the effect of the factor $[n_B(\omega) + 1]$, the results of Fig. 14(b) are obtained. This figure is directly comparable to Fig. 14(a). The temperature factor has little effect on the spectra at Q = 0.6 and 1.0, but there is a major change in the broad, high temperature response at Q = 1.8, emphasising the softening at higher temperatures.

8. CONCLUSION

Recent studies of the excitations of liquid ^{4}He have exploited increased neutron fluxes and improved instrumental resolution to show up effects hitherto unobserved and to quantify others only qualitatively studied previously. The most important of these new investigations concerns the temperature dependence of S(Q, ω). These new results have led to a reappraisal of theoretical interpretations of the phonon-roton excitations, such as that of Glyde and Griffin[28]. Although space restrictions do not permit a detailed comparison of the neutron observations with theories for ^{4}He, there are a number of experimental properties of S(Q, ω) which must be described by a successful theory :

(1) The existence of a sharp phonon-maxon-roton excitation at low temperatures, with significant anomalous dispersion up to about 0.5$Å^{-1}$.
(2) The coexistence, at higher energies, of a broad continuum response with structure at low temperature mirroring the two-phonon density of states ; at higher temperatures and pressures, this structure is "washed-out".
(3) The disappearance of the sharp component at the lambda point for medium to large wavevectors, while the phonon peak at small wavevectors broadens but persists into the normal phase.

There have been, of course, many theoretical treatments[5] of the ^{4}He excitation spectrum. While some are successful in reproducing some particular feature, to our knowledge none adequately describes all of the above observations. It is hoped that this work will stimulate a productive confrontation between theory and experiment which will lead to a more profound understanding of the excitations in liquid helium.

ACKNOWLEDGEMENTS

The author is grateful to the many colleagues who contributed to this work, with particular thanks to L.K.H. Andersen, H.R. Glyde, D. Puschner, R. Scherm and E.C. Svensson. The support of the Science and Engineering Research Council (U.K.) and of the North Atlantic Treaty Organisation (NATO) is gratefully acknowledged. The Institut Laue-Langevin is thanked for the provision of the experimental facilities used in this work.

REFERENCES

1. R.A. Cowley and A.D.B. Woods, Can. J. Phys. 49: 177 (1971).
2. A.D.B. Woods and R.A. Cowley, Rep. Prog. Phys. 36: 1135 (1973).
3. O.W. Dietrich, E.H. Graf, C.H. Huang, and L. Passell, Phys. Rev. A 5: 1377 (1972).
4. E.H. Graf, V.J. Minkiewicz, H. Bjerrum Møller and L. Passell, Phys. Rev. A10: 1748 (1974).
5. H.R. Glyde and E.C. Svensson, Solid and Liquid Helium, in : "Methods of Experimental Physics, Vol 23B; Neutron Scattering", D.L. Price and K. Sköld, eds., Academic Press Inc. (1987).
6. W.G. Stirling, in: "75th Jubilee Conference on Helium-4", J.G.M. Armitage, ed., p109, World Scientific, Singapore (1983).
7. W.G. Stirling, Proc. Int. Conf. Phonon Phys. 2nd, J. Kollår, N. Kroó, N. Menyhård and T. Siklós, eds., p. 829, World Scientific, Singapore (1985).
8. W.G. Stirling, Proceedings of LT19, Brighton, U.K., August 1990 : to be published.
9. E.F. Talbot, H.R. Glyde, W.G. Stirling and E.C. Svensson, Phys. Rev. B38: 11229 (1988).
10. W.G. Stirling and H.R. Glyde, Phys. Rev. B41: 4224 (1990).
11. L.K.H. Andersen, W.G. Stirling, R. Scherm, A. Stunault, B. Fåk, A.J. Dianoux and H. Godfrin, to be published.
12. V. Sears, Nucl. Inst. and Methods 123: 521 (1975).
13. H.J. Maris, Rev. Mod. Phys. 49: 341 (1977)
14. E.C. Svensson, P. Martel, and A.D.B. Woods, Phys.Lett. 55A: 151 (1975).
15. W.G. Stirling, J.R.D. Copley, and P.A. Hilton, in : "Neutron Inelastic Scattering 1977", Vol. II, p. 45, IAEA, Vienna, (1978).
16. R.C. Dynes and V. Narayanamurti, Phys. Rev. B12: 1720 (1975).
17. C.H. Aldrich III, C.J. Pethick, and D. Pines, J. Low Temp. Phys. 25: 691 (1976).
18. J. Maynard, Phys. Rev. B14: 3868 (1976).
19. C.A. Murray, R.L. Woerner, and T.J. Greytak, J. Phys. C 8: L90 (1975).
20. A.D.B. Woods, P.A. Hilton, R. Scherm, and W.G. Stirling, J.Phys. C10: L45 (1977).
21. R.J. Donnelly, J.A. Donnelly and R.N. Hills, J.Low Temp. Phys. 44: 471 (1981).
22. E.C. Svensson, P. Martel, V.F. Sears, and A.D.B. Woods, Can.J.Phys. 54: 2178 (1976).
23. E. Manousakis and V.R. Pandharipande, Phys. Rev. B 33: 150 (1986).
24. A.D.B. Woods and E.C. Svensson, Phys. Rev. Lett. 41: 974 (1978).
25. F. Mezei, Phys. Rev. Lett. 44: 1601 (1980).
26. L.D. Landau and I.M. Khalatnikov, Zh. Eksp. Teor. Fiz. 19: 637 (1949).
27. F. Mezei and W.G. Stirling, in: "75th Jubilee Conference on Helium-4", J.G.M. Armitage, ed., p111, World Scientific, Singapore (1983).
28. H.R. Glyde and A. Griffin, to be published; and this conference.

APPENDIX

Table 1 Pressure-dependence of the parameters of the phonon dispersion relation.

Pressure	C_o (ms^{-1})	γ (Å^2)	Q_A (Å^{-1})	Q_B (Å^{-1})
s v.p. Maris[13]	238.3	1.112	0.542	0.332
Stirling[6]	238.3	0.978	0.548	0.384
5 bar	272.0	2.618	0.430	0.180
15 bar	323.2	10.70	0.225	0.050
20 bar	343.5	8.23	0.151	0.040

Table 2 Temperature and pressure-dependence of the roton parameters.

T(K)	P(bar)	Δ(K)	Q_R (A^{-1})	m*(^{4}He)
0.05	s.v.p.	8.608(±.01)	1.92(±.002)	0.136(±.005)
0.90		8.605	1.924	0.150
0.75[20]		8.618(±.01)	1.926(±.005)	0.126(±.03)
"Low"[21]		8.616	1.930	0.153
0.90	15.2 (±1)	7.71 (±02)	2.019(±.003)	0.109(±.005)
1.28[3]	15.06	7.66	1.998	0.14
0.90	24.0(±1)	7.30(±.02)	2.051(±.003)	0.112
1.25[3]	24.26	7.08	2.028	0.127

MOMENTUM DISTRIBUTIONS IN LIQUID HELIUM

P.E. Sokol

Department of Physics
The Pennsylvania State University
University Park, PA 16802

W.M. Snow

Intense Pulsed Neutron Source and Materials Science Division
Argonne National Laboratory
Argonne, IL 60439

INTRODUCTION

Liquid ^{4}He is one of the most intensively studied systems in the history of physics. The weak attractive forces between the atoms and their large zero point motion combine to stabilize the liquid phase of helium under its saturated vapor pressure even at absolute zero. The effects of the Bose statistics of the helium atoms are important in the liquid, and at a sufficiently low temperature liquid ^{4}He undergoes a second order phase transition to a superfluid phase. The efforts over many decades to understand the superfluidity of liquid ^{4}He have produced innovative ideas and techniques of broad significance in both theory and experiment. Today liquid helium enjoys a unique status as a model system for the study of several aspects of many body physics. Ironically, it has attained this status despite the fact that a detailed microscopic theory of the liquid does not exist. A glance at the current state of our understanding of one of the simplest properties of liquid ^{4}He, its momentum distribution, will serve to bring this irony into focus.

Bose condensation provides the microscopic basis for our present theoretical understanding[1] of superfluidity in bulk liquid ^{4}He. According to this view, the superfluid phase is described by a macroscopic wave function

$$< \psi >= \sqrt{n_0}\, exp(i\phi) \tag{1.1}$$

The square of the amplitude of $< \psi >$ is the fraction of atoms in the condensate, and its phase is coherent on a macroscopic scale. The gradient of this phase is proportional to the superfluid velocity, the new hydrodynamic degree of freedom which appears in the well-known two fluid model. Quantized vortices arise as a result of the quantization of superfluid circulation required by the single-valued nature of the macroscopic wave function. Even some of the characteristics of the phonon-

Excitations in Two-Dimensional and Three-Dimensional Quantum Fluids
Edited by A.G.F. Wyatt and H.J. Lauter, Plenum Press, New York, 1991

maxon-roton spectrum of elementary excitations may be an indirect consequence of the presence of a condensate.[2] None of these properties of superfluid ^{4}He, however, depends in detail on the exact magnitude of the condensate. Only its existence is necessary. In this way, many of the properties of the liquid which make it a model system are insulated from our ignorance of the condensate and the detailed microscopic physics which determine its size.

The momentum distribution $n(p)$ of superfluid ^{4}He, however, is different in this respect. In the absence of a detailed microscopic theory, the momentum distribution is the only measurable property of the liquid which can reveal the size of the condensate. It is this characteristic of the momentum distribution which makes its measurement an issue of fundamental interest. In a sample of liquid ^{4}He at rest, a condensate would produce a δ-function spike in the momentum distribution centered at $p = 0$ with an intensity proportional to the fraction of atoms in the condensate, n_0.

The most direct method for measuring the momentum distribution in liquid ^{4}He is through the use of deep inelastic neutron scattering (DINS). Deep inelastic scattering experiments are now performed in many fields of physics in an effort to gain information on momentum distributions.[3] The basic idea behind all such measurements is quite simple. In a deep inelastic scattering event, the energy and momentum transferred from the scattering probe to a particle in the target are very high compared to the characteristic energies and momenta of the particles in the system. In this limit, the dynamic structure factor $S(Q,\omega)$ may be related to the momentum distribution through the Impulse Approximation (IA). This direct relation between $S_{IA}(Q,\omega)$ and $n(p)$, coupled with the interest in the Bose condensate, has been a major source of inspiration for the numerous inelastic neutron scattering experiments performed on helium.

Most high resolution measurements to date have been confined to the relatively low Q (4-10 Å^{-1}) neutrons available at reactors. In the most recent of these studies,[4,5] a definite change in the scattering is observed upon entering the superfluid phase. Unfortunately, sizeable deviations from the IA obscure the detailed features of the momentum distribution. Experiments at significantly larger $Q > 50$ Å^{-1} have also been performed at spallation neutron sources.[6,7] However, in these measurements the instrumental resolution was comparable to the intrinsic width of the scattering, making the extraction of information on the momentum distribution difficult. Direct comparisons of past experimental results with the accurate calculations of $n(p)$ now available have shown large discrepancies.[8-11] This lack of agreement between theory and experiment is a clear indication that there is room for improvement on the measurements of momentum distributions in helium. Finally, it is important to realize that no distinct feature characteristic of the superfluid, like a narrow peak in $n(p)$ at $p = 0$, has ever been observed in any of these experiments.

Since the experiments mentioned above were performed, several simultaneous advances have been made. The availability of more powerful computational techniques and facilities for the calculation of $n(p)$ has lead to very accurate theoretical predictions. In addition, accurate theoretical calculations for deviations from the IA, called final state effects, have been performed in both the normal and superfluid phases. Finally, the development of spallation neutron sources has enabled us to obtain high quality, interpretable experimental results with good statistical accuracy. Taken together these advances have made possible, for the first time, a quantitative comparison between theory and experiment for the momentum distribution of liquid helium.

In this paper we will present a selection of recent results from high resolution neutron scattering studies of $S(Q,\omega)$ in liquid ^{4}He in both the normal and superfluid phases. For the first time, we obtain excellent agreement between theoretical calculations of $n(p)$ and neutron scattering data. In addition, we present a new method for estimating the The estimates for the condensate behave as expected on physical grounds as a function of temperature and density.

We begin with a brief review of the current microscopic theoretical studies of the momentum distribution in liquid ^4He. Complete and detailed descriptions of the theoretical work have recently appeared elsewhere.[12–15] Therefore, we shall limit our discussion to the most recent examples.

Liquid helium provides a very challenging system for theoretical studies. The attractive part of the helium potential is quite weak, as evidenced by the stability of the liquid even at T=0. This leads to a large overlap of the atomic wavefunctions which make quantum statistics an important factor in determining the properties of the ground state. However, while the attractive interactions are quite small the hard core repulsion between atoms is very large. This makes liquid helium a very strongly interacting system, in some sense even more strongly interacting than nuclear matter.[12] Therefore, it has been extremely difficult to develop a comprehensive microscopic theory of the dense liquid phase using perturbative techniques.[16] Some specific results can be obtained using analytic techniques. For example, a major success of these theories is the prediction of the relationship[2] between the broken Bose symmetry that characterizes the ground state when a condensate is present and the superfluid properties. In addition, it is known that the presence of a condensate induces a weak singularity in the momentum distribution of the non-condensate atoms. In general, though, it has not been possible to develop a comprehensive microscopic theory of the ground or excited states of the liquid.

In view of the difficulty associated with analytic theories, the most detailed and comprehensive results for $n(p)$ have come from numerical calculations. Variational[13] and Greens Function Monte Carlo (GFMC)[15] techniques have been applied extensively to calculating the properties of the ground state at T=0. These first principles calculations, using realistic potentials, have obtained good agreement with several measured properties for the ground state of the liquid, such as the total energy and static structure of the liquid.

The most prominent feature in the ground state $n(p)$ is the appearance of a condensate, a δ function singularity at $p = 0$. Fig. 1 shows two recent calculations of $n(p)$ in the ground state, using GFMC[15] and variational[13] techniques. Both calcula-

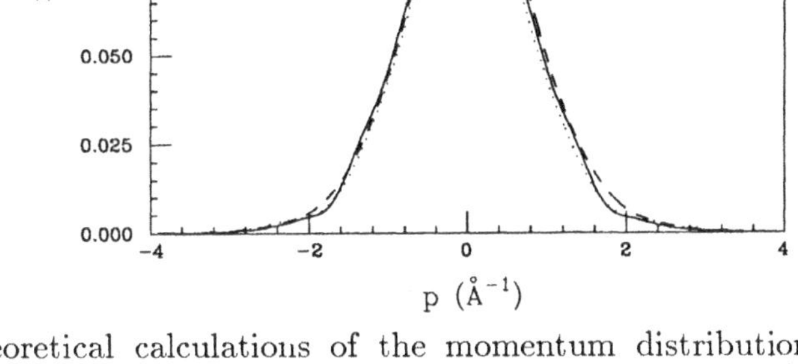

Fig. 1. Theoretical calculations of the momentum distribution. The ground state momentum distributions have been calculated using GFMC[15] (solid line) and variational[13] (dotted line) techniques. The momentum distribution in the normal liquid (dashed line) has been calculated using PIMC[14] techniques.

tions show a condensate δ function (not visible in the variational calculation). The intensity in this δ function, which is known as the condensate fraction n_0, is ≈ 9 % of the total intensity.

In a non-interacting gas, where the wavefunctions of the particles are free particle states and p is a 'good' quantum number, n_0 is just the fraction of atoms in the zero momentum state. This is not true in the liquid where the interactions modify the states so that they are no longer free particle states and p is no longer a 'good' quantum number. Every particle in the liquid now participates in the condensate and it is not appropriate to speak of a particular fraction of the atoms occupying the condensate, or $p = 0$ state. In the interacting liquid, the appearance of the condensate signals the development of off-diagonal long range order (ODLRO) at the one-particle level.[17,18] The condensate fraction is the magnitude of the microscopic order parameter for this phase.

The remainder of the momentum distribution, excluding the condensate δ function, is known as the uncondensed momentum distribution and exhibits several interesting features. At small p, which corresponds to long range interactions among the atoms, the effects of statistics are the dominant factor. For example, singular behavior[19,20] due to the coupling of the condensate to long wavelength collective excitations (phonons) appears. This singular behavior can be obtained exactly at small p where the phonons are well defined and goes as $1/p^2$ in the collisionless regime and $1/p$ in the hydrodynamic regime. The variational $n(p)$ in Fig. 1 explicitly shows this behavior at small p. It is not present in the GFMC results, although this is presumably due to the relatively small size of the samples used in the numerical calculations. Alternately, the large p behavior of $n(p)$ is determined primarily by the short range repulsive interaction between atoms and the statistics play little role.

The great majority of the numerical studies of the liquid have concentrated on the ground state due to the intrinsic limitations of the GFMC and variational approaches. However, Path Integral Monte Carlo (PIMC)[14] methods have recently been applied to study the liquid properties at finite temperatures. These calculations yield similar results to the ground state calculations at low temperatures i.e. a condensate fraction of approximately 9%. However, they have the distinct advantage that they can provide results at finite temperature. For example, the condensate fraction can be obtained as a function of temperature where a rapid increase of n_0 is observed upon entering the superfluid with very little variation with temperature thereafter.

The momentum distribution may also be calculated in the normal liquid using PIMC. For example, Fig. 1 shows $n(p)$ at 3.33 K, well above the superfluid transition.[14] The momentum distribution is broad and featureless with a nearly Gaussian form, the familiar classical result. The width of the momentum distribution is determined by the quantum zero point motion of the liquid and is much larger than the width expected for classical particles. However, aside from this the shape of the momentum distribution in the normal liquid shows little effect due to quantum statistics.

DEEP INELASTIC NEUTRON SCATTERING

Inelastic neutron scattering at large momentum transfer Q provides the most direct means to obtain information on $n(p)$. In this limit, the scattering is due to single atoms and the final state of the scattering particle is assumed to be a free particle state. This is the well known Impulse Approximation (IA). In the IA, $S(Q,\omega)$ is directly related to the atomic momentum distribution, $n(\vec{p})$

$$\lim_{Q\to\infty} S(Q,\omega) = S_{IA}(Q,\omega) = \frac{1}{\rho}\int n(\vec{p}_i)\;\delta\Big[\omega - E(p_f) + E(p_i)\Big]\frac{d^3\vec{p}_i}{(2\pi)^3} \tag{3.1}$$

where $\vec{p}_i$ is the initial momentum of an atom, $\vec{p}_f = \vec{p}_i + \vec{Q}$ is the final momentum of the recoiling atom, and $E(p) = p^2/2M_{He}$ is its kinetic energy. The delta function in

Equation (3.1) represents the conservation of energy and momentum for the scattering of a neutron from a single atom.

It is also possible to express $S(Q,\omega)$ in terms of a Compton profile $J(Y,Q)$

$$(Q/M_{He})S(Q,\omega) \equiv J(Y,Q) \tag{3.2}$$

$$Y \equiv (M_{He}/Q)(\omega - \frac{Q^2}{2M_{He}}) \tag{3.3}$$

which, in the IA limit, becomes the longitudinal momentum distribution[21]

$$\lim_{Q\to\infty} J(Y,Q) \to J_{IA}(Y) \tag{3.4}$$

where Y is the component of the momentum along the direction of $\vec{Q}$. $J_{IA}(Y)$ is simply related to the momentum distribution

$$J_{IA}(Y) = \frac{1}{(2\pi)^2\rho}\int_{|Y|}^{+\infty} pn(p)dp = \int_{-\infty}^{+\infty}\int_{-\infty}^{+\infty} n(p_x,p_y,Y)dp_x dp_y \tag{3.5}$$

where the z-axis is chosen to lie along the direction of $\vec{Q}$. $J_{IA}(Y)$ exhibits several features which are characteristic of the IA. It is symmetric about $Y=0$ and depends on Q only through the scaling variable Y. In bulk liquid ^{4}He, both of these conditions are well satisfied[22] for Q's greater than 15Å^{-1}.

In principle, the observed scattering at high Q provides direct information on $n(p)$. However, in practice the momentum and energy transfers currently attainable are not sufficient to reach the IA limit. Deviations from the IA, due to interactions of the scattering particle with its neighbors, significantly affect the scattering. These deviations are known as final state effects (FSE) since they modify the ideal plane wave final state of the scattering particle necessary for the application of the IA.

Final state effects have been the subject of several theoretical studies[23] and considerable controversy over both the form and the importance of FSE exists. Recently we have presented[24] a detailed comparison of several theories for FSE in liquid helium. The comparison was performed at a momentum transfer of $Q = 23$ Å^{-1}. We find that, while several theories provide a good description of FSE in the normal liquid phase, most do not accurately describe FSE in the superfluid phase. However, we do find that one of the available theories provides a good description of FSE in both the normal and superfluid phases. This theory, which is due to Silver, represents FSE as a convolution of a final state broadening function with the IA result for $J(Y)$. We shall use this theory later in comparisons of the theoretical predictions with the experimental results.

Given knowledge of FSE corrections, measurements of $J(Y)$ can be used to directly determine $n(p)$. For example, if $n(p)$ is Gaussian then $J(Y)$ is also a Gaussian with the same second moment. Furthermore, the condensate, which appears as the three-dimensional delta function in $n(p)$, is a one dimensional delta function in $J(Y)$.

While there is a direct relationship between $J(Y)$ and $n(p)$, it is important to note that it is not a one-to-one correspondence. Features that are prominent in the momentum distribution may not be prominent in $J(Y)$. To illustrate this consider the two recent calculations of the ground state $n(p)$ discussed earlier and shown in Fig. 1. Both calculations predict a condensate fraction, which appears as a delta function with n_0=9.2 % . Both calculations also predict quite similar behavior at intermediate and large p. However, they differ markedly at small p. The variational calculation exhibits singular behavior due to coupling of long wavelength density fluctuations to the condensate which are not apparent in the GFMC calculation, presumably due to finite size effects in the calculation.

While $n(p)$ for these two calculations is quite different, the corresponding $J_{IA}(Y)$, shown in Fig. 2a, is remarkably similar. The singular behavior, which is the dominant feature in the variational $n(p)$ at small p, is quite small in $J(Y)$. When FSE and instrumental broadening are taken into account the (now small) differences between the predicted scattering for the two calculations all but disappear, as shown in Fig. 2b. The predictions of the two very different $n(p)$'s are now nearly indistinguishable! In principle, the differences between the two different $n(p)$'s we began with are still present in Fig. 2b. In practice, a measurement of the scattering would need fantastically good statistical accuracy to ever hope to observe these differences.

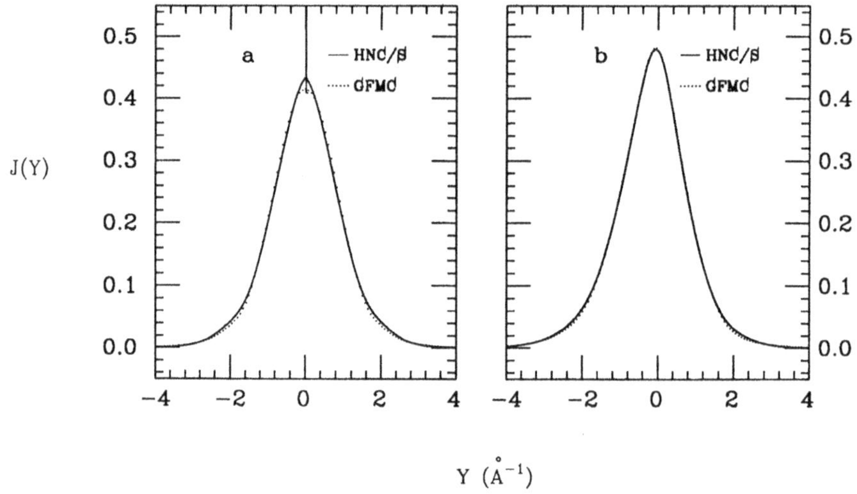

Fig. 2. Illustration of the relationship of $n(p)$ and $J(Y)$. a) shows the $J(Y)$ corresponding to the two ground state $n(p)$'s in Fig. 1. Both calculations have a condensate fraction of 9 %. However, the variational calculation (solid) shows the predicted $1/p$ singular behavior. This is lacking in the GFMC (dotted) result. b) shows the effect of instrumental resolution and FSE on the $J(Y)$ shown in a). The small differences between the predicted scattering for the two different calculations is now almost entirely gone.

This insensitivity to the details at small p is a direct consequence of the IA. The Compton profile, which is proportional to the measured scattering, is the momentum distribution in the direction of the momentum transfer, with the longitudinal components averaged over. Since this is the integral of $pn(p)$ features at small p, such as singular behavior, will be suppressed due to the p in the integrand. Alternately, features at large p will be enhanced, for exactly the same reason.

In principle, we can invert this procedure and obtain $n(p)$ directly from $J(Y)$. In practice, converting from an experimentally determined $J(Y)$ to $n(p)$ presents several difficulties. First, the instrumental resolution and final state effect broadening must be removed. Deconvolution, particularly on data with statistical noise, is an unstable and ill-defined procedure. In addition, numerical differentiation substantially increases the error associated with the data and can only be avoided if the data is smoothed or the results are otherwise biased in some way. Finally, in order to obtain $n(p)$ the differentiated data must be divided by Y, and the results of this division are particularly susceptible to any statistical fluctuations at small Y.

Because of these difficulties, we have chosen not to convert the results of our measurements to $n(p)$. Instead, we first convert the theoretical predictions for the momentum distributions to $J(Y)$. We then convolute the theoretical predictions with both the instrumental resolution function and the FSE broadening function and compare the result directly to the data. In this way, the errors in the data due to counting statistics can be used to place meaningful constraints on the shape of the underlying momentum distribution, and all of the problems mentioned above are avoided.

COMPARISON TO THEORY

Most previous measurements of liquid helium aimed at extracting information on $n(p)$ have been carried out using reactor based instruments[25-30] where the maximum momentum transfers obtainable, with reasonable resolution, are on the order of 5-15 $Å^{-1}$. Unfortunately, obvious deviations from the IA predictions, such as asymmetry in the peak and oscillations in the width as Q is varied, are quite significant at these Q's. In addition, accurate theoretical descriptions of FSE, particularly in the superfluid, are lacking at these low Q's.

In this paper we will concentrate on recent measurements carried out using spallation neutron sources. These sources, with their high flux in the epithermal region, allow measurements at much larger Q's with relative resolutions comparable to the lower Q measurements at reactors. The experiments carried out at higher Q's have the advantage that FSE are more amenable to theoretical treatment.

To illustrate the results obtainable at spallation sources we will discuss recent measurements using the high resolution PHOENIX spectrometer at the Intense Pulsed Neutron Source at Argonne National Laboratory. We have completed an extensive series of measurements at several temperatures and densities in both the normal and superfluid phases. In this paper, we will present results for a representative sample of the measurements.

Fig. 3 shows the measured scattering at a Q of 23 $Å^{-1}$, converted to $J(Y)$, at a constant density of 0.147 g/cm^3. The scattering shown in Fig. 3 is in qualitative agreement with the predictions of the IA. It is centered at and symmetric about $Y = 0$. In addition, measurements[22] at a variety of Q's have shown that the shape of $J(Y)$ is independent of Q above approximately 15 $Å^{-1}$. Thus, at least at this qualitative level the scattering is well described by the IA.

The scattering in both the normal liquid and the superfluid phase is broad and featureless. In the normal liquid, $J(Y)$ is nearly Gaussian, the classical result. There is a distinct change in the observed scattering upon entering the superfluid phase: the scattering is visibly more peaked near $Y = 0$ at T=2.0 K than it is at T=2.3 K. As the temperature is lowered further the observed scattering sharpens further, but no distinct condensate peak is observed. This behavior of the scattering data is consistent with the appearance of a condensate peak in the superfluid phase which is broadened by instrumental resolution and FSE. However, due to the finite statistical accuracy of the results it is also consistent with a wide variety of momentum distributions, some which do not contain a condensate. Therefore, while the increase in scattering intensity at small Y is consistent with the appearance of a condensate, it does not provide conclusive evidence for a condensate fraction.

As can be seen from Fig. 3, the agreement between theory and experiment is excellent for all of the temperatures. This excellent agreement has also been observed at higher densities in both phases. Singular behavior other than a condensate δ-function could be responsible for the increase in the scattering at small p observed in the superfluid. However, as seen in the comparison with the variational $n(p)$, this would have to be a very singular behavior, more singular than $1/p$, to obtain agreement with the experimental results. Thus, while the experimental results can not rule out a ground state $n(p)$ which does not contain a condensate, they do provide strong evidence for a very narrow feature containing 10±2% of the total area. The

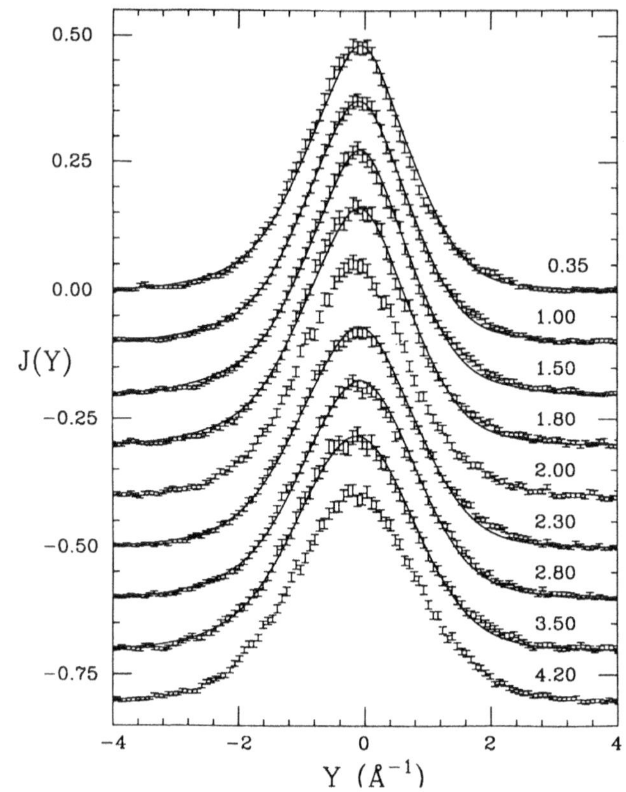

Fig. 3. The measured scattering $J(Y, Q)$ in liquid ^{4}He at a density of 0.147 g/cm^3 for temperatures of 0.32, 1.0, 1.5, 1.8, 2.0, 2.3, 2.8, 3.5, and 4.2 K. The solid lines show the theoretical calculations of momentum distributions using the PIMC[14] and GFMC[15] methods converted to $J(_{IA}(Y)$ and broadened by the instrumental resolution function and the FSE broadening function of Silver[23]. The GFMC calculations are used for comparison with the 0.35 K results, while the PIMC calculations are used, where available, at temperatures above 1 K. The agreement between theory and experiment is excellent in both the normal and superfluid phases.

fact that the scattering data is in excellent agreement with the PIMC calculations and the GFMC calculation which yields a condensate fraction of 9.2 % suggests that this very narrow feature is indeed the Bose condensate first predicted by London.[31]

EXTRACTION OF THE CONDENSATE FRACTION

Due to its importance in understanding the superfluid phase, considerable emphasis has been placed on determining the condensate fraction. However, a direct determination of the condensate fraction has been frustrated by the failure to observe a distinct condensate peak. Since the width of FSE broadening follows the very slow decrease of the He-He cross section for higher Q's, it is very unlikely that a condensate peak could be resolved in any deep inelastic neutron scattering experiment in the foreseeable future. In addition, previous comparisons between the observed scattering and the theoretical predictions were not in agreement, and no value for the condensate could be inferred based on the theories.

The excellent agreement between theoretical calculations of $n(p)$ and the scattering data is strong (though not conclusive) evidence for the presence of a condensate. However, it is important to realize that there are a number of physical systems, such as ^{3}He-^{4}He mixtures and ^{4}He in confined geometries, for which accurate theoretical calculations of $n(p)$ are more difficult obtain. For these systems, it may not be possible to obtain strong evidence for the presence of a condensate based on a direct comparison to theory.

Therefore, it is important to develop alternate techniques to estimate the condensate fraction. These all, in one form or another, involved modeling the momentum distribution in the superfluid to extract a value for the condensate fraction. A wide range of values for n_0 (2-17 %) were obtained, depending on the particular model used. One central feature of all these models was that they implicitly assumed the condensate exists. All the measurements can also be viewed as consistent with some increase in $n(p)$ at small p, but with no condensate.[33]

We will examine one of the techniques to estimate the condensate fraction, due to Sears et. al.[4], which has been used in most recent measurements aimed at extracting the condensate. [4,5,22,28] In the Sears method, the momentum distribution in the superfluid phase is modeled as a delta function due to the condensate, a singular term induced by the condensate, and a uncondensed component which is assumed to have the same shape $n(p)$ in the normal liquid. Using this technique, values for the condensate fraction of $\sim$ 10-13 % , when extrapolated to T=0, were originally obtained. These results were in resonable agreement with theoretical predictions. Recently, however, Griffin[34] pointed out that the original calculation regarding the changes induced in the uncondensed component of the momentum distribution, was incorrect. When the correct form for these changes is used, the inferred values of n_0 are 4-5 % , considerably below the theoretical estimates.

The Sears method suffers from some important limitations. For example, the Sears method assumes that the shape of the nonsingular part of the momentum distribution in the superfluid is the same shape as the full momentum distribution in the normal liquid just above the superfluid transition. Although this appears to be a reasonable assumption, there is no evidence that it is exactly true, and one would expect that it is only approximately true. In addition, the necessity of measuring the momentum distribution in the normal liquid phase introduces other uncertainties. Any errors in the determination of the momentum distribution in the normal liquid will contribute to uncertainties in the determination of the condensate from the superfluid measurement.

We have developed an alternative method for estimating the size of the condensate fraction from deep inelastic neutron scattering data. The basic idea of our approach is to use a model for $n(p)$ which explicitly incorporates into its form all that is known about the momentum distribution of superfluid helium from many body theory. We use this model $n(p)$, converted to $J(Y)$ and broadened by instrumental resolution and FSE, to obtain fits to the superfluid data, and we extract estimates of the condensate without using any information from the momentum distribution in the normal liquid. The fitting function has been chosen in such a way that each free parameter is primarily determined by a different feature of the momentum distribution.

The fitting function we use takes the following form in $n(p)$

$$n(p) = n_0(\delta(p) + f(p)) + A_1 exp(\frac{-p^2}{2\sigma_1^2}); p \leq p_c \tag{5.1}$$

$$n(p) = A_2 exp(\frac{-p^2}{2\sigma_2^2}); p \geq p_c \tag{5.2}$$

where $f(p)$ is the term due to the condensate-induced singularity[35,36]

$$f(p) = \frac{mc}{2}\frac{1}{p}\left(\frac{exp(cp/k_BT)+1}{exp(cp/k_BT)-1}\right) \tag{5.3}$$

The model $n(p)$ is positive definite, normalized to unit area, continuous at p_c, and is symmetric about and centered at $p = 0$. Enforcing these physical conditions on our model $n(p)$ yields two linear equations in the amplitudes A_1 and A_2 in terms of various functions of the remaining parameters. The resulting model has four free parameters: the widths of the two Gaussians σ_1 and σ_2, the cutoff momentum p_c, and the condensate fraction n_0. The condensate fraction and its uncertainty is obtained by allowing the fitting parameters to vary within their allowed values.

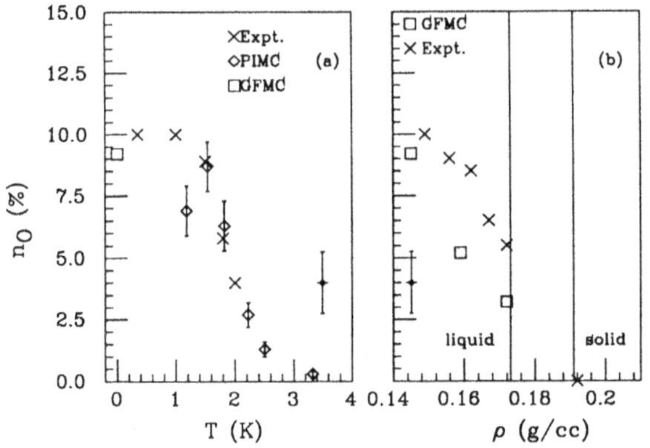

Fig. 4. Figure 4a shows the condensate fraction estimated using the model $J(Y)$ as a function of temperature for the constant density slice at 0.147 g/cm^3. Figure 4b shows the condensate fraction estimate as a function of density for the constant temperature slice at 0.75 K. The preliminary result for the size of errors in the condensate fraction estimate is indicated with the error bars in the lower right hand corner of Fig. 4a and the lower left hand corner of Fig. 4b

The fitting parameters are all related to various unknown aspects of the momentum distribution. p_c is the momentum at which the condensate-induced singularity becomes negligible in comparison to the nonsingular part of the momentum distribution. The point where this occurs is not known. Our approach to this problem is to choose values for p_c in the range $0.1 \leq p \leq 0.7$ Å^{-1} in the zero sound region where the above expression for the condensate-induced singularity is correct.[36] σ_2 is determined in the course of the fit primarily by the width of the momentum distribution in the intermediate to large Y region. σ_1, on the other hand, is not tightly constrained by the scattering data. It represents the unknown effect near $Y = 0$ of the condensate on the nonsingular part of the momentum distribution.

Figure 4 shows our estimates of the condensate fraction along with the results of theoretical calculations. Fig. 4a shows the temperature dependence results for the constant density slice at 0.147 g/cm^3. The estimates of the condensate fraction on the 0.147 g/cm^3 slice start at 10 % for the lowest temperature measurement, which is in good agreement with the GFMC value of 9.2 % at T=0 within the errors. The estimates decrease with increasing temperature and tend to zero at the superfluid transition, in qualitative agreement with the PIMC values.

The weak temperature dependence of the condensate at low temperatures seen in Fig. 4a is consistent with the results of many body theory and the finite temperature extensions of the variational calculations. The condensate wave function must vanish at the second order superfluid-normal fluid phase transition, since it is the microscopic order parameter for the superfluid phase. Unfortunately, the errors in the estimates are too large to place any useful constraints on the critical exponent which controls the behavior of the condensate close to the superfluid transition.

Fig. 4b shows the density dependence results at T=0.75 K. The estimates of n_0 are systematically somewhat higher than the theoretical results using the GFMC method, but show the same density dependence within the experimental errors. The estimates of the condensate fraction on the T=0.75 K slice decrease with increasing density. Notice that the estimate for the condensate remains finite close to the superfluid-solid phase transition. Since the superfluid-solid phase transition is first order, there is no reason for the condensate to vanish continuously near the critical density. In addition, we see no evidence for the presence of a condensate in the solid phase.

CONCLUSIONS

We have now reached a stage where there is excellent agreement between the theoretical results and the experimental observations for all aspects of the momentum distribution in liquid ^{4}He. In particular, the agreement between theory and experiment settles the long-standing question regarding the magnitude, and even the presence, of a Bose condensate in the superfluid. The experimental results provide convincing evidence for a Bose condensate containing 10 % of the atoms.

A better understanding of the strengths and weakness's of DINS as applied to determinations of $n(p)$ in quantum systems has also evolved. For example, the insensitivity of the observed scattering to some of the singular behavior in $n(p)$ is now understood. We have tried to convey an appreciation of where the measurements can provide a definite test of theories and where they are not sensitive to particular details.

We have also developed a method for the estimation of the condensate from scattering data which produces results for superfluid ^{4}He which are in agreement with theoretical calculations. With this method, it will be possible to estimate the condensate fraction of helium in more complicated environments without accurate theoretical calculations of $n(p)$.

Unfortunately, the original goal for much of the work in liquid helium, a direct observation of the condensate fraction, has not come to pass. In view of our current understanding of FSE in helium, it is unlikely that this goal will ever be reached in deep inelastic neutron scattering experiments. While the current experimental results do not definitively prove the existence of a condensate, they do provide such overwhelming evidence that we can now consider the problem, for the bulk liquid at least, solved.

This work was supported by the NSF under grant DMR-8704288 and by OBES/DMS support of the Intense Pulsed Neutron Source at Argonne National Laboratory under DOE grant W-31-109-ENG-38.

References

[1] P. Nozières and D. Pines, **The Theory of Quantum Liquids**, Volume 2, Addison-Wesley (1989), chapter 6.

[2] A. Griffin, Can. J. Phys. **65**, 1368 (1987).

[3] An overview of such measurements can be found in **Momentum Distributions**, edited by R. N. Silver and P. E. Sokol, Plenum Press (1989).

[4] V. F. Sears, E. C. Svensson, P. Martel and A. D. B. Woods, Phys. Rev. Lett. **49**, 279 (1982).

[5] H. A. Mook, Phys. Rev. Lett. **51**, 1454 (1983).

[6] S. Ikeda and N. Watanabe, Phys. Lett. A **121**, 34 (1987).

[7] R. S. Holt, L. M. Needham, and M. P. Paoli, Phys. Lett. A **126**, 373 (1989).

[8] M. Puoskari and A. Kallio, Phys. Rev. B **30**, 152 (1984).

[9] E. Manousakis, V. R. Pandharipande, Q. N. Usmani, Phys. Rev. B **31**, 7022 (1985); E. Manousakis and V. R. Pandharipande, ibid., 7029.

[10] M. H. Kalos, M. A. Lee, P. A. Whitlock, and G. V. Chester, Phys. Rev. B **24**, 115 (1981).

[11] P. Martel, E. C. Svensson, A. D. B. Woods, V. F. Sears, and R. A. Cowley, J. Low Temp. Phys. **23**, 285 (1976).

[12] P.E. Sokol, R.N. Silver, and J.W. Clark, in ref. 3.

[13] E. Manousakis, in ref. 3.

[14] D.M. Ceperley, in ref. 3.

[15] R. Panoff and P. Whitlock, in ref. 3.

[16] J.W. Clark and M.L. Ristig, in ref. 3.

[17] O. Penrose and L. Onsager, Phys. Rev. **104**, 576 (1956).

[18] C.N. Yang, Rev. Mod. Phys. **B34**, 694 (1962).

[19] P.C. Hohenburg and P.C. Martin, Ann. Phys. (N.Y.) **34**, 291 (1965).

[20] J. Gavoret and P. Nozières, Ann. Phys. (N.Y.) **28**, 349 (1964).

[21] G. B. West, Phys. Rep. **18C**, 263 (1975).

[22] P. E. Sokol, Can. J. Phys. **65**, 1393 (1987).

[23] R. N. Silver, in Proceedings of the 11th International Workshop on Condensed Matter Theories, Oulu, Finland, 1987, Plenum Press; Rapid Communications, Physical Review B, March 1 (1988) and references therein.

[24] P.E. Sokol, T.R. Sosnick, W.M. Snow, and R.N. Silver, in ref 3; P.E. Sokol, T.R. Sosnick, W.M. Snow, and R.N. Silver, to be published in Phys. Rev. B.

[25] R.A. Cowley and A.D.B. Woods, Phys. Rev. Lett. **21**, 787 (1968).

[26] R.A. Cowley and A.D.B. Woods, Can. J. Phys. **49**, 177 (1971).

[27] H.A. Mook, Phys. Rev. Lett. **32**, 1167 (1974).

[28] H.A. Mook, Phys. Rev. B **37**, 5806 (1988).

[29] E.C. Svensson, V.F. Sears, A.D.B. Woods and P. Martel, Phys. Rev. B **21**, 3638 (1980); V.F. Sears, E.C. Svensson, A.D.B. Woods and P. Martel, Atomic Energy of Canada Limited Report No. AECL-6779 (unpublished).

[30] A.D.B. Woods and V.F. Sears, Phys. Rev. Lett. **39**, 415 (1977).

[31] F. London, Nature **141**, 643 (1938).

[32] H.A. Mook, R Scherm, and M.K. Wilkinson, Phys. Rev. A **6**, 2268 (1972).

[33] H. W. Jackson, Phys. Rev. A **10**, 278 (1974).

[34] A. Griffin, Phys. Rev. B **32**, 3289 (1985).

[35] E. M. Lifshitz and L. P. Pitaevskii, **Statistical Physics, part 2**, p. 108.

[36] A. Griffin, private communication.

S(Q,ω) FOR LIQUID ^{4}He: WHAT MORE DO WE NEED TO KNOW?

E.C. Svensson

AECL Research
Chalk River Laboratories
Chalk River, Ontario, K0J 1J0, Canada

INTRODUCTION

Since I completed my review article[1] on the temperature dependence of the dynamic structure factor $S(Q,\omega)$ of liquid ^{4}He for the proceedings of the Hiroshima Symposium, there have been two very important developments: Stirling and Glyde[2] have provided us with a much more complete set of results for the temperature dependence of $S(Q,\omega)$ at low Q (0.4 Å^{-1}), and Glyde and Griffin[3] have proposed a new and very appealing interpretation of $S(Q,\omega)$ covering the whole range of wave vectors Q and frequencies $\omega = 2\pi\nu$ for both the superfluid and normal phases of liquid ^{4}He. In the Glyde-Griffin (GG) picture, the main peaks in $S(Q,\omega)$ at low Q ($\lesssim$ 0.7 Å^{-1}, phonon region), which have been known[4] since 1965 to change relatively little on passing through the superfluid transition temperature, T_λ, correspond to collective zero-sound (ZS) modes as first suggested by Pines.[5] In contrast, the sharp peaks in $S(Q,\omega)$ at larger Q ($\gtrsim$ 0.8 Å^{-1}), which have been known[6,7] since 1978 to be present only below T_λ, correspond to single-particle (SP) excitations. A very schematic representation[8] of the GG interpretation is shown in Fig. 1. The shaded regions are simply meant to indicate that the SP modes may disappear at low Q while the ZS modes are expected to disappear at high Q. Hybridization between the ZS and SP modes, which GG propose is responsible for the existence of a smooth continuous dispersion relation for the

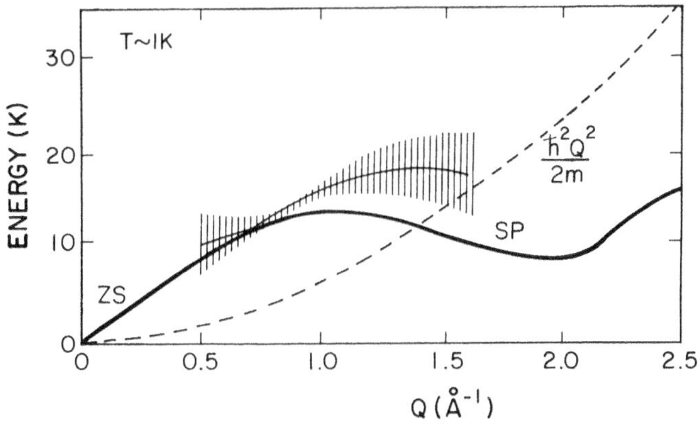

Fig. 1. Schematic representation of the Glyde-Griffin interpretation of the dispersion relation for superfluid ^{4}He. Figure adapted from Ref. 8.

Excitations in Two-Dimensional and Three-Dimensional Quantum Fluids
Edited by A.G.F. Wyatt and H.J. Lauter, Plenum Press, New York, 1991

single excitations, might cause the shaded region to actually be pushed away from the solid curve.

As discussed in detail by Glyde and Griffin[3], the Bose broken symmetry and the associated macroscopic Bose condensate fraction $n_o(T)$ which exist below T_λ allow the single-particle excitations, which are poles of the single-particle Green's function, to also become poles of the density response function, giving rise to corresponding peaks in $S(Q,\omega)$. These peaks will have a weight determined by $\Lambda^2(Q,\omega,T) \propto n_o(T)$, where $\Lambda(Q,\omega,T)$ is a Bose broken symmetry vertex function. (For additional details see Refs. 3, 9 and 10.) This gives a natural explanation of the fact that, as I emphasized in Ref. 1, the weights of the sharp peaks in $S(Q,\omega)$ for $Q \geq 0.8$ Å^{-1} go rapidly to zero as one approaches T_λ from below while the values of the widths and frequencies indicate that the corresponding excitations remain well defined with only very moderate softening of the frequencies. The single-particle excitations should of course continue to exist above T_λ but since $n_o = 0$ they are then no longer density fluctuations, and hence do not contribute to $S(Q,\omega)$. To add a bit of historical perspective, I point out that we were on the verge of recognizing this explanation of the experimental observations at the time of the Hiroshima Symposium as evidenced by the fact that I wrote in my paper[1] that "It almost appears as if these excitations continue to exist above T_λ but then no longer have any weight in $S(Q,\omega)$, i.e. are no longer density fluctuations." while Griffin wrote in his[11] that "If taken seri-

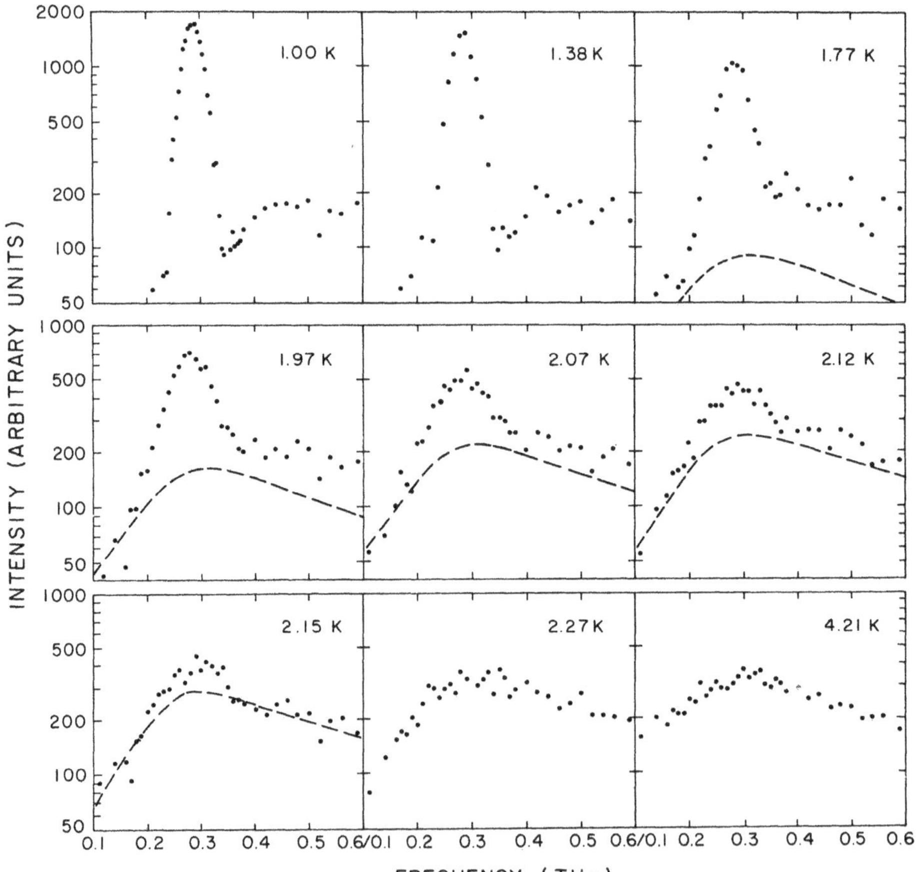

Fig. 2. Resolution-broadened dynamic structure factors of liquid ^{4}He for $Q = 1.13$ Å^{-1} at SVP. The dashed curves show the "normal-fluid" components in the Woods-Svensson model (Ref. 7). Figure from Ref. 7.

ously, the Woods-Svensson picture would seem to imply that the phonon-roton resonance still exists above T_λ but is no longer a density fluctuation.".

In this paper, I will focus mainly on those experimental results which I feel bear most directly on the GG interpretation of $S(Q,\omega)$ for liquid ^4He. There is certainly much support for this interpretation but there are also instances where the experimental results do not seem to exhibit features that the GG picture would lead one to expect. As we proceed, I will indicate what additional studies I feel would be most valuable. This paper should be viewed as a sequel to Ref. 1, where a great deal of relevant background information can be found. I will make repeated references to specific Figs. therein, so the reader may find it valuable to have a copy at hand. To facilitate intercomparison of results, it may be useful to note that 1 THz is equivalent to 48.0 K or 4.14 meV and that 1 bar is equivalent to 0.1 MPa or 0.987 atm.

RESULTS AND DISCUSSIONS

Largely because of their historical significance, I begin this section by showing, in Fig. 2, our results[6,7] of 1978 for the temperature dependence of $S(Q,\omega)$ at the maxon wave vector (Q_M = 1.13 Å^{-1}) at saturated vapour pressure (SVP). These were the first results to show a qualitative change in $S(Q,\omega)$ on passing through T_λ (T_λ = 2.17 K at SVP). We see that the sharp peak, which dominates the scattering at the lowest temperature (note the logarithmic scale), steadily broadens and becomes weaker as the temperature is raised but changes very little in frequency. It remains a clearly identifiable feature in $S(Q,\omega)$ at 2.15 K (just 0.02 K below T_λ) but then disappears as we go above T_λ, leaving us with just the single broad peak characteristic of the scattering by nonsuperfluid ^4He. Similar behaviour was also observed[7] for Q values of 0.80, 1.30, 1.40 and 1.926 Å^{-1}. The sharp peak in $S(Q,\omega)$ in

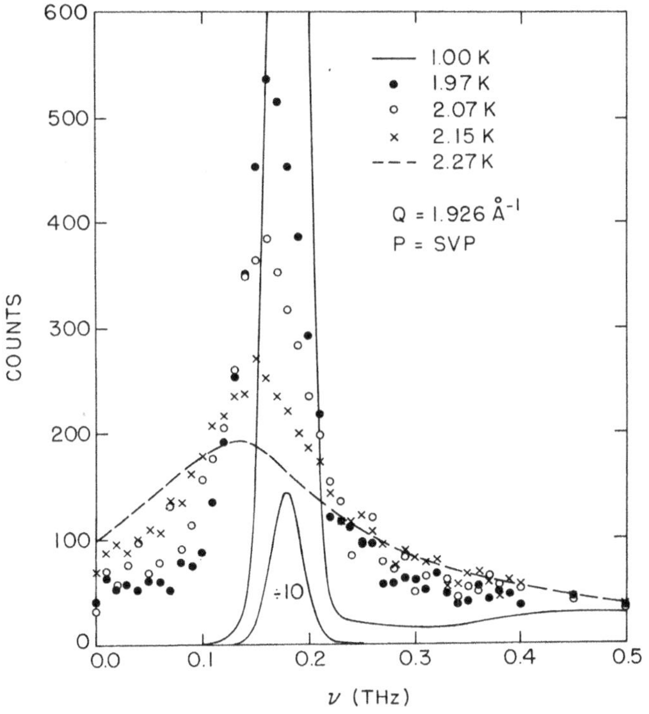

Fig. 3. Results of Woods and Svensson (Ref. 12) for the resolution-broadened dynamic structure factor of liquid ^4He for Q = 1.926 Å^{-1} at SVP.

this range of Q "is thus unique to the superfluid phase and hence a signature of the presence of the Bose condensate".[1] These were the results which set us on the road to the eventual understanding in terms of SP excitations provided by the GG picture. Unfortunately, it took 12 years to get there. The dashed curves in Fig. 2 show the normal-fluid components from the two-fluid Woods-Svensson (WS) model[7] for $S(Q,\omega)$ which will be discussed later.

Results for the roton wave vector (Q_R = 1.926 Å^{-1} at SVP) at selected temperatures from the study of Woods and Svensson[12] are shown in Fig. 3. Solid and dashed curves are smooth curves drawn through the distributions for 1.00 and 2.27 K. Again we see the very strong and sharp peak observed at 1.00 K broadening and decreasing dramatically in intensity as we approach T_λ, and then apparently disappearing from $S(Q,\omega)$ as we go above T_λ. There is substantially more softening of the excitation frequency than at the maxon position (Fig. 2) but still only a very moderate amount, about 13% over the whole range of superfluid temperatures. It is perhaps even more instructive to consider the results in Fig. 3 in terms of what happens when one lowers the temperature. Since there is relatively little change above T_λ, $S(Q,\omega)$ for $T = T_\lambda$ is still given rather closely by the dashed curve. Then as we cool further a sharp peak suddenly appears and grows very rapidly in intensity while continuously sharpening and moving slowly to higher frequency. Intensity is very rapidly removed from the regions to either side of the sharp peak, especially rapidly on the low frequency side, somehow being transferred into the sharp peak. By the time we reach 1.00 K, there is no intensity left at all on the low frequency side of the sharp peak. Below T_λ, the total intensity, $S(Q) = \int S(Q,\omega)d\omega$, only changes[13-15] by ~5% at the roton position, and barely at all at the maxon position (Fig. 2). The marked change in $S(Q,\omega)$ caused by cooling only 0.02 K below T_λ emphasizes the necessity of very high temperature accuracy and stability in the experiments.

The fact that the sharp peak in $S(Q,\omega)$ in the maxon-roton region is unique to the superfluid phase has been confirmed by the study of Svensson et al.[16] and Talbot et al.[17] for maxons and rotons at P = 20 bars, by the study of Stirling and Glyde[2] for rotons at SVP, and by the very recent work of Andersen et al.[18] (see also Stirling[19]) for a wide range of Q at SVP. Composite plots, like in Fig. 3, for maxons and rotons at 20 bars are shown by Figs. 13 and 14 in Ref. 1. The maxon at 20 bars gives a very nice visual demonstration of the sharp peak being unique to the superfluid phase since at that pressure, in contrast to the situation for SVP shown in Fig. 2, the sharp peak is centred at a frequency well below that of the broad peak seen above T_λ. In actual fact one can, however, follow the sharp maxon peak much closer to T_λ at SVP since the peak starts out about a factor of two weaker[20] at low temperature at the higher pressure (see Fig. 2 of Ref. 1) and also broadens much more rapidly with increasing temperature (see Fig. 8). The fact that sharp peaks in $S(Q,\omega)$ in the maxon-roton region are only observed below T_λ strongly supports the contention[3] that these peaks correspond to SP excitations visible in $S(Q,\omega)$ only because of the Bose broken symmetry. The variations with temperature of the intensities, frequencies and widths of these sharp peaks add additional support as we shall discuss later.

The temperatures of most importance for the understanding of liquid ^{4}He are undoubtedly $T = 0$ and $T = T_\lambda + \delta$. In practice, δ must be rather small but large enough that you are sure you are not below T_λ. $\delta \sim 0.1$ K is probably fine for studies in the maxon-roton region. As regards $T = 0$, what one requires in practice is that $S(Q,\omega)$ has effectively reached its low temperature limit. We have already noted that the intensity of the SP peak would be expected to have a temperature dependence essentially the same as n_o. The temperature dependence of n_o is not very well known,[21,22] but over most of the range of superfluid temperatures (although not in the strict limits $T \rightarrow 0$ and $T \rightarrow T_\lambda$, see Ref. 14) it is probably sufficiently close to that of the superfluid fraction ρ_s/ρ, which is well known,[23] that, for purposes of the present discussion of the temperature dependence of $S(Q,\omega)$, we can use the latter. If we require $\rho_s/\rho \geq 0.99$, we find that we need to reach lower temperatures which range from 1.05 K at SVP to 0.87 K at P = 25 atm, values which should be aimed for in any new studies.

Results[12] for $S(Q,\omega)$ at 1.00 and 2.27 K, effectively the two crucial temperatures noted above, for five Q values are shown in Fig. 4. The intensities for the upper four panels may be compared by using the listed monitor (MON) values, but those for the lowest panel cannot be directly compared with the others since higher experimental resolution was used for

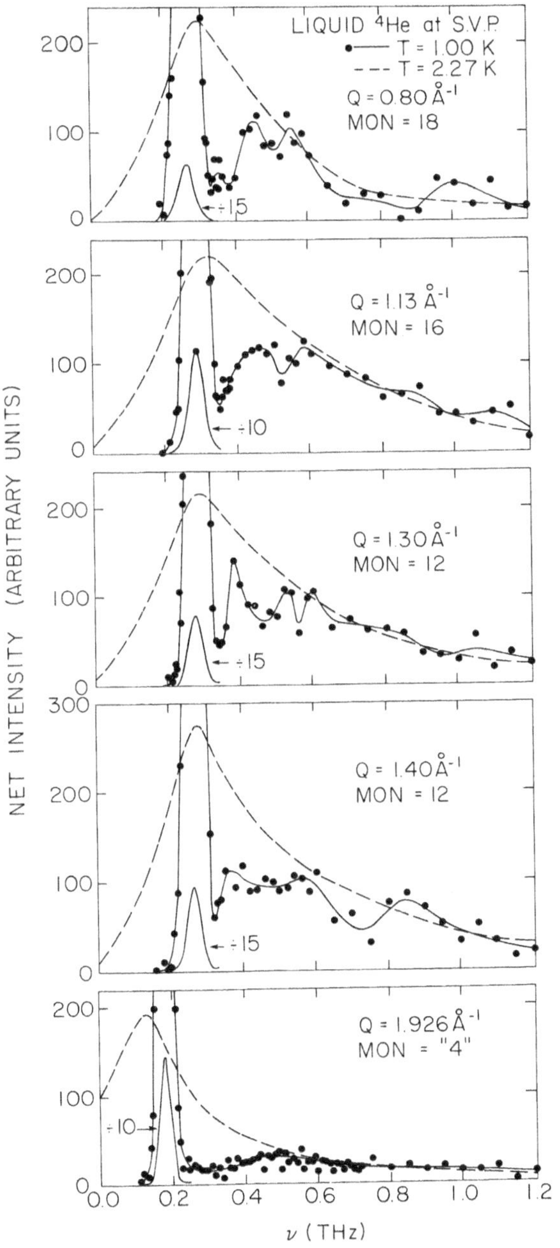

Fig.4. Results of Woods and Svensson (Ref. 12) for the resolution-broadened dynamic structure factor of liquid ^{4}He at SVP and 1.00 and 2.27 K.

these measurements. For clarity, the results for 2.27 K, where $S(Q,\omega)$ consists of just a single broad asymmetric peak, are shown simply by smooth curves (dashed lines). The solid lines are simply curves drawn through the experimental data points for 1.00 K as guides to the eye. At this temperature, $S(Q,\omega)$ consists of a very sharp (intrinsic FWHM known[24] to be $\lesssim 4 \times 10^{-4}$ THz for the roton) and strong single-excitation peak (shown reduced by a factor of 10 or 15 to indicate the experimental resolution) and, primarily at higher frequencies, a broad component which exhibits marked structure, some of which appears to be essentially as sharp as the experimental resolution. I will refer to the broad component as $S_M(Q,\omega)$. Traditionally[25] this component has been thought of as arising entirely from multiexcitation scattering but Glyde and Griffin[3] have proposed that it also contains, for at least some range of Q in the region of the maxon, a contribution from a broad remanent ZS mode as indicated schematically in Fig. 1. One of the important questions we want to consider is whether or not there is any evidence in the experimental results for such a ZS contribution.

First we note that the structure in $S_M(Q,\omega)$ varies quite dramatically with Q. Calculations[26-31] have shown that there should be marked structure in $S_M(Q,\omega)$ and that, in particular, one should be able to identify various features as arising from combinations of maxons and rotons, i.e., roton + roton (2R), maxon + roton (M+R), maxon + maxon (2M), etc. The calculations of Götze and Lücke[28] and Manousakis and Pandharipande[29] are the most detailed. In Fig. 5 we compare the results of Manousakis and Pandharipande for $Q = 0.825$ Å^{-1} (dashed curve) with the experimental results for $Q = 0.80$ Å^{-1} from Fig. 4. Arrows indicate the experimental values of the 2R, M+R and 2M frequencies taken from the single-excitation peaks in Fig. 4. There can be little doubt that the three peaks in the experimental $S_M(Q,\omega)$ correspond primarily to these three two-excitation processes. The three highest frequency peaks in the calculated $S_M(Q,\omega)$ correspond[29] to the same three processes. Manousakis and Pandharipande have noted that, as seen in Fig. 5, their values of the two-excitation frequencies are too high and have attributed this to the fact that they did not include certain self-energy corrections which would have made their numerical calculations much more complicated. At the time of their study, there was little incentive to proceed to this higher level of complication but now that we have the high-quality results shown in Fig. 4, and other excellent results for $S_M(Q,\omega)$ from the work[18,19,32,33] of Stirling and collaborators, such extra effort would be merited. It seems highly probable that the lowering of the frequencies that would result from this improvement would, together with appropriate broadening by the experimental resolution, bring the calculated $S_M(Q,\omega)$ into very

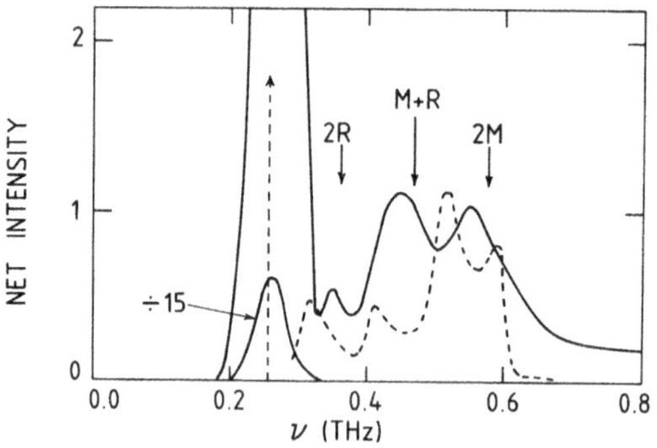

Fig. 5. Comparison of the experimental $S(Q,\omega)$ (solid curve) for liquid ^{4}He at $Q = 0.8$ Å^{-1} and $T = 1.00$ K from Fig. 4 with the calculated $S(Q,\omega)$ (dashed curve) of Manousakis and Pandharipande (Ref. 29) for $Q = 0.825$ Å^{-1}. The normalization of the two sets of results is arbitrary. Arrows indicate the experimental values of the 2R, M+R and 2M frequencies (see text).

good agreement with experiment over the frequency region ($\nu \lesssim 0.6$ THz) where two-excitation processes are dominant. The observed intensity at higher frequencies comes from higher-order processes which are not included in the calculations of Manousakis and Pandharipande. The lowest frequency peak in their $S_M(Q,\omega)$ arises from two-phonon (2P) processes, i.e., from two excitations on the nearly linear low-Q part of the dispersion relation (Fig. 1). If this peak is actually present, it is merged with the single-excitation peak because of the finite experimental resolution. It would be worth a try to pull it out by using very high resolution.

From the comparison between experiment and theory discussed above, my conclusion is that the observed $S_M(Q,\omega)$ for $Q = 0.8$ Å^{-1} can very likely be completely accounted for by multiexcitation processes. There appears to be no evidence for any residual broad ZS scattering at this wave vector, at least not in the region below 0.6 THz (~30 K), and it seems unlikely that ZS-SP hybridization would push any separate component of the scattering (of either ZS or SP character) to an even higher frequency (see Fig. 1). A transfer of intensity to high frequency could of course take place in a stepwise fashion via hybridization with the multiexcitations (2R, M+R, 2M ...) as would be expected to occur and has been considered by Fukushima et al.[30] While a detailed theoretical investigation of such multi-hybridization would be very valuable, the present evidence seems to indicate that essentially all of the intensity in the broad peak observed at 2.27 K for $Q = 0.8$ Å^{-1}, which in the GG picture we interpret as a ZS peak, has by 1.00 K been transferred into either the sharp single-excitation peak or the multiexcitation scattering. Since the temperature dependence of $S(Q,\omega)$ for $Q = 0.8$ Å^{-1} is very similar[1,7] to that at larger Q, I feel we should interpret the single-excitation peak here as having predominantly SP character. Where then is the "peak" of predominantly ZS character that one would expect to be a consequence of the ZS-SP hybridization?

At larger Q, there is not as good agreement between the experimental $S_M(Q,\omega)$ and the calculations of Manousakis and Pandharipande.[29] Their $S_M(Q,\omega)$ for $Q = 1.125$ Å^{-1} is dominated by the M+R peak with the 2R and 2M process just giving shoulders on its sides. Our results (Fig. 4) for $Q = 1.13$ Å^{-1} indicate that the 2R and M+R peaks have indeed merged, but they still show evidence of a separate 2M peak as do the results of Stirling.[32,33] The poorer agreement between theory and experiment at this higher Q could be evidence for some ZS scattering hidden in the experimental $S_M(Q,\omega)$ but, since we observe an easily identifiable peak (the 2M peak) which is not there in the theory, I am inclined to believe that the explanation is simply that the theory is not as good as at lower Q. I cannot see any feature in the experimental $S_M(Q,\omega)$ in Fig. 4 which would indicate the presence of a ZS mode at $T = 1.00$ K behaving like one might expect, i.e., broadening and getting weaker as Q increases, and perhaps beginning to decrease in frequency, as indicated schematically in Fig. 1. In proposing that the broad component of $S(Q,\omega)$ consisted of both the usual multiexcitation scattering and a ZS component, Glyde and Griffin[3] focussed specifically on the results[17] for $Q = 1.13$ Å^{-1} at $P = 20$ bars. Although the detailed features of $S(Q,\omega)$ are different at this pressure, I still don't see anything in $S_M(Q,\omega)$ at 20 bars and low T that requires a ZS component for its explanation. Hence I see no convincing evidence in the experimental $S_M(Q,\omega)$ for $Q \gtrsim 0.8$ Å^{-1} at low T for the ZS component that would be expected to be one of the consequences of ZS-SP hybridization. Perhaps the ZS mode is just very weak, or perhaps the hybridization has different consequences than we expect and we are not looking in the right region of (Q,ω) space?

I feel that it is absolutely essential that we understand $S_M(Q,\omega)$ in complete detail. The agreement in Fig. 5, and the ways that are already known[29] to improve it, shows that we are in fact well on our way to doing so. The improved calculations should be a very high priority for the immediate future. We also need additional measurements, at better resolution than used for the results in Fig. 4, with better statistical accuracy, at many more Q values, and for the full range of pressures. It is well known[34] (see p. 349 in Ref. 35 for additional references) that the high-frequency tail of $S(Q,\omega)$ should exhibit a $\nu^{-7/2}$ frequency dependence. Wong[34] found that this appeared to be the case (for $\nu \gtrsim 0.5$ THz) for the earlier results of Woods et al.[36] for $Q = 0.8$ Å^{-1}, at least if he arbitrarily assumed an extra flat background under the tail. A smooth $\nu^{-7/2}$ dependence for the tail does not, however, seem to be consistent with the results in Fig. 4 since there is evidence of structure for at least 0.8 and 1.4 Å^{-1}, and the decreases in intensity between 0.6 and 1.2 THz are much less than expected from a $\nu^{-7/2}$ dependence. An additional background (of unknown origin) under the tail would give better agreement with the expected decrease, but we have no reason to suspect this. It seems more likely that we have to go to even higher ν to reach the $\nu^{-7/2}$ region. The measurements[12] actually extended to ν values that ranged from 1.50 THz at 0.8 Å^{-1} to 2.35 THz at 1.926 Å^{-1}, but the intensity becomes so low beyond the range shown in Fig. 4 that the frequency dependence cannot be accurately determined. The measurements[12] do indicate real intensity out to 1.8-2.0 THz at 1.926 Å^{-1}, in substantial agreement with the calculations of Götze and Lücke[28]. The measurements of Talbot et al.[17] for $Q = 1.13$ and 2.03 Å^{-1} at 20 bars indicated that the tail of $S(Q,\omega)$ was independent of temperature for $\nu \gtrsim 0.7$ THz (see Figs. 13 and 14 in Ref. 1). This does not appear to be the case for the results in Fig. 4. Additional work is needed to check if the apparent structure in the 1.00 K results is real and at what frequency, if any, the tail becomes independent of temperature.

Having not found the expected evidence for a ZS component in $S_M(Q,\omega)$ at 1.00 K, we might ask whether or not any or all of the broad peaks for 2.27 K (i.e. $T_\lambda + \delta$) in Fig. 4 really represent zero sound? And if they don't, what do they represent? Glyde and Griffin[3] (again referring specifically to results[17] for 20 bars) propose that the broad peak at the maxon position is a ZS peak but that the broad (and much lower frequency) peak at the roton position is not. They argue that the latter peak can be understood as simply the scattering from thermally excited rotons. They do not consider intermediate Q values. The results in Fig. 4 give no indication of a change in the character of the normal-fluid scattering over the range 0.8-1.4 Å^{-1}. All of the peaks have much the same width and are centred at much the same frequency. We do not have results from this study[12] for Q values between 1.40 and 1.926 Å^{-1} but we know from the studies of Woods et al.[37-39] at 4.2 K and SVP that there is a smooth continuous change of $S(Q,\omega)$ in this region, which also continues to higher Q. Both the width and the mean energy of the broad peak exhibit[37] a minimum at approximately the roton wave vector. While we know from these[37-39] and other[40-43] studies that the normal-fluid peak sharpens up quite dramatically when we go to Q values smaller than those in Fig. 4, and that the identification with ZS is then quite certain, it is not at all clear from the experiments where (i.e., at how large a Q value) we must cease to think of these peaks as corresponding to zero sound. The results certainly don't seem to give any indication of a ZS mode broadening and dying out as suggested by the shaded area in Fig. 1. If this happens, the ZS mode must merge into something else so smoothly that we see no obvious evidence of it happening. Another case of hybridization? We urgently need a theory which correctly describes $S(Q,\omega)$ for normal liquid ^{4}He over the whole range of Q. If we don't know what "entity" the normal-fluid peak corresponds to at some particular Q, we are on rather shaky ground when we try to hybridize that "entity" with the SP mode that suddenly appears below T_λ.

In Fig. 6 we summarize results for the two crucial temperatures for liquid ^{4}He at SVP. The solid curve shows the frequencies of the sharp excitations at $T \leq 1.2$ K from a compilation of results by Donnelly et al.[44] Solid circles show the mean frequencies for $T = 2.3$ K and $Q \leq 0.7$ Å^{-1} from a study by Woods et al.[40] while open circles show the values at larger Q from the dashed curves in Fig. 4. The x's show the positions of maximum intensity of these dashed curves. The straight lines show frequencies for first-sound ($\nu = cQ/2\pi$) and zero-sound ($\nu = c_0Q/2\pi$) propagation (see caption). As known from the detailed study by Stirling,[45] the solid curve lies above the first-sound line (dashed curve) below $Q = 0.55$ Å^{-1} (anomalous dispersion region). The results for 2.3 K show[40] first-sound propagation at very low Q ($\lesssim$ 0.2 Å^{-1}) and then a transition to zero-sound propagation, with a velocity very close to that predicted by Etters[46] (dash-dot curve), at larger Q. We have to be rather cautious in interpreting the results for the normal phase for $Q \gtrsim 0.8$ Å^{-1} since it is not totally clear which of the normal-fluid peaks (dashed curves) in Fig. 4 correspond to zero sound. And also, how does one extract the best values of the (ZS) frequency from such broad asymmetric peaks? Until we have a theory which gives a correct quantitative description of $S(Q,\omega)$ for normal ^{4}He over the full range of Q, we can't satisfactorily answer these questions. Even with these uncertainties, however, the indications are that the results for the normal phase exhibit sound-propagation behaviour (i.e., follow a linear dispersion relation) to considerably higher Q than those for the superfluid phase. If we use the mean frequencies (circles), the results follow the zero-sound line to at least 0.8 Å^{-1} and are not much below it at 1.13 Å^{-1}. In contrast, the results for the superfluid phase begin to fall rapidly away from the sound-velocity line beyond 0.55 Å^{-1}. I interpret this difference in behaviour for the two phases as rather strong support for the GG picture. When the SP modes appear below T_λ, the hybridization forces what was a pure ZS branch to bend over much more rapidly and merge smoothly with the SP branch.

It would be very valuable to have a much more complete set of results for $T_\lambda + \delta$ at SVP than shown in Fig. 6 and also sets of complete results for

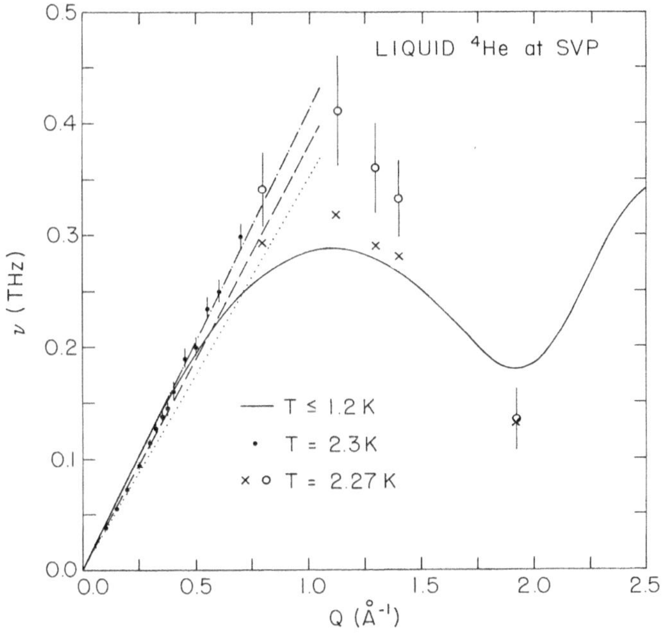

Fig. 6. Comparison of the dispersion relation (solid curve) for superfluid ^{4}He at $T \leq 1.2$ K from Ref. 44 with the values of the mean frequency at half height (circles) and the position of maximum intensity (x's) of the peaks observed for normal liquid ^{4}He just above T_λ. Open circles and x's are from the results shown in Fig. 4 and solid circles are from Ref. 40. The dashed and dotted curves show values of $cQ/2\pi$ for 1.2 and 2.3 K, respectively, while the dash-dot curve shows values of $c_0Q/2\pi$ for 2.3 K from Ref. 46.

$T = 0$ and $T_\lambda + \delta$ over the full range of densities. Much effort is going to have to be devoted to measuring and understanding $S(Q,\omega)$ for normal ^4He since this is the "baseline" material we start with just before further cooling brings in the condensate and the SP modes. Except for the extensive measurements of Woods et al.[37-39] at 4.2 K and SVP, most of the work on normal ^4He has been in the low Q region where the interest has primarily been comparison with sound velocities. In the hope of stimulating new theoretical work on zero-sound in normal ^4He, I show in Fig. 7 the results of our earlier studies[40-42] for three different thermodynamic conditions. Here we see more clearly than in Fig. 6 the transition from first-sound to zero-sound propagation at 2.3 K and SVP, and the very good agreement with the predicted[46] zero-sound velocity. There is similar behaviour at 2.5 K and 10 atm, but there are no calculations of c_0. For 4.2 K and 25 atm, there appears to be first-sound propagation to at least $Q = 0.5$ Å^{-1}. Again there are no calculations of c_0. If as much effort had been devoted to understanding zero sound in normal ^4He as has been devoted to such studies for liquid ^3He (see Ref. 35) we would undoubtedly be much more advanced in our understanding of normal ^4He, and better poised to exploit the new GG interpretation of $S(Q,\omega)$.

I now want to turn to a consideration of models for $S(Q,\omega)$ and values of the single-excitation parameters, i.e., the frequency, $\omega(Q,T)$, the intrinsic half width, $\Gamma(Q,T)$ and the intensity, $Z(Q,T)$. Returning first to Fig. 4, we note again the enormous qualitative difference between 1.00 K and 2.27 K. In order to be able to say that we fully understand liquid ^4He, we have to ultimately be able to calculate correctly, from first principles and for all Q values, $S(Q,\omega)$ for these two crucial temperatures (i.e., $T = 0$ and $T_\lambda + \delta$), and for all temperatures in between. We are still a long way from achieving this goal and, on the shorter term, it would be very valuable to have a model for correctly interpolating between the results for $T = 0$ and those for $T_\lambda + \delta$, i.e. for describing the temperature variation of $S(Q,\omega)$ in the superfluid phase. One absolutely essential feature that any such model must possess is that the intensity of the sharp peak in $S(Q,\omega)$ in the maxon-roton region go to zero as $T \to T_\lambda$, a fact now well established by experiment.[2,6,7,16-19] Another is that, in the limit $T \to 0$, there be essentially no intensity on the

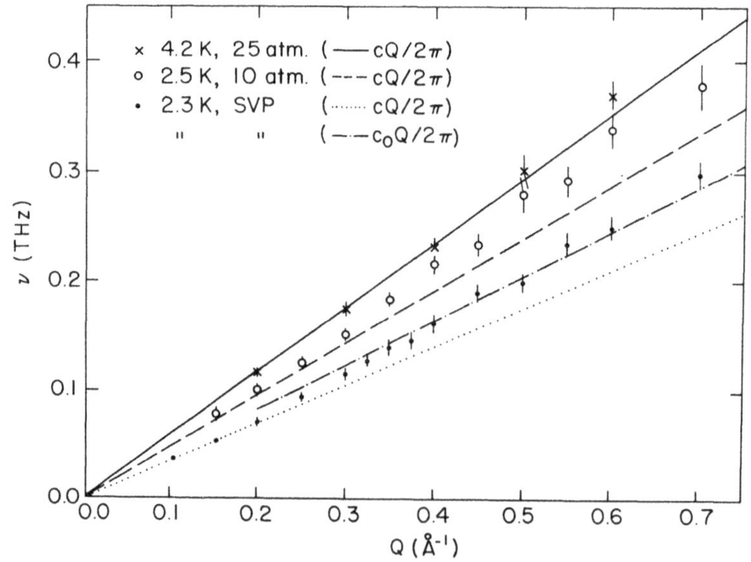

Fig. 7. "Dispersion relations" for normal ^4He at low Q from the studies of Ref. 40 (solid circles), Ref. 41 (x's) and Ref. 42 (open circles). Straight lines show the corresponding results for first-sound ($\nu = cQ/2\pi$) and zero-sound ($\nu = c_0Q/2\pi$, from Ref. 46) propagation.

low frequency side of the sharp peak. The first (and still the only) model to incorporate both these essential features was the two-fluid model of Woods and Svensson[7] (WS model). In this model, $S(Q,\omega)$ for any $T < T_\lambda$ is envisaged as consisting of "superfluid" and "normal-fluid" components, namely

$$S(Q,\omega) = \frac{\rho_S}{\rho} S_S(Q,\omega) + \frac{\rho_N}{\rho} S_N(Q,\omega). \tag{1}$$

Here ρ_S and $\rho_N = \rho - \rho_S$ are the macroscopic superfluid and normal-fluid densities which go to zero in the limits $T \to T_\lambda$ and $T \to 0$, respectively. The basic shape of the quantity $S_N(Q,\omega)$ is determined by the experimental results for $T_\lambda + \delta$ (dashed curves in Fig. 4) but for $T < T_\lambda$ a thermal-population factor is introduced to slightly adjust the shape at low frequency as needed to satisfy detailed balance. The dashed curves in Fig. 2 show examples of the normal-fluid components, $(\rho_N/\rho)S_N(Q,\omega)$, thus obtained. The excess intensity at each temperature (essentially all the observed intensity at 1.0 K where $\rho_N = 0.007\rho$) consists of a sharp (single-excitation) peak and a broad (multiexcitation) component at high frequency (see Fig. 5 in Ref. 1). The sharp peaks are then analyzed to obtain the single-excitation parameters. The WS model, a purely empirical model, was found[7,12] to work rather well for all five Q values (Fig. 4) that were studied at SVP. It was later[16,17] found to also work reasonably well for maxons and rotons at P = 20 bars though somewhat less well than at SVP, especially for rotons. As discussed in detail in Ref. 1, there are numerous reasons to expect the description given by (1) to be less good at certain Q values and pressures than at others. The good overall description given by the WS model indicates that the intensity of the sharp peak in $S(Q,\omega)$ in the maxon-roton region varies with temperature rather closely as ρ_S/ρ (see Fig. 6 in Ref. 1). Note that since one normalizes, for each Q, to the observed intensity of the sharp peak at the lowest temperature, it is only the temperature dependence that is being determined. The ρ_S/ρ temperature dependence for $Z(Q,T)$ given by the WS analysis, which as we have noted earlier implies at least approximately an $n_0(T)$ temperature dependence is, I believe, strong support for the GG interpretation of these sharp peaks as corresponding to SP excitations and hence having an expected $\Lambda^2(Q,\omega,T) \propto n_0(T)$ temperature dependence. Since there has been considerable objection[11] to the use of ρ_S and ρ_N in (1), I suggest here an alternative,

$$S(Q,\omega) = \frac{n_0(T)}{n_0(0)} S_S(Q,\omega) + \frac{[n_0(0) - n_0(T)]}{n_0(0)} S_N(Q,\omega). \tag{2}$$

I have not tried this model since there is, at present, inadequate information about $n_0(T)$ but I strongly suspect that it will give at least as good a description of $S(Q,\omega)$ as (1), and very similar single-excitation parameters.

We should note here that ZS-SP hybridization, which might be expected to be quite Q dependent, could well alter the shape of $S_N(Q,\omega)$ much more drastically with decreasing temperature than assumed in the applications to date of the WS model. The comparisons shown in Fig. 4 and the results[7,12] for intermediate temperatures shown partly in Figs. 2 and 3, together with the fact that the WS model works well for all five Q values, seem to suggest, however, that any hybridization does not vary greatly over this range of Q, or very drastically affect the shape of $S_N(Q,\omega)$. Glyde and Griffin[3] have proposed a simple hybridization model in which a sharp SP mode interacts with a broad ZS mode but have not worked out the consequences in any quantitative detail. Until we have a much more detailed understanding of ZS-SP hybridization in superfluid ^{4}He, I am prepared to tentatively accept the values of the single-excitation parameters given by the WS model. As noted earlier, it is still the only model which incorporates the two essential features of the data and, if we want to characterize the SP modes throughout the whole superfluid phase, we have nothing else to use at present.

Intrinsic widths for rotons and maxons obtained[7,12,17] by using the WS model are shown in Fig. 8. The roton widths are in excellent agreement, right up to T_λ, with the values given by the Landau-Khalatnikov (LK) theory[47] (see Refs. 1 and 17 for details). At SVP, the maxon widths (and those[7] for 0.8, 1.3 and 1.4 $Å^{-1}$) are also in excellent agreement with the LK widths indicating that we have the same decay mechanism (for the SP excitations) over a wide range of Q at SVP. This is undoubtedly a 4-excitation process, with the temperature dependence of $\Gamma(Q,T)$ and $\omega(Q,T)$ being determined mainly by the changing population of the thermal rotons needed for this process to occur. At 20 bars, the maxon width is very much larger than the roton width indicating some additional decay mechanism, probably the decay of a maxon into two rotons which is now allowed since the maxon frequency is greater than twice the roton frequency. The widths for $Q = 0.4$ $Å^{-1}$ at SVP from the new work of Stirling and Glyde[2] are also shown in Fig. 8 (triangles). These widths were obtained simply by fitting a harmonic oscillator (HO) function[2,17] to the main peak in $S(Q,\omega)$. They should tentatively be viewed as widths for ZS whereas the other values are widths for SP excitations. It is known from the work of Mezei and Stirling[48] that the widths in the low Q (ZS) region depend markedly on Q.

Frequencies for the maxon and roton SP excitations in superfluid ^{4}He obtained by using the WS model are plotted against ρ_N/ρ in Fig. 9. To within the experimental uncertainty, there is a linear dependence on ρ_N/ρ in all cases which indicates that the frequency shifts are directly proportional to the density of thermal rotons. Note that the frequencies decrease by only very moderate amounts (21% maximum) over the full range of superfluid temperatures. The results in Figs. 8 and 9 show that the modes which correspond

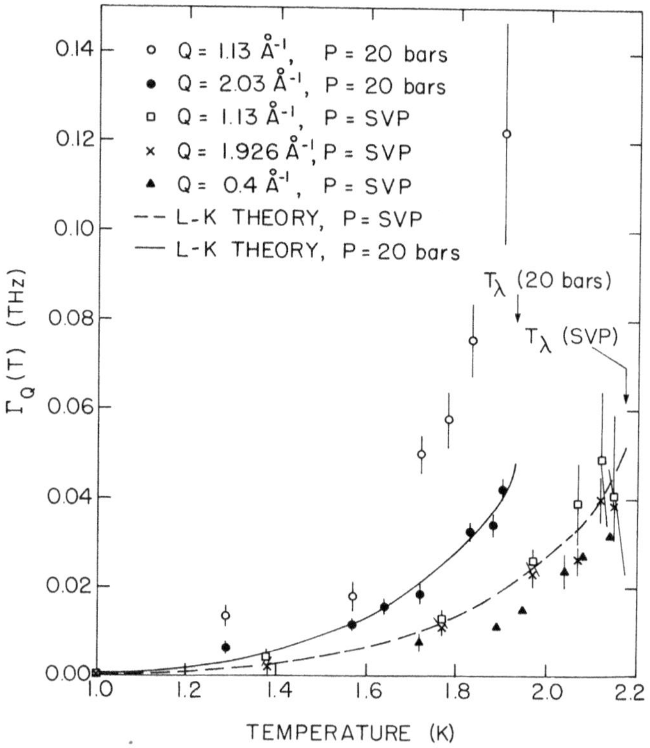

Fig. 8. Intrinsic widths for maxons ($Q = 1.13$ $Å^{-1}$) and rotons ($Q = 1.926$ and 2.03 $Å^{-1}$) in superfluid ^{4}He at SVP (Refs. 7 and 12) and 20 bars (Ref. 17) from analysis in terms of the WS model, and those for $Q = 0.4$ $Å^{-1}$ from Ref. 2. Curves show the Landau-Khalatnikov widths (see text).

to the sharp peaks in $S(Q,\omega)$ in the maxon-roton region remain well defined excitations as long as they remain visible, i.e., essentially right up to T_λ. At SVP, the ratio $2\Gamma(Q,T)/\omega(Q,T)$ at $T = T_\lambda$, taken from the dashed curves in Figs. 8 and 9, is 0.5 for the roton and 0.3 for the maxon. There is certainly no indication of any soft-mode behaviour as one approaches T_λ, as has been proposed by Ruvalds[49] and as is common behaviour as one approaches a phase transition characterized by a broken symmetry. It was this behaviour of well-defined modes just disappearing that led me to speculate in Ref. 1 that these modes probably continued to exist above T_λ but then couldn't be seen in $S(Q,\omega)$. Now that we have the theoretical understanding provided by Glyde and Griffin,[3] the behaviour of the frequencies and widths shown in Figs. 8 and 9 together with the fact that the intensities of the corresponding peaks vary essentially as $n_0(T)$, as discussed earlier, seem to me to in fact demand that we identify these modes as SP excitations.

The dotted curve in Fig. 9 shows 2Δ, twice the roton frequency, taken from the solid curve in the right panel. The indication is that the maxon frequency will fall below 2Δ for $T \leq 1.21$ K. Hence, if the large maxon widths at 20 bars (Fig. 8) are indeed caused by the decay of a maxon into two rotons, the maxon width might be expected to decrease below ~1.2 K. It would be very valuable to have a detailed study of the widths of maxons, and also rotons, at low T for pressures around 20 bars similar to the excellent study of Mezei[24] for rotons at SVP. The point at which the maxon frequency exceeds 2Δ can be moved to higher or lower T by varying the pressure.

I have not included in Figs. 8 and 9 results[17] for maxons and rotons at 20 bars and for[2] rotons at SVP obtained by analysis of $S(Q,\omega)$ in terms of a model usually referred to[17] as the simple-subtraction (SS) model. An example of the use of the SS model is shown in Fig. 10. In this model one separates $S(Q,\omega)$ at the lowest temperature into a single-excitation part and a multi-excitation part, $S_M(Q,\omega)$, as shown for example by the dotted curves. One then assumes that $S_M(Q,\omega)$ is independent of temperature and subtracts it from $S(Q,\omega)$ for each higher temperature, including $T > T_\lambda$, to obtain the single-excitation part for that temperature. The difference in the bottom panel of Fig. 10 is thus the SS model result for the single-excitation at 2.27 K

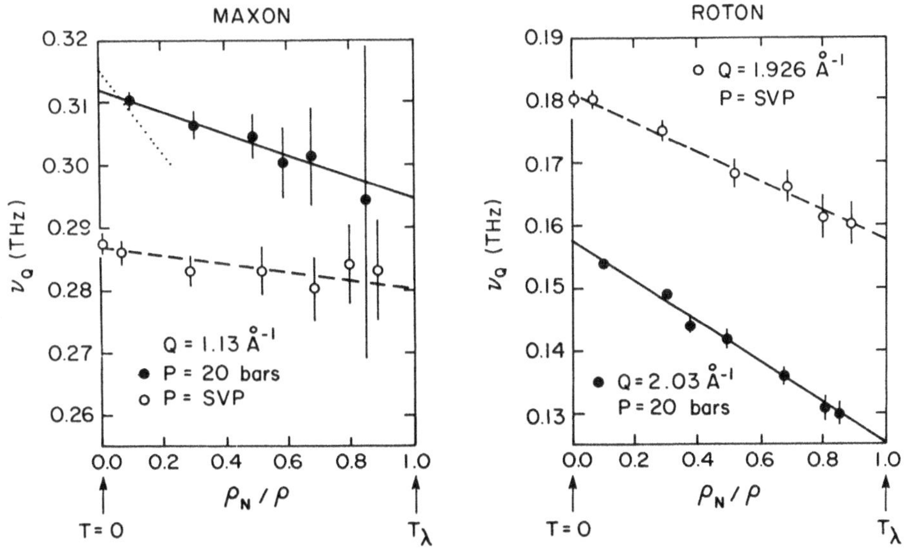

Fig. 9. Frequencies for maxons and rotons in superfluid ^{4}He at SVP (Refs. 7 and 12) and 20 bars (Ref. 17) from analysis in terms of the WS model. The solid and dashed curves are linear fits. The dotted curve shows twice the roton frequency for P = 20 bars.

(i.e., just above T_λ). I have strongly criticized the SS model in Ref. 1 which the reader should consult for details. It gives a "single-excitation" peak at all temperatures and, since one is subtracting a constant $S_M(Q,\omega)$ from an $S(Q,\omega)$ whose integrated intensity, $S(Q)$, varies by $\lesssim 5\%$ with temperature, it forces the "single-excitation" weight to be roughly independent of temperature. This is totally inconsistent with what the experiments have been telling us for 12 years and with the GG interpretation of $S(Q,\omega)$ wherein the sharp peaks in the maxon-roton region correspond to SP excitations and vary in weight as $n_0(T)$, disappearing at T_λ. Note that the difference in Fig. 10 shows marked structure and possibly regions of unphysical negative intensity in the tail. Even the main "bump" has a very unusual lineshape. In my opinion, it is totally misleading to fit such a difference distribution with say an HO lineshape and claim you have extracted single-excitation parameters. Except in the limit of low T, I feel that the parameters obtained by analysis[2,17] in terms of the SS model are essentially meaningless. They are certainly not the parameters for the sharp SP peak in $S(Q,\omega)$ which is the entity of prime importance that we want to follow throughout the superfluid phase.

Thus far I have focussed mainly on the region $Q \geq 0.8$ Å^{-1} where $S(Q,\omega)$ exhibits a sharp peak which disappears at T_λ. I now want to consider what happens at lower Q. In Fig. 11 we see the frequency shifts for 0.2-0.6 Å^{-1} from the work of Cowley and Woods[25] and those for 0.4 Å^{-1} from the work of Stirling and Glyde.[2] In all cases the frequency steadily decreases with increasing T in the superfluid phase until we get to a temperature close to T_λ when it suddenly jumps. It then appears to increase steadily and rather rapidly, and by an amount which increases with increasing Q, until somewhat above T_λ after which it falls rapidly. The drop at high T is almost certainly attributable to the central Rayleigh mode which appears in $S(Q,\omega)$ above T_λ and gets increasingly stronger as T increases (see Ref. 39 for an example). This mode was ignored in the analysis[2,25] used to obtain the results for $T > T_\lambda$ in Fig. 11. Note that the data are not sufficiently dense or accurate to firmly establish that the maximum frequency occurs above T_λ as suggested by the curves. It could well be that the maximum occurs essentially at T_λ for all Q values. It is rather important to know. New measurements for this range of Q at higher resolution and for more closely spaced temperatures near T_λ would be exceedingly valuable.

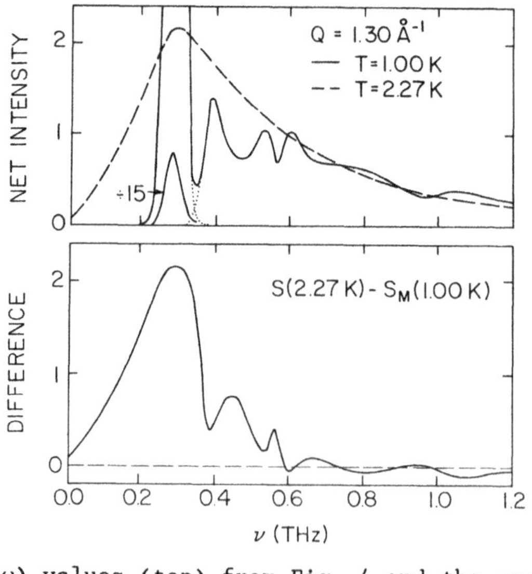

Fig. 10. $S(Q,\omega)$ values (top) from Fig. 4 and the result (bottom) of subtracting the multiexcitation part for 1.00 K from $S(Q,\omega)$ for 2.27 K.

The behaviour for $Q \leq 0.6$ Å^{-1} near T_λ in Fig. 11 (and also the behaviour near T_λ of the corresponding widths - see Fig. 7 in Ref. 2 and Fig. 17 in Ref. 25) undoubtedly holds very important clues to the understanding of liquid ^{4}He. How might we for example understand the jump in frequency below T_λ? As $n_0(T)$ decreases rapidly as we approach T_λ, taking away the SP modes and rapidly decreasing the strength of the ZS-SP hybridization, is some downward pressure on the ZS mode being rapidly removed causing it to rebound? Or might there even be both ZS and SP modes within what now appears from experiment to be a single peak, with a transfer of intensity between the two and changes in their frequencies as n_0 and the hybridization change with temperature? Does hybridization necessarily push the ZS and SP modes apart or might it pull them together? We can ask numerous such questions that we can't satisfactorily answer until we have better measurements and a better theoretical understanding. The somewhat anomalous behaviour near T_λ in Fig. 11 is probably telling us that we are not dealing with a pure ZS mode at 0.4 Å^{-1}, and possibly not even at 0.2 Å^{-1}. For comparison with the results at low Q, the frequency shifts (for the SP modes) at higher Q obtained[7,12,17] using the WS model are also shown in Fig. 11. Note that at SVP there is a hint of a rise in frequency just below T_λ for Q values up to 1.4 Å^{-1}, but no indication of a rise at 1.926 Å^{-1}. Is this evidence for ZS-SP hybridization in the maxon region but not in the roton region, as suggested by Glyde and Griffin[3]? The apparent slight rise in the SP frequencies for Q values in the range 0.8-1.4 Å^{-1} and the fact that the (ZS?) frequencies for normal ^{4}He lie above the SP frequencies for superfluid ^{4}He in this region of Q (Fig. 6) suggest that ZS-SP hybridization is pushing the modes apart here. At 20 bars there is no indication of a frequency rise for either the maxon or the roton though the errors for the maxon are sufficiently large near T_λ (Fig. 9) that a rise cannot be ruled out.

Griffin and Svensson[8] have pointed out that, with the new GG interpretation, the quantity Z(Q), the intensity of the sharp peak in $S(Q,\omega)$ at low T, should be viewed as the superposition of ZS and SP components. This

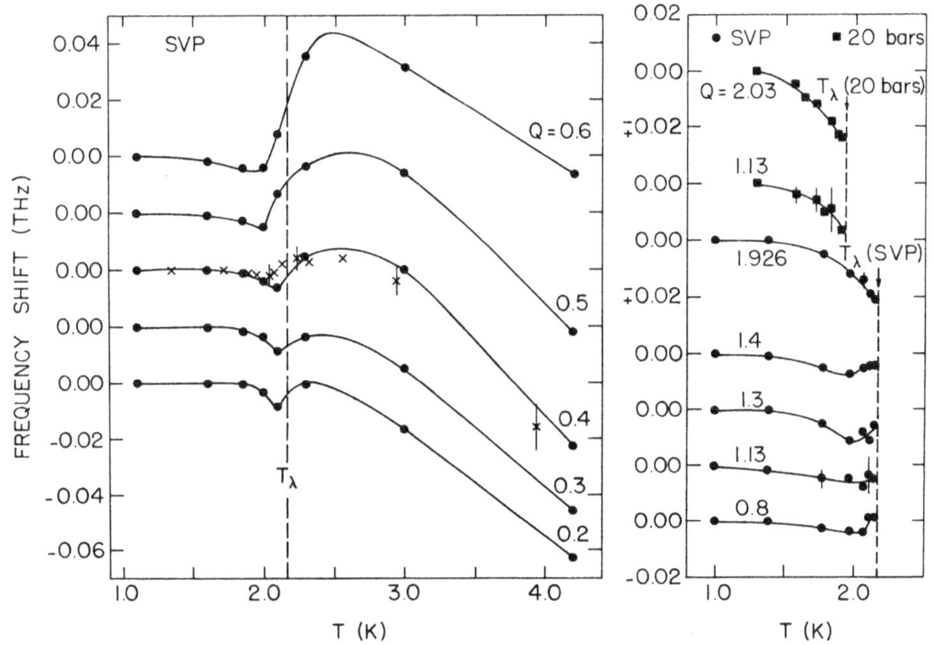

Fig. 11. Frequency shifts at small Q at SVP from Ref. 25 (circles) and Ref. 2 (x's) and at larger Q at SVP (circles) from Refs. 7 and 12 and at 20 bars (squares) from Ref. 17. Curves are simply guides to the eye.

is illustrated in Fig. 12 together with a similar separation for $H_I(Q)$, the contribution of the sharp peak to the total first moment. The fact that these separations give such natural explanations of the previously puzzling shapes of the Z(Q) and $H_I(Q)$ curves is further support for the GG interpretation. Although the separations in Fig. 12 have been drawn in a rather arbitrary way, there is not enormous leeway in how to do them, so they probably give us some additional indication of where and how the change from ZS to SP character occurs along the single-excitation branch (Fig. 1). Griffin and Svensson[8] have also speculated that the SP weight at the maxon position may be a reflection of the value of n_0. We see from Fig. 12 that this is indeed ~0.1, consistent with accepted values[21,22] of n_0 for T ~ 1.0 K at SVP. In Fig. 12, I have also included values[20] of Z(Q) for maxons and rotons at low and high pressures at 1.2 K. We see that Z(Q) for the maxon is about 50% smaller at the higher pressure suggesting that n_0 is also about 50% smaller. This is consistent with the decrease in n_0 over the same range of pressure inferred from other studies.[22,50] The SP peaks in $S(Q,\omega)$ for superfluid ^{4}He clearly contain a great deal of information on the temperature and density dependence of n_0, to be exploited by future experimental and theoretical studies. With hindsight, one now realizes that the work[6,7] of 1978 which first showed the sharp maxon-roton peak in $S(Q,\omega)$ disappearing at T_λ was a clear experimental demonstration of the existence of a finite Bose condensate in superfluid ^{4}He, some four years before conclusive proof of this was obtained[51] from (much more difficult) inelastic neutron scattering measurements at large Q.

CONCLUDING REMARKS

We clearly need a lot of new work, both theoretical and experimental, to reach our ultimate aim of understanding the complete $S(Q,\omega)$ for liquid ^{4}He for all temperatures and densities. I have made numerous specific recommendations as we have gone along and I will only reiterate here that I feel the highest priority right now should be to understand $S(Q,\omega)$ for the normal

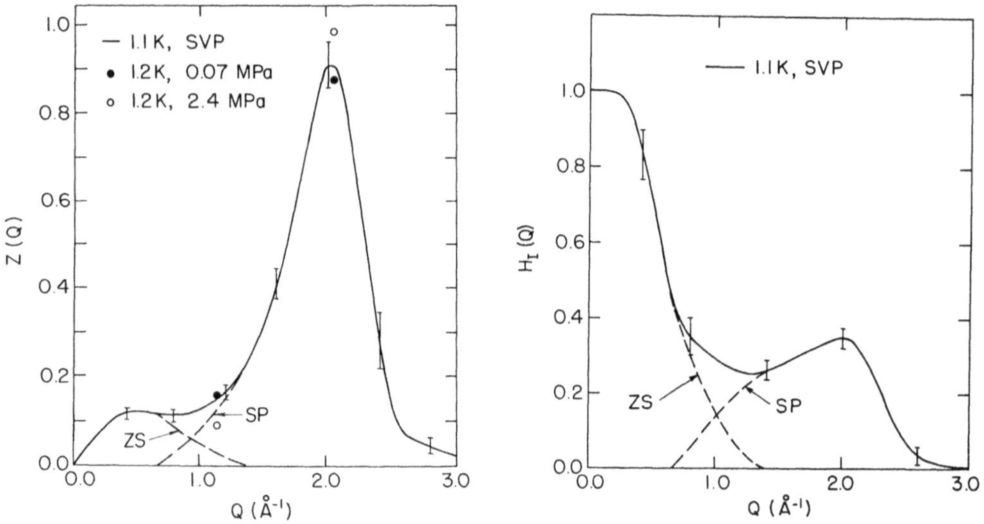

Fig. 12. Left panel: The intensity, Z(Q), of the sharp peak in $S(Q,\omega)$ at T = 1.1 K and SVP (solid curve) from Ref. 25 and the values for low (solid circles) and high (open circles) pressures from Ref. 20. Dashed curves show a plausible separation into ZS and SP parts based on Ref. 8. Right panel: The contribution (solid curve) of the sharp peak in $S(Q,\omega)$ to the total first moment from Ref. 25 and a plausible separation into ZS and SP parts.

phase. Although in attempting to develop a model for describing the temperature variation of S(Q,ω) below T_λ one could decide to just use the experimentally determined S(Q,ω) for $T_\lambda + \delta$ without really understanding what they correspond to, this is not a very satisfactory approach. While, in the hope of stimulating new work to fill in the gaps in our understanding, I have deliberately focussed on the areas where our knowledge is most deficient, we really do know a great deal about S(Q,ω) for liquid ^{4}He, with much recent progress. At the end of Ref. 1, I wrote that I thought we were "poised for a breakthrough in our understanding of liquid ^{4}He". The new GG interpretation represents that breakthrough. We have leapt another major hurdle on our way to a true fundamental understanding of liquid ^{4}He, but we still have some way to go as evidenced by the many important questions we can ask but not answer. We certainly now appreciate much more clearly the fundamental role the condensate plays in determining what we see in the superfluid phase, though it may play only a rather minor role in determining the intrinsic properties (e.g., the frequencies and lifetimes of the SP modes) of some of the things it allows us to see. Much of the confusion, uncertainty and disagreement prevalent at the time of the Hiroshima Symposium is now behind us. Most satisfying to me is the now rather widespread acceptance of the fact that the sharp peak in S(Q,ω) in the maxon-roton region really does disappear at T_λ, something I have firmly believed and lobbied for since 1978. For the first time in many years, theory has caught up to experiment and given us ideas, and incentive, for new experimental studies. There is a lot of excitement in the field, and I anticipate further major progress in the near future.

ACKNOWLEDGEMENTS

I have benefitted greatly from many discussions about liquid ^{4}He with many colleagues, and I want to particularly acknowledge the numerous stimulating and very enjoyable recent discussions with Allan Griffin.

REFERENCES

1. E.C. Svensson, in "Elementary Excitations in Quantum Fluids", ed. by K. Ohbayashi and M. Watabe (Springer-Verlag, Heidelberg, 1989) p. 59.
2. W.G. Stirling and H.R. Glyde, Phys. Rev. B **41**, 4224 (1990).
3. H.R. Glyde and A. Griffin, Phys. Rev. Lett. **65**, 1454 (1990).
4. A.D.B. Woods, Phys. Rev. Lett. **14**, 355 (1965).
5. D. Pines, in "Quantum Fluids", ed. by D.F. Brewer (Wiley, New York, 1966), p. 257.
6. E.C. Svensson, R. Scherm and A.D.B. Woods, J. de Phys. **39**, C6-211 (1978).
7. A.D.B. Woods and E.C. Svensson, Phys. Rev. Lett. **41**, 974 (1978).
8. A. Griffin and E.C. Svensson, Physica B **165&166**, 487 (1990).
9. H.R. Glyde, these proceedings.
10. A. Griffin, these proceedings.
11. A. Griffin, in Ref. 1, p. 23.
12. A.D.B. Woods and E.C. Svensson, unpublished work from the study reported in Ref. 7.
13. E.C. Svensson, V.F. Sears, A.D.B. Woods and P. Martel, Phys. Rev. B **21**, 3638 (1980).
14. E.C. Svensson, V.F. Sears and A. Griffin, Phys. Rev. B **23**, 4493 (1981).
15. E.C. Svensson and A.F. Murray, Physica **108B**, 1317 (1981).
16. E.C. Svensson, W.G. Stirling, E. Talbot and H.R. Glyde, Jpn. J. Appl. Phys. **26** (Suppl. 26-3), 33 (1987).
17. E.F. Talbot, H.R. Glyde, W.G. Stirling and E.C. Svensson, Phys. Rev. B **38**, 11229 (1988).
18. L.K.H. Andersen, W.G. Stirling, R. Scherm, A. Stunault, B. Fåk, A.J. Dianoux and H. Godfrin, to be published.
19. W.G. Stirling, these proceedings.

20. E.C. Svensson, P. Martel, V.F. Sears and A.D.B. Woods, Can. J. Phys. **54**, 2178 (1976).
21. E.C. Svensson, in "Proceedings of the 1984 Workshop on High-Energy Excitations in Condensed Matter" (Los Alamos National Laboratory publication LA-10227-C, 1984), Vol. II, p. 456.
22. P.E. Sokol and W.M. Snow, these proceedings.
23. J. Maynard, Phys. Rev. B **14**, 3868 (1976).
24. F. Mezei, Phys. Rev. Lett. **44**, 1601 (1980).
25. R.A. Cowley and A.D.B. Woods, Can. J. Phys. **49**, 177 (1971).
26. H.W. Jackson, Phys. Rev. A **4**, 2386 (1971) and **8**, 1529 (1973).
27. F. Iwamoto, K. Nagai and K. Nojima, in "Proc. 12th Int. Conf. on Low Temp. Phys.", ed. by E. Kanda (Keigaku Publ. Co., Tokyo, 1971) p. 189.
28. W. Götze and M. Lücke, Phys. Rev. B **13**, 3825 (1976).
29. E. Manousakis and V.R. Pandharipande, Phys. Rev. B **33**, 150 (1986).
30. K. Fukushima, N. Koyama and T. Sugiyama, Prog. Theor. Phys. **61**, 367 (1979).
31. K. Fukushima and F. Iseki, Phys. Rev. B **38**, 4448 (1988) and in Ref. 1, p. 144.
32. W.G. Stirling, in "Proc. 2nd Int. Conference on Phonon Physics", ed. by J. Kollar, N. Kroo, N. Menyhard and T. Siklos (World Scientific, Singapore, 1985) p. 829.
33. W.G. Stirling, Physica B **165&166**, 501 (1990).
34. V.K. Wong, Phys. Lett. **61A**, 454 (1977).
35. H.R. Glyde and E.C. Svensson, in "Neutron Scattering", ed. by D.L. Price and K. Sköld, Methods of Exp. Phys., Vol. 23, Part B (Academic Press, New York, 1987) p. 303.
36. A.D.B. Woods, E.C. Svensson and P. Martel, in "Neutron Inelastic Scattering 1972" (IAEA, Vienna, 1972) p. 359.
37. A.D.B. Woods, E.C. Svensson and P. Martel, in "Low Temperature Physics - LT14", ed. by M. Krusius and M. Vuorio (North Holland, Amsterdam, 1975) Vol. 1, p. 187.
38. A.D.B. Woods, E.C. Svensson and P. Martel, in "Proc. of the Conf. on Neutron Scattering", ed. by R.M. Moon (CONF-760601-P2, Natl. Tech. Inf. Serv., Springfield, VA, 1976) Vol. II, p. 1010.
39. A.D.B. Woods, E.C. Svensson and P. Martel, Can. J. Phys. **56**, 302 (1978).
40. A.D.B. Woods, E.C. Svensson and P. Martel, Phys. Lett. **57A**, 439 (1976).
41. E.C. Svensson, W.G. Stirling, A.D.B. Woods and P. Martel, in Ref. 38, p. 1017.
42. A.D.B. Woods, P. Martel and E.C. Svensson, in "Neutron Inelastic Scattering 1977" (IAEA, Vienna, 1978) Vol. II, p. 37.
43. K.S. Pedersen and K. Carneiro, Phys. Rev. B **22**, 191 (1980).
44. R.J. Donnelly, J.A. Donnelly and R.N. Hills, J. Low Temp. Phys. **44**, 471 (1981).
45. W.G. Stirling, in "75th Jubilee Conf. on Helium-4", ed. by J.G.M. Armitage (World Scientific, Singapore, 1983) p. 109.
46. R.D. Etters, Phys. Rev. Lett. **16**, 119 (1966).
47. L.D. Landau and I.M. Khalatnikov, Zh. Eksp. Teor. Fiz. **19**, 637 (1949).
48. F. Mezei and W.G. Stirling, in Ref. 45, p. 111.
49. J. Ruvalds, Phys. Rev. Lett. **27**, 1769 (1971).
50. J.H. Root and E.C. Svensson, Physica B (in press).
51. V.F. Sears, E.C. Svensson, P. Martel and A.D.B. Woods, Phys. Rev. Lett. **49**, 279 (1982).

RAMAN SCATTERING FROM SUPERFLUID HELIUM

Kohji Ohbayashi

Faculty of Integrated Arts and Sciences
Hiroshima University
Hiroshima 730, Japan

1. INTRODUCTION

Various excitations in quantum fluids can be studied by light scattering. Quasielastic light scattering probes excitations of small wave vector such as first and second sounds. By inelastic light scattering, excitations of larger momentums can be studied under the restriction that the total momentum is practically zero. To use inelastic Raman scattering to investigate excitations in quantum fluids was proposed by Halley.[1] The first experiment was done by Greytak and Yan.[2] Subsequently, Iwamoto[3] and Ruvalds and Zawadowski[4] proposed the theory of two-roton bound state to explain the experimental spectrum based on the quasi-particle interaction model. Soon after, the prediction was experimentally confirmed by the Greytak's group.[5,6] The success attracted attention because it yielded accurate estimation of the roton-roton interaction energy which is difficult to obtain by other methods. Although the the concept of the two-roton bound state was criticized,[7–9] the model based on quasi-particle interactions was widely accepted for analyses of both neutron and Raman scattering experiments. I refer the readers to the conference proceedings for the works along this line.[10]

In recent years, present author's group restarted experiments on Raman scattering from superfluid ^{4}He . We found a few results which claim necessity of reinvestigation of above conventional interpretations. Three years ago, we had a symposium at Hiroshima as a satellite meeting of LT-18 conference. The proceedings of the symposium was published last year after a long delay.[11] A few theorists reviewed the theoretical problems of Raman scattering.[12–14] I wrote an experimental review.[15] Contents of this report partly overlap with those reported there. But, by adding new experimental data except for the polarization property, following four problems are discussed; (a) Atomic process and polarization property, (b) Two-roton Raman scattering, (c) Fine structure, and (d) Temperature dependence of roton parameters.

2. EXPERIMENTAL METHOD

The block diagram of the apparatus which we have been using in recent years is shown in Fig. 1. Incident light source is an argon ion laser operated at 514.5 nm. The laser beam out of the source is vertically polarized and a half-wave plate is used to rotate the polarization direction. The polarizer 1 is for determination of the incident beam polarization direction. The laser beam is focussed into the sample with the lens 1. Scattered light is collected with

Excitations in Two-Dimensional and Three-Dimensional Quantum Fluids
Edited by A.G.F. Wyatt and H.J. Lauter, Plenum Press, New York, 1991

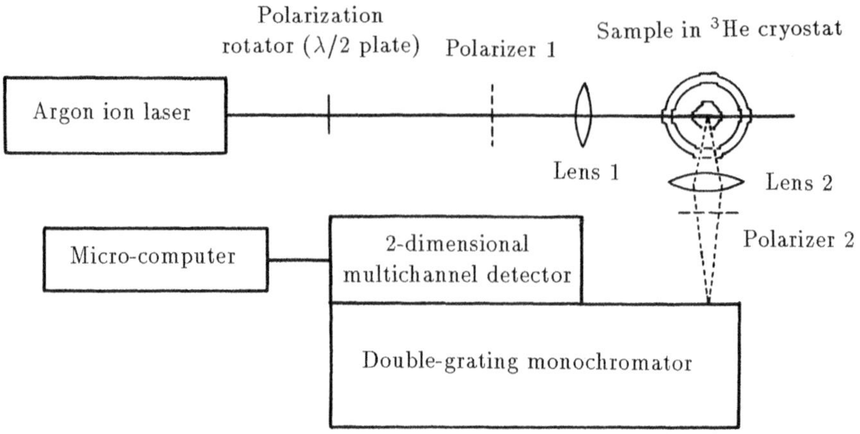

Figure 1. The experimental system for Raman scattering from liquid ^{4}He .

the lens 2 of $f/3.7$ at 90 degree to the direction of the incident beam. For the polarization analysis polarizer 2 is used with its easy axis vertical.

Spectra are analyzed by a double grating spectrometer. The whole spectrum imaged on the surface of the exit slit is detected by a multichannel detector. Fixing the grating direction in the multichannel detection improves reproducibility and linearity of energy shift. The entrance slit is set to a very narrow width to realize a good instrumental resolution. Other three slits are wide open. After measuring the central component, their widths are adjusted so that the central component is cut just enough not to affect the Raman spectrum. The energy shift of the Raman spectrum is determined by the separation of the position from the central component. It was calibrated by the rotational Raman spectrum of N_2 and O_2 gases. In our former work, we used a 1-dimensional diode array multichannel detector. Recently a 2-dimensional multichannel detector of photon-counting type has been introduced, which enables the detection practically free from thermal noises. Accumulated data in the detector are transferred to a micro-computer to store in floppy disks and to analyze.

The sample is cooled with a ^{3}He circulation type cryostat down to 0.6 K. The temperature of the sample is measured with a germanium resistor, which was calibrated with a precision pressure gage in reference to the vapor pressure of ^{4}He and ^{3}He . To control temperature at a higher value, the circulation rate of ^{3}He gas was reduced. An electric feedback circuit was used to control the temperature within ± 0.015 K of the desired value.

3. ATOMIC PROCESS AND POLARIZATION PROPERTY

Raman scattering occurs via two atoms. The incident light excites the first atom at $\boldsymbol{r}_i$, the second atom at $\boldsymbol{r}_j$ is virtually excited by the first atom via inter-atomic interactions, and then the latter atom deexcites accompanying an emission of light. The scattered field is the sum of the emissions from all the atoms. This atomic process is characterized by the diatomic polarizability tensor $\hat{\alpha}(\boldsymbol{r}_i, \boldsymbol{r}_j)$, which depends on the positions of the two atoms. The fact that $\hat{\alpha}(\boldsymbol{r}_i, \boldsymbol{r}_j)$ is not well known makes the calculation of Raman spectrum difficult. We can make phenomenological guesses. If the two atoms are far apart, the second atom is excited by the dipole field of the first atom. This is the dipole-induced-dipole (DID) Raman scattering. If two atoms are situated closely, the direct interactions due to overlap of electronic wave

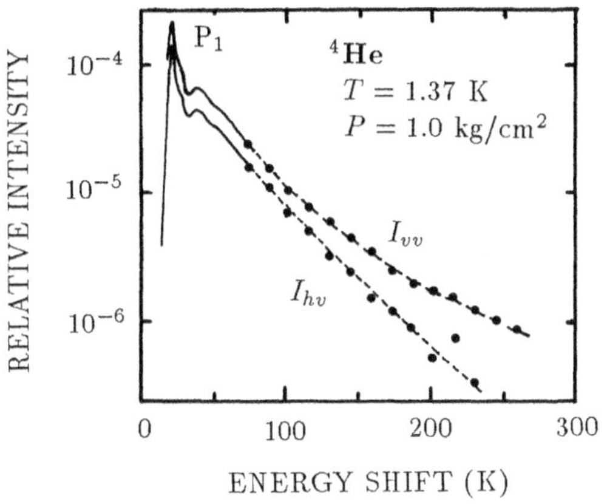

Figure 2. Polarization properties of Raman spectrum of superfluid ^{4}He

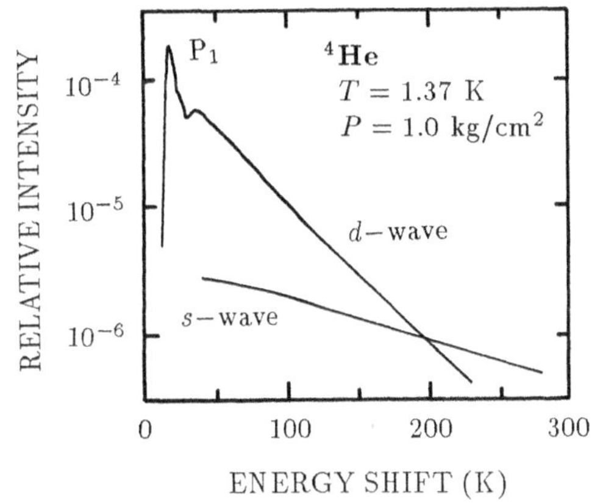

Figure 3. Raman spectra of the d-wave $I_d(\omega)$ and the s-wave $I_s(\omega)$ of superfluid ^{4}He .

functions dominate the scattering. This is the EO Raman process. The change from the DID process to the EO is continuous. They are not distinct.

The elementary process is indirectly probed by the polarization property. By a kinematical theory, Iwamoto obtained the polarization property of Raman spectrum $I_R(\omega)$ as[3]

$$I_R(\omega) = (\boldsymbol{\epsilon}_o \cdot \boldsymbol{\epsilon}_i)^2 I_s(\omega) + [3/4 + (\boldsymbol{\epsilon}_o \cdot \boldsymbol{\epsilon}_i)^2/4] I_d(\omega). \tag{1}$$

The diagonal part of the polarization tensor $\hat{\alpha}(\boldsymbol{r}_i, \boldsymbol{r}_j)$ contributes to the s-wave scattering $I_s(\omega)$. The d-wave scattering $I_d(\omega)$ is related to the off-diagonal part of the tensor. The DID process contributes only to $I_d(\omega)$, while the EO process both to $I_s(\omega)$ and $I_d(\omega)$. Thus presence of the s-wave scattering clearly shows the importance of the EO interaction. Experimentally, the s-wave and d-wave contributions are separately estimated. By taking $\boldsymbol{\epsilon}_o$ vertical, we consider the spectra of two selections of $\boldsymbol{\epsilon}_i$; I_{vv} for the case of $\boldsymbol{\epsilon}_i$ vertical and I_{hv} for $\boldsymbol{\epsilon}_i$ horizontal. From Eq. 1,

$$I_{vv}(\omega) = I_s(\omega) + I_d(\omega), \qquad I_{hv}(\omega) = (3/4) I_d(\omega). \tag{2}$$

A typical result of our measurements[16,17] is shown in Fig. 2 for superfluid ^{4}He at 1.37 K under a pressure of 1 kg/cm^2. This experiment was done using a photomultiplier detector and a double grating monochromator of the instrumental FWHM of 3.7 K. The solid curves are drawing through many data points, which cannot be plotted as distinct points. The broken lines are fit by eye to the closed circle data points. Calculated $I_s(\omega)$ and $I_d(\omega)$ using Eq. 2 from the curves in Fig. 2 are plotted in Fig. 3. The s-partial wave scattering is clearly confirmed. As the energy shift increases, the s-partial wave becomes dominant. Near the two-roton peak P_1, $I_s/I_d \leq 10^{-2}$. The spectrum has practically d-wave symmetry. This is consistent with the early work by Woerner and Greytak[18] which confirmed d-wave symmetry of the two-roton peak. The fact does not necessarily mean that the two-roton Raman scattering occurs by the DID process. We need a theory which would explain whole the spectral features shown in Fig. 3.

A few theoretical discussions of polarization properties were made analytically.[3,19-21] Kleban and Halley made a semi-empirical microscopic calculation and found considerable s-wave contribution.[22] They predicted that the s-wave appears at large energy shift and estimated the ratio of integrated intensities; $\mathcal{I}_s/\mathcal{I}_d = \int I_s(\omega)d\omega / \int I_d(\omega)d\omega$. Their numerical calculation was $0.11 \leq \mathcal{I}_s/\mathcal{I}_d \leq 1.45$. Later Campbell et al.[23] made another microscopic calculation and obtained $0.08 \leq \mathcal{I}_s/\mathcal{I}_d \leq 1.88$. Numerical estimation using the experimental data shown in Fig. 3 is $\mathcal{I}_s/\mathcal{I}_d = 0.09 \pm 0.03$, which marginally agrees with the theoretical prediction. We made polarization measurements both on superfluid and normal ^{4}He . The integrated intensity ratios of all the measurements lie within $0.06 \leq \mathcal{I}_s/\mathcal{I}_d \leq 0.16$. Normal liquid ^{3}He also show similar value.[24] From these observations, the feature shown in Fig. 3 has common origin related to diatomic interactions.

4. TWO-ROTON RAMAN SCATTERING

In Fig. 2, P_1 is the two-roton Raman scattering peak. We have been measuring the two-roton spectrum at various pressures to determine the pressure dependence of the roton-roton interaction. A part of the early results was reported at LT-18 conference[25] and also at the satellite Hiroshima symposium.[15] Recently, we have improved the measurements using a two-dimensional photon counting detector and also reduced the tail of the central component to a negligible order by observing $I_{hv}(\omega)$ spectrum instead of former $I_{vv}(\omega)$. Two typical spectra measured at 0.65 K are plotted in Fig. 4 for pressures of svp and 20 kg/cm^2.[26] With an increase of pressure, the intensity decreases and the peak position shifts to a lower energy. However, the energy shift does not exactly follow the pressure dependence of the roton minimum energy $\Delta_0(n)$ measured by neutron scattering.[27-30] The two-roton peak appears at an energy lower than $2\Delta_0(n)$ at svp but shifts to a higher energy at 20 kg/cm^2.

The two-roton Raman intensity is mainly proportional to the density of state $D_2(\omega)$ of the two rotons created by light. If created two rotons are not interacting,

$$D_2(\omega) \propto \frac{1}{\sqrt{\omega - 2\Delta_0}}\Theta(\omega - 2\Delta_0). \tag{3}$$

Here, $\Theta(x)$ is a unit function which is unit for $x > 0$ and zero for $x < 0$. The peak is expected at $2\Delta_0$ from this. However, the experimental spectrum $I_e(\omega)$ is a convolution of this function with instrumental profile function $F_I(\omega)$ shown by broken curves in Fig. 4;

$$I_e(\omega) \propto \int_{-\infty}^{+\infty} F_I(\omega - \omega')D_2(\omega')d\omega'. \tag{4}$$

As the result of this integral, the peak shifts to a higher energy. In Fig. 5, the observed two-roton peak positions are compared with those expected from the neutron $2\Delta_0(n)$ using Eqs. 3 and 4. The negative (positive) shift means that smaller (larger) energy is needed to

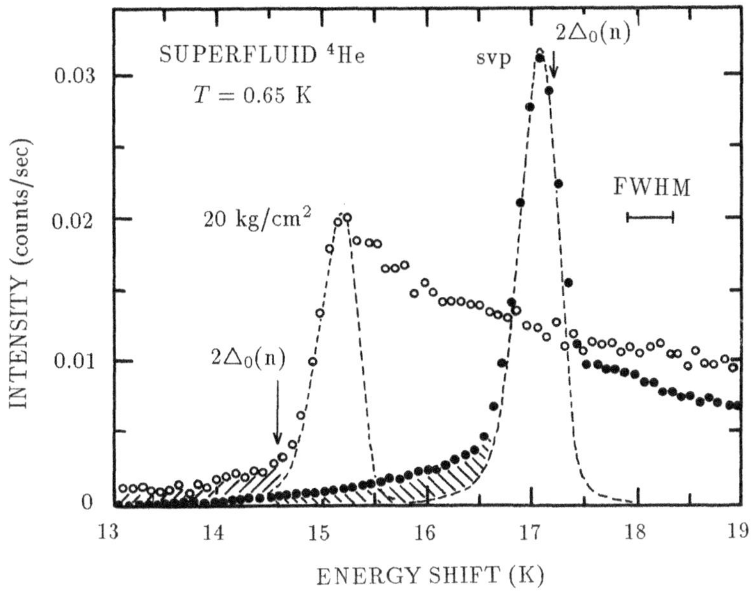

Figure 4. Two-roton Raman spectra of superfluid ^{4}He at svp and at 20 kg/cm^2.

create the roton pair than to create two separated single rotons. The energy difference may be interpreted as roton-roton interaction energy. Bedel, Pines, and Zawadowski[31] predicted that the roton-roton interaction observed by Raman scattering (d-wave of total momentum zero) changes from attractive to repulsive as pressure increases. The result of Fig. 5 is in good agreement with their prediction at least qualitatively.

It is interesting to test whether the energy shift and the change of the spectral shape with pressure can be consistently explained by the roton-roton interaction model proposed by Iwamoto[3] and Ruvalds and Zawadowski[4] (the IRZ model), which was experimentally verified to fit to the svp data by Greytak's group.[5,6] $D_2(\omega)$ of the IRZ model for interacting two rotons with total momentum zero is

$$D_2(\omega) \propto 2\pi\sqrt{E_i}\delta(\omega - 2\Delta_0 + E_i) + \frac{\sqrt{\omega - 2\Delta_0}}{\omega - 2\Delta_0 + E_i}\Theta(\omega - 2\Delta_0). \tag{5}$$

The δ-function represents a single bound state with binding energy E_i for attractive roton-roton interaction. The second term represents unbound states. The first term is a sharp peak with negative shift and the second term a broad peak with positive shift. If the interaction between rotons is repulsive, the bound state disappears and the density of unbound states is given by the same expression as the second term of Eq. 5. In this case E_i is the repulsive energy between two rotons.

In our previous paper, we pointed out that the two-roton spectrum at svp cannot be fitted by Eq. 5 well.[25] It seemed to contradict to Murray et al.[6] Recently, we have confirmed the Raman spectrum due to two-phonon process below the two-roton peak. In Fig. 4, the broken curves are the instrumental profile plotted so that the height and the position of the peak agree with the two-roton peak. As shown by the hatched regions in the figure, responses exist at the left of the two-roton peak. Because there is no branch other than phonons there, these are regarded as scattering by two-phonon processes. To compare with Eq. 5, we took into account the phonon contribution as an adjustable constant. The best fit was obtained with $E_i = 0.2$ K and $2\Delta_0(\mathrm{r}) = 17.28$ K as shown by the solid curve in Fig. 6 (a). The hatched region is the phonon contribution. The inclusion of the phonon contribution makes the fit good and yields roton energy consistent with the neutron result and that obtained by Murray et al.[6] Although they did not take into account the phonon contribution, it was implicitly subtracted included in the Brillouin and Raman background. One experimental discrepancy is that we did not observed *residual* width of the two-roton spectrum claimed by them.

For 20 kg/cm^2, the spectrum must be analyzed with the IRZ model with repulsive interaction because of the positive peak shift. A similar analysis to that of svp, however, showed considerable deviation even at the best fit. Considering uncertainty of Raman and neutron measurements, the repulsive energy of maximum allowance consistent with the peak position is $E_i = 0.29 \pm 0.16$ K. Calculated curves for three repulsive interaction energies are shown in Fig. 6 (b). The deviations at the lower energy side of the peak show that the IRZ model does not give a good fit.

5. FINE STRUCTURE

In our previous publications, we have claimed existence of fine structures.[15,16,32] Former results were measured at 1.3 K. In this paper, spectra measured at 0.65 K are shown in Fig. 7 for (a) svp and for (b) 20 kg/cm^2. The features we notice are appearance of peaks such as P_2, P_3, P_4, and P_5. P_4 and P_5 are not clear in Fig. 7 (b). But they were confirmed at 1.3 K with a better statistical accuracy.[15] Also the change of the slope is noticed at energies labeled as c, d, and e.

If only two-phonon processes are considered, a peak or an anomaly is expected at $2\Delta_0$ and at $2\Delta_1$, where Δ_1 is the maximum of maxon energies. Hirashima and Iwamoto proposed possibility of three-phonon Raman processes.[33,34] Kinematically, multiphonon processes of the zero total momentum are possible. And abrupt change of the spectrum is expected near the energies

$$\omega = \mathrm{n}\Delta_0 + \mathrm{m}\Delta_1 \qquad (\ \mathrm{n,m} = \text{positive integer}\ ). \tag{6}$$

Those energies for two-phonon and three-phonon processes are shown in Fig. 7. The four-roton energy is also shown.

The features in the svp spectrum show good correspondence to those energies. P_1, c, d, P_3, P_2, and the shoulder of P_2 at the higher energy side are respectively observed near the energies $2\Delta_0$, $3\Delta_0$, $2\Delta_1$, $2\Delta_0 + \Delta_1$, $4\Delta_0$, and $\Delta_0 + 2\Delta_1$. Iwamoto predicted a cusp at $2\Delta_1$.[3] Hirashima and Iwamoto analyzed the three-roton process in detail and found that the binding energy may be of the order of 1 K.[34] The feature just at the lower energy side of $3\Delta_0$ may be interpreted as the three-roton process.

On the contrary, the features observed for the spectrum at 20 kg/cm^{2}do not show good correspondence with the extremal energies of Eq. 6 as shown in Fig. 7 (b).

Halley[12] discussed that there are two sources of higher order Raman processes; the quasiparticle interactions such as discussed by Hirashima and Iwamoto and non-linearlity in the relation between the density fluctuations and the boson operators of quasiparticles. How the latter affects the Raman spectral structure is not yet clarified. The model that takes into account the former effect describes the svp data well. As the pressure of the sample increases, however, deviations become significant.

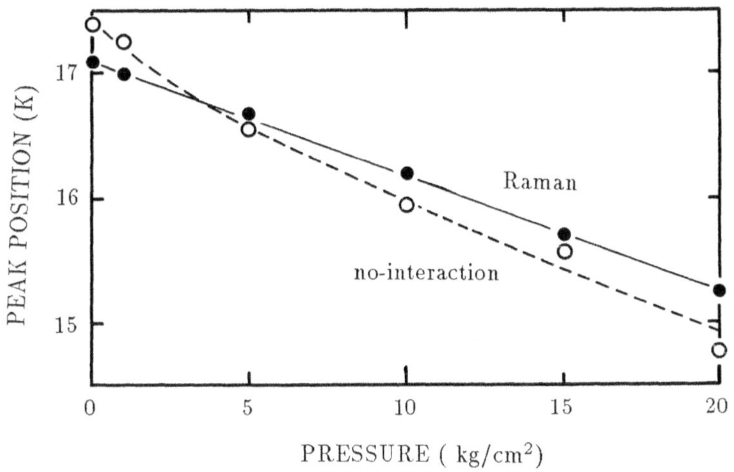

Figure 5. Comparison of the two-roton Raman peak positions with those expected from neutron data assuming no interaction between the two rotons.

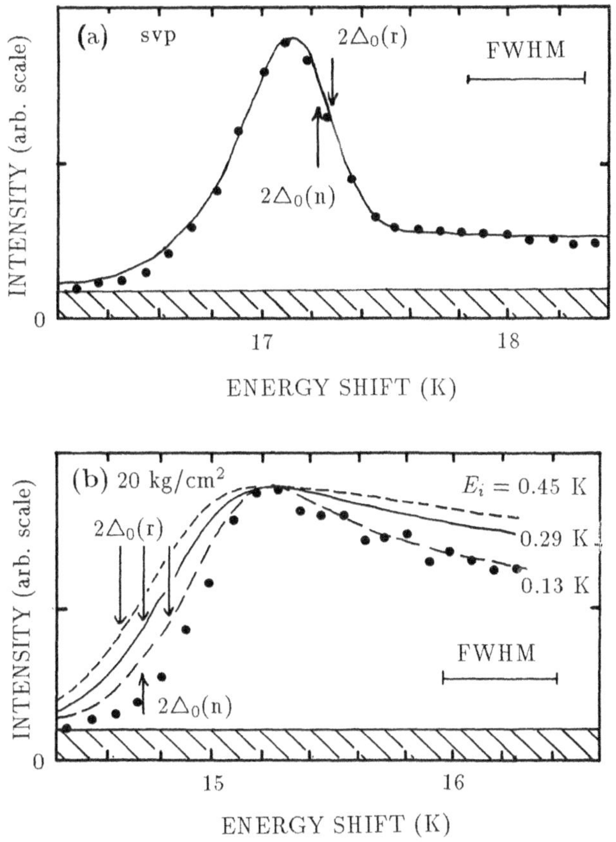

Figure 6. Comparison of the two-roton spectral shape at $T = 0.65$ K with the IRD model; (a) svp and (b) 20 kg/cm^2. The closed circles are data points and the curves are calculated values.

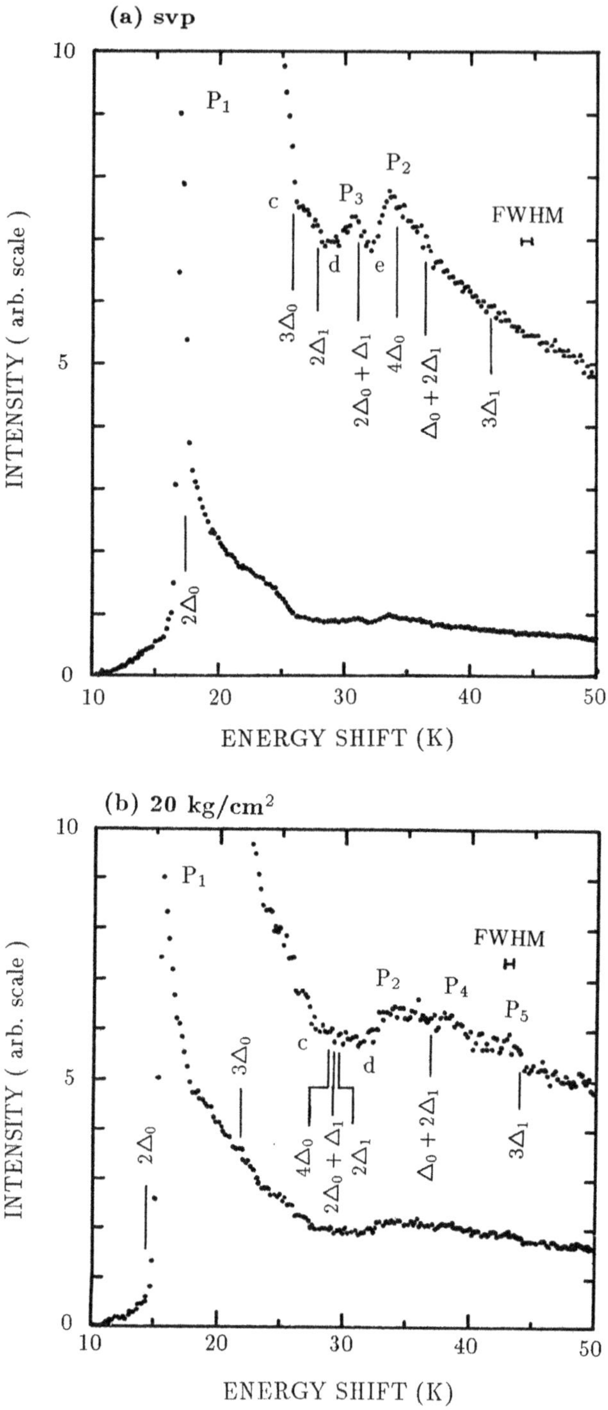

Figure 7. Raman spectra of superfluid ^4He measured at 0.65 K with the instrumental FWHM of 0.75 K. Pressures are (a) svp and (b) 20 kg/cm^2. Fine structures are clearly observed.

6. TEMPERATURE DEPENDENCE OF ROTON PARAMETERS

In spite of extensive experimental and theoretical studies, the temperature dependence of $S(k,\omega)$ determined by inelastic neutron scattering is not well understood.[35] Woods and Svensson proposed a two-component model that the one-phonon peak has a weight proportional to macroscopic superfluid density $\varrho_s(T)$.[36] Talbot et al. claimed the simple multiphonon subtraction model.[30] Recent work by Stirling and Glyde[37] reported different behavior at different wave vectors. At finite temperatures, the roton parameters obtained by neutron scattering is model dependent. Theoretically, Griffin proposed that the Bose-Einstein condensation does not cause a drastic change to $S(k,\omega)$ and only abrupt narrowing is expected as we pass through T_λ from above.[38,39] The Raman spectrum of ^{4}He is given by

$$I_R(\omega) \propto \int\int t(\boldsymbol{q})t(\boldsymbol{q}')H(\boldsymbol{q},\boldsymbol{q}',\omega)d\boldsymbol{q}d\boldsymbol{q}', \tag{7}$$

where $t(\boldsymbol{q})$ is the light-helium coupling factor and $H(\boldsymbol{q},\boldsymbol{q}',\omega)$ is related to $S(k,\omega)$ as

$$H(\boldsymbol{q},\boldsymbol{q}',\omega) = \int_{-\infty}^{\infty} S(\boldsymbol{q},\omega)S(\boldsymbol{q}',\omega-\omega')d\omega'. \tag{8}$$

The temperature broadening of $S(k,\omega)$ broadens $I_R(\omega)$ by a factor of two. Thus temperature dependence of $S(k,\omega)$ may be studied by Raman scattering also.

To explore the temperature dependence, we measured Raman spectrum of ^{4}He at various temperatures under a pressure of 5 kg/cm^2.[40] The spectra measured at the three representative temperatures are shown in Fig. 8: (a) The lowest temperature 0.65 K at which the spectrum is broadened only by the instrumental profile. (b) The temperature 1.75 K below T_λ where temperature broadening is not so small but the spectrum still keeps the feature apparently similar to that at the lowest temperature. (c) The temperature 2.45 K above T_λ at which the spectrum is almost featureless.

Trials of analysis with a few models showed that the one-component model with lorentzian temperature-broadening fits most excellently at all temperatures. Because the width of the peak in Fig. 8 (a) is determined only by the instrumental width, being the temperature width is negligible, convolution of it with a lorentzian function was calculated for various widths 2Γ to find the best fit a spectrum at higher temperatures. Γ is the width of a single roton. In order to superpose a best fit calculation with an experimental plot, the calculated curve must be shifted along the energy axis. We interpret the difference as the temperature variation of the roton energy. Thus the analysis enabled simultaneous determination of the roton width and the roton energy. As shown in 8 (b) and (c) by broken curves, the calculated curves fit excellently. In these figures, the peak heights of both experimental and calculated plots are normalized to show the fit clearly. Actual peak height decreased with an increase of temperature as shown in Fig. 9. The decrease can be explained also by the convolution. As the width increases, the peak height of the convoluted spectrum decreases. In Fig. 9, the broken curve is the decrease of the calculated peak height due only to the increase of the width. The fit with experimental data is good. This shows that total integarated intensity does not show a feasible temperature variation. The fact that both the spectral shape and the intensity do not show a marked change above and below the T_λ demonstrates that there is no qualitative change of the Raman spectrum across it. It is consistent with Griffin's proposal.[38,39]

The temperature dependence of the roton width Γ is shown in Fig. 10. The temperature dependence of the single roton energy is shown in Fig. 11.

Because the roton energy is given as the function of temperature with no ambiguity, we can make a comparison of the temperature dependence of the roton width with the well

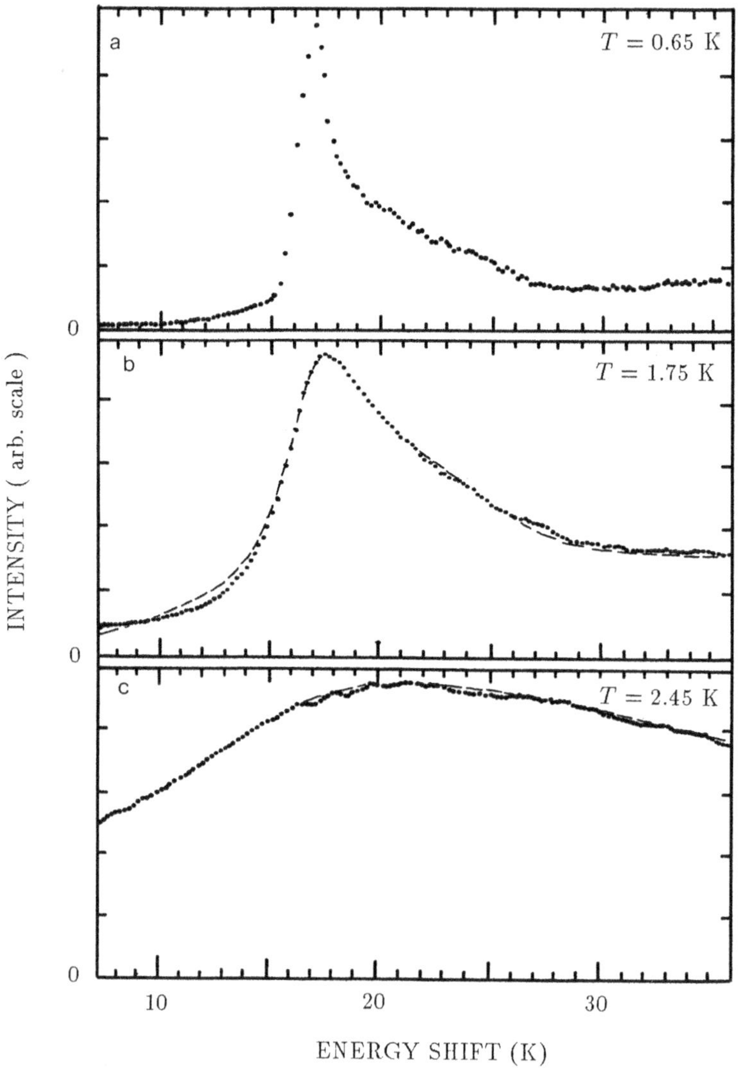

Figure 8. The Raman spectra of liquid ^{4}He under pressure of 5 kg/cm^{2}measured at three different temperatures. The broken lines are the calculated curves.

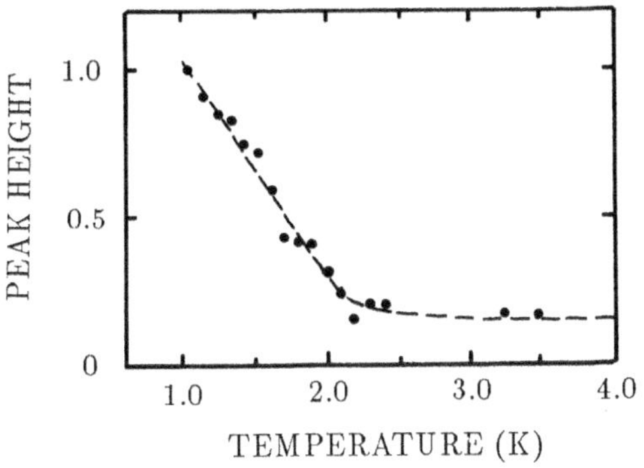

Figure 9. Temperature dependence of the peak height of the two-roton Raman scattering. The closed circles are data points. The broken curve is the calculated variation due to lorentzian broadening. It is normalized to pass the data point at 1 K.

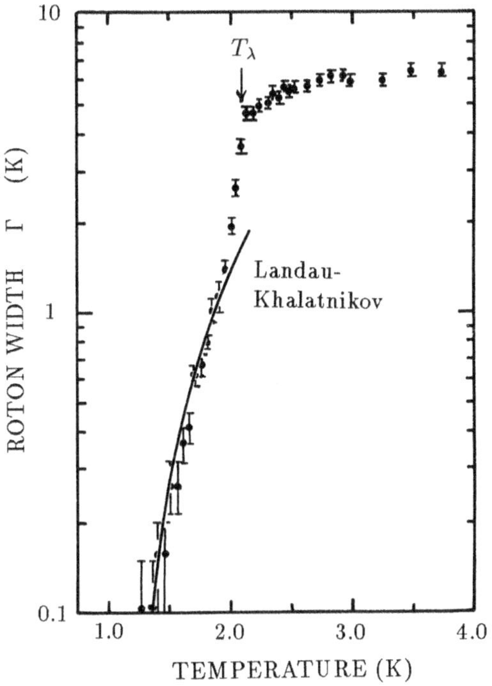

Figure 10. The temperature dependence of the roton width determined by Raman scattering. The solid curve is the Landau-Khalatnikov theory.

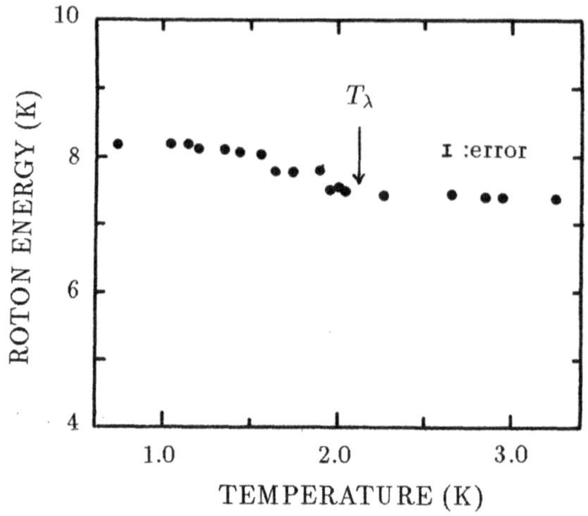

Figure 11. The temperature dependence of the roton energy determined by the Raman scattering

known Landau-Khalatnikov theory as shown by the solid curve in Fig. 10. Above about 1.80 K, the observed width show significant upward deviation from the theory. This suggests that an additional broadening mechanism other than the roton-roton interactions appears near T_λ. A possible candidate for it is vortex excitations. Theories suggest that vortex-excitations exceed roton excitations in their number near T_λ.[41] And the deviation may be detecting the vortex-excitations through roton-vortex interactions.

ACKNOWLEDGEMENTS

Discussion and collaboration with M. Watabe, K. Nagai, M. Udagawa, and N. Ogita are gratefully acknowledged.

REFERENCES

1. J. W. Halley, Phys. Rev. **181**: 338 (1969).
2. T. J. Greytak and J. Yan, Phys. Rev. Lett. **22**: 987 (1969).
3. F. Iwamoto, Prog. Theor. Phys. (Japan) **44**: 1135 (1970).
4. J. Ruvalds and A. Zawadowski, Phys. Rev. Lett. **25**: 333 (1970).
5. T. J. Greytak, R. L. Woerner, J. Yan, and R. Benjamin, Phys. Rev. Lett. **25**: 1547 (1970).
6. C. A. Murray, R. L. Woerner, and T. J. Greytak, J. Phys. **C 8**: L90 (1975).
7. A. D. B. Woods and R. A. Cowley, Rep. Prog. Phys. **36**: 1135 (1973).
8. P. Kleban and R. Hastings, Phys. Rev. **B 11**: 1878 (1975).
9. P. Kleban, Phys. Rev. **B 19**: 3511 (1979).
10. J. Ruvalds and T. Regge, eds., "Quantum Liquids", North Holland, Amsterdam (1978).
11. K. Ohbayashi and M. Watabe, eds.,"Elementary Excitations in Quantum Fluids", Springer-Verlag, Berlin (1989).
12. J. W. Halley, Theory of Light Scattering from ^{4}He , *in*: ibid.
13. F. Iwamoto, Effect of the Three-Excitation States on the Raman Scattering from superfluid ^{4}He , *in*: ibid.

14. J. Ruvalds, Reflections on Excitations in Superfluid Helium, *in*: ibid.
15. K. Ohbayashi, Raman Scattering of Liquid ^{4}He , *in*: ibid.
16. K. Ohbayashi and M. Udagawa, Phys. Rev. **B 31**: 1324 (1985).
17. M. Udagawa, H. Nakamura, M. Murakami, and K. Ohbayashi, Phys. Rev. **B 34**: 1563 (1986).
18. R. L. Woerner and T. J. Greytak, J. Low Temp. Phys. **13**: 149 (1973).
19. M. J. Stephen, Phys. Rev. **187**: 279 (1969).
20. A. L. Fetter, J. Low Temp. Phys. **6**: 487 (1972).
21. S. Nakajima, Prog. Theor. Phys. (Japan) **45**: 353 (1971).
22. P. Kleban and J. W. Halley, Phys. Rev. **B 11**: 3520 (1975).
23. C. E. Campbell, J. W. Halley and F. J. Pinski, Phys. Rev. **B 21**: 1323 (1980).
24. M. Udagawa, T. Imada, and K. Ohbayashi, J. Phys. **C 20**: 1063 (1987).
25. K. Ohbayashi, T. Akagi, N. Ogita, M. Watabe, and M. Udagawa, Jpn. J. Appl. Phys. **26**, Supplement 26-3: 5 (1987).
26. N. Ogita, M. Watabe, M. Udagawa, and K. Ohbayashi, "Proceedings of LT-19 Conference", (1990).
27. A. D. B. Woods, P. A. Hilton, R. Scherm, and W. G. Stirling, J. Phys. **C 10**: L45 (1977).
28. W. G. Stirling, New High-Resolution Neutron Scattering Investigations of Excitations in Liquid Helium-4, *in*: "Proc. 2nd. Int. Conf. on Phonon Physics," J. Kollar, N. Kroo, N. Menyhard, T. Siklos, eds., World Scientific, Singapore (1985).
29. O. W. Dietrich, E. H. Graf, C. H. Huang, and L. Passell, Phys. Rev. **A 5**: 1377 (1972).
30. E. F. Talbot, H. R. Glyde, W. G. Stirling, and E. C. Svensson, Phys. Rev. **B 38**: 11229 (1988).
31. K. S. Bedel, D. Pines, and A. Zawadowski, Phys. Rev. **B 29**: 102 (1984).
32. K. Ohbayashi, M. Udagawa, and M. Watabe, Can. J. Phys. **65**: 1571 (1987).
33. D. S. Hirashima and F. Iwamoto, Prog. Theor. Phys. (Japan), **75**: 744 (1986).
34. D. S. Hirashima and F. Iwamoto, Phys. Rev. **B 40**: 10891 (1989).
35. E. C. Svensson, Temperature Dependence of $S(Q, \omega)$ for Liquid ^{4}He , *in*: "Elementary Excitations in Quantum Fluids," K. Ohbayashi and M. Watabe,eds., Springer-Verlag, Berlin (1989).
36. A. D. B. Woods and E. C. Svensson, Phys. Rev. Lett. **41**: 974 (1978).
37. W. G. Stirling and H. R. Glyde, Phys. Rev. **B 41**: 4224 (1990).
38. A. Griffin, Can J. Phys. **65**: 1368 (1987).
39. A. Griffin, Elementary Excitations in Bose-Condensed Liquids and Gases at Finite Tempratures, *in*: "Elementary Excitations in Quantum Fluids", K. Ohbayashi and M. Watabe, eds., Springer-Verlag, Berlin (1989).
40. K. Ohbayashi, M. Udagawa, H. Yamasita, M. Watabe, and N. Ogita, "Proc. LT-19 Conf." (1990).
41. G. A. Williams, Phys. Rev. Lett. **59**: 1926 (1987).

INTERPRETATION OF INELASTIC LIGHT SCATTERING MEASUREMENTS IN ^{4}He

J. W. Halley

School of Physics and Astronomy
University of Minnesota
Minneapolis, MN 55455

M. S. Korth

Physics Department
University of Minnesota
Morris, MN 56267

We describe calculations of the inelastic light scattering spectrum for frequencies of the order of up to several times the roton frequency. The calculations are based on the two body approximation in a theoretical description of the fluid due to Pandiharipande and Manousakis. We use the results to determine the role of two mechanisms for the development of multiquasiparticle structure in the spectrum which we proposed earlier. We find that the mechanism arising from quasiparticle anharmonicities plays a much larger role in producing these structures in the spectrum than does the mechanism arising from nonlinearities in the relation between the density and quasiparticle operators. We find that the three quasiparticle couplings on which the theory rests depend strongly on the magnitudes and directions of the quasiparticles in a way which strongly favors processes involving colinear momenta. This has striking qualitative effects on the details of the predicted spectra.

INTRODUCTION

Second order inelastic light scattering was first observed in superfluid helium four twenty one years ago in experiments by Greytak and Yan[1]. The experiment was preceded by a prediction [2]which, while qualitatively in accord with the experiments, did not get the details right and this provided the occasion for a great deal of interesting theoretical work in the next few years. The most important new aspect of the subsequent theories was

Excitations in Two-Dimensional and Three-Dimensional Quantum Fluids
Edited by A.G.F. Wyatt and H.J. Lauter, Plenum Press, New York, 1991

the suggestion [3-4] that the two rotons excited by the light interact attractively and form a bound state. Further experiments by the Greytak group [5-6] at higher frequency resolution gave evidence of the expected structure below the frequency corresponding to two free rotons and these experiments were fit quite successfully to the theory of reference 4. With these successes in hand many felt that the fundamental processes involved in this experiment were well understood and interests turned elsewhere. That this satisfaction was illusory has been discussed elsewhere [7]. More recently, experiments by Ohbayashi and his group [8-10] have provided data which can clarify the nature of the coupling to the light, the role of the excluded volume constraint and the accuracy of theories of quasiparticle dynamics. In this paper, we focus on one aspect of those results, the apparently rather rich structure of the spectrum which these workers reported in the relatively high frequency region between three and about six times the roton frequency. We wish to address the question of what mechanisms could be causing such structure and to evaluate the effectiveness of a modern theoretical description of the fluid in accounting for them.

In the next section we briefly review some general features which a theory of the light scattering process must possess including two possible mechanisms for the multiquasiparticle processes reported in the work by the Japanese group. In the third section describe how we have calculated the light scattering spectrum in this region and give results. The last section contains a discussion and conclusions.

FUNDAMENTAL THEORETICAL CONSIDERATIONS

We start with a rather general expression for the interaction H_{eff} of the electromagnetic field with the fluid :

$$H_{eff} = \int d\vec{r} \int d\vec{r'} \rho(\vec{r}) \vec{E}(\vec{r}) \cdot \alpha(\vec{r}, \vec{r'}) \cdot \vec{E}(\vec{r'}) \rho(\vec{r'}) \tag{1}$$

where $\rho(\vec{r})$ is the operator

$$\rho(\vec{r}) = \sum_i \delta(\vec{r} - \vec{r_i}) \tag{2}$$

($\vec{r_i}$ is the position of the ith helium atom), $\vec{E}(\vec{r})$ is the microscopic electric field and $\alpha(\vec{r}, \vec{r'})$ is a polarizability tensor giving the local dipole moment of the medium at the point $\vec{r}$, given that there is a field at the point $\vec{r'}$. From this coupling one finds that the extinction coefficient $h(\omega)$ is given up to constants which will not interest us here by

$$h(\omega) \propto \int < H_{eff}(0) H_{eff}(t) > e^{-i\omega t} dt \tag{3}$$

$$\propto \int d\vec{r_1} \int d\vec{r_2} \int d\vec{r_3} \int d\vec{r_4} \hat{\epsilon}_0 \cdot \alpha(\vec{r_1}, \vec{r_2}) \cdot \hat{\epsilon}_n \hat{\epsilon}_0 \cdot \alpha(\vec{r_3}, \vec{r_4}) \cdot \hat{\epsilon}_n < \rho(\vec{r_1}, 0) \rho(\vec{r_2}, 0) \rho(\vec{r_3}, t) \rho(\vec{r_4}, t) > e^{-i\omega t} dt$$

$$\propto \int d^3q \int d^3q' t(\vec{q}) t(\vec{q'}) S_2(\vec{q}, \vec{q'}, \omega)$$

Here $\hat{\epsilon}_0$ and $\hat{\epsilon}_n$ are respectively the polarizations of the incoming and outgoing light in the scattering process. The factors $t(\vec{q})$ are defined as

$$t(\vec{q}) = \int d^3r e^{i\vec{q}\cdot\vec{r}} \hat{\epsilon}_o \cdot \alpha(\vec{r}) \cdot \hat{\epsilon}_n \tag{4}$$

$$t(\vec{q}) = t_d(q) \hat{\epsilon}_o \cdot (3\hat{q}\hat{q} - \mathbf{1}) \cdot \hat{\epsilon}_n + t_s(q) \hat{\epsilon}_o \cdot \epsilon_n \tag{5}$$

in which $t_d(q) = 4\pi \int r^2 j_2(qr)\alpha_d(r)dr$ and $t_s(r) = 4\pi \int r^2 j_0(qr)\alpha_s(r)dr$. With these developments it then follows that the extinction coefficient can be written in the form

$$h(\omega) \propto I_s(\omega)(\epsilon_o \cdot \epsilon_n)^2 + I_d(\omega)[3/4 + 1/4(\hat{\epsilon}_o \cdot \hat{\epsilon}_n)^2]. \tag{6}$$

Here

$$I_d(\omega) = \int 4\pi q^2 dq \int 4\pi q\prime^2 t_d(q) t_d(q\prime)(4/25) S_2^d(q, q\prime, \omega) \tag{7}$$

and

$$I_s(\omega) = \int 4\pi q^2 dq \int 4\pi q\prime^2 t_s(q) t_s(q\prime) S_2^s(q, q\prime, \omega) \tag{8}$$

The functions $S_2^{s,d}(q, q\prime, \omega)$ are the l=0,2 Legendre projections of the four point correlation function $S_2(\vec{q}, \vec{q\prime}, \omega)$:

$$S_2^{s,d}(q, q\prime, \omega) = (2l+1)/2 \int P_l(cos\gamma_{\hat{q},\hat{q}\prime}) S_2(\vec{q}, \vec{q\prime}, \omega) dcos\gamma_{\hat{q},\hat{q}\prime}, \quad l = 0, 2 \tag{9}$$

The integrated s-wave scattering is weaker[8] than the d-wave scattering by about a factor of 10 and has been discussed elsewhere[7]. We will not consider it further here.

As we have also discussed elsewhere[7], the hardest theoretical part of the problem is to be sure that excluded volume constraints have been correctly included in the calculation of the four point density correlation function in these expressions. One approach to a calculation including these effects is described in references 11 and 12 . In the language of the present formulation, this work assumes that the (up to orthogonalization corrections which we will not discuss here)

$$\rho_B(\vec{k}) = \sqrt{Z}(\vec{k})(\alpha^{\dagger}_{\vec{k}} + \alpha_{-\vec{k}}) \tag{10}$$

where the operators $\alpha_{\vec{k}}$ enter a quasiparticle Hamiltonian of the sort written earlier. The "density" $\rho_B(\vec{k})$ is related to the spatial Fourier transfrom of the particle density through the relation

$$\rho_B(\vec{k}) = \sum_i^N e^{i\vec{k}\cdot\vec{r}_i}(1 + \sum_{j\neq i} \vec{k} \cdot \vec{r}_{ij}\eta(r_{ij})). \tag{11}$$

Without the term involving $\eta(r)$, this would just be the spatial Fourier transform of the particle density. The added term is intended to take account of "backflow" in the variational wavefunction describing the quasiparticle excitations in a way somewhat like the one used in the original variational calculations of Feynman and Cohen[13]. The function $\eta(r)$ was determined variationally in an approximate way in references 12 where the the quasiparticle excited state was assumed to be equal to $\rho_B(\vec{k})$ times a Jastrow-like ground state and the energy was minimized with respect to parameters in the function $\eta(r)$. In order to correctly describe the observed quasiparticle spectrum, it was necessary to assume that these quasiparticles interact according to the quasiparticle Hamiltonian

$$H = \sum_{\vec{q}} \epsilon_{\vec{q}} \alpha^{\dagger}_{\vec{q}} \alpha_{\vec{q}} + \sum_{\vec{q},\vec{k}} g_3(\vec{q}, \vec{k}) \alpha^{\dagger}_{\vec{k}} \alpha_{\vec{q}} \alpha_{\vec{k}+\vec{q}} + h.c. \tag{12}$$

$$+ \sum_{\vec{q},\vec{k},\vec{l}} g_4(\vec{q}, \vec{k}, \vec{l}) \alpha^{\dagger}_{\vec{q}} \alpha^{\dagger}_{\vec{k}} \alpha_{\vec{l}} \alpha_{\vec{q}+\vec{k}-\vec{l}} + \cdots$$

The operators $\alpha_{\vec{q}}$ are boson destruction operators. In the approach of references 12 it is possible to calculate the couplings g_3 and g_4 microscopically and they report results for g_3. A full calculation along these lines takes explicit account of the hard core of the helium atoms and should predict the light scattering spectrum correctly. We report some preliminary results of this sort in the next section.

In this paper we will focus attention on problem of understanding the multi-particle structures seen reported in references 9 and 10 . In earlier work [7] we pointed out two possible mechanisms for such structures. We refer to the two mechanisms as 1) quasiparticle anharmonicity and 2) nonlinearities in the density-quasiparticle relation. The terms involving quasiparticle interactions in the quasiparticle Hamiltonian above (involving g_3,g_4,...) are what is meant by quasiparticle anharmonicities. They lead to multiquasiparticle processes in the light scattering spectrum through splitting of one or both members of the quasiparticle pair initially excited by the light. The mechanism for quasiparticle nonlinearities arising from nonlinearities in the density quasiparticle relation can be understood in terms of the expression (11) from which one easily shows that

$$\rho(\vec{k}) = \rho_B(\vec{k}) - \sum_{i \neq j} e^{i\vec{k}\cdot\vec{r}_{ij}} i\vec{k}\cdot\vec{r}_{ij}\eta(\vec{r}_{ij}) \tag{13}$$

The last term can be written in turn in terms of two factors of the density ρ. Finally, iterating the preceding equation by substituting ρ_B for ρ in the second term one obtains

$$\rho(\vec{k}) = \rho_B(\vec{k}) + \sum_{\vec{q}} \rho_B(\vec{q})\rho_B(\vec{k}-\vec{q})M(\vec{q},\vec{k}) + \ldots \tag{14}$$

in which

$$M(\vec{q},\vec{k}) = -i\int \frac{d^3r}{V}(\vec{k}\cdot\vec{r})\eta(\vec{r})e^{i\vec{q}\cdot\vec{r}}. \tag{15}$$

This, together with the relation $\rho_B(\vec{k}) \propto (\alpha^\dagger_{\vec{k}} + \alpha_{-\vec{k}})$ generates terms describing three and more quasiparticle excitations when it is used in the basic expression for the extinction coefficient written down in Eq. (3). This mechanism for multiquasiparticle excitation does not require that the quasiparticles interact at all. As discussed in reference 14 , these nonlinear terms in the density-quasiparticle relation are likely to be required in order to satisfy the excluded volume constraint. In the present formulation, based on the formalism of references 11 and 12 one sees the connection to the excluded volume constraint in a very physical way, because the terms involving η in the relation of ρ_B to ρ are thought physically to arise to account for "backflow" which occurs in order that the particles not overlap during the motion associated with the excitation. We report a preliminary version of a calculation incorporating both these mechanisms of multiparticle excitation in the light scattering spectrum in the next section.

CALCULATION OF THE LIGHT SCATTERING SPECTRUM

We use the two body approximation of reference 11 to calculate the light scattering spectrum. In this approximation, one approximates the ground state function of liquid helium by a wave function of Jastrow variational form:

$$|\Psi_0\rangle = \prod_{1\leq l<j\leq N} f_J(r_{ij}) \tag{16}$$

In practise, in the two body approximation, the only properties of the ground state which are needed are those described by the radial distribution function $g(r)$ and for this function, we follow references 11 and 12 and use experimental values [15]. The variationally calculated $g(r)$

is in reasonably good agreement with experiment. The quasiparticle states are described in zeroth approximation in the scheme of references 11 and 12 by

$$|n>=\rho_B(\vec{k}_1)\rho_B(\vec{k}_2)...\rho_B(\vec{k}_n)|\Psi_0> + orthogonalization\ terms \tag{17}$$

where $\rho_B(\vec{k})$ was defined in (11). The energy of the state $|1>$ is denoted $e_B(\vec{k})$. We used the "short range" form of the backflow function discussed in 11

$$\eta(r) = Aexp(-[(r-r_0)/w_0]^2) \tag{18}$$

with the parameters reported by reference 11 to minimize the energy of the one quasiparticle states: $A = 0.15; r_0 = 0.8\sigma; w_0 = 0.44\sigma$ where $\sigma = 2.55\mathring{A}$ is the hard core radius of helium. In the two body approximation, all the three body integrals which enter the calculation of matrix elements are dropped. As we will see, this leads to a single quasiparticle spectrum which is not in extremely good agreement with experiment. Nevertheless, we can learn a great deal about the amplitude of multiparticle signatures in the light scattering spectrum from such a calculation.

To describe the scattering in terms of this approach we define the quantity $D_2(\vec{q},\vec{q}',\omega)$ by

$$D_2(\vec{q},\vec{q}',\omega) =< \Psi_0|\rho(\vec{q}')\rho(\vec{q}')\left(\frac{1}{H-E_0-\omega-i\epsilon}\right)\rho(\vec{q})\rho(\vec{q})|\Psi_0> \tag{19}$$

It is easy to show that

$$2ImD_2(\vec{q},\vec{q}',\omega) = S_2(\vec{q},\vec{q}',\omega) \tag{20}$$

as defined in the preceding section so that calculation of $D(\vec{q},\vec{q},\omega)$ will give us the needed function. As in reference 12 , we let $H = H_0 + H_I$ where H_0 is diagonal in the states (17) and make a perturbation expansion in H_I. If we expand the operators $\rho(\vec{q})$ in terms of the operators $\rho_B(\vec{q})$ using (14) and take the first term, we get a description in terms of a two quasiparticle bare propagator and its interactions. (A discussion of the contribution of the next term in the expansion (14) appears below.) We select a subset of the perturbation series corresponding to ignoring interaction between the quasiparticles but including all their self energy corrections . Then we obtain

$$D_2(\vec{q},\vec{q}',\omega) = -|<\Psi_0|\rho_B(\vec{q})\rho_B(-\vec{q})|\vec{q},-\vec{q}>|^2\delta_{\vec{q},-\vec{q}'}\times \tag{21}$$

$$\int\frac{d\omega'}{2\pi i}\frac{1}{(e_B(\vec{q})+\Sigma(\vec{q},\omega-\omega')-(\omega-\omega')-i\epsilon)}\frac{1}{(e_B(-\vec{q})+\Sigma(-\vec{q},\omega')-\omega'-i\epsilon)}$$

where Σ is the self energy. The state $|\vec{q},-\vec{q}>$ is the normalized and orthogonalized two quasiparticle state defined by (17) . In the calculations reported here ,we have approximated the matrix element whose square appears in front of the integral as $<\Psi_0|\rho_B(\vec{q})|\vec{q}>\equiv N_{\vec{q},\vec{q}}$. It is important to note that this last approximation can contribute to violations of the hard core constraint discussed in the introduction and we hope to improve on it. The integration in (21) is done by use of Cauchy's theorem. We define the complex numbers $z_i', i = 1...,n$ as the solutions to the equation

$$z_i' = e_B(-\vec{q}) + \Sigma(-\vec{q}, z_i') - i\epsilon \tag{22}$$

giving

$$D_2(\vec{q},\vec{q}',\omega) = \tag{23}$$

$$\sum_{i=1}^{n}\frac{|<\Psi_0|\rho_B(\vec{q})|\vec{q}>|^2\delta_{\vec{q},-\vec{q}'}}{e_B(\vec{q})+e_B(-\vec{q})+\Sigma(-\vec{q},z_i')+\Sigma(\vec{q},\omega-z_i')-\omega-2i\epsilon}$$

It is quite feasible to evaluate the spectrum using this expression. However, for numerical simplicity we make the additional assumptions that 1) we can replace z_i' in the arguments of the self-energies of the last equation by its real part and 2) the contribution of the pole i to the sum in the last expression is dominantly near $\omega = e_{sc}^i(\vec{q}) + e_{sc}^i(-\vec{q})$ where $e_{sc}^i(\vec{q})$ is a solution to

$$e_{sc}^i(\vec{q}) = e_B(\vec{q}) + Re\Sigma(\vec{q}, e_{sc}^i(\vec{q})) \tag{24}$$

With these further assumptions the expression for $D_2(\vec{q},\vec{q},\omega)$ becomes

$$\sum_{i=1}^{n} \frac{|N_{\vec{q},\vec{q}}|^2}{2e_{sc}^i(\vec{q}) + 2iIm\Sigma(\vec{q}, e_{sc}^i(\vec{q})) - \omega - 2i\epsilon} \tag{25}$$

This has an obvious physical interpretation . To evaluate it we need to compute the self-energy function arising from the perturbation expansion.

To evaluate the self energy we use the lowest order expression

$$Im\Sigma(\vec{k},\omega) = -\frac{\pi}{2}\sum_{\vec{l},\vec{m}} |g_3(\vec{k},\vec{l},\vec{m})|^2\delta(\omega - e_B(\vec{l}) - e_B(\vec{m})) \tag{26}$$

(A better approximation would be to use the the solutions to (24) in place of e_B in the self energy and solve the last equation and (24) together.) To evaluate the real part of the self energy we use the Kramers-Kronig relation, as in reference 12. At large momenta, it is not hard to show that, because all the bare spectra become free particle like, the real part of the self-energy acquires a divergence so that it increases with the square root of the upper limit of the Kramers-Kronig integral. We made a preliminary exploration of the effects of this on our calculations by studying the self-energy as a function of momentum cutoff and of frequency at high frequencies. We conclude that the divergence is not strongly affecting our results, but its existence is a reason for treating the application of these approximations with caution. To evaluate (26) we need the values of the coupling parameters g_3 which we obtain by evaluating expressions given in reference 11 . It is here that we confine ourselves to the two body (TB) approximation of that paper. We show values of $g_3(\vec{k},\vec{l},\vec{m})$ (for $\vec{k} = \vec{l} + \vec{m}$) for a few values of $\vec{k}$ in Figure 1.

To finally evaluate the light scattering spectrum we need expressions for the function $t_d(q)$ which describes the coupling of the light to the fluid. We made a fit to the anisotropic part of the pair polarizability of helium as reported by Dacre[16] which takes the following analytical form:

$$\alpha_d(r) = \frac{A}{(Br^3 + C)}(1 - e^{-Dr^3}) \tag{27}$$

in which $A = 11.4264\text{Å}^3$, $B = 6.748\text{Å}^{-3}$, $C = 2.0$ and $D = 0.108\text{Å}^{-3}$. The corresponding values of the function $t_d(q)$ are compared with the dipole induced dipole form first used by Stephen [17] in Figure 2. It turns out that details of the spectrum in the high frequency region are quite sensitive to this function as shown below.

We show solutions to the equation (24) in Figure 3. There is a very sharp quasiparticle branch and many broader multiparticle branchs. The quasiparticle spectrum differs from the experimental one [18] by as much as 20 % . In particular, the maxon energy is 12.5 ° K compared to the experimental value of 13.8 ° K and the roton energy is 10.8 ° K compared to the experimental value of 8.7 ° K. This is consistent with the results of 12 for the TB approximation. Using these and the other approximations discussed, we obtain the d-wave light scattering spectrum shown in Figure 4. The positions of the peaks cannot correspond to those in experiment because of the deficiencies in the quasiparticle spectrum just noted.

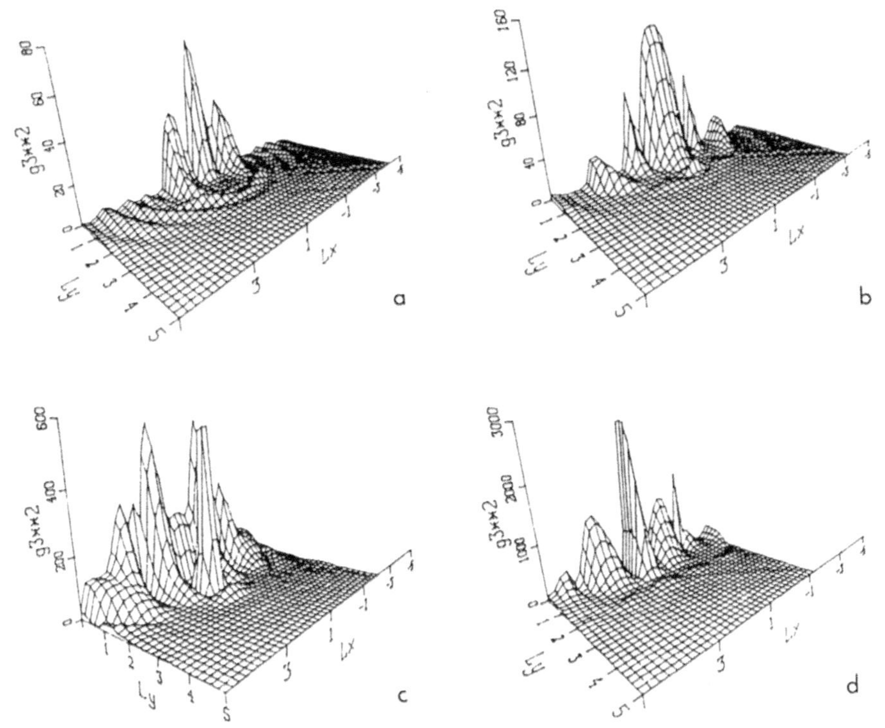

Figure 1. The coupling functions $|g_3(\vec{k},\vec{l},\vec{m})|^2$ for various values of $|\vec{k}|$. $\vec{k}$ lies along the x axis. a) $|\vec{k}|$=1.08 $\AA^{-1}$, b) $|\vec{k}|$ = 1.92 $\AA^{-1}$, c) $|\vec{k}|$ = 3.00 $\AA^{-1}$ d) $|\vec{k}|$ = 3.96 $\AA^{-1}$.

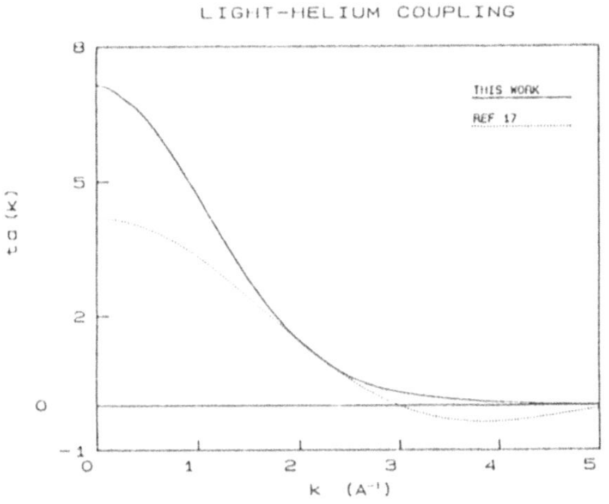

Figure 2. The coupling matrix element $t_d(q)$ from equation (27) compared with that of reference 17.

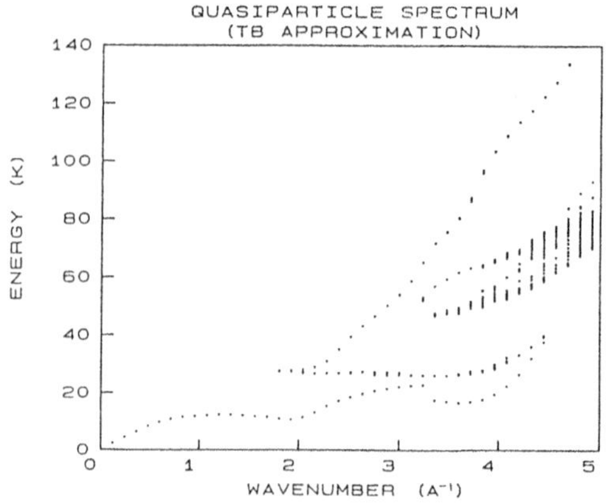

Figure 3

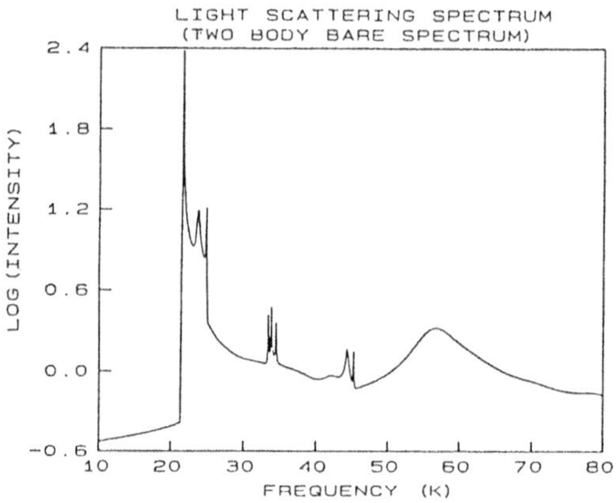

Figure 4

The solutions to equation (24) and the light scattering spectrum using the bare spectrum from the two body approximation of references 11 and 12 .

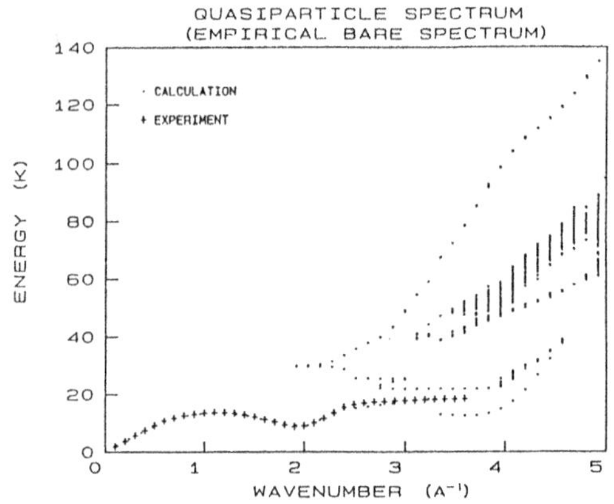

Figure 5

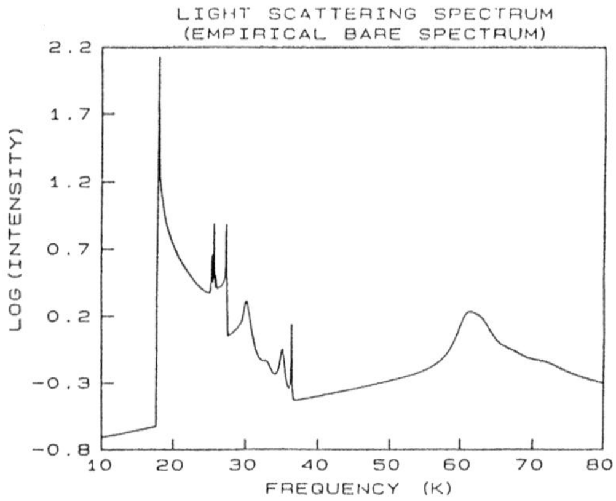

Figure 6

Same as Figures 3 and 4 but with the empirical bare spectrum. The plusses (+) in Figure 5 show the quasiparticle spectrum from inelastic neutron experiments.

The correct way to improve the quasiparticle spectrum is to go to a better approximation, such as the extended two body approximation of reference 12. However, to get some idea how well such a calculation is likely to improve the result, we have done the calculation of the spectrum again using a function $e_B(\vec{q})$ which is chosen to give a quasiparticle spectrum $e^1_{sc}(\vec{q})$ which is in good agreement with the experiments as shown in Figure 5 which shows the solutions to equation (24) as well as the experimental quasiparticle spectrum. The light scattering spectrum using this $e_B(\vec{q})$ is shown in Figure 6. The peaks now correspond to frequencies like those which would be identified as multiparticle peaks in the experimental spectrum. In Figure 7 we show the predicted light scattering spectrum using the Stephens dipole-induced -dipole function $t_d(q)$ and the empirical bare quasiparticle spectrum in order to illustrate the effects of variations in the function $t_d(q)$.

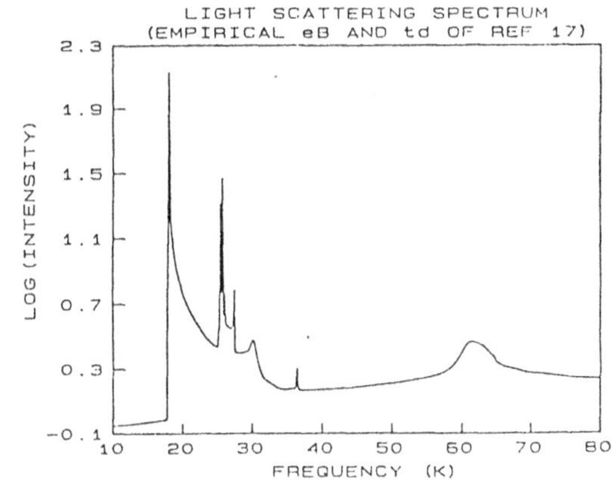

Figure 7. Same as Figure 6 but with the function $t_d(q)$ of reference 17.

Finally we consider the light multiquasiparticle spectrum arising from the second of the mechanisms discussed in the introduction. The next term in the expansion (14) gives a contribution proportional to the three quasiparticle propagator in the spectrum. We have evaluated the ratio of the frequency integral of this contribution to the frequency integral of the contribution from the two quasiparticle propagator, which with our approximations takes the form

$$I_3/I_2 = \frac{\sum_{\vec{k},\vec{q}} |t_d(\vec{k})|^2 N_{\vec{k},\vec{k}} |M(\vec{k},\vec{q})|^2 N_{\vec{q},\vec{q}} N_{\vec{k}-\vec{q},\vec{k}-\vec{q}}}{\sum_{\vec{k}} |t(\vec{k})|^2 |N_{\vec{k},\vec{k}}|^2} \tag{28}$$

We find that this ratio is about .02 with our approximations so that the second type of multiparticle excitation mechanism appears to be making a small contribution to the light scattering spectrum within this model. Furthermore, analysis of the wave vector dependence of the summands in (28) suggests that this conclusion will hold across the frequency spectrum.

DISCUSSION

The structure of the coupling constants $g_3(\vec{k},\vec{l},\vec{m})$ revealed in Figure 2 is quite interesting and intricate and is partly unexpected. When $|\vec{k}|$ is the maxon wave vector ($\approx 1\AA^{-1}$) we find three large peaks in $g_3(\vec{k},\vec{l},\vec{m})$ when $\vec{l}$ and $\vec{m}$ are approximately collinear with $\vec{k}$. The positions of two of the peaks correspond to processes of decay into one maxon and one roton. The other peak corresponds to a process of decay into two phonons. The latter is not anticipated on simple intuitive grounds. A small peak can be associated with the process of decay into two rotons. When $|\vec{k}|$ is the roton wave vector, we find 5 peaks associated with collinear processes. The peak nearest the origin is associated with decay into two maxons. The other two peaks are associated with decays into (1) one phonon + an excitation in the plateau region above the roton region of the quasiparticle spectrum and one roton + an excitation associated with the second (lower) plateau in the quasiparticle spectrum which occurs above 3.4 $\AA^{-1}$ in our calculations (see below). Even greater complexity appears in the g_3 values at higher values of $|\vec{k}|$ which are shown for example in the other two parts of Figure 2. We note especially the importance of forward scattering processes in all cases and the appearance of sharp structure associated with the momenta well above the roton wavevector.

With regard to the self-consistent spectra we will focus for simplicity on Figure 5, which is qualitatively very similar to Figure 3. The lowest, quasiparticle branch is quite similar to experimental results [18] for wavevectors below about 3.4 $\AA^{-1}$. Above this wavevector, there is a sharp drop in the predicted quasiparticle energy which is not observed, though the resolution at these wavenumbers might be low enough to accomodate it (see Figure 7 of reference 18 .) At higher energies, the predicted multiphonon energies are qualitatively consistent with the neutron results except that the calculated multiphonon spectrum begins at a higher wavevector (about 2 $\AA^{-1}$) than the experimental one. Within our approximations, the neutron scattering function will have multiparticle weight below this momentum because of the imaginary part of the self energy. We have not made a detailed comparison of the calculated neutron scattering function with experiment, as was done in reference 12 using the extended two body approximation.

With regard to the calculated light scattering spectrum, we will discuss Figure 6. At a broad qualitative level, the spectrum is similar to the experimental one (Figure 8) except for the predicted peak at 60 $^{\circ}K$. This peak arises from a double excitation of the onset in the calculated self-consistent multiphonon particle spectrum at 2 $\AA^{-1}$ as seen in Figure 5. As just discussed, this feature of the multiparticle spectrum is not observed in neutron scattering experiments and appears to be a defect of the two body approximation, since it seems also not to appear in the results of reference 12 . Looking at the remaining structure one finds some features that definitely correspond to identifications like those proposed in references 8 and 10 . For example, the peaks at 31.6 °K and at 36.4°K can be unambiguously identified with two rotons plus one maxon and two maxons plus one roton respectively. On the other hand, other structure in our calculated spectrum cannot be identified in this way. For example, the peaks near 25 ° K appear to arise from double excitation of the low frequency (≈ 12.6 ° K) excitations which are predicted (see Figure 5) to lie at momenta above 3.4 $\AA^{-1}$ in our calculation. Because these excitations are not confirmed experimentally, there may naturally be some scepticism about their reality. Within the model they arise because at around 3.4 $\AA^{-1}$ there is a large increase in the matrix element for collinear decay into a roton and a maxon, leading to a corresponding drop in the self-consistent quasiparticle energy. This will be discussed in more detail elsewhere[19].

Finally, we note that Figure 7 shows that matrix element effects are indeed important in the details of the light scattering spectrum in the frequency region of interest here.

CONCLUSIONS

We have described a calculation of the light scattering spectrum in the two body approximation of references 11 and 12 . Some details of the treatment of the calculation of the two quasiparticle propagator differ from that of those authors. The coupling functions $g_3(\vec{k},\vec{l},\vec{m})$ are very strong functions of the magnitudes and directions of their vector arguments. In particular, processes in which the wave vectors are all parallel are strongly favored. To facilitate comparison with experiment, we have calculated the light scattering spectrum arising from these coupling functions, using an empirical bare quasiparticle spectrum which reproduces the experimental spectrum self-consistently at low momenta. Within the model, a new, sharp quasiparticle branch appears starting at 3.4 $\mathring{A}^{-1}$ and 12.6 ^{o} K. In calculating the light scattering spectrum, we used a matrix element $t_d(q)$ consistent with current understanding of the pair polarizability of helium. The resulting spectrum looks qualitatively like the experimental one except for a feature at 60 ^{o} K which is likely to be an artifact of our approximations. Other features in our calculated spectrum in the multiquasiparticle region arise from the high momentum branch in the spectrum mentioned above. It is not obvious that this branch is an artifact and it will be interesting to see if it occurs in more sophisticated theories and experiments at higher momentum transfer.

ACKNOWLEDGEMENT

We wish to thank the Minnesota Supercomputer Institute, the University of Minnesota Graduate School and the University of Minnesota at Morris for support.

REFERENCES

1. T. Greytak and J. Yan, Phys. Rev. Lett. **22**,987 (1969)

2. J. W. Halley, Bull. Am. Phys. Soc. **13**, 398 (1968);
Phys. Rev. **181**, 338 (1969)

3. F. Iwamoto, Prog. Theo. Phys.**44**, 1135 (1970)

4. J. Ruvalds and A. Zawadowski, Phys. Rev. Lett. **25**, 333 (1970);
A. Zawadowski, J. Ruvalds, and J. Solana, Phys. Rev. **A5**, 399 (1972)

5. T. J. Greytak, R. Woerner, J. Yan and R. Benjamin,
Phys. Rev. Lett. **25**, 1547 (1970)

6. C. A. Murray, R. L. Woerner, and T. J. Greytak, J. Phys. **C 8**, L90 (1975)

7. J. W. Halley, in "Elementary Excitations in Quantum Fluids", K. Ohbayashi
and M. Watabe, eds., Springer-Verlag Series in Solid State Sciences 79, p. 106 (1989)

8. M. Udagawa, H. Nakamura, M. Murakami and K. Ohbayashi,
Phys. Rev. **B 34**, 1563 (1986)

9. K. Ohbayashi and M. Udagawa, Phys. Rev. **B31**,1324(1985)

10. K. Ohbayashi and A. Ikushima, J. Phys. **C 7**,L206 (1974)

11. E. Manousakis and V. R. Pandharipande, Phys. Rev **B30**, 5062 (1984)

12. E.Manousakis and V. R. Pandharipande, Phys. Rev. **B31**,7029(1985); **B33**,150(1986)

13. R. P. Feynman and M. Cohen, Phys. Rev. **102**, 1189 (1956)

14. P. Kleban and R. Hastings, Phys. Rev. **B11**, 1878 (1975)

15. H. N. Robkoff and R. B. Hallock, Phys. Rev. **B24** , 159 (1981)

16. P. Dacre, Molecular Physics **45** , 17 (1982)

17. M. J. Stephen, Phys. Rev. **187**, 279 (1969)

18. R. A. Cowley and A. D. B. Woods, Can. J. Phys. **49**, 177 (1971)

19. J. W. Halley and M. S. Korth, unpublished

DISCUSSION ON EXCITATIONS IN BULK ^4He

Chairmen : J W Halley, W G Stirling

Halley began the discussion of the theory of excitations in Bulk ^4He with some remarks about the nature of the theoretical problem of describing liquid ^4He. The main point is that the helium-helium potential has a hard core, e.g. $V(r) \propto (\sigma/r)^{12}$. This means that its Fourier transform V(Q) is extremely large (essentially infinite) at all wavelengths. A theory which expands in powers of V(Q) (as in the formulation of the Glyde-Griffin theory presented this morning) is unlikely to be reliable. The hard core problem is what stimulated Feynman and later Feenberg and others including Krotschek, Campbell, Pandharipande and Manousakis to take a different approach. The variational ground state and excited states which they find take full account of the hard core problem, so they can use microscopic atomic potentials. They then can do a convergent pertubation theory in the interactions of the approximate variational quasiparticle states. The achievements of this approach are impressive and numerous: correct ground state energy, correct phonon-roton spectrum, multiparticle terms $S(q,\omega)$, and a condensate fraction $\approx$ 10%. A weakness is that the theory has not been fully extended to finite temperatures.

Maris : The polarization potential theory of Pines had sometimes seemed sensitive to anomalous dispersion. Halley responded that he had meant to refer in his remarks to theories of the Feenberg school, starting with microscopic potentials, and not to the polarization potential theory.

Krotschek pointed out that a link has now been established between perturbation theory in the (large) interatomic potential with a hard core and the correlated basis function (CBF) approach. The link, which involves summation of parquet diagrams in the pertubation theory, leads to a theory of the CBF form.

Glyde : The main results of the dielectric theory of Glyde and Griffin do not depend on a summation of a pertubation series in V(Q). The theory can be reformulated in terms of the t-matrix.

Platzman : There seemed to be some confusion between two problems: there is a well-defined problem of finding the ground state and lowest excited states of the fluid which can be approached variationally. Secondly there is the problem of finding the response of the system to a given probe, such as a neutron or a photon. The results of solving the second problem are likely to be complicated.

Campbell : Thermodynamics could provide some tests of the ideas proposed by Glyde and Griffin: For example, if there are really two long wavelength modes 'clamped' together at long wavelengths, then this will have consequences in the specific heat.

Griffin presented a transparency and used it to discuss the possible relation of his theory to the CBF theories. He wrote the expression:

$$\rho_{\mathbf{k}}|\varphi_0\rangle + \sum_{\mathbf{q}} A_{\mathbf{q},\mathbf{k}} \rho_{\mathbf{q}+\mathbf{k}} \rho_{-\mathbf{q}} |\varphi_0\rangle$$

for the variational excited state and suggested that the second term might be related to the mixing of single particle with collective excitations which is emphasized in his theoretical ideas.

Campbell : The term to which Griffin referred had been studied extensively and was related to three body correlations in the fluid. It has nothing directly to do with the condensate, unlike the one-particle excitations in the Glyde-Griffin theory.

Ferrell : How will we know when we have arrived at a full understanding of a complicated many body system like ^{4}He? It's not a problem like quantum electrodynamics in which one has a well defined answer.

Golub : If the roton is a manifestation of short range order, how does it disappear at T_λ? Platzmann replied that the question is an example of the confusion he discussed earlier: Above T_λ the roton is still an excited state, but its signature in the neutron scattering response function changes.

Scherm and Platzmann both commented that roton like spectra are observed in systems without condensates, including ^{3}He and quasi two dimensional metals in large magnetic fields (quantum Hall effect).

Putterman commented that someone had announced resolution of the problem of the role of the condensate in at least ten conferences in his memory. He asked if it was really an important parameter if it took so much trouble to define its role. What, he asked, if the single particle condensate were small but the pair condensate amplitude were large. Would it change the properties of the fluid?

Sokol : The microscopic details depend on the existence of the condensate but not on its magnitude.

Griffin : The new theories were now revealing the role of the macroscopic order parameter which was not previously understood.

G Williams showed a picture from a simulation of the vortices in the 3 dimensional xy model and suggested that such a model could be used to understand how rotons are damped as T approaches T_λ.

Turning to the current experimental situation, Stirling proposed four topics for discussion:

(1) Do the new neutron scattering experiments show a difference in the nature of the phonon excitations and the maxon-roton excitations?

(2) What then is the relation between the Raman scattering and neutron scattering results, which are apparently inconsistent?

(3) Will there be an observable 'cross-over' in the excitation spectrum at about 0.6 to 08$Å^{-1}$?

(4) Are there other ways to test the Glyde-Griffin model of the excitation spectrum?

Svensson pointed out that at small Q, above T_λ, it was necessary to take explicit account of the Rayleigh mode in the neutron spectra. Referring to the talk on the theory of Raman scattering by Halley, Maris found the ideas of hybridization and possible "extra" modes to be interesting; this might possibly provide an explanation of unexplained effects in propagation experiments. Donnelly pointed out that the existence of another excitation branch was ruled out by thermodynamics. The danger of thinking in terms of non-interacting modes was emphasised by Platzmann; what is important is the response function relevant to a particular type of observation. Griffin added his support to Maris' comments, adding that hybridisation around the maxon region is particularly interesting.

Glyde then proposed that the workshop consider two important questions:

(1) What are the key experimental observations to be explained?

(2) Why does the maxon, but not the phonon, scattering disappear as T tends to T_λ?

Referring to the neutron observations of peaks within the multiphonon continuum, Wyatt asked how maxon-roton resonances etc. could be distinguished from multiple neutron scattering. Stirling and Svensson described how the experimental configuration was designed to minimise these effects, following the calculations of Sears, and Svensson emphasised the importance of calculating the magnitude of the multiple scattering.

Lauter wondered how the single-particle excitations were expected to behave at small Q.

Dahm returned to the problem of a possible second (roton) minimum at higher Q. Stirling responded that there was no generally accepted observation by neutron scattering, in this region. Experiments at wavevectors of about 3 to 4$Å^{-1}$ are difficult as the instrumental resolution tends to worsen with increasing Q.

Stringari then proposed a physical quantity, the calculation of which could give further insight into the validity of the Glyde-Griffin model. He suggested that the matrix element

$$< o \mid \hat{\rho}_q \mid q >$$

be calculated between the ground state <o | and the phonon-roton state $|q>$, at zero temperature. Here $\hat{\rho}_q$ is the non-condensate component of the density operator. In the small wavevector limit the contribution of $\hat{\rho}_q$ to the total density matrix element can be shown to be crucial; according to the Glyde-Griffin model this should be very small in the roton region.

Sokol questioned the statistical significance of the 'bumps' appearing in the (neutron) multiphonon spectra. Stirling responded that several different experiments had given identical results. This was supported by Scherm, who presented neutron results taken at very low temperature, employing the time-of-flight technique.

EXCITATIONS IN LIQUID HELIUM AT NEGATIVE PRESSURES

Humphrey J. Maris

Department of Physics
Brown University
Providence, RI 02912, U.S.A.

INTRODUCTION

In this article we will discuss the behavior of liquid helium at negative pressures, with particular emphasis on the dispersion curve for phonons and rotons. We will show that even though the amount of experimental work so far performed on helium in a state of tension is quite small, it is possible to deduce the main features of the variation of the dispersion curve with pressure in surprising detail. Furthermore, the results of the analysis provide a new perspective on the behavior of helium for positive pressure.

DISPERSION CURVE AS A FUNCTION OF PRESSURE

The dispersion curve for elementary excitations in superfluid helium-4 has been studied as a function of pressure by several groups[1]. The main features of the variation of the dispersion curve in the range between $P = 0$ and 25 Bars (the freezing pressure) are as follows:

1) The variation of the sound velocity c with pressure has been carefully measured by Abraham et al[2]. They determined the quantity $c(P) - c(0)$ with random errors of about $\pm\, 3 \times 10^{-5} c(0)$. To obtain $c(P)$ itself they had to use the value of $c(0)$ previously determined by Whitney and Chase[3] which has a much larger uncertainty of $\sim 4 \times 10^{-4}$. The sound velocity increases with increasing pressure and for $P = 0$

$$\frac{\rho}{c}\frac{dc}{d\rho} = 2.84 \tag{1}$$

2) The variations of the dispersion curve in the maxon and roton regions are not known with comparable precision. Brooks and Donnelly[1] have given simple expressions that fit the neutron scattering data for the maxon energy Δ_M, the roton energy Δ, the roton effective mass μ, and the momentum p_o at the roton minimum as a function of density. Of these quantities Δ_M and p_o increase with increasing P, and Δ and μ decrease.

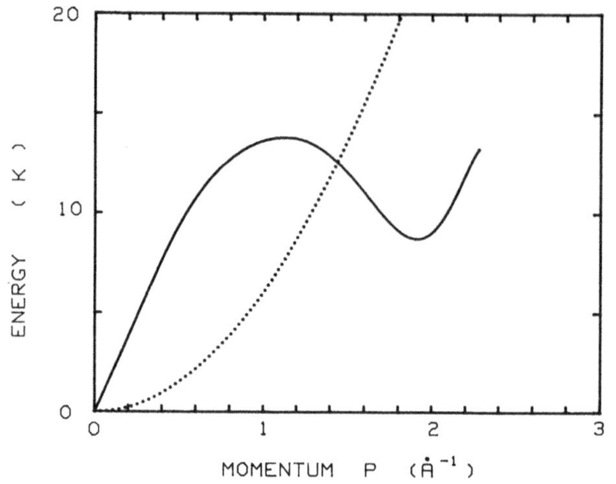

Fig. 1. A comparison of the free-atom dispersion curve (dotted curve) Eq. (2), with the dispersion curve of superfluid helium-4 at zero pressure.

At first sight a "natural explanation" of these variations is the following. In the limit of zero density one could argue that the excitations in helium must be simply

$$\epsilon = p^2/2m \tag{2}$$

where m is the bare mass of a helim atom. To get to the dispersion relation (2) starting from the measured dispersion relation at zero pressure the roton minimum must disappear as the density decreases. For this to happen the roton minimum must become shallower as ρ is lowered below the density ρ_o corresponding to $P = 0$. Thus, μ must increase as ρ decreases. By comparison of the free-atom dispersion with the liquid dispersion curve (see Fig. 1), it is also clear that Δ_M and Δ must decrease and increase, respectively, with decreasing ρ. In addition, as $p \rightarrow 0$ the linear term in the function $\epsilon(p)$ (corresponding to the sound velocity) must vanish, and so c has to decrease with decreasing ρ. Thus, these simple arguments give the correct sign for the variation of the quantities c, Δ_M, Δ, and μ with pressure; no prediction is made about the variation of p_0.

Despite this apparent success the "explanation" cannot possibly be correct. To see this consider what happens when a negative pressure is applied to the liquid. For the moment let us ignore the possible formation of bubbles in the liquid under the action of the negative pressure, i.e., assume that the liquid remains homogeneous. As the pressure is made more negative the density will decrease, but it will not decrease continuously to zero. The point is that the energy E of any condensed system will vary with molar volume V as shown qualitatively in Fig. 2. Consequently, the pressure -dE/dV will always have a maximum negative value P_c at some finite molar volume V_c. Hence, if a pressure more negative than P_c is applied the liquid is unstable, and so the density cannot be smoothly decreased to zero.

Since P has a minimum value at $V = V_c$ it follows that the bulk modulus B = -V dP/dV vanishes at this point. If we assume for the moment that E can be expanded as a simple power series near to V_c, P and B will vary approximately as

$$P - P_c \propto (V_c - V) \tag{3}$$

$$B \propto (V_c - V) \tag{4}$$

Since the density remains finite the sound velocity varies as

$$c \propto (V_c - V)^{1/2}$$

$$\propto (P - P_c)^{1/4} \tag{5}$$

Hence, the sound velocity goes to zero at the density ρ_c corresponding to the volume V_c, not when $\rho \rightarrow 0$ as assumed in the explanation given above.

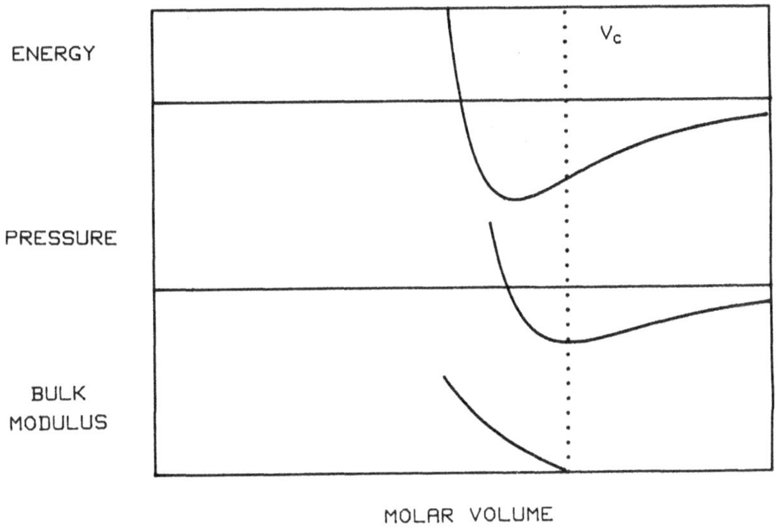

Fig. 2. Qualitative form expected for the variation of the energy, pressure, and bulk modulus with molar volume. V_c is the critical volume discussed in the text.

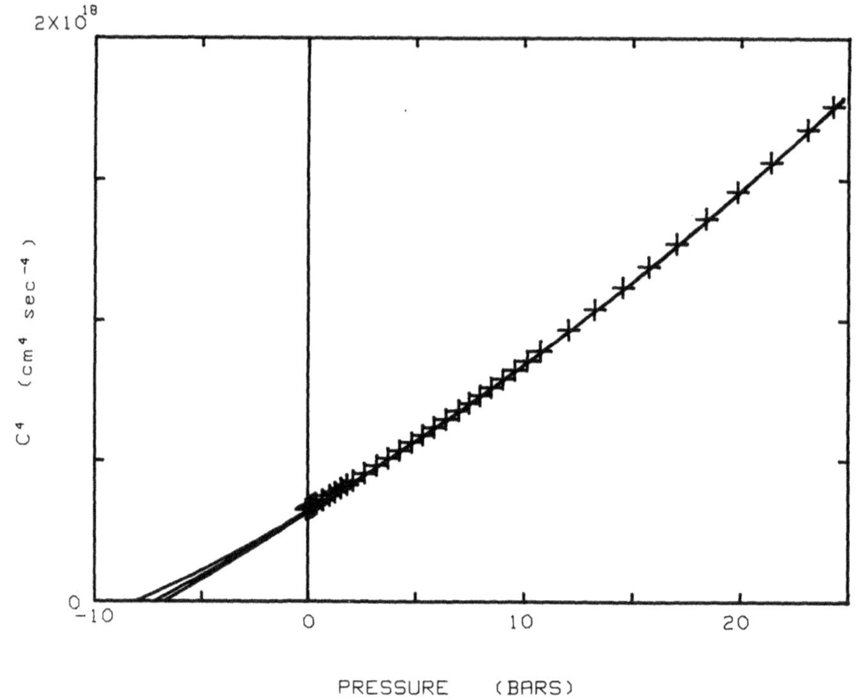

Fig. 3. Three least-squares fits of c^4 versus P. The fitting functions are given in Eq. 6. In the range of positive pressures one cannot distinguish the three fits on this plot. For negative pressure the curves are Eqs. 6(b), 6(a), and 6(c) from top to bottom.

ESTIMATE OF CRITICAL PRESSURE

To determine P_c, V_c, etc, the most attractive method is to perform an extrapolation of the sound velocity data into the negative pressure regime,[4,5] and to use this extrapolation to find the point at which c becomes zero. Once $c\,(P)$ is determined it is straightforward [5] to obtain by integration the dependence of other quantities such as E or B on pressure or density. Assuming that c does indeed vary as $(P - P_c)^{1/4}$ near P_c it is natural to perform the extrapolation by fitting c^4 to some simple function of P. This is shown in Fig. 3. The fits are obtained using the functions

$$c^4 = a_o + a_1 P + a_2 P^2 \tag{6a}$$

$$= b_o + b_1 P + b_2 P^2 + b_3 P^3 \tag{6b}$$

$$= \frac{d_o + d_1 P}{1 + e_1 P} \tag{6c}$$

One can see that the variations between the three fits are fairly small. The values of P_c obtained are - 7.21, - 8.07, and - 6.74 bars, respectively. A fit of the Padé form of Eq. (6c) but going to second order in P gives [5] P_c = - 8.87 bar, and we will use this value in most of the following discussion. The other critical parameters based on this fit are $\rho_c = 0.110$ gm.cm^{-3} and $V_c = 36$ cm^3.

DISPERSION IN PHONON REGION

According to the above discussion the dispersion curve for small momentum becomes linear, rather than quadratic, for small p as P approaches P_c. The dispersion can thus be approximated by

$$\epsilon = p^2/2m^* \tag{7}$$

where m^* is an effective mass. One can obtain a value for m^*, and a prediction for the dispersion for small p for all values of the pressure by the following method. Consider a phenomenological model for the energy density H in the liquid valid for disturbances which vary slowly in space and time.

Let

$$H = U(\rho) + \frac{1}{2}\rho v^2 + \lambda \mid \nabla\rho \mid^2 + \frac{1}{2}\beta\dot{\rho}^2 \tag{8}$$

where $U(\rho)$ is the internal energy per unit volume for a homogeneous and static system, and v is the velocity. The last two terms with coefficients λ and β add to the usual energy density in fluid mechanics contributions in space and time-derivatives which are of the lowest order possible. Based on the energy density (8) it is straightforward to calculate the phonon dispersion relation. The result is for small p

$$\epsilon = p\left(\frac{c^2 + 2\lambda\rho p^2/\hbar^2}{1 + \rho\beta p^2/\hbar^2}\right)^{1/2} \tag{9}$$

where c is the sound velocity $\left(= \sqrt{\partial^2 U/\partial\rho^2}\right)$. By comparison of (7) and (9) we obtain

$$m^* = \frac{\hbar}{\sqrt{8\lambda\rho_c}} \tag{10}$$

One can estimate λ by the following method[5]. From the value of c in the negative pressure region one can calculate U as a function of ρ. Then one can use Eq. (8) to calculate the density profile and energy of the interface between liquid helium and a vacuum. Since this is a static problem the coefficient β does not enter into the results. Then the coefficient λ can be determined by requiring that the calculation give the experimentally-measured value for the liquid surface energy.

The result is

$$\lambda = 9.1 \times 10^{-7} gm^{-1} cm^7 sec^2 \tag{11}$$

which in turn gives

$$m^* = 0.18 m_4 \tag{12}$$

where m_4 is the ^{4}He mass.

This value can only be considered approximate because of the simple form taken for the energy density (8), and because we have neglected the possible variation of λ with density. As far as we can see, λ and β should remain finite at the critical point, and thus must vary slowly compared to c. As a test of this idea we have assumed that λ and β are constants and have determined values for them by comparing Eq. (9) with the measured phonon dispersion for positive pressures. For small p, Eq. (9) can be written

$$\epsilon = cp(1 + \alpha_2 p^2 + ...) \tag{13}$$

with

$$\alpha_2 = \rho\left(\frac{\lambda}{c^2} - \frac{\beta}{2}\right) \tag{14}$$

Thus, a plot of α_2/ρ as a function of $1/c^2$ should give a straight line with slope λ and intercept $-\beta/2$. This is shown in Fig. 4 using the experimental data of Rugar and Foster[6] for α_2. A least-squares fit gives $\lambda = 10.5 \times 10^{-7}$ cgs which is in good agreement with the estimate made above (Eq. 11). The good fit to the data obtained when it is plotted as in Fig. 4 provides support for the assumption that λ and β are slowly varying compared to c.

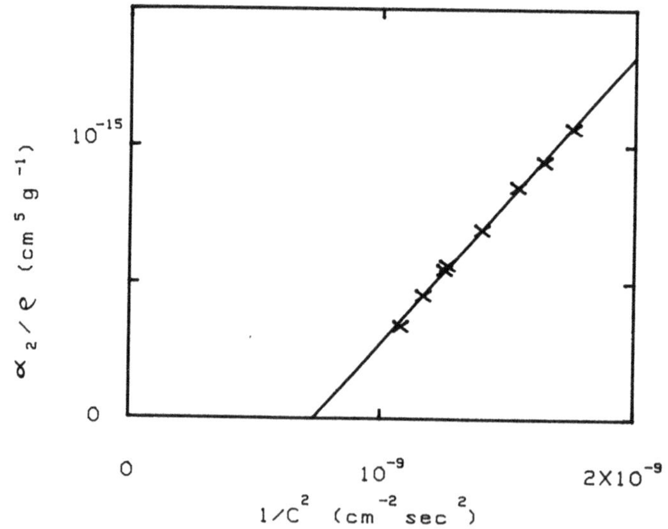

Fig. 4. Determination of the parameters λ and β from the phonon dispersion measurements of Rugar and Foster (ref. 6).

The value obtained for β is 1.55×10^{-15} gm^{-1} cm^5. The term $1/2\ \beta\dot{\rho}^2$ represents an extra internal energy that the liquid has when its density is changing. This term can be understood on the basis of a simple model[5]. Let us focus on a particular helium atom moving in the potential V due to the neighboring atoms in the liquid. We ignore the motion of the neighbors and simply view them as the source of the potential V. If the density of the liquid is constant, V will be constant and the atom will be in the ground state ψ_0. If ρ changes very slowly, V will also change slowly, and to a good approximation the wave function ψ will always be close to the ground state wavefunction for the instantaneous value of the potential. The small difference between ψ and ψ_0 means that the atom will have an excess energy δE relative to the energy in the instantaneous ground state. This excess energy can be shown to be[5]

$$\delta E = \hbar^2 \sum_n \frac{|\dot{V}_{no}|^2}{E_{no}^3} \tag{15}$$

where $\dot{V}_{no}$ is the matrix element of the time-derivative of V between the ground state and state n, and E_{no} is the difference in energy of these states. This gives for the coefficient β the result

$$\beta = \frac{2\hbar^2\rho}{M} \sum_n \frac{1}{E_{no}^3} \left| \frac{d\dot{V}_{no}}{d\rho} \right|^2 \tag{16}$$

Based on a reasonable form for V, one can make a rough estimate of β. The result is in reasonable agreement with the value obtained above from the analysis of α_2.

When α_2 is positive a graph of ϵ versus p curves upwards. This is referred to as anomalous dispersion, since in most crystalline solids the curvature is in the opposite direction. The existence of anomalous dispersion was predicted[7] from an analysis of phonon transport data, and was confirmed by specific heat measurements[9]. At high pressures (above about 20 bars) the dispersion becomes "normal", i.e., α_2 becomes negative. The discussion we have given provides a simple explanation for the existence of anomalous dispersion, and for the change of sign of α_2 at high pressures. One can see from Eq. (14) that λ and β make contributions to α_2 which have opposite signs, and that because of the variation of the sound velocity with pressure at sufficiently low pressure the dispersion will become anomalous.

In Fig. 5 we show the small momentum part of the dispersion relation for several pressures. These results are again within the approximation that λ and β are constant.

CRITICAL BEHAVIOR AT P_c

In the above discussion we have argued that at some negative pressure the sound velocity will become zero, and that the pressure dependence of c near to this critical pressure will be as $(P - P_c)^\nu$ with $\nu = 1/4$. This value of ν is based on the assumption that at the critical volume V_c the energy of the system has a simple inflection point, i.e.,

$$E(V) = E_c + (V - V_c)E_1 + (V_c - V)^3 E_3 + \ldots\ldots \tag{17}$$

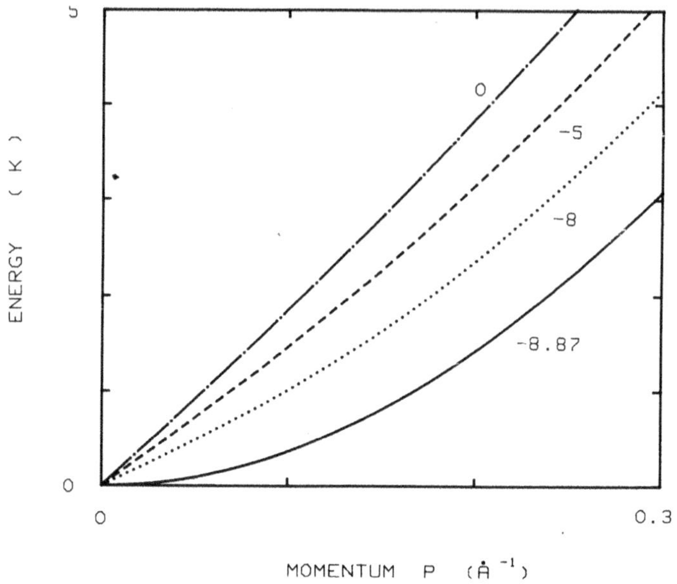

Fig. 5. Dispersion curve for phonons of small momentum in liquid ^{4}He. The different curves are labelled by the pressure in bars.

If the cubic term is replaced by a term in $(V_c - V)^\gamma$, then one obtains instead

$$\nu = \frac{\gamma - 2}{2(\gamma - 1)} \tag{18}$$

It is interesting to see if an analysis of the sound velocity data without imposing any theoretical prejudice leads to $\nu = 1/4$. One can see from Fig. 3 that a plot of c^4 versus P curves upwards, and this suggests that possibly ν is larger than $1/4$. To make a more quantitative analysis we used a method described by Baker[10]. We calculated $dlogc/dP$ from the data and made a fit of this quantity to a Padé function of the form

$$\frac{dlogc}{dP} = \frac{c_o + c_1 P + ...c_m P^m}{1 + d_1 P + ...d_m P^m} \tag{19}$$

This function will have a pole at P_c and the residue at the pole is the exponent ν. The results of fits of this type for ^{4}He and ^{3}He are shown in Table I. In these fits the sums of the squares of the deviation of the data from the fit decrease by a very small amount (0.03% for ^{4}He) on going from $m = 1$ to $m = 2$. Both for ^{3}He and ^{4}He the analysis indicates a value of ν in the range 0.31 to 0.33, i.e., significantly larger than $1/4$.

These results for ν imply that a plot of energy as a function of volume does not have a simple inflection point at the volume V_c. We have constructed a simple model that

Table 1. Values of P_c and the sound velocity exponent ν for ^{3}He and ^{4}He. The order of the Padé approximant is m. The ^{3}He data are not sufficiently accurate to make a meaningful fit to a second-order Padé.

	m=1		m=2	
	P_c	ν	P_c	ν
^{3}He	-3.1	0.324	—	
^{4}He	-9.49	0.331	-9.42	0.312

provides an explanation of this interesting behavior. We assume that we can divide the energy of liquid helium into two contributions, a potential energy term U which is an analytic function of V and a term K_{ZP} arising from the zero-point energy. Thus,

$$E(V) = U(V) + K_{ZP}(V) \tag{20}$$

If we know the function $U(V)$ we can calculate an approximate value for K_{ZP} by first finding the bulk modulus $B = Vd^2U/dV^2$. Then we calculate the sound velocity $c = (B/\rho)^{1/2}$, and from this obtain in some approximation (see example below) the dispersion curve for elementary excitations. Finally, we find $K_{ZP}(V)$ as the zero-point energy of this spectrum, and obtain $E(V)$ from (20).

This calculation completes one renormalization cycle. We now repeat the calculation but using for the bulk modulus $B = Vd^2E/dV^2$. The idea is to continue going around the loop until the results do not change, and then examine how the sound velocity varies with volume or pressure as it approaches the critical point.

To carry out this procedure we need an approximate expression for K_{ZP} in terms of c. For this we have taken Eq. (9) with $\beta = 0$, i.e.,

$$\epsilon = p(c^2 + 2\lambda\rho p^2/\hbar)^{1/2} \tag{21}$$

(Note that it is essential to retain the term involving λ to get the correct form for ϵ at small p and close to P_c, but that setting $\beta = 0$ is unimportant). We then take as the zero point energy

$$\begin{aligned} K_{ZP} &= \frac{2\pi V}{h^3}\int_0^{p_m} \epsilon p^2\,dp \\ &= \frac{4\pi V}{\hbar^4}\left(\frac{\lambda\rho}{2}\right)^{1/2} p_m^5\left[\frac{2y^5}{15} + \left(\frac{1}{5} - \frac{2y^2}{15}\right)\left(1+y^2\right)^{3/2}\right] \end{aligned} \tag{22}$$

where p_m is a cut-off at high momentum, and

$$y = \frac{\hbar c}{p_m(2\lambda\rho)^{1/2}} \tag{23}$$

With this simple form for K_{ZP}, we have determined the critical behavior of the sound velocity. It is straightforward to show (either analytically or by actually performing the iteration numerically) that near P_c

$$c \propto (P - P_c)^{1/3} \tag{24}$$

This result for the exponent ν is in extremely good agreement with the experimental results. As a simple way of showing this we plot in Fig. 6 c^3 as a function of P for ^{3}He and ^{4}He. This gives linear plots even out to pressures as far from the transition as $\pm$ 10 bars.

The result $\nu = 1/3$ implies that the energy as a function of volume near the V_c does not have a simple inflection point. Instead, one has

$$E \approx E_c + a_1(V_c - V) + a_4(V_c - V)^4 + ... \tag{25}$$

Despite the remarkable success of the theory in predicting the critical behavior of c, much more work remains to be done to put the calculation on a solid basis. The calculation of the zero-point energy does not properly take account of short-wavelength excitations (rotons). In addition, it is implicitly assumed that K_{ZP} is well-approximated by an integral over a spectrum of non-interacting excitations, i.e., anharmonic effects are included in a highly-simplified way.

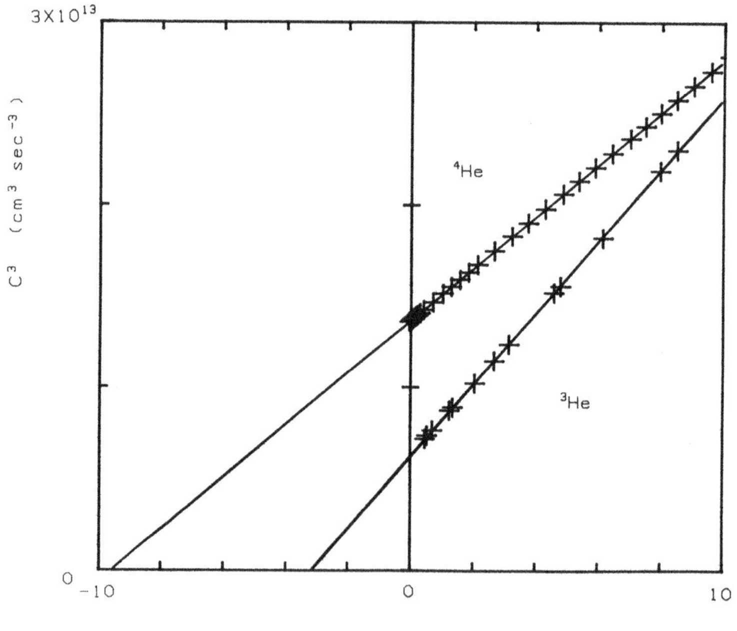

Fig. 6. Plot of the cube of the sound velocity versus pressure for ^{3}He and ^{4}He.

SUMMARY

We have shown by simple physical arguments that one can deduce a surprisingly complete picture of the excitations in liquid helium at negative pressure. Liquid ^{3}He and ^{4}He become unstable at long wavelengths (a "macroscopic instability") at critical pressures P_c of around -3 and -9 bars, respectively. We have given a simple theory of the behavior of the liquid in the vicinity of the critical pressure, and this theory agrees well with the critical exponent of the sound velocity at P_c estimated by extrapolation of data in the positive pressure range. As we have described, the understanding of the excitation spectrum at negative pressures also provides a simple explanation of anomalous phonon dispersion.

ACKNOWLEDGMENTS

I thank Quan Xiong for many discussions about this work. The research was supported in part by the National Science Foundation through grant DMR-8820469.

REFERENCES

1. For a review, see J. S. Brooks and R. J. Donnelly, J. Phys. Chem. Ref. Data 6, 51 (1977).
2. B. M. Abraham, Y. Eckstein, J. B. Ketterson, M. Kuchnir, and P. R. Roach, Phys. Rev. A 1, 250 (1970).
3. W. M. Whitney and C. E. Chase, Phys. Rev. Lett. 9, 243 (1962).
4. H. J. Maris and Q. Xiong, Phys. Rev. Lett. 63, 1078 (1989).
5. Q. Xiong and H. J. Maris, J. Low Temp. Phys. 77, 347 (1989).
6. D. Rugar and J. S. Foster, Phys. Rev. B 30, 2595 (1984).
7. H. J. Maris and W. E. Massey, Phys. Rev. Lett. 25, 220 (1970).
8. H. J. Maris, Rev. Mod. Phys. 49, 341 (1977).
9. N. E. Phillips, C. G. Waterfield, and J. K. Hoffer, Phys. Rev. Lett. 25, 1260 (1970).
10. G. A. Baker, in "The Padé Approximant in Theoretical Physics," G. A. Baker and J. L. Gammel, eds. (Academic, New York, 1970), p. 1.

NEUTRON SPIN ECHO OBSERVATION OF SPONTANEOUS PHONON DECAY IN SUPERFLUID ^{4}He

F. Mezei, C. Lartigue, and B. Farago*

Hahn-Meitner-Institut, Pf. 390128, D-1000 Berlin 39
*Institut Laue-Langevin, 156X, F-38042 Grenoble Cedex

INTRODUCTION

Due to the peculiar shape of the phonon-roton dispersion curve in ^{4}He, the spontaneous decay of one excitation into a number of others is forbidden for wavenumbers above $q_c \simeq 0.55$ Å^{-1}. Therefore, $q > q_c$ excitations can only decay by scattering on another, thermal excitation, and consequently, in the ground state $T \to 0$ their lifetime becomes infinite, as put to evidence by ballistic photon/roton experiments[1]. In contrast the $q \leq q_c$ phonons must have a finite lifetime at $T \to 0$ and we shall present the first quantitative observation of this phenomenon.

EXPERIMENTAL RESULTS

The generalized Neutron Spin Echo (NSE) approach[2] allows the observation of the linewidth of elementary excitations by neutron scattering with a resolution of about 1 μeV FWHM (this value corresponds to 1.32 nsec lifetime) compared to some 20-40 μeV resolution limit of conventional high resolution neutron scattering techniques (triple axis spectroscopy (TAS).) In a previous study[3] the temperature dependence of the phonon linewidth was investigated combining NSE and TAS techniques. It was found by extrapolation that the temperature dependent part of the linewidth below 1 K becomes smaller than the above NSE resolution limit. However, the residual linewidth at $T \to 0$ could not be unambiguously established due to the poor statistical accuracy of the data.

In the present improved series of experiments we have observed a significant q dependent linewidth below $q \leq 0.45$ Å^{-1}, as shown in Figure 1. This value of q corresponds well to the kinematical cut-off wavenumber $q_c^{(2)} \simeq 0.43$ Å^{-1} (as determined from the dispersion relation) of the decay process of one phonon into two others (three phonon process) due to the third order anharmonic coupling. The dashed line in the figure shows the theoretical results of Maris[4], who found that higher order processes give negligible contributions compared to this three phonon process. He used the observed pressure dependence of the sound velocity in order to estimate the anharmonic coupling. Even if we take into account that the q resolution in this experiment was about 8% FWHM, there is a substantial discrepancy between the theoretical and experimental results.

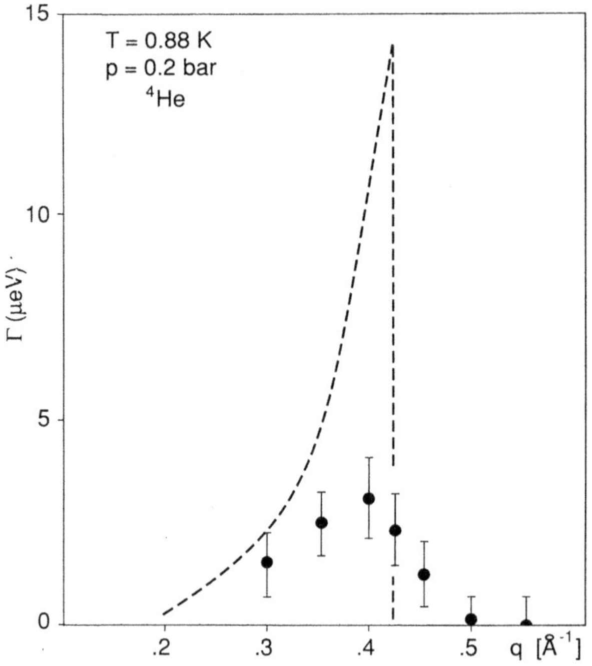

Fig. 1. Measured and calculated HWHM phonon linewidth in liquid He.

References

1. A. F. G. Wyatt, in this volume and references therein.
2. F. Mezei, The Principle of Neutron Spin Echo, in: "Neutron Spin Echo", F. Mezei, ed., Springer Verlag Heidelberg (1980).
3. F. Mezei and W. G. Stirling, Neutron Scattering study of the lifetime of phonons in superfluid ^{4}He, 75th Jubilee Conf. on ^{4}He, J.G.M. Armitage, ed., World Scientific, Singapore (1983).
4. H.J. Maris, Phonon-phonon interactions in liquid helium, Rev. Mod. Phys. 49:341 (1977).

DYNAMICS OF ROTONS

David C. Samuels and Russell J. Donnelly

Department of Physics
University of Oregon
Eugene, Oregon 97403

ABSTRACT

We consider the interaction of rotons with quantized vortex lines in the theory of mutual friction. A good account is given of mutual friction at low temperatures and under pressure. We also consider collisions of rotons with rotons in the context of a recent experiment by Forbes and Wyatt.

1. INTRODUCTION

At temperatures below approximately 1.8K, it is reasonable to treat superfluid helium as a dilute gas of elementary excitations. We describe these elementary excitations as point particles with a Hamiltonian

$$H = \epsilon(p) - \vec{p}\cdot\vec{v}_s \qquad (1)$$

where $\epsilon(p)$ is the free particle Hamiltonian, $\vec{p}$ is the excitation momentum, and $\vec{v}_s$ is the local superfluid velocity at the position of the excitation. The free particle Hamiltonian is given by the dispersion curve shown in Figure 1. Rotons with $p > p_0$ have their group velocity $u = d\epsilon/dp$ parallel to their momentum and are called R^+ rotons; those for $p < p_0$ have their group velocity antiparallel to their momentum and are called R^- rotons. We discuss the dispersion curve in more detail in section 3.

The trajectory of an excitation is determined by solving Hamilton's equations numerically. In this paper we present the results of this procedure in two specific cases; the scattering of rotons by the velocity field of a quantized vortex line, and the scattering of two roton beams incident at right angles.

2. ROTON - VORTEX INTERACTION

The first objective of our investigation was to simulate roton - vortex collisions [1] to gain insight into this complicated interaction. The vortex line is considered to be rigid and aligned along the z-axis with a circulation $\vec{\kappa} = (h/m)\hat{z}$ where m is the mass of the helium atom. The velocity field of the vortex is azimuthal in the x-y plane and is given by $\vec{v}_s \geq (\kappa/2\pi r)\hat{\phi}$ where the unit vector $\hat{\phi}$ is in the azimuthal direction in the x-y plane and r is the radial distance from the core.

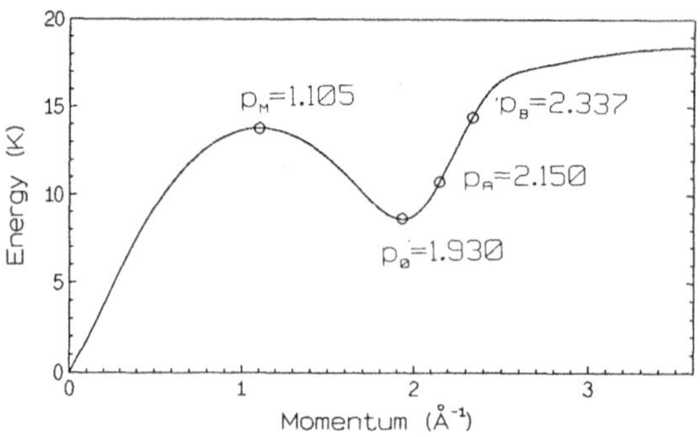

Fig. 1. Dispersion curve for elementary excitations in helium II. The minimum is described by an energy Δ and a momentum p_0. The momenta labeled p_M, p_A, and p_B are explained in section 3.

Hall and Vinen described the interaction of excitations with quantized vortices by defining two microscopic parameters D and D′, [2]

$$\vec{f}_{ex} = D\vec{v}_R + D'\vec{\kappa} \otimes \vec{v}_R \tag{2}$$

where $\vec{f}_{ex}$ is the net force per unit length on the vortex line, and $\vec{v}_R$ is the average excitation velocity. We calculate the parameters D and D′ from the trajectories of rotons interacting with the velocity field of the vortex.

We illustrate in Fig. 2(a) the trajectories of a single R^+ roton of momentum 2.0 $Å^{-1}$ incident upon a vortex line. This relatively slow roton (~72 m/sec) experiences minor deflections for positive impact parameters, but truly major deflections for negative impact parameters. These rotons make a transition from one side on the roton minimum to the other with no change in energy. These dramatic reversals of group velocity do not necessarily result in a large momentum exchange because the momentum vector may be deflected only a small amount. This type of "species change" may also be found in an early paper by Goodman (see especially his Fig. 1) [3].

The corresponding momentum exchange parallel to the initial trajectory is shown in Fig 2(b). One can see that while the maximum momentum exchange occurs for close collisions, features such as the reversal of the group velocity extend so far from the vortex line (~150 Å in this example) that they add up to a substantial net momentum transfer to the line.

Using the coordinate system described above and assuming a macroscopic drift of rotons with mean velocity in the +x direction, we can show by methods developed by Rayfield and Reif [4] that

$$D = -2\pi \int_0^\infty dp\, p^4 \frac{df}{d\epsilon} u \int_0^\pi d\theta (\sin^3\theta)\sigma \tag{3}$$

where f is the distribution function of rotons with zero net velocity, σ is the cross section,

$$\sigma = \int_0^\infty db \frac{\Delta p}{2p} \tag{4}$$

Θ is the polar angle and Δp is the momentum transfer parallel to the original trajectory.

The simulation results of Fig. 2(b) suggest that the integrand of σ can be closely approximated by the simple function:

$$\Delta p = \begin{Bmatrix} 0 & b < b_{crit} \\ |p - p'| & b_{crit} < b < 0 \\ 0 & b > 0 \end{Bmatrix} \tag{5}$$

where b is the impact parameter of the roton, b_{crit} is the critical impact parameter dividing scattered from unscattered rotons, and p′ is the momentum of a roton on the opposite side of the roton minimum with the same energy as the original roton. The critical impact parameter is simply the distance from the vortex where the $\vec{p}\cdot\vec{v}_s$ potential reduces the roton's kinetic energy to the roton minimum. The roton's group velocity then reverses direction and it backs out of the vortex potential.

Therefore

$$\sigma \approx \frac{(p-p')}{2p} b_{crit} = \frac{|p-p'|\kappa\sin\theta}{4\pi(\epsilon-\Delta)} \tag{6}$$

and

$$D_1 = -2\pi\int_0^\infty dp\, p^4 \frac{df}{d\epsilon} u \frac{3}{32}\frac{|p-p'|\kappa}{(\epsilon-\Delta)} \tag{7}$$

where we define D_1 as the part of D that is due to scattering from the velocity field of the vortex.

To find D_1, we can perform the integration in (7) numerically, or (at low temperatures) use the Landau parabolic approximation

$$\epsilon \cong \Delta + (p-p_0)^2/2\mu \tag{8}$$

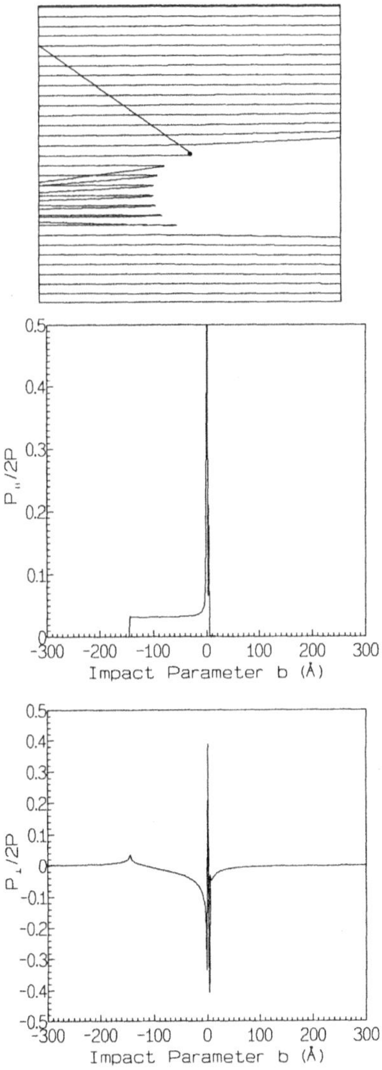

Fig. 2. (a) Trajectories of rotons incident upon a vortex line (indicated by a dot) with clockwise circulation. The momentum of the rotons is p = 2.0 Å^{-1} and the impact parameter b is stepped in intervals of 20Å. The almost total reversals of direction for negative impact parameters corresponds to the species change $R^+ \to R^-$.

(b) Parallel momentum exchange for rotons with orbits such as in (a). This function can be approximated by the rectangular function in Eq(4), ignoring close collisions.

(c) Transverse momentum exchange from (a). Note that the total transverse momentum exchange, the integral under this curve, very nearly vanishes.

where μ is the roton effective mass and Δ is the energy at the roton minimum. Assuming a simple Bose distribution for rotons, and the Landau parabolic approximation, we calculate

$$D_1 \cong \frac{3\pi}{4}\kappa\sqrt{\frac{2\pi\mu}{kT}}\frac{p_0^4 e^{-\Delta/kT}}{h^3} \tag{9}$$

Eq. (9) is a new result in the theory of mutual friction.

For close collisions (< 10 Å) we cannot expect reasonable physical results from the simulation for at least three reasons. First, the vortex may deform in a close collision and send waves along the core. Second, there are localized roton states near the line which are not taken into account in a single excitation model [5]. Finally, the (still unknown) quantum mechanical structure of the vortex lines and the roton must be involved in these collisions. Since the integral of the perpendicular momentum transfer is small at large distances (figure 2c), the core interaction is dominate and D′ cannot be described by our classical calculation.

Hillel and Vinen [6] have produced an ingenious account of the consequences of close collisions. They assume roton interactions between the incoming roton and the localized rotons near the core will result in the absorption of the roton. Taking into account this absorption and subsequent (non-isotropic) re-emission of rotons, they find the corresponding microscopic friction parameters

$$D_2 = 2\rho_n v_G \sigma_{core} \tag{10}$$

$$D' = \rho_n \kappa / 2 \tag{11}$$

where $v_G = \sqrt{2kT/\pi\mu}$ is the average group velocity of rotons and σ_{core} is the core cross section which we take to be the core diameter.

The microscopic parameters D and D′ are not measured in an experiment. Instead, two related parameters, B and B′, are obtained. The connection between the parameters B and B′ and D and D′ is intricate but direct. Taking $D = D_1 + D_2$ from (9) and (10), D′ from (11), and using the formulae of Barenghi et al [2] gives a good account of B and B′ at low to moderate temperatures (i.e. T < 1.8 K).

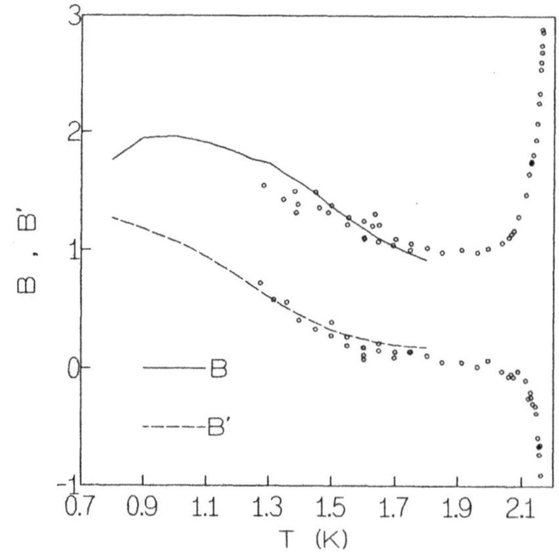

Fig. 3. Values of B and B′ deduced from D and D′ compared to experimental values taken from Ref. 2.

We can test our theory below 1 K by comparing with the drag on vortex rings measured by Rayfield and Reif [4]. They determine a drag coefficient α, which after allowance for drag by phonons and ^{3}He, is interpreted in terms of a roton-line cross section:

$$\alpha = \frac{3\pi^2}{8}\frac{\kappa}{\hbar^3}p_0^4 e^{-\beta\Delta}\bar{\sigma}_0 \tag{12}$$

and quote the surprisingly large result $\bar{\sigma}_0 = 9.5 \pm 0.7 Å$. Vortex ring drag formulae of section 3 of Ref. 1 can be shown to yield $D \cong 2\alpha/\kappa$. Interpreting our result (9) in the same way as (12) gives the contribution to the cross section from distant collisions

$$\bar{\sigma}_0 = \kappa\sqrt{2\mu/kT\pi} \tag{13}$$

which yields $\bar{\sigma}_0 = 8.5Å$ at T = 0.67 K. The contribution from the core in D_2 is about 2Å and raises $\bar{\sigma}_0$ to ~ 10.5Å in substantial agreement with experiment.

The observed quantities B and B′ have been reported only above 1.3 K[1]. They depend on the frequency of second sound used in their observation in a complicated way and the means to get accurate values are tedious [7]. It would be better to have the microscopic parameters D and D′ available at all temperatures and pressures. Unfortunately, our single excitation model cannot be extended above 1.8 K.

Data available today limit the calculation of B and B′ at higher pressures. No reliable tabulation of viscosity as a function of temperature and pressure exists. If, however, we use the roton viscosity formula of Ref. 8 and the low temperature core parameter of Ref. 2 we can compute B and B′ at T = 1.4 K from P = 0 to 24 bar. We find B is nearly independent of pressure, falling about 25% over the pressure range from a value of ~1.4 at P=0. The data of Mathieu, Marachal and Simon [9] are roughly constant at ~1.5 over the same range in pressure. This result is surprising since D and D′ are a function of ρ_n which increases by a factor of 3 over this pressure range at T = 1.4 K [10].

This research gives, we believe, the basis for understanding the mutual friction coefficients D and D′, and B and B′, at temperatures below 1.8K and under pressure.

3. ROTON - ROTON SCATTERING

Our attention was drawn to the roton-roton interaction problem by a recent report by Forbes and Wyatt [11] of the interaction between two crossed roton beams. The experiment indicated a velocity dependent interaction, which, counter to intuition, increases as the velocity of the rotons increase, in a restricted momentum range.

To model the superfluid velocity field of a roton, we apply the dipole concept of Onsager, and Feynman [13], using the classical dipole field model of Donnelly and Roberts [14]. The $\vec{p}\cdot\vec{v}_s$ interaction potential for this model is given by

$$U(\vec{p}_1,\vec{p}_2,\vec{r}) = \frac{1}{4\pi\rho}\left[\frac{(\vec{p}_1\cdot\vec{r})(\vec{p}_2\cdot\vec{r})}{r^5} - \frac{\vec{p}_1\cdot\vec{p}_2}{r^3}\right] \tag{14}$$

where ρ is the total helium density and $\vec{r}$ is the separation vector of the rotons. The $(4\pi\rho)^{-1}$ is due to strength of the dipole moment formulated quantum mechanically with consideration given to the back flow nature of the roton. This potential is quite complicated. The magnitude and even the sign of the potential change as the momentum vectors of the roton vary during an interaction, leading to alternating

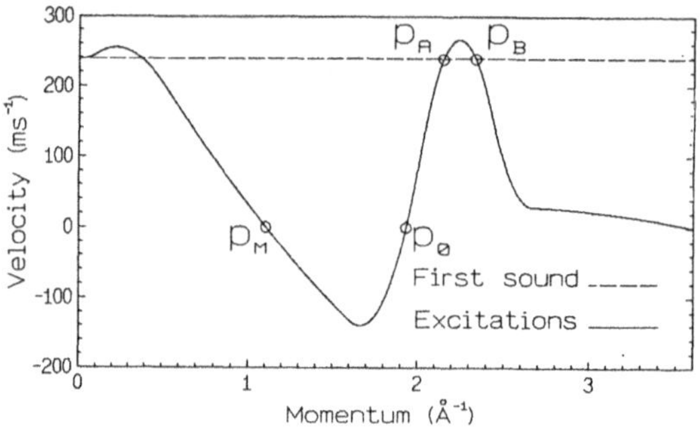

Fig. 4. The group velocity of an excitation is plotted with critical momenta labeled. p_M is the maxon peak, p_0 is the roton minimum and excitations in interval of p_A to p_B have group velocities greater than the velocity of first sound.

repulsive and attractive potentials. The velocity of a roton is then given by the slope of this dispersion curve $v(p) = \partial\epsilon/\partial p$ and is shown in Fig. 4.

In Fig. 1 and Fig. 4, four critical momenta are indicated, which separate the dispersion curve into regions. The first two critical momenta are the maxon peak, p_M, and the roton minimum p_0. The momenta marked as p_A and p_B border an interval where the roton group velocity is greater than the velocity of first sound. The consequences of this region are our concern here.

To calculate the roton trajectories, we again solve Hamilton's equations using the interaction potential (14). The Landau parabolic approximation to the dispersion curve is not useful here since the interactions involve energies that are beyond the validity of such an approximation. We solve the equations of motion numerically, using a Runge-Kutta-Fehlberg method.

For simplicity, we consider two perfectly collimated roton beams crossing at right angles, which is an idealized construction of the geometry of the Forbes and Wyatt experiment. One beam is labeled the 'target' beam and the other is the 'probe' beam. The probe beam is modeled by a rectangular array of initial positions for the roton, perpendicular to the beam direction. Each trajectory in this array is calculated separately as the roton interacts with the target beam. The cross section of the probe beam is defined as the area of the initial rectangular array which contains trajectories where the roton satisfies one of three conditions during its trajectory. These conditions are: (I) the probe roton is accelerated above the speed of sound, (II) the probe roton undergoes a species change (i.e. $R^+ \rightarrow R^-$), and (III) either roton has an energy greater than 2Δ. Condition (III) is due to the theory of Pitaevskii [15], which predicts that rotons that reach a value of 2Δ must decay into phonons by some process. Condition (II) is a process that primarily produces large angle scattering.

Condition (I) is the process which we believe describes the anomalous experimental results of Forbes and Wyatt. This condition is suggested by observation of the rise in the attenuation of the probe roton beam by the target beam, shown in Fig. 5(a). As the probe momentum increases from 2.0 to 2.2 Å^{-1}, the attenuation by the 1.95 Å^{-1} target beam increases. The attenuation of the probe beam defined by Forbes and Wyatt is the natural logarithm of the ratio of the attenuated beam to the unattenuated beam:

$$-\ln\left(\frac{\phi(p_1)}{\phi_0(p_1)}\right) = n(p_2)\sigma(p_1, p_2)d \qquad (15)$$

where $\phi(p_1)$ and $\phi_0(p_1)$ are the attenuated and unattenuated probe beams respectively. The right hand side of Eq. 15 contains the perpendicular cross section, $\sigma(p_1, p_2)$, which is a function of probe and target momenta, the density of target rotons, $n(p_2)$, produced by the target heater (which is not known experimentally) and a characteristic length, d, of the interaction region.

Forbes and Wyatt present a dispersion curve to which they correlate roton velocity to momentum. This curve has an inflection point at a momentum of 2.25 $Å^{-1}$, corresponding to a group velocity equal to the velocity of first sound. The experimental values of the beam attenuation rise as the roton momentum approaches this value, labeled p_* in Fig. 5(a). The more recent dispersion curve of Donnelly, Donnelly and Hills [16] (Figs. 1 and 4) has a considerable region with velocities greater than the velocity of first sound, as described earlier. This difference in dispersion curves does not change our interpretation of the experimental results.

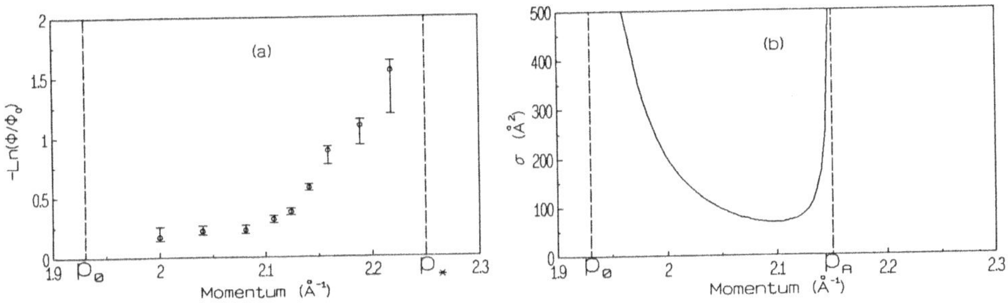

Fig. 5. In Fig. 5(a) the attenuation of a probe beam of rotons by a fixed target beam as a function of probe roton momentum is shown, as measured by Forbes and Wyatt [11]. The dashed lines indicate the location of p_0 and p_*. In Fig. 5(b) the calculated perpendicular roton-roton cross section is plotted for a target roton of 21 m/s. The dashed lines indicate the location of p_0 and p_A.

We present our cross section results in Fig. 6 for two different target momenta: a R^+ target with momentum equal to 1.95 $Å^{-1}$ and a R^- target with momentum equal to 1.90 $Å^{-1}$. Both rotons correspond to the roton which Forbes and Wyatt select for their target which has a velocity of 21 m/s. Forbes and Wyatt select target velocity, a method that will not uniquely select target momentum. Notice the qualitative difference between interactions of like and unlike species. Singularities in the cross section occur for both target and probe momenta where the velocity is zero, p_M and p_0 in Fig 3. Condition (II) is the primary contribution to the cross section near these two momenta. Near the probe momentum interval of p_A to p_B, both target cross sections rise exponentially. Condition (I) is responsible for this rise. As the probe momenta reach the end of the known dispersion curve, the excitation velocities go to zero and the cross sections begin to rise again. This effect is due to Condition (III).

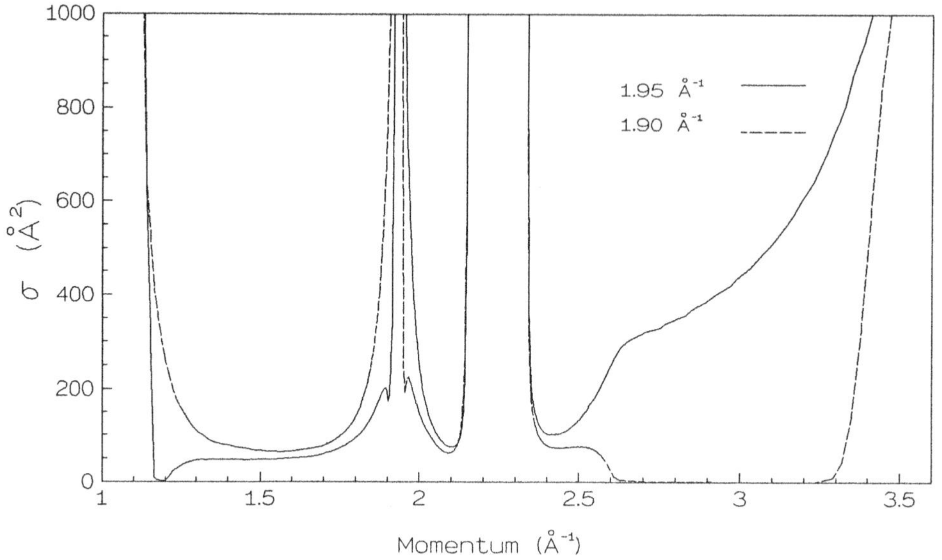

Fig. 6. Perpendicular roton-roton cross sections are plotted for two target momenta as a function of probe momentum from the maxon peak to the end of the known dispersion curve. The two target momenta have the same magnitude of velocity equal to 21 m/s.

In Fig. 5(b) we present an average of the R^+ and R^- cross sections in the momentum interval of the Forbes and Wyatt experiment. Assuming a Bose distribution of energy, the statistical weighting of the excitations by energy shows that the 1.90 $Å^{-1}$ and the 1.95 $Å^{-1}$ rotons are equally statistically significant. Comparing to the experimental results in Fig. 6, there is good qualitative agreement with the experimental values. Since we have an idealized geometry which cannot adequately approximate the angular sensitivity of the experimental apparatus, and that the target roton density, $n(p_2)$, in Eq. 15 is unknown, direct comparison of our numerical results with the experimental values is not possible. Despite the inability to compare quantitatively, it is clear that with the implementation of Condition (I), we can explain the anomalous rise in the experimental results.

To illustrate the individual contribution of each condition to the roton-roton cross section, we present two representative cross section diagrams in Fig. 7. For each calculation, the target roton has a momentum value of 1.95 $Å^{-1}$ and its dipole moment points to the left (which defines the symmetry axis). Probe momenta of 1.99 $Å^{-1}$ and 2.14 $Å^{-1}$ are used in Fig. 7(a) and 7(b), respectively, with dipole moments pointing into the page. In Fig. 7(a), all three conditions are present with the axis of symmetry clearly visible. In the region defined by Condition (II), two large holes exist. Rotons trajectories in this region suffer angular scattering but avoid the three cross section conditions. In Fig. 7(b) only Conditions (I) and (II) contribute to the cross section. Two island regions of Condition (I) have formed linked to the larger core by a region of Condition (II). As the probe momentum nears p_A, these islands grow together and eventually connect to the core. The shapes of these cross sections emphasize the complexity of the interaction potential.

We have investigated the possibility of a lifetime effect for rotons that travel faster than the velocity of first sound. For the effect to be noticeable for this model, lifetimes on the order of 10^{-11} seconds or less would be necessary. Estimates by Jäckle and Kehr [17] for the lifetime of a single roton with momentum just above p_A are on the order of 10^{-8} seconds. The discrepancy in these values may be

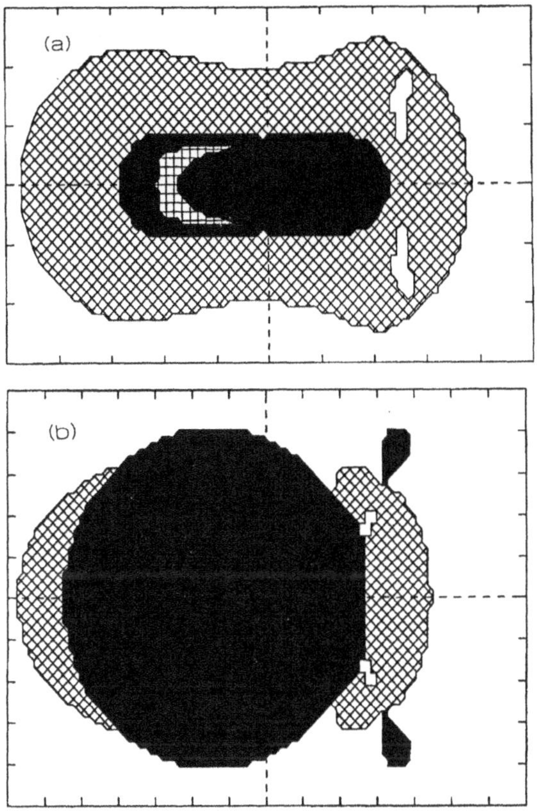

Fig. 7. Presented are calculated shapes of perpendicular roton-roton cross sections for two probe momenta: Fig. 7(a) 1.99 $Å^{-1}$ and Fig. 7(b) 2.14 $Å^{-1}$ with a target momentum of 1.95 $Å^{-1}$. The solid fill corresponds to Condition (I). The slanted and orthogonal cross hatch fills correspond to Condition (II) and (III) respectively, as explained in the text. Axes marks are separated by 3 Å and 1 Å in (a) and (b), respectively.

attributed to the difference in the type of interactions. The process of R + R ⇔ R + P + R, where R and P are roton and phonon respectively, is different than the process R ⇔ R + P, considered by Jäckle and Kehr.

The formalism for the classical roton-roton model is based on the theory of action at a distance or instantaneous field propagation. In this model of an excitation gas and background superfluid in helium II, the field propagation velocity is the velocity of first sound. Cursory development of retarded equations of motion has been performed but initial results indicate some difficulty with implementing this approach. Phenomena such as tachyonic behavior of rotons and phonon shock waves, which are physically reasonable, are not readily introduced into our model.

The experimental results observed by Forbes and Wyatt are explained by a decay mechanism in the region where the roton velocity is greater than the velocity of first sound. Further experiments of the type presented by Forbes and Wyatt will shed light on the single excitation dispersion curve in the roton region and perhaps will lead to a better understanding of the structure of the roton. The calculations presented indicate the need for the further development of a decay theory for multi-roton interactions in which roton velocities reach or exceed the velocity of first sound.

4. CONCLUSION

The roton's complicated dispersion curve and equally complicated interaction potential make roton dynamics a natural arena for the use of computer simulations. We have just begun to apply simulation techniques to this problem. The procedures described in Section 3 may eventually be extended to describe the roton viscosity, a quantity about which we know little theoretically or experimentally. When these simulations move beyond simple one or two particle interactions to include the full many-body behavior of the roton gas we can expect to learn even more, especially about the statistical distribution of rotons.

The essential limitation of this approach is its classical nature. This is not as severe a restriction as it may seem. Assuming a simple distribution of rotons, the greatest population should occur near the roton minimum (p_0). Due to the small (or zero) group velocity for these rotons, the cross sections of any roton interaction will tend to have an infinite value at p_0. Large cross sections for the rotons that are expected to be the most significant part of the population will lead to interactions at very large distances relative to the wavelength of the rotons (approximately 3Å near the roton minimum). If interactions of this type dominate the behavior of the roton gas, a classical description is a good approximation.

ACKNOWLEDGEMENT

This research is supported by the NSF Low Temperature Physics Program under grant DMR 88-15803.

REFERENCES

1. D. C. Samuels and R. J. Donnelly, Phys. Rev. Lett. **64**, 1385 (1990).
2. C. F. Barenghi, R. J. Donnelly and W. F. Vinen, J, Low Temp. Phys. **52**, 189 (1983).
3. S. C. Goodman, Phys. Fluids **14**, 1293 (1971).
4. G. W. Rayfield and F. Reif, Phys. Rev. **136**, A1194 (1964).
5. W. I. Glaberson, D. M. Strayer and R. J. Donnelly, Phys. Rev. Lett. **21**, 1740 (1968).
6. A. J. Hillel and W. F. Vinen, J. Phys. C. **16**, 3267 (1983).
7. C. E. Swanson, W. T. Wagner, R. J. Donnelly and C. F. Barenghi, J. Low Temp. Phys. **66**, 263, (1987).
8. P. H. Roberts and R. J. Donnelly, J. Low Temp. Phys. **15**, 1 (1974)
9. P. Mathieu, A. Marechal and Y. Simon, Phys. Rev. B **14**, 3753 (1976).
10. J. S. Brooks and R. J. Donnelly, J. Phys. Chem. Ref. Data **6**, 51 (1977).
11. A.C. Forbes and A.F.G. Wyatt, Phys. Rev. Lett. **64**, 1393 (1990).
12. *Quantum Fluids and Solids*, edited by S.B. Trickey, E.D. Adams and J. W. Duffy (Plenum, New York, 1977).
13. R.P. Feynman and M. Cohen, Phys. Rev. **102**, 1189 (1956).
14. P.H. Roberts and R.J. Donnelly, J. Low Temp. Phys. **15**, 1 (1974).
15. L.P. Pitaevskii, Sov. Phys. JETP **12**, 155 (1961).
16. R.J. Donnelly, J.A. Donnelly and R.N. Hills, J. Low Temp. Phys. **44**, 471 (1981).
17. J. Jäckle and K.W. Kehr, Phys. Lett. **39A**, 327 (1972).

SPECTROSCOPIC STUDY OF SCATTERING OF UCN BY LIQUID He^4

E. Gutsmiedl, R. Golub (+) and J. Butterworth (*)

Technical University of München
James-Franck-Str
D-8046 Garching
and ILL Grenoble (France)
(+) Technical University of Berlin
and ILL Grenoble (France)
(*) ILL Grenoble

Abstract

We report on our first recent measurements of the energy spectrum of upscattered Ultra-Cold-Neutrons in liquid He^4 at temperatures around 1 Kelvin. The measurements were made on the Superthermal UCN Source at the ILL [1]. This source produces UCN's in liquid He^4 by down-scattering (phonon-emission) of 10 Å neutrons. The UCN's produced in this way are stored in the He^4 container, since they are totally reflected from the walls. The UCN's can gain energy through scattering by Phonons [3] and Rotons [3], and hence leave the container. We have measured the spectrum of these upscattered neutrons by using a pseudo statistical chopper [4]. The first preliminary measurements show a discrepancy with the spectrum expected on the basis of phonon and roton scattering.

Excitations in Two-Dimensional and Three-Dimensional Quantum Fluids
Edited by A.G.F. Wyatt and H.J. Lauter, Plenum Press, New York, 1991

In this paper we want to report our first recent measurements of the energy spectrum of upscattered Ultra-Cold-Neutrons (UCN) in liquid Helium 4 at temperatures around 1 Kelvin. The UCN's are created in the liquid He^4 itself by downscattering of incoming 10 Å neutrons through one-phonon emission [1].

The He^4 container is exposed to 10 Å neutrons from a cold neutron guide at the ILL [2]. After some time (200 seconds) the beam is shut off by a Li shutter (see fig. 1).

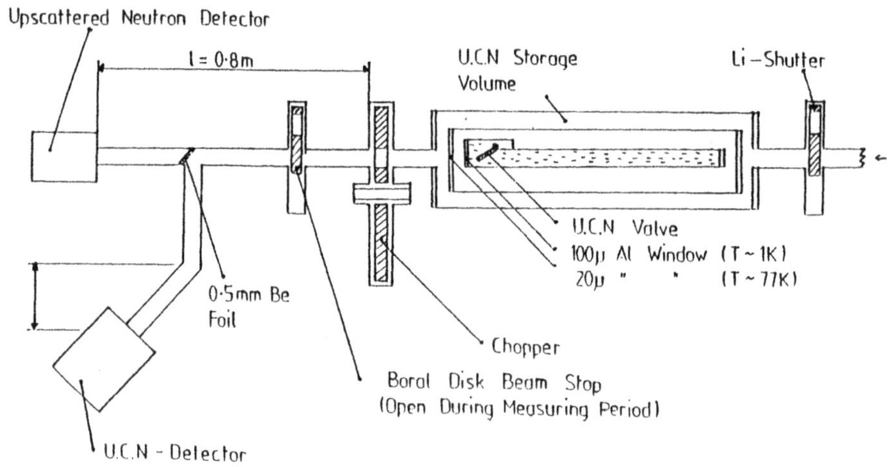

Figure 1. Experimental Setup

UCN's, which were created in the He^4 container are trapped in this containment due to their property of being totally reflected from material walls at all angles of incidence. The UCN's can now gain energy again during their random walk in the bottle through scattering by Phonons [3], and Rotons [3]. After that process, these neutrons can leave the He^4 container. We have investigated these neutrons with a pseudostatistical time-of-flight apparatus [4], in order to measure the energy spectrum of the upscattered neutrons. The theory of the phonon/roton scattering [3] predicts a curve with a peak lying in the region of 20 Å (final wavelength of the upscattered neutron) for a temperature of 1.25 Kelvin. The data of the experiment show two peaks (see fig. 2). One peak is at 15 Å and one is at 45 Å for temperature 1.25 K. We were later able to identify the first peak as partially due to a contamination through the closed cold neutron shutter. After our experiment with the He^4 we removed the Helium and warmed up the Cryostat. We then made a time-of-flight measurement with the closed cold beam shutter. The measurements showed a clear peak at 15 Å. We improved then the shielding of the incoming beam shutter and measured again. The result was, that the background peak at 15 Å was now completely removed.

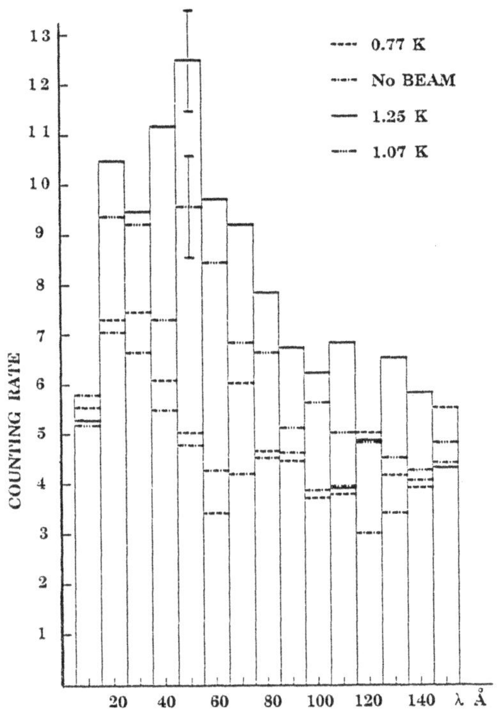

Figure 2. Raw Spectra of the Upscattered Neutrons

The remaining peak in the data indicates, that there is still a possible discrepancy between the data and the phonon/roton uspcattering theory [3], (see fig. 3).

One possible explanation could be, that we have used in the theory the hydrodynamic limit of the Grüneisen constant $U_0 = 2.84$ [5], which determines the strength of the matrix element of the three phonon interaction. At higher phonon wavenumber the Grüneisen constant may be momentum dependent [5], as recently calculated by Halley and Korth [8].

Therefore we have to improve the theory of the phonon/roton scattering and compare it to our data again.

The phonon/roton upscattering theory has in the total neutron cross section a strong temperature dependence. The cross section shows in the 2-phonon regime (0.5 - 0.9 Kelvin) a T^7-dependence. This kind of behaviour was observed in earlier measurements [6], [7].

We intend to improve our experimental technique by rebuilding the cryostat in order to get more precise measurements of the upscattered neutrons.

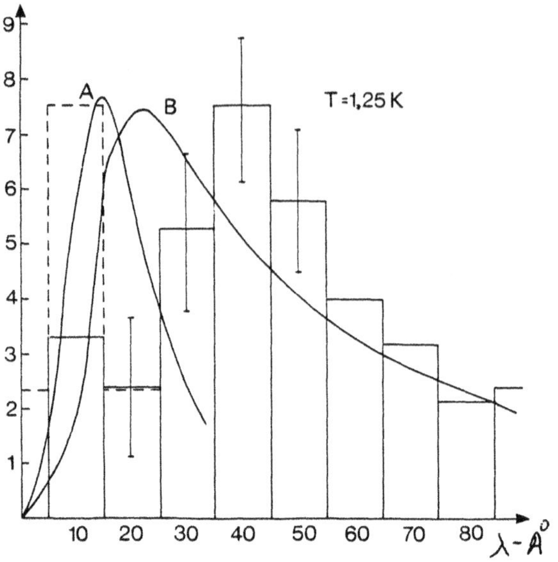

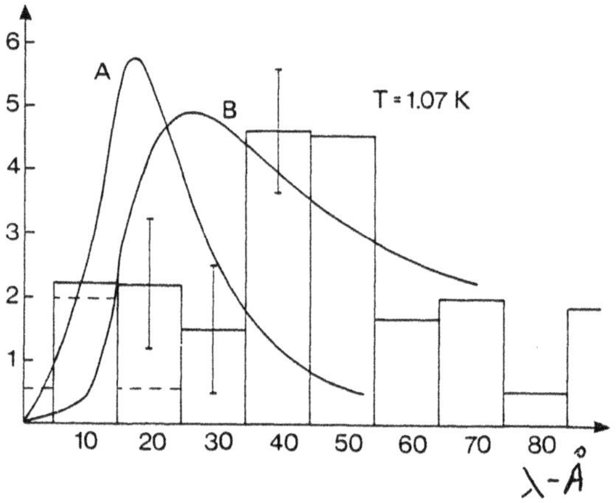

Figure 3. Background Subtracted Spectra
A - Phonon-Roton Theory (3)
B - Theory Corrected for Transmission of an Ideal Lossless Guide

Dashed Histogram - Expected One Phonon Contribution Normalized to Measured Spectrum

Literature

[1] R. Golub, J.M. Pendlebury Phys. Lett. 53A.133 (1975)

[2] H. Blank, B. Maier "The Yellow Book": Guide to the neutron research facilities at the ILL p. 35

[3] R. Golub Phys. Lett. 72A, 387 (1979)

[4] U. Freudenberg, W. Gläser Nucl. Instr. Meth. A243, 429 (1986)

[5] H.J. Maris Phys. Rev. A vol 8, n° 4, 1980 (1973)

[6] R. Golub et al. Z. Phys. B51, 187 (1983)

[7] A.I. Kilvington et al. Phys. Lett. A125, 416 (1987)

[8] J.W. Halley and M.S. Korth, paper presented at this conference.

ROTON-ROTON SCATTERING

A.F.G.Wyatt and A.C.Forbes

Department of Physics
University of Exeter
Stocker Road, Exeter, EX4 4QL, UK

1 INTRODUCTION

The combination of quantum evaporation (QE) and time of flight measurements enables roton-roton scattering to be studied as a function of wavevector[1]. Hitherto scattering has been has been seen with thermal populations[2] which confines the wavevector to values very near that of the roton minimum $q = 1.924Å^{-1}$[3], and averages over all angles between the two scattering rotons. Preliminary results using QE indicate a strong wavevector dependence of the scattering cross-section[1].

Very little is known about rotons apart from their dispersion curve and lifetimes. We need a good microscopic picture of them if we are to understand their scattering. The difficulty in describing rotons stems in part from their short wavelengths, ~3Å which is similar to both atomic dimensions and atomic separations in the liquid ^4He. This last property makes liquid ^4He, in essence, a strongly interacting condensed Bose system. It is hoped that scattering measurements will lead to a deeper understanding of their nature, as it has done for particles in other branches of physics.

We know that rotons have very long lifetimes when they are not scattered by other excitations, that is, they do not spontaneously decay[4,5]. This means that beams of rotons can be created in liquid ^4He which is cold enough, T<0.1K. The beams can be collimated into reasonably narrow pencils and can travel distances of order centimetres without decay. If two such beams are crossed then they should interact due to roton-roton scattering.

In the experiment described here one roton beam is monitored by quantum evaporation. It is seen to be attenuated when a second broad beam crosses it, approximately at right angles. As the second beam is increased in intensity so the probe beam is reduced. The rotons in the probe have their energy and momentum changed by scattering. Simulation studies show that if their propagation direction is changed by a few degrees then they will not be detected. This means that the experiment is sensitive to small angle scattering and hence to relatively weak interactions.

A broad energy spectrum of rotons is injected into the liquid in a short pulse from heater h_1, see figure 1. As rotons with different wavevectors travel at widely different velocities, so their time of flight can be used to distinguish between them. The time dispersion must be larger than both the injected pulse length and the detectors time constant. When this is so, the signal detected as a function of time can be interpreted as a spectrum of wavevectors. We shall analyse the relative attenuation of the signal at various times in the received signal.

The second beam of rotons, that cause the scattering of the probe beam, is also injected as a pulse from heater h_2. The delay between the injection of these rotons from h_2 and the injection of the probe rotons from h_1, selects the wavevectors of the rotons with which the probe rotons scatter. In this paper we present results where this delay is relatively long, 30μs, which means that the probe rotons are scattering from slow rotons with $q_2<1.95Å^{-1}$. This is in contrast to the previously published results[1] where the delay was negative, -10μs, which meant that the probe rotons scattered from rotons with a wide spectrum of wavevectors, $q_2 \sim 2.0 - 2.3Å^{-1}$.

The experimental arrangement is described in the next section and the measured results in the following section, 3. The analysis and discussion of the results which leads to relative values of the cross-section as a function of wavevector is described in section 4 and is followed by the conclusion in section 5.

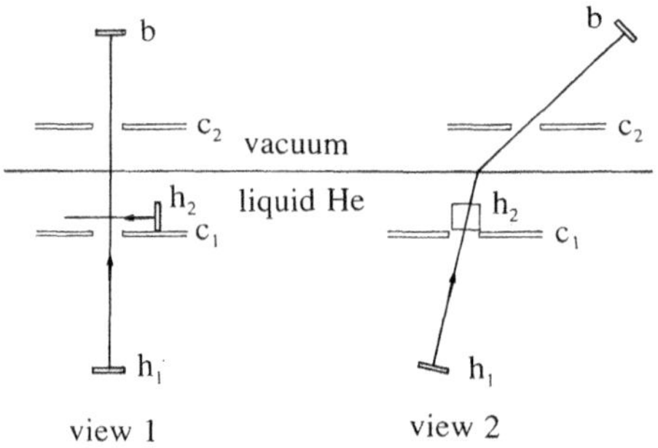

Fig. 1. The experimental arrangement is shown schematically from 2 perpendicular view directions. The probe rotons are injected by heater h_1 and the scattering rotons by heater h_2. c_1 and c_2 are collimators and b is a superconducting Zn bolometer.

2 THE EXPERIMENTAL ARRANGEMENT

The experimental arrangement is shown schematically in figure 1. The thin metal film heater h_1 injects both probe R^+ rotons and phonons. These are collimated into a beam by a small hole in a mica sheet and then are incident, at a mean angle of 12° to the normal, on the free surface of the liquid He. Here atoms are quantum evaporated, those atoms evaporated by R^+ rotons are detected by the bolometer b at an angle of 47° to the normal. Atoms evaporated by phonons are not detected by the bolometer as they are at angles much closer to the normal. The scattering of the phonons is discussed elsewhere[7,8].

Another thin metal film heater h_2, injects the scattering R+ rotons. The interaction point is defined by the intersections of the centre lines from h_1 and h_2. This point is 5.0 mm from h_1 and 1.5 mm from h_2. The collimator c_1 is 4.5 mm from h_1 and the hole is 0.8 mm diameter. The collimator for the atoms, c_2, is 1.0 mm diameter and 4.5 mm from the bolometer. The heaters are 1 mm^2.

The R^+ rotons are defined as those with $q>1.924Å^{-1}$ and only these rotons are observed. Their spectrum is shown in figure 2 [9] together with their group velocity as a function of wavevector[10]. The group velocity varies from zero at $q = 1.924Å^{-1}$ to $\sim 238ms^{-1}$ at $q = 2.24\ Å^{-1}$.

In figure 3 is shown the sequence of events. At t = 0, typically a $10\mu s$, 2.0mW heating pulse is applied to h_2. Then after a delay of $30\mu s$, measured from the start of the h_2

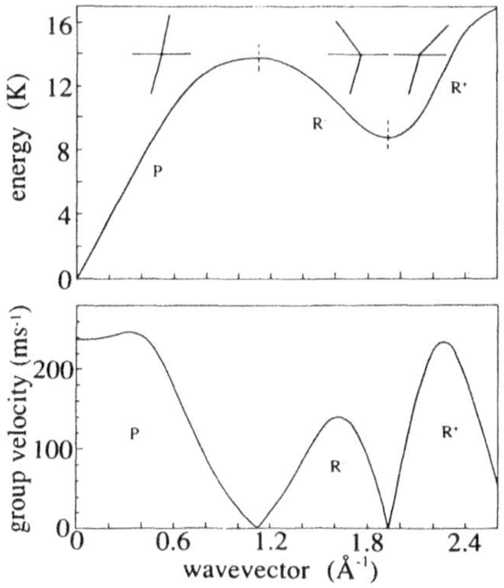

Fig. 2. The dispersion curve for ^4He and the group velocity for phonons, P, R^- rotons and R^+ rotons.

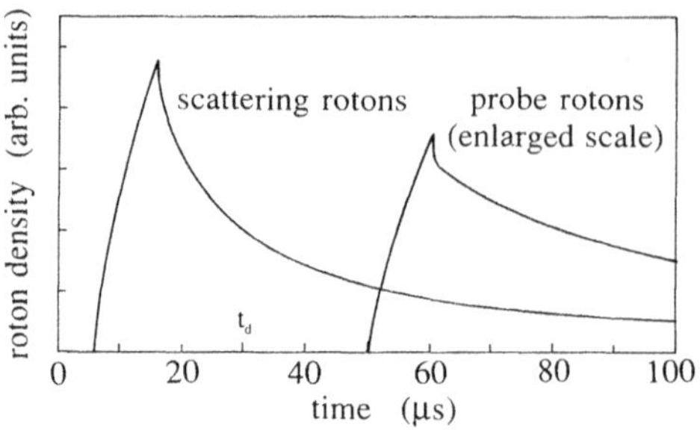

Fig. 3. The roton number densities for the probe and scattering rotons as a function of time at the interaction point for a delay $t_d = 30\mu s$.

pulse, a 10μs, 3.1mW heating pulse is applied to h_1 which injects in the probe rotons. The fastest probe rotons take 20.6μs to arrive at the interaction point and so they arrive at 50.6μs after the start of the h_2 pulse. The fastest h_2 rotons that scatter these probe rotons at the interaction point have velocity $1.5\ 10^{-3}/50.6 \times 10^{-6} = 29.6\mu s^{-1}$ and have $q_2 = 1.95Å^{-1}$. Scattering occurs when the two beams overlap, this can be seen in figure 3 to be for times >50.6ms. It can be seen from this figure that the scattering roton number density is developing slowly as the probe beam passes through the interaction point.

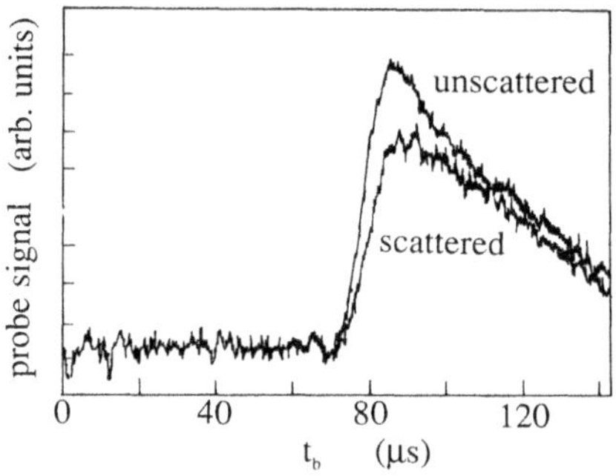

Fig. 4. Examples of the probe beam signals with and without scattering rotons. The heater pulses are, h1; 10μs, $2.0mWmm^{-2}$, and h_2 10μs, $3.1mWmm^{-2}$. The probe pulse is delayed 30μs. Time is measured from the start of the probe heater pulse.

3 RESULTS

The received signals on the bolometer are due to the atoms quantum evaporated by the probe rotons. Typical unscattered and scattered signals are shown in figure 4. The signal can be seen to decrease more at the beginning than at the end of the received pulse. The fastest roton-atom signal are due to rotons with $q\sim2.3Å^{-1}$ and later signals are due mainly to rotons with lower wave numbers. This signal-time dependence of the attenuation is the first indication that the scattering is dependent on the wavevector of the probe roton.

The probe signal is measured for a range of scattering roton pulse widths. The effect of increasing the h_2 pulse width is to increase the scattering roton density at the interaction point, because the pulse widths are small compared to the dispersion for propagation times of >50μs; i.e. the scattering density is increased with only a negligible change in the spread of scattering roton wavevectors. The probe signal is found to decrease expontially with the scattering roton density as can be seen in figure 5, and the rate of decrease depends on the time of the received probe signal.

4 ANALYSIS AND DISCUSSION

The probe beam of rotons passes through a slowly developing cloud of scattering rotons. We know from scattering simulations, that the scattered rotons have a negligible chance of being detected via quantum evaporation. So any scattering event will simply attenuate the probe beam. We also ensure that the scattering roton number density is many times greater than the decrease in probe roton number density, at the interaction point. This means that the scattering roton density is not depleted by the scattering with the probe rotons. This point is confirmed by the results shown in figure 5 which we discuss below.

From the above picture the flux ϕ is given by

$$\phi = \phi_o \exp(-n_2\sigma D) \qquad (1)$$

where ϕ_o is the unscattered flux, n_2 is the scattering roton number at the interaction points, σ is the cross section for scattering between the two rotons and D is the linear extent of the scattering roton cloud in the direction of the probe beam. Subscript 1 is used for probe rotons and 2 for scattering rotons.

Because σ is a function of the wavevectors of both rotons in a scattering event, ie. $\sigma(q_1,q_2)$ then equation 1 only strictly applies for scattering between two mono energetic beams of R+ rotons. However, as all the scattering rotons have nearly the same wavevector then we shall use equation 1 for each instant of the probe pulse where the dispersion causes essentially only a narrow range of wavevectors to be scattering at any time.

From equation 1 we find

$$-\ln(\phi/\phi_o) \propto \langle n_2 \rangle \qquad (2)$$

where $\langle n_2 \rangle$ is the average number density over the time and space determined by the probe rotons. To verify the behaviour expressed in equation 2 the bolometer signal $s(t)$ at a particular time is taken to be $\propto\phi(t)$, as it would exactly be if the detector had zero time constant. In figure 5 we plot $-\ln(\phi(t)/\phi_o(t))$ at a fixed t, against pulse width which as we have discussed earlier, varies the scattering number density $\langle n_2 \rangle$.

It can be seen in fig 5 that $-\ln(\phi(t)/\phi_o(t))$ is linearly proportional to the pulse width but that the gradient depends on the time, t_b, of the probe signal. At later times the gradient is less. The linear dependence at all times shows that the scattering roton density is not being depleted by the early scattering events. The change in the gradient with time t_b is mainly due to the scattering cross-section changing with the wavevector of the probe roton.

In order to extract the variation in cross-section we must take out the variation of $\langle n_2 \rangle$ over the duration of the probe rotons passing through the interaction point. The dominant time smearing effect in the experiment is due to the finite pulse width of the probe roton injection, 10μs, which

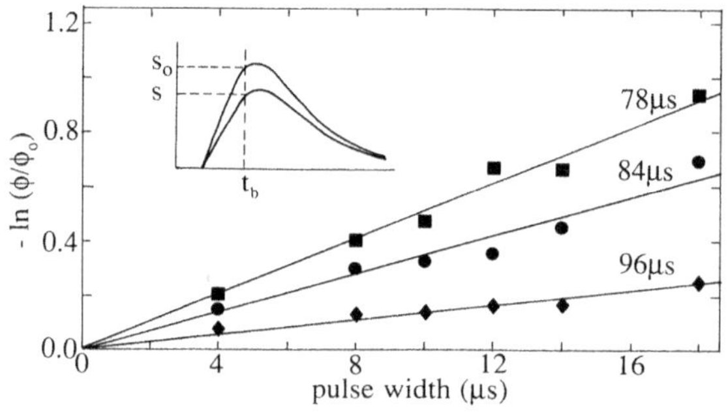

Fig. 5. The logarithm of the ratio of the scattered probe flux ϕ to the unscattered probe flux ϕ_o is shown as a function of the width of the scattering pulse to heater h_2, for various times, t_b, of the probe signal. The probe pulse is 10μs, 2mWmm^{-2} and the scattering pulse 3.1mWmm^{-2}, the delay is 30μs. Inset is shown how s_o and s are defined, we take $\phi/\phi_o = s/s_o$, and t_b is measured from the start of the probe heater pulse.

is greater than the detector time constant, $\tau_c \sim 5\mu s$, and path length variations due to the finite collimation. We consider the effect of the pulse length in detail.

Consider the signal from the bolometer at a particular time t_b, and imagine the heater h_1 giving out the same spectrum of rotons continuously. The rotons emitted by the heater at time t_h give a signal at time t_b if their wavevector q_1 is given by

$$t_b - t_h = \frac{L_1}{v_g(q_1)} + \frac{L_a}{v_a} \equiv t(q_1) \tag{3}$$

where $v_g(q_i)$ is the group velocity, L_1 and L_a are the roton and atom path lengths respectively. v_a is the velocity of the evaporated atom and this also depends on q, through the equation

$$\hbar\omega(q_1) = E_B + m_4 v_a^2 / 2 \tag{4}$$

where $\omega(q_1)$ is the energy of the probe roton, E_B is the binding energy, 7.16K, and m_4 is the mass of a ^4He atom.

In figure 6 we show q as a function of $L_1/v_g + L_a/v_a$ ($\equiv t(q_1)$. On this diagram we can show the range of wavevectors that contribute to the signal at t_b due to the finite heater, h_1, pulse length. Slower rotons from the beginning of the pulse give a signal at the same time as faster rotons from the end of the pulse. If the heater is on between $t_h = 0$ and $t_h = t_p$ then the wavevectors on the curve and between the lines $t_b = t(q_1)$ and $t_b - t_p = t(q_1)$, contribute at t_b; i.e. $q_{L1} < q_1 < q_{L2}$ and $q_{u2} < q_1 < q_{u1}$. When the line $t_b - t_p = t(q_1)$ does not intersect the curve but $t_b = t(q_1)$ does then, then the range is $q_{L1} < q_1 < q_{u1}$. (Subscripts L and u denote the lower and uppper wavevector ranges respectively).

In terms of these ranges the energy flux on the bolometer is proportional to[11].

$$\varphi(t_b) = \int_{q_{L1}}^{q_{L2}} \hbar\omega \dot{N}_1(q_1) dq_1 + \int_{q_{u2}}^{q_{u1}} \hbar\omega \dot{N}_1(q_1) dq_1 \tag{5}$$

where $\dot{N}_1(q_1)dq_1$ is the number of rotons emitted per second by the heater, with wavevectors in the range q_1 to $q_1 + dq_1$. Other experiments have shown that the roton spectrum can be approximately characterised by an effective temperature, T_e, in which case.

$$\dot{N}_1 \propto \frac{q^2}{\exp\left(\frac{\hbar\omega(q)}{kTe}\right) - 1} \tag{6}$$

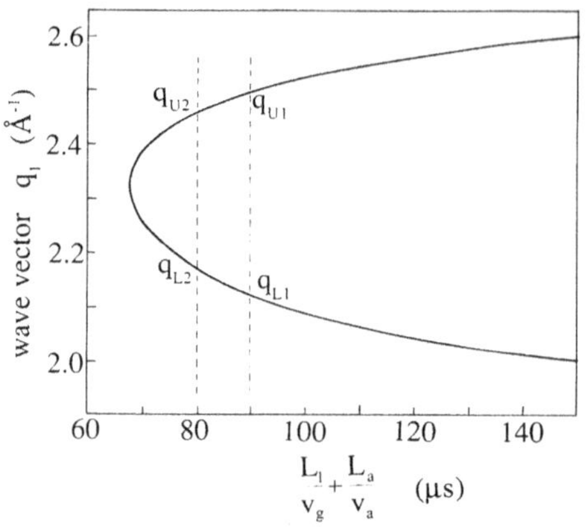

Fig. 6. The probe wavevector is shown as a function of the transit time for the probe roton and atom through the liquid ^{4}He and vacuum respectively (see text). The roton probe flux at the bolometer at a time 90μs contains wavevectors which lie between the dashed lines i.e. q_{L1} - q_{L2} and q_{u1} - q_{u1}, due to the 10μs probe heater pulse.

T_e is typically 1.5K and so the contribution form the upper range of wavevectors in equation 5 is relatively small.

Having found the wavevector range for the probe rotons that contribute at a particular time to the signal we now go to the interaction point in the liquid He. The probe rotons in these ranges do not arrive at the interaction point at the same time, for example, rotons in the lower range arrive in the interval

$$\frac{l_1}{v_g(q_{L1})}+t_d < t_i < \frac{l_1}{v_g(q_{L2})}+t_d+t_{p_1} \qquad (7)$$

where t_d is the time delay between the h_1 and h_2 pulses and l_1 is the distance from h_1 to the interaction point. This interval is less than t_{p1}. Some examples are shown in figure 7.

During this time the scattering roton number density is changing. Because the h_2 pulse is also finite, at any instant t_i, there is a range of scattering roton wavevectors at the interaction point. The ranges of q_2 that contribute at t_i are given by

$$t_i = \frac{l_2}{v_g(q_2)}$$

$$\text{and} \quad t_i - t_{p2} = \frac{l_2}{v_g(q_2)} \qquad (8)$$

where l_2 is the distance between h_2 and the interaction point and t_{p2} is the h_2 pulse length. The number density at t_i is given by the number flux divided by group velocity hence

$$n_2(t_i) \propto \int \frac{\dot{N}_2(q_2)}{v_g(q_2)} dq_2 \tag{9}$$

where the integration ranges are given by the wavevectors that solve the equations 8.

The average scattering number density is given by

$$\langle n_2 \rangle = \frac{\int n_1(q_1) n_2(t_i) dq_1}{\int n_1(q_1) dq_1} \tag{10}$$

where the integration ranges are the same as in equation 5 and t_i and q_1 are related by

$$\frac{l_1}{v_g(q_1)} = t_i - t_d \tag{11}$$

This value of $\langle n_2 \rangle$ relates to one value of t_b; it is given by equation 3 with $t_h = 0$. In the range of times which show a clear attenuation, 68 - 102μs (measured from start of h_1 pulse), $\langle n_2 \rangle$ only varies by 12%.

Finally we find the relative values of cross-section from

$$\sigma \propto -\frac{\ln(\phi / \phi_0)}{\langle n_2 \rangle} \tag{12}$$

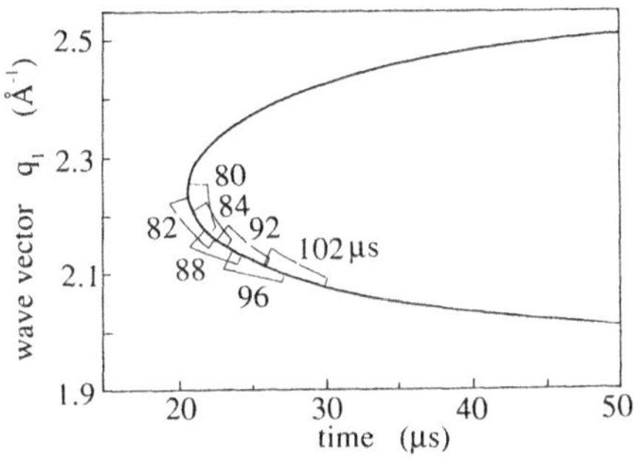

Fig. 7. The probe roton wavevector is shown as a function of propagation time from the heater to the interaction point in the liquid. The ranges are labelled with the times that the signal arrives at the bolometer. The range is finite due to the 10μs probe heater pulse width.

In figure 8 is shown the relative σ against the average value of roton wavevector that contributes at a particular signal time, t_b. It can be seen that σ rises considerably between 2.1 and 2.25Å^{-1}.

The absolute value of σ depends on the absolute value of the scattering number density which we do not know. Only a small fraction of the h_2 heater output is in the form of R^+ rotons and as yet we do not have a model for roton generation. The results should be treated as preliminary as there may well be systematic errors which only further experiments can indicate.

The rise in $\sigma(q_1,q_2 = 1.95Å^{-1})$ with q_1 should be steeper than figure 8 indicates, as each measurement represents an average value over a range of wavevectors due mainly to the finite pulse length of the probe pulse and the time constant of the detector.

5 CONCLUSION

This first experiment shows that direct excitation scattering experiments are now feasible. We have found that the probe pulse is selectively attenuated by a second roton beam and we have interpreted this as a wavevector dependent scattering cross-section. We believe that this technique opens up many other interesting measurements which will increase our understanding of the excitations of liquid ^{4}He.

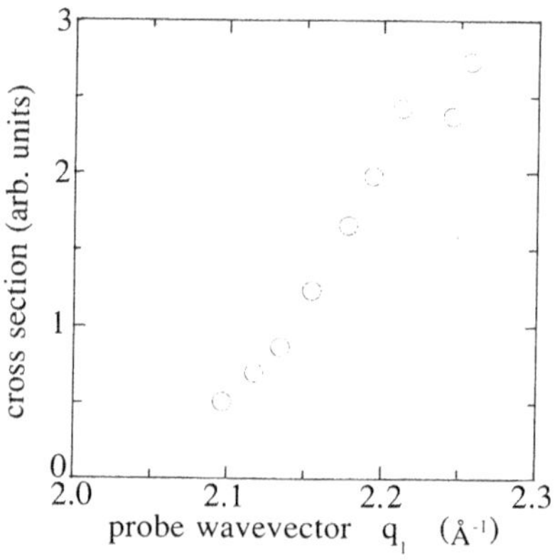

Fig. 8. The cross-section for roton-roton scattering is shown as a function of average probe wavevector. The scattering rotons have q_2~1.95Å^{-1}.

6 REFERENCES

1. A C Forbes and A F G Wyatt
Phys. Rev. Letts. 64 1393 (1990).

2. See for example F Mezei
Phys. Rev. Letts. 44 1601 (1980) and the references therein.

3. The one exception is reference 2 where the neutron line width was not found to be significantly different to that at the roton minimum.

4. F R Hope, M J Baird and A F G Wyatt
Phys. Rev. Letts. 52 1528-31 (1984)

5. M Brown and A F G Wyatt
J.Phys: Condens. Matter 2 (1990) 5025-46.

6. This was inadvertently quoted as 1.95Å^{-1} in reference 1.

7. A C Forbes and A F G Wyatt, "Phonons 89", S Hunklinger, W Ludwig and G Weiss, eds., World Scientific, Singapore (1990) p1047.

8. A C Forbes and A F G Wyatt. Proc LT19.

9. See for example A D B Woods and R A Cowley. Reports in progress in Physics 36 1135-1231 (1973) and references therein.

10. M Brown priv. comm.

11. Equation 5 only applies for QE at normal incidence. Finite collimation and a non normal angle of incidence will give a significantly different signal shape.

THEORY OF ROTON-ROTON SCATTERING

Fumiaki Iwamoto

Institute of Physics, College of Arts and Sciences
University of Tokyo, Komaba, Tokyo 153

1. INTRODUCTION

Direct observation of roton-roton scattering by Forbes and Wyatt[1] is the first example of the scattering between controlled beams of excitations in the condensed matter, and supplies information of elementary processes without thermal average. They found a dramatic increase of the scattering cross section as the probe roton momentum* increases from 2.0 to 2.2Å^{-1}.

In this article we report a theoretical study[2] stimulated by their experiment. Theoretically, the roton-roton scattering presents an interesting scattering problem to us with the following characteristics.

(i) Galilei non-invariance: An excitation in a medium is not the same as a particle in the vacuum. The energy-momentum relation of a roton is not simple $E = p^2/2m$, but the dispersion curve $E = \omega(p)$ found by neutron scattering experiment. Although the total momentum $\mathbf{P}$ of a pair of rotons is a constant of motion due to the uniformity of the medium, the concept of center of mass frame does not apply to the present case, because a roton of momentum $\mathbf{p}$ moves with its group velocity $\mathbf{v}(\mathbf{p}) = \mathrm{grad}_{\mathbf{p}}\omega(p)$, and the 'center of mass velocity', $\{\mathbf{v}(\mathbf{P}/2 + \mathbf{r}) + \mathbf{v}(\mathbf{P}/2 - \mathbf{r})\}/2$, depends not only on the total momentum $\mathbf{P}$ but also on the relative momentum $2\mathbf{r}$.

(ii) Invalid Born approximation: Although a roton has an appreciable momentum, its group velocity is small near the minimum or the maximum† of the dispersion curve. The simple Born approximation does not give correct result for such slow excitations even if the interaction may be weak. Correct transition matrix element follows after summing up the interaction up to the infinite order.

(iii) Non-conservation of particle number: A one-roton state of momentum from about 0.6Å^{-1} to around 3.2Å^{-1} is completely stable. However, a many-roton state, in general, couples with the states of different number of rotons.

*We use unit where $\hbar = 1$ and $k_B = 1$. Numerical value of momentum or wave number is given in Å^{-1}, and of energy or frequency in °K.

†The name 'roton' is used rather comprehensively in this article, not restricted around the minimum of the dispersion curve.

Excitations in Two-Dimensional and Three-Dimensional Quantum Fluids
Edited by A.G.F. Wyatt and H.J. Lauter, Plenum Press, New York, 1991

(iv) Ambiguous interaction: At present, study of the roton dynamics is still in a primitive stage. We have to follow a phenomenological approach; we infer from the experiment the strength and the shape of the interaction on one hand, interpret the experimental results and try to make predictions on the other hand. In such approaches certain arbitrariness exists in the definition of the non-interacting one-roton energy and the corresponding interaction. We can choose suitable model Hamiltonian to clarify the physical mechanism in each situation.

Because the total momentum P of the two rotons in the experiment by Forbes and Wyatt is already in the plateau region ($P \approx 2.8\text{Å}^{-1}$), the hybridization model is suitable for the present purpose. The model has been widely used by many people (Pitaevskii[3], Ruvalds and Zawadowski[4], Smith et al.[5], Nagai[6], Fukushima et al.[7], ...) for the interpretation of the plateau spectrum and the analysis of the two-branch structure found by neutron scattering[8].

In the next section we give general expression of the roton-roton scattering cross section based on the hybridization model. In section 3 the integral equation of the transition matrix is explicitly solved by assuming the interaction to be separable. In section 4 the behavior of the cross section is discussed. We can derive several definite properties on the cross section. Because the behavior of the cross section found by the experiment can be reproduced qualitatively in a wide range of parameters, a shape-independent treatment is introduced. Further discussion is given in section 5.

2. HYBRIDIZATION MODEL

In this section we derive the roton-roton scattering cross section from the hybridization model making use of the kinematical restrictions imposed on the interaction.

In the energy-momentum plane two-roton density of states forms a high-density band[9] due to the non-monotonic nature of the dispersion curve. The band extends in energy from $2\Delta_0$ to $2\Delta_1$, and in momentum up to $2Q_0$ at the energy $2\Delta_0$ and up to $2Q_1$ at $2\Delta_1$. The roton minimum (maximum) energy and its momentum are denoted by Δ_0 and Q_0 (Δ_1 and Q_1). It contains a logarithmically diverging ridge at the energy $\Delta_0 + \Delta_1$ from the momentum $Q_0 - Q_1$ to $Q_0 + Q_1$. It was proposed by Pitaevskii[3] and by Ruvalds and Zawadowski[4] that the non-interacting roton energy is a steeply increasing function beyond Q_0 joining smoothly to the free atom energy $p^2/2m$ at large momenta, and the observed spectra result after the interaction with the two-roton high-density band.

In the following we use notations $\mathbf{p}_1$ and $\mathbf{p}_2$ for the momenta of a pair of rotons, $\mathbf{P}$ for their total momentum, and $\mathbf{r}$ for half the relative momentum. A prime or zero put on the right shoulder distinguishes different pairs. The hybridization model is represented by the following Hamiltonian,

$$\mathcal{H} = \sum_{\mathbf{P}} \omega(p)\, b_{\mathbf{P}}^{\dagger} b_{\mathbf{P}} + \frac{1}{\sqrt{\Omega}} \sum_{\mathbf{P},\mathbf{r}} \{ V_3(\mathbf{P},\mathbf{r})\, b_{\mathbf{P}}^{\dagger} b_{\frac{\mathbf{P}}{2}-\mathbf{r}}\, b_{\frac{\mathbf{P}}{2}+\mathbf{r}} + \text{h.c.} \}$$
$$+ \frac{1}{2\Omega} \sum_{\mathbf{P},\mathbf{r},\mathbf{r}'} V_4(\mathbf{P},\mathbf{r},\mathbf{r}')\, b_{\frac{\mathbf{P}}{2}+\mathbf{r}}^{\dagger}\, b_{\frac{\mathbf{P}}{2}-\mathbf{r}}^{\dagger}\, b_{\frac{\mathbf{P}}{2}-\mathbf{r}'}\, b_{\frac{\mathbf{P}}{2}+\mathbf{r}'}, \tag{2.1}$$

where $b_{\mathbf{p}}^{\dagger}$ and $b_{\mathbf{p}}$ are the creation and annihilation operators of a non-interacting roton of momentum $\mathbf{p}$, and Ω is the volume of the system. We take the non-interacting roton energy $\omega(p)$ to coincide with the observed roton energy up to around the momentum

2.4Å^{-1}, then to deviate smoothly entering into the high-density band somewhere at 2.5~2.6Å^{-1}. Because the observed one-roton state is stable, the interaction $V_3(\mathbf{P},\mathbf{r})$ vanishes as long as $\omega(P)$ coincides with the observed one-roton energy.

We investigate the interacting eigenstate $|\lambda\rangle$ of total momentum $\mathbf{P}$ as a superposition of the non-interacting one-roton state and two-roton states,

$$|\lambda\rangle = C_\lambda b^\dagger_{\mathbf{P}} |\ \rangle + \sum_{\mathbf{r}}{}' \psi_\lambda(\mathbf{r})\, b^\dagger_{\frac{\mathbf{P}}{2}+\mathbf{r}}\, b^\dagger_{\frac{\mathbf{P}}{2}-\mathbf{r}} |\ \rangle, \tag{2.2}$$

$|\ \rangle$ being the ground state. A prime on the sum means that the $\mathbf{r}$-sum is done over half momentum space to avoid overcounting of the identical states. The coupled equations of the amplitudes are

$$\{E-\omega(P)\}C_\lambda = \frac{2}{\sqrt{\Omega}}\sum_{\mathbf{r}}{}' V_3(\mathbf{P},\mathbf{r})\psi_\lambda(\mathbf{r}), \tag{2.3}$$

$$\{E-\omega_2(p_1,p_2)\}\psi_\lambda(\mathbf{r}) = \frac{2}{\sqrt{\Omega}}V_3^*(\mathbf{P},\mathbf{r})C_\lambda + \frac{2}{\Omega}\sum_{\mathbf{r}'}{}' V_4(\mathbf{P},\mathbf{r},\mathbf{r}')\psi_\lambda(\mathbf{r}'), \tag{2.4}$$

E being the energy of the state $|\lambda\rangle$, and $\omega_2(p_1,p_2)$ the non-interacting two-roton energy, $\omega(p_1)+\omega(p_2)$. By eliminating C_λ, an equation of $\psi_\lambda(\mathbf{r})$ follows,

$$\{E-\omega_2(p_1,p_2)\}\psi_\lambda(\mathbf{r}) = \frac{1}{\Omega}\sum_{\mathbf{r}'}{}' U(\mathbf{P},\mathbf{r},\mathbf{r}')\psi_\lambda(\mathbf{r}'), \tag{2.5}$$

with the E-dependent interaction,

$$U(\mathbf{P},\mathbf{r},\mathbf{r}') = 2V_4(\mathbf{P},\mathbf{r},\mathbf{r}') + 4V_3^*(\mathbf{P},\mathbf{r})\frac{1}{E-\omega(P)}V_3(\mathbf{P},\mathbf{r}'). \tag{2.6}$$

For a scattering state, the quantum number λ is specified by the incident roton momenta $\mathbf{p}_1^0$ and $\mathbf{p}_2^0$ or by $\mathbf{P}$ and $\mathbf{r}^0$, and $E=\omega_2(p_1^0,p_2^0)$. The amplitude is expressed as

$$\psi_\lambda(\mathbf{r}) = \delta_{\mathbf{r},\mathbf{r}^0} + \frac{1}{E-\omega_2(p_1,p_2)+i\epsilon}\frac{1}{\Omega}T_\lambda(\mathbf{r}), \tag{2.7}$$

with the transition T-matrix defined by

$$T_\lambda(\mathbf{r}) = \sum_{\mathbf{r}'}{}' U(\mathbf{P},\mathbf{r},\mathbf{r}')\psi_\lambda(\mathbf{r}'). \tag{2.8}$$

From Eqs. (2.7) and (2.8) the T-matrix obeys an integral equation,

$$T_\lambda(\mathbf{r}) = U(\mathbf{P},\mathbf{r},\mathbf{r}^0) + \frac{1}{\Omega}\sum_{\mathbf{r}'}{}' U(\mathbf{P},\mathbf{r},\mathbf{r}')\frac{1}{E-\omega_2(p_1',p_2')+i\epsilon}T_\lambda(\mathbf{r}'). \tag{2.9}$$

The cross section σ for a probe roton of momentum $\mathbf{p}_1^0$ to be scattered by a scattering roton of momentum $\mathbf{p}_2^0$ propagating at right angle to $\mathbf{p}_1^0$ is defined by dividing the transition probability by the probe roton flux v_1/Ω,

$$\sigma = \frac{2\pi}{v_1/\Omega}\sum_{\mathbf{r}}{}' \left|\frac{1}{\Omega}T_\lambda(\mathbf{r})\right|^2 \delta(E-\omega_2(p_1,p_2)), \tag{2.10}$$

v_1 being the probe roton group velocity, $|d\omega(p_1^0)/dp_1^0|$.

So far we have used only the momentum conservation. Now we study other kinematical restrictions on the interaction. From the isotropy of the medium the interactions should have the following properties[10],

$$V_3(\mathbf{P},\mathbf{r}) = V_{3P}(p_1,p_2), \tag{2.11}$$

$$V_4(\mathbf{P},\mathbf{r},\mathbf{r}') = \sum_m V_{4Pm}(p_1,p_2;p_1',p_2') \exp\{im(\phi-\phi')\}. \tag{2.12}$$

In place of $\mathbf{r}$, which is (r,θ,ϕ) in the spherical coordinate with $\mathbf{P}$ as z-axis, magnitudes p_1, p_2 of momenta of the two rotons together with ϕ are taken as variables. Bose statistics restricts the interaction $V_{3P}(p_1,p_2)$ to be symmetric with respect to p_1 and p_2, and $V_{4Pm}(p_1,p_2;p_1',p_2')$ to be symmetric (antisymmetric) with respect to p_1 and p_2 for even m (odd m). Accordingly, the interaction V_{4Pm} is weak for odd m as long as p_1 and p_2 are nearly equal. Time reversal invariance requires $V_{4Pm}(p_1,p_2;p_1',p_2') = V^*_{4P-m}(p_1',p_2';p_1,p_2)$. We frequently use the relation,

$$\frac{1}{\Omega}\sum_{\mathbf{r}}' \rightarrow \int \frac{d\mathbf{r}}{2(2\pi)^3} = \int_{\mathcal{D}} \frac{p_1 p_2}{P}\frac{dp_1\,dp_2\,d\phi}{2(2\pi)^3}, \tag{2.13}$$

in which the trigonal inequality, $|p_1-p_2| \leq P \leq p_1+p_2$, defines the integration domain $\mathcal{D}$.

Expanding U in a similar way as V_4,

$$U(\mathbf{P},\mathbf{r},\mathbf{r}') = \sum\nolimits_m U_{Pm}(p_1,p_2;p_1',p_2') \exp\{im(\phi-\phi')\}, \tag{2.14}$$

we find that the E-dependence appears only at $m=0$,

$$U_{Pm}(p_1,p_2;p_1',p_2') = 2V_{4Pm}(p_1,p_2;p_1',p_2') + 4\delta_{m,0}V^*_{3P}(p_1,p_2)\frac{1}{E-\omega(P)}V_{3P}(p_1',p_2'). \tag{2.15}$$

The T-matrix is also expanded as

$$T_\lambda(\mathbf{r}) = \sum\nolimits_m T_{Pm}(p_1,p_2;p_1^0,p_2^0) \exp\{im(\phi-\phi^0)\}. \tag{2.16}$$

Then, the integral equation (2.9) becomes

$$T_{Pm}(p_1,p_2;p_1^0,p_2^0) = U_{Pm}(p_1,p_2;p_1^0,p_2^0) + \int U_{Pm}(p_1,p_2;p_1',p_2')\frac{1}{E-\omega_2(p_1',p_2')+i\epsilon}T_{Pm}(p_1',p_2';p_1^0,p_2^0)\frac{p_1'p_2'}{P}\frac{dp_1'\,dp_2'}{8\pi^2}. \tag{2.17}$$

The cross section is expressed as

$$\sigma = \frac{2\pi}{v_1}\sum_m \int |T_{Pm}(p_1,p_2;p_1^0,p_2^0)|^2\,\delta(E-\omega_2(p_1,p_2))\frac{p_1p_2}{P}\frac{dp_1 dp_2}{8\pi^2}. \tag{2.18}$$

The formulas derived in this section are almost exact (components of more than two rotons are neglected). There are four constants of motion, $\mathbf{P}$ and m. The remaining two variables, p_1 and p_2, do not separate. For further study we introduce approximations in the following sections.

3. SEPARABLE APPROXIMATION

To solve the integral equation of T-matrix explicitly, we employ in this section a separable interaction, which is no more complicated than the constant interaction. Because the cross sections for $m \neq 0$ channels are not expected to show such energy dependence as found in the experiment by Forbes and Wyatt, we confine ourselves to the $m = 0$ channel and suppress suffices such as P and m.

The separable approximation to the interaction (2.15) is

$$U(p_1, p_2; p_1', p_2') = h(p_1, p_2)\, g(E)\, h(p_1', p_2'), \tag{3.1}$$

its energy dependence being contained in

$$g(E) = 2g_4 + \frac{4g_3^2}{E - \omega(P)}. \tag{3.2}$$

We normalize the form factor $h(p_1, p_2)$ such that $h(Q_0, Q_0) = 1$.

The integral equation (2.17) for the T-matrix can be solved to give

$$T(p_1, p_2; p_1^0, p_2^0) = h(p_1, p_2)\frac{g(E)}{1 - g(E)F(E)}h(p_1^0, p_2^0), \tag{3.3}$$

where $F(E)$ is the propagator of a pair of free rotons,

$$F(E) = \int \frac{h^2(p_1, p_2)}{E - \omega_2(p_1, p_2) + i\epsilon} \frac{p_1 p_2}{P} \frac{dp_1 dp_2}{8\pi^2}. \tag{3.4}$$

By expanding the denominator of Eq. (3.3) in powers of $g(E)$, we are convinced that Eq. (3.3) contains all the higher order terms of the interaction. The following quantity defined by the imaginary part of the free propagator,

$$\begin{aligned} \rho(E) &= -\frac{1}{\pi} \operatorname{Im} F(E) \\ &= \int h^2(p_1, p_2)\, \delta(E - \omega_2(p_1, p_2))\, \frac{p_1 p_2}{P} \frac{dp_1 dp_2}{8\pi^2}, \end{aligned} \tag{3.5}$$

represents the density of two-roton states in unit volume if we take the form factor $h(p_1, p_2) = 1$. The real part is expressed with $\rho(E)$ as

$$\operatorname{Re} F(E) = \mathcal{P} \int \frac{\rho(E')}{E - E'} dE'. \tag{3.6}$$

Although its connection to the asymptotic form of the wave function is not so clear as in the usual scattering problem, we define the 'phase shift' δ by the following equation,

$$\cot \delta = \frac{1}{\pi\rho(E)} \{\operatorname{Re} F(E) - \frac{1}{g(E)}\}. \tag{3.7}$$

Then, we have from the relations (3.3) $\sim$ (3.5),

$$T(p_1, p_2; p_1^0, p_2^0) = -\frac{h(p_1, p_2)h(p_1^0, p_2^0)}{\pi\rho(E)} e^{i\delta} \sin\delta, \tag{3.8}$$

and we have the cross section in a form,

$$\sigma = \frac{2}{v_1} \frac{h^2(p_1^0, p_2^0)}{\pi\rho(E)} \sin^2\delta. \tag{3.9}$$

4. BEHAVIOR OF CROSS SECTION

Even in the separable approximation, theory still contains many indefinite quantities, i.e., $\omega(P)$, g_3, g_4, and $h(p_1, p_2)$. It seems premature to determine all these quantities from the experiment. The situation makes us reminiscent of the nucleon-nucleon scattering in the late forties. Yukawa, exponential, Gauss, or even square-well potential can fit the experiment as far as low energy scattering is concerned. Then, shape-independent treatment was invented, and the analysis was done with only two parameters — scattering length and effective range. In this section we pursue an analogous treatment assuming the incident energy E to be near $2\Delta_0$. When E approaches $\Delta_0 + \Delta_1$, several modifications are required.

As a guide we show in Fig. 1 density of two-roton states at $P = 2.8\text{Å}^{-1}$ computed from Eq. (3.5) by taking $h(p_1, p_2) = 1$ and using the experimental dispersion curve, which is cut at momentum 2.4Å^{-1}, by Cowley and Woods[8]. It has a finite jump at $2\Delta_0$ and a logarithmic peak at $\Delta_0 + \Delta_1$. Figure 1 also shows the corresponding $\text{Re}\, F(E)$ computed from Eq. (3.6). Reflecting the analytic properties of $F(E)$, $\text{Re}\, F(E)$ is logarithmically divergent at $2\Delta_0$ and has a finite jump at $\Delta_0 + \Delta_1$.

Although the form factor $h(p_1, p_2)$ can modify the behavior of these functions, their irregularities remain the same. We can express $\rho(E)$ near $2\Delta_0$ with a step function as

$$\rho(E) = \theta(E - 2\Delta_0)\rho_r(E), \tag{4.1}$$

where $\rho_r(E)$ is a regular function near $2\Delta_0$ and its value ρ_0 at $2\Delta_0$ is

$$\rho_0 = \rho_r(2\Delta_0) = \frac{\mu_0 Q_0^2}{4\pi P}. \tag{4.2}$$

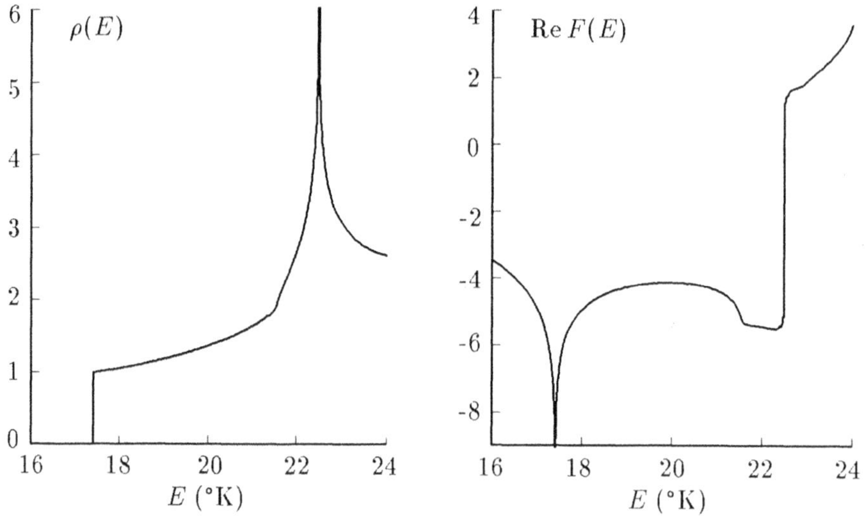

Fig. 1. Two-roton density of states $\rho(E)$ and the corresponding real part $\text{Re}\, F(E)$ of the free two-roton propagator at $P = 2.8\text{Å}^{-1}$, ordinates in unit of $\rho_0 = \mu_0 Q_0^2/4\pi P$.

The roton mass μ_0 is defined by the second derivative of the dispersion curve at its minimum, $1/\mu_0 = d^2\omega(p)/dp^2 \mid_{p=Q_0}$. Equation (3.6) shows that $\mathrm{Re}\, F(E)$ has a functional form near $2\Delta_0$,

$$\mathrm{Re}\, F(E) = \rho_r(E) \ln \frac{|E - 2\Delta_0|}{2\Delta_0} - f_1(E), \tag{4.3}$$

with a regular function $f_1(E)$. Putting Eqs. (4.1) and (4.3) into Eq. (3.7), we have

$$\cot \delta = \frac{1}{\pi} \ln \frac{|E - 2\Delta_0|}{2\Delta_0} - f(E), \tag{4.4}$$

where

$$f(E) = \frac{1}{\pi \rho_r(E)} \{ \frac{1}{g(E)} + f_1(E) \}. \tag{4.5}$$

Gross behavior of the function $f(E)$ is dominated by a hyperbola,

$$\frac{1}{\pi \rho_0 g(E)} = \frac{1}{\pi \rho_0} \{ \frac{1}{2g_4} - \frac{(g_3/g_4)^2}{E - \omega(P) + 2g_3^2/g_4} \}, \tag{4.6}$$

which is an increasing function of E. Also, near $2\Delta_0$, quantities appearing in Eq. (3.9) can be replaced as $h(p_1^0, p_2^0) = 1$, $\rho(E) = \rho_0$, and $v_1 = q_1/\mu_0$ ($q_1 = |p_1^0 - Q_0|$) to give

$$\sigma = \frac{8P}{q_1 Q_0^2} \sin^2 \delta \,. \tag{4.7}$$

Using the above formulas we can see behaviors of the phase shift δ and the cross section σ. Various sets of parameters, $g_3, g_4, \omega(P)$, can give rise to a resonance ($\cot \delta = 0$) several degrees above $2\Delta_0$. Its peak value is $8P/q_1 Q_0^2$ from Eq. (4.7). We can see from Eq. (3.9) that the cross section diverges as $1/v_1$ for a very slow probe roton. The divergence is weakened when the scattering roton is also slow because of the effect of $\ln |E - 2\Delta_0|$ term in Eq. (4.4). It is to be noted that the right-hand side of Eq. (4.4) can vanish at $E = E_P < 2\Delta_0$. It is the Pitaevskii branch[3] representing the plateau energy. Indeed, when we look for the bound state solution of the coupled Eqs. (2.3) and (2.4), we can easily see that the binding energy is determined by the condition

$$\mathrm{Re}\, F(E) - \frac{1}{g(E)} = 0\,, \tag{4.8}$$

which means the vanishing of the right hand side of Eq.(3.7).

For a practical analysis of the cross section near $2\Delta_0$, we expand the function $f(E)$ in Eq. (4.4) at $2\Delta_0$ and retain up to its linear term,

$$\cot \delta = \frac{1}{\pi} \ln \frac{|E - 2\Delta_0|}{2\Delta_0} - A(P) - B(P)(E - 2\Delta_0)\,. \tag{4.9}$$

In place of the quantities, $\omega(P), g_3, g_4$, and $h(p_1, p_2)$, we take $A(P)$ and $B(P)$ as phenomenological parameters. Undoubtedly, the parameter $B(P)$ is positive keeping the E-dependence of Eq. (4.6). Also, strong P-dependence is expected in these parameters. It comes mainly from $\omega(P)$ and g_3, strength of $V_3(\mathbf{P}, \mathbf{r})$, which vanishes while $\omega(P)$ coincides with the observed one-roton energy and increases as $\omega(P)$ deviates from the observed one-roton energy. As an example we draw in Fig. 2 the cross section from Eq. (4.7) by taking the parameters in Eq. (4.9) arbitrarily. It is to be remarked that simple straight lines for Eq. (4.5) can lead to various types of cross section.

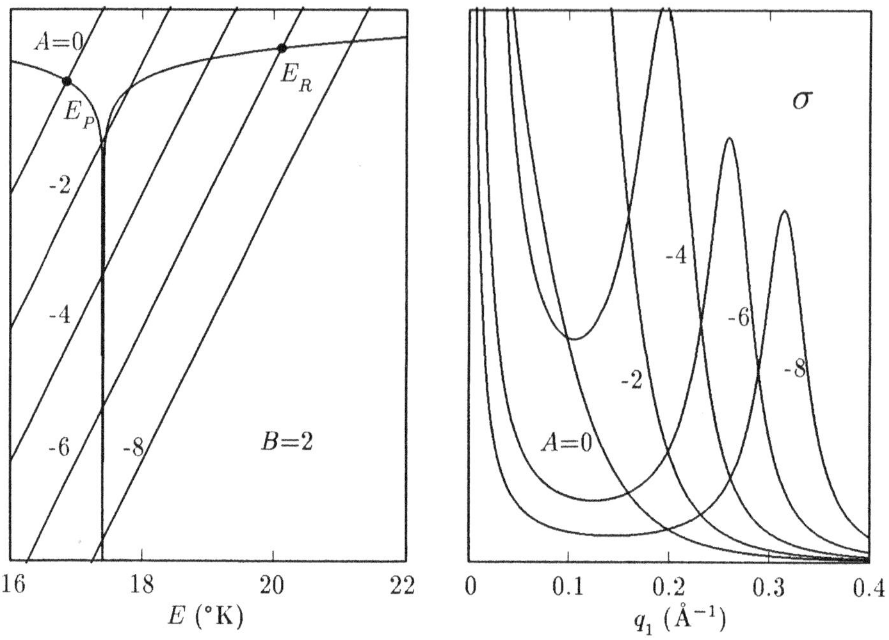

Fig. 2. $\cot\delta$ is the difference of curve $(1/\pi)\ln(|E - 2\Delta_0|/2\Delta_0)$ and straight line $A(P) + B(P)(E - 2\Delta_0)$, the parameters A, B being chosen arbitrarily. E_P and E_R are the plateau and the resonance energies (Left). Right shows the corresponding cross section for a probe roton of momentum $Q_0 + q_1$ incident on a slow scattering roton.

5. DISCUSSION

As we have seen in the previous sections, almost all features of the roton-roton scattering stem from the fact that the dispersion curve has a minimum at a finite momentum. This property leads to the finite step in the two-roton density of states and the logarithmic term in $\mathrm{Re}\, F(E)$. The irregularity, together with the energy dependent interaction arising from the hybridization, can give various types of behaviors to the cross section in a narrow energy range above $2\Delta_0$. As to the irregularity at $\Delta_0 + \Delta_1$, similar treatment is possible and gives a cusp to the cross section just at $E = \Delta_0 + \Delta_1$.

At first sight it is somewhat puzzling that the cross section diverges as $1/v_1$ at finite momentum by considering the usual scattering formula $(4\pi/k^2)\sin^2\delta$, which restricts the cross section to be of the order of the square of the de Broglie wave length. However, in the present case only the angular momentum around the total momentum $\mathbf{P}$ is conserved and the remaining two degrees of freedom represented by p_1 and p_2 do not separate. Accordingly, it is not correct to compare our cross section with the one for a given partial wave in an ordinary central potential. In the present case, the upper bound of the cross section is given from Eq. (3.9) as $2h^2(p_1^0, p_2^0)/\pi v_1 \rho(E)$, which becomes $8P/q_1 Q_0^2$ near $2\Delta_0$.

We have put aside the scattering of $m \neq 0$ channels from sections 3 and 4. Apart from the $1/v_1$-behavior for a very slow probe roton, the cross sections of $m \neq 0$ channels do not appear with strong energy dependence near $2\Delta_0$ because the hybridization does not show up in these channels.

It will be really interesting if the experiment can be performed by changing the angle between the incident beams of rotons from 90° so that the total momentum P changes, because strong P-dependence is also expected.

REFERENCES

1. A. C. Forbes and A. F. G. Wyatt, Phys. Rev. Lett. **64**, 1393 (1990).
2. K. Nagai and F. Iwamoto, *to be published.*
3. L. P. Pitaevskii, Sov. Phys. JETP **9**, 830 (1959).
4. J. Ruvalds and A. Zawadowski, Phys. Rev. Lett. **25**, 333 (1970); A. Zawadowski, J. Ruvalds, and J. Solana, Phys. Rev. **A5**, 399 (1972).
5. A. J. Smith, R. A. Cowley, A. D. B. Woods, W. G. Stirling, and P. Martel, J. Phys. **C10**, 543 (1977).
6. K. Nagai, J. Phys. **C11**, L759 (1978); *Elementary Excitations in Quantum Fluids* edited by K. Ohbayashi and M. Watabe (Springer Verlag 1989) p.140.
7. K. Fukushima, N. Koyama and T. Sugiyama, Prog. Theor. Phys. **61**, 367 (1979); *Elementary Excitations in Quantum Fluids* edited by K. Ohbayashi and M. Watabe (Springer Verlag 1989) p.144.
8. R. A. Cowley and A. D. B. Woods, Canad. J. Phys. **49**, 177 (1971).
9. F. Iwamoto, K. Nagai, and K. Nojima, Proceedings of the 12th International Conference on Low Temperature Physics, Kyoto, 189 (1970).
10. I. A. Fomin, Sov. Phys. JETP **33**, 637 (1971).

THEORY OF THE PHONON-ROTON SPECTRUM OF LIQUID ^{4}He

C.E. Campbell

School of Physics and Astronomy
University of Minnesota
Minneapolis, Minnesota 55455, U.S.A.

INTRODUCTION

The theory of the excited states of boson fluids, and particularly liquid ^{4}He, has been the subject of attention for nearly fifty years, beginning with Landau's observation in 1941 that the unusual properties of liquid ^{4}He at temperatures well below the λ point could be understood in terms of a phonon and roton spectrum of excited states.[1] Subsequently Landau asserted without proof that the phonons and rotons were part of a single, continuous spectrum as a function of momentum,[2] having essentially the same form as is presently widely accepted.

This phenomenological theory was given some microscopic support by various field-theoretical treatments of the non-ideal Bose gas, beginning with the work of Bogoliubov.[3-5] The vacuum for these field theories is the non-interacting Bose gas ground-state. The chief effect of the interaction on the excitation spectrum was shown to be the renormalization of the free-particle spectrum at long wavelengths, from the non-interacting $\hbar^2k^2/2m$, into the form $\hbar c k$. While these theories do not produce a roton structure, i.e., a local minimum in the spectrum at a finite wave-number, they do in fact confirm an essential feature of Landau's observation that the low energy excited states which contribute to the low temperature properties of a boson fluid must contain a linear (phonon) form, a feature which he also showed to be a minimum requirement for superfluidity.

Further developments in this field theoretic approach are reviewed in the contributions of Glyde and Griffin to this volume. We will make a few additional comments in the Discussion section at the end of this paper.

The direct application of these field theoretic methods to liquid ^{4}He was thought to be precluded by the fact that this system is a very strongly interacting, high density system, indeed the highest density of any quantum fluid in the sense that the repulsive core in the helium interaction occupies 18% of the fluid

Excitations in Two-Dimensional and Three-Dimensional Quantum Fluids
Edited by A.G.F. Wyatt and H.J. Lauter, Plenum Press, New York, 1991

volume at equilibrium density, in contrast to 13% for liquid ^{3}He, and only 1% for nuclear matter. It is certainly not at all clear that conclusions drawn from analysis of the interacting Bose gas in the weak interaction limit can be applied to this system, although Landau's observations point to the same conclusions. Beginning in 1953, Feynman provided the nexus in a series of elegant arguments applicable to *nearly* any boson fluid, including liquid ^{4}He.[6-9] The most important point in Feynman's analysis of the situation is that the phonons (density fluctuations) are the lowest energy excitations in the system, which is certainly not the case for a many-body system satisfying statistics other than Bose statistics. While it is commonly asserted that this analysis depends only on the Bose statistics, it must be that the phase coherence of the underlying ground state, i.e. the existence of the Bose condensate, is also necessary to the argument. However, the *magnitude* of the condensate fraction plays no role in the argument; it is only necessary that the number of particles in the condensate be macroscopic. It is a distinguishing feature of the Landau-Feynman approach that there is no overt reference to the condensate fraction, in contrast with the field-theoretic approaches. Similarly there is no direct experimental evidence that the magnitude of the condensate fraction governs the physics of the fluid, (except possibly for direct probes of the condensate through deep inelastic scattering [see the contribution to these proceedings by Sokol]).

Nozieres and Pines make the interesting observation that interacting Bose fluids are related to the ideal Bose gas in a way which is reminiscent of the relationship between Fermi liquids and the ideal Fermi gas, namely the ground state of the interacting system evolves continuously from the ideal system as the interaction potential is turned on adiabatically. We should add that this is true until and unless the potential moves into a range where a phase transition occurs, such as the transition to a superconducting state (for the fermion system) or a solid. Of course the boson system differs from the fermion system in that, in the boson system, the role of the potential is to deplete the condensate, and one might imagine that this would happen in such a way that the condensate vanishes for some potential which is in a region where the system has not yet solidified, i.e., the system goes through a transition from ODLRO to non-ODLRO, whereas the Fermi system has no ODLRO until the potential moves into an appropriately attractive region. This adiabatic modification also describes the evolution of the quasiparticle states. Turning the argument around in the boson systems, as one turns off the interaction, the sound velocity decreases to zero at zero interaction strength, leaving only the quadratic term which is the free-particle spectrum.

In the next sections we show how the excited state structure of liquid ^{4}He is an evolution from that of the ideal Bose gas, albeit quantitatively a distant evolution because of the strength of the interaction. In this way we are exhibiting the natural consequences of Feynman's argument as it pertains to the excited states of an interacting boson liquid, including liquid ^{4}He, as they have been worked out quantitatively in subsequent years. In particular we see explicitly how this spectrum depends only on the *diagonal*-structural information in the ground state, and has no reference to the off-diagonal order present in the system.

It should be emphasized that we are specifically focussing on the excited state spectrum, and not on the response of the fluid to neutron and photon probes. Nevertheless, it is our good fortune that these probes which couple to the density are so well coupled to the elementary excitations which emerge from the theory. *A priori* it was not obvious that there would not be some other low lying excitations which couple weakly or not at all to these probes. It should also be noted that, while the excited states are properties of a system at $T = 0$, there are resonant modes at finite temperature which are not present at $T = 0$ and thus in the strictest sense are not excited states of the system (the best example being second sound modes), and those modes which are present at $T = 0$ will generally continue to exist at finite temperatures but with a shift in frequency (energy) due to coupling to thermally excited modes. In liquid ^{4}He it is fortunate that less that *0.25%* of the possible modes are excited at $T = 1K$, so that to a very good approximation the $T = 0$ excited states are being measured by neutron and light scattering.

THE EXCITED STATES OF THE IDEAL BOSE GAS

We begin by examining the complete set of excited states of the ideal Bose gas, followed in the next section by the obvious generalization to the complete set of states of the interacting Bose fluid. The single excitation state of the ideal Bose gas, corresponding to the excitation energy $\hbar^2k^2/2m$ and momentum $\hbar\mathbf{k}$, can be written in the form

$$|\mathbf{k}\rangle = \frac{\rho_\mathbf{k}\Psi_0}{\sqrt{N}}$$

where N is the number of particles and Ψ_0 is the *non-interacting* ground state. The state as written is normalized if Ψ_0 is normalized. The operator ρ_k is the density fluctuation operator, which can be written in first quantized form

$$\rho_\mathbf{k} = \sum_{j=1}^{N} e^{i\mathbf{k}\cdot\mathbf{r}_j}$$

or in second quantized form

$$\rho_{\vec{k}} = \sum_{\vec{p}}^{N} a^\dagger_{\vec{k}-\vec{p}}\, a_{\vec{p}}$$

where $a^\dagger$ and a are free-particle creation and annihilation operators, respectively. Similarly, the doubly excited state has the form

$$|\mathbf{k},\mathbf{q}\rangle = \frac{[\rho_\mathbf{k}\,\rho_\mathbf{q} - \rho_{\mathbf{k}+\mathbf{q}}]\,\Psi_0}{\sqrt{N^2}}$$

with an additional factor of $\sqrt{2}$ in the denominator when $\boldsymbol{k} = \boldsymbol{q}$. The effect of the second term in the brackets is to orthogonalize this doubly excited state to the single excited state of wave number $\boldsymbol{k} + \boldsymbol{q}$. This pattern generalizes to an arbitrary number of excitations totalling less than or equal to the number of particles:

$$|n_{\mathbf{k}_1}, n_{\mathbf{k}_2}, \cdots, n_{\mathbf{k}_\alpha}\rangle = \frac{\left[\prod_{i=1}^{\alpha} (\rho_{\mathbf{k}_i})^{n_i} - \text{orthog}\right]\Psi_0}{\sqrt{N^M\left[\prod_{i=1}^{\alpha} n_i!\right]}}$$

where the total number of excitations $M = \sum n_i \leq N$, and *orthog* represents a polynomial in the density fluctuation operators of order $M - 1$ with simple integer coefficients which has the effect of orthogonalizing the state to all states of total number of excitations $M' < M$ and the same total wave number. It is interesting to note that, to leading order in N/M, the orthogonalization terms in this equation do not affect the normalization nor, more importantly, the energy if it is calculated as an expectation value.

THE EXCITED STATES AND EXCITATION ENERGIES OF THE BOSE FLUIDS INCLUDING LIQUID ^{4}He

In the case of an interacting Bose fluid such as liquid ^{4}He, it is simple to see that the system can be described by a complete set of states similar to those shown above for the ideal Bose gas, but with the only, albeit essential, difference that the function Ψ_0 is now the *ground state of the fully interacting hamiltonian:* $H\Psi_0 = E_0\Psi_0$. While this basis is complete, the fact that the "vacuum" Ψ_0 is correlated means that the basis is no longer orthogonal or normalized. Normalization is a simple matter. The normalization is completely expressible in terms of the liquid structure function $S(k)$, with the result

$$|n_{\mathbf{k}_1}, n_{\mathbf{k}_2}, \cdots, n_{\mathbf{k}_\alpha}\rangle = \frac{\left[\prod_{i=1}^{\alpha} (\rho_{\mathbf{k}_i})^{n_i}\right]\Psi_0}{\sqrt{N^M\left[\prod_{i=1}^{\alpha} \{n_i!\, S(k_i)^{n_i}\}\right]}}$$

where the "orthog" has also been dropped, since it does not affect the normalization or Hamiltonian expectation value in these states. It is essential to the completeness of this correlated basis set that the ground state wave function Ψ_0 is real and positive, which is indeed the situation for the exact ground state of an interacting boson system.[10] Note that the simplest state is identical to the Feynman state

$$|\mathbf{k}\rangle = \frac{\rho_{\mathbf{k}}\Psi_0}{\sqrt{NS(k)}}$$

with expectation value giving the well-known Bijl-Feynman excitation energy

$$\varepsilon_k^{BF} = \langle k | H - E_0 | k \rangle = \frac{\hbar^2 k^2}{2m\, S(k)}$$

The expectation value in the general state is then of the form of a Bose excitation spectrum:

$$E = E_0 + \sum_i n_i\, \varepsilon_{k_i}^{BF}$$

with however the possibly important caveat that the total number of "excitations" *M* not exceed *N*. (It has been suggested that this restriction might be related to the lambda transition.[11])

The introduction of this correlated basis was originally motivated by the observation that the strong, short-range dynamical interactions should be essentially the same in low excited states as in the ground state, and thus are largely accounted for in trial states of these forms.[12,13,10] The minimal requirement that an *ab initio* theory of liquid ^{4}He must satisfy is the inclusion of the strong short-range correlations induced by the strong, short-range repulsive interaction potential. Of course the single excitation state appeared earlier in Feynman's work, and is precisely the structure of the of the singly occupied excited state in Bogoliubov's theory. The strong correlations enter the theory phenomenologically in the Feynman ansatz through the use of the experimental liquid structure function, whose structure is dominated at all but long wavelengths by the short-range structure of the fluid.

The analysis until this point shows that going beyond the single excitation state to the correlated basis provides a straightforward extension to multiple Feynman excitations, and at the same time saturates the number of excited states because of the completeness of the basis. However, we know from experiment, as well as the arguments of Feynman and Cohen, that the Bijl-Feynman excitation energy is only qualitatively correct, except at long wavelengths. What is left out of the above simple analysis is the fact that this set of states which give multiple Bijl-Feynman excitations are no longer orthogonal, nor is the Hamiltonian diagonal in this basis. A systematic approach for obtaining the correct eigenvalues and eigenfunctions from the above analysis is achieved by orthogonalizing and diagonalizing this correlated basis. The matrix elements of the Hamiltonian and the unit operator (the latter needed for orthogonalization) are easily shown to be simple functions of the $M + M'$ order structure functions of the ground state. Thus, to go beyond the Bijl-Feynman theory, one is immediately faced with the task of determining higher order ground state correlation functions than available from experiment, which again are dominated by the short-range correlations in precisely the wave number regimes where the Bijl-Feynman spectrum falls short. While this task has only been carried out partially and approximately, the structure of the results is immediately clear. In particular, any state of wave number $\mathbf{k}$ must be expressible as an Nth order polynomial in the density fluctuation operators acting upon the ground state:

$$\Psi_k = \Bigg[A_k^{(1)} \rho_k + \sum_{\substack{q_1, q_2 \\ q_1+q_2=k}} A^{(2)}_{q_1, q_2} \rho_{q_1} \rho_{q_2} +$$

$$\sum_{\substack{q_1, q_2, q_3 \\ q_1+q_2+q_3=k}} A^{(3)}_{q_1, q_2, q_3} \rho_{q_1} \rho_{q_2} \rho_{q_3} + \dots + \sum_{\substack{q_1, \dots, q_N \\ q_1+\dots+q_N=k}} A^{(N)}_{q_1, \dots, q_N} \rho_{q_1} \cdots \rho_{q_N} \Bigg] \Psi_0$$

The "Fermi-liquid-like" philosophy suggests that, for this expression to represent the phonon-roton spectrum, one would expect that, as the interaction potential is turned on from zero, the value of $A_k^{(1)}$ would evolve from $1/\sqrt{N}$ to a value of the same order, while the other expansion constants would evolve from zero. Moreover, to be a practical theory, it should be possible to truncate this series after a few terms. We shall show below that it appears to be unnecessary to go beyond third order polynomials to find convergence to very good agreement with experiment.

The remaining excited state structures are deduced in a similar manner, all having the form of an Nth order polynomial in the density fluctuation operator. For example, the bi-excitation state has the same form as above, but with a different dependence of the coefficients on the number of particles. It can be expressed in the form

$$\Psi_{k-p,p} = F_{k-p,p} \Psi_0$$

where

$$F_{k-p,p} = B^{(0)}_{k-p,p} \rho_{k-p} \rho_p + B_k^{(1)} \rho_k + \sum_{\substack{q_1, q_2 \\ q_1+q_2=k}} B^{(2)}_{q_1, q_2} \rho_{q_1} \rho_{q_2} +$$

$$\sum_{\substack{q_1, q_2, q_3 \\ q_1+q_2+q_3=k}} B^{(3)}_{q_1, q_2, q_3} \rho_{q_1} \rho_{q_2} \rho_{q_3} + \dots + \sum_{\substack{q_1, \dots, q_N \\ q_1+\dots+q_N=k}} B^{(N)}_{q_1, \dots, q_N} \rho_{q_1} \cdots \rho_{q_N}$$

which is distinguished from the expression for the single excitation state by the fact that the constant $B^{(0)}$ is one order of magnitude higher in N than the coefficients $B^{(2)}$ of the other bi-linear terms.

Application to the liquid helium phonon-roton spectrum

Beginning with the work of Feynman and Cohen, there are numerous calculations of energies for the phonon-roton spectrum using the quadratic form, and at least one calculation which goes to third order. The problem faced in these calculations is the calculation of the constants $A^{(2)}$, and in one case, $A^{(3)}$. In each of these cases, the task requires knowledge about third and higher order correlation functions of the ground state, information which is not available from experiment, unlike the simple Feynman ansatz where all that is needed is the experimentally available liquid structure function $S(k)$ (supplemented by the theoretically known long wave-length structure of this function at $T = 0$). Consequently one must have a good theory of the ground state wave function to obtain quantitative accuracy, and indeed even qualitative trends as a function of pressure and temperature.

Here we will review briefly only two of these calculations; the contribution of Halley and Korth elsewhere in this volume is based upon yet another calculation, due to Manousakis and Pandharipande.

The present "state-of-the-art" in semi-analytic theories of the ground state of liquid ^{4}He is in terms of the extended Jastrow wave function:

$$\Psi_0 = \prod_{i<j}^{N} e^{\frac{1}{2}u_2(r_{ij})} \prod_{i<j<k}^{N} e^{\frac{1}{2}u_3(r_i, r_j, r_k)}$$

where the functions u_2 and u_3 are determined by functional variation.[14,15] It turns out that the long and intermediate wavelength structure of u_2 is essentially determined by the zero-point motion of the Bijl-Feynman phonons, while u_3 is determined almost entirely by the additional zero-point motion which occurs because of the bi-linear contributions to the excitation spectrum, e.g. Feynman-Cohen backflow.[16]

Figure 1 shows three theoretical calculations of the phonon-roton spectrum and the experimental curve, all at equilibrium density. Each of these theoretical calculations employed the

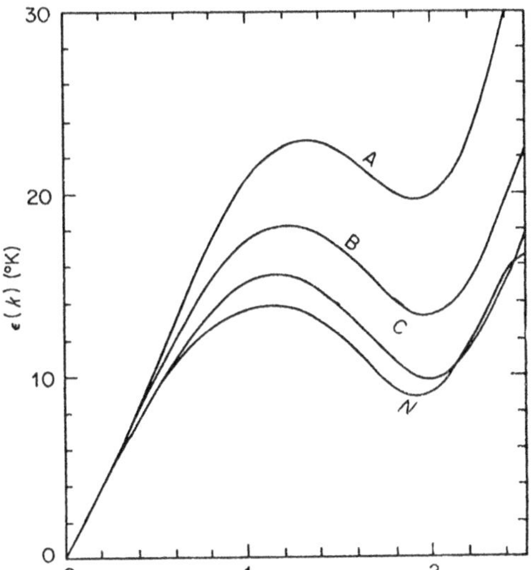

Fig. 1. Elementary excitation spectrum for liquid ^{4}He at equilibrium density. A: Bijl-Feynman form. B: bi-linear form. C: tri-linear form. N: experiment.

optimum Jastrow wave function, i.e., with $u_3 = 0$. The simple Feynman ansatz shows the familiar quantitative failure outside of the phonon region. The bi-linear calculation, obtained by using second order Brillouin-Wigner perturbation theory to determine $A^{(2)}$,[17,16] is a significant improvement over the Feynman ansatz, and indeed is better than the Feynman-Cohen result (which is in fact also a bi-linear form but with $A^{(2)}$ of the dipolar back-flow form together with a single variational parameter[16]). The third theoretical result by Lee and Lee includes the tri-linear terms with $A^{(3)}$ also determined by second order Brillouin-Wigner perturbation theory. The agreement of this result with experiment is well within the systematic errors due to the approximations used as well as the uncertainty of the interaction potential. Most importantly, it appears that the general procedure is converging to the experimental result as terms are added to the polynomial expression for the elementary excitation wave function given above.

At this level of agreement with experiment the accuracy of the interaction potential becomes an important quantitative issue. Each of the three theoretical results discussed above were obtained with the Lennard-Jones potential, which is analytically convenient but is less accurate than the HFDHE2 potential of Aziz *et al.*[18] Results obtained at the bi-linear level are compared in Figure 2, where it is seen that there is some improvement with the improved potential.[19] Unfortunately the Lee and Lee tri-linear calculation has not been repeated with the Aziz potential. It is tempting to suggest that the difference between the Lee and Lee results and experiment can be mostly accounted for by the difference between the two bi-linear results shown in Figure 2, though this is somewhat dangerous since there are approximations in the Lee and Lee calculation which should be of similar magnitude.

The two calculations including the bi-linear terms shown in Fig. 2 also include the effects of the extended Jastrow factor u_3. A comparison with Fig. 1 then suggests that the additional three body factor in the wave function has little effect on the calculated excitation spectrum. However, as the density increases, the dependence of the calculated excitation spectrum on density is qualitatively and quantitatively affected when u_3 is included, producing considerably improved agreement with the experimental trends. Figures 3-5 show the density dependence of three parameters which characterize the roton region of the excitation spectrum: the roton energy Δ, the roton wave number k_0, and the curvature at the spectrum at the roton minimum, given by an effective mass μ:

$$\varepsilon_k = \Delta + \frac{\hbar^2 (k - k_0)^2}{2\mu}$$

These figures emphasize the substantial improvement which is obtained by the combination of the three-body factors in the wave function and the bi-linear terms in the excitation energy. Each of these contribute substantially to the improvement.[17]

Because of the complexity of the calculation, the Lee and Lee results were obtained at only one density. We now know that their theory can be simplified by making use of the optimization of the

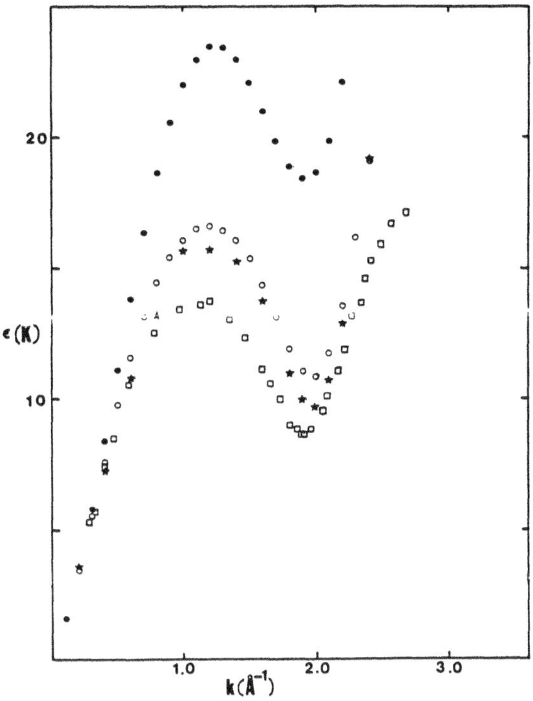

Fig. 2. Elementary excitation spectrum for liquid ^{4}He at equilibrium density. Solid circles: BF spectrum. Bi-linear result with LJ potential (open circles) and Aziz potential (star). Squares: experiment.

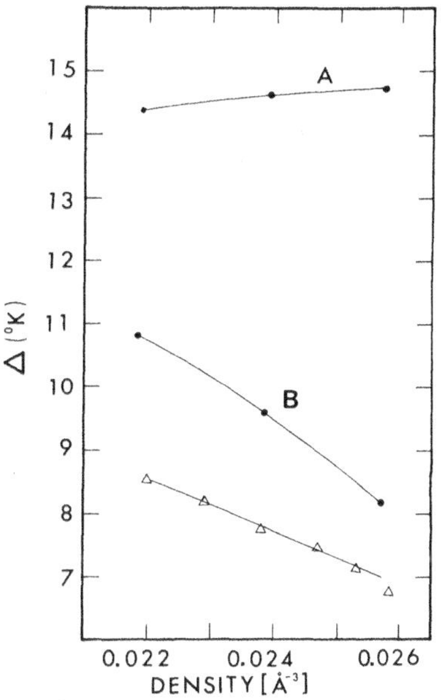

Fig. 3. A: Bijl-Feynman.
B: bi-linear.
Δ: experiment.

three-body factor in the ground state wave function; it would be very interesting to repeat their calculation with those simplifications together with the improved helium interaction, and to obtain results for the entire density regime.

It is interesting to note that the above analysis also provides a qualitative and semi-quantitative account of the phonon dispersion. Family used our extended Jastrow results and the same expression for the excitation spectrum (including bi-linear terms) to show that this *ab initio* theory (with no adjustable parameters) gives both the anomalous dispersion at low pressures, and the transition to normal dispersion with increasing pressure, which occurs in the theory at a pressure of about 15 atm.[20] Earlier Berdahl had pointed out the importance of these bi-linear phonon terms (three-phonon vertices) in the structure of the three-body distribution function at long wavelengths, and employed Landau's quantum hydrodynamics to relate this long-range structure to the Grueneisen constant.[21] One of the lessons of that work is that, while hydrodynamics predicts the long wavelength structure

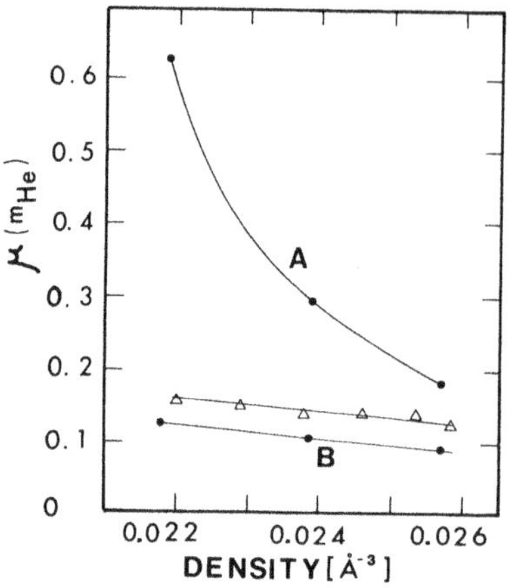

Fig. 4. Roton curvature parameter as a function of density Labelled as in Fig. 3.

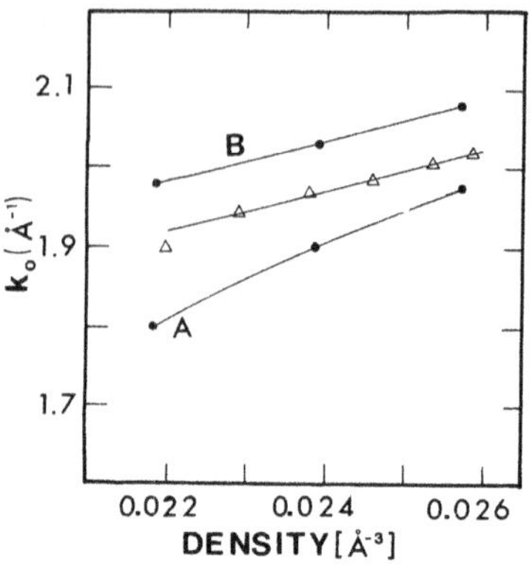

Fig. 5. Roton wavenumber as a function of density. Labelled as in Fig. 3.

of the three-body distributions in a boson fluid, the simple Jastrow function (i.e., $u_3 = 0$) does not give the correct result, while the inclusion of the u_3 function does give the correct behavior. Thus it is not surprising that the three-body factor in the ground state is essential for obtaining the correct dispersion in the long wavelength excitation spectrum.

DISCUSSION

We hope we have demonstrated here that a very successful theory of the phonon-roton spectrum emerges from the natural generalization of Feynman's ansatz for the structure of the excited state wave function, which goes hand-in-hand with, and requires a similarly sophisticated analysis of the ground state wave function. This approach differs from the field theoretic approach to the weakly interacting Bose gas in the way in which the condensate fraction appears to play a role. The evidence which has been presented here demonstrates that the excitation spectrum is determined entirely by the on-diagonal, multi-body distribution functions of low index: At the bi-linear level, the result depends only on the two-body distribution function (or equivalently $S(k)$) and the three- and four-body distribution functions, while the inclusion of tri-linear terms brings in a dependence on the five- and six-body distribution functions. The fact is that the condensate fraction plays no overt quantitative role in this theory, although it does play a surpassing qualitative role, namely, the analysis would be qualitatively incorrect if the condensate fraction vanished; i.e., this approach is valid as long as the number of particles in the condensate is macroscopic. The fact that the existence of a non-zero condensate fraction is required to understand the properties of liquid ^4He is, however, more than enough reason to be very interested in the value of the condensate fraction, and how it is sensitive to temperature, density and inhomogeneity. Fortunately it is a straightforward matter to calculate the condensate fraction (as well as the entire free particle occupation number and its fourier transform, the off-diagonal one-body density matrix).within a Jastrow theory. All such calculations have given a condensate fraction near 10% in the ground state at equilibrium density.

It is interesting to contemplate the question of the properties of a boson quantum fluid which has no Bose condensate in its ground state. A clear set of examples is the class of non-ideal bose systems in one dimension, where quantum fluctuations destroy both solid order (i.e., diagonal long-range order) and off-diagonal long-range order, i.e., these systems are all Bose quantum fluids with no condensate at $T = 0$. Lieb and Liniger solved exactly for the ground and excited states of one such system, where the interaction is a repulsive δ function.[24] While they find the usual elementary excitation spectrum, analogous to the free-particle spectrum except that it has a phonon nature at long wavelengths, it is most noteworthy that they also find a second set of excitations which fall below this branch. We conjecture that this second, lower branch appears primarily because of the *absence* of the condensate fraction. Moreover, Lieb speculates that there may be a second branch in some three-dimensional systems. Indeed, that may be the case, and it might also correspond to interactions for which the system is in a quantum fluid state, but the condensate fraction vanishes. The construction of such an interaction potential is an interesting intellectual challenge. One such example should occur in two-dimensions, namely the two-dimensional bose system with repulsive logarithmic interactions.[10] All that appears to be known about this system is that its ground state has no condensate, which seems to make it unique among two-dimensional boson systems (n.b., $T = 0$).

Returning to the question of liquid ^{4}He, we have not discussed the role of the analysis presented here in accounting for the properties of the inelastic neutron scattering cross section, $S(k, \omega)$. As can be seen in the contribution by Halley and Korth elsewhere in this volume, it is a relatively straightforward matter to incorporate the same physics in the determination of response functions such as this, with results in comparable agreement with the experimental results.

The methods discussed herein are open to systematic improvement in several obvious ways: improved calculations of the ground state wave function; improvements upon second-order Brillouin Wigner perturbation theory; and the inclusion of higher multi-density fluctuation terms in the expressions for the excitation wave function. Comparisons of the extended Jastrow ground state results[14] to the "exact" Green Function Monte Carlo simulation results of Kalos *et al*[22] suggest that there is very little to be gained by further improvements in this regard (although formal internal consistency at the tri-linear level would require the inclusion of four-body factors in the ground state wave function). Similarly, a careful analysis of the structure of the three-excitation vertices (closely related to the bi-linear contributions in the excitation spectrum) by Krotscheck indicates that there is little to be gained by going beyond the second order BW structure.[15] Concerning the third possibility for improvement, i.e. the inclusion of high order polynomials in the excitation wave function, it is likely that the furthest that one need proceed is the tri-linear terms in the fashion of Lee and Lee, but with a somewhat improved calculation using more recent methods and wave functions. It should be noted that Manousakis and Pandharipande have done a second-order BW perturbation theory in which the density fluctuation operator is augmented to include Feynman-Cohen backflow, so that the bi-linear state which is mixed in through the perturbation theory in fact includes cubic and quartic density fluctuation operator terms.[23] The resulting excitation spectrum lies even closer to the experimental result than the Lee and Lee curve shown in Fig. 1, though it was done with the improved helium interaction and thus cannot be directly compared to the Lee and Lee result.

We should note that, with the continuing progress in quantum Monte Carlo simulation methods, there should soon be an "exact" or nearly exact excited state wave function and excitation energy for a system of a few hundred to a thousand particles.

Finally, we return to the question of field theoretical methods and their applicability to the strongly correlated system of liquid ^{4}He. In recent years, Jackson, Lande and Smith have succeeded in solving a subset of the boson parquet theory, which is arguably the most sophisticated field theoretic calculation to have been carried out for an interacting Bose fluid.[25,26] This method involves the self-consistent, simultaneous summation of rings and ladders, which is made manageable by using a local approximation in certain key places in the parquet equations. The results are most remarkable: they are identical to the solution of the Jastrow Euler-Lagrange theory in the HNC/0 approximation, which itself is the simplest self-consistent solution for the Jastrow approximation to the ground state (i.e., $u_3 = 0$ in the expression for the extended Jastrow wave function shown above).[26] At this level, all of the essential physics of the ground state already appears, including both the correct short- and long-range correlations. This then brings us back to the question of where the condensate fraction appears overtly in the theory. The answer is that it plays no overt role in the boson parquet theory as long as it is not depleted. Unfortunately, the corresponding theory of the low excited states is not yet available.

REFERENCES

1. L. D. Landau, Zh. Eksp. Teor. Fiz. 11:592 (1941); J. Phys. Moscow 5:718 (1941).
2. L. D. Landau, J. Phys. Moscow 11:91 (1947).
3. N. Bogoliubov, J. Phys. USSR 11:23 (1947); for further references and discussion, see references 4 and 5.
4. D. Pines, "The Many Body Problem," W. P. Benjamin, New York (1962).
5. P. Nozieres and D. Pines, "The Theory of Quantum Liquids, Volume II," Addison-Wesley, New York (1990).
6. R. P. Feynman, Phys. Rev. 91:1301 (1953).
7. R. P. Feynman, Phys. Rev. 94:262 (1954).
8. R. P. Feynman and M. Cohen, Phys. Rev. 102:1189 (1956).
9. R. P. Feynman, Prog. Low Temp. Phys. 1:17 (1955).

10. C. E. Campbell, <u>in</u>: "Progress in Liquid Physics," C. A. Croxton, ed., Wiley, New York (1978).
11. H. W. Jackson, private communication.
12. H. W. Jackson and E. Feenberg, <u>Rev. Mod. Phys.</u> 34:686 (1962).
13. E. Feenberg, "Theory of Quantum Fluids," Academic Press, New York (1969).
14. C. C. Chang and C. E. Campbell, <u>Phys. Rev. B</u> 13:3779 (1976).
15. E. Krotscheck, <u>Phys. Rev. B</u> 33:3158 (1986).
16. C. E. Campbell and F. J. Pinski, <u>Nucl. Phys.A</u> 328:21 (1979).
17. C. C. Chang and C. E. Campbell, <u>Phys. Rev. B</u> 15:4238 (1977).
18. R. A. Aziz, V. P. S. Nain, J. S. Carley, W. L. Taylor, and G. T. McConville, <u>J. Chem. Phys.</u> 70:4330 (1979).
19. L. R. Whitney, Ph. D. Dissertation, University of Minnesota (1981).
20. F. Family, <u>Physica</u> 107B:699 (1981).
21. P. Berdahl, <u>Phys. Rev. A</u> 10:2378 (1981).
22. M. H. Kalos, <u>Nucl. Phys. A</u> 328:153 (1979); P. M. Whitlock, D. M. Ceperley, G. V. Chester and M. H. Kalos, <u>Phys Rev. B</u> 19:5598 (1979).
23. E. Manousakis and V. R. Pandharipande, <u>Phys. Rev. B</u> 30:5062 (1984).
24. E. Lieb and W. Liniger, <u>Phys. Rev.</u> 130:1605 (1963); E. Lieb, <u>Phys. Rev.</u> 130:1616 (1963).
25. A. D. Jackson, A. Lande and R. A. Smith, <u>Phys. Rept.</u> 86:55 (1983).
26. A. D. Jackson, A. Lande and R. A. Smith, <u>Phys. Rev. Lett.</u> 54:1469 (1985).

TOWARDS A UNIFIED DESCRIPTION OF BULK EXCITATIONS AND THE TWO FLUID THEORY OF SUPERFLUIDS IN TERMS OF A SINGLE CLASSICAL EQUATION OF MOTION

S. Putterman, P.H. Roberts, C.A. Jones* and A. Larraza†

Physics Department, University of California, Los Angeles USA;
*University of Newcastle upon Tyne, England
†U.S. Naval Postgraduate School, Monterey, Calif.

INTRODUCTION

A successful theory of Helium 4 would have to unify the two fluid behavior and the phonon-roton spectrum with the quantization of circulation. A complete microscopic theory would also be able to relate the material constants that appear in these equations to the potential of interaction which exists between He atoms. When such a theory is achieved it will for instance be able to describe the critical behavior near the lambda temperature as well as the propagation and attenuation of second sound. Towards the goal of achieving this theory we would like to outline our picture of how such a theory might be constructed.

ONE FLUID THEORY OF A PHONON SUPERFLUID

First we consider the phonons and the two fluid theory. In this regard we have shown that the Euler equations of a one component isotropic fluid yield the two fluid equations[1,2] as a result of the symmetry breaking that characterizes nonlinear field equations with behavior on multiple time scales. The Euler equations at T=OK are given by four equations in four unknowns:

$$\partial\rho/\partial t + \vec{\nabla}\cdot\rho\vec{v} = 0 \qquad (1)$$

$$\rho\,\frac{D\vec{v}}{Dt} = -\vec{\nabla}\,p(\rho) \qquad (2)$$

where ρ is the density p is the pressure and v is the velocity. According to the equation of state pressure is a function of density. Our assertion that the two fluid equations follow from 1,2 is motivated by the fact that these equations possess wave solutions with both long and short wavelengths. The motion typical of the superfluid component is the long wavelength solution while the motion characteristic of the normal fluid is the short wavelength solution. Although the nonlinearities in 1,2 cause these solutions to interact they nevertheless possess an approximately conserved quantity; the adiabatic invariant of the short waves. This quantity is the momentum of the normal fluid relative to the superfluid rest frame.

The derivation of the two fluid theory from the one fluid Euler equations first requires the separation of time scales by setting the

$$\rho = \rho_0(\vec{r},t) + \int \rho_1(\vec{k},\vec{r},t)\exp\left[i\theta(\vec{k},\vec{r},t)\right]d^3k + c.c. \qquad (3)$$

where ρ_0 is the long wavelength motion and ρ_1 represents the high frequency motion. Substitution of 3 into 1,2 requires consideration of three key nonlinear effects:

A) the scattering of one sound wave by another to produce sum and difference frequencies and wave numbers,

B) the refraction of short wavelength sound waves by long wavelength background variations,

C) the reaction of the background so that total energy and momentum are conserved during the refraction.

Substitution of 3 and a similar decomposition for v into 1,2 yields coupled equations for the high frequency component and the background (i.e. low frequency component). The equation for the noise (i.e. high frequency component) is found at leading order to be:

$$\left(\frac{\partial n}{\partial t}\right)_{k,r} + \frac{\partial \omega}{\partial k_i}\left(\frac{\partial n}{\partial r_i}\right)_{k,t} - \left(\frac{\partial \omega}{\partial r_i}\right)_k \left(\frac{\partial n}{\partial k_i}\right)_{r,t} = I(n) \qquad (4)$$

where

$$\omega = \omega_0(k) + \vec{v}_0\cdot\vec{k}; \quad \omega_0 = c_0 k \qquad (5)$$

$$n(k,r,t) \equiv \epsilon_k/\omega_0 \qquad (6)$$

$$\epsilon_k = (c_0^2/\rho_0)|\rho_1|^2 + \rho_0|v_1|^2 \qquad (7)$$

so that c_0 is the speed of sound. The left hand side of 4 describes effect B. We will not write out the collision integral in all its detail since we will mainly require its conserved moments:

$$\int \vec{k} I(n) d^3k = 0 \qquad (8)$$

$$\int \omega I(n) d^3k = 0 \qquad (9)$$

Inserting the expansion 3 into 1,2 and collecting together the long wavelength terms yields the equation for the reaction of the background (effect C):

$$\partial\rho_0/\partial t + \vec{\nabla}\cdot\left[\rho_0\vec{v}_0 + \left(\int \rho_1\vec{v}_1{}^* d^3k + c.c.\right)\right] = 0 \qquad (10)$$

$$\frac{\partial \vec{v}_0}{\partial t} + (\vec{v}_0\cdot\vec{\nabla})\vec{v}_0 = -\vec{\nabla}\left[\Phi_0 + (U_n/c_0)dc_0/d\rho_0\right] \qquad (11)$$

where,

$$U_n = \int n\omega_0 d^3k = \int \epsilon_k d^3k; \quad \Phi_0 = \int^{\rho_0} dp/\rho \qquad (12)$$

Equations 4,10,11 describe how the mixture of slow and fast components evolve in time and they are in fact more general than the two fluid equations. They reduce to the Landau two fluid equations when supplemented with the approximation of local equilibrium:

$$n = n_{eq}(\beta(\omega_0 - \vec{k}\cdot\vec{w})) \tag{13}$$

where $\beta,\vec{w}$ vary slowly with r,t. Applying 8,9 to 4,10,11 yields precisely the two fluid equations with

$$\vec{w} = \vec{v}_n - v_s,\ v_0 = v_s,\ \rho_0 = \rho.$$

So we see that a phonon theory of superfluid He4 can be obtained from the Euler equations plus an input of ρ and c_0.

EXTENSION OF THE ONE FLUID PHONON THEORY TO INCLUDE THE ROTON MINIMUM

As is well known the excitation spectrum of He4 possess a roton dip at a wavelength scaled to an angstrom. This effect can easily be included within the framework of the above theory by including in the pressure a nonlocal dependence upon the density. For instance if the pressure is derivable from an internal energy density given by:

$$U(\rho) = (\gamma/\rho)(\nabla\rho)^2 + \frac{1}{2m^2}\int \rho(r)V(r-r')\rho(r')dr' \tag{14}$$

then the new dispersion law for small oscillations is given by

$$\omega_0(k)^2 = 2\gamma k^4 + (2\pi)^3\rho_0 V(k)k^2/m^2 \tag{15}$$

where m is the mass of a He4 atom and $V(r-r')$ is the two particle potential. For reasonable V the dispersion law matches the phonon-roton curve except that at high k, $\omega \sim k^2$ rather than flattening out. Solving the Euler equations with a p determined by 14 should again yield the two fluid theory but with more realistic equations of state.

UNIFICATION OF THE PHONON-ROTON THEORY WITH QUANTIZED VORTICES

There still remains the question of the quantization of circulation:

$$\oint \vec{v}_s\cdot d\vec{\ell} = nh/m \tag{16}$$

While this condition is consistent with the Euler equations they do not require it. However 16 can be unified with 14 and 1 by use of a macroscopic wave function applicable at T=OK:

$$\psi = \sqrt{\rho}\,\exp i\phi \tag{17}$$

where:

$$v = (\hbar/m)\vec{\nabla}\phi$$

Now 1,2,16,14 are contained in[3]:

$$i\hbar\frac{\partial\psi}{\partial t} = -\frac{\hbar^2}{2m}\nabla^2\psi + \bar{\mu}(\rho)\psi + \frac{1}{m}\int |\psi^2(r')|V(r-r')\psi(r)d\vec{r} \tag{18}$$

where

$$c_0^2 = \rho_0(\partial\bar{\mu}/\partial\rho_0)/m \tag{19}$$

The unification of the macroscopic quantization with the small amplitude excitation spectrum enriches the theory so that it now includes a new type of excitation first predicted by Onsager. These are the famous quantized vortex rings. These excitations have the property that at high momentum, energy goes as the square root of the momentum. Namely these excitations have the property observed in the phonon roton spectrum at momentum transfers above the roton minimum. So now it is essential to study the nature of the quantized vortex ring solutions to 18 and also how these solutions mix with the usual normal mode solutions. In this direction a simpler form of 18 has been considered.[4] It is the case given by:

$$i\hbar\frac{\partial\psi}{\partial t} = -\frac{\hbar^2}{2m}\nabla^2\psi + w_0|\psi|^2\psi \qquad (20)$$

so that

$$c_0^2 = w_0\rho_0/m .$$

The Ginzburg-Pitaevsky equation above has a vortex ring solution for radii much greater than the wavelength of the roton minimum. Starting from these rings smaller quantized objects were created on a computer[4] (always according to 20). It was found that the energy and momentum of these objects decreased until a point was reached where the circulation adiabatically disappeared but a solitary condensation remained. Below this point the localized state continued to shrink until at some minimum energy, greater than zero, it expanded into a larger and larger wave-like object whose energy momentum relation obeyed that which applies to sound waves. That is one can follow the continuous transition of a quantized vortex into a sound-like wave. This is all quite suggestive of the picture first proposed by Onasager.

As can be seen from 20 the G.P. equation is a unification of a phonon spectrum with macroscopic quantization. This of course misses the role of the roton minimum. One would like to see the above analysis carried out for the more realistic equation 18. While this has not been accomplished it may be appropriate to speculate on how this will turn out. For 18 the vortex ring spectrum will cross the $\omega \sim k^2$ spectrum characteristic of the small oscillations. At this point level repulsion will split the spectrum into two brances; one which will flatten out at high k and attach onto the roton minimum and another which will be pushed to high frequency. This is quite close to what is observed. These remarks therefore suggest that a multiple scales analysis of 18 which includes normal modes and vortex rings might provide a rather complete picture of He4. The main roadblock to this analysis will probably involve the mathematical description of the mutual interaction of the localized and delocalized modes; i.e. the rings and small oscillations.

CLASSICAL SECOND SOUND

According to the above analysis some superfluid concepts such as second sound are strictly classical in their origin. Thus one is led to ask if there are classical systems in which this phenomenon can be observed. We have proposed that classical systems which are filled with large amounts of wave energy should display second sound.[5] The idea here is that the amplitude of excitation should be such that the nonlinear interactions dominate the effects due to viscosity. To the extent that this applies the classical system will mimic the two fluid theory of superfluids. The properties of these classical systems will be determined by the power spectrum of the turbulence[6] and not the Planck or equipartition spectra. So we are saying that classical systems which are sufficiently far off equilibrium will mimic superfluidity.

Applications which we have in mind are wind driven waves on the surface of the ocean[7)] and even drift wave turbulence in plasmas. In the case of plasmas we might hope to use our understanding of superfluids to understand the problems of confinement. After all, second sound causes energy to spread out at a rapid rate that is characteristic of hyperbolic processes.

REFERENCES

1. S.J. Putterman, P.H. Roberts, Phys. Lett. 89A, 444 (1982).
2. S.J. Putterman, P.H. Roberts, Physica 117A, 369 (1983).
3. K.A. Brueckner, K. Sawada, Phys. Rev. 106, 1117, 1128 (1957).
4. C.A. Jones, P.H. Roberts, J. Phys. A15 (1982) 2599;
   J. Grant, P.H. Roberts, J. Phys. A7, (1974) 260;
   C.A. Jones, S.J. Putterman, P.H. Roberts, J. Phys. A19 (1986) 2991;
   P.H. Roberts, J. Grant, J. Phys. A4, 55 (1971).
5. A. Larraza, S.J. Putterman, Phys. Rev. Lett. 57, 2810 (1986).
6. S.J. Putterman, P.H. Roberts, Phys. Reports 168, 209 (1988).
7. Phys. Rev. A41, 3144 (1990).

DISSCUSSION ON BULK ^{4}He DYNAMICS

Chairmen : R J Donnelly, S Stringari

Glyde to Campbell : Many properties of liquid ^{4}He including the dispersion curve for the elementary excitations you describe, can be calculated with impressive precision using the method to Correlated Basis Functions (CBF). Can this method be extended to finite temperatures near T_λ? How would you distinguish clearly between a description of superfluid and normal ^{4}He in the CBF method? Would you expect the dispersion curve to be substantially different in normal and superfluid ^{4}He?

Griffin : comment on Campbell's paper. The finite temperature microscopic theory which Glyde and I have described is firmly based on the dynamical role of a Bose broken symmetry. As shown especially by Hohenberg and Martin as well as Bogoliubov in the early 1960's, at low frequencies, this theory leads precisely to the fluid hydrodynamics of Landau at all temperatures below T_λ. As successful as they are for certain properties, it is not clear if the variational wavefunctions described by Campbell are to be viewed as coherent macroscopic wavefunctions which lead directly to superfluid hydrodynamics. They imply a finite Bose condensate but the latter seems to be almost an accidental feature, as Campbell himself has noted. Further work should try to clarify the crucial question concerning the role of the condensate in such variational approaches.

Campbell : My presentation was concerned with the eigenstates of the Hamiltonian, and therefore in a technical sense should not be generalized to $T > 0$, of course, once one knows the complete set of eigenvalues and eigenstates, they may in principle be used to calculate properties at finite temperature. The most general method to accomplish that is a theory of the full density matrix of the system. Reatto and Chester carried out such a theory at low temperatures some years ago, and Campbell, Kürten, Rietig and Sengen have reformulated their approach as a variational method which has been applied successfully to the liquid-gas critical point of the fluid. These theories give the correct temperature dependence of the condensate fraction at low temperatures, as well as the correct temperature dependence of S(q), including critical opalescence at the critical point. It also correctly gives the T = 0 transition to quasi off-diagonal long range order in two-dimensional helium. We have shown in principle how the theory produces the λ transition, but mathematical difficulties associated with the calculation of the entropy have thus far prevented us from doing an actual calculation through the λ point. It is quite clear, however, that the λ point corresponds to the disappearance of the condensate fraction within this theory.

While our presentation focussed on excited states and did not address the CBF theory of $S(q,\omega)$, Halley's talk included a review of how the CBF theory has been used to calculate this quantity as well as the response function to second order light scattering. The theoretical $S(q,\omega)$ as obtained most recently by Manousakis and Pandharipande, shows excellent agreement with the measured phonon-roton region, and semi-quantitative agreement with the multi-excitation features of the spectrum. To our knowledge, the CBF theory is the only theory to give such a good agreement with experiment. This theory has not, however, been carried out at any finite temperature, let alone near and above the λ transition. The recent experimental analysis of $S(q,\omega)$ through the λ point, as well as the very stimulating theoretical conjectures of Glyde and Griffin, provide new impetus to carry out the CBF theory through the λ point. The structure of the CBF theory at low temperatures makes it clear the $S(q,\omega)$ would evolve away from its T = 0 form in a continuous manner due to thermal population and excitation interactions. This is also indicated by the calculated low temperature behaviour $S(q)$, which is the frequency integral of $S(q,\omega)$.

In answer to a question we posed in the first session, Glyde agreed that his theory with Griffin includes T = 0. It follows that a T = 0 theory of the phonon-roton spectrum should, according to their theory, show evidence of the role of both the zero sound mode and the single particle mode which combine to produce the phonon-roton spectrum. The CBF theory shows no evidence of this effect. To our knowledge, the Glyde-Griffin theory has not actually been implemented in a quantitave calculation of this spectrum at any temperature, including T = 0. We look forward to seeing such real calculations applied to helium.

Concerning questions of the role of the condensate, our conjecture in this presentation was that it is qualitatively of surpassing importance, but not of quantitative significance. We had no intention of indicating that the finite Bose condensate "seems to be almost an accidental feature", and we welcome the opportunity to correct any misimpressions our oral presentation might have given. We attempted to make the point, as we have published elsewhere, that the finite Bose condensate is very difficult to avoid in any boson system which has a quantum <u>fluid</u> ground state, and that, should one be able to devise a model in two or three dimensions which has no condensate, then it would by contrast, further clarity the role of the condensate. The qualitative role of the condensate in liquid helium is certainly to prohibit any excitation with energy below that of the density fluctuations, and thus to permit the presence of Landau superfluidity. Again we should point out that it is an easy matter to demonstrate within our theory, that the condensate vanishes at high temperatures. The theory most easily distinguishes between normal fluid and superfluid through the off diagnol long range order. Of course, at T = 0 we find, trivially that $\rho_s = \rho$, because within CBF the system is in its exact ground state at T = 0. If one asks what the Jastrow approximation (and its improvements) say about the superfluid density at T = 0, the answer is the same: $\rho_s = \rho$. This was first demonstrated by the fact that the ground state Jastrow function is unchanged upon rotation until the rotation rate exceeds a critical value ω_c whereupon the ground state includes a single vortex line. The finite temperature theory also give a normal fluid density with the proper temperature dependence at low temperatures where it has been calculated. The single excited states (or any excited state) of ρ_k, polynominal degree $O(N^o)$, does not destroy the coherence of the state. This may be true until the degree equals N, but this is purely conjecture.

Hallock : Some years ago Cumming's and his collaborators proposed a prescription for the determination of the condensate fraction from measurements of the pair correllation function. Although this prescription has been criticised, its application to temperature and pressure dependent data obtained by both neutron and X-ray scattering experiments yields condensate fraction values in general agreement with the value obtained by high momentum transfer neutron scattering through a completely different analysis. It would be surprising if this were an accident. What is the theoretical basis of the Cumming's prescription, and if weak, will the theorists please resolve the issue.

Griffin : it is weak.

Glyde : Cumming et al have confirmed the claim and there is some other claim of its validity.

Campbell : Concerning the contribution by Maris about the critical negative pressure at which the sound velocity vanishes, I wish to note that, in the late 1970's Jackson, Jennings, Castillejo and Smith found this effect in their Jastrow-Euler Langrange analysis of liquid ^4He. Subsequently Campbell, Kürten, Ristig and Senger calculated the locus of this spinodal line from T = 0 to the critical temperature. However, none of this work analysed the approach to this line as a function of pressure.

BULK PROPERTIES OF ^{3}He-^{4}He MIXTURES

R.M.Bowley

Department of Physics
University of Nottingham
Nottingham NG7 2RD, England

INTRODUCTION

In the space available it is impossible to review all aspects of research into bulk properties of helium mixtures, and so I shall omit all reference to the thermodynamics of mixtures, a subject which has been thoroughly explored and is reviewed in the articles by Ebner and Edwards[1] and by Ghozlan and Varoquaux[2]. Another excellent review, but with a more theoretical slant, is that of Baym and Pethick[3]. Rather, I will concentrate on several new ideas have emerged since these reviews were published. For example: it has become clear that, with the advent of large magnetic fields and new cooling techniques, it is possible to polarise dilute mixtures by brute force methods. This has opened up a whole new field of research on degenerate spin polarised Fermi systems, a topic pioneered by Bashkin and Meyerovich[4]. Another new idea: three phonon decay processes are allowed in ^{4}He at low pressures,and these processes affect phonon transport in dilute ^{3}He-^{4}He mixtures[5,6]. Another new field was opened up by the discovery of spin waves in mixtures[7]. These are the topics which I will concentrate on in this review. The transport properties of dilute mixtures near the lambda line, a fascinating subject, will be dealt with elsewhere in the proceedings by Horst Meyer[8] and Richard Ferrell[9].

First I will review the current state of our understanding of the properties of ^{3}He-^{4}He mixtures, and then I will review experiments on spin waves and second sound waves, collective excitations which reveal a lot about the interactions between the ^{3}He quasiparticles. This will enable me to discuss most of the new experiments and developments of the last ten years, and to point out where more research is needed.

^{3}He quasiparticles of in superfluid ^{4}He are described by the dispersion relation

$$E_p = -E_3 + \frac{p^2}{2m_0^*}(1+\gamma p^2) \qquad (1)$$

where **p** is the wavevector m_0^* is the effective mass for a single

Excitations in Two-Dimensional and Three-Dimensional Quantum Fluids
Edited by A.G.F. Wyatt and H.J. Lauter, Plenum Press, New York, 1991

quasiparticle, and γ describes the deviation of the dispersion from that of a free particle. (Units are such that $\hbar = 1$.) This form of dispersion was first proposed by Brubaker et. al[10]. The value of γ obtained from an analysis of second sound velocities[11] is -0.074 +/-0.004 A^2 at a pressure of zero bar. Other estimates of γ can be found from specific heat measurements, such as those of Greywall[12] and of Owers-Bradley et al.[13]

The value of the effective mass m_0^* at zero bar quoted by Ghozlan and Varoquaux[2], of 2.255 m_3, was obtained by them by fitting all available experimental data. Analysis of the inertial mass[11], obtained from more recent second sound measurements, gave a value of 2.27 m_3, and analysis of accurate specific heat measurements gave 2.25 m_3 [13]. Other values have been obtained in the past but these seem to be the most reliable and are mutually consistent.

The interaction between quasiparticles is usually modelled by a **q** dependent interaction, **q** being the momentum transferred between quasiparticles. The form favoured by Ghozlan and Varoquaux[2] is

$$V(p,p',q) = V_0 + V_1 q^2 + V_2 \frac{(p.q)\,(p'.q)}{q^2} \tag{2}$$

where **p** and **p'** are the momenta of the two quasiparticles. The term proportional to V_2 describes the backflow interaction between the ^{3}He. The quantities V_0 and V_1 are not known with great accuracy. Different measurements give widely differing values. This is somewhat embarrassing because the temperature at which the ^{3}He becomes superfluid depends very sensitively on the interaction; we would all like to know how cold the mixture has to become before we get a new superfluid.

The **q** independent part of the interaction is expressed in one of two ways. Some people prefer to describe V_0 using the scattering length $a = m_0^* V_0 / 4\pi\hbar^2$; others prefer to express it as $V_0 = \xi m_4 s^2 / n_4$. A scattering length of -1 Å corresponds to a value of ξ of -0.072. Ghozlan and Varoquaux quote a value which corresponds to a scattering length of -1.1 Å. This is a value which lies in the middle of other estimates. For example the analysis of spin wave or spin echo experiments in dilute mixtures allows us to get a value for V_0 from the Leggett-Rice effect and a separate value for V_0 from the ^{3}He-^{3}He scattering rate. This one experiment[14] yields a scattering length of either -0.77 Å or -1.5 Å.

An estimate of V_0 can be made from measurements of the velocity of second sound, and the latest value[11] is about -1.3 Å. This value is not very reliable partly because it depends sensitively on the temperature scale which is used, and partly because the analysis is correct to lowest order in the interaction. A better estimate could be made by including terms of order a^2 in the formula for the second sound velocity; such a calculation will be worth doing once the temperature scale is determined more precisely.

Estimates of the interaction can also be made from the analysis of scattering rates associated with thermal conductivity and viscosity; the best current estimate by Fu and Pethick[15] gives $a = -0.93$ Å. These authors analyzed, to third order in perturbation theory, the viscosity and thermal conductivity measurements that had been made in concentrated solutions.

This wide spread in values of the interaction is unacceptable. We need to know the interaction more accurately so as to predict the superfluid transition temperature. An accurate measurement of V_0 would also help to test Baym's model[16] of the interaction.

PHONON ^3He INTERACTION

The behaviour of mixtures above about 0.1 K is affected by the presence of phonons. The theory of the interaction between phonons and ^3He and the transport properties was developed in a series of papers by Baym Saam and Ebner[17]. At the time the theory was developed it was not appreciated that the phonon dispersion is anomalous at low pressures, and that three phonon decay processes are allowed. These processes change the kinetics of the phonons, and modify the relaxation rates which appear in the formulae for the damping of second sound and the thermal conductivity. It has a particularly large effect at low pressures and in dilute mixtures. As the pressure is raised so the region of anomalous dispersion contracts, and above about 18 bar it disappears completely. Increasing the concentration of ^3He means that phonon ^3He scattering rates become larger than the three phonon decay rates which therefore become less important.

These ideas were first put forward by Adamenko and Tsyganok[5] (and discovered later by Bowley and Sheard[6]). The inclusion of the three phonon decay process enables us to account successfully for the concentration dependence of the thermal conductivity, measured by Rosenbaum et.al.[18], and the temperature dependence of the velocity and absorption of first and second sound. These latter topics are reviewed in the excellent paper by Adamenko and Rudavskii[19].

A more detailed of test the theory comes from the temperature and frequency dependence of the damping of second sound. This was first studied in detail by Greywall[20]. He measured second sound resonances

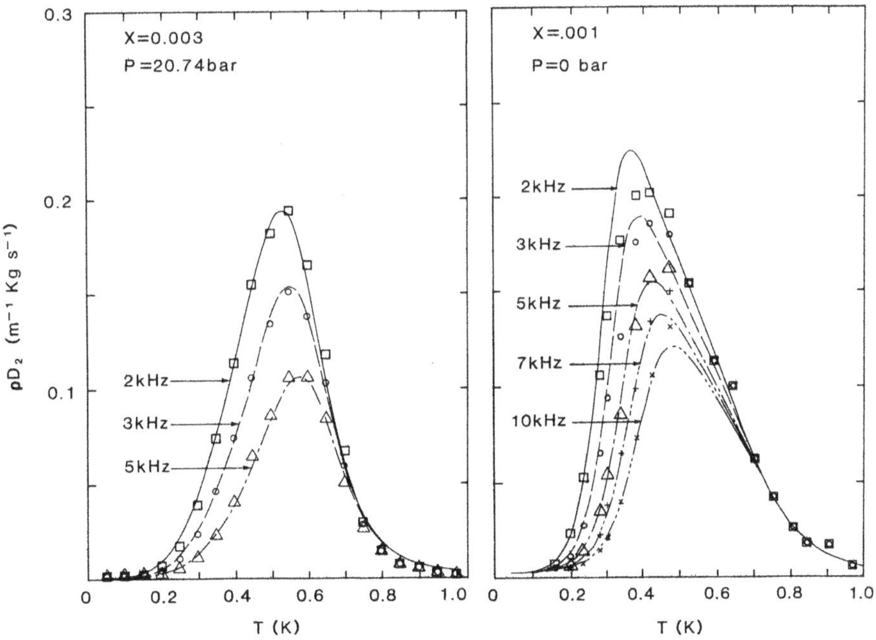

Figure 1. *The attenuation of second sound as a function of temperature for a 0.3% solution at 20.74 bar, (Fig. 1(a)) and for a 0.1% solution at 0 bar (Fig. 1(b)).*

in a cylindrical cell, and could obtain the damping of second sound from the widths of the resonances. The frequency range which can be covered lies between 1 kHz and 10 kHz. For low frequencies compared to the scattering rate the behaviour is hydrodynamic and independent of frequency. As the temperature is lowered the scattering rate decreases and becomes smaller than the frequency; consequently, the attenuation becomes a function of frequency. At 20 bar the theory is simpler, because we can ignore three phonon decay processes and use the Baym Saam and Ebner[17] theory to fit the data. The best fit, obtained by McHale et.al.[21], is shown in figure 1a above. The theory is in excellent agreement with experiment, but it gives ^{3}He phonon scattering rates which are larger than expected. The values found by McHale et.al. agree with the experimental value for the scattering rate obtained by Husson and de Bruyn Ouboter [22], but are about 50% larger than the theoretical estimates made by them at 20 bar (see table I of reference 22). More work could be done to unravel the source of this discrepancy for pressures above 18 bar where three phonon processes can be ignored.

At zero bar the three phonon decay process is important and alters the shape of the curves. Even with the three phonon process it is not possible to get a very good agreement with the data; the best fit found by using the theory of McHale et.al.[21] is shown in figure 1b. Although the agreement is good below 0.35 K and above 0.5 K, it is poor in the region in between where the damping reaches a maximum. The theory overestimates the damping here. The cause of this problem remains obscure. This experiment has been repeated by Church Owers-Bradley and Main[23] for a wider frequency range with the same poor fit around 0.4 K (figure 1b shows their data which is identical to that of Greywall[21] where they overlap).

CHANGES ON SPIN POLARISATION

Dilute mixtures (say less than 0.1%) can be spin polarised by applying a strong magnetic field of about 10 tesla, and by cooling to 10 mK or below. This changes thermodynamic properties like the specific heat, the osmotic pressure and the velocity of second sound. These changes are quite small, at most of order 20%. In contrast transport properties are dramatically altered.

Since the scattering between the ^{3}He is mainly "s-wave", for small momentum transfers, it involves quasiparticles of opposite spin. As the system becomes spin polarised so the scattering rate decreases and the mean free path of the quasiparticles increases, leading to increases in the thermal conductivity, the viscosity and the longitudinal diffusion constant. These changes are expected to be spectacular provided one can get to temperatures well below the degeneracy temperature of the mixture. For a 0.1% solution T_F is about 26 mK and a 10 tesla field produces a Zeeman splitting of 15.6 mK. Substantial effects should be seen if the mixture could be cooled to say 5 mK. Until recently the lowest temperature that had been reached[24] was just under 10 mK, so the polarisation is only about 40%, and the system was not completely degenerate. However even under these circumstances it is possible to see changes in transport properties.

Greywall and Paalanen[24] were the first to observe a change in the transport properties on spin polarisation. The studied the velocity and damping of second sound in a cylindrical cavity by measuring the frequency and the width of the resonances. They reported that the widths increased on applying a magnetic field. Their experiment was repeated by Owers-Bradley et.al.[25] with greater precision. The damping is caused by bulk losses due to thermal conduction and viscosity, to surface losses due to thermal conduction and shear viscosity and to reflection losses off the transducers used to excite the second sound. In order to simplify the analysis Owers-Bradley et.al.[25] used the theoretical prediction of Hampson et.al.[26] for the ratio of thermal conductivity to viscosity, so that only the viscosity was left as a free parameter to fit to the data.

A typical set of resonance widths is shown in figure 2, where the open circles show data taken in a field of 11 tesla and the crosses show data in zero field. The increase in the width is clearly visible. By fitting the theory to the data it is possible to extract values of the viscosity, and these are in excellent agreement with the theory of Hampson et al[26]. The best fit to the theory, which uses the Born approximation, yields values of V_0 and V_1 of

$$V_0 = -0.067 m_4 s^2/n_4$$

$$V_1 = 0.025\ m_4 s^2/n_4 k_0^2$$

where k_0 is the Fermi momentum of a 5% solution. These values are quite plausible, and the Born approximation is expected to be fairly accurate for dilute solutions, for corrections are of order $k_F a$ in the degenerate limit. For a 0.05% solution this correction is about 7% using a = -1Å.

Direct measurements of the viscosity and thermal conductivity are more difficult. Ordinary vibrating wire viscometers are impractical in high magnetic fields because of the voltages induced in the wires, which are moving in a strong magnetic field. A more promising experiment is Poiseuille flow through a tube.

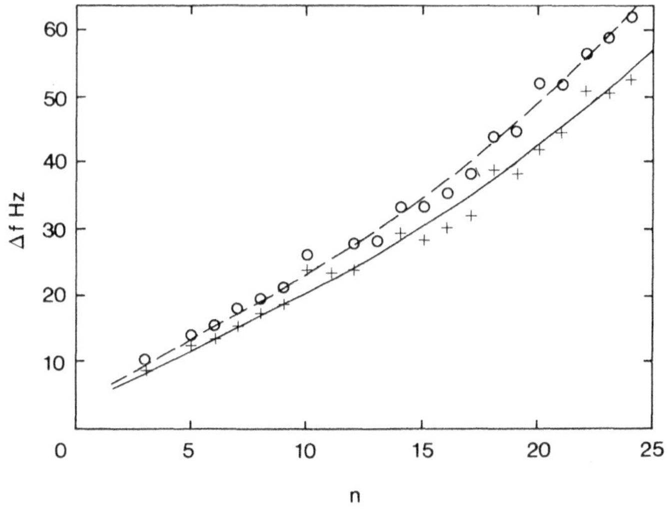

Figure 2 *The widths of various resonances in fields of 11 and 0 Tesla*

Thermal conductivity measurements are tricky, partly because of the lack of a reliable thermometer in a large magnetic field. One possibility is to use the second sound cell as a thermometer. This has the great advantage in that it is a direct measurement of a property of the mixture. In contrast a glass capacitance thermometer, for example, will make poor thermal contact with the ^3He because of the large Kapitza resistance.

The second sound cell could be calibrated by measuring the frequency of say the 15th resonance for several standard temperatures, and by fitting this to the formula for the temperature dependence of the second sound velocity. Since one can measure the frequency of the 15th resonance with great accuracy (better than 1 in 10^5), it should be possible to measure small temperature differences quite accurately.

However there is a complication. Suppose a heat current is set up between two points separated by a distance L. If the mixture is held in a closed cell with non magnetic walls, one can ensure that there is no longitudinal spin current or mass current. In the steady state the chemical potential of the spin up and spin down ^3He must be equal along the length L and this implies that there is a concentration gradient of ^3He. Consequently, the densities of the ^3He at either end of the length L are different, and so the calibration of the thermometers must allow for this change in density. This problem only has a minor effect for the classical regime where the second sound velocity is almost independent of concentration. In the degenerate regime it is important. This difficulty can be overcome and a second sound cell could, in principle, be used as a thermometer.

Another difficulty is that the phonon contribution to thermal conductivity can be quite large, even down to 20 mK. The reason is that the phonons travel perhaps 20-30 times as fast as the ^3He. The phonon contribution can be made much smaller by confining the mixture to a collection of very narrow channels (say 0.05 mm in diameter) acting in parallel. But this will not be a problem if temperatures as low as 5mK can be reached.

Certainly experiments that directly measure the thermal conductivity would be of great interest, but they do appear to be quite tricky. Perhaps this is the reason why they have not yet been attempted.

SPIN WAVES

The existence of propagating transverse spin waves in a degenerate Fermi liquid was first proposed by Silin[27]. Direct experimental observation of spin waves were made in liquid ^3He by Masuhara et. al.[28], and in ^3He -^4He mixtures by Owers-Bradley et.al.[7], and later by Ishimoto et.al.[29] and by Akimoto et.al[30]. These experiments give information on the parameter

$$\lambda = \frac{1}{(1+F_0^a)} - \frac{1}{(1+F_1^a/3)} \qquad (3)$$

It is possible to measure λ as a function of concentration and hence to obtain information about the interaction between the quasiparticles.

In the Hartree Fock approximation we have

$$F_0^a = \frac{-N(0)}{2k_F^2}\int_0^{2k_F} dq\, qV(q) \tag{4}$$

$$\frac{F_1^a}{3} = \frac{-N(0)}{2k_F^2}\int_0^{2k_F} dq\, qV(q)\,(1-q^2/2k_F^2) \tag{5}$$

where N(0) is the density of states for a single spin and k_F is the Fermi wavevector. If we knew V(q) we could calculate the Landau parameters and obtain λ in the Hartree Fock approximation. Roughly we find

$$\lambda \approx \frac{N(0)}{4k_F^4}\int_0^{2k_F} dq\, q^3V(q) \tag{6}$$

By varying the concentration we vary the Fermi wavevector k_F, and so we integrate over different ranges of q in the equation above. Because V(q) is negative for small q and positive for large q, we expect λ to be negative for low concentrations and positive for larger ones. Estimates based on the model potential of Murdock Mountfield and Corruccini[31] indicate that λ changes sign for a 3% solution at 0 bar pressure. The experiment of Owers-Bradley et. al.[7] for a 5% solution gave $\lambda = 0.03$, and the experiment of Ishimoto et. al.[29] for a 1% solution gave $\lambda = -0.045$ confirming this analysis. Further experiments at other concentrations would be very interesting and may help to specify V(q) more precisely. If this happens there will be incentive for someone to go beyond the Hartree Fock approximation, and calculate λ to second order in the interaction.

More recently Bedell[32] has pointed out that when $\lambda = 0$ new features appear in the dispersion and attenuation of spin waves and in the Leggett Rice effect[33]. In particular he predicts that the spin wave dispersion is linear in the wavevector k as opposed to the usual k^2 dependence. The best way of finding the region where $\lambda = 0$ is to choose a concentration close to 3%, and to vary the pressure so as to sweep through the region where $\lambda \approx 0$. Such an experiment may be quite difficult to perform.

MAGNETIC EFFECTS WHEN SPIN POLARISED

What happens to the spin waves as the Fermi system becomes spin polarised? Continuous wave nmr experiments allow us to measure the frequency shift and the broadening of the nmr line, and hence to get the parameter λ and the transverse relaxation time $\tau_\perp$. How do $\tau_\perp$ and λ vary with spin polarisation experimentally? To obtain a large spin polarisation one needs to use a dilute mixture (0.1% or less is ideal), and for such mixtures the nmr signal is weaker. But if one uses a strong magnetic field the signal improves. Another way of improving the signal is to create larger transverse magnetisation by tipping the spins further. Spin echo experiments by Gully and Mullin[34] were the first study of the magnetic properties of spin polarised mixtures.

In these experiments the spins are tipped away from the z direction (the direction of the polarising field) by an angle θ_1 at time t = 0, and then they precess in a field magnetic field gradient G. A second tipping pulse, which rotates the spins through an angle θ_2, is applied at time $t=\tau$

and an echo forms at $t = 2\tau$. In the usual spin echo experiment θ_2 is taken as 180° and the height and phase of the echo is measured as a function of τ and θ_1. The presence of an internal field leads to a non-linear equation for the spin current and this gives rise to a complicated response, first analyzed by Leggett and Rice. If the sequence θ_1-τ-θ_2 is used with θ_2 not equal to 180°, then multiple spin echoes are expected, that is echoes at times $t = n\tau$ where n is 3,4,5... . Multiple spin echoes have been observed in ^{3}He by Hirayoshi et al.[35] and in mixtures by Bedford et al.[36].

If both θ_1 and θ_2 are small then the system behaves as if spin waves were generated. For example, the recently reported experiments by Candela et al.[7] used pulsed nmr and the tipping angle is small. The Fourier transform of the free induction decay then shows the frequency shift and the damping from spin waves.

Leggett and Rice[33] used the quantity μ to describe this non-linear term in the spin current, where $\mu M_0 = -\omega_0\lambda\tau_\perp$. Here ω_0 is the Larmor frequency $\tau_\perp$ is the transverse relaxation time and M_0 is the equilibrium magnetisation. If the magnitude of the magnetisation remains spatially uniform, the equation for the spin current simplifies, and Leggett found that the echo heights h vary as

$$(1+\mu^2 M_0^2\cos^2(\theta_1))\log_e(h)+\frac{1}{2}(\mu M_0\sin(\theta_1))^2[h^2-1]=-2\gamma^2G^2D_\perp\tau^3/3 \quad (7)$$

This equation was used by Gully and Mullin[34] to fit their data taken in a 0.037% mixture in a field of 8.9 Tesla; from the fit they obtained values of μM_0 and the transverse diffusion constant.

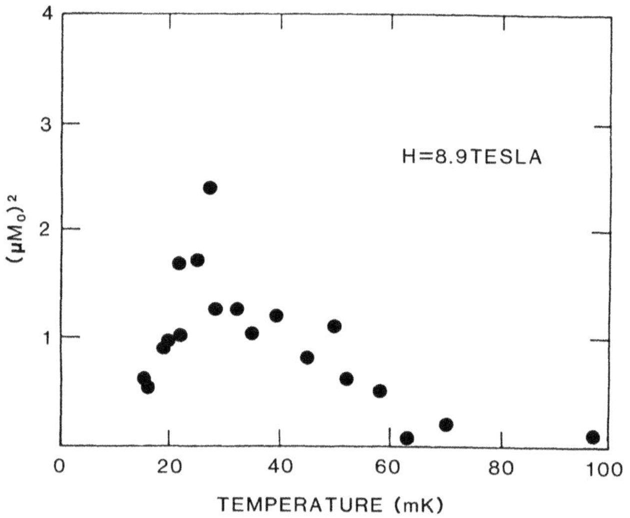

Figure 3 *The values $\mu^2 M_0^2$ obtained by Gully and Mullin[34] by fitting to the theory of Leggett and Rice.*

It was expected that μM_0 would increase monotonically as the temperature decreases, but this is not what they observed. Their data showed a peak in the temperature dependence of $\mu^2 M_0^2$ at about 25 mK. Moreover, the diffusion coefficient was much smaller than expected at the lowest temperatures used.

Ivanova and Meyerovich[37] pointed out that ratio $\mu M_0/D_\perp$ is independent of the relaxation time, and hence it should give a good estimate of the interaction. For dilute solutions at very low temperatures the interaction is nearly constant, and from fitting Gully and Mullin's data Ivanova and Meyerovich estimated the scattering length to be between -0.5 Å and -0.7 Å, a fairly low value. If the interaction is taken to vary with q in a more reasonable way then the value of $-V_0$ is increased, but only by about 20%. The goodness of the fit to the data lead Ivanova and Meyerovich[37] to trust Gully and Mullin's data. But we are still left with the puzzle: why does μM_0 in Gully and Mullin's experiment appear to reach a maximum at 25 mK?

One possibility is that there is a defect in the experiment. Gully and Mullin state that, below 25 mK, the chamber was cooled differently, and that there was a heat leak down the filling capillary. Perhaps heat currents are set up and these affect the data.

Another possibility is that the transverse relaxation time $\tau_\perp$ is much smaller than expected below 25 mK. Meyerovich[38] was the first to notice that, in degenerate spin polarised mixtures, the phase space for final states is opened up for transverse processes, processes in which spin up and spin down quasiparticles scatter into two spin up unoccupied states. This leads to an increase in the scattering rate and a reduction in the tranverse relaxation time. The theory was worked out in detail by Jeon and Mullin[39], and numerical calculations based on their theory show that the transverse relaxation time is reduced, but only for temperatures well below the Fermi temperature. In contrast the longitudinal relaxation time is expected to increase on spin polarisation.

Does this new idea help us to explain the data of Gully and Mullin? Unfortunately it doesn't because, in Gully and Mullin's experiment, the Fermi temperature is about 13.7 mK, whereas the lowest temperature used is about 17 mK. The system is not degenerate, there is not expected to be any opening up of phase space, and the transverse relaxation time is not reduced.

There are now two objectives to current research in this field. The first is to measure μM_0 as a function of temperature for a spin polarised mixture, and to confirm or refute the findings of Gully and Mullin[34]. The second objective is to measure both the longitudinal and the transverse diffusion constants, so as to test the theory of Jeon and Mullin[39]. This is a more difficult task, for it requires solutions at temperatures well below both the Fermi temperature and the magnetic energy $\hbar\gamma B$. Recent progress in designing experiments to measure both diffusion coefficients has been reported by Nunes et. al.[40] for a 0.035% mixture at 5 mK in a field of 9 Tesla.

Candela et.al.[14] have measured $D_\perp$ and μM_0 by analyzing spin wave resonances in a spherical cavity for a 0.182% mixture in a field of 8 tesla. They found that μM_0 increased monotonically as the temperature was reduced down to 8 mK, in clear distinction to Gully and Mullin's results.

The value of the ratio $\mu M_0/D_\perp$ which they measured agrees with theory and gives for the scattering length a = -0.77 Å. (Remember that this ratio is independent of the transverse relaxation time[37].) Candela et.al.[14] assumed that the interaction is a constant, but if they had taken it to be of the form $V(q) = V_0 + V_1q^2$, then λ would be larger. Simple integration gives

$$\lambda \approx N(0)\,(V_0+8V_1k_F^2/3) \tag{8}$$

for degenerate mixtures. Using the best current estimate of V_1 would change the value scattering length to -0.84 Å. Everything now is consistent, but only if we ignore quantities which depend on the transverse relaxation time, such as the diffusion constant. The value of the interaction which Candela et al. obtain from fitting their transverse diffusion coefficient is about -1.5 Å.

In other words it is not possible at present to fit both μM_0 and D with a single value of the interaction.

The cause of this problem is not clear. It may be that we need to include both V_0 and V_1 and to go beyond the Born approximation when calculating the scattering rate. This has been shown to be important for concentrated mixture by Fu and Pethick[15]; perhaps it is still important for these dilute mixtures. Alternatively we may need to include the dipolar backflow term.

It now seems certain that there was a flaw in the experiment of Gully and Mullin. What is the flaw? The theory of Leggett applies only to situations where the magnitude of the magnetisation is uniform in space, as is normally the case in low magnetic fields. In very high magnetic fields there are larger non uniformities in the field, and it is tempting to use a large field gradient to swamp these non uniform fields. Gully and Mullin used a gradient of 26 gauss/cm or 0.26 T/m. The length, d, of their cell was about 0.7 cm, so there is a difference in magnetic field across the cell of about 18 gauss. A typical value for the rf field for the split coil, which they use, is about 5 Gauss, which is much less than the change in magnetic field across the cell,Gd. To ensure that the spins are tipped over uniformly, we need to use an rf field much larger than this. If this is not done, the spins at different parts of the cell are tipped about different axes, leading to large variations in the direction of the magnetisation.

This is the explanation proposed by Bedford et. al.[36] to account for several unexpected effects seen in the spin echo experiments done at Nottingham. For example, we have seen multiple spin echoes for a nominal θ_2-τ-180° sequence where theory predicts that only single echoes should occur. The explanation is that the spins are only tipped over through 180° at the centre of the cell; at either end the tipping angle is much less than 180° and so these regions give rise to the multiple spin echoes.

Further experiments to measure both $D_\perp$ and μM_0 are needed ,in which care is taken that cell is sufficiently small so that the rf field is larger than Gd.

Finally let us consider the case where **M** is non uniform. The equation of motion for **M** given by Leggett and Rice can be rewritten in terms of $\mathbf{M} = \boldsymbol{\ell}M$ where $\boldsymbol{\ell}$ is a unit vector in the direction of **M**. The equation for the spin current then becomes[41]

$$J = -D_{\parallel}(\nabla M)\ell - \frac{D_{\perp}}{1+\mu^2 M^2}(M\nabla\ell + \mu M^2 \ell x \nabla^2 \ell) \tag{9}$$

where $D_{\parallel}$ is the longitudinal spin diffusion coefficient. When this is substituted in the Leggett-Rice equation of motion for **M**, we get

$$\frac{\partial M}{\partial t} = D_{\parallel}\nabla^2 M - \frac{D_{\perp}M(\nabla\ell)^2}{1+\mu^2 M^2} \tag{10}$$

Provided $(\nabla\ell)^2$ remains uniform in space M remains spatially uniform for all time. However, if $(\nabla\ell)^2$ is not spatially uniform, variations in M will occur leading to longitudinal spin currents.

Suppose we want to measure $D_{\parallel}$, how should we do it? If we use a system in which Gd is larger than the rf tipping field we will set up a non zero $\nabla\ell$. After a short time the region in the centre of the cell will lose most of its magnetisation (i.e. M becomes small) and longitudinal spin diffusion occurs. The precise evolution of the magnetisation is complicated because the equations of motion are non linear, but it can be calculated numerically. Suppose we have already obtained μM_0 and $D_{\perp}$ from fitting to the Leggett-Rice effect. We would then be in a position to compute the evolution of the system as a function of $D_{\parallel}$ and to find the value gives the best fit to the data.

Theory predicts different behaviour for the transverse and longitudinal relaxation times. In the degenerate regime the relaxation time increases as T^{-2} as the temperature is lowered. The prediction is that at low temperatures (that is temperatures much less than the magnetic separation of the Fermi surfaces) the transverse relaxation time increases less rapidly than T^{-2} as the temperature is lowered, eventually tending to a constant at very low temperatures. In contrast the longitudinal relaxation time increases more rapidly than T^{-2} as the spin polarisation of the system increases. This difference between the two relaxation times (or diffusion coefficients) should be observable in concentrated as well as dilute solutions. All that is essential is that the temperature is small (say 3 mK) compared to the magnetic energy (15.6 mK for a 10 tesla field).

ACKNOWLEDGEMENTS

This work has been done in close collaboration with John Owers-Bradley. I have benefitted greatly from his experience and insight.

References

1. A. Ghozlan and E. Varoquaux Annales de Physique **4**:239 (1979)
2. C. Ebner and D. O. Edwards Physics Reports **2**:77(1971)
3. G. Baym and C. J. Pethick in The Physics of Liquid and Solid Helium, part II ed. by K.H.Bennemann and J.B.Ketterson (Wiley, New York,1978)
4. E. P. Bashkin and A. E. Meyerovich Advances in Physics **30**:1,(1981)
5. I. N. Adamenko and V. I. Tsyganok, Zh.Eskp.Teor.Fiz. **87**:865(1984): **88**:1641,(1985)
6. R. M. Bowley and F. W. Sheard J.Phys.C:Solid State Phys. **18**:L867(1985)

J. R. Owers-Bradley ,H. Chocolacs,R. M. Mueller. C. Buchal, M. Kubota and F. Pobell Phys Rev.Lett. **51**:2120 (1983);R. M. Bowley and J. R. Owers-Bradley J.Low Temp. Phys. **63**:331,(1986)
8. H. Meyer these proceedings
9. R. Ferrell these proceedings
10. N. R. Brubaker, D. O. Edwards, R. E. Seligmann and R. A. Sherlock, Phys.Rev.Lett. **25**:715(1969); J.Low Temp.Phys.**3**:619,(1970)
11. R. M. Bowley J.Low Temp Phys. **61**:291(1985);**71**:319(1988)
12. D. S. Greywall Phys.Rev.Lett.**41**:177(1978)
13. J. R. Owers-Bradley, P. C. Main, R. M. Bowley, G. J. Batey and R. J. Church J.Low Temp.Phys. **72**:201(1988)
14. D. Candela, D. R. McAllaster, L-J. Wei and G. A. Vermeulen Preprint
15. H. H. Fu and C. J. Pethick Phys.Rev.B **14**:3837(1976)
16. G. Baym J.Low Temp Phys. **18**:335(1975)
17. G. Baym and C. Ebner Phys.Rev.**164**:235(1967);G. Baym and W. F. Saam Phys.Rev. **171**:172(1968); G. Baym,W. F. Saam and C. Ebner Phys. Rev. **173**:306(1969)
18. R. L. Rosenbaum, J. Landau and Y. Eckstein J.Low Temp.Phys **16**:131(1974)
19. I. N. Adamenko and E. Ya. Rudavskii Sov.J.Low Temp.Phys **13**:1,(1987)
20. D. S. Greywall Phys.Rev.Lett. **42**:1758,(1979)
21. G. McHale, M. A. Jones and R. M. Bowley J.Phys. C:Solid state Phys. **21**:5757(1988)
22. L. P. J. Husson and R. de Bruyn Ouboter Physica **122B**:201(1983)
23. R. J. Church J. R. Owers-Bradley and P. C. Main unpublished
24. D. S. Greywall and M. A. Paalanen Phys Rev.Lett. **46**:1292,(1981)
25. J. R. Owers-Bradley, P. C. Main, R. J. Church, T. M. M. Hampson, G. McHale and R. M. Bowley Phys.Rev.Lett. **61**:1619,(1988)
26. T. M. Hampson, R. M. Bowley, D. Brugel and G. McHale J.Low Temp.Phys. **73**:333(1988)
27. V. Silin Zh.Eksp.Teor.Fiz.**33**,726(1958)[Sov.Phys.JETP **6**:945,(1958)]
28. N. Masuhara, D. Candela, D. O. Edwards, R. F. Hoyt, H. N. Scholtz and D. S. Sherrill Phys. Rev. Lett. **53**:1168(1984); D. Candela, M. Masuhara, D. S. Sherrill and D. O. Edwards, J. Low Temp. Phys. **63**:369 (1986)
29. H. Ishimoto, H. Fukuyama, N. Nishida, Y.Miura, Y. Takano, T. Fukuda, T. Tazaki and S.Ogawa Phys.Rev.Lett. **59**:904(1987)
30. H. Akimoto, T. Kawae, G. Oh, O. Ishikawa, T.Hata, T.Kodama, and T.Shigi, Jpn.Jour. Appl. Phys. **S26-3**:65(1987)
31. E. S. Murdock, K. R. Mountfield and L. R. Corruccini J.Low Temp. Phys. **31**:581(1978)
32. K. S. Bedell Phys.Rev.Lett **62**:167(1989)
33. A. J. Leggett and M. J. Rice Phys Rev.Lett. **20**:586(1968) and **21**:506 (1968); A. J. Leggett, J.Phys C**3**,448(1970)
34. W. J. Gully and W. J. Mullin Phys Rev.Lett. **52**:1810(1984)
35. Y. Hirayoshi, G. Eska, G. Bonfait, A. S. Greenberg, K. Neumaier and D. Thoulouze Proceedings of LT17:1229(1984); D. Einzel, G. Eska, Y. Hirayoshi, T. Kopp and P. Wolfle, Phys Rev.Lett. **53**:2312(1984)
36. A. S. Bedford, R. M. Bowley, D. Wightman and J. R. Owers-Bradley Proceedings of LT 19
37. K. D. Ivanova and A. E. Meyerovich Sov.Phys. JETP **64**:964,(1986)
38. A. E. Meyerovich Phys.Lett.**107A**:177(1985)
39. J. W. Jeon and W. J. Mullin Phys.Rev.Lett. **62**:2691(1989); J.Phys. (Paris) **49**:1691(1988)
40. G. Nunes jr.,C. Jin, A. M. Putnam and D. M. Lee Bull.Am.Phys.Soc. **35**,1078,(1990)
41. L. Levy Phys. Rev. B**31**:7077(1985)

TRANSPORT PROPERTIES OF VERY DILUTE SUPERFLUID MIXTURES OF ^{3}He IN ^{4}He

H. Meyer, J. Tuttle, and F. Zhong

Department of Physics
Duke University, Durham, N.C. 27706

INTRODUCTION

The theory of transport properties of superfluid helium mixtures, based on the two fluid model, has been developed more than thirty years ago [1-3]. One of the tests of this theory was by heat conductivity experiments [4], performed with dilute mixtures of ^{3}He in the ^{4}He with $1.4 \times 10^{-4} \leq X \leq 1.3 \times 10^{-2}$, where X is the ^{3}He mole fraction. For the range $X \leq 10^{-2}$ the predictions become particularly simple and were confirmed by Ptukha's experiments[4]. However more recent experiments with more dilute mixtures [5-7] showed the heat conductivity κ_{eff} to depart from predictions for $X \leq 5 \times 10^{-4}$. Furthermore the thermal relaxation times τ also showed an anomalous behavior [6,7]. This stimulated further measurements of κ_{eff}, τ and also of the thermal diffusion ratio k_T^* which we report in this paper.

Measurements on very dilute mixtures offer a number of experimental challenges and these will be briefly discussed below. In fluid layers with a height of up to ~ 3 mm, one measures the temperature drop between thermometers embedded in the two metal pieces with parallel flat faces that form the horizontal boundaries of the fluid. Then, one of the most important aspects in the experiment is a careful determination of the thermal boundary resistance, which becomes comparable and even larger than the bulk fluid resistance as X decreases. Close enough to the superfluid transition, which is of interest in the mixture investigations, the boundary resistance has an anomalous behavior, which we have investigated in some detail for nearly pure ^{4}He ($X \simeq 2 \times 10^{-9}$).

We shall first present briefly the predictions of the bulk transport properties for superfluid mixtures and the theory for the anomalous boundary resistance for ^{4}He near the superfluid transition. After a short experimental part we present results and discussions.

THEORY

The complete predictions for ^3He-^4He mixtures that result from the two-fluid hydrodynamics have been described by Khalatnikov [3]. For dilute mixtures with $X \leq 10^{-2}$, where the assumption of ideal mixtures is a good approximation, the equations simplify considerably and certain predictions on measurable quantities are made which can be tested, and where the internal consistency can be checked. We restrict ourselves to the thermal conductivity $\kappa_{eff} = Q/\text{grad}\, T$ and to $k_T^* = -T\, \text{grad}\, c/\text{grad}\, T$ where Q is the heat flux and c is the mass concentration where $c \simeq 3X/4$ for dilute mixtures. The predictions are in the limit $Q \rightarrow 0$ and are given by

$$\kappa_{eff} = \kappa + \frac{\rho k_B D_{iso} m_4}{m_3^2} \frac{\left(k_T^*\right)^2}{X} \tag{1}$$

$$k_T^* = \frac{\sigma_4 m_3}{k_B} + \frac{m_3}{m_4} X \tag{2}$$

Here ρ is the mass density, m_3 and m_4 are the isotopic masses per particle, κ is the diffusive conductivity and the second term in Eq. 1 is the contribution to the heat flow from mass counterflow. Also D_{iso} is the mass diffusion coefficient of isolated ^3He impurities in the ^4He and which is independent of X. Finally σ_4 is the entropy per gr of ^4He, accurately measured as a function of T [8]. For high enough temperatures at small enough X, one has $\kappa_{eff} \gg \kappa$ and also $\sigma_4 m_3/k_b \gg m_3 X/m_4$. Then $\kappa_{eff} X$ and k_T^* are only functions of T.

The thermohydrodynamic equations by Khalatnikov [3] can be used to calculate the thermal relaxation times τ after the heat current Q across a fluid layer of thickness h is turned on or off [9]. The relation for τ also becomes simple for low enough X. One then obtains

$$\tau = \left(\frac{h}{\pi}\right)^2 \frac{\rho k_B m_4 \left(k_T^*\right)^2}{m_3^2 \kappa_{eff} X} \tag{3}$$

If $\kappa_{eff} \gg \kappa$, one can substitute Eq. 1 into Eq. 3 and one obtains

$$\tau = \left(\frac{h}{\pi}\right)^2 D_{iso}^{-1} \tag{4}$$

Hence τ, like k_T^* is only dependent on T. Furthermore D_{iso} can be determined from both κ_{eff} and from τ. Hence, the internal consistency for the determination of D_{iso} can be checked. The complete expressions for k_T^* and τ when the approximations made in Eqs. 2 and 3 break down have been presented elsewhere[3,9] and have been used below in the comparison of theory and experiments for higher X and lower T.

We now briefly discuss the thermal transport across the solid-liquid boundaries in the conductivity cell. It is well known [10] that there is a boundary resistance resulting from the phonon mismatch between the solid and the liquid phases which we denote by $R_b^o(T)$, and which is a smooth function of temperature. An additional boundary resistance ΔR_b in pure superfluid ^{4}He arises as T_λ is approached [11,12]. It results from the boundary layer of the fluid of a thickness comparable with the correlation length ξ. In that layer the superfluid density ρ_s decreases from its bulk value to zero at the boundary. Then the heat transport by counterflow of the bulk fluid has to be converted into the smaller diffusive transport in this boundary layer. An order of magnitude calculation [11] for $Q \rightarrow 0$ then shows that $\Delta R_b \approx \xi/\kappa$. One assumes that the diverging κ in the superfluid phase is the same as κ in the normal phase at the same reduced temperature $|\varepsilon|$ where $\varepsilon = (T-T_\lambda)/T_\lambda$. Since $\xi = \xi_o\, \varepsilon^{-\upsilon}$ with $\upsilon = 0.67$ and $\xi_o = 1.4\times10^{-8}$ cm, and $\kappa = \kappa_o\, \varepsilon^{-\psi}$ with $\psi \simeq 0.44$ and $\kappa_o = 1.28\times10^{-2}$ mW/cmK, one obtains

$$\Delta R_b \approx \frac{\xi_o}{\kappa_o} |\varepsilon|^{-0.23} \approx 1\times10^{-3} |\varepsilon|^{-0.23} \text{ cm}^2\text{K/Watt} \qquad 5)$$

A Renormalization group calculation [13] for $Q \rightarrow 0$ has determined ΔR_b more exactly, and still more recently Ferrell [14] has developed another theoretical approach. The weakly diverging ΔR_b has been detected in the linear regime, namely the limit $Q \rightarrow 0$ by Duncan et al. [15] and also in the weakly Q-dependent regime at higher Q by Dingus et al. [16] and Zhong et al. [17]. So far theory has not dealt with this non-linear aspect. However the surprising discovery [15,16,17] has been a more strongly diverging and stronger Q-dependent anomalous thermal resistance as T_λ is approached still further. There is no explanation so far for this additional resistance.

EXPERIMENTAL PART

Simultaneous measurements of κ_{eff}, k_T^* and of τ were carried out in a specially designed conductivity cell (cell B). When a current was passed through the fluid layer, temperature changes across this layer were measured as well as the ^{3}He concentration changes which were recorded from nearly isothermal density measurements. The density of fluid at the top boundary was probed by a capacitor situated above the top plate and immersed in fluid connected with the main bath via small channels. The arrangement, the experimental procedures and the analysis are described in Ref. 17. The parallel surfaces bounding the fluid were of gold plated copper, and the fluid layer thickness was h = 1.81 mm. In a different conductivity cell of conventional design (cell A), spacers of variable length could be introduced between parallel copper plates which were lapped but not gold plated. The spacing used so far was for mixture experiments was 1.47 mm. Presently measurements are in progress with h=0.97 mm.

The total temperature drop measured across the cells when a heat current was switched on is given by $\Delta T = \Delta T_b + \Delta T_{liq}$. Here ΔT_b includes the drop across the copper end pieces and the liquid-solid boundary resistance, while ΔT_{liq} is the drop across the bulk fluid layer. In a series of experiments, measurements were first conducted with nearly pure ^{4}He ($X = 2\times10^{-9}$) where $\Delta T_{liq} = 0$, to determine ΔT_b as a function of the reduced temperature $|\varepsilon|$ and for various heat currents. Metered amounts of ^{3}He were then added to the fluid to form mixtures of progressively increasing X. From measurements with very diluted mixtures, it was found that $\Delta T_b/Q$ was nearly independent of X, except possibly very close to T_λ where the anomalous behavior was found. Hence $\Delta T_b/Q$ for ^{4}He was used in the determination of $\Delta T_{liq}/Q$ in mixtures. The ratio $\Delta T_b/\Delta T_{liq}$ is found to decrease with increasing X.

A heat current Q across a fluid mixture layer induces a concentration difference between the cold (c) and warm(w) boundaries $\Delta X = (X_c - X_w) \propto k_T^* Q \kappa_{eff}^{-1}$. When ΔX becomes comparable with the average fluid concentration X, non-linear effects will become noticeable, for instance the measured conductivity $Q/\Delta T_{liq}$ will be an increasing function of Q. The heat flux Q_c above which a sharp increase in non-linear behavior takes place is found[6,17] to be roughly proportional to $X^{0.8}$. This means that for measurements restricted to the linear regime, the maximum permissible heat currents become severely limited as X decreases. This in turn implies a decrease in the accuracy of the measurements. Because in addition κ_{eff} increases with decreasing X, the conventional germanium thermometry which resolves $\delta T \sim 3\text{x}10^{-7}$K becomes inadequate for measurements in mixtures with $X < 10^{-6}$. More sensitive thermometry with nano degree resolution developed recently[15,18] would be able to probe mixtures with $X > 10^{-8}$, assuming that the trend observed in κ_{eff}[5,6,7,17] continues for $X < 10^{-6}$.

RESULTS AND DISCUSSIONS

We will first discuss the transport measurements in mixtures where it is understood that the results were obtained in the linear regime for $Q < Q_c$. Secondly, we briefly discuss the observations for $Q > Q_c$. We then present results of the anomalous boundary resistance in ^{4}He which is only detected for $\varepsilon \le 10^{-3}$, at least at heat currents up to 100 μW/cm^2.

Mixtures, linear regime

In Figure 1, we show the conductivity results taken with cell B with h_B = 1.81 mm, plotted versus X along paths at

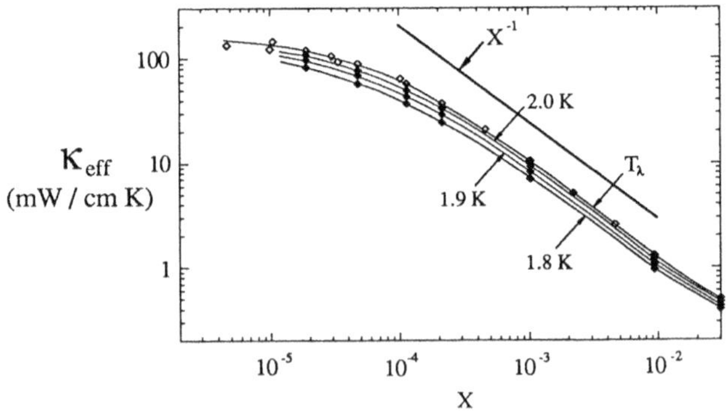

Fig. 1 The conductivity κ_{eff} versus X along paths of constant reduced temperature $|\varepsilon|$. Solid diamonds: isotherms. Open diamonds: lambda line $T_\lambda(X)$. The solid curves are a guide to the eye through the data (symbols). The straight line indicates $\kappa_{eff} \propto X^{-1}$.

constant T and along the lambda line T_λ. (For $X<10^{-2}$, T_λ is constant within 14 mK) As can be seen, for $X \leq 10^{-3}$ the curves show a systematic departure from the $\kappa_{eff} \propto X^{-1}$ prediction indicated by a straight line. To avoid overcrowding, we have not presented the results of cell A ($h = 1.47$ mm) that show an even stronger departure from predictions [6], nor the data by Putkha [4] with $h = 6$ mm that show no departure down to $X = 1.4 \times 10^{-4}$, her most dilute mixture.

Similarly Fig 2 shows the results for k_T^* versus X along paths of constant $|\varepsilon|$. Here the symbols label the interpolated data and the dashed curves are a guide to the eye. The solid lines are the predictions from theory. The dilute mixture approximation, Eq. 2 is valid only for $X \leq 10^{-2}$. Here again, the largest systematic differences between the data and the predictions occur for $X \leq 10^{-3}$.

Fig. 3a shows the data[19] for τ in the same representation as in Figs. 1 and 2. Here again the dashed lines are a guide to the eye and the solid ones are the predictions for Eq. 3 where the data of κ_{eff} were extrapolated for $X \leq 10^{-3}$ assuming the predictions $\kappa_{eff} \propto X^{-1}$ to hold. Again there are substantial discrepancies as X decreases below 10^{-3}. We note that these are observed over the whole temperature range, not just near $T_\lambda(X)$. If by contrast we substitute the measured values of κ_{eff} into Eq. 3, we obtain a much improved agreement with the data for τ as shown in Fig.3b. Other discrepancies are noteworthy. For instance we find that the diffusion coefficient D_{iso}, as determined from conductivity and from τ measurements via Eqs. 1 and 4 respectively, to be different by a factor of namely ≈ 1.5, as can be seen in Fig. 4. Furthermore the

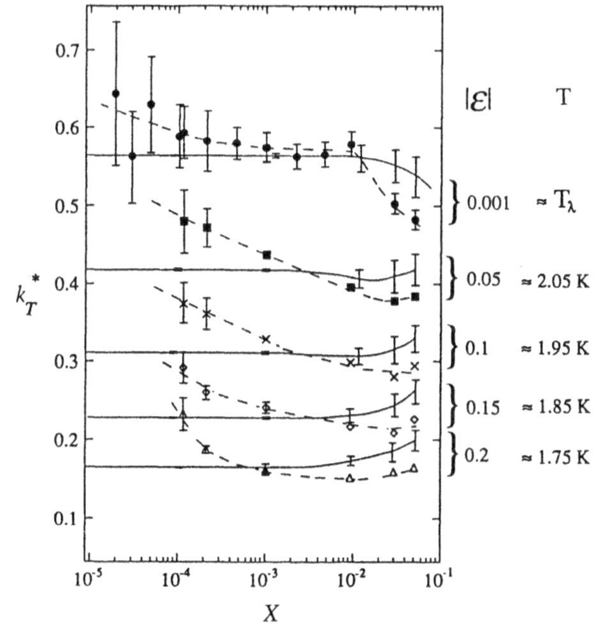

Fig. 2 The effective thermal diffusion ratio k_T^* versus X along paths $|\varepsilon|$ = constant. The dashed curves are a guide to the eye through the data (symbols). The solid lines are the predictions.

relaxation times measured with cells A and B do not scale as h^2, contrary to predictions. By contrast, the τ measured in the normal phase for nearly pure ^{4}He scale as h^2, and are in quantitative agreement with the predictions that use the measured thermal diffusivity $\kappa/\rho C_p$.

<u>Mixtures, non-linear regime</u>

We now discuss measurements as a function of the heat current. As described above, there is a sharp increase of the observed conductivity $Q/\Delta T_{liq}$ as Q increases beyond Q_c. For a given functional dependence of κ_{eff} on X and for a thickness h of the layer, $Q_c(X)$ and the non-linear $Q/\Delta T_{liq}$ have been predicted by Behringer [19] and semiquantitatively confirmed by the experiments [17]. In this calculation, $k_T{}^*$ was assumed to be given by Eq 2. For the heat currents up to 100 $\mu W/cm^2$ used in our experiments, non-linear behavior is observed for $X \leq 5x10^{-5}$. Behringer's theory [19] furthermore shows that if κ_{eff} followed the predicted X^{-1} behavior, the observed $Q/\Delta T_{liq}$ should not show a dependence on Q although the concentration profile in the fluid layer is not linear above a certain Q.

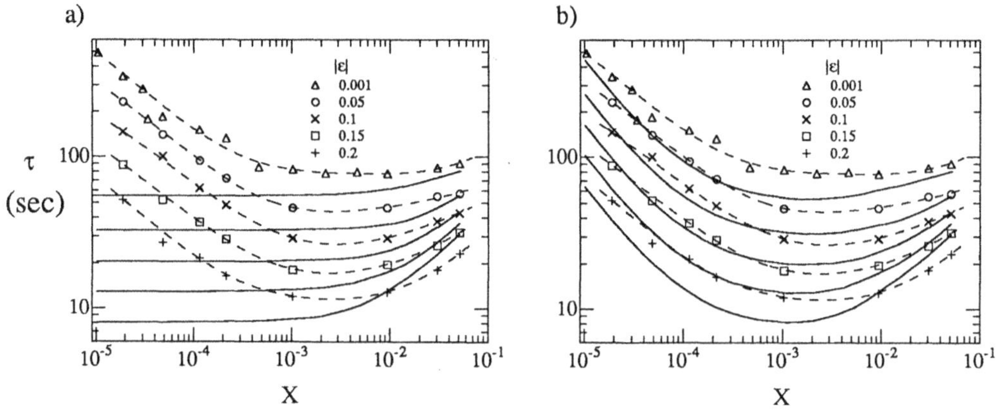

Fig. 3 The relaxation time τ versus X along paths $|\varepsilon|$ = constant. The dashed curves are a guide to the eye through the data (symbols).a) The solid lines are the predictions as explained in the text, where κ_{eff} follows Eq. 1 for $X < 10^{-2}$. <u>b)</u> The solid lines are the predictions using the measured κ_{eff}.

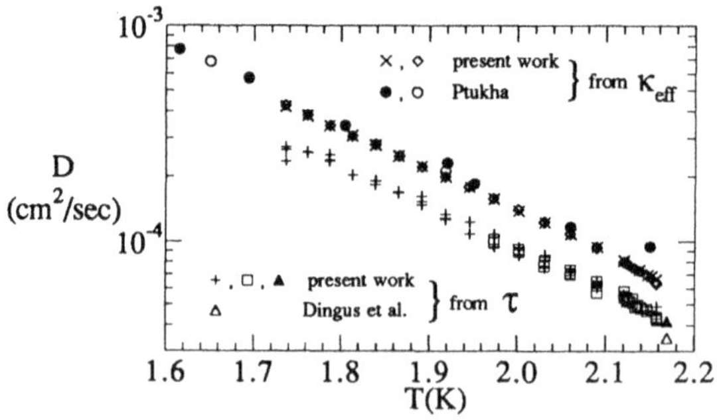

Fig. 4 The mass diffusion coefficient versus T as determined from κ_{eff} and from τ data, respectively. The open and closed triangles result from τ data for cells A and B for $T=T_\lambda$ and for $10^{-3}<X<10^{-2}$.

The observed rise in $Q/\Delta T_{liq}$ is therefore internally consistent with the departure of κ_{eff} from the X^{-1} prediction.
Hence a number of unanswered questions remain in the transport properties of dilute ^3He-^4He mixtures. One might ask whether the fluid dimensions matter, since the experiments from cells A and B are qualitatively but not quantitatively internally consistent. A tentative observation, based on Ptukha's and our data so far seems to be: the smaller the fluid layer thickness, the larger is the departure from the predicted $\kappa_{eff} \propto X^{-1}$. This is surprising because the layer thickness is still much larger than certain relevant length parameters, such as ξ and the average distance between the ^3He impurities. So far two approaches have been proposed to explain the departure of κ_{eff} from the two-fluid predictions [21,22] and we must wait for futher progress in both experiments with different layer heights, and for further theoretical insights.

Boundary resistance

The boundary resistance results were obtained both as a function of temperature at constant heat current, and also as a function of Q at constant $|\varepsilon|$. The temperature dependence of the background resistance shows the expected behavior of increasing with decreasing T. This is seen in Fig 5b as a rise on a semilogarithmic plot of $R_b(|\varepsilon|)$ when $|\varepsilon|$ increases beyond $\sim 10^{-2}$. For smaller $|\varepsilon|$, upon approaching T_λ, R_b passes through a minimum and first shows a weak, then a sharper increase as $|\varepsilon|$ further decreases beyond a critical value $\varepsilon_c(Q)$. Finally at a temperature where the warmer fluid boundary changes from superfluid to normal, there is a general breakdown of the bulk superfluid phase, shown by a very large sudden increase of ΔT. This temperature is marked by an arrow in Figs.5a and 5b. The regime where R_b diverges weakly is identified with that described by Eq.5. The predicted effect is small, however, and has been carefully investigated by Duncan and Ahlers with nanoK resolution [15] for $Q \to 0$. In our experiments the currents needed for an adequate detection are above those appropriate to the linear regime. This might be one reason why our

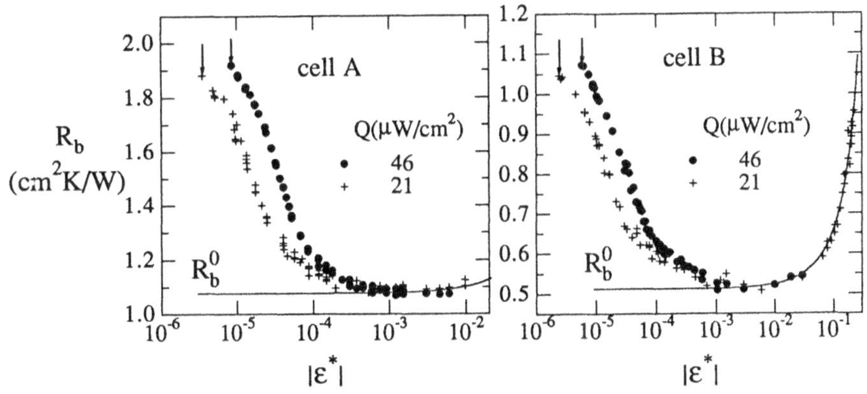

Fig. 5 The boundary resistance R_b in superfluid ^4He versus the reduced temperature $|\varepsilon|$ for two heat currents. a) Cell A with h = 1.47 mm and b) Cell B with h = 1.81 mm. The vertical arrows mark the superfluid breakdown point

results for $\Delta R_b(|\varepsilon|)$ have been consistently higher than the experiments of ref 15 and the predictions [11-13]. It should be noted, however, that Li and Lipa [23] did not detect the weakly divergent ΔR_b in spite of also using nanoK resolution thermometry. Our sensitivity is sufficient for a systematic study of the anomaly at $Q \geq 8\ \mu W/cm^2$. In Fig.5a and b we show R_b taken with cells B (gold-plated surfaces) and A (copper surfaces). We note that at its minimum value near $|\varepsilon| \simeq 10^{-3}$ R_b^o is nearly twice for cell A as for cell B. The same ratio is found likewise for the maximum anomaly ΔR_b^{max} at the superfluid breakdown. In Fig. 6 we plot ΔR_b^{max} for both cells and also for cell C (with h=0.97 mm and copper surface) as a function of Q and note that it is nearly constant. Our ΔR_b^{max} and R_b^o for cell B are ~ 20% larger than that reported by Duncan and Ahlers, who also used gold-plated surfaces. Observations with cell C give R_b^o=0.70 and ΔR_b=0.85 cm^2K/W, the latter consistent with the results for cell A with the same surface. These observations suggest that the magnitude of ΔR_b^{max} might depend on the nature of the surface. In Fig.7 we show a plot of

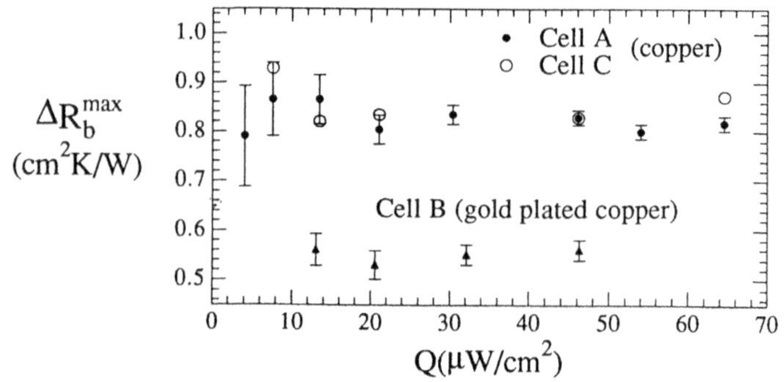

Fig. 6 The maximum boundary anomaly ΔR_b^{max} at the breakdown versus Q for cells A, B and C.

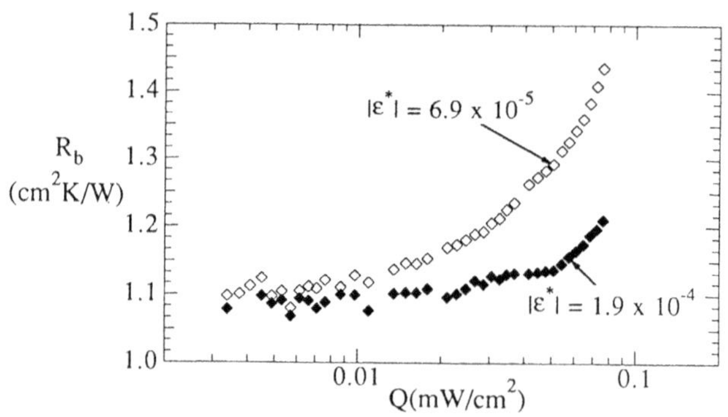

Fig. 7 The boundary resistance R_b versus Q along paths at constant $|\varepsilon|$, where ε is the average temperature of the fluid layer, as taken with cell A.

R_b versus Q for several reduced temperatures with cell A and demonstrate the weak non-linearity at small currents which becomes strong at higher currents and is a function of $|\varepsilon|$.

In conclusion, the strongly divergent boundary resistance anomaly of ^{4}He near the superfluid transition is a complex and interesting phenomenon that remains to be clarified. The results with cells A and C will be published in more detail in the near future. Further experiments are planned.

Acknowledgments

This research was supported by grant DMR 8820479 of the National Science Foundation. The authors acknowledge stimulating discussions with R.P. Behringer, G. Ahlers, R.V Duncan and R.A. Ferrell and the experimental help of D. Murphy.

References

1. F. London Superfluids Vol. II, (Dover Publications, New York, 1954), p. 190 and ff.
2. I. M. Khalatnikov and V. N. Zharkov, Zh. Eskp. Teor. Fiz. **32**, 1108 (1957). [Sov. Phys. JETP, **5**, 905, (1957)].
3. I. M. Khalatnikov, Introduction to the Theory of Superfluids (Benjamin, New York, 1965), p. 149 and ff.
4. T. P. Ptukha, Zh. Eksp. Teor. Fiz. **40**, 1583 (1961) [Sov. Phys. JETP **13**, 1112 (1961)].
5. M. Tanaka, A. Ikushima and K. Kawasaki, Phys. Lett. **61A**, 119 (1977).
6. M. Dingus, F. Zhong, J. Tuttle, and H. Meyer, J. Low Temp. Phys. **65**, 213 (1986).
7. T. C. P. Chui and J. A. Lipa, Jpn. J. Appl. Phys. **26**, (1987) Supplement 26-3.
8. A. Singsaas and G. Ahlers, Phys. Rev. **B29**, 4951 (1984).
9. R. B. Behringer and H. Meyer, J. Low Temp. Phys. **46**, 435 (1982).
10. J. M. Pfotenhauer and R.J. Donnelly, Adv. Heat Transfer **17** 65 (1985).
11. A. Onuki, Prog. Theor. Phys. Supp. **79**, 191 (1984).
12. V. L. Ginzburg and A. A. Sobaynin, Usp. Fiz. Nauk. **120**, 15 (1973) [Sov. Phys. Usp. **19**, 773 (1976)].
13. D. Frank and V. Dohm, Phys. Rev. Lett. **62**, 1864 (1989).
14. R. Ferrell, Proc. International Low Temperature Conference LT19, Brighton Sussex 1990.
15. R. Duncan, G. Ahlers and V. Steinberg, Phys. Rev. Lett. **58**, 377 (1987). R. Duncan and G. Ahlers, Phys. Rev. B (1990) To appear.
16. M. Dingus, F. Zhong and H.Meyer, J. Low Temp. Phys. **65**, 185 (1986).
17. F. Zhong, J.Tuttle and H.Meyer,J Low Temp. Phys.**49**,9(1990)
18. T. C. P. Chui and J. A. Lipa in Proc. LT17, 1984, edited by U. Eckern, A. Schmid, W. Weber and H. Wuehl (North Holland, Amsterdam, 1984), p. 932.
19. J.Tuttle, F. Zhong and H. Meyer, J. Low Temp. Phys (to be published)
20. R.P. Behringer, J.Low Temp. Phys. **62**, 15 (1986).
21. T. C. P. Chui and J. A. Lipa, Phys. Rev. **A40**, 4306 (1989).
22. R. A. Ferrell. Proc. NATO Advanced Research Workshop, Excitations in 2D and 3D Quantum Fluid, Exeter, 1990.
23. Q.Li, T.C.P. Chui and J.A. Lipa, Bull., Am. Phys. Soc. **33**, 1373 (1988).

ANOMALOUS THERMAL CONDUCTIVITY IN VERY DILUTE ^3He-^4He SOLUTIONS

Richard A. Ferrell

Center for Theoretical Physics and Center for Superconductivity Research, Department of Physics and Astronomy, University of Maryland, College Park, MD 20742

ABSTRACT

The measurements of Meyer's group[1] at Duke University have revealed an unexpected deviation from Khalatnikov's theory of very dilute ^3He-^4He solutions. The theoretical prediction is that κ_{eff}, the effective thermal conductivity in the superfluid phase, should diverge as the reciprocal of the X, ^3He mole fraction. In contrast, Meyer and coworkers have found κ_{eff} leveling off at a finite value and apparently becoming nearly independent of X. This finding was corroborated by one measurement by Lipa and Chui at the very low value of $X = 10^{-7}$. We have been able to account phenomenologically for this extremely puzzling behavior by means of a "3-fluid model," which postulates that there is an additional normal component that interacts both with the ^3He and with the gas of excitations in the liquid ^4He (i.e., with the usual normal component). By fitting the parameters to the observed plateau, the 3-fluid model predicts that κ_{eff} should rise again for even lower concentrations, in the range $X < 10^{-7}$. Although difficult, this test ought to be feasible with the advanced thermometric techniques that are now available. The search for the predicted rise at $X < 10^{-7}$ and the development of "quasi-Khalatnikov" behavior at the ultra-low concentrations will be a crucial test for the 3-fluid hypothesis.

I. INTRODUCTION

Dingus, Zhong, Tuttle, and Meyer[1] have reported an anomalous and unexpected behavior of the transport properties of ^3He-^4He mixtures at very low concentration.[2] The increase in κ_{eff}, the effective thermal conductivity, upon decreasing X, the mole fraction of ^3He, falls markedly below the theoretically predicted X^{-1} dependence. A similar discrepancy has been reported by Chui and Lipa.[3] To date there has been, in our opinion, no satisfactory explanation of these remarkable observations. We are therefore motivated in this paper to explore a phenomenological explanation based on the qualitative and intuitive idea that, at the very low concentrations under consideration, the ^3He atoms must remain free and mobile. It follows that the observed discrepancy must simply be revealing the setting in of some addition drag force acting on the ^3He atoms. To implement this qualitative idea we hypothesize a third component of the ^4He host fluid, over and above the usual super- and normal components. The consequences of this postulate are worked out in Section V, where we see that a crucial prediction is that, in the ultra-low mole fraction regime, $X < 10^{-7}$, a "quasi-Khalatnikov" behavior of κ_{eff} should set in. In this region κ_{eff} would again be inversely proportional to X, but with a greatly reduced magnitude.

Because of the drastic and unconventional nature of the 3-fluid hypothesis we first set the stage by reviewing, in a rather didactic fashion, the most basic aspects of the problem. We hope, in this way, to present a compelling case for the subsequent introduction of the 3-fluid hypothesis. Section II is devoted to the underlying thermodynamics of very dilute ^3He-^4He solutions, while the steady state response to a heat current is studied in Section III. In Section IV the analysis of the diffusive relaxation of a concentration inhomogeneity yields strong evidence that the observed phenomenon is a true bulk, rather than surface, process. Following the working out of the 3-fluid hypothesis in Section V, we conclude in Section VI with a brief summary.

II. THERMODYNAMICS

Because of the surprising discrepancy between the predictions of Khalatnikov's theory and the observed behavior of very dilute ^3He - ^4He solutions,[1] it is well to review the very most basic principles that are applicable to such solutions. At low concentrations the ^3He atoms can be regarded as a perfect gas whose entropy is related to T, V, and N, the temperature, volume, and number of atoms, by

$$dS = Nc_V \frac{dT}{T} + Nk_B \frac{dV}{V} , \tag{2.1}$$

where k_B is Boltzmann's constant and c_V is the constant volume specific heat per atom. On a, perhaps, even more basic level, Eq. (2.1) is equivalent, by virtue of the perfect gas law, $PV = Nk_BT$, to the first law,

$$dU = Nc_VdT = TdS - PdV , \tag{2.2}$$

where P and U are the pressure and energy content of the gas, respectively. Integration of Eq. (2.1) yields

$$S = Nc_V \ln T + Nk_B \ln V + \Sigma(N) , \tag{2.3}$$

In order for S to be an extensive variable, the constant of integration is required to depend upon N according to

$$\Sigma(N) = - Nk_B \ln N + Nk_B \Sigma_o , \tag{2.4}$$

where Σ_o is independent of N. Substitution of $c_V = \frac{3}{2} k_B$ and Eq. (2.4) into Eq. (2.3) yields the familiar expression

$$\frac{S}{Nk_B} = \frac{3}{2} \ln T + \ln V - \ln N + \Sigma_o$$

$$= \frac{3}{2} \ln T - \ln n + \Sigma_o , \tag{2.5}$$

where $n = N/V$ is the atomic density. Eq. (2.5) can, of course, be derived directly from statistical mechanics. The ln V term comes from the number of different locations that can be occupied by any given atom; the ln N term comes from the indistinguishability of the atoms and is related to the Gibbs paradox.

By setting N equal to N_3, the number of ^{3}He atoms dissolved in a much larger number $N_4 \gg N_3$ of ^{4}He atoms, Eq. (2.5) describes S_3, the entropy of mixing. With an energy of mixing of ε_3 per ^{3}He atom, the total Gibb's function for the system is

$$G(T,P,N_4,N_3) = G(T,P,N_4,0) + N_3\varepsilon_3 - TS_3 \; , \qquad (2.6)$$

neglecting interaction between the ^{3}He atoms. The dependence of the chemical potential of the ^{4}He atoms on the mole fraction

$$X = \frac{N_3}{N_4} \qquad (2.7)$$

is obtained from

$$\xi_4(T,P,X) = \frac{\partial G}{\partial N_4} = \xi_4(T,P,0) - T\,\frac{\partial S_3}{\partial N_4} \; . \qquad (2.8)$$

For $X \ll 1$ the interaction between the ^{3}He atoms can also be neglected in calculating the total volume of the solution,

$$V = N_4 v_4 + N_3 v_3 \; , \qquad (2.9)$$

where the partial volumes v_4 and v_3 are constants. Thus,

$$\frac{\partial \ln V}{\partial N_4} = \frac{v_4}{N_4 v_4 + N_3 v_3} = \frac{N_4^{-1}}{1 + \frac{v_3}{v_4} X} \; . \qquad (2.10)$$

Substituting Eq. (2.10) into Eqs. (2.5) and (2.8) yields

$$\frac{\partial S_3}{\partial N_4} = N_3 k_B \frac{\partial \ln V}{\partial N_4} = k_B \frac{X}{1 + \frac{v_3}{v_4} X} \simeq k_B X \qquad (2.11)$$

and

$$\xi_4(T,P,X) = \xi_4(T,P,0) - k_B T X \; . \qquad (2.12)$$

The negative gradient of ξ_4 constitutes an accelerating force for the superfluid. For the latter to remain in steady state motion, in the absence of any pressure gradient, requires

$$0 = -\,\mathrm{grad}\,\xi_4 = \sigma_4 k_B\,\mathrm{grad}\,T + k_B T\,\mathrm{grad}\,X \; , \qquad (2.13)$$

or

$$\frac{1}{T}\,\mathrm{grad}\,T = -\,\sigma_4^{-1}\,\mathrm{grad}\,X \; . \qquad (2.14)$$

Here we have used

$$\left(\frac{\partial \xi_4^{(0)}}{\partial T}\right)_P = -\,\sigma_4 k_B \; , \qquad (2.15)$$

where σ_4 is the entropy per ^{4}He atom of the pure fluid, measured in units of k_B. Near the lambda point $\sigma_4 \simeq 0.6$, so that the proportionality coefficient in Eq. (2.14) is $\sigma_4^{-1} \simeq 1.7$. This inescapable, and quite rigorous, linear relationship connecting the temperature and concentration gradients has recently been experimentally verified over several decades of low concentration.[4] Because of Eq. (2.14), any measurement of grad T can be regarded as essentially a direct measurement of grad X (which is generally a less accessible observable). For this reason, most of the remaining work in this paper will be couched in terms of the linear response of grad X to an imposed heat current as well as the transient behavior of grad X that occurs when the heat current is turned on or off.

From Eq. (2.6) we can also obtain the chemical potential of the ^{3}He atoms,

$$\xi_3(X) = \frac{\partial G}{\partial N_3} = \varepsilon_3 - T\,\frac{\partial S_3}{\partial N_3}\,, \tag{2.16}$$

where

$$\frac{\partial S_3}{\partial N_3} = \left(\frac{3}{2}\ln T + \Sigma_0 - 1\right)k_B - k_B \ln n_3\,. \tag{2.17}$$

Substitution of Eq. (2.17) into Eq. (2.16) and further differentiation yields the inverse susceptibility, or "thermodynamic stiffness"

$$\chi^{-1} = \frac{\partial \xi_3}{\partial n_3} = k_B T\,\frac{\partial \ln n_3}{\partial n_3} = \frac{k_B T}{n_3} = \frac{k_B T}{n_4}\,X^{-1}\,. \tag{2.18}$$

The chemical potential gradient that is required to sustain a gradient in X is therefore

$$\mathrm{grad}\ \xi_3 = \frac{k_B T}{X}\,\mathrm{grad}\ X\,. \tag{2.19}$$

III. HEAT FLOW

Below the lambda point the normal fluid, consisting of a gas of interacting excitations, carries the entropy density $n_4\sigma_4 k_B$. A drift velocity v_n results in a heat current density

$$Q = n_4\sigma_4 k_B T v_n\,. \tag{3.1}$$

As a result, the individual ^{3}He atoms, at low concentration, will experience a drag force

$$F_3 = \mu_3^{-1} v_n = (n_4\sigma_4 k_B T)^{-1}\,\frac{Q}{\mu_3}\,, \tag{3.2}$$

where μ_3 is the mobility for an isolated ^{3}He atom. The effect of F_3 on the ^{3}He distribution is illustrated in Fig. 1. Consider that the system is bounded by horizontal plates, separated by height h. For convenience in visualization we imagine that the heat current Q is introduced at the upper plates and flows in the negative z-direction. This heat current produces a downwards drag force, of magnitude given by Eq. (3.2), which is balanced by the chemical potential gradient

$$\frac{d\xi_3}{dz} = -\,F_3\,. \tag{3.3}$$

Substitution from Eq. (2.19) yields

$$X^{-1}\,\frac{dX}{dz} = \frac{d\ln X}{dz} = -\,\frac{F_3}{k_B T}\,, \tag{3.4}$$

which, integrated, gives the familiar barometric profile

$$X(z) = \text{const.} \exp\left(-\frac{F_3 z}{k_B T}\right) , \tag{3.5}$$

as sketched in Fig. 1.

It is useful to carry out an alternative integration of Eq. (3.4). Written in differential form, Eq. (3.4) becomes

$$dX = -\frac{F_3}{k_B T} X \, dz . \tag{3.6}$$

Integrating from $z = 0$ to $z = h$ and imposing ^3He conservation, which requires $\langle X \rangle$, the mole fraction averaged over the entire profile to equal X_0, the $Q = 0$ unperturbed value, we obtain

$$X(h) - X(0) = -\frac{F_3 h}{k_B T} \langle X \rangle = -\frac{F_3 h}{k_B T} X_0 . \tag{3.7}$$

Dividing by h gives the average gradient as

$$\left\langle \frac{dX}{dz} \right\rangle = -\frac{F_3}{k_B T} X_0 , \tag{3.8}$$

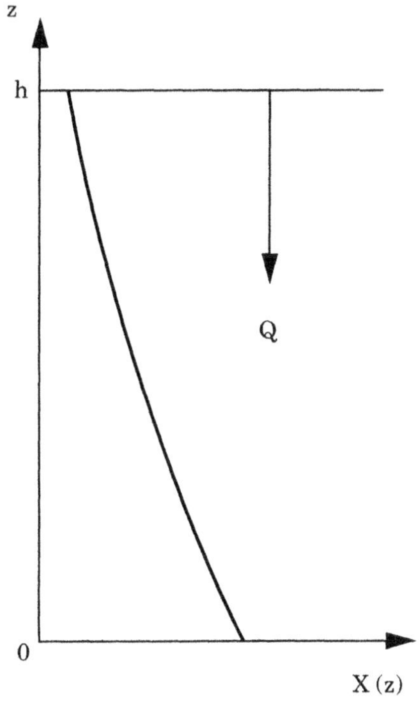

Fig. 1. Barometric profile for the space-dependent mole fraction, X(z) for a heat current Q flowing downwards in a ^3He-^4He solution of thickness h.

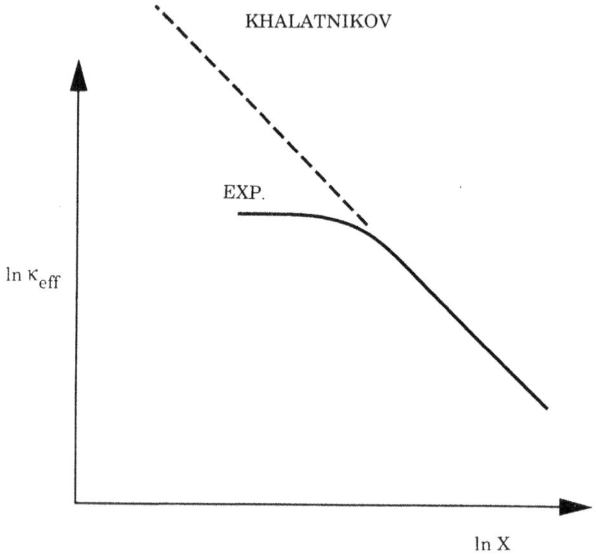

Fig. 2. Observed deviation at low mole fraction of κ_{eff}, the effective thermal conductivity, from Khalatnikov theory.

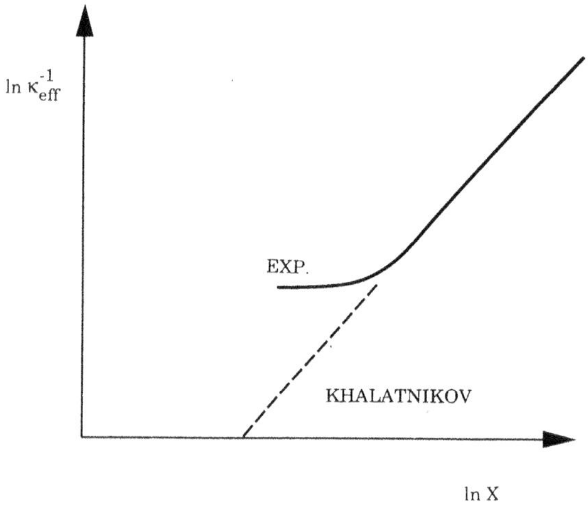

Fig. 3. Deviation at low mole fraction of the temperature response function, κ_{eff}^{-1}, from Khalatnikov theory.

which can be expressed in terms of the linear response function

$$R \equiv - Q^{-1} \langle\frac{dX}{dz}\rangle . \tag{3.9}$$

Substituting Eqs. (3.2) and (3.8) into Eq. (3.9) yields

$$R = \frac{(k_BT)^{-2}}{n_4\sigma_4} \frac{X_0}{\mu_3} . \tag{3.10}$$

The concentration response in Eq. (3.9) is readily translated into the temperature response by means of Eq. (2.14). It is conventional to describe the linear relationship between Q and the resulting temperature gradient in terms of the effective thermal conductivity

$$\kappa_{eff} \equiv \frac{Q}{\langle dT/dz\rangle} = \frac{\sigma_4}{TR} = n_4\sigma_4^2 k_B^2 T \frac{\mu_3}{X_0} , \tag{3.11}$$

a formula that follows from Khalatnikov's more complete treatment of the problem, in the limiting case $X_0 \ll 1$. The dashed line in Fig. 2 illustrates the inverse proportionality to X_0 that is predicted by Eq. (3.11). The unexpected deviation of κ_{eff} from the Khalatnikov ideal behavior that is observed experimentally[1] is indicated by the solid curve.

There is some conceptual advantage in dealing with this problem in terms of R, or, equivalently, in terms of κ_{eff}^{-1}, the concentration dependence of which is illustrated in Fig. 3. From this point of view, the proper relationship of cause to effect ensues from the fact that the primary agent is the heat flow, Q, which is controlled experimentally. The observations consist in measuring the gradient that is established in the presence of Q. Thus, as indicated by the dashed line in Fig. 3, the Khalatnikov prediction is not that there is a divergence, but, rather, that the observable effect produced by Q should vanish as $X \to 0$. This is, of course, eminently reasonable in view of the superfluid property of pure ^{4}He, viz. $Q \neq 0$ in the absence of any gradient. The observed deviation suggests that some effect is setting in at very low concentration that hinders the disappearance of the effect. This is the motivation for the 3-fluid model that is presented in Section V below.

IV. MASS DIFFUSION

The only unknown factor in R, the linear response function of Eq. (3.10), is the drag coefficient, μ_3^{-1}. It is therefore fortunate that this transport coefficient can be measured in another way, independently of the steady-state heat flow experiment. By studying the transient behavior of the solution when the heat current is suddenly turn off or on, Dingus et al.[1] have been able to infer the value of μ_3^{-1}. Suppose that at time $t = 0$ the heat current that produces the concentration profile depicted in Fig. 1 is suddenly switched off. The chemical potential gradient, previously balanced by the drag exerted by the normal fluid, now drives the ^{3}He current density

$$\underset{\sim}{J}_3 = - n_3 \mu_3 \,\text{grad}\, \xi_3 . \tag{4.1}$$

Substituting Eq. (2.19) gives

$$\underset{\sim}{J}_3 = - n_4 k_B T \mu_3 \,\text{grad}\, X . \tag{4.2}$$

Conservation of ^{3}He determines the change in the local value of the mole fraction as

$$\frac{\partial X}{\partial t} = - \frac{1}{n_4} \operatorname{div} \underset{\sim}{J}_3 = k_B T \, \mu_3 \, \nabla^2 X \, , \tag{4.3}$$

which will be recognized as the standard diffusion equation, with the diffusion coefficient identified as

$$D = k_B T \, \mu_3 \, , \tag{4.4}$$

in accord with Einstein's relation.

The initial condition for the diffusion process, for a profile only slightly perturbed from equilibrium, is proportional to

$$H(z,0) = 1 - \frac{2z}{h} \, . \tag{4.5}$$

The normal modes are

$$\cos k_\ell z = \cos \frac{\ell\pi}{h} z \, , \tag{4.6}$$

associated with the relaxation rates

$$\gamma_\ell = Dk_\ell = \frac{\pi^2 D}{h^2} \ell^2 \, , \tag{4.7}$$

where ℓ is a positive odd integer. It is convenient to introduce the dimensionless time

$$\tau = \frac{\pi^2 D}{h^2} t \, , \tag{4.8}$$

based on the fundamental. The expansion of the relaxing concentration profile in terms of the normal modes is consequently

$$H(x,t) = \frac{8}{\pi^2} \sum_{\ell \text{ odd}} \ell^{-2} e^{-\ell^2 \tau} \cos \frac{\ell \pi z}{h} \, . \tag{4.9}$$

Of special interest is the transient behavior of the excess mole fraction at the lower boundary, as given by

$$F(\tau) \equiv H(0,t) = \frac{8}{\pi_2} \sum_{\ell \text{ odd}} \ell^{-2} e^{-\ell^2 \tau} \, , \tag{4.10}$$

and as illustrated by Fig. 4. Truncation of the series after the first term engenders a fractional error of the order of

$$3^{-2} \frac{e^{-9\tau}}{e^{-\tau}} = \frac{1}{9} e^{-8\tau} \, , \tag{4.11}$$

which is less than two-tenths of one-percent for $\tau \geq 0.5$. The dashed line in Fig. 4 shows $(8/\pi^2)e^{-\tau}$, the contribution of the fundamental to $F(\tau)$.

For the interval $0 \leq \tau \leq 0.5$ the function $F(\tau)$ is more conveniently evaluated by Poisson summation, which is equivalent to reflecting $H(z,0)$ into the region $- h \leq z \leq 0$ as

$$H(z,0) = 1 - \frac{2|z|}{h} , \tag{4.12}$$

and then continuing this function periodically outside the symmetric interval $-h \le z \le h$. With the Green's function

$$G(z-z';t) = (4\pi Dt)^{-1/2} \exp - \frac{(z-z')^2}{4Dt} \tag{4.13}$$

we find

$$H(z,t) = \int_{-\infty}^{\infty} dz' \, G(z-z';t) \, H_p(z',0) , \tag{4.14}$$

where $H_p(z,0)$ is the periodic extension of Eq. (4.12). For $z = 0$ and for τ not too large, the Green's function samples $H_p(z',0)$ only near the origin. This permits replacing $H_p(z',0)$ by $H(z',0)$, giving

$$F(\tau) \simeq \int_{-\infty}^{\infty} dz \, G(z;t) \, H(z,0)$$

$$= 1 - \frac{4}{h} \int_{0}^{\infty} dz \, z \, G(z;t)$$

$$= 1 - \frac{4}{h} \frac{2Dt}{(4\pi Dt)^{1/2}} = 1 - \frac{4\sqrt{\tau}}{\pi^{3/2}} , \tag{4.15}$$

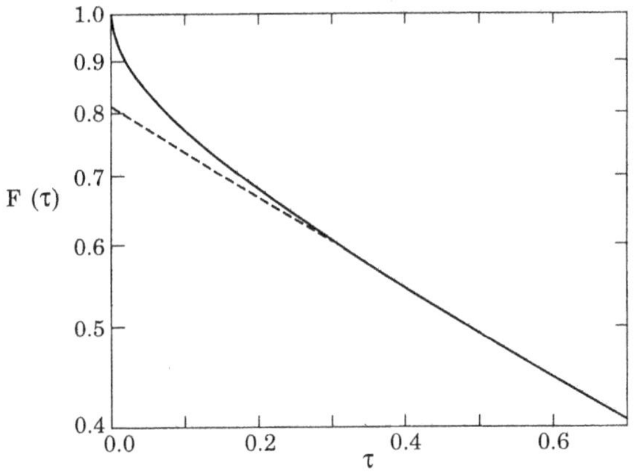

Fig. 4. Diffusive relaxation of the boundary temperature versus reduced time τ. The dashed line indicates the contribution of the fundamental mode to the relaxation process.

with an error of less than 0.2 percent in the interval $0 \le \tau \le 0.5$. The abrupt drop in $F(\tau)$ at $\tau = 0$ is evidently a feature of the transients reported by Dingus et al.[1] This is strong evidence that the relaxation that is taking place in the dilute solutions is indeed a bulk diffusion process rather than some surface phenomenon.

The relaxation of a strongly perturbed concentration distribution, produced by larger heat currents than envisaged in the treatment just above, offers the possibility of further experimental verification of the bulk nature of the diffusion process. According to Eq. (3.5), the mole fraction has generally the spatial dependence exp $(-2\alpha z/h)$, where

$$\alpha = \frac{hF_3}{2k_BT} , \tag{4.16}$$

and F_3 is expressed in terms of the heat current, by Eq. (3.2). The ratio of X to its spatial average (unaffected by the heat current) is

$$\left.\frac{X(z,t,\alpha)}{\langle X(z,t,\alpha)\rangle}\right|_{t=0} = \frac{e^{-\alpha\frac{2z}{h}}}{\langle e^{-\alpha\frac{2z}{h}}\rangle} = \frac{e^{\alpha H(z,0)}}{\alpha^{-1}\sinh\alpha} , \tag{4.17}$$

with $H(z,0)$ given by Eq. (4.5). For $|\alpha| \ll 1$, Eq. (4.17) reduces to

$$X/\langle X\rangle \simeq 1 + \alpha H , \tag{4.18}$$

as studied above. But for larger values of α, because of the lack of symmetry about the midplane at $z = h/2$, the function $F(\tau)$ has to be generalized to

$$F(\tau,\alpha) \equiv \frac{X(0,t,\alpha) - X(h,t,\alpha)}{X(0,0,\alpha) - X(h,0,\alpha)} \tag{4.19}$$

in order to describe the relaxation of the <u>difference</u> in concentration at the two boundaries. Expanded in normal modes,

$$F(\tau,\alpha) = \sum_{l=1}^{\infty} A_l(\alpha)\, e^{-l^2\tau} , \tag{4.20}$$

where, for $\alpha \neq 0$, the even integers also contribute, because of the lack of symmetry. The effect of the heat current on the strength of the fundamental is represented by

$$A_1(\alpha) = \frac{8}{\pi^2}\,\frac{\alpha\coth\alpha}{1 + \frac{4}{\pi^2}\alpha^2} . \tag{4.21}$$

For small values of α, $A_1(\alpha)$ drops slowly below its $\alpha = 0$ value of $8/\pi^2$, as

$$A_1(\alpha) \simeq \frac{8}{\pi^2}\left[1 - \left(\frac{4}{\pi^2} - \frac{1}{3}\right)\alpha^2\right] , \tag{4.22}$$

but, for $\alpha \gg 1$, approaches zero as

$$A_1(\alpha) \simeq \frac{2}{\alpha} . \tag{4.23}$$

In other words, a straight line fit to $F(\tau,\alpha)$ in the large τ range, such as illustrated by the dashed line in Fig. 4, should result in a greatly reduced $\tau = 0$ intercept, for $\alpha >> 1$. The shape of the curve will, of course, also be different, with the abrupt drop at $\tau = 0$ becoming even more pronounced.

In the range of very low mole fractions, $X << 1$, it is expected that any interaction between the ^3He atoms should become negligible and that their transport properties should be completely characterized by the mobility, μ. This is an individual particle parameter, by its very nature independent of X. From Einstein's relation, Eq. (4.4), the diffusion coefficient should, in turn, be X-independent, as indicated in Fig. 5 by the horizontal dashed line labelled "Khalatnikov." By Eq. (4.7), the mean time for relaxation of the fundamental,

$$\gamma_1^{-1} = \frac{h^2}{\pi^2 D}, \tag{4.24}$$

should also be independent of X. This is found experimentally, however, not to be the case. The mean times rise with decreasing X and the inferred values of D therefore drop, as indicated in Fig. 5 by the portion of the curve labelled "EXP." Substitution of Eq. (4.4) into Eq. (4.24) gives

$$\gamma_1^{-1} = \frac{h^2}{\pi^2 k_B T} \mu^{-1}, \tag{4.25}$$

so that the observed rise in γ_1^{-1} can be translated into the plot of μ^{-1} versus X that is shown by the portion of the curve in Fig. 6 that is labelled "EXP." The flat X-independent behavior that is predicted by the Khalatnikov independent particle approach is indicated by the horizontal dashed line in Fig. 6. When the values of μ^{-1} determined from γ_1^{-1} are substituted into Eq. (3.11) the resulting values for κ_{eff} were found to agree well with those determined from the steady-state measurements. It is consequently clear that the failures of the Khalatnikov approach that are illustrated in Figures 2 and 3, on the one hand, and in Figures 5 and 6, on the other, have a common origin. It would seem that some effect is setting in at very low concentrations to invalidate, or, at least, modify,

Fig. 5. Deviation at low mole fraction of the diffusion coefficient from Khalatnikov theory.

the simple concept of a drag coefficient. Any proposed explanation of the discrepancy will have to predict the course of the curve in Fig. 6 at ultra-low concentrations, in the unexplored region beyond the range of the concentrations so far measured, as indicated by the question mark. This is the subject of the next section.

V. THE THREE-FLUID HYPOTHESIS

The drop in the diffusion coefficient that sets in at very low concentration, as illustrated in Fig. 5, might suggest some trapping mechanism that would tend to immobilize the ^{3}He atoms. This hypothesis does not seem tenable, however, in view of Eqs. (2.16) and (2.17), which can be written as

$$\xi_3(X) = \Phi + k_B T \ln X , \tag{5.1}$$

where Φ is an X-independent quantity. Thus, the chemical potential is an unbounded and monotonically decreasing function of decreasing concentration which would be expected to result in the "evaporation" of any trapped ^{3}He atoms, in the low concentration limit. It, therefore, seems more promising to consider a mechanism for decreasing the mobility of the ^{3}He atoms, without actually immobilizing them completely. The latter idea has been proposed by Chui and Lipa,[5] and is illustrated by the dot-dashed line in the plot in Fig. 7 of $(X\kappa_{eff})^{-1}$ versus X. The horizontal dashed line labelled "Khalatnikov" is the conventional X-independent version of the drag coefficient; the rising portion of the solid curve labelled "EXP." indicates the observed discrepancy. The flat part of the curve, to the left of this, labelled "3-FLUID," illustrates the prediction of the three-fluid hypothesis that we want to propose in what follows.

Because of the evidence of variation in the drag coefficient, we are led to postulate the existence of a third component in the liquid Helium, over and above the usual super- and normal-fluid components of the customary two-fluid model. Our phenomenological approach does not require us to identify the microscopic nature, or origin, of this hypothetical

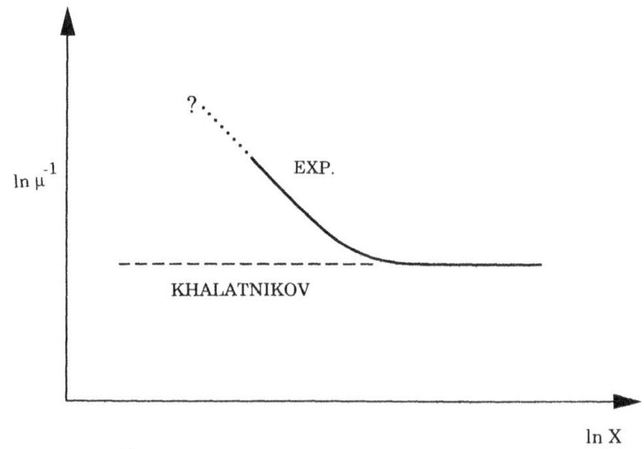

Fig. 6. Deviation at low mole fraction of the drag coefficient from Khalatnikov theory.

third component, although it is reasonable that it might have something to do with vortices or vortex rings, or some kind of mesh of vortices. We assume that the third component has negligible entropy and that it does not contribute to the heat current. Our phenomenology consists in assigning separate mobilities, μ and $\bar{\mu}$, to the motion of an isolated ^{3}He atom in the usual normal fluid and in the additional normal fluid, respectively. We also assume a mutual friction acting between the two normal fluids, linear in their relative velocity, with coefficient $n_4 f$ for the force density. Without loss of generality we can take the ^{3}He gas to be at rest, so that the sum of the drag forces exerted, on the average, on a ^{3}He atom by the two normal fluids moving at velocities v_n and $\bar{v}_n$ is

$$F_3 = \mu^{-1} v_n + \bar{\mu}^{-1} \bar{v}_n , \qquad (5.2)$$

a generalization of Eq. (3.2). If $\bar{v}_n \neq v_n$, the mutual drag force per unit volume acting between the two normal fluids is, as stated above, $n_4 f$ $(v_n - \bar{v}_n)$. For the third component to be in steady state equilibrium, this frictional force between the two normal fluids needs to be balanced by the drag from the ^{3}He gas at its density, n_3. Equilibrium therefore requires

$$n_3 \bar{\mu}^{-1} \bar{v}_n = n_4 f \ (v_n - \bar{v}_n) , \qquad (5.3)$$

or

$$\frac{X}{f\bar{\mu}} \bar{v}_n = v_n - \bar{v}_n , \qquad (5.4)$$

which yields

$$\bar{v}_n = \frac{v_n}{1 + X/f\bar{\mu}} . \qquad (5.5)$$

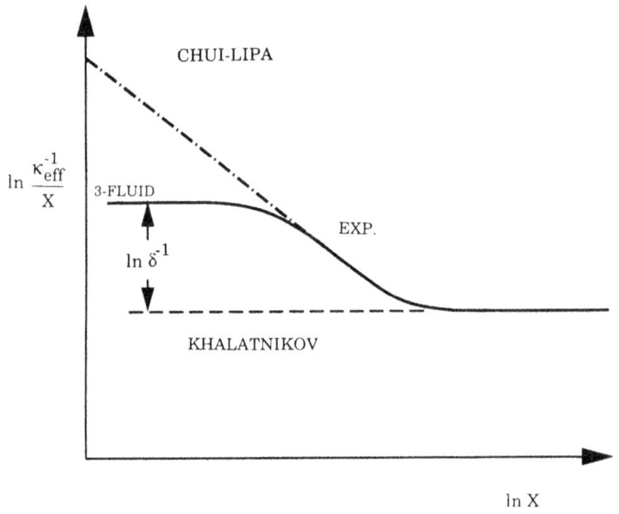

Fig. 7. Three-fluid model interpolating between the Khalatnikov regi[r] and the postulated ultra-low mole fraction "quasi-Khalatnik[o] regime.

The physical content of the three-fluid hypothesis is immediately evident from Eq. (5.5). In the limit, $X \to \infty$, the velocity of the postulated third component vanishes. It then has no observable effect, there being no relative motion between it and the ^3He gas, so that the conventional Khalatnikov theory applies. In the other limit, however, as $X \to 0$, $\bar{v}_n = v_n$ and the two normal fluids move together, resulting in an enhanced drag coefficient $\mu^{-1} + \bar{\mu}^{-1}$ and giving a modified "quasi-Khalatnikov" regime at the ultra-low concentrations.

Substituting Eq. (5.5) and factoring out v_n gives the effective X-dependent drag coefficient

$$\mu(X)^{-1} = \frac{1}{\mu} + \frac{1}{\bar{\mu} + X/f}$$

$$= \mu^{-1} E(x) \ , \qquad (5.6)$$

where the concentration dependent enhancement factor has the scaled form

$$E(x) = \frac{1 + \dfrac{X}{f(\mu + \bar{\mu})}}{\dfrac{\bar{\mu}}{\mu + \bar{\mu}} + \dfrac{X}{f(\mu + \bar{\mu})}} = \frac{1 + x}{\delta + x} \ . \qquad (5.7)$$

Here we have introduced the scaled mole fraction

$$x = \frac{X}{X_c} \ , \qquad (5.8)$$

with the crossover value

$$X_c = f(\mu + \bar{\mu}) \ . \qquad (5.9)$$

The characteristic shape parameter is the ratio

$$\delta = \frac{\bar{\mu}}{\mu + \bar{\mu}} \ . \qquad (5.10)$$

The Khalatnikov and "quasi-Khalatnikov" regimes are characterized by the two limiting values

$$E(x) \simeq \begin{cases} 1, & x >> 1 \\ \delta^{-1}, & 0 \le x << 1 \ . \end{cases} \qquad (5.11)$$

The experimental observations indicate a very small value for δ, so that for $x >> \delta$,

$$E(x) \simeq 1 + \frac{1}{x} \ , \qquad (5.12)$$

which gives the rise shown by the "EXP." portion of the curve in Fig. 7.

Once the general qualitative idea of a "quasi-Khalatnikov" regime at the ultra-low concentrations is accepted, it is natural to attempt, purely phenomenologically, to find some simple interpolation formula that connects it with the usual Khalatnikov regime of moderately low concentrations. Eq. (5.7) is precisely such a formula, and predicts that the response function should have the scaled behavior

$$\kappa_{eff}^{-1} \propto x \frac{1 + x}{\delta + x} . \tag{5.13}$$

On a log-log plot, this has the slope

$$\frac{d \ln \kappa_{eff}^{-1}}{d \ln x} = \frac{x}{1 + x} + \frac{\delta}{\delta + x}$$

$$\simeq x + \frac{\delta}{x} , \tag{5.14}$$

where the final expression assumes $\delta << 1$, and is valid in the intermediate region $\delta << x << 1$. The log-log plot of κ_{eff}^{-1} should, therefore, always have a non-vanishing positive slope, whose minimum value of $2\sqrt{\delta}$ occurs at $x = \sqrt{\delta}$. Examination of the experimental data[1] suggests $\delta = 0.6 \times 10^{-3}$, with the crossover mole fraction chosen as $X_c = 1.3 \times 10^{-4}$. A log-log plot[6] of log κ_{eff} versus log X, based on Eq. (5.13) and these parameters, is exhibited in Fig. 8. The data[1] are shown by the crosses, and it is evident that an extension of the measurements to concentrations smaller than $\delta \cdot X_c \simeq 10^{-7}$ would be very useful. This would be a crucial test of the basic hypothesis on which the 3-fluid model rests, namely that there must be a "quasi-Khalatnikov" region for $X < \delta \cdot X_c$ where κ_{eff} rises as X^{-1}.

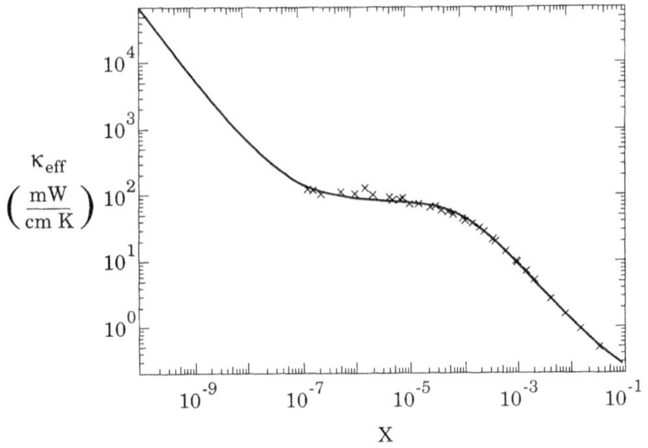

Fig. 8. Effective thermal conductivity versus mole fraction: comparison of the 3-fluid model prediction with the data of reference 1.

IV. SUMMARY

In the first three sections of this paper we have given a great deal of attention to elementary and well-understood aspects of dilute solutions. The justification for this abuse of the reader's patience has been the extremely puzzling findings reported in references 1 and 2. These discrepancies require a reexamination of the fundamental principles, and lead to the 3-fluid model that is developed in Section V. As emphasized there, a vital test of this model will be to extend the measurements of κ_{eff} into the ultra-low mole fraction region, $X < 10^{-7}$.

If κ_{eff} does not, eventually, rise and exhibit a "quasi-Khalatnikov" behavior, the 3-fluid model will have to be discarded unequivocally. It is a pleasure to acknowledge discussions with Drs. A. Onuki and H. Meyer as well as support by the U. S. National Science Foundation under Grant for Basic Research DMR-89-01723.

REFERENCES

1. M. Dingus, F. Zhong, J. Tuttle, and H. Meyer, J. Low Temp. Phys. 65, 213 (1986). See also M. Dingus, F. Zhong, and H. Meyer, Phys. Rev. Letters 54, 2347 (1985).
2. For a general review and complete bibliography of this subject, see H. Meyer, these proceedings.
3. J.A. Lipa and T.C.P. Chui, Phys. Rev. Letters 58, 1340 (1987); Proceedings of the 18th International Conference on Low Temperature Physics, Kyoto, 1987 [Jpn. J. Appl. Phys. Suppl. 26-3, 13 (1987)].
4. F. Zhong, J. Tuttle, and H. Meyer, J. Low Temp. Phys. 49, 9 (1990).
5. T.C.P. Chui and J.A. Lipa, Phys. Rev. B40, 4306 (1989).
6. I am indebted to Prof. Meyer for providing this fit to the data of reference 1.

THERMAL COUNTERFLOW IN SUPERFLUID ^3He-^4He MIXTURES

H. van Beelen

Kamerlingh Onnes Laboratorium, Rijksuniversiteit Leiden
Postbus 9506 2300 RA Leiden – The Netherlands

INTRODUCTION

The two-fluid model provides an adequate description of the hydrodynamic behaviour of superfluid ^3He-^4He mixtures. The ^3He component is incorporated in the normal fraction. The additional thermodynamic variable is accounted for by the conservation law of ^3He.

The experimental evidence in support of the above statement is perhaps somewhat scarce, particularly for mixtures with higher concentrations of ^3He. The experiments reported in literature mainly concern the linear regime, i.e. acoustic phenomena or measurements of the damped oscillations of a torsional pendulum. The flow properties of mixtures have only been investigated incidentally. From the extensive studies on capillary flow of pure helium II it is known that the superfluid turbulence in the flow tube exhibits many interesting features, which show up at certain combinations of the transport velocities of the superfluid and the normal fraction[1], and it seems of interest to investigate the effect of the admixture of ^3He on the production process of the superfluid vorticity.

The conservation of ^3He implies, however, that superfluid mixtures can not simply be considered as He II with an enhanced normal fluid fraction. Because the ^3He component flows with the normal fluid a steady state cannot establish itself in the majority of flow experiments. Large nonuniformities in the ^3He distribution can thereby build up. These tend to remain after the flow is stopped, as they can be balanced by temperature differences while also the relevant diffusion process is extremely ineffective. The complications due to the occurrence of this so-called heat-flush effect explain perhaps the almost complete lack of experimental data on turbulent flow in superfluid mixtures.

In this contribution the analysis of a thermal counterflow experiment[2], concerning mixtures with molar ^3He concentrations of up to 11% in the temperature range between 1 K and 2 K, will be discussed.

Excitations in Two-Dimensional and Three-Dimensional Quantum Fluids
Edited by A.G.F. Wyatt and H.J. Lauter, Plenum Press, New York, 1991

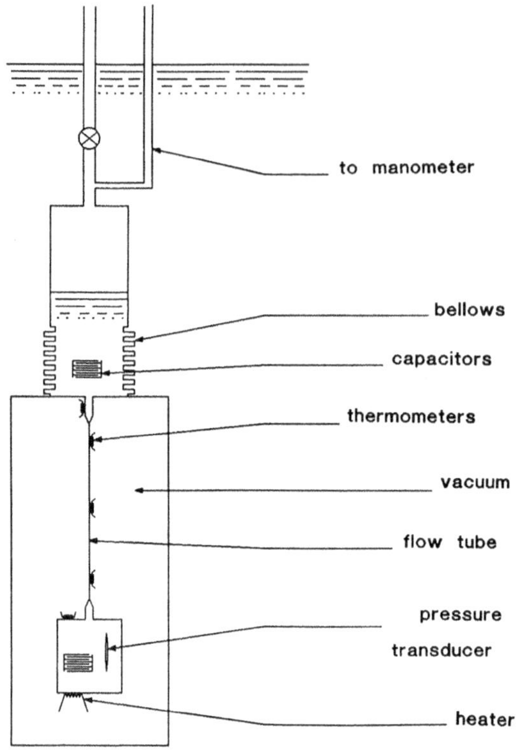

Figure 1. The thermal counterflow device.

THERMAL COUNTERFLOW

Even in the simplest thermal counterflow experiment at small constant heat input, a concentration difference, accompanied by an associated temperature difference, will steadily build up between the reservoirs on either side of the flow tube. In the experimental setup depicted in Fig.1 it is observed that the steady increase is largely accommodated just outside the ends of the tube, the temperature gradient along the tube remaining very small[2]. At the same time, the ^{3}He concentration and the temperature distribution further down inside each reservoir become increasingly nonuniform. These effects rule out the use of a straightforward analysis of the measuring data. Even though the choice of sufficiently large reservoirs can ensure the establishment of quasi-steady flow conditions, the determination of the quasi-steady flow velocities in the flow tube becomes ambiguous. The same applies to the concentration gradient along the tube and with that of the gradient of the partial chemical-potential of ^{4}He, which becomes unequal to zero in the turbulent-flow regime.

In the earlier analysis of our experiments[2] we overcame the second problem by deriving the mass concentration gradient $\partial c/\partial z$ from the temperature gradient $\partial T/\partial z$, which was measured directly along the flow tube (see Fig.1). We thereby made use of the property that under quasi-steady circumstances the divergences of the currents of

^{3}He and entropy ($\rho c\vec{v}_n$ and $\rho s\vec{v}_n$ respectively) will both remain negligibly small. From this it follows that the two gradients are simply proportional and point in the same direction, according to:

$$\frac{1}{c}\frac{\partial c}{\partial z} = \frac{c_m}{s_4 T}\frac{\partial T}{\partial z} \tag{1}$$

in which c_m is the specific heat of the mixture and s_4 the partial entropy per unit mass of ^{4}He in the solution. This observation explains the experimental finding that during laminar counterflow the temperature gradients along the tube remain even smaller than the corresponding ones in ^{4}He, since in laminar flow the contributions of the two gradients to the chemical-potential gradient $\partial\mu_4/\partial z$ together balance the contribution of Poiseuille's pressure gradient $\partial P/\partial z$, as:

$$\frac{\partial \mu_4}{\partial z} = \frac{1-\tilde{c}}{\rho}\frac{\partial P}{\partial z} - s_4\frac{\partial T}{\partial z} - c\left(\frac{\partial(\mu_3-\mu_4)}{\partial c}\right)_{P,T}\frac{\partial c}{\partial z} \tag{2}$$

(the contribution $-(\rho_n/2\rho)\partial(\vec{v}_n - \vec{v}_s)^2/\partial z$ is here neglected). In this relation μ_3 is the partial chemical potential per unit mass of ^{3}He in the solution and $\tilde{c} \equiv -\frac{c}{\rho}(\frac{\partial \rho}{\partial c})_{P,T}$. Unfortunately, the quasi-steady increase of T obstructs in practice the accurate determination of the small temperature gradients encountered in laminar flow.

With respect to the first problem there are at least five independent ways to obtain experimental values for the flow velocity of the normal fluid in the flow tube. Two of these values follow from the independent capacitive measurements of the rates of change of the ^{3}He concentrations in the two reservoirs of Fig.1, via the application of the ^{3}He conservation law:

$$\rho c A v_n = -V_1\left(\frac{d\rho c}{dt}\right)_1 = V_2\left(\frac{d\rho c}{dt}\right)_2 \tag{3}$$

with A the cross-sectional area of the flow tube and V_1 and V_2 the ("effective") volumes of the two reservoirs. A third value is derived from the measured rate of increase of the temperature in the reservoir near the heater, by applying the balance equation for the entropy of that reservoir:

$$\frac{\rho s_4 T A}{1-\tilde{c}} v_n = \dot{Q} - \rho c_m V_1\left(\frac{dT}{dt}\right)_1 \tag{4}$$

where $\dot{Q}$ is the rate of heat supply. Another value is deduced from the observed rate of change of the vapour pressure above the free liquid level in the downstream reservoir, using:

$$\left(\frac{d\rho c}{dt}\right)_2 = \frac{dP_v}{dt} / \left(\frac{\partial P_{sat}}{\partial \rho c}\right)_T \tag{5}$$

together with eq.(3), taking literature values for the saturated vapour pressures of the mixtures[3]. We thereby assumed that equilibrium conditions between liquid and vapour are maintained quasi- steadily, owing to the good thermal contact with the surrounding helium bath that is provided by the heat exchanging bellows.

It was found that the four values so obtained differed rather strongly. The values deduced from the changes in ^{3}He concentration near the heater were found to be systematically larger than the ones deduced from the corresponding changes in the downstream reservoir, the differences amounting to an almost constant ratio of 1.8. The other two values were always found to lie somewhere in between.

These systematic differences must clearly be attributed to the heat-flush effect, the various detectors only measuring local values instead of the averaged values appearing in the relevant balance equations. It is a priori not at all obvious which value of v_n one should choose. In ref.2 we argued that there is some reason to prefer the values obtained from the detectors in the "isothermal" downstream reservoir, even though this choice implies the occurrence of a huge nonuniformity in the region near the heater.

A conclusive solution of this dilemma could be provided by the fifth method, that is to derive the normal fluid velocities in the regime of laminar flow from the measured flow pressures by means of Poiseuille's law:

$$\Delta P_f = -\frac{8\pi\eta\ell}{A} v_n \tag{6}$$

with η the shear viscosity of the normal fluid and ℓ the length of the tube. The strength of this method lies in the fact that the pressure inside the reservoirs will remain uniform, no matter how nonuniform the distributions of the concentration and the temperature become.

These laminar values can then be used to calibrate the other four detectors, as the various quasi-steady rates of change are found to vary nicely proportionally with the rate of heat input, as do the observed pressure drops. These calibrations can subsequently be used in the regime of turbulent flow, attained at larger heat inputs, if one makes the plausible assumption that the flow patterns in the reservoirs, particularly the one in the downstream reservoir, will not be affected by the state of turbulence in the very narrow tube. Then the proportionality with v_n will be maintained.

Unfortunately we could not make use of the latter method because of a lack of reliable data on the values of the shear viscosity of the mixtures at the time of the measurements. Values reported in literature, obtained from the damping of a torsional pendulum[4] and from isothermal capillary-flow data[5] differ markedly, possibly also due to the presence of the almost uncontrollable heat-flush effect during the measurements. In ref.2 we therefore reversed the procedure and deduced viscosity values from the proportionality found between the laminar flow pressures and our "choice" of normal fluid velocities. The viscosity values so obtained turned out to be significantly larger than the literature values.

It was then decided to determine the shear viscosity of the mixtures in a separate experiment, in which special care would be taken to avoid the occurrence of the heat-flush effect. This can be accomplished by containing the mixture inside a closed copper capillary immersed in a He-II bath, ensuring almost ideally isothermal conditions.

DETERMINATION OF THE SHEAR VISCOSITIES

The simple isothermal tube configuration can be used as an organ-pipe for fourth sound, provided that the tube is sufficiently long. In that case the frequencies of the first few harmonic modes are small enough for their viscous penetration depths λ_v to be much larger than the radius of the tube:

$$\lambda_v = \left(\frac{2\eta}{\rho_n\omega}\right)^{\frac{1}{2}} >> \frac{A}{\pi}^{\frac{1}{2}} \tag{7}$$

From our experience with superfluid ^{4}He under similar circumstances we know that, in the temperature range between 1 K and 2 K, the damping of the free fourth-sound oscillations is completely dominated by the contribution from the viscous motion of the normal fluid[6]. Because one can expect the additional damping in a mixture due to diffusion to be extremely small, this will also hold in the case of mixtures. From a measurement of the damping rates one can thus deduce values of the shear viscosities, using the relation between these two quantities that follows from the set of linearized two-fluid equations proposed by Landau and Khalatnikov[7]:

$$\alpha_v = \frac{\rho_s}{(1-\tilde{c})^2}\left[\frac{\rho_n}{\rho_s} + \tilde{c} - \frac{c^2}{u_{40}^2}\left(\frac{\partial(\mu_3-\mu_4)}{\partial c}\right)_{P,T}\right]^2 \frac{A}{16\pi\eta}\,\omega^2 \tag{8}$$

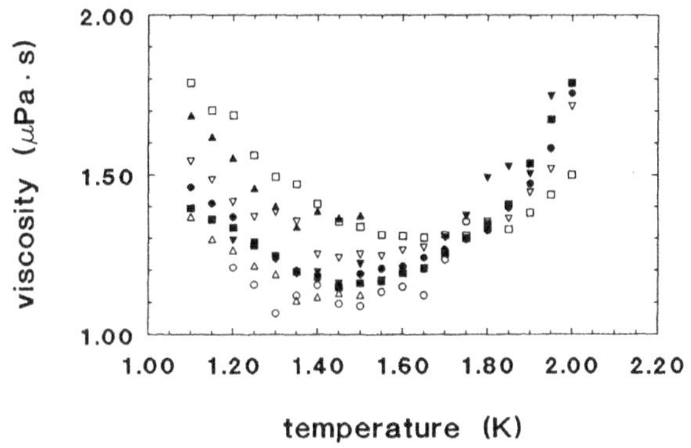

Figure 2. The shear-viscosity data for various molar concentrations, as deduced from the measured damping rates of fourth sound.
□: ^{4}He; ▲: 0.6%; ▽: 1.2%; ●: 3.1%; ■: 5.25%; △: 7.4%; ▼: 8.3%; ○: 11.2%.

The various thermodynamic parameters appearing in the relation can be obtained from literature[8,9]. Values for the speed u_{40} of isothermal fourth sound are directly provided by the measured frequencies.

We measured the damping rates by means of a capacitive pressure transducer connected to one end of a very long (308 m) and narrow (i.d. 170μm, o.d. 1.2 mm) copper tube[10]. The resulting shear viscosities are presented in Fig.2. They are in good agreement with the torsional oscillator results reported very recently by Wang, Howald and Meyer[11].

THE LAMINAR-COUNTERFLOW DATA REVISITED

The new viscosity values are about 30% smaller than those obtained in ref.2. It must thus be concluded that the true counterflow velocities are about 50% larger than the values indicated by the detectors in the downstream reservoir.

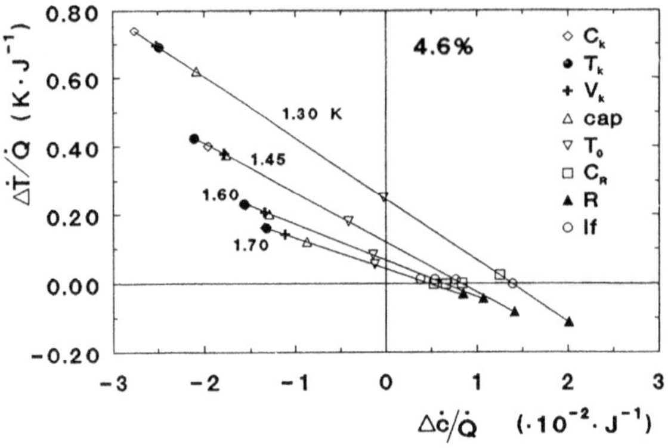

Figure 3. The reduced values for the quasi-steady rates of change of the temperature and the ^{3}He mass concentration, computed for various locations in the device.

Such a large correction has, of course, a number of important consequences for the quantitative analysis of the data. It leads, for instance, to a considerable reduction of the nonuniformities in the reservoirs when computed as in ref.2 from the condition that in the laminar case the partial chemical potential of ^{4}He as well as the rate of change of the pressure should be spatially constant throughout the entire liquid. This condition establishes a linear relationship between the simultaneous local rates of change of the temperature and of the concentration throughout the liquid, reading as:

$$s_4 \Delta \dot{T} + c \left(\frac{\partial(\mu_3 - \mu_4)}{\partial c} \right)_{P,T} \Delta \dot{c} = K \tag{9}$$

The value of the parameter $K/\dot{Q}$ is determined by the value of $v_n/\dot{Q}$, and thus changes with the new calibration. It is computed by applying eq.(9) to the *mean* rates of change in the reservoir near the heater, expressing these mean rates in terms of $\dot{Q}$ and v_n by means of the balance equations(3) and (4).

Equation (9) allows the computation of the rates of the temperature change at the concentration detectors and of the concentration change at the thermometers. In combination with the other conservation law represented by eq.(3) it also allows the computation of the mean rates of change in the downstream reservoir. Finally, using the data for the variations of the saturated vapour pressure with concentration and temperature from ref.3, one can obtain the corresponding values at the interfaces.

It turns out that the new values obtained for the local rates at both detectors of the upstream reservoir (C_k and T_k in the examples shown for a mixture of 4.6% in Fig.3) never differ by more than 15% from their mean values (V_k), calculated from the counterflow velocities by straightforward application of the appropriate balance equations to that reservoir. In our earlier computations (which correspond to parallel shifts of the straight lines in Fig.3), these differences even exceeded a factor two in some cases. In addition, our new computations reveal that the temperatures of the capacitor (C_R) in the downstream reservoir and of the vapour-liquid interface (If) both stay nicely

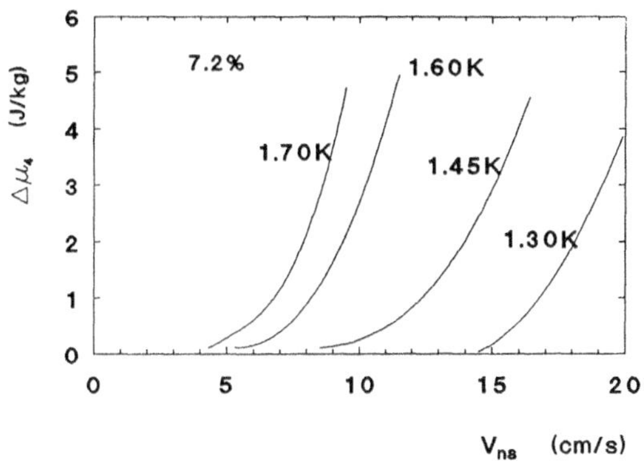

Figure 4. Curves of the chemical-potential difference across the flow tube (i.d. = 100 μm, $\ell = 79$ mm) for a 7.2% mixture.

close to that of the bath, so that the concentration at the interface also remains close to the measured value (see the examples in Fig.3). Unexplained remains, of course, why this measured value is so much smaller than the computed mean value (R) of the reservoir. It suggests that more than 50% of the inflow of ^{3}He is deposited on the wall of the bellows. It seems rather remarkable that this deposit will not come loose from the wall again until some contact gas has been admitted to the vacuum can, by which the flow of normal fluid through the tube is reversed and overall equilibrium is restored.

THE REGIME OF TURBULENT COUNTERFLOW

The analysis of our data on turbulent counterflow[12] is, of course, also affected by the new velocity values. They do not alter the values of the chemical-potential gradients $\partial\mu_4/\partial z$, as these are deduced by means of eqs.(1) and (2) from the observed temperature gradients and pressure drops. But the change of the velocity scale alters the values of the vortex-line density L_0 deduced from them.

As an example, Fig.4 shows the chemical-potential difference across the glass flow tube (of inner diameter 100 μm and length 79 mm) versus the counterflow velocity inside it, for a mixture with a molar ^{3}He concentration of 7.2%, at various temperatures. Fig.5 depicts the corresponding vortex-line densities derived under the assumption of an isotropic vortex tangle according to:

$$BL_0 = 3\frac{1-\tilde{c}}{\kappa}\frac{\rho_s}{\rho_n}\frac{\Delta\mu_4}{\ell v_n} \tag{10}$$

in which κ is the quantum of circulation ($\kappa \approx 10^{-7}\mathrm{m^2s^{-1}}$). The values of the Hall-Vinen parameter B of the mixtures are unknown, but in He II B is found to be of order unity, depending only weakly on temperature.

For comparison, curves obtained with pure He-II in the flow tube are also shown in Fig.5. Although the accuracy with which the above results could be obtained is rather poor, they allow us to conclude that, in spite of its appearing otherwise in the experiments, the turbulent behaviour of thermal counterflow in these mixtures is even quantitatively (not merely qualitatively) closely similar to that of He II. This conclusion is corroborated by the results obtained for the other concentrations. The presence of the ^{3}He seems to have surprisingly little influence on the production process of superfluid turbulence.

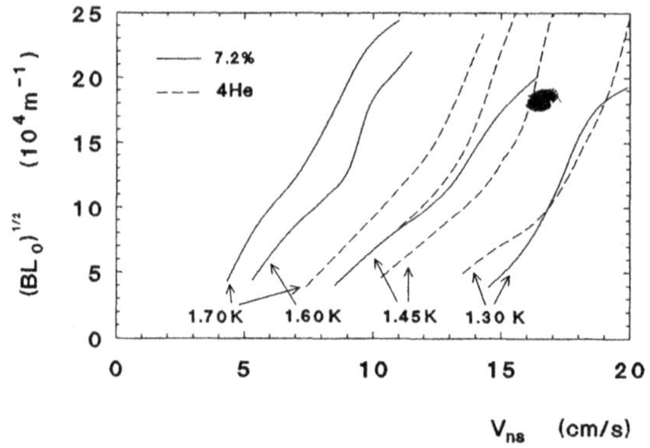

Figure 5. The vortex-line densities deduced from the curves in Fig.4 by means of eq.(10). For comparison, the corresponding results obtained for pure ^{4}He are shown by the dashed lines.

REFERENCES

1. H. van Beelen, W. van Joolingen and K. Yamada, On a balance equation for superfluid vorticity in capillary flow of Helium II, Physica B 153:248 (1988).
2. R.F. Mudde and H. van Beelen, Counterflow in superfluid ^{3}He-^{4}He mixtures part I: the laminar case, Physica B 162:197 (1990).
3. S.G. Sydoriak and T.R. Roberts, Vapour pressure of ^{3}He- ^{4}He mixtures, Phys. Rev. 118:901 (1960).
4. J.G Dash and R.D. Taylor, Density and viscosity of normal fluid in dilute solutions of ^{3}He in ^{4}He, Phys. Rev. 107:1238 (1957).
5. F.A. Staas, K.W. Taconis and K. Fokkens, Viscosity of liquid ^{3}He-^{4}He mixtures in the Helium II region above 1.0 K. Physica, 26:669 (1960).
6. H.P. Sierat, H. Versnel and H. van Beelen, Attenuation of fourth sound in He II between 0.5 K and 2.05 K, Physica B 153:159 (1988).
7. I.M. Khalatnikov, "An Introduction to the theory of superfluidity", W.A. Benjamin, New York-Amsterdam (1965).
8. V.I. Sobolev, B.N. Esel'son, O.S. Novovitskaya and L.A. Pogorelov, Density of the normal component of ^{3}He-^{4}He solutions at elevated pressures, Sov. J. Low. Temp. Phys. 5:269 (1979).

9. K.W. Taconis and R. de Bruyn Ouboter, "Progress in Low Temp. Phys.", Vol.IV, Ch.2, North-Holland, Amsterdam (1964).
10. J.J. van Son and H. van Beelen, Isothermal fourth sound in superfluid ^3He-^4He mixtures, Proceedings LT19, Brighton, Sussex (1990).
11. S. Wang, C. Howald and H. Meyer, Shear viscosity of liquid ^4He and ^3He-^4He mixtures, especially near the superfluid transition, J. Low Temp. Phys., 79:151 (1990).
12. R.F. Mudde, "Thermal counterflow in ^3He-^4He mixtures", Ph.D. Thesis, Leiden (1989).

ELEMENTARY EXCITATIONS IN FERMI LIQUIDS:

NEUTRON SCATTERING BY ^{3}HE AND ^{3}He-^{4}He MIXTURES

Reinhard Scherm, Kurt Guckelsberger, Anne Stunault

Physikalisch Technische Bundesanstalt (PTB)
Braunschweig F.R.G.

Björn Fåk

Institute Laue-Langevin (ILL) Grenoble, France

INTRODUCTION

In 1947, Landau predicted in his phenomenological theory of superfluid He the existence of rotons[1]. It was 1957 that at a small reactor in downtown Stockholm neutrons scattered for the first time off liquid He at 1.4 and 4.2 K which showed conclusively that rotons were real[2]. The papers presented on the first day of this conference show that liquid ^{4}He is still a subject of high interest and lively research. The question: what is the roton ? still excites long discussions.

The Fermi liquid ^{3}He, by no means of less interest, has drawn much less experimental attention: because of its enormous absorption cross section for thermal neutrons, scattering experiments were for a long time considered as hopeless. Only after 1974 experiments had been performed by two groups, one at the ILL[3] the other at Argonne National Laboratory[4] in the US.

Here, we report on recent experiments where the two former groups, now at PTB, ILL, the Studsvic Research Center (Sweden) and University Keele (GB) finally collaborated at the time of flight spectrometer IN6 at the ILL. This instrument, designed with ^{3}He experiments in mind, offers very high intensity at very low background. With more than 300 detectors, covering a continuous angular range, the neutron scattering function $S(Q,\omega)$ can be obtained simultaneously in a wide Q-range (we use $\hbar$ =1 and k_B = 1 such expressing momentum in nm^{-1} and energy in Kelvin). By careful optimisation of the instrument, cryostat and sample cell we minimized background and parasitically scattered neutrons.

PURE ^{3}He

Fig. 1 shows selected spectra of pure ^{3}He obtained at 120 mK and zero pressure. The experiment[5], on which had been reported previously, has now been reanalysed with a refined data treatment[6]. At the lowest Q's, the

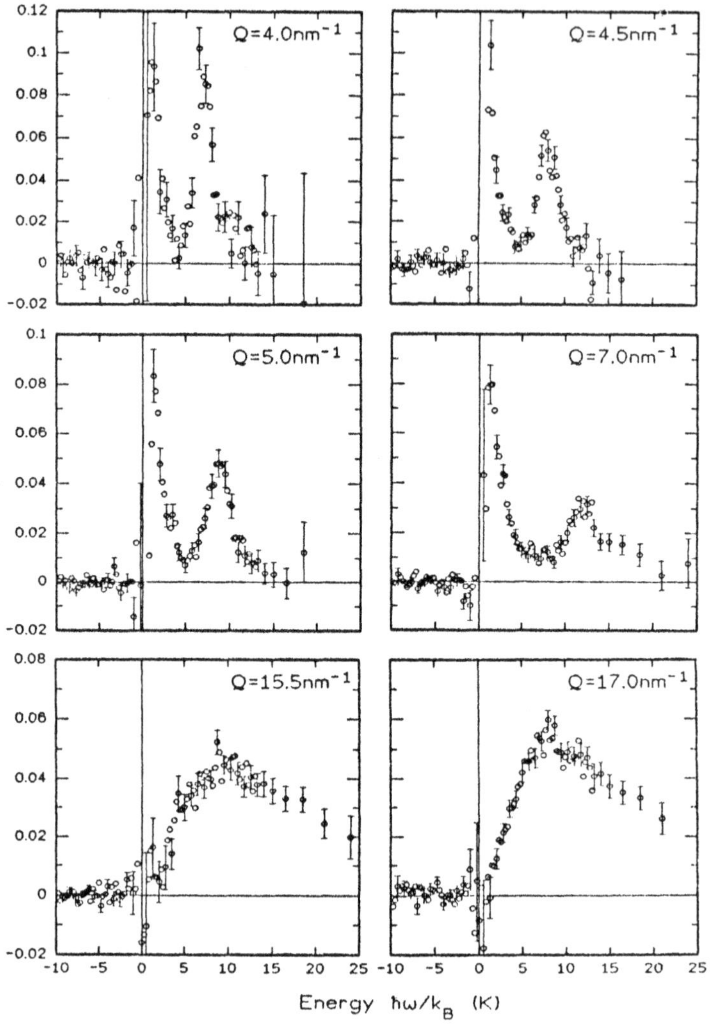

Fig. 1 The neutron scattering function S(Q,ω) of liquid ^{3}He at 120 mK and at saturated vapour pressure at a few selected Q-values. A refined treatment[6] of the original data[5] yields 35 spectra with $\Delta\omega$ =0.25 K and ΔQ = 0.5 nm^{-1} bins. The instrumental resolution is 0.9 K.

zero sound (zs) peak at higher energy is now completely separated from the low energy spin fluctuation peak or one particle - one hole (1p-1h) excitation sometimes called "paramagnon". The zs mode shows much stronger anomalous dispersion than in ^{4}He which vanishes for both isotopes at higher pressures[5].

In contrast to superfluid ^{4}He the zs mode is strongly damped at finite momenta even at low temperatures. Glyde and Khanna[7] proposed the decay of zs into two (1p-1h) pairs to be the dominant damping mechanism. This has been confirmed quantitatively in a recent calculation by Holas and Singwi [8].

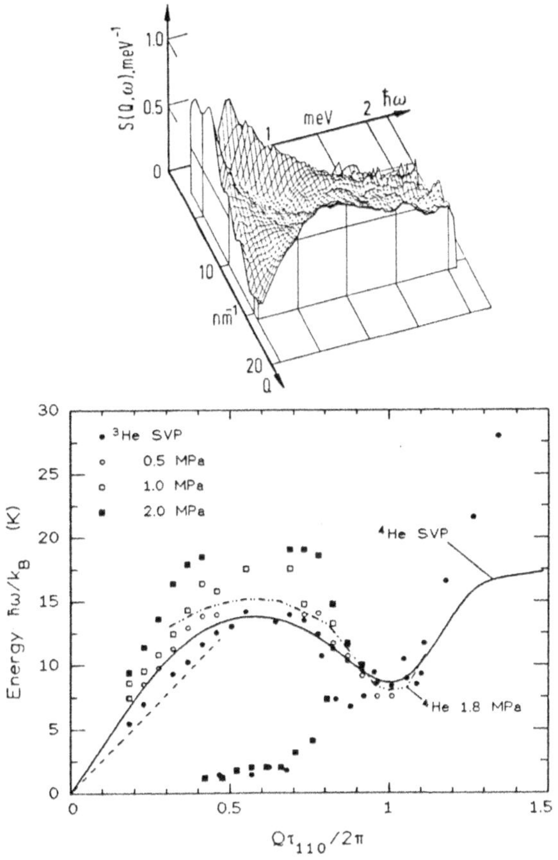

Fig. 2 above: General view of $S(Q,\omega)$ of ^3He at SVP and 120 mK shown for $Q <= 2k_F$. Below: peak energies ω of both, ω_F the particle hole excitatation and zero sound mode of ^3He at SVP. The dashed straight line extrapolates the ultrasound velocity. Solid line: ω_B (ph-r) of ^4He at SVP. Momentum Q is rescaled to density. The agreement between both isotopes fails at higher pressures.

Fig. 2 shows the peak energy ω_F, where $S(\omega)$ peaks at constant Q i.e. the "ridge" of the rather broad $S(Q,\omega)$ mountains. We observe a striking resemblance to the ω_B of the sharp phonon-roton (ph-r) branch of ^{4}He. If one rescales the momentum Q with the lattice spacing d_{110} of a bcc crystal of identical number density, the normal fluid ^{3}He exhibits a roton like minimum at $Qd_{110}/2\pi = 1$ just as superfluid ^{4}He. This is a consequence of the local short range order which is essentially the same in both liquids independent of quantum statistics.

^{3}He-^{4}He MIXTURES

Dilute mixtures of ^{3}He in liquid ^{4}He allow to extend the study of Fermi liquids towards small Fermi momentum $k_F = (3\pi^2 n_3)^{1/3}$ where the picture of a Fermi gas is still relevant. Neutron measurements have been made at 1% and ≈5% concentration at different temperatures and pressures[9].

Fig. 3 shows the ph-r mode of ^{4}He besides the much weaker "mode" of the fermion excitations of ^{3}He. The ph-r energies are only slightly shifted against those of pure ^{4}He at identical particle density. Therefore, the 5% mixture had been compressed at a pressure of 160 kPa. The phonon-rotons are broadened by the ^{3}He impurities. This linewidth Γ significantly increases at $Q \approx 18$ nm^{-1} slightly below the roton momentum, just where both modes seem to intersect.

A very dilute solution of ^{3}He in ^{4}He is commonly treated as a gas of nearly free fermion quasiparticles moving in a background of superfluid bosons. The backflow of ^{4}He around the moving ^{3}He gives rise to the effective mass m* of the quasiparticle. At low Q, the single quasiparticle then behaves just like a free gas with the Landau-Pomeranchuk (LP) spectrum:

$$\varepsilon_k = k^2/2m^* \qquad (1)$$

The structure of the solvent is reflected in the maximum of the structure factor $S_B(Q)$ and in the roton minimum of $\omega_B(Q)$. Similarily, we expect the energy ε_k of a single ^{3}He quasiparticle to be lowered at higher momenta. In parametrized form to lowest order:

$$\varepsilon_k = \frac{k^2}{2m^*}\left(1 + \gamma k^2 \right) \qquad \gamma < 0 \qquad (2a)$$

or expressed as a k-dependent effective mass

$$\varepsilon_k = \frac{k^2}{2m^*(1 + \beta k^2)} \qquad \beta > 0 \qquad (2b)$$

A neutron excites a ^{3}He quasiparticle from its state $|k\rangle$ into $|k+Q\rangle$ with the corresponding excitation energy $\omega_Q = \varepsilon_{k+Q} - \varepsilon_k$. At T=0 $|k\rangle = |0\rangle$, hence the excitation spectrum ω_Q of one impurity atom of course coincides with the single particle spectrum ε_Q. At finite ^{3}He densities ω_Q generally is different from ε_k. The scattering function is[10]:

$$S(Q,\omega) = 2/N \ \Sigma_k \ f_k(1 - f_{k+Q}) \ \delta(\omega - \varepsilon_{k+Q} + \varepsilon_k) \quad (3)$$

The Fermi function is $f_k = \{1,0\}$ depending on whether the state $|k\rangle$ is occupied or not. Equ. 3 can easily be understood intuitively:

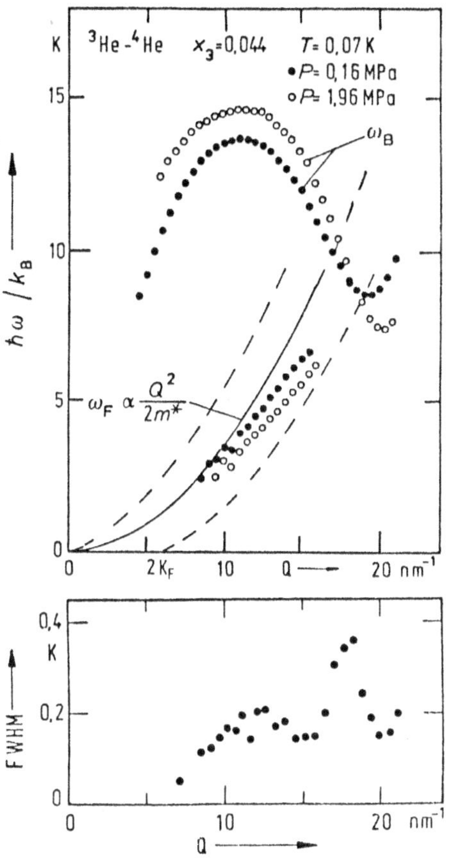

Fig. 3 ^3He-^4He Mixtures at 70 mK: 4.7% at 0.16 MPa, 4.2% at 1.96 MPa. The dispersion ω_B of the boson mode (ph -r) and the dispersion ω_F of the fermion mode , i.e. the position of Gaussians fitted to the measured peaks are shown.----- is the boundary of the zero temperature Lindhard function for free quasiparticles. The linewidth 2Γ (FWHM) is the broadening of the boson mode due to the presence of ^3He.

if the initial state $|k\rangle$ is occupied, then f_k = 1 and if the final state $|k+Q\rangle$ is empty then f_{k+Q} = 0. Then $1*1*\delta(..)$ contributes to the sum Σ_k.

COMPUTATION

For calculations, we replace the sum by an integral over $d^3\underline{k}$ and f_k by the Fermi-Dirac function

$$f_k = \{ \exp(-[\varepsilon_k-\mu(T)]/T) +1 \}^{-1} \qquad (4)$$

The chemical potential $\mu(T)$ has to be calculated from the implicite expression $\Sigma_k f_k$ = N, which is the conservation of the number of particles N. We computed $S(Q,\omega)$ numerically by calculating $f_k(1-f_{k+Q})$ for a given Q and all k sorting the results with respect to the energy ($\omega = \varepsilon_{k+Q} - \varepsilon_k$) and calculated its mean energy

$$\langle\omega\rangle_Q = \int \omega\, S(Q,\omega)d\omega / \int S(Q,\omega)d\omega \qquad (5)$$

The expression (2) fails at higher momentum. While (2b) levels off at $\varepsilon_\infty = 1/2m^*\beta \approx 42$ K, Eq.(2a) even becomes negative for $k > 1/\sqrt{-\gamma} \approx 34\ nm^{-1}$. For that reason, the integration is only carried out up to a rather arbitrarily chosen upper limit $k_c = 19$ or $24\ nm^{-1}$. At $Q > 16\ nm^{-1}$ and particularily at higher temperatures, this integration limit k_c strongly influences the shape of $S(Q,\omega)$ and therefore $\langle\omega\rangle_Q$.

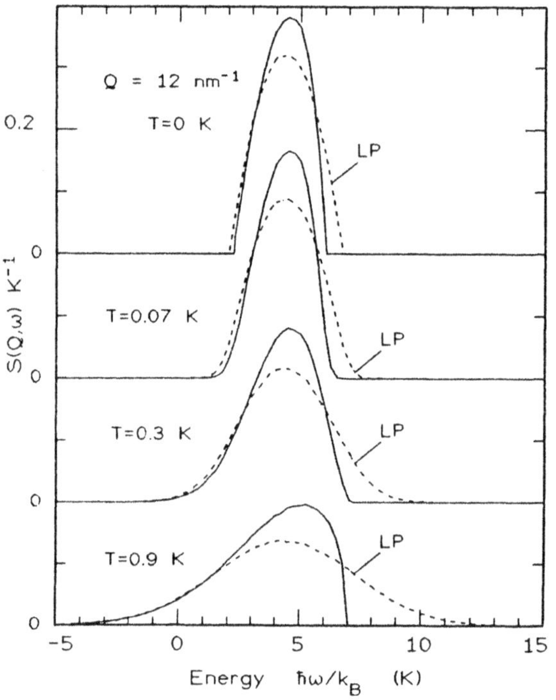

Fig. 4 Model calculations of $S(Q,\omega)$ for 4.7% ^3He in ^4He at $Q = 12\ nm^{-1}$ and various temperatures. Models: ε_k from Eq. (2) with $m^* = 2.22$ and $\gamma = -0.00085$; LP: single particle spectrum Eq. (1) with $m^* = 2.53$. The latter parameter is chosen such that $\varepsilon(k=12)$ coincides for both models.

Some results are shown in Fig. 4. With a LP single particle spectrum Eq. (1) one finds of course the Lindhard function (which can be derived analytically) a distribution which, for $Q>2k_F$, is symmetric around its mean $\langle\omega\rangle_{LP} = \varepsilon_Q$. With a nonparabolic dispersion ε_k however, the lineshape of $S(\omega)$ becomes asymmetric and the mean $\langle\omega\rangle_Q$ is now shifted to lower frequencies.

In Fig. 5, one sees both the single particle dispersion ε_Q and the mean excitation energy $\langle\omega\rangle_Q$ calculated for a solution containing 4.7 at% ^3He at 70 mK. These are compared to the experimental fermion energies as obtained from a Gaussian fit of $S(Q,\omega)$ measured at 70 mK from two solutions of 1% and 4.7% ^3He of identical particle density, hence at different pressures.

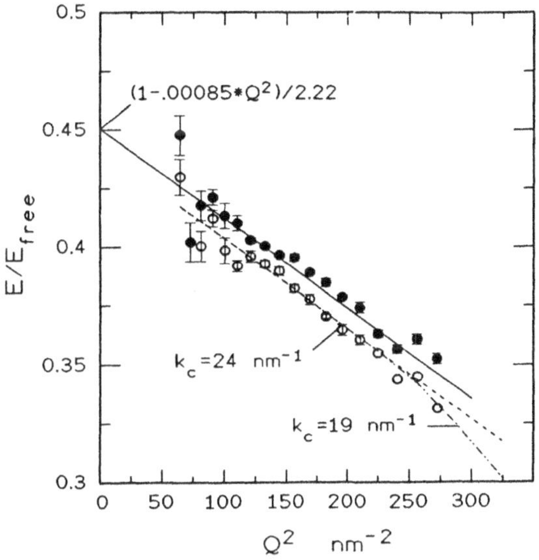

Fig. 5 The fermion energy E normalized to the free ^{3}He energy $E_{free} = Q^2/2m_3$. Experimental: o 1% ^{3}He at P = 10 kPa o 5% ^{3}He at P = 160 kPa. Calculated: _______ single particle energy;----- mean fermion energy ⟨ω⟩ for 4.7% ^{3}He and k_c=24 nm^{-1}; -..-..- k_c = 19 nm^{-1}.

CONCLUSION

A unique single particle energy ε_k describes very well the variation of $\langle\omega\rangle_Q$ with concentration at constant density. The parameters m* and γ agree with those deduced from specific heat and second sound velocity measurements[11,12]. One has to bear in mind that the latter experiments probe rather low Q's while our data cover only the high momenta $>2k_F$ i.e. $8 < Q < 18\ nm^{-1}$.

TABLE I

Parameters for the single particle energy

$$\varepsilon_k = k^2/2m^*(\ 1 + \gamma k^2\ )$$

	Thermodynamic[11]	n-scattering
Q-range		$8 < Q$ 18 nm^{-1}
m^*/m_3	2.225	2.22
γ	- 0.00076	- 0.00085 nm^2
k_c		24 nm^{-1}

References

1. L. Landau, J. of Phys. **XI**, 91-92 (1947)
2. H. Palevski, K. Otnes, K.E. Larsson, R. Pauli,R. Stedmann, Phys. Rev. **108**, 1346-1347 (1957)
3. R. Scherm, W. G. Stirling, A. D. B. Woods, R. A. Cowley, G. J. Coombs, J. Phys. **C7**, L341-L345 (1974);see also P. A. Hilton, R. A. Cowley, R. Scherm, W. G. Stirling, J. Phys. **C13**, L295 (1980)
4. K. Sköld and C. Pelizzari in "Recent Developpements in Condensed Matter Physics", edited by DeVreese **Vol 1** (Plenum, New York, 1981)
5. R. Scherm, K. Guckelsberger, B. Fåk, K.Sköld, A. J. Dianoux, H. Godfrin and W. G. Stirling, Phys. Rev. Lett. **59** 217 (1987)
6. A. Stunault, in preparation
7. H. R. Glyde and F. C. Khanna, Can. J. Phys. **58**, 343 (1980)
8. A. Holas and K. S. Singwi, Phys. Rev. **B40**, 167 (1989)
9. B. Fåk, K. Guckelsberger, M. Körfer, R. Scherm and A. J. Dianoux, Phys. Rev. **B41**, 8732 (1990)
10. S. Lovesey "Theory of Neutron Scattering from Condensed Matter",**Vol 1** p 81, Clarendon Press, Oxford (1984)
11. R. M. Bowley, this conference
12. A. Ghozlan and E. Varoquaux, Ann. Phys. Fr. **4** 239 (1979)

THE EFFECT OF FILM THICKNESS AND THE SUBSTRATE ON

SUPERFLUID ^{3}He FILM FLOW

J.P. Harrison, A. Sachrajda[†], S.C. Steel[*] and P. Zawadzki[†]

Physics Department, Queen's University
Kingston, Ontario, Canada

ABSTRACT

The self-emptying beaker technique was used to study the superfluid properties of thin film liquid ^{3}He. Superfluid film flow was observed only below the bulk transition temperature at 0.93 mK. In a separate experiment the minimum film thickness was determined as a function of ^{3}He level in the beaker. Thus, the flow rate, or critical current density, was determined as a function of temperature and film thickness. Extrapolation allowed a determination of the critical temperature (T_c) and zero temperature critical current density $J_c(0)$ as a function of film thickness. T_c agreed with the predicted superfluid-normal phase boundary at $2d/\xi(T) \cong \pi$ where d is the film thickness and $\xi(T)$ is the coherence length. $J_c(0)$ was an order of magnitude smaller than expected for dissipation by pair-breaking. When a ^{4}He monolayer was adsorbed on the surface of the beaker, it suppressed the diffuse scattering of ^{3}He quasiparticles at the substrate boundary, as also observed by Freeman et al. There was no measurable suppression of T_c even for films as thin as 100 nm. With or without the ^{4}He monolayer there was always an abrupt drop in the flow-rate when the film thinned to about 200 nm. This may be associated with the transition between the expected thick film B-like phase and thin film A-like phase.

INTRODUCTION

The ^{3}He film is a two or three dimensional quantum liquid depending upon the relative sizes of the quantum length scale parameter and the film thickness. There is considerable interest in following the quantum liquid through the cross-over from 3-D to 2-D behaviour. For the normal Fermi liquid the length scale parameter is the Fermi wavelength which is of the order of the interatomic spacing in pure ^{3}He and inter-^{3}He atom spacing in dilute ^{3}He-^{4}He mixtures. These dimensions are of the order of 1 nm. By contrast the length scale for the superfluid state of liquid ^{3}He is the coherence length which is of the order of 100 nm. Fortuitously this is comparable with the thickness of the van der Waals film that adheres to a substrate a millimetre or so above a free surface of bulk liquid ^{3}He. Therefore a study of superfluidity of the van der Waals film can be used to investigate superfluidity in liquid ^{3}He in the region of the dimensional crossover. Both the film thickness and the coherence length can be varied: The relationship between film thickness (d) and height of the film above the free surface (h) can be expressed as:

$$d = \alpha\, h^{-1/3}$$

where α is a constant which depends upon the substrate. The coherence length is given by [2,3]:

$$\xi(T) = \xi(0)(1-T/T_c)^{-1/2}$$

where $\xi(0) \sim 50$ nm. Thus, for example, with $\alpha = 10^{-8}\ m^{4/3}$, $T \ll T_c$ and h = 0.2 mm then $d/\xi(T) \sim 3$, corresponding to a 3-D superfluid. On the other hand, with $T \sim 0.9\ T_c$ and h = 1 mm then $d/\xi(T) < 1$, approaching a 2-D superfluid. Hence with readily available experimental parameters the crossover can be achieved.

Soon after the discovery of superfluid ^3He [4] and the basic understanding of the A and B phases in terms of spin-triplet pairing [5], theoreticians started to study the superfluid in confined geometry, both slab (2-D) and cylinder (1-D)[2,3,6-11]. The order parameter is complex. For the slab geometry there are three components, two parallel to the walls and one perpendicular. As d approaches $\xi(T)$ the order parameters are suppressed for diffuse boundary conditions. For specular boundaries only the perpendicular order parameter is suppressed. The predicted phase diagrams are shown in Figure 1. For the diffuse slab, i.e., both confining walls diffusely scatter ^3He quasiparticles, the superfluid state ceases for $D/\xi(T) = \pi$ where D is the slab thickness. Between the bulk ^3He

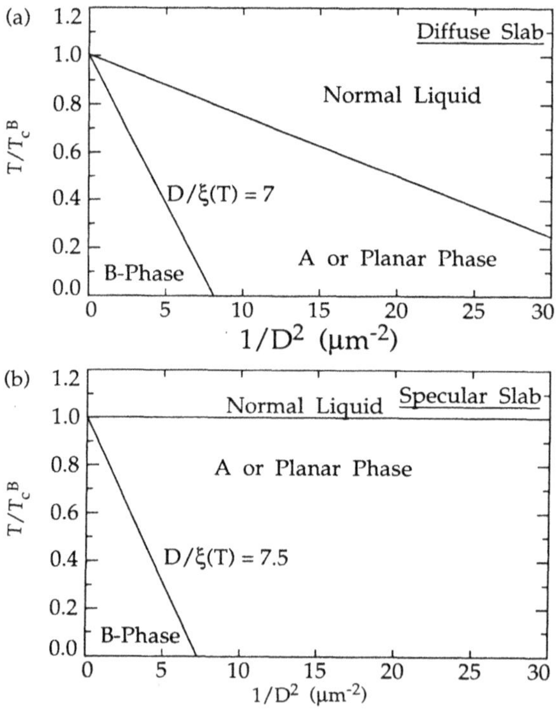

Fig. 1. Predicted phase diagrams for ^3He confined to a slab of thickness D for (a) diffusely, and (b) specularly scattering walls. T_C^B is the bulk superfluid transition temperature.

for which $D/\xi(T) >> 1$ and the superfluid/ normal transition there is a predicted phase transition between B-like and A-like phases at $D/\xi(T) = 7$. There is no suppression of the superfluid state by specular walls; however as for the diffuse slab there is a predicted phase transition from the B-like to the A-like phase at $D/\xi(T) = 7.5$. Prior to this work these phase transitions had never been observed. Unfortunately there have been no calculations for the case of one specular and one diffuse wall which might be expected to correspond to the van der Waals film on a rough substrate.

In summary, as the ^3He film is thinned we might expect first to see a phase transition from the B-like phase characteristic of the bulk to the A-like phase and then to see a transition to the normal state if the substrate diffusely scatters the ^3He quasiparticles. The B→A transition is not expected to influence the critical current density if dissipation is limited by pair-breaking.

Earlier experiments have confirmed the suppression of superfluidity in thin films and narrow pores [12-16] and that for 350 nm films between mylar sheets the equilibrium phase is A-like[17]. Further, two monolayers of ^4He on the mylar sheets removed the suppression of superfluidity by the confining walls, suggesting that the ^4He coating switched the substrate from diffuse to specular.

Superfluid film flow out of a glass beaker was used by Daunt and Mendelssohn 50 years ago to study superfluidity and dissipation in liquid ^4He. Five years ago Sachrajda et al[18] found that there was also ^3He film flow in a similar geometry. This paper presents a brief report on a follow-up study on the dependence of the superfluid film flow on the film thickness and the substrate. There has been a major improvement in the quality of the substrates when compared with our earlier study [15]. As a result there is now agreement with the Ginzburg-Landau prediction for the superfluid/normal transition in the film. Also, there is a distinct flow-rate transition that we identify as the B-like to A-like phase transition as the film thins.

PREPARATION AND CHARACTERIZATION OF SUBSTRATES

A film flow experiment measures the flow rate of superfluid out of a beaker with the inner level higher than the outer level. The flow is driven by the gravitational head but is limited by the critical mass-flow current density of the superfluid. Thus, the flow is limited by the film with the thinnest cross section which, in normal circumstances, will be at the rim of the beaker. For our previous study we used a beaker with a flat-topped rim [15]. This had the disadvantage of having small radius of curvature edges and consequent film thinning [19]. Further, it was not a good rim geometry for obtaining a high polish. The principal change made for this study was in the design of the rim of the beaker. However, the experiments were far more extensive; five different beakers were used and in some cases the beaker surfaces were preplated with one or three monlayers of ^4He atoms.

The new beakers all had rounded rims with radius of curvature ~ 0.5 mm. The dimensions were 2.5 mm i.d., 4.5 mm o.d. by 20 mm high. The rims could be uniformly polished, both mechanically and electrically. The rounded rims provided a meniscus geometry that was analysable, although with more diffuculty than the case of a vertical wall. The 0.5 mm radius of curvature still produces film thinning, but only ~ 20% which is manageable. At the beginning it was necessary to understand the nature of the ^3He meniscus as the level of ^3He in the beaker is varied. This

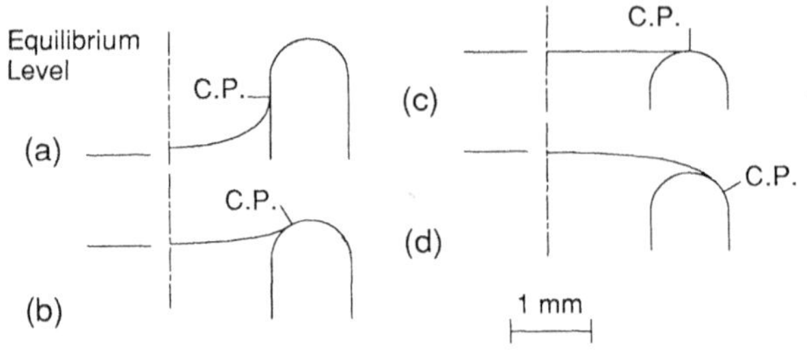

Fig. 2. Shape of the liquid meniscus within the round rim beaker for several free surface levels (shown on the left) relative to the rim of the beaker. Figure 2(d) is the upper limit for the meniscus.

understanding is summarized in Figure 2 which shows the meniscus shape for four different levels of ^3He in the beaker. Figure 2(d) shows the upper limit for a stable meniscus. Beyond that limit a viscous liquid will run over the rim. For our case the limiting level was 220 μm above the rim. The labels C.P. show the contact points for the meniscii. As the level rises, the film is thinnest at the top of the rim until the level is at the rim (Figure 2(c)); beyond that the film is thinnest at the contact point. The basis for these results will be described elsewhere [20].

The five beakers were two copper, two stainless steel and one glass. All metal beakers were mechanically polished with progressively finer polishing powder. The copper beakers were given an additional electropolish. The glass beaker rim was fused and consequently only approximated a semicircular cross-section. Table 1 is a summary of the beakers. Electron microscope pictures were taken of the metal beaker rims. However our experience soon taught us that for film flow the electron micrograph

Table 1. The Beakers used in this Film Flow Study

Material	Code	Preparation	$\alpha_4(10^{-10}m^{4/3})$	Label
Copper	Theory		60	
	RC1	M+E	105 ± 10	Rough
	RC2	M+E	75 ± 5	Smooth
	W/P		126	
Stainless Steel	Theory		50 - 60	
	RS1	M(<50nm)	59 ± 6	Smooth
	RS3	M(1200 Grit)	95 ± 20	Rough
	Keller	M(900 Grit)	100	
Glass	RG3	Fused		

M: Mechanically Polished ; E: Electropolished
W/P: Williams and Packard

was not a good way to describe the surface. Instead we use the van der Waals thickness constant α as our characterization parameter. We have found that the smoother the surface the smaller is α and the smaller is the superfluid film flow rate.

The film thickness constant (α_4) was determined by measuring Atkins oscillations of superfluid ^4He over the rim of the beaker. Initially these were performed in the film flow chamber but there was a systematic error due, perhaps, to the u-tube geometry of the film flow apparatus. Therefore a separate chamber was designed specifically for the Atkins oscillation experiments. The earlier measurements were then recalibrated for the systematic error[20].

The results for the van der Waals parameter α_4 of the surfaces are summarized in Table 1. The multiplying factor of 10^{-10} was chosen so that the numbers that appear within the table also correspond to the film thickness in nm at a height of 1 mm above a free surface of ^4He. For comparison the theoretical parameters calculated from Dzyaloshinskii et al theory[21] and corrected by a factor 0.8 as suggested recently by Cheng and Cole[22] are given, as are other earlier measurements for polished copper and stainless steel substrates.

The two copper beakers were prepared in the same way except that RC2 was given a longer electropolish. There was a significant difference in the parameters α_4 for the two copper substrates and, as will be described later, in the film flow over the substrates. It is expected that the smoother the substrate, the closer will be α_4 to the theoretical value. Roughness causes small "puddles" that fill with helium by surface tension and add to the calculated thickness of a smooth substrate. The table shows that RC1, used for most of our work, was better than the Williams and Packard polished copper substrate but not as smooth as RC2. Further, α_4 for substrate RC2 was very close to that expected for a flat surface. This was very encouraging.

The substrate RS1 was polished very carefully and again produced a thickness constant close to ideal. RS3 was purposely left "rough" to simulate the copper beaker RC1 and confirm the influence of surface roughness on superfluid film flow. The polishing of RS3 was stopped at 1200 grit. This surface should have been comparable with Keller's stainless steel surface which was polished down to 900 grit and indeed the results show this to be so. The glass substrate was not tested. However many other metal beakers substrates were measured and the results confirmed that the smoother the surface, as judged from the extent of the polishing and the electron micrographs, the smaller was the parameters α_4.

For future reference the smoother and rougher beakers will be labelled smooth and rough, remembering that these are relative terms. For ^3He, the thickness parameter α_3 is simply related to α_4 by[15],

$$\alpha_3 = \left(\frac{4}{3}\right)^{1/3} \alpha_4$$

THE FILM FLOW EXPERIMENT

The measurement of superfluid ^3He film flow over the rim of the beakers was performed in essentially the same way as described in our earlier report[15]. Therefore it will be described only briefly here, with emphasis on the more careful preparation of the meniscus prior to entering the superfluid state.

The bottom of the beaker was open to a 12 cm³ reservoir of liquid ^{3}He that, during a run, was cooled to below 1 mK by a $PrNi_5$ refrigerator via a sintered silver powder heat exchanger. The level of ^{3}He in the beaker was determined by a capacitance level detector fitted in a separate tower above the ^{3}He reservoir. The beaker was surrounded by a closed copper cell which also contained a capacitance level detector to determine the ^{3}He level outside the beaker.

Prior to cooling below 1 mK, the ^{3}He level was raised in the beaker to its full height (see Figure 2(d)) by allowing some to spill over. This was done by condensing extra ^{3}He gas into the ^{3}He reservoir with the reservoir at 80 mK. The low temperature stage (refrigerant plus ^{3}He reservoir and experiment towers) was then precooled to 10 mK with the dilution refrigerator. The negative expansion coefficient of liquid ^{3}He[23] caused a continual spillover of ^{3}He during this precool. Computer simulations,taking into account the measured flow rate of normal liquid ^{3}He as a function of gravitational head at 80 mK, the temperature dependence of the viscosity and of the expansion coefficient, and the temperature versus time profile of the precool demonstrated that the free surface would always have been within a few micrometres of the equilibrium upper limit at the end of the precool.

The demagnetization of the refrigerant to below 1 mK took approximately 3 hours. During this time the ^{3}He at the point of contact of the meniscus with the rim would have been held almost immobile by the high viscosity of the normal fluid below 10 mK. Thus, just before the onset of superfluidity we estimate that the volume expansion from the final 10 to 1 mK cooldown left the ^{3}He level about 10 μm above the equilibrium upper limit.

On cooling below 1 mK superflow started, usually with a large flow rate until the level dropped from 20 to 100 μm. There followed an abrupt change in the flow rate, down by about an order of magnitude. For a period of 20 hours or longer the level detectors were monitored while ^{3}He flowed smoothly over the rim of the beaker.

There were two thermometers, a lanthanum diluted cerium magnesium nitrate (LCMN) magnetic thermometer for use below 1 mK and a melting curve thermometer to monitor the precool and to calibrate the LCMN thermometer. The time constant of the LCMN thermometer was about 2 hours. However, this caused no problem because the warm-up rate of the low temperature stage was less than 1 μK per hour at 0.4 mK.

When sufficient flow data had been accumulated, heat was added to the low temperature stage to raise the temperature in steps in order to calibrate the LCMN thermometer against the melting curve thermometer. The $PrNi_5$ was then remagnetized, the beaker topped up with ^{3}He and the whole procedure repeated for a different measurement temperature. The measurement temperature was varied by varying the $PrNi_5$ magnetizing field. Film flow was studied over the temperature range 0.4 mK to the bulk transition temperature at 0.93 mK.

FILM FLOW RESULTS

Most of the experimental effort went into the copper beaker RC1. At the time it was believed that this was a smooth beaker because the trial polishes on similarly prepared beakers produced excellent surfaces as judged from electron micrographs. However after the film flow experiments with RC1, it was characterized by electron micrsocopy and Atkins oscillations and found to be relatively rough. This was good and bad. Being relatively rough the ^3He film was some 50% thicker than on a similar smooth beaker and there was substantial film flow. However, a strict comparison with theory is marred by some uncertainty in the profile of the substrate. Nevertheless we believe that, taking all of the results into account, the results for this beaker are significant. First the results for this rough copper breaker will be presented. Then the effect of preplating the beaker surface with one atomic layer of ^4He will be described. Finally, the results for a similarly rough stainless steel beaker and the smooth copper and stainless steel beakers will be compared with those for the rough copper beaker.

Pure ^3He over the Rough Copper Beaker

The primary data from a film flow measurement is a plot of the levels inside and outside the beaker as a function of time. The slope of these graphs is a measure of the flow rate. When multiplied by the effective area this gives the volume flow rate. Figure 3 shows a representative flow rate curve; it is a plot of dh/dt as a function of h where h is the difference between the meniscus level and the equilibrium upper limit of the meniscus. The closed circles are for flow over the bare copper beaker at 0.43 mK. The flow rate decreases rapidly as the level drops because the film thins. It is expected that as the film thins, the critical temperature, the superfluid density and the critical current density will

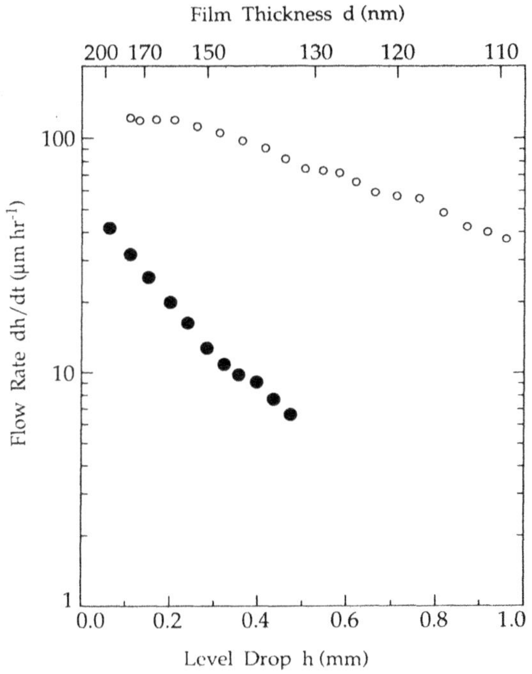

Fig. 3. Flow rate as a function of the level drop below the equilibrium upper limit. The closed circles are for the bare substrate and open circles for the ^4He preplated substrate. In both cases the temperature was 0.45 mK.

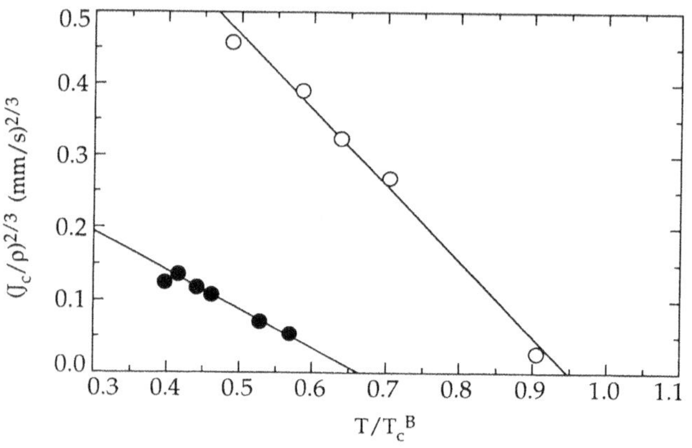

Fig. 4. The critical current density as a function of reduced temperature for ^3He films 150 nm thick. The closed circles are for the bare substrate and the open circles for the ^4He preplated substrate.

all decrease, as also does the flow cross-section of course. The curve shown is one of seven that were measured in the temperature range 0.37 mK to 0.68 mK. The flow rate decreases with increase in temperature. The smallest film thickness along the flow path can be calculated as a function of h from the measured van der Waals parameter $\alpha_3 = 1.10\alpha_4$, the density and surface tension of liquid ^3He, and the known geometry of the round rim beaker[20]. This calculated film thickness is shown along the upper scale. By taking vertical cuts down the family of flow rate curves the critical current density j_c can be determined as a function of temperature for different film thicknesses. One representative plot is shown in Figure 4. The ordinate is $j_c^{2/3}$ to reflect the predicted $(1 - T/T_c^F)^{3/2}$ dependence of the critical current density if limited by pair-breaking[8,24]. T_c^F is the transition temperature in the film. From this plot, two numbers can be derived: the critical current density at zero temperature and the transition temperature for that film thickness. It was found that the critical current density for any particular film thickness was about an order of magnitude smaller than expected for pair-breaking[8,24]. However as shown in Figure 5, the thickness dependence of the transition temperature in the film was in excellent agreement with the predictions of Ginzburg-Landau theory for a diffuse boundary, i.e. $D/\xi(T) = \pi$.* In Figure 5, T_c^F/T_c^B is plotted as a function of D^{-2} to bring out the dependence of $\xi(T)$ upon $(1-T/T_c^B)^{-1/2}$.

The Effect of a ^4He Monolayer

Freeman et al[17] discovered that two monolayers of ^4He atoms, preplated onto a mylar surface, transformed that surface into a specular

* In view of the lack of a calculation for a slab with one diffuse and one specular surface, we treat the film as a slab with two diffuse boundaries and double the thickness of the film. That is, we treat the free surface of the film as a mirror surface for the ^3He quasiparticles.

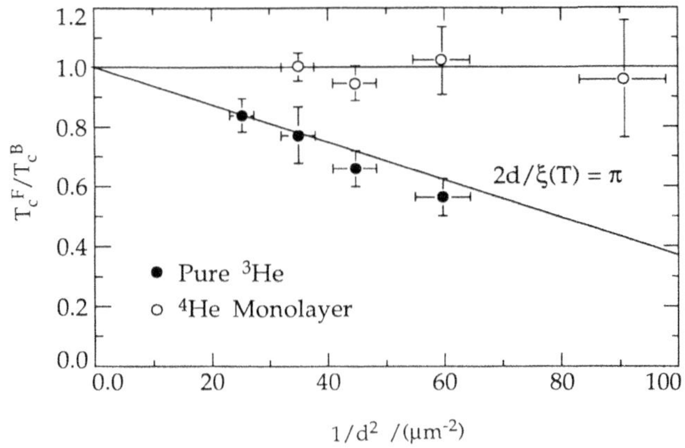

Fig. 5. The superfluid/normal transition temperature, plotted as T_C^F/T_C^B as a function of $1/d^2$ where d is the film thickness. The line $2d/\xi(T) = \pi$ is the Ginzburg-Landau theory prediction for the transition for a slab with diffuse boundaries. The closed circles are for the bare substrate and the open circles for the ^4He preplated substrate.

boundary for ^3He quasiparticles, so suppressing the decrease of the transition temperature in thin film superfluid ^3He. With this in mind we preplated the surface of the rough copper beaker RC1 with one monolayer of ^4He atoms and repeated all of the above experiments. The change in the results was dramatic, as is seen from the open circles in Figures 3, 4 and 5. At the same temperature the flow rate is significantly higher and falls much less rapidly as the level drops. For the same film thickness the critical current density is larger. Finally, and most striking, is the absence of any decrease in T_c^F as the film thins.

The Flow Transition

In all of the film flow runs there was always an abrupt flow rate transition when the level had dropped by between 20 and 100 micrometres, depending upon the temperature, beaker, and whether there was preplating with ^4He. The universality of the transition and systematic variation with the above parameters precludes any artifact such as a particular blemish on one of the rims. At first it was attributed to overfilling of the beaker due to the negative thermal expansion of liquid ^3He. However, computer simulations and a study of the flow of normal ^3He over the rim convinced us otherwise. We now believe that the flow rate transition is genuine and corresponds to the predicted transition from the B-like phase, characteristic of bulk and thick film ^3He, to the A-like phase characteristic of thin film ^3He. The flow rate transition does not coincide with the phase boundary shown in Figure 1. This, however, is not surprising because the transition is first order[25] and subject to superthinning.

Other Beakers

Film flow over the smooth copper and stainless steel beakers was much the same and quite different from the rough copper beaker. Following cooldown below 1 mK there was the initial high flow rate followed by an abrupt stop to the flow. The total level drop was only about 30 μm for the smooth copper and 75 μm for the smooth stainless steel. However, when these level drops are converted to a film thickness, the resultant thickness corresponds to that expected for the superfluid to normal phase transition for a diffuse substrate. It appears that the ^{3}He film remains in the B-like phase all the way to the normal boundary. The difference is the significantly thinner film on the smooth substrate. Note that the superfluid to normal phase boundary is second order and not subject to superthinning.

When preplated with two monolayers of ^{4}He atoms there was both B-like and A-like flow over the smooth copper beaker rim. That is, the suppression of T_c was inhibited by the prescence of the ^{4}He monolayers.

The results for the bare rough stainless steel beaker were similar to those for the bare rough copper beaker, apart from the larger flow before the flow transition that was also observed for the smooth stainless steel beaker. One other difference was that following the flow transition it took about an hour for the A-like flow rate to build up. However, once started the A-like flow continued for many hours.

The only experiment performed with the glass beaker was with preplated ^{4}He on the surface. The results were similar to those for the rough copper beaker except that the flow transition was not abrupt. This was attributed to an unevenness of about 50 μm in the rim level around the beaker.

CONCLUSION

This study of film flow over several round rim beakers has produced a systematic picture of superfluidity in thin film ^{3}He in the region $d \sim \xi(T)$. Results for four metal beakers, rough and smooth copper, rough and smooth stainless steel, have shown a suppression of T_c in agreement with Ginzburg-Landau theory for slab geometry with diffuse boundaries. Preplating the beaker substrate with just one monolayer of ^{4}He atoms appears to switch the substrate from diffuse to specular for ^{3}He quasiparticle scattering. This does make sense. It is known that a monolayer of ^{3}He on a substrate is solid-like and behaves as a Curie-Weiss ferromagnet at low temperatures[26]. It has been proposed that such a magnetic layer would scatter ^{3}He quasiparticles diffusely[27]. Replacing the first monolayer of ^{3}He by a monolayer of ^{4}He removes the magnetic surface layer. The prescence of the ^{4}He weakens the van der Waals bond to the substrate and leaves the first layer of ^{3}He at liquid density and hence almost non magnetic Fermi liquid. Therefore one monolayer of ^{4}He is sufficient to switch off completely the magnetic scattering.

There is a flow rate transition that we identify with the predicted B-like to A-like phase transition as the ^{3}He film is thinned. Our experiment measures the film "on the fly"-- as the level drops the film thins. Therefore at present we cannot control the rate at which the film thins. The transition is first order and therefore subject to superthinning. Most of the film flow that we measure occurs after the flow rate transition and would therefore correspond to flow of the A-like phase.

The measured critical current density for the A-like phase is an order of magnitude smaller than expected for dissipation limited by pair breaking. However, the critical current density of the B-like phase may be about right; the speed with which this region of the phase diagram is traversed is too fast for a good estimate of the temperature. It has been suggested that other dissipation mechanisms in the A-like phase are quite likely-e.g. motion of texture[28].

The main influence of the substrate, rough or smooth, copper or stainless steel, seems to be its influence on the film thickness. It is satisfying that the film thickness determined from Atkins oscillations gives good agreement of the measured ^{3}He superfluid transition temperature with the prediction of Ginzburg-Landau theory. But then, both film flow and Atkins oscillations average over a region of the substrate. The most dramatic influence of the substrate is whether the first monolayer of helium is ^{3}He or ^{4}He.

Finally, it is noted that with a preplating of ^{4}He there was film flow over the rough copper beaker with the level 0.2 mm down, and the temperature 0.84 mK. In this circumstance $d/\xi(T) \sim 1$, the cross-over point from 3-D to 2-D for the superfluid film.

ACKNOWLEDGMENTS

We would like to acknowledge helpful discussion with U. Erb, A.L. Fetter, M. Freeman, R.B. Hallock, R. Harris-Lowe, J. Kurkijarvi, D.D. Osheroff, J.D. Reppy, J.A. Sauls, N. Schopohl, Z. Tesanovic and R.R. Turkington. The patience and skill of T. Bovingdon and J. Thompson in providing many tiny beakers, and S. Kurylowicz and J. Odell in providing endless liquid helium is also acknowledged. This work has been supported by the Natural Sciences and Engineering Research Council of Canada.

REFERENCES

1. I.E. Dzyaloshinskii, E.M. Lifshitz and L.P. Pitaevskii, Advances in Phys. 10, 165 (1961).
2. V. Ambegaokar, P.G. deGennes and D. Rainer, Phys. Rev A9, 2676 (1974).
3. A.L. Fetter and S. Ullah, J. Low Temp. Phys. 70, 515 (1988).
4. D.D. Osheroff, R.C. Richardson and D.M. Lee, Phys. Rev. Lett. 28, 885 (1972).
5. A.J. Leggett, Rev. Mod. Phys. 47, 331 (1975).
6. L.H. Kjaldman, J. Kurkijarvi and D. Rainer, J. Low Temp. Phys., 33, 577 (1978).
7. G. Barton and M.A. Moore, J. Low Temp. Phys. 21, 489 (1975).
8. K.W. Jacobsen and H. Smith, J. Low Temp. Phys, 67, 83 (1987).
9. Y.-H. Li and T.-L. Ho, Phys. Rev B38, 2362 (1988).
10. Z. Tesanovic and O.T. Valls, Phys. Rev. B34, 7610 (1986).
11. J. Hara and K. Nagai, J. Low Temp. Phys. 72, 407 (1988).
12. M.T. Manninen and J.P. Pekola, J. Low Temp. Phys. 52, 497 (1983).
13. J.P. Pekola, J.C. Davis and R.E. Packard, J. Low Temp. Phys. 71, 141 (1988).
14. V. Kotsuba, K.D. Hahn and J.M. Parpia, Phys. Rev. Lett. 58, 804 (1987).
15. J.G. Daunt, R.F. Harris-Lowe, J.P. Harrison, A. Sachrajda, S.C. Steel, R.R. Turkington and P. Zawadzki, J. Low Temp. Phys. 70, 547 (1988).
16. J.C. Davis, A. Amar, J.P. Pekola and R.E. Packard, Phys. Rev. Lett. 60, 302 (1988)
17. M.R. Freeman, R.S. Germain, E.V. Thunenberg and R.C. Richardson, Phys. Rev. Lett. 60, 596 (1988).
18. A.J. Sachrajda, R.F. Harris-Lowe, J.P. Harrison, R.R. Turkington and J.G. Daunt, Phys. Rev. Lett. 55, 1602 (1985).

19. G.A. Williams and R.E. Packard, Phys. Rev. Lett. 32, 587 (1974).
20. S.C. Steel, P. Zawadzki, J.P. Harrison and A. Sachrajda, (to be published).
21. E.S. Sabisky and C.H. Anderson, Phys. Rev A7, 790 (1973)
22. E. Cheng and M. Cole, Phys. Rev. B38, (1988).
23. D.F. Brewer and J.G. Daunt, Phys. Rev. 115, 843 (1959).
24. D. Vollhardt, K. Maki and N. Schopohl, J. Low Temp. Phys. 39, 79 (1980), H. Kleinert, J. Low Temp. Phys. 39, 451 (1980).
25. T.-L. Ho, private communication.
26. A.I. Ahonen, T. Kodama, M. Krusius, M.A. Paalanen, R.C. Richardson, W. Schoepe and Y. Takano, J. Phys. C. 9, 1665 (1976).
27. J.A. Sauls, private communication.
28. N. Schopohl, private communication.

† Present address: N.R.C., Ottawa, Canada
* Present address: Kamerlingh Onnes Laboratorium, Leiden, The Netherlands

DISCUSSION ON BULK ^3He AND $^3He/^4He$ MIXTURES

Chairmen : S. Balibar, H. Meyer

Discussion on Scherm's talk on neutron study of elementary excitations in ^3He and ^3He - ^3He Mixture.

Stringari to Scherm : You compared ^3H-^4He mixtures to pure ^4He by rescaling q with total density. Could you not rather use a scaling on the compressibility?

Scherm : The momentum of the roton minimum is closely related to the peak in the structure factor $S(q,\omega)$ and I believe that this is essentially a result of the total density. We had the choice to compare the mixture and ^4He at the same ^4He density or at the same overall density and we chose the latter. Your suggestion to rescale on the compressibility can be applied to the energy. This is equivalent to normalizing the sound velocities.

Edwards : Does the roton minimum remain at the same q when adding ^3He to ^4He?

Scherm : Yes, the roton minimum appears at the same $q_o.a$ (here a is the inter-atomic distance) in ^4He at different pressures as well as in ^3He-^4He mixtures between 0 and 20 bar. We see of course the change of the density with $X(^3He)$ via $q_o.a$.

Griffin : I would like to encourage you to continue these experiments on $S(q,\omega)$ in ^3He-^4He mixtures to temperatures through T_λ. In particular it is interesting to find out whether or not the maxon-roton peak in $S(q,\omega)$ disappears above T_λ. The dielectric formalism has been worked out for mixtures (see the paper by Griffin at this workshop). Seeing how the particle-hole excitations of the ^3He modify the temperature - dependent line shape found in pure ^4He would be a nice test of an extended Glyde - Griffin picture for superfluid mixtures.Campbell : Concerning the question of whether number density rescaling is appropriate in the analysis of the spectrum in the roton region in mixtures, it should be noted that theoretical calculations of

$S(q,\omega)$ indicate that the structure of $S(q,\omega)$ at intermediate wavelengths (including the roton region) and at short wavelengths depends only on the number density, not on the mass or on statistics. This gives some theoretical support for Scherm's choice of density rescaling of q for mixtures for his comparison of the neutron scattering experiments for mixtures to ^4He.

Platzman to Scherm and Sokol: What happens at high momentum transfers?. There must be a distinction between the mass of the quasipaticles and the mass of the recoiling particle ($q^2/2m$).

Sokol : We have carried out preliminary (unpublished) measurements of ^3He-^4He mixtures. These measurements are at $q>10Å^{-1}$. We see two single particle peaks, one centred at the ^4He recoil energy and the other centred at the ^3He recoil energy.

Discussion of Harrison's talk on the effect of substrate on ^3He flow.

Gasparini : How does the ^3He in the film experience the solid boundary and how can a substrate with a ^4He monolayer avoid bringing ζ_s and the transition temperature T_c down to zero?

Harrison : Liquid ^3He has a complex order parameter. Essentially in the B phase and for diffuse boundaries, the three components in the order parameter go to zero at the boundaries. In this way, the diffuse boundaries induce decreases of ζ_s, T_c and j_c, the critical flow. For specular boundaries, the perpendicular component of the order parameter still dips to zero at the boundary. However the components parallel to the surface do not. This holds until the thickness becomes smaller than the fermi wavelength.

Vinen: The same effect of a diffuse or a specular boundary is seen in superconductors, a normal metal-superconducting interface brings the order parameter to zero and lowers the transition temperature.

Bozler : How do you know that you have added a ^4He monolayer, when you want to plate your substrate.

Harrison : The total area of the cell, including the sintered heat exchanger, is quite large, about $50m^2$, as measured by adsorption isotherms. Thus the total helium gas required is quite easy to measure to form a monolayer of the ^4He.

Golub : Do the beaker surfaces receive any other treatment after polishing?

Harrison : Just before they are mounted, they are rinsed in ethanol.

Balibar : Why should one monolayer make the difference that it does?

Harrison : It was predicted by J Sauls (North Western University) that magnetic scattering by the first monolayer of

Harrison : It was predicted by J Sauls (North Western University) that magnetic scattering by the first monolayer of

the ^3He would cause the diffuse scattering of the quasiparticles. This first monolayer has a solid-like density and Curie-Weiss magnetism when there is an overlay of liquid ^3He to provide the required exchange interaction. One monolayer of ^4He is sufficient to remove that Curie-Weiss magnetism.

Gasparini : When determining T_c for a phase transition, it is more reliable to use a thermodynamic parameter such as superfluid density or the heat capacity, rather than the critical velocity.

Harrison : That is true, and indeed the superfluid density in confined ^3He has been used by Freeman et al at Cornell University to demonstrate the depression of T_c in confined geometry. However our business is film flow and that is what we have used to determine our values of T_c in the films.

Saunders : How are you sure that you have only one monolayer and not a superfluid layer of ^4He, for your preplating experiments.

Harrison : As noted above, we are fairly certain of the surface area in our cell. Also there was no major increase in the time constant of our Lanthanium Ceriun Magnesium nitrate thermometer, which is very sensitive to ^4He coverage. Therefore we suspect that it had very little ^4He on it. We would expect the ^4He to go preferentially to the copper surfaces and to start plating the magnetic thermometer only after a monolayer had formed on the copper. Also we know from later experiments that when there is a superfluid layer, the ^4He is more mobile and re-settles on the surfaces with a time constant of several days. This caused an initial heat leak of about 3nW. During the monolayer experiment, the heat leak was 1nW.

Gasparini : The lack of shift in T_c in the case of specular surfaces seems hard to understand in the light of the fact that if one makes the ^3He film arbitrarily thin - strictly 2D - one would certainly expect a lower T_c.

Discussion on Meyer's and Ferrell's talks on the transport properties of very dilute ^3He-^4He mixtures and the 3-fluid model.

Platzman to Ferrell : How can you tell from your theory when the conductivity will turn up again at X still smaller than 10^{-7}?

Ferrell : This predicted increase $X \leq 10^{-1}$ results from the two-parameter fit to the experimental results to $X \simeq 10^{-6}$. Now at $X = 10^{-7}$, Chui and Lipa's data point does not show this turn-up. Only experiments at smaller X could settle this point. Now, according to Chui and Lipa's theory (Phys. Rev. B 1989) the conductivity will remain flat with decreasing X, which means that the drag coefficient will increase with decreasing X, while my theory predicts that it will turn over to a constant valve.

Lauter : Can we hear above prospective experiments at $X < 10^{-7}$?

Meyer : There have been striking advances by Lipa and by Ahlers and their collaborators with a temperature resolution of better than 50 times that of germanium.thermometry. So one can certainly expect the range of measurements in the linear regime down to say $X \simeq 10^{-8}$. What puts a limit on the lowest concentration is the maximum heat flux Q_c before entering the non-linear regime. Since Q_c is roughly proportional to X, this limits the sensitivity with which one detects temperature changes.

Leiderer : What determines the relaxation times in your experiments? Should it not depend on the specific heat, and how long is τ?

Meyer : The specific heat at low enough X is eliminated from the expression for τ, which then depends only on the mass diffusion coefficient. There is a strong coupling between the thermal and the concentration relaxation and their time constants are the same. (See our paper for their magnitude).

Ferrell : The relaxation time should only be a function of T and not of X at low enough X, but the experiments show otherwise.

Laheurte : What do you think about gravity and surface adsorption effects on the concentration profiles that you assume? Is the average ^3He concentration which is measured at the top and bottom of the cell for very low concentration really known, in other words, are gravity effects important?

Meyer : I should note that our experiments are always started with almost pure ^4He, with a ^3He concentration $X(^3He) \sim 2\times10^{-9}$ to do boundary resistance measurements. Therefore the surfaces are preplated with layers of solid ^4He. It would be surprising if upon successive additions of ^3He to produce dilute mixtures there would be an appreciable substitution of ^4He by ^3He. Anyway, ^3He would less likely to be absorbtion that ^4He on solid surfaces. Concerning the gravity effect that might introduce a concentration gradient : this effect is entirely negligible. In our cell with a fluid layer height of 2mm, I can estimate that the relative difference between top and bottom, $\delta X/X \approx 10^{-6}$. This situation is very different from that near the tricritical point in helium mixtures where gravity effects are indeed very important.

Lauter : Could your anomalous Kapitza resistance be caused by a melting of your solid ^4He layer on your plates at the superfluid transition?

Meyer : This would very much surprise me that ^4He would melt away at precisely that point. Rather I think that this melting should not be a sharp process but rather one spread over a certain temperature range above 3K. When "cleaning" our cell from absorbed gas, we have to heat it well above 4K.

Golub : Using a collimated neutron beam as a probe and ultraviolet scintillations in the liquid as a detector, it may be possible to measure distributions of ^3He at concentrations of $X \leq 10^{-8}$. By measuring in the presence of applied heat currents, we could address some of the questions raised in the talks of Ferrell and Meyer.

Griffin : The experiments at Duke suggest that the Khalatnikov-Landau theory does not work at very low concentrations. Is this a major event in the last few years?

Ferrell : Well, one can also have the suspicion that Meyer and his colleagues are playing a practical joke on the theorists, and later tell us that all is in order.

COLLECTIVE EXCITATIONS AND FERMI-LIQUID EFFECTS IN ^{4}HE AND ^{3}He-^{4}He FILMS

E. Krotscheck and C. J. Tymczak

Center for Theoretical Physics, Department of Physics
Texas A&M University, College Station, Texas 77843

INTRODUCTION

This contribution reports on recent progress in the development of a uniform picture for non-uniform quantum many-body systems. We give a brief account of the optimized Hypernetted-Chain (HNC/EL) theory and its implementation for non-uniform systems. The theory is applied to films of ^{4}He atoms adsorbed to a substrate and to ^{3}He-^{4}He mixture films. We discuss the structure of collective excitations and their transition from the thin, surface-mode dominated to the thick, zero-sound dominated films. We calculate properties of ^{3}He-impurities in the films and study the interaction of such atoms with both the background and other impurities.

GROUND STATE STRUCTURE

The study of the structure and the excitation spectrum of films of helium atoms adsorbed to a substrate provides an extremely interesting theoretical problem in the sense that state of the art many-body theory is required for a satisfactory description. Part of this contribution will, therefore, be devoted to a discussion of the considerations needed for a comprehensive understanding of the relevant physics, and for an interpretation of the the experimentally accessible information. The basic structure of the problem is most clearly illuminated in ^{4}He at zero temperature. The simplicity of the system makes the implementation of the theory technically straightforward and the physical interpretation most transparent.

Helium atoms are known to interact via a potential that consists essentially of a repulsive hard-core and an attractive long-ranged part[1]. The interplay between the hard-core repulsion and the long-ranged attraction determines the saturation density and the binding energy of the system. The hard-core like short-ranged repulsion plays an important role in a restricted geometry like, for example, a system of atoms adsorbed to a substrate: Due to the geometric restrictions, the atoms will form a layer-structure on the substrate, which would not be present for weakly interacting particles in an external field.

We will see in this meeting a number of contributions[2,3] that address the experimental evidence for the formation of layers in adsorbed helium films. In particular neutron-scattering experiments[3] have

Excitations in Two-Dimensional and Three-Dimensional Quantum Fluids
Edited by A.G.F. Wyatt and H.J. Lauter, Plenum Press, New York, 1991

recently provided a wealth of information; the interpretation of the data is a challenge for theorists. It is worth noting here that neutron scattering experiments normally probe the long-ranged structure of the system, nevertheless one might be able to see some evidence of a layer formation. We find therefore an intriguiging interplay between long- and short- ranged correlations in adsorbed helium films.

Most many-body theorists agree today that a self-consistent treatment of long- and short ranged correlations is an important requirement for *any* satisfactory microscopic description of strongly interacting quantum systems[4]. Adsorbed films of helium atoms are the most prominent example for a system where the simultaneous importance of short- and long-ranged correlation is immediately evident. "Mean-field" or "density-functional" theories fail badly[5,6], even though they have some apparent success in describing the surface of bulk quantum liquids[6]. It is worth mentioning here that there is also evidence for the importance of short-ranged structure in quantum-liquid clusters[7] qhich can not be accounted for by the non-local density functional theory[8] that will be presented in this meeting.

The only satisfactory approach known today is a truly microscopic theory that takes both short- and long-ranged effects systematically into account. The quality of results will then depend on nothing but the investment of computational effort, large-scale numerical simulations[9] providing the highest level of accuracy at the expense of a loss of flexibility in the application. In this contribution we will not attempt to reach the best possible level of accuracy for the description of the ground-state structure, but rather devise a theory that is flexible enough to allow us the study of collective excitations, impurity states, and impurity quasiparticle interactions.

We start the microscopic description of a many-body system in an external potential with an empirical Hamiltonian

$$H = \sum_{i=1}^{N} \left\{ -\frac{\hbar^2}{2m}\nabla_i^2 + U_{ext}(\mathbf{r}_i) \right\} + \sum_{1 \le i < j \le N} v(|\mathbf{r}_i - \mathbf{r}_j|), \tag{1}$$

where $U_{ext}(\mathbf{r})$ is an external potential, and $v(|\mathbf{r}_i - \mathbf{r}_j|)$ is the interaction between individual particles. Our goal is to predict the behavior of N of particles in a given volume Ω from no other knowledge but the fundamental interactions. ^{4}He is one of the systems where this problem can partly be solved by large-scale numerical simulations[9]. But there is a large number of experimental data on, for example, low-lying excitated states, whose energetics, structure, and lifetime are most adequately discussed in the language of conventional many-body theory. The most successful approach for this purpose available today is the optimized hypernetted chain theory.

One starts[10] with a variational *ansatz* for the ground-state wave function of a system of N particles with coordinates $\mathbf{r}_1, \ldots, \mathbf{r}_N$:

$$\Psi_0(\mathbf{r}_1, \ldots, \mathbf{r}_N) = \exp \frac{1}{2} \left\{ \sum_{1 \le i \le N} u_1(\mathbf{r}_i) + \sum_{1 \le i < j \le N} u_2(\mathbf{r}_i, \mathbf{r}_j) + \ldots \right\} \Phi_0(\mathbf{r}_1, \ldots, \mathbf{r}_N). \tag{2}$$

The one-body function $u_1(\mathbf{r})$ describes the spatial structure of the system, the two-body function $u_2(\mathbf{r}_i, \mathbf{r}_j)$ describes the short- and long-ranged correlations between pairs of particles, and $\Phi_0(\mathbf{r}_1, \ldots, \mathbf{r}_N)$ is the ground state of a weakly interacting model system; its main purpose is to guarantee antisymmetry of the wave function for Fermions. The correlation functions and the model wave function are determined by minimization of the energy-expectation value E_0:

$$\frac{\delta E_0}{\delta u_1(\mathbf{r})} = 0, \qquad \frac{\delta E_0}{\delta u_2(\mathbf{r}_i, \mathbf{r}_j)} = 0. \tag{3}$$

If one uses approximations for the evaluation of the energy functional, it is essential that such approximations lead to approximate Euler equations which have the same formal structure and mathematical properties as the Euler equations of the exact problem; the variational problems (3) may otherwise be meaningless. The existence of physically meaningful solutions of the variational problems (3) has been shown to be intimately related to an adequate, self-consistent treatment of both short- and long ranged correlations. It turns out that the so-called HNC hierarchy of approximations has such meaningful variational minima at each level of its implementation[11]; it includes both the right physics and the appropriate mathematical structure. Unfortunately, the simplest level of the HNC/EL hierarchy provides only a qualitative description of the relevant physics, the quantitative predictions can generally be considered to have an error margin of 30 %. Methods to improve the accuracy of the theory are well known[12,13], but have not yet been implemented for non-uniform systems.

To show how short- and long ranged correlations are treated in the HNC/EL theory, we start with the basic equations for the simplest case, the homogeneous Bose liquid. These equations are most conveniently formulated in terms of the pair correlation function $g(r)$, and the static structure function $S(k) = 1 + \rho \int d^3r[g(r) - 1]e^{i\mathbf{k}\cdot\mathbf{r}}$. One finds[11]

$$S(k) = \left\{\sqrt{1 + \frac{4m\rho}{\hbar^2 k^2}\tilde{V}_{ph}(k)}\right\}^{-\frac{1}{2}} \tag{4}$$

where

$$V_{ph}(r) = g(r)v(r) + \frac{\hbar^2}{m}|\nabla\sqrt{g(r)}|^2 + [g(r) - 1]w_I(r). \tag{5}$$

is the "particle-hole interaction", note that

$$V_{ph}(|\mathbf{r} - \mathbf{r}'|) = \frac{\delta^2 E_0}{\delta\rho_1(\mathbf{r})\delta\rho_1(\mathbf{r}')}, \tag{6}$$

in particular $\rho\tilde{V}_{ph}(0+) = mc^2$, where c is the zero sound velocity. $w_I(r)$ is the so-called "induced interaction", its analytic form is irrelevant for our further discussions. In Eq. (4) we recover the RPA expression for the static form factor. The HNC/EL theory supplements the RPA with a microscopic theory of the particle- hole interaction $V_{ph}(r)$. Some algebraic manipulations also permit to write the equations in a "Bethe-Goldstone" like form.

$$\frac{\hbar}{m}\nabla^2\sqrt{g(r)} = [v(r) + w_I(r)]\sqrt{g(r)} \tag{7}$$

which shows that the short-ranged correlations are also treated properly by the theory.

We see from Eq. (4) that the theory has, as required, no *uniform* solution with negative compressibility, $\rho\tilde{V}_{p-h}(0+) < 0$, c.f. Eq. (6). The theory has also a unique signature for symmetry breaking with a finite wave number: Then, the term under the square-root Eq. (4) becomes negative at *finite* wave numbers. If an instability of either kind occurs, one must allow for the appropriate symmetry breaking in the wave function (1) by including a one-body function $u_1(\mathbf{r})$ and breaking the translational invariance of $u_2(\mathbf{r}_i, \mathbf{r}_j)$. The theory then has solutions if the system is, *within the assumed geometry*, stable against small perturbations.

To derive an algorithm for the numerical solution of the generalizations of the equations (4-6) in geometries with broken symmetry is is highly non-trivial even for the simplest types of symmetry breaking. This is simply due to the amount of data required: If roughly 100 data points are needed to tabulate a typical pair distribution function in the uniform case, one needs typically 100^3 data points for the plane surface or spherical geometry. An important breakthrough in the development of microscopic theories was the development of an efficient algorithm to solve the generalizations of Eqs. (4-6) for uniform systems[14]. We will not try to describe the technical details of the algorithm, it suffices to mention that it is based on a normal-mode decomposition which gives the collective excitated states of the system as a natural byproduct.

The physical system considered in our application of the basic formalism is a number of ^{4}He atoms adsorbed to a substrate which is described by an external field $U_{ext}(z)$. The substrate potential $U_{ext}(z)$ is obtained by averaging Lennard-Jones interactions over a half space. Given the substrate potential, the two-body interaction, and the surface coverage $n = \int dz \rho_1(z)$, the physical problem is completely defined. We have solved the HNC/EL equations for the ^{4}He numerically[15] for a sequence of surface coverages n. Some typical examples of density profiles are shown in Fig. 1. Most remarkable are the strong density-oscillations of the ^{4}He film, which are due to the geometric core-exclusion between the ^{4}He particles.

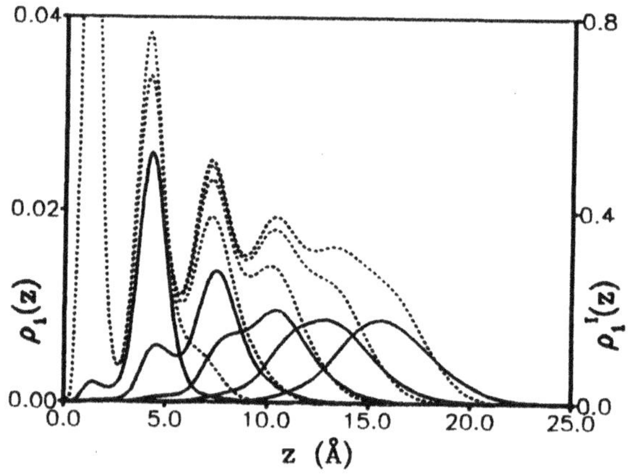

Fig. 1. The density $\rho_1(z)$ of the background film of ^{4}He atoms (dotted lines) and of the ^{3}He impurity, $\rho_1^I(z)$ (solid lines) are shown for surface coverages of n = 0.15, 0.20, 0.25, 0.30, and 0.35 ^{4}He atoms/Å^2. The ^{4}He densities of all films shown are virtually identical within the first layer, whose density maximum is about 0.08 atoms/Å^3. The ^{3}He impurity density is normalized such that $\int dz \rho_1^I(z) = 1$. The figure is taken from Ref. 16

COLLECTIVE EXCITATIONS

We have already mentioned above that the solution of the optimization problem for the ground state implies the calculation of the collective excitations of the system. Instead of giving the technical details of the optimization procedure and the relationship to a microscopic theory of excited states[17], we will provide here an alternative route for the derivation of the excitation spectrum, which is more general in the sense that it does not rely on the assumption of a variational wave function, and perhaps more familiar to the technically less inclined reader.

Following Feynman[18], we write a trial function of the excited state as

$$\Psi_F(\mathbf{r}_1, \ldots \mathbf{r}_N) = F(\mathbf{r}_1, \ldots \mathbf{r}_N)\Psi_0(\mathbf{r}_1, \ldots \mathbf{r}_N), \tag{8}$$

where $F(\mathbf{r}_1, \ldots \mathbf{r}_N) = \sum_{i=1}^{N} f(\mathbf{r}_i)$.

Assuming that $\Psi_0(\mathbf{r}_1, \ldots, \mathbf{r}_N)$ is *either* the *exact* ground state wave function *or* is of the Jastrow-Feenberg form (2) with optimized one- and two-body correlations, the energy difference between the (variational) ground state E_0 and the trial excited state $E_F \langle \Psi_F \mid H \mid \Psi_F\rangle / \langle \Psi_F \mid \Psi_F \rangle$ is

$$\begin{aligned} \hbar\omega \equiv E_F - E_0 &= \frac{1}{2}\frac{\langle \Psi_0 \mid [F,[T,F]] \mid \Psi_0\rangle}{\langle \Psi_0 \mid F^2 \mid \Psi_0\rangle} = \frac{\hbar^2}{2m} \sum_{1\le i\le N} \frac{\langle \Psi_0 \mid |\nabla f(\mathbf{r}_i)|^2 \mid \Psi_0\rangle}{\langle \Psi_0 \mid F^2 \mid \Psi_0 \rangle} \\ &= \frac{\hbar^2}{2m} \frac{\int d^3r \rho_1(\mathbf{r}) |\nabla f(\mathbf{r})|^2}{\int d^3r \rho_1(\mathbf{r})|f(\mathbf{r})|^2 + \int d^3r d^3r' f(\mathbf{r})[\rho_2(\mathbf{r},\mathbf{r}') - \rho_1(\mathbf{r})\rho_1(\mathbf{r}')]f(\mathbf{r}')}. \end{aligned} \tag{9}$$

This energy difference is an upper bound for the exact or the variational excitation energy. For any trial excitation function $f(\mathbf{r})$, the transition density is

$$\delta\rho(\mathbf{r}) = \rho_1(\mathbf{r})f(\mathbf{r}) + \int d^3r'[\rho_2(\mathbf{r},\mathbf{r}') - \rho_1(\mathbf{r})\rho_1(\mathbf{r}')]f(\mathbf{r}'). \tag{10}$$

The lowest upper bound for the excitation energy is obtained by minimizing the energy difference $\hbar\omega$ with respect to the excitation function $f(\mathbf{r})$. This leads to the eigenvalue equation

$$H_1\sqrt{\rho_1(\mathbf{r})}f(\mathbf{r}) = \hbar\omega \int d^3r' S(\mathbf{r},\mathbf{r}')\sqrt{\rho_1(\mathbf{r})}f(\mathbf{r}') \tag{11}$$

with the coordinatate space representation of the static structure function

$$S(\mathbf{r},\mathbf{r}') = \delta(\mathbf{r}-\mathbf{r}') + \frac{\rho_2(\mathbf{r},\mathbf{r}') - \rho_1(\mathbf{r})\rho_1(\mathbf{r}')}{\sqrt{\rho_1(\mathbf{r})\rho_1(\mathbf{r}')}}, \tag{12}$$

and the non-local kinetic energy operator

$$H_1 = -\frac{\hbar^2}{2m}\frac{1}{\sqrt{\rho_1(\mathbf{r})}}\nabla\cdot[\rho_1(\mathbf{r})\nabla]\frac{1}{\sqrt{\rho_1(\mathbf{r})}}. \tag{13}$$

This is the desired generalization of the Feynman dispersion relation[16]. Note that in the uniform limit $\rho_1(\mathbf{r}) = \rho_0 = const$ and $S(\mathbf{r},\mathbf{r}') = S(|\mathbf{r}-\mathbf{r}'|)$, Eq. (11) leads directly to the Feynman dispersion relation $\hbar\omega(k) = \hbar^2k^2/2mS(k)$. Having calculated the excitation energies and the transition densities, we can finally construct the dynamic response function

$$S(\mathbf{r},\mathbf{r}';\omega) = \delta\rho_\omega(\mathbf{r})\delta\rho_\omega(\mathbf{r}'). \tag{14}$$

The two first sumrules

$$\sqrt{\rho_1(\mathbf{r})}H_1\sqrt{\rho_1(\mathbf{r}')} = \sum_\omega \hbar\omega S(\mathbf{r},\mathbf{r}';\omega) \tag{15}$$

and

$$\sqrt{\rho_1(\mathbf{r})}S(\mathbf{r},\mathbf{r}')\sqrt{\rho_1(\mathbf{r}')} = \sum_\omega S(\mathbf{r},\mathbf{r}',\omega) \tag{16}$$

are trivial consequences of the structure of the eigenvalue problem (11) and the definition (14). Note that the sum over the excitation energies has to be interpreted as an integral if the spectrum is continuous.

The solution of an equation equivalent to Eq. (11) is part of the algorithm, therefore we also get the spectrum of collective excitations as a part of the theory. In the plane surface geometry, Eq. (11) can be decoupled as a function of the momentum $\mathbf{q}_{\|}$ parallel to the surface, i.e. the eigenfunctions $f_\omega(\mathbf{r})$ and the corresponding transition densities $\rho_\omega(\mathbf{r})$ have the form $f_\omega(\mathbf{r}) = f_{\epsilon,\mathbf{q}_{\|}}(z)e^{i\mathbf{q}_{\|}\cdot\mathbf{r}_{\|}}$, where z the distance of a particle from the surface. The index ϵ labels the quantum number related to the motion in the direction perpendicular to the surface. This quantum number is discrete for energies $\hbar\omega + \mu = \epsilon + \hbar^2 q_{\|}^2/2m + \mu < 0$, otherwise it is continuous. A numerical calculation will, of course, always provide a descrete spectrum, however, if the problem is discretized on a sufficiently large mesh, reliable information on the level density can be obtained.

Normally, the dynamic structure function is discussed in momentum space. In our geometry, the dynamic structure function is diagonal in the momentum $\mathbf{q}_{\|}$ parallel to the surface, and for the sake of simplicity we consider also only diagonal momenta k perpendicular to the surface. Thus, we define the *diagonal* dynamic structure function in momentum space

$$S(k,\mathbf{q}_{\|};\omega) = \int dz dz' d^2 r_{\|} e^{i[(z-z')k+\mathbf{q}_{\|}\cdot\mathbf{r}_{\|}]} S(\mathbf{r},\mathbf{r}';\omega), \tag{17}$$

where $\mathbf{r}_{\|}$ is the distance between $\mathbf{r}$ and $\mathbf{r}'$ parallel to the surface. Experimentally most interesting is, of course, the case of grazing angles $k = 0$.

Figs. 2a-c show contour plots of our calculated static structure functions $S(0,q_{\|};\omega)$ for a sample of three films, typically having two, three, and four layers. Darker areas correspond to higher intensity. When discussing these results, one has to keep in mind that the HNC approximation yields a structure function whose peak is somewhat too low. More importantly, our theory is based on the Feynman ansatz, which leads in the bulk liquid to a phonon-roton spectrum that is about a factor of two too high.

A number of observations apply:

(a) The lowest-lying collective excitation is the ripplon. In all cases considered here, the ripplon dispersion relation lies below the continuum which is determined by the line $-\mu + \hbar^2 q_{\|}^2/2m$.

(b) We see also a number of dispersion branches that lie, for small momenta, roughly parallel to the ripplon dispersion relation. As the films become thicker, the density of these branches increases. One can think of these modes as of ripplons with a transverse momentum superimposed. These dispersion branches occasionally disappear into or reappear out of the continuum.

(c) At higher energies, we see in all cases in the continuum peaks of the structure function that have the general shape of a phonon-roton dispersion relation. In fact, one can apparently distinguish between the dispersion branches corresponding to different layers: The double-layer film shows one phonon-roton-like dispersion branch, whereas the four-layer film lets us identify three distinct phonon-roton branches. The first layer in our calculation is apparently so highly compressed that its excitation energy is well above the scale shown here. In other areas of the $(q_{||}, \omega)$-plane is very little strength.

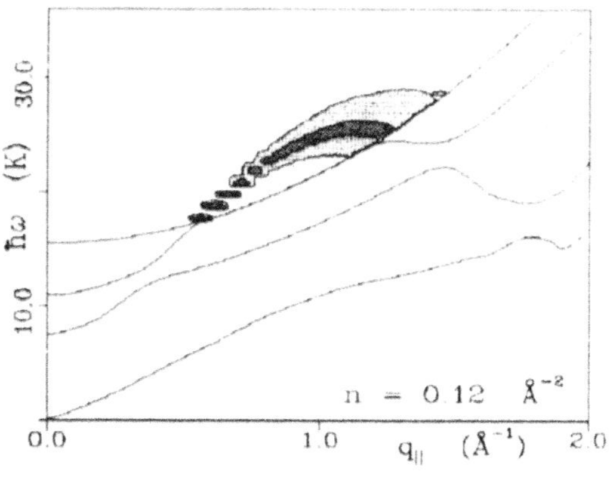

Fig. 2a

(d) In the momentum range of the roton minimum, we see a typical example of a level-crossing between different excitation modes. Indeed, one can quite easily identify volume and surface modes by looking at the shape of their transition density. We have shown in Ref. 15 that below the roton minumum the lowest-lying excitation is a clear surface mode, whereas after the level-crossing the lowest-lying excitation is a volume mode.

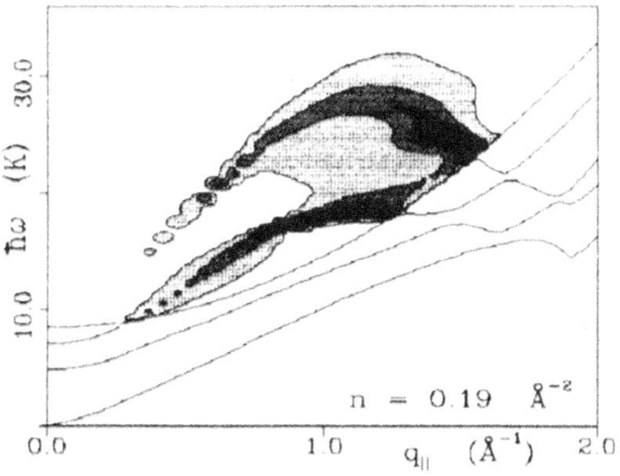

Fig. 2b

Most remarkable is for the very thick film the high density of collective modes in the area of the roton minimum. Of course, in reality these modes will be smeared out such that one should only see a broad regime of excitations. As the film thickness increases, these modes should become denser and more strongly localized around the normal liquid phonon-roton spectrum.

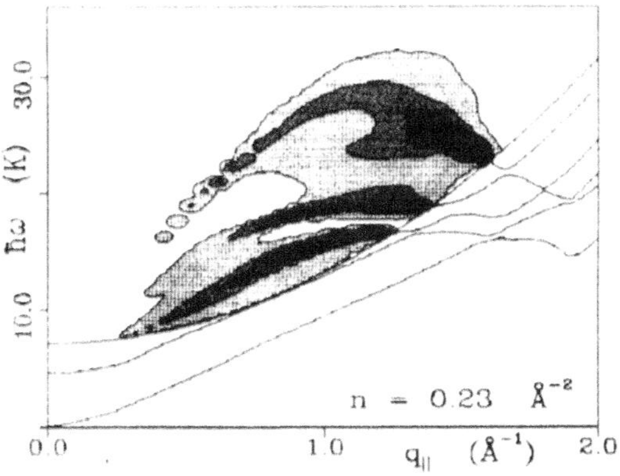

Fig. 2c

Since a perpendicular component of the momentum can never be excluded in neutron-scattering experiments, one has to expect that the dynamic structure function becomes partly modulated by the contribution from the transverse neutron momentum. While one should expect for high-lying modes that one sees, again, essentially the kinetic energy, the Fourier analysis of the transverse component of the dynamic form factor should let us analyze the geometric shape of the transition density. Figs. 3a-3c show plots $S(k, \mathbf{q}_{\|}, \omega)$ for a film with surface coverage 0.23 $Å^{-2}$ for the values of parallel momentum $q_{\|} = 1.071$, 1.357 and 1.643 $Å^{-1}$ as a function of the perpendicular momentum k. The discrete contours parallel to the k-axis below the continuum show the strength of the discrete modes. The third discrete mode appearing slightly below the continuum for $q_{\|} = 1.357$ $Å^{-1}$ has a peak at *finite* wavenumber, one would identify this mode as a volume excitation.

Going to higher momenta (Fig. 3c), we see that the mode corresponding to a *finite* wave number is the lowest discrete excitation, whereas the surface mode appears to be the second excitation. A second surface-type excitation (i.e. one that has considerable weight for $k = 0$ appears to be directly below the continuum. As the parallel momentum increases, we also see a more and more pronounced kinetic energy branch.

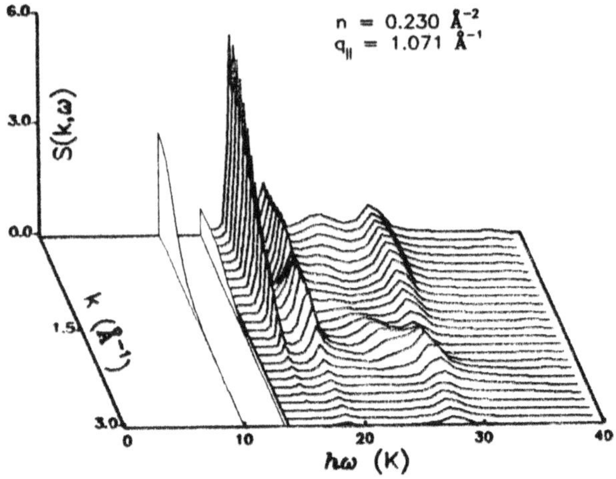

Fig. 3a

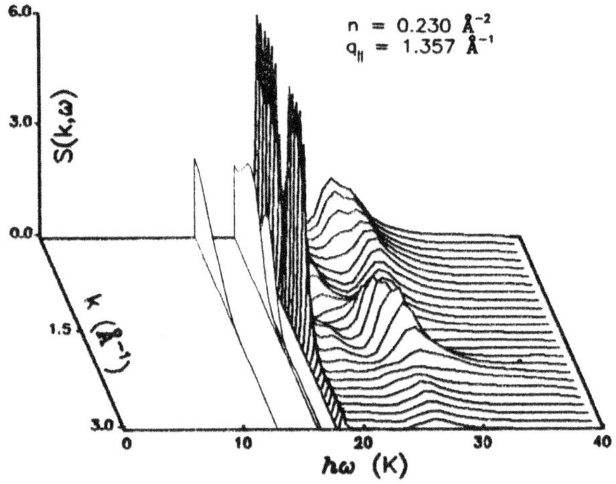

Fig. 3b

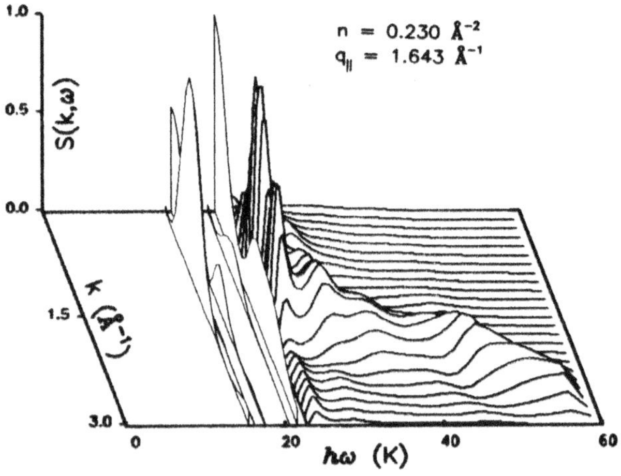

Fig. 3c

Neutron-scattering experiments may reveal a variety of interesting features that might even permit a direct resolution of the layer structure. Present data[3] show unfortunatly a much more diffuse picture, and it is not yet clear whether the rich structures outlined above can be experimentally resolved. One should, therefore, also look for other experimental probes for adsorbed helium films.

Useful probes for the surface structure are ^{3}He impurities since these can be distinguished from the ^{4}He background by both their magnetic moment and statistics. Experimental efforts focus on

(a) Measurements of the binding energy and the specific heat of ^{3}He impurities[19], and

(b) Measurement of magnetic properties of ^{3}He impurities adsorbed to ^{4}He surfaces[20,21].

Besides probing the structure of the underlying ^{4}He film, these experiments also provide the exciting possibility to study quasi-twodimensional Fermi systems. The experimental efforts require the development of three levels of theoretical tools:

(a) Single-impurity theories aim at the calculation of the impurity binding-energy and effective mass. The static correlations between particles determine if and how many surface bound states exist, and whether the impurity penetrates into the substrate. The (complex) self-energy of the impurity gives access to the specific heat and the mobility.

(b) Mixture films in the dilute limit correspond to a system of *two* static impurities[16]. If the concentration of the impurities is small, one can ignore higher-order correlations between the impurities and dynamic effects due to momentum-dependent correlations.

(c) A two-component theory is needed for large impurity concentrations. This enables us to study the structure of quantum-liquid interfaces.

Technically, impurities are added to the system by includint impurity-background and impurity-impurity correlatios $u_1^I(\mathbf{r}^I)$, $u_2^{IB}(\mathbf{r}_i^I, \mathbf{r}_j)$, and $u_2^{II}(\mathbf{r}_i^I, \mathbf{r}_j^I)$ extending the wave function. For Fermion impurities, the wave function is multiplied with a Slater determinant to ensure the antisymmetry with respect to exchange. The impurity-background and impurity-impurity correlations are again determined by optimization.

The interaction of a single ^{3}He atom with the ^{4}He background is normally described by a complex self-energy $\Sigma(q, E)$ which contains the information on both the energy/momentum relationship and the damping of the impurity motion. For energies, all of that information can be condensed in the so-called *hydrodynamic effective mass* m_H. This effective mass can be calculated[23] by either including a momentum dependence in the correlation functions or in CBF perturbation theory.

The interaction between two impurities is conveniently described by an effective interaction $V_{eff}^{II}(\mathbf{r}, \mathbf{r}')$. If the impurity atoms are in the ground state, the motion perpendicular to the surface can be integrated out. In the limit of low impurity concentration, the effective interaction can be interpreted as Landau's quasiparticle interaction in two dimensions. The variational theory provides tool for the microscopic calculation of this interaction which includes both the direct Van der Waals force, local screening, and the exchange of ripplons and phonons through the ^{4}He background. The quasiparticle interaction depends only on the relative momentum and spin of the two impurity atoms:

$$f(i,j) = f_c(q_{\|}) + f_\sigma(q_{\|})\sigma_1 \cdot \sigma_2 \tag{18}$$

Given the interaction (18), we can calculate the Landau parameters F_m^s and F_m^a of the two-dimensional Fermi liquid,

$$N(0)f_{c,\sigma}(|\mathbf{q}_1 - \mathbf{q}_2|) = \sum_{m=0}^{\infty} F_m^{(}s,a)\cos(m\phi) \tag{19}$$

where ϕ is the angle between $\mathbf{q_1}$ and $\mathbf{q_2}$, and $N(0) = m^*/\pi\hbar^2$ the density of states at the Fermi surface. From these, we obtain the magnetic susceptibility $\chi(0)$, in units of the susceptibility of the free two-dimensional Fermi gas, χ_{30},

$$\chi(0)/\chi_{30} = (m_H/m_3)(1 + F_1^s/2)/(1 + F_0^a) \tag{20}$$

and the total effective mass

$$m^* = m_H(1 + F_1^s/2). \tag{21}$$

We found it convenient to use the parametrization

$$f(i,j) = -\frac{1}{2}V_0\left[1 - \cos(q/q_0)\right]\left[1 + \sigma_i \cdot \sigma_j\right]. \tag{22}$$

to fit the quasiparticle interaction in the relevant momentum regime. With that parametrization, we find the analytic expression

$$F_m^{s,a} = -V_0 N(0)\left[\delta_{m0} - J_{2m}(2k_F/q_0)\right] \tag{23}$$

for the Landau-parameters in terms of the strength V_0 and the range parameter q_0 of the effective interaction and the two-dimensional Fermi momentum k_F.

We have carried out extensive calculations[16,22,23] for the static ground state of one ^{3}He impurity on a family of ^{4}He backgrounds ranging from a double layer ($n = 0.12\text{Å}^{-2}$) to a system of about five helium layers ($n = 0.35\text{Å}^{-2}$). The solid lines in Fig. 1 show the impurity density in comparison with the background density. We see that the ^{3}He particle is, as expected, inside the outermost ^{4}He layer for our calculation of a double-layer system. As the thickness of the background increases, the ^{3}He particle is pushed outward into the low-density regime of the film.

The interaction of the ^{3}He impurity with the ^{4}He background gives rise to the "hydrodynamic effective mass". This contribution to the total effective mass has been calculated using second-order CBF perturbation theory for the impurity self-energy in terms of the emission and re-adsorption of ripplons and phonons. Our results are compared in Fig. 4 with the experimental results of Refs. 19. The calculated hydrodynamic mass is somewhat lower than the experimental mass. However, the experiment has been done for a finite ^{3}He coverage of 0.3 layers. Therefore we must include the effects of the quasiparticle interaction between the ^{3}He atoms.

Figure 4 (solid line) shows the total effective mass m^* calculated from Eq. (21). For one active layer of ^{4}He, the hydrodynamic mass is about 1.8 m_3, in reasonable agreement with both the data of Refs. 7 and the conclusion of Ref. 8 from their thinnest ^{4}He film. In this regime the correction due to the quasiparticle interaction becomes quite sizable with increasing ^{3}He concentration, but it is difficult to make a quantitative statement due to both the high density of the background and the rapid variation of the effective mass.

For all cases with thicker ^{4}He films, the accuracy of our theoretical prediction is quite satisfactory. The microscopic theory predicts a hydrodynamic mass that is consistent with the estimate $m_H/m_3 =$

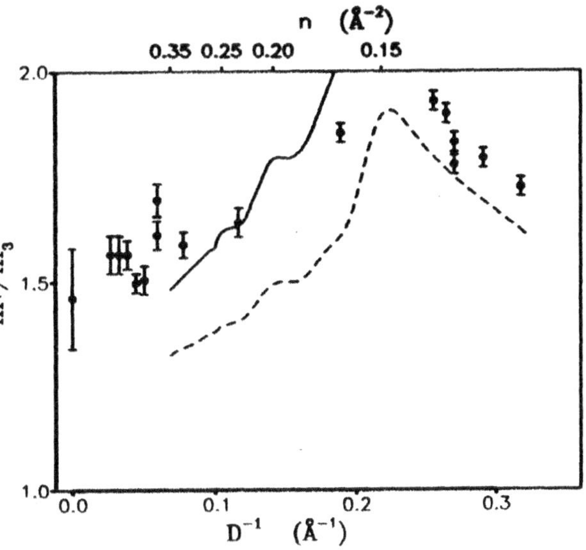

Fig. 4. The calculated effective mass m^* is shown in units of the bare ^{3}He mass m_3 as a function of the *inverse film thickness* D^{-1} (dashed line). The circles with error bars are the experimental data of Refs. 7. The upper scale shows the surface coverage corresponding to the film thickness D. The dashed line shows the hydrodynamic effective mass m_H for 0.3 layers of ^{3}He. This figure is taken from Ref. 12.

1.26 ± 0.15 given in Ref. 19. The agreement with the latter data is improved when the corrections due to the quasiparticle interaction are included. The quasiparticle interaction between ^{3}He atoms in the surface gives a density-dependent correction to the effective mass. The contribution is about 10% for the case of 0.3 layers of ^{3}He.

A comparison of the theoretical and the experimental magnetic susceptibility ratio $\chi(0)/\chi_{30}$ is shown in Fig. 5. Here, we have taken a hydrodynamic mass of $1.26 m_3$, which gives the best fit to experimental data[20]. This choice is consistent with our calculation of the effective mass since the experimental data refer to a much smaller ^{3}He coverage. We find a quite satisfactory agreement between theory and experiment. The agreement is not quite so good for smaller ^{4}He coverages, where our theoretical description is intrinsically less accurate.

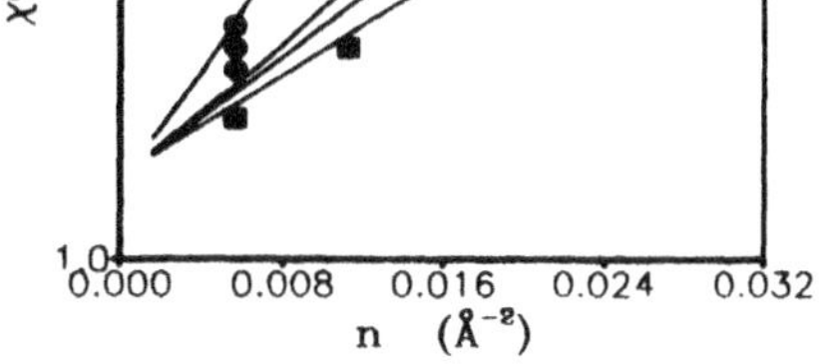

Fig. 5. Magnetic susceptibility χ/χ_{30} of the ^{3}He film as a function of areal density n in atoms/Å^2, for ^{4}He coverages between 0.15 atoms/Å^2 (uppermost curve), and 0.30 atoms/Å^2 (lowest curve). Solid squares: experimental data[20] for a surface coverage of 9.5 ^{4}He layers. Circles: experimental data[20] for a ^{3}He density of 0.088 layers, for ^{4}He coverages of 2.8, 3.4, 4.75, an' 5 24 atoms/Å^2. This figure is taken from Ref. 11

The good agreement between our calculations and the experiments has been obtained with a *static, momentum-dependent* quasiparticle interaction. Measurements of the type of Ref. 20 appear to be extemely useful to enhance our theoretical understanding of the quasiparticle interaction in liquid ^{3}He. The analytic expression (23) should provide a useful analytic tool.

Acknowledgements

The work was supported, in part, by the National Science Foundation under Contract PHY-8806265, the Robert A. Welch foundation under Grant A-1111, and the Texas Advanced Research Program under Grant No. 010366-012. C. J. T. acknowledges support through a Texas A&M Regent's fellowship.

References

1. R. A. Aziz, V. P. S. Nain, J. C. Carley, W. L. Taylor, and G. T. McConville, J. Chem. Phys. **70**, 4330 (1979).
2. Contributions by F. Gasparini, H. Godfrin, R. B. Hallock, and H.-J. Lauter to these proceedings.
3. H.-J. Lauter, these proceedings.
4. A. D. Jackson, *Ann. Review Nucl. Part. Sci***33** (1983) 105.
5. G. Ji and M. Wortis, Phys. Rev. **B34**, 7704 (1986).
6. S. Stringari and J. Treiner, Phys. Rev. **B36**, 8369 (1987).
7. S. A. Chin and E. Krotscheck, Phys. Rev. Lett (submitted).
8. S. Stringari, these proceedings.
9. D. Ceperley, private communication.
10. E. Feenberg, *Theory of Quantum Fluids* (Academic, New York, 1969).
11. C. E. Campbell and E. Feenberg, Phys. Rev. **188**, 396 (1969).
12. C. E. Campbell, Phys. Lett. **44A**, 471 (1973); C. C. Chang and C. E. Campbell, Phys. Rev. **B15**, 4238 (1977).
13. E. Krotscheck, Phys. Rev. **B33**, 3158 (1986).
14. E. Krotscheck, G.-X. Qian, and W. Kohn, Phys. Rev. **B31**, 4245 (1985).
15. J. L. Epstein and E. Krotscheck, Phys. Rev. **B37**, 1666 (1988).
16. J. L. Epstein, E. Krotscheck, and M. Saarela, Phys. Rev. Lett **64**, 427 (1990).
17. E. Krotscheck, Phys. Rev. **B31**, 4258 (1985).
18. R. P. Feynman, Phys. Rev. **94**, 262 (1954).
19. X. Wang and F. M. Gasparini, Phys. Rev. **B38**, 11245 (1988); B. K. Bhattacharyya, M. J. DiPirro, and F. M. Gasparini, Phys. Rev. **B30**, 5029 (1984).
20. J. M. Valles, Jr., R. H. Higley, R. B. Johnson, and R. B. Hallock, Phys. Rev. Lett. **60**, 428 (1988).
21. R. B. Hallock, these procedings.
22. E. Krotscheck, M. Saarela, and J. L. Epstein, Phys. Rev. **B38**, 111 (1988).
23. E. Krotscheck, M. Saarela, and J. L. Epstein, Phys. Rev. Lett. **61**, 1728 (1988).

NMR MEASUREMENTS OF ^{3}He IN ^{3}He-^{4}He MIXTURE FILMS

Robert B. Hallock†

Laboratory for Low Temperature Physics
Department of Physics and Astronomy
University of Massachusetts, Amherst, Massachusetts 01003 USA

INTRODUCTION

^{3}He atoms atop a thin film of ^{4}He constitute a conceptually simple but remarkably rich system in which the ^{3}He, depending on the coverage, can be studied from the extremes of a nearly non-interacting 2D Fermi gas, to a 3D interacting liquid. In such a system, the Fermi temperature depends on coverage and, for the 2D case, may be readily tuned. Several important questions may be asked of this system: (1) how does the behavior of the ^{3}He change as the ^{3}He coverage is increased and the interactions among the atoms increase, (2) how does the ^{3}He interact with the ^{4}He film which supports it, (3) for ^{3}He coverages which approach and exceed a monolayer, how does the ^{3}He evolve from a 2D to a 3D system, (4) are there phase transitions in the system as a function of either ^{3}He coverage or temperature, (5) what are the dynamical properties of the ^{3}He as a function of these parameters?

Early NMR experiments on the mixture film system were performed by Brewer and his co-workers[1] on a Vycor substrate. More recently, detailed measurements of the specific heat for mixture films were carried out by Gasparini and his colleagues[2-3] on a Nuclepore substrate. These later experiments showed the 2D Fermi gas behavior of the ^{3}He for the dilute case, found ^{3}He states within the ^{4}He film and observed an anomaly in the temperature dependence of the heat capacity which was interpreted as evidence for a liquid-gas (or puddling) transition among the ^{3}He.

Third sound has also been used to study the ^{3}He-^{4}He mixture film system. In a series of experiments in this laboratory it was established that the fundamental configuration of the film was bilayer[4], that the Lekner[5] picture of dilute ^{3}He was correct[6] and that the third sound attenuation as a function of ^{3}He coverage possessed unexplained structure[7]. Temperature dependence of the third sound attenuation was also seen[7,8]; thermally activated behavior was observed for $T > 200$ mK for ^{3}He coverages above 0.6 layers.

In an effort to better understand this remarkable system and its dynamics, we are conducting NMR experiments on the mixture film. For these experiments we chose to preplate our substrates with at least two atomic layers of ^{4}He to ensure that there is no solid ^{3}He in the mixture film system and hence no extra complication to the interpretation of the NMR signal. Our first experiments[9] showed evidence for ^{3}He - ^{3}He interactions, but saw no evidence for a liquid-gas transition.

THE APPARATUS AND PROCEEDURE

The apparatus[9] used for this work is illustrated in Fig. 1. Inside the experimental chamber is located a substrate for the NMR studies which consists of 400 Nuclepore filters with 0.2 μm pores and a total surface area of approximately 1.8 m^2. Each filter is 13 mm in diameter with a 3 mm hole in the center and was pressed onto a 3 mm diameter copper post located on the axis of the NMR coil. This helps to ensure thermal contact to each filter. Also available in the same experimental chamber is a third sound resonator made from a single sheet of Nuclepore. Film thickness oscillations at the resonator's first harmonic are driven by an evaporated thermal resistance heater at one end of the resonator and detected in the Nuclepore pores between silver capacitor plates evaporated at the other end of the Nuclepore. The changing capacitance modulates a tunnel diode oscillator which provides $\sim$ 0.004 Å amplitude resolution. Capillary condensation is not present in the Nuclepore substrate. Detailed third sound and capacitance studies[10] have shown Nuclepore of characteristic pore size 0.2 μm to remain free of capillary condensation for ^{4}He coverages below 8 - 9 atomic layers. The temperature was measured by use of a Speer 100 Ω carbon resistor calibrated to a typical accuracy of 3 mK against the ^{3}He melting curve by use of a ^{3}He melting curve thermometer on the apparatus.

The description of the amounts of helium in the mixture system is sometimes confusing with various authors choosing different conventions. For ^{3}He coverages we will use d_3 in atomic layers [with one layer defined to be 1 atom/(3.93 Å^2)] . Thus, by this convention we have one atomic layer of ^{3}He when there are 6.45×10^{14} atoms/cm^2 on the surface. We express the distance from the Nuclepore substrate to the surface of the ^{4}He film as d_4 in atomic layers with one such layer defined to be a distance of 3.6 Å. Third sound measurements provide a measure of d_4. We will also discuss coverage of ^{4}He in the apparatus in terms of D_4 where we use D_4 as a measure of the amount of adsorbed ^{4}He. Our convention is D_4 = 1 layer = 12.82 μmoles/m^2. An assumption about the amount of ^{4}He in the "solid" adjacent to the substrate allows a connection between these two measures: our measurements

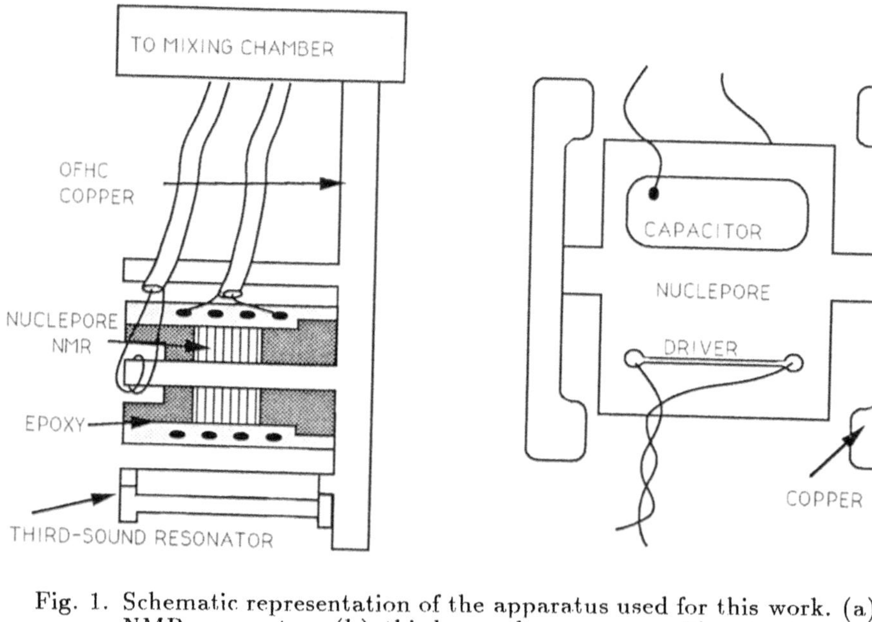

Fig. 1. Schematic representation of the apparatus used for this work. (a) NMR apparatus, (b) third sound resonator. The third sound resonator is located below the NMR apparatus.

show that $d_4 = D_4 - 1.33$ layers. Finally, in terms often used by the theorists, $n_4 = D_4/(3.6)^2$ relates ^{4}He coverage in units of $\AA^{-2}$. Thus, $D_4 = 1$ layer $= 0.0772\ \AA^{-2} = n_4 = 7.72 \times 10^{14}$ atoms/cm^2. In our experiments ^{4}He is added in known amounts, thus n_4 is determined. Differences in descriptions among different authors center on different assumptions about the density of ^{4}He in the film adjacent to the substrate. In some cases, authors describe coverages in terms of the amount of liquid present. The translation to n_4 removes this ambiguity. However, different substrates may have different Van der Waals coefficients which may result in difficult intersubstrate comparisons.

For the NMR experiments we report here, measurements of the magnetization and the relaxation times T_1 and T_2 were carried out as a function of temperature and coverage of ^{3}He ($0.005 < d_3 < 4.0$ layers) on the Nuclepore for several coverages of ^{4}He. We have also initiated a detailed study of the ^{4}He coverage dependence of these quantities for $d_3 = 0.1$.

Thermal equilibrium values of the ^{3}He magnetization, $M = \mu_m(N_+ - N_-)$ and the relaxation time T_2 were measured by means of 90-τ-180 spin-echo pulse sequences in a 2-T (Larmor frequency = 62.9 Mhz) magnetic field with H_o oriented perpendicular to the average Nuclepore pore axes. The height $E(\tau)$ of the spin-echo was measured for several values of τ. Extrapolation of $\ln E(\tau)$ vs τ to $\tau = 0$ provided a measure of the magnetization. For $0.3 < \tau < 1.5$ msec $\ln E(\tau)$ was linear in τ for most coverage conditions and the slope yields $T_2 \sim 7$ msec with d_3 and T dependence to be described later. For $\tau > 1.5$ msec, dependent on d_3, deviations from linear behavior are seen. Calibration of the magnetization is made by comparison to Curie-law data from the lowest ^{3}He coverages $d_3 \sim 0.01$ layer studied. To determine T_1 we use 90 - τ_{long} - 90 - τ - 180 spin echo pulse sequences and observe T_1 to be well described by a single exponential[11] over more than two decades in τ_{long}.

The magnetization data are normalized by $M_o = N\mu_m^2 H_o/k_B T_F(m_3)$ the magnetization at $T = 0$ of an area A of a weakly polarized ($\mu H_o \ll k_B T$) 2D ideal Fermi gas (2DIFG) of N particles of bare mass m_3. In two dimensions $T_F(m) = \pi\hbar^2 N/k_B mA$; thus M_o is independent of ^{3}He coverage.

RESULTS AND DISCUSSION

Some of our most important observations concerning the magnetization are summarized[12–14] in Figures 2a and 2b where we show $M(T)/M_o$ vs d_3 for the lowest temperature data available at several values of d_4. These data are best discussed with a focus of attention on specific regions of ^{3}He coverage. For $0.1 < d_3 < 0.65$ the magnetization is degenerate at the temperatures shown. For these data the linear increase in M with d_3 is due to interactions among the ^{3}He which apparently increase with coverage. Extrapolation to $d_3 = 0$ yields values for the hydrodynamic mass, m_H, of the ^{3}He. These values for m_H may be compared with recent predictions due to Krotscheck[15] and Treiner[16] and their co-workers. In both cases, theory predicts and increase in m_H with a decrease in d_4 consistent with our measurements[13,14,17]. Krotscheck's[15] calculation of the dependence of the effective mass on coverage shows the same behavior and is consistent with the measurements of Bhattacharyya et al.[3]. At the moment we are conducting experiments to be described in more detail below with $d_3 = 0.1$ as a function of ^{4}He coverage in an effort to observe structure in m_H vs d_4 which has been predicted[15]. In this d_3 regime, the temperature dependence of the magnetization is a monotonic increasing function of 1/T and can be fit to a 2D Fermi gas model $M(T) = M_o(m/m_3)[1\text{-}\exp(T_F(m)/T)]$ as shown in Figure 3 for $d_3 < 0.6$.

For the coverage region $0.7 < d_3 < 0.9$ we observe a dramatic step in the magnetization which is smooth and very well defined. The midpoint of this step is observed to depend weakly on d_4. At the low coverage end of this step in M we observe an anomalous temperature dependence in the magnetization which is illustrated for a typ-

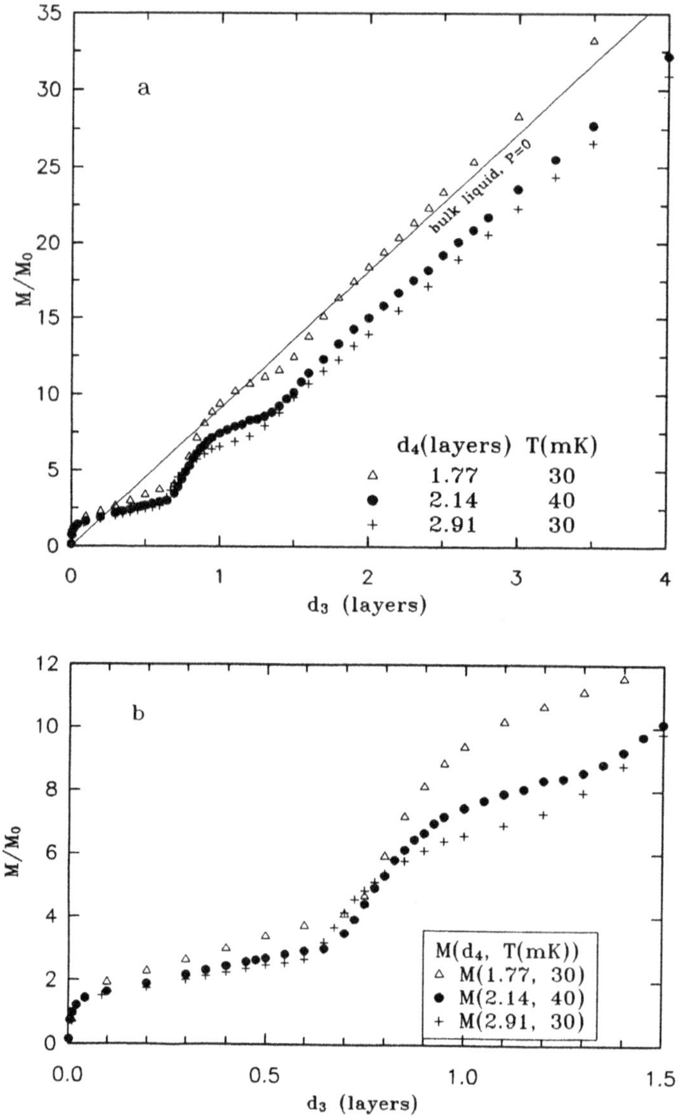

Fig. 2. Magnetizaton of the sample vs ^{3}He coverage, d_3, for three ^{4}He coverages: $d_4 = 1.77$ (triangles, 30 mK, upper data), 2.14 (solid circles, 40 mK) and 2.91 layers (plusses, 30 mK). Two views are shown, our complete ^{3}He coverage and an expanded view which more clearly shows the structure in the vicinity of the first step.

ical case in Figure 3. There is some evidence for a second step (Fig. 2) which appears to begin at $d_3 \sim 1.4$ layers. The presence of these steps suggests the occupation of higher discrete energy levels for the ^{3}He in the mixture film system. We have applied a simple two-level 2D Fermi liquid model[12] to the data to extract values ϵ_{12} for the energy difference between the two levels and find[12,13] $\epsilon_{12} = 1.80$, 1.72 and 1.67K for the three coverages $d_4 = 1.77$, 2.14 and 2.91 layers, respectively. ϵ_{12} is a weak function of ^{4}He coverage, decreasing as d_4 increases.

A simple physical picture which begins with an ideal 2D Fermi gas provides insight into the origin of the step at $T = 0$. Consider for the moment a collection of non-interacting 2D levels which can be populated by ideal non-interacting Fermi particles. For such a system, the magnetization (normalized by M_o) is a constant independent of the number of particles in the system. This is illustrated by the solid line in Figure 4. As we continue to add particles T_F increases; ultimately T_F reaches the energy of the second state. Further addition of particles begins the occupation of a second 2D level which also has a magnetization independent of coverage; thus the system magnetization doubles and the magnetization displays a step. In the actual non-ideal system several additional processes are at work: individual ^{3}He atoms interact with the ^{4}He substrate and become "dressed" (the hydrodynamic mass; m_3 goes to $m_H > 1$); interactions among the ^{3}He quasiparticles increase the effective mass due to "Fermi" effects and the magnetization and energy levels change with coverage (and the magnetization is no longer independent of d_3). Thus, as shown schematically by the dashed lines in Figure 4, the magnetization has "tilted plateaus" and shows a step.

The "dressing" of the ^{3}He mass by the ^{4}He and interactions with other ^{3}He can be conveniently thought of by a simplistic but picturesque analogy with a person

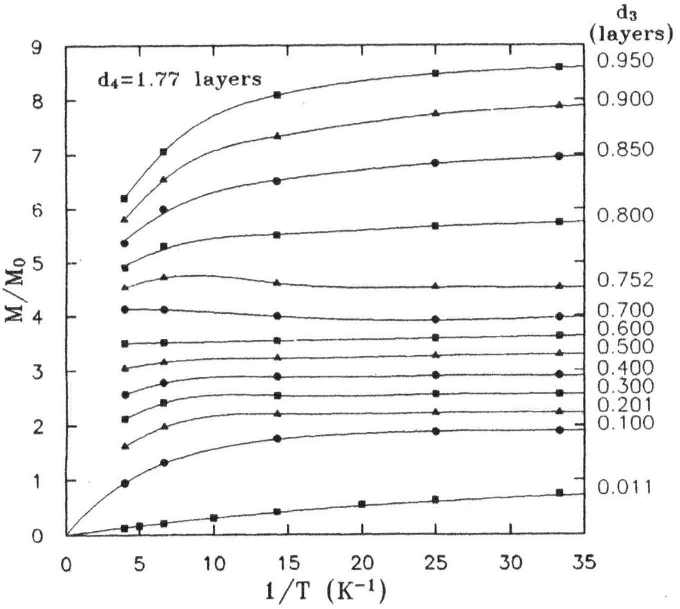

Fig. 3. Magnetization vs. 1/T at fixed ^{4}He coverage 39.7 μmol/m^2 for the d_3 shown at the right. Note $dM/dT > 0$ for $d_3 \sim 0.65$. The curves for $d_3 \leq 0.2$ are 2D Fermi fits; the other curves are guides to the eye.

and a lake. The bare mass corresponds to a single person in winter on a perfectly frictionless frozen lake; there is no interaction to impede the motion of the person as he slides along the lake. The hydrodymanic mass represents the effect the person feels due to the interaction with the now liquid surface of the lake in the summer when he becomes a swimmer and interacts with it. The Fermi effects are in analogy to what happens to the swimmer when additional swimmers are introduced into the lake; there is interaction among the swimmers and their motion is further complicated.

In somewhat different language[12], the ^{3}He in the film is viewed as a Fermi liquid with quasiparticles that are free to occupy the momentum states parallel to the ^{4}He surface and the stationary states of the ^{3}He wave functions perpendicular to the substrate. The energy spectrum of states is thus continuous in two degrees of freedom and discrete in the third. The available Fermi space is reduced from a sphere in the three dimensional case to a set of Fermi disks. The discrete energy level of each Fermi disk will depend on the ^{3}He coverage, the ^{4}He film thickness, the Van der Waals potential of the substrate and the ^{4}He surface profile.

Recent work[13,14] has shown the relaxation times T_1 and T_2 also have structure and that this structure is strongly correlated with that in the magnetization. An example of this for data taken at $d_4 = 1.77$ is shown in Figure 5. That T_1 decreases with an increase in ^{3}He coverage is perhaps not surprising. An increase in coverage should increase the collision rate among the quasiparticles and thus[18] drive down T_1. At present we do not understand the mechanism which causes the increase in T_1 at the steps in the magnetization.

The theoretical challenge is to come to grips with these observations and properly describe the system. Guyer[19] and colleagues have developed a density functional theory for the ^{3}He film in the absence of a substrate and shown that in such a model a sequence of states emerges as the film thickens. In this model there are no explicit ^{3}He - ^{3}He interactions, no interactions with the ^{4}He and none with the substrate. The Fermi energy is a function of the amount of ^{3}He present and Fermi disks are present. As a result, stepped structure is present in the magnetization, and structure is predicted for the specific heat as a function of ^{3}He coverage. Krotscheck[15] and separately Treiner[16] and their colleagues have considered the problem in the presence of interactions. Epstein et al.[15] consider the dilute ^{3}He limit and in a microscopic calculation include interactions among the ^{3}He and between the ^{3}He and ^{4}He. They are able to obtain values for m_H and Fermi parameters. Pavloff and Treiner[18] have considered the surface of bulk ^{4}He in a density functional approach which has no explicit ^{3}He - ^{3}He interactions. They have also recently extended the calculations to the case of helium films[16] and obtained values for m_H. The experiments show that for our system, the step structure washes out rapidly with ^{3}He coverage. This may be due to inhomogeneity in the Nuclepore substrate[12]. Pavloff and Treiner[16] find that the ^{4}He film can support two surface states separated in energy by 1.8 - 2.0 K in general agreement with the value $\epsilon_{12} \sim 1.7$K which we deduce[12,13] by application of an approximation to the Havens-Sacco and Widom[20] theory to our data. Further work will be required to establish whether this two-surface-state picture is an appropriate description of our experiments.

In an effort to more completely test some of these recent theoretical predictions, we are currently involved in a program to study the magnetization vs d_4 for $d_3 = 0.1$ (in an effort to measure the d_4 dependence of m_H) and to study the evolution of M as a function of d_3 for a much greater d_4 coverage than we have used in the past (in an effort to compare the evolution of the magnetization vs d_3 for high d_4 compared to low d_4). Preliminary results are available; $M(d_3 = 0.1)/M_o \sim m_H$ is observed to decrease[17] with and increase in d_4 in a manner generally consistent with the trends predicted by Krotscheck[15] and Treiner[16].

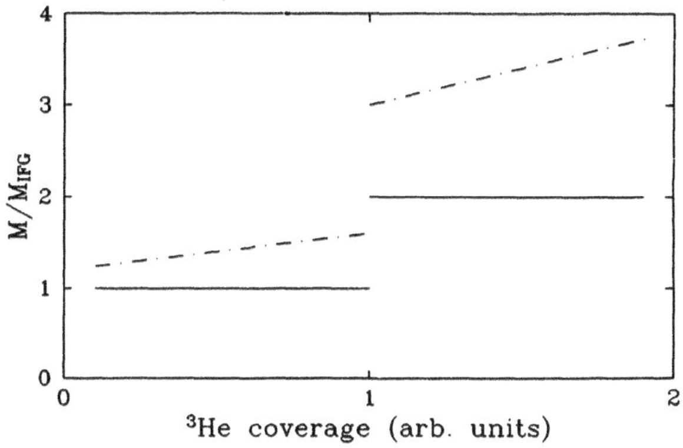

Fig. 4. Magnetization vs d_3 for the case of an ideal 2D Fermi gas (solid lines) and for the non-ideal case discussed in the text (dashed lines).

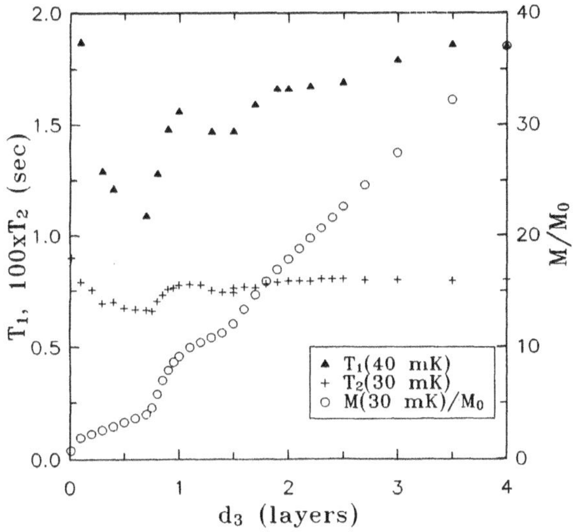

Fig. 5. NMR relaxation times T_1 and T_2 and the magnetization as a function of d_3 for $d_4 = 1.77$; note the registry between the structure in the relaxation times and the magnetization.

For this dilute $d_3 = 0.1$ case we have also been conducting detailed measurements of T_1 and T_2 for various d_4 as a function of temperature. These measurements have found[17,21] an unexpected temperature dependence for T_1. Shown in figure 6 is the T_1 vs ^{4}He coverage for various temperatures[21,22]. The most pronounced feature in the data is a peak near $n_4{}^* = 0.23\,\AA^{-1}$, the coverage at which the third sound velocity reaches its maximum [presumably for $n_4 > n_4{}^*$ the ^{4}He film displays superfluidity; for $n_4 < n_4{}^*$ the film modes are heavily damped]. For $n_4 < n_4{}^*$ T_1 is at most weakly linearly dependent on temperature as might be expected[23] in two dimensions. For $n_4 > n_4{}^*$ we observe $T_1{}^{-1} = A + B/\sqrt{T}$ for $T < 250$ mK with B a linearly increasing function of n_4 above $n_4{}^*$. This behavior is shown for several n_4 values in Fig. 7. For $T > 250$ mK, subtraction of the $\sqrt{T}$ dependence shows[17,21] that $T_1{}^{-1} \sim \exp(-\epsilon/T)$ with $\epsilon \sim 2$K. We observe ϵ to decrease with an increase of n_4. This behavior is consistent with that predicted by Krotscheck[24] and contrary to that seen by Bhattacharyya et al.[3].

We find the $\sqrt{T}$ behavior difficult to understand. Interactions of the ^{3}He with other ^{3}He, with film excitations and with any substrate impurities should all contribute to the relaxation rate. The magnetization data indicates that the system is fully degenerate. A system of degenerate Fermi particles in 3D is expected to have $T_1 \sim T^{-2}$; in 2D the details of the temperature dependence are not well known. For a classical system in 2D Cowan[23] has shown T_1 to be weakly linear in temperature. In considering ^{3}He interactions with the ^{4}He film, the riplon density $\sim T^{5/3}$, for rotons, $n_R \sim \exp(-\Delta/T)$ where for thin films $\Delta \sim 5 - 8$K; none of these yield $T_1 \sim \sqrt{T}$.

The thermally activated process $T_1{}^{-1} \sim R\exp(-\epsilon/T)$ suggests the presence of an unoccupied second level or the existence of a film excitation with a small energy gap in the system. The activation energy deduced from the behavior of T_1 for this dilute case, $\epsilon \sim 2$K, is consistent with the values of ϵ_{12} found from the magnetization

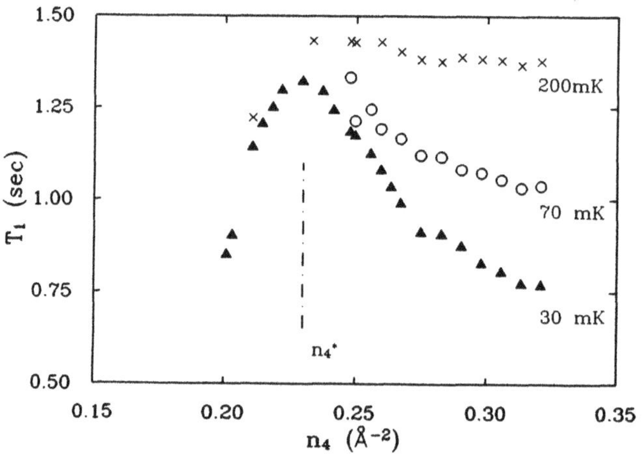

Fig. 6. T_1 vs n_4 for several temperatures. The peak in T_1 occurs at a coverage $n_4{}^*$ for which the third sound velocity attains its maximum.

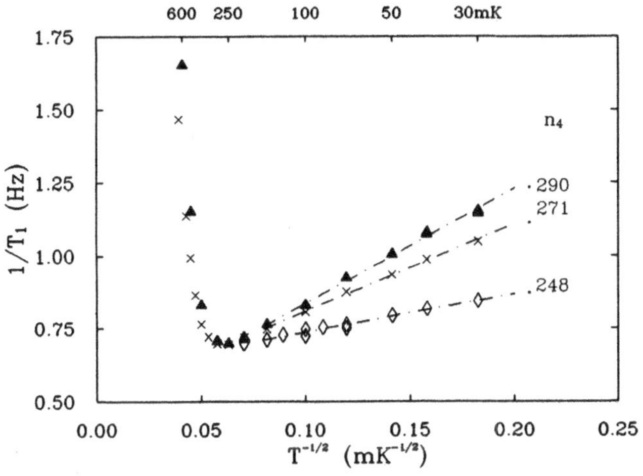

Fig. 7. Temperature dependence of T_1 for several values of the ^{4}He coverage. Note the dependence on $\sqrt{T}$ for $T < 250$ mK.

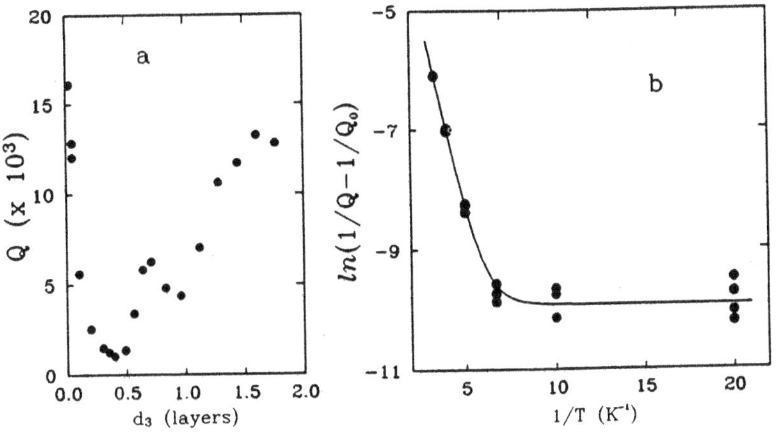

Fig. 8. Third sound Q behavior as seen in a glass resonator. (a) The Q as a function of d_3 showing structure, (b) The excess dissipation due to the ^{3}He showing thermally activated behavior. For these data $d_4 = 5.3$ and $d_3 = 1.28$.

step data; however, the coefficient R for the two cases differs by a factor of 200. ϵ determined from T_1 appears to have a stronger ^4He coverage dependence than does ϵ_{12} determined from the magnetization for higher ^3He coverages. This might be expected since the energy of an unoccupied state is determined by the ^4He surface profile and the Van der Waals potential, whereas, the energetics of heavily occupied states will be strongly influenced by the ^3He itself.

Although taken at a d_4 value greater than any of the d_4 values for which detailed NMR measurements vs d_3 have been made to date, earlier work[7] with third sound in a glass resonator done in this laboratory found strong structure in the Q of the third sound; pronounced peaks in the Q were seen for $d_4 = 5.3$ and 5.7 layers at $d_3 \approx 0.7$ and 1.5 layers (Fig. 8). The strong structure evident at T = 100 mK is reduced at T = 250mK. It is tempting to conclude that this structure in the third sound Q is directly related to the structure in the magnetization. The fact that the calculations due to Treiner show the surface states to be rather insensitive to a thickening ^4He film lend credence to the notion that the Q structure and the steps in the magnetization may be related. Further evidence for this comes from the predictions of Guyer whose theoretical work[19,25] shows that in the presence of discrete levels, the Q is expected to show such structure. Preliminary work with the third sound resonator in the present apparatus is suggestive, but additional measurements are necessary. In the glass resonator experiments, the third sound Q also showed interesting temperature dependence. For $d_3 > 0.6$ the dissipation due to the presence of the ^3He showed thermally activated behavior. The excess damping due to the ^3He obeys $1/Q - 1/Q_o \sim \exp(-C/T)$ with $C \sim 1.5K$. C was found to be weakly dependent on both d_3 and d_4. Here Q_o is the third sound Q for the pure ^4He underlayers in the resonator. Interestingly, for low ^3He coverages, $d_3 < 0.3$, no thermally activated behavior was seen in the excess damping.

In conclusion, the ^3He - ^4He mixture film is a remarkably rich system. In addition to the experiments with only liquid ^3He in the film system described here, there have been a number of interesting experiments reported recently[26] on films for which at least some of the ^3He is a solid. The interconnection between those results and the ones we have reported here is emerging.

It is a pleasure to acknowledge my co-workers in these experiments over the past few years and to thank them for allowing the use in some cases of unpublished data: N. Alikacem, R.H. Higley, B.R. Johnson, D.T. Sprague and J.M. Valles, Jr. Conversations with H. Godfrin, R.A. Guyer, E. Krotscheck, W.J. Mullin and J. Treiner have also been of benefit. Permission from J. Treiner to quote m_H values prior to publication is appreciated. In addition, I am greatful for detailed comments and suggestions from N. Alikacem and D.T. Sprague which were incorporated into this manuscript. Our recent work was supported by the National Science Foundation through grants DMR 85-17939, 88-20517.

REFERENCES

† Recent work done in collaboration with N. Alikacem, R.H. Higley, and D.T. Sprague.

1. D.F. Brewer, D.J. Creswell and A.L. Thompson, in Proceedings of the Twelfth International Conference on Low Temperature Physics, Kyoyo, 1970, ed. E. Kanda (Keigaku, Tokyo, 1970), p. 157.

2. M.J. Dipirro and F.M. Gasparini, Phys. Rev. Lett. **44**, 269 (1980).

3. B.K. Bhattacharyya, M.J. Dipirro and F.M. Gaspirini, Phys. Rev. B **30**, 5029 (1984).

4. F.M. Ellis, R.B. Hallock, M.D. Miller and R.A. Guyer, Phys. Rev. Lett. **46**, 146 (1981). See also J.C. Noiray, D. Sornette, J.P. Romagnan and J.P. Laheurte, Phys. Rev. Lett. **53**, 2421 (1984).

5. J. Lekner, Philos. Mag. **22**, 669 (1970).

6. J.M. Valles, Jr., R.M. Heinrichs and R.B. Hallock, Phys. Rev. Lett. **56**, 1704 (1986).

7. F.M. Ellis and R.B. Hallock, Phys. Rev. B **29**, 497 (1984); R.M. Heinrichs and R.B. Hallock, Proceedings of the Seventeenth International Conference on Low Temperature Physics, ed. U. Eckhern et al. (North-Holland, Amsterdam, 1984), p. 59; R.M. Heinrichs, Ph.D. Dissertation (Univ. of Massachusetts, 1985), unpublished.

8. R.B. Hallock, Can. J. Phys. **65**, 1517 (1987).

9. J.M.Valles, Jr., R.H. Higley, B.R. Johnson and R.B. Hallock, Phys. Rev. Lett. **60**, 428 (1988).

10. J.M. Valles, Jr., D.T. Smith and R.B. Hallock, Phys. Rev. Lett. **54**, 1528 (1985); D.T. Smith, K.M. Godshalk and R.B. Hallock, Phys. Rev. B **36**, 202 (1987).

11. For the smallest values of τ_{long} ($\tau_{long} < 0.2$ sec) we observe a slight nonlinearity.

12. R.H. Higley, D.T. Sprague and R.B. Hallock, Phys. Rev. Lett. **63**, 2570 (1989).

13. R.H. Higley, D.T. Sprague and R.B. Hallock, Phys. Rev. B, (to be published).

14. R.H. Higley, D.T. Sprague and R.B. Hallock, Proceedings of the Nineteenth International Conference on Low Temperature Physics, ed. D. Betts et al. (in press).

15. J.L. Epstein, E. Krotscheck and M. Saarela, Phys. Rev. Lett. **64**, 427 (1990).

16. N. Pavloff and J. Treiner (preprint); J. Treiner, private communication.

17. D.T. Sprague, N. Alikacem and R.B. Hallock (to be published).

18. A. Abragam, The Principles of Nuclear Magnetism, (Oxford Univ. Press, London, 1961), p. 324.

19. R.A. Guyer, this volume and R.A. Guyer, K.R. McCall and D.T. Sprague, Phys. Rev. B **40**, 7417 (1989).

20. S.M. Haven-Sacco and A. Widom, J. Low Temp. Phys. **40**, 357 (1980).

21. D.T. Sprague, N. Alikacem and R.B. Hallock, Proceedings of the Ninteenth International Conference on Low Temperature Physics, ed. D. Betts et al., (postdeadline, in press)

22. The T_1 values from our most recent run represented by figure 6 are generally consistent with earlier runs except for the case of the run whose data is shown in figure 5 where T_1 appears a bit longer. The echo height was the same in the Curie region for the case of $d_3 = 0.1$ for each of these runs; there is no error in the ^3He coverage. We don't presently understand this quantitative discrepancy; it does not affect any of the trends of the data within a single run. In all cases in a given run the data are reproducible and heating effects are absent.

23. B.P. Cowan, J. Low Temp. Phys. **50**, 135 (1983).

24. E. Krotscheck, Phys. Rev. B **32**, 5713 (1985).

25. R.A. Guyer, private communication.

26. D.S. Greywall and P.A. Busch, Phys. Rev. Lett. **65**, 64 (1990); C.P. Lusher, J. Saunders and B.P. Cowan, Proceedings of the Nineteenth International Conference on Low Temperature Physics, ed. D. Betts et al. (in press).

FERMION OCEANS

R.A. Guyer

Department of Physics and Astronomy
University of Massachusetts
Amherst, MA 01003

INTRODUCTION

Fermion oceans are neutral, self bound fermi liquid films; they are present in the systems described at this workshop by Godfrin, Hallock, Laheurte, Saunders, etc.[(1)]. These self bound films stand in contrast to the charged, unbound fermi gases, the fermion atmospheres, that are present in the systems described at this workshop by Leiderer, Dahm, Platzmann, Lea, etc.[(2)].

Phase Diagrams

The behavior of fermion oceans is influenced importantly by the circumstance of their experimental realization. In Fig. 1 we show a global ^{4}He - ^{3}He film "phase diagram" on which the possible ^{4}He - ^{3}He films can be located in composition, (amount of ^{4}He in layers, amount of ^{3}He in layers = (D_4, D_3)[(3)]. In Fig. 2 we show sketches of film structure at points A1, B1, and C1 on Fig. 1. These films were encountered in the experiments done along lines A, B, and C in Fig. 1:

A. Greywall[(4)] and Godfrin and co-workers[(5)] on grafoil at ($D_4 = 0$, D_3).
B. Saunders, Luscher, and Cowan[(6)] at ($D_4 = 1.0$, D_3).
C. Higley et al. at ($D_4 = 2^+$, D_3) on Nuclepore.

The meaning of a point on Fig. 1 is not independent of the substrate that supports the film. In Fig. 3 the complicated phase diagram of ^{3}He on grafoil is shown. The evolution of the system along A for ^{3}He on grafoil, i.e. at ($D_4 = 0$, D_3) is followed on this diagram. In Fig. 2 a sketch of the structure of the ^{3}He film at A1 (see Figs. 1 and 3) is shown. This film has 3 components, a very dense solid first layer, a solid second layer, and a liquid layer system.

Single Particle States

Interesting features in the behavior of fermion oceans have their source in the influence of the depth of the ocean (film thickness) on the states available to the particles within it. As an example, consider a free standing[(8)] film having 2×10^{15} ^{3}He atoms in volume 10Å x 1 cm x 1 cm. These atoms are in a potential of their own making, i.e. in a "square well" potential of approximate width 10Å and approximate depth -2.5 K. This potential is responsible for the single particle states in which the atoms reside; e.g.

$$\psi_\alpha \propto \phi_m(z)\, e^{i \vec{q} \cdot \vec{\rho}}\, \chi(\sigma) \qquad (1)$$

(1)

having energy

$$\varepsilon_\alpha = e_m + \frac{\hbar^2 \vec{q}^{\,2}}{2m_3} \quad . \qquad (2)$$

Here m is the index for the single particle states that describe particle motion perpendicular to the substrate, $\vec{q}$ is the two dimensional wave vector for motion parallel to the substrate, e_m the energy associated with $\phi_m(z)$ and σ is the spin; $\alpha = m \vec{q} \sigma$. See Fig. 4. As more ^3He atoms are added to the film, i.e. D_3 increases, the film thickness increases. The single particle states in which the particles reside, e_m and ϕ_m, evolve; e.g. $\phi_1(z) \sim \cos \pi z/D_3$, $e_1 \propto D_3^{-2}$. As the particles are fermions the way in which they choose to occupy these states also evolves. See Figs. 4 and 5 for illustration of the evolution of the single particle energies and of the $\vec{k}$-space the particles inhabit. This latter is a sequence of planes of two

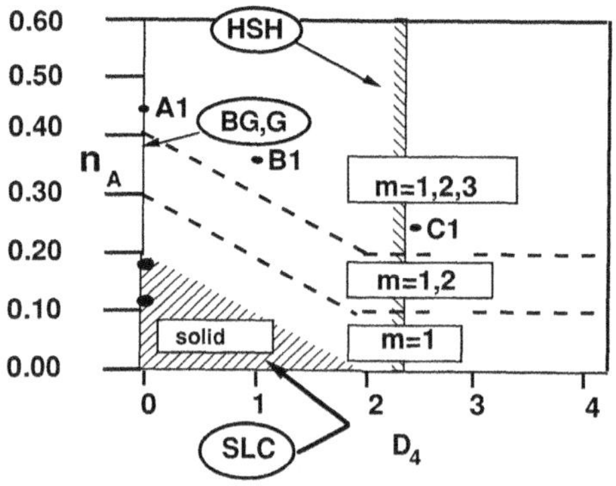

Figure 1. Global ^{4} - ^3He Phase Diagram. Possible ^4He - ^3He films are described by the co-ordinate pair (D_4, D_3) that indicates the total composition (layers of $^4He = D_4$, layers of ^3He - D_3). The experiments in references 4 - 7 are done along lines A(4, 5), B(6), and C(9) on this diagram. Further elaboration of the diagram requires substrate specific information like that in Fig. 3.

dimensional wave vectors $\vec{q}$ at $e_1(D_3)$, $e_2(D_3)$, ... The number of ^3He atoms determines, D_3, e_1, e_2, ... and the filling of the states in each plane, i.e. the radius q_F^m of the fermi surface of each disk(9). We have $(q_F^m)^2 = 2\pi^2 N_m/A$ where N_m is the number of particles in state m, A = (1 cm)^2,

$$e_F = e_m + \frac{\hbar^2 (q_F^m)^2}{2m_3} \quad .$$

A measurement of the magnetization of this system in an external field of strength H_o would yield

$$M = M_o\, \bar{m}\,(D_3) \quad , \qquad (3)$$

where

$$M_o = \mu^2 H_o \frac{2m_3 A}{\hbar^2}$$

and $\bar{m}$ is a step function of D_3. The magnetization evolves in a sequence of quantized steps. This result has its source in:

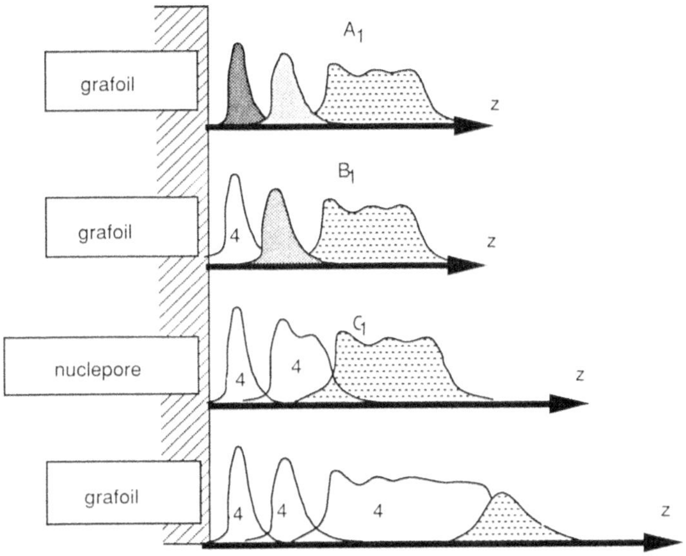

Figure 2. Film Structure. The mixture films at A1, B1, and C1 on Fig. 1 have in common a fermion ocean atop 2 solid ^{3}He layers, A1, one solid ^{4}He and one solid ^{3}He layer, B1, atop 2$^+$ (2 solid) ^{4}He layers, C1.

(1) The additive nature of thermodynamic quantities, $M = \sum_{m=1}^{\bar{m}} M_m$, where M_m is the contribution to the magnetization of the particles in disk m.

(2) $M_m \propto N_m/\varepsilon_F^{(m)}$ and $\varepsilon_F^{(m)} \propto \frac{\hbar^2 N_m}{2m_3 A}$; i.e. the response of each disk is inversely proportional to the number of particles in it. Thus M_m is independent of N_m; a measurement of the magnetization simply "counts" the number of occupied disks.

The features, (1) and (2) above, lead to step structure in the specific heat and to step or oscillatory behavior in all thermodynamic quantities. There is every reason to expect that oscillatory features, related to the nature of the single particle states, will also be present in transport properties. Magnetization measurements (Higley et al.[7]), specific heat measurements (Gasparini et al.[10], Greywall[4]), third sound Q measurements (Ellis and Hallock[11]) and T_1, T_2 measurements (Higley, et al.[12]) confirm these expectations.

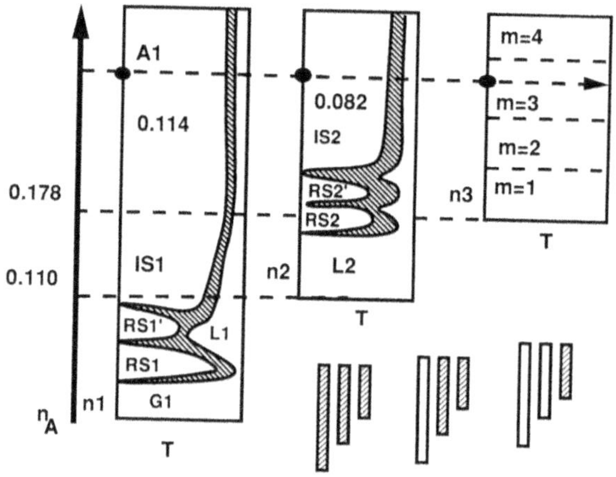

Figure 3. Phase Diagram of ^{3}He on Grafoil. As the total number of ^{3}He atoms in the film increases, the total areal density increases ($N_A = {}^3$He atoms $/(A)^2$, 1 liquid layer $\simeq$ 0.07 atoms/$(A)^2$). With increase in N_A the areal density in the first layer, second layer, and liquid layer system, n_1, n_2, and n_3 respectively, changes; n_1 and n_2 saturate at 0.114 and 0.08^+ part for suitably large n_A.

THEORETICAL MODEL

Let us construct a phenomenological model that will describe both the structure and dynamics of a fermion ocean. We write

$$E[n^o_\alpha, \delta n_\alpha, p; H_o] = F_\perp[n^o_\alpha] + \frac{1}{2} \sum_{\alpha\alpha'} \mathcal{F}_{\alpha\alpha'} \, \delta n_\alpha \, \delta n_{\alpha'}$$

$$- \sum_\alpha \sigma\mu H_o n_\alpha + \sum_\alpha \Lambda_\alpha \, \delta n_\alpha \, p + \frac{1}{2} \frac{1}{\chi_s} p^2 \tag{4}$$

$$- \mu H_o p.$$

(1) $F_\perp[n^o_\alpha]$ is a density functional model of the Helmholtz free energy of the equilibrium film[13,14]; the single particle energies $\varepsilon^o_\alpha = \delta F_\perp/\delta n^o_\alpha$ determine the occupation probabilities, n^o_α, etc.

(2) The Landau interaction functional, $\mathcal{F}_{\alpha\alpha'}$, describes the consequence of fluctuations in n_α away from n^o_α.

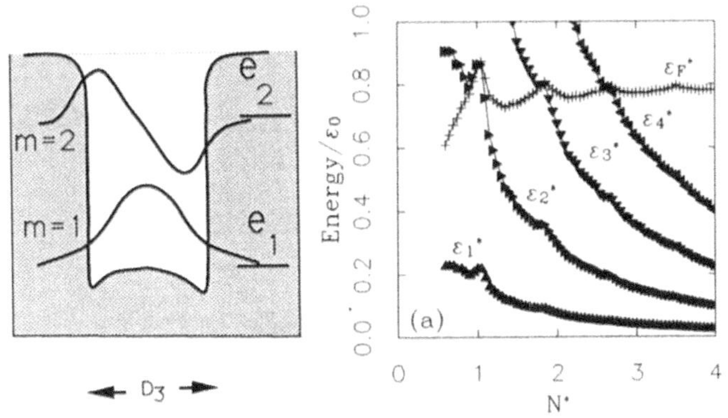

Figure 4. Wavefunctions and Energies. In a film of thickness D_3 the single particle states are approximately the states cos $m\pi z/D_3$ of a particle in a box of width D_3. See a. These states evolve with $D_3 = N^*$ as shown on the right, $e_1(D_3)$, ... The crosses show the evolution of the fermi energy with D_3. This evolution corresponds to evolution of the number of disks; number of m states, occupied by particles.

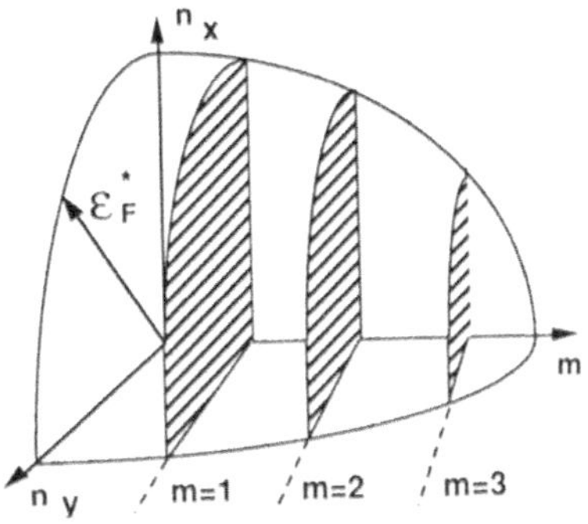

Figure 5. k- Space. For this sytem the states occupied by particles are in disks in each of a sequence of planes at $e_1(D_3)$, $e_2(D_3)$ .... . The number of occupied planes, disks, increases with D_3, although the fermi energy changes very little, because the single particle energies decrease as D_3 increases and more planes fit within the fermi sphere.

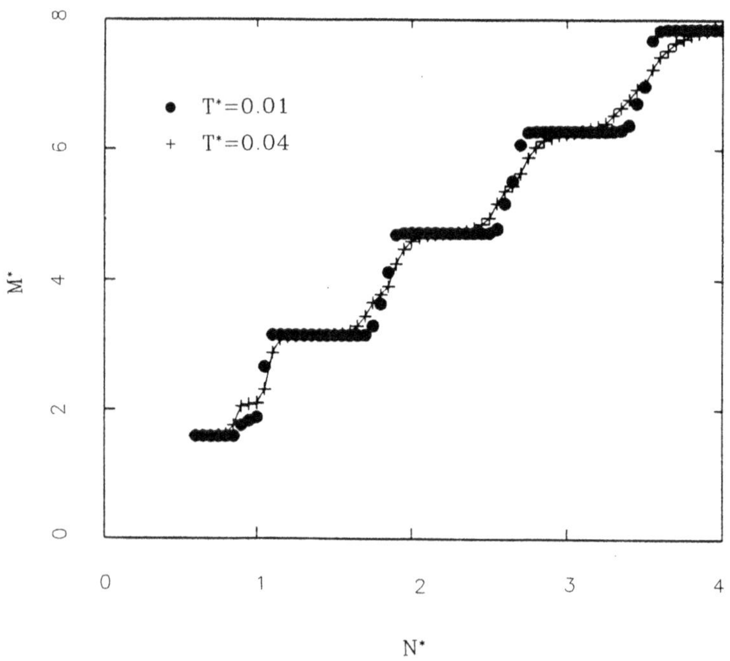

Figure 6. Magnetization vs. $D_3 = N^*$.

(3) An external field H_o couples to the spins of the particles in the ocean and to the polarization, p, of a system to which the ocean may be coupled. As an example of p consider the second solid ^3He layer in Fig. 2, B1.

(4) The polarization energy of the p system, $p^2/2\chi_s$, may have its souce in a microscopic coupling to the ocean. Expect χ_s (D3). (See the discussion of an RKKY interaction by Guyer[15].)

(5) A coupling constant, here Λ_α, between the ocean and the system described by p can be found in microscopic analysis similar to that in reference 15.

To illustrate some of the interesting content of the model let us consider the case Λ = 0, e.g. a ^3He ocean atop ^4He as in Higley et al.

Structure

To describe the structure of this film we would study $F_\perp[n^o_\alpha]$ as has been done by Guyer, McCall, and Sprague[13] and others[14].

Thermodynamics

To discuss the thermodynamic response of this system we use the formal apparatus of Haven-Sacco and Widom[16] which relates the thermodynamic quantities to the Landau interaction functional. To wit

$$\chi = \left(\frac{\hbar\gamma}{2}\right)^2 \sum_\alpha \sum_\beta \sigma\sigma' G_{\alpha\beta}$$

$$\rho = \frac{1}{m} \sum_\alpha \sum_\beta p^z_\alpha p^z_\beta G_{\alpha\beta} \tag{5}$$

$\alpha = m\vec{q}\,\sigma$, $\beta = m'\vec{q}\,'\sigma'$. The function G is related to F through the system of equations

$$G_{\alpha\beta} = -h_\alpha \delta_{\alpha\beta} - h_\alpha h_\beta A_{\alpha\beta}$$

$$A_{\alpha\beta} = F_{\alpha\beta} + \sum_\gamma F_{\alpha\gamma} h_\gamma A_{\gamma\beta}\ , \tag{6}$$

with $h_\alpha = \partial n_\alpha/\partial\varepsilon_\alpha$. A theory of χ **may** start with a model for F. For example

$$F_{\alpha\alpha'} = a_{mm'} + \delta_{mm'}\delta_{\sigma\sigma'}(c_m + d_m \cos\phi); \tag{7}$$

here ϕ is the angle between $\vec{q}$ and $\vec{q}\,'$, $a_{mm'}$, c_m and d_m are a set of D3 dependent constants. Using Eqs. (6) $A_{\alpha\alpha'}$ can be found in the form

$$A_{\alpha\alpha'} = A_{mm'} + \delta_{mm'}\delta_{\sigma\sigma'}(B_m + D_m \cos\phi)$$

with for example [17]

$$C_m = \frac{c_m}{1 + \frac{1}{2} g_m c_m}$$

and

$$D_m = \frac{d_m}{1 + \frac{1}{4} g_m d_m} \qquad (8)$$

where g_m is the density of states at the fermi surface of disk m, $g_m = Am_m^*/\pi\hbar^2$. We find

$$\chi = \sum_{m=1}^{\bar{m}} \chi_m \qquad (9)$$

and

$$\rho = \sum_{m=1}^{\bar{m}} \rho_m ,$$

where

$$\chi_m = \left(\frac{\hbar\gamma}{2}\right)^2 \frac{g_m}{1+\frac{1}{2} g_m c_m} \qquad (10)$$

and

$$\rho_m = \frac{p_{F,m}^2}{2m_3} \frac{g_m}{1 + \frac{1}{2} g_m d_m}$$

or

$$\frac{m_m^*}{m_3} = 1 + \frac{1}{2} g_m d_m .$$

The simplicity of these results, an effective mass for each disk, an enhanced susceptibility for each disk, is due to the features of the interaction functional that are probed by χ and ρ. The elastic response and the specific heat probe $a_{mm'}$ and are not additive over disks as they are in a non-interacting model mentioned above or as are χ and ρ.

At low D_3, so that only the m = 1 state is required(18), we can combine the results above to find

$$\frac{\chi_1}{\chi_o} = \frac{m_3^*(0)}{m_3} \left[1 + \frac{1}{2} g_1(d_1 - c_1)\right]$$

where $m_3^*(0)/m_3 = 1.3$ and $g_1(d_1-c_1)/2 \simeq 3 D_3$ from the experiment of Valles et al.(19). This result is in rather nice agreement with the qualitative and quantitative results of a microscopic theory due to Krotschek and co-workers(20).

Application of Eq. (10) to the experimental results of Higley et al. suggests that for the m = 2 state behaves much like the m = 1 state;

$$m_3^*(0)_2 \simeq m_3^*(0)_1$$

and

with a ferromagnetic coupling, Λ, leads to an enhancement of the susceptibility contribution of each disk. This result, in qualitative accord with experiment, has not been put to a quantitative test. One barrier to undertaking such a test is the complexity of the resulting theoretical model for χ.

T1

It is difficult to make a theory of T_1 in this system. The principal reason for this is that the spectrum of possibilities that one might consider is broad while the amount of reliable data is very limited. Thus, the informed choices that are necessary to guide theory toward a correct explanation cannot be made. However, to illustrate some of the physics that is operating in these systems we will describe some elements that may appear in a model for T_1. We take T_1 to be due to the dipole-dipole coupling between ^{3}He atoms. At low D_3, when only the m = 1 state is occupied, T_1^{-1} will be proportional to the average strength of the dipolar interaction. As D_3 increases, still only m = 1, the average strength of this interaction increases and T_1 should decrease. Of course the strength of the dipolar coupling is really an expectation value that involves $|\phi_1(z)|^2$. At fixed small D_3 as D_4 increases we expect $|\phi_1(z)|^2$ to describe a more tightly bound particle. Again, the strength of the dipolar coupling should increase and T_1 decrease. These expectations are consistent with experiment. At fixed D_4 as D_3 is increased the m = 2 state is populated; the magnetization doubles. If the m = 2 magnetization relaxes independently of the m = 1 magnetization [there is no cross relaxation] then the decay will be the sum of two exponential decays; a fast, m = 1, decay and a slow, m = 2, decay. Should the m = 1 and m = 2 states be strongly coupled then a single exponential will result. The possibilities we are describing may well appear against a background of other processes and not be at all apparent. Much more, careful, data is required to illuminate this issue.

I am pleased to acknowledge valuable conversations with K.R. McCall, D.T. Sprague, and Nadir Alikacem.

References

1. An ocean is a self bound fluid film. Its density is approximately uniform having a value near that of bulk material; cf an atmosphere.

2. An atmosphere is a fluid film of non-self-bound material, a gas, that owes its existence to an external force that holds it in place. It is non-uniform having a density determined by the number of particles in it and the strength of the external field.

3. A layer is a measure of the areal density in terms of the number of particles required to fill a surface to an areal density that corresponds to bulk particle separation. For bulk ^{4}He the particle density is n = 1 particle/(3.6Å)^3. Thus $n_A = 1/(3.6\text{Å})^2 = 7.7 \times 10^{14}$ part/cm^2 is a "layer" of ^{4}He. For ^{3}He one layer corresponds to 6.6×10^{14} part/cm^2.

4. D. Greywall, Phys. Rev. B**41**, 1842 (1990).

5. H. Godfrin, R.R. Ruel, and D.D. Osheroff, Phys. Rev. Lett. **60**, 305 (1988), H. Franco, R.E. Rapp, and H. Godfrin, Phys. Rev. Lett. **57**, 1161 (1981).

6. J. Saunders, C. Luscher, and B.P. Cowan, preprint of LT-19 paper.

7. R.H. Higley, D.T. Sprague, and R.B. Hallock, Phys. Rev. Lett. **63**, 2570 (1989). See also the paper by Hallock in this volume.

8. A free standing film, i.e. a film free of a substrate potential to hold it "down", can be usefully studied as a simple approximation. Such a film is unstable to modes that let it become a lump of material. The most modest substrate potential will stabilize a free standing film.

9. This discussion illustrates matters of principle that are robust in the presence of various computational details, as for example in references 13 and 14.

10. B.K. Bhattacharyya, M.J. Dipirro, and F.M. Gasparini, Phys. Rev. B**30**, 5029 (1984).

11. F.M. Ellis and R.B. Hallock, Phys. Rev. B**29**, 497 (1984).

12. R.H. Higley, D.T. Sprague, and R.B. Hallock, preprint and submissions to LT-19.

13. R.A. Guyer, K.R. McCall, and D.T. Sprague, Phys. Rev. B**40**, 7417 (1989).

14. N. Pavloff and J. Treiner preprint. A number of density functional theories of finite pieces of liquid have been developed. See for example Stringari and Treiner, J. Chem. Phys. **87**, 5021 (1987), R.A. Guyer, J. Low Temp. Phys. **69**, 49 (1986), G. Ji and M. Wortis, Phys. Rev. B**34**, 7709 (1986). Also the thesis by Ji and reference 13.

15. R.A. Guyer, Phys. Rev. Lett. **64**, 1919 (1990). The model used in this paper admits many possibilities for solid layer-ocean interaction. Here Λ_α captures the essence of what is possible.

16. S.M. Havens-Sacco and A. Widom, J. Low Temp. Phys. **40**, 357 (1980).

17. The equation for $A_{mm'}$ is much more complicated than that for C and D.

18. The ocean picture is not required for understanding of some of the features being discussed here.

19. J.M. Valles, R.H. Higley, B.R. Johnson, and R.B. Hallock, Phys. Rev. Lett. **60**, 428 (1988).

20. E. Krotschek, M. Saarela, and J.L. Epstein, Phys. Rev. Lett. **61**, 1728 (1988).

STUDIES OF DISSIPATIVE SUPERFLOW THROUGH POROUS VYCOR GLASS

J. D. Reppy* and A. Tyler

Physics Department, Schuster Laboratory
University of Manchester
Manchester, M13 9PL, U.K.

INTRODUCTION

The work reported here has been largely motivated by the recent interest in problem of dissipation or flux flow in the "Vortex Liquid" state of high T_c superconductors[1]. Early persistent current measurements for ^{4}He in porous Vycor glass[2] demonstrated that a possibly analogous state of dissipative flow exists for an extended temperature interval below the superfluid transition as observed in the limit of low frequency and amplitude. Additional motivation is found in the development of a new theoretical description for superfluid helium contained within a 3D connected porous medium in terms of the explicit dynamics of vortices of the superfluid system[3]. We will be interested in testing the predictions of such models for the regime of dissipative phase slip near the transition temperature as well as for thermally assisted phase slip at lower temperatures. A superfluid Helmholtz resonator has been developed for these measurements and our preliminary results are presented here.

BACKGROUND

The physics of superfluid ^{4}He contained within porous media has been a rich field of research for at least three decades. An excellent review of early work on this subject has been given by D. F. Brewer[4]. Much of the early work was motivated by the desire to study the influence of finite size or restricted geometry on the superfluid. It was expected that ^{4}He contained within a sufficiently small volume would display a shifted transition temperature as well as a rounding of the heat capacity signature. These expectations appeared to be confirmed by experiment[4]. The onset of superflow was seen to decrease as the pore size was reduced or, in the case of adsorbed ^{4}He films, when the film thickness was made thinner. Even more convincing evidence came from measurements of films adsorbed on porous media where progressive rounding of the heat capacity peak was seen as the film thickness was reduced. The measurements of Brewer's group at Sussex for helium contained within the 50Å channels of porous Vycor glass showed this effect particularly clearly[4,5]. For the full pore case the heat capacity was similar to slightly rounded version of the infinitely sharp anomaly seen for bulk ^{4}He. Studies of films[5] adsorbed in Vycor showed a broadening of the peak and a shift to lower temperatures as the film thickness was reduced all in keeping with the idea of finite size effects.

In contrast to the rounded heat capacity peaks seen in the early ^{4}He experiments in Vycor, ^{4}He adsorbed on a clean graphite surface showed remarkable sharp heat capacity peaks associated with various types of ordering transitions in the ^{4}He-graphite system[6]. This suggested that the rounded behaviour seen in the Vycor and for other glassy substrates might be a manifestation of non-homogeneity induced by a heterogeneous substrate. Indeed, attempts were made to explain the superfluid transition in 2D helium films in terms of an heterogeneity induced droplet model[7].

Thus it was surprising when in the mid seventies it was discovered that the superfluid density in full-pore Vycor displayed a sharp transition and further that the superfluid density displayed a power-law temperature dependence as T_c was approached with a critical exponent almost identical to that of pure bulk ^{4}He[8]. These results were difficult to understand in terms of the common theoretical views of the time. However, it was suggested in their initial paper by Kiewiet, Hall & Reppy[8] that the 3D connectivity of the superfluid within the Vycor might be responsible for the critical behaviour similar to 3D bulk superfluid. An immediate consequence of this conjecture was that a film within Vycor would be expected to also display 3D critical behaviour since the connectivity is the same as for the full pore case. A series of experiments[9] with progressively thinner films with transitions as low as 75mK has confirmed this expectation.

For a time the absence of an observable heat capacity singularity at the superfluid transition presented a difficulty for this point of view. However, in 1978 P. Hohenberg pointed out[10] that if ^{4}He in Vycor was in the same universality class of the bulk fluid then the size of the heat capacity anomaly associated with the superfluid transition could be estimated on the basis of 2-scale factor universality (hyper universality)[11]. When this was done for the full pore case, using the measured value for the superfluid density, the anomaly is found to be 10^4 times smaller per unit volume than for the bulk fluid. This estimate is well below the experimental resolution of current heat capacity experiments. However in the case of thin films adsorbed in Vycor it has proved possible to resolve a sharp heat capacity anomaly at the superfluid transition as demonstrated by Moses Chan and his group at Penn State University[12]. The size of the observed peaks are in accord with estimates based on hyper-universality[13].

Thus at the present time, the experimental evidence supports the original contention that the superfluid transition for ^{4}He in Vycor is 3D in character. However, for other porous media such as the fractal silica aerogels[14], there is clear indications of new and different critical behaviour possibly induced by the long range correlated structure of these fractal media.

Alternative explanations have been offered for the temperature and film thickness dependence of the superfluid density in Vycor. First by Kotsubo and Williams[15] and subsequently by Wang and Yu[16]. Their approach applies a finite size cut off to the Kosterlitz-Thouless renormalization similar to the finite size cut-off in the non-zero frequency theory of Ambegaokar, Halperin, Nelson and Siggia (AHNS)[17]. This produces a broadening of the static K-T transition and, in addition, a rounding or "foot" on the superfluid density wave as it approaches zero. This approach has been criticised since it neglects the 3D connectivity of the ^{4}He-Vycor system and is incapable of producing the observed power-law behaviour as T_c is approached. In an attempt to remedy this defect and yet retain the essence of the vortex pair model, Minoguchi and Nagaoka[3] (MN), and independently Machta and Guyer[3] (MG), considered the problem of vortices on a 3D connected lattice or "jungle gym" as MG have termed it. In addition, MG have shown that their model is equivalent to a 3D XY Villain model which is believed to be in the same universality class as the bulk 3D superfluid. If true, then the vortex approach has been brought full circle and there should be no conflict with the earlier claims[8,9] for a 3D transition in the ^{4}He-Vycor system.

Although this approach as advanced by MN, MG, and more recently by Gallet and Williams[18] may not be able to provide new information on the static 3D transition there should be, within these models, a rich dynamics associated with the motion of the vortices in the presence of time varying superflow. Indeed, in beautiful series of experiments, Mulders and Beamish[19] have demonstrated frequency dependent behaviour in the 10 to 100MHz range for the ^{4}He-Vycor film system. Cho, Gallet and Williams[20] have shown that the observed broading with increasing frequency can be understood in terms of the 3D vortex models.

Our motivation in pursuing the series of experiments presented here is to examine the dynamics of superflow at low frequency in Vycor and in particular to make a start at understanding dissipative processes involved in this flow possibly in terms of the recent 3D vortex models.

It has been known from persistent current measurements[2] that the critical velocity for superflow in Vycor is rather different in character from that observed in larger pore size porous media. In these experiments[2] it was seen that the critical velocity decreased rapidly as temperature was raised, apparently heading for zero while the superfluid density still remained finite. A similar behaviour was also observed for persistent currents in unsaturated ^{4}He films adsorbed on 3D connected media[21]. When the full-pore Vycor critical velocity data was analysed in terms of a thermally activated model for phase slip, it was found that the energy barrier could be fit by a log function of the superfluid velocity as one might expect for vortex pair barrier model[2]. A barrier of the $\mathrm{Log}(1/v_s)$ form is only a weak function of superfluid velocity and hence the critical velocity is not sharply defined. What is observed is a band of velocities for which phase slip occurs at only a slowly decreasing rate. Schematically this is shown in Figure 1. The persistent current critical velocities as determined by Kukich[2] are shown as well as the superfluid density from torsional oscillator measurements[22].

At low temperatures long lived persistent currents are allowed with velocities up to several meters per second for temperatures approaching zero. However as the temperature is raised the region of relatively stable superflow (very slowly decaying supercurrents) becomes restricted to decreasingly small velocities. If, on the other hand, one performs a superflow measurement at sufficiently low velocity, say 1μ/sec, and at sufficiently high frequency ~100hz, then low dissipation measurements can be made to within 10^{-3}K of the transition temperature before dissipative effects become dominant. Measurements satisfying these criterion are easily performed using torsional oscillator techniques and have become a standard technique for measuring the superfluid density in the ^{4}He-Vycor and other helium porous media systems[22].

The aim of the present experiment is to investigate those regions of the velocity temperature plane that are not easily accessible to the persistent current or torsional oscillator approach. Our technique involves pressure driven ac flow through a plate of Vycor glass. The ac pressure is provided by a systematic double chamber Helmholtz oscillator. The two chambers of the oscillator are nearly equal in volume ($10\mathrm{cm}^3$). In addition to the Vycor superleak connecting the two chambers we have

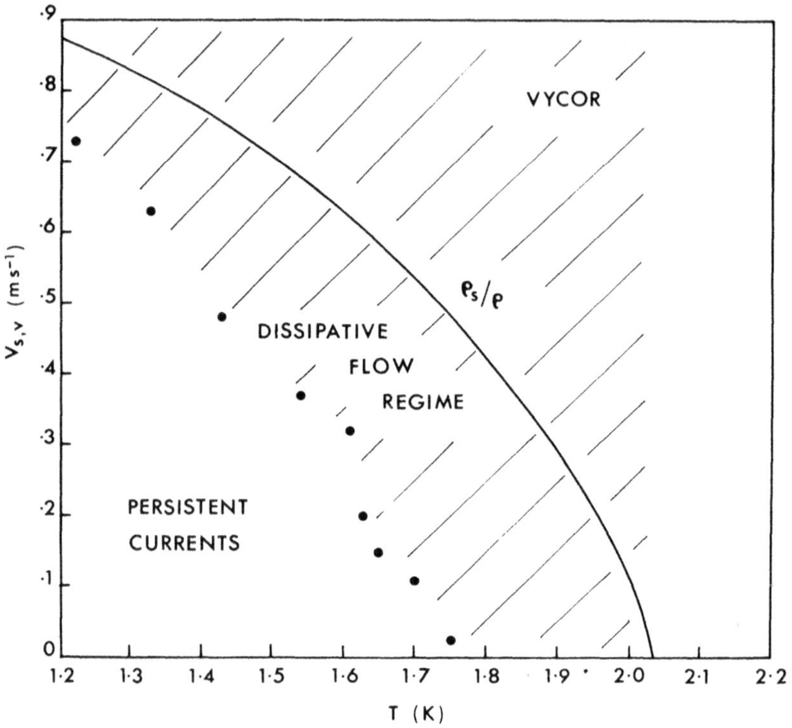

Figure 1. Dissipative and persistent current regimes. Closed circles, critical velocities from Kukich[2].

introduced a second parallel channel which allows operation above the T_c of the Vycor, as in an earlier acoustic resonator developed by Tyler and Vavasour[22]. The Q of our oscillator varies from about 10^2 just above the Vycor T_c to several thousand at temperatures near 1K.

The Helmholtz oscillations are driven by piezoelectric transducers mounted on the slightly flexible end walls of the two chambers. These transducers also serve to measure the ac pressure within each cavity. On one end wall we have mounted a conventional parallel plate capacitor, which can be used both as a dc strain gauge as well as a driver or motion detector for calibration of the piezo electric transducers.

In Figure 2 we show an overall view of the resonance frequency of this Helmholtz oscillator at low drive level for a temperature from 2.3K down to 1K. Above the Vycor T_c all of the flow takes place through the open channel and the resonant frequency essentially tracks the temperature variation of the first sound velocity. In this analysis we are neglecting the flexibility of the end walls of the Helmholtz oscillator; the reader is referred to work of Brooks, Sabo, Schubert and Zimmerman[23]. At the λ-point the frequency drops to a minimum. At the Vycor transition additional superflow begins and the frequency of the oscillator shifts rapidly upward reflecting the temperature dependence of

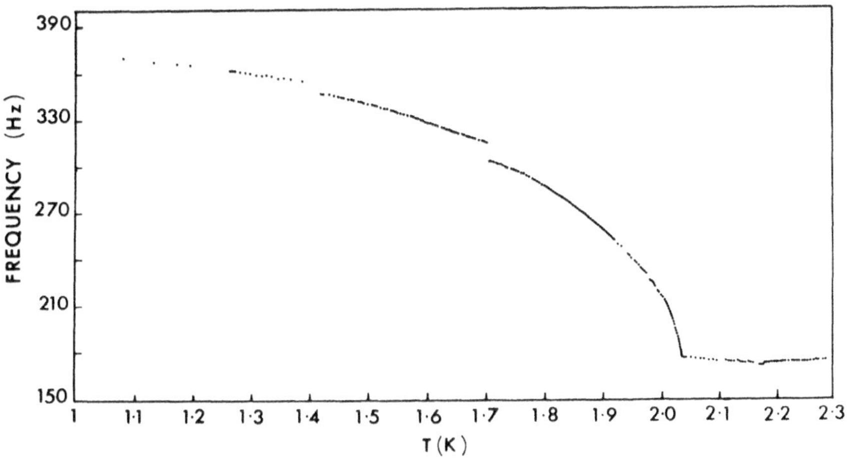

Figure 2. Helmholtz resonator frequencies at 35mV rms drive.

the superfluid density in the ^{4}He-Vycor system. The oscillating superflow through the Vycor is an effective generator of second sound. There are two breaks in the frequency caused by mode coupling with second sound resonances in the open chambers. Since this mode coupling complicates the analysis it is desirable to choose design parameters that minimize the influence of such mode crossings. A reduction in the area of the Vycor could be used to bring the frequency below the lowest lying second sound mode at all temperatures of interest. There still is the unavoidable problem of fourth sound resonances in the Vycor sample itself at temperatures very near the transition, where the frequencies are low. As has always been seen in torsional oscillators, these will cause mode coupling as the superfluid density in the Vycor goes to zero at T_c. However in our case the small size of the Vycor sample and our relatively low operating frequency restricts the mode crossing to a very small temperature interval of the order of 10^{-4}K below the transition.

In Figure 3 we show an enlarged view of the data near the Vycor transition temperature for two different excitation levels differing by a factor of 14.7. Both curves shown an excess dissipation peak below T_c. It is clear that for these modest drives, the higher drive level produces a broader dissipation feature with a peak at a lower temperature, and the frequency (and hence ρ_s) is slightly suppressed. Note that at the lower drive level the dissipation feature has a temperature width of only one part in 10^3 of T_c. From the very small frequency rounding, this is the sharpest transition yet observed for helium in Vycor and is believed to be due to the macroscopy uniformity of this particular Vycor sample. An advantage of our technique is the possibility of using very small Vycor samples (volume ($<10^{-2}$cm^3) and thus reducing the problem of large scale inhomogeneities.

In a second series of measurements we have reduced the area of the Vycor sample. This allows us to reach much higher driving pressures and thus investigate a wider range of dissipative flow states. In the next two figures we show data obtained at various fixed temperatures by varying the drive level over four orders of magnitude. For these

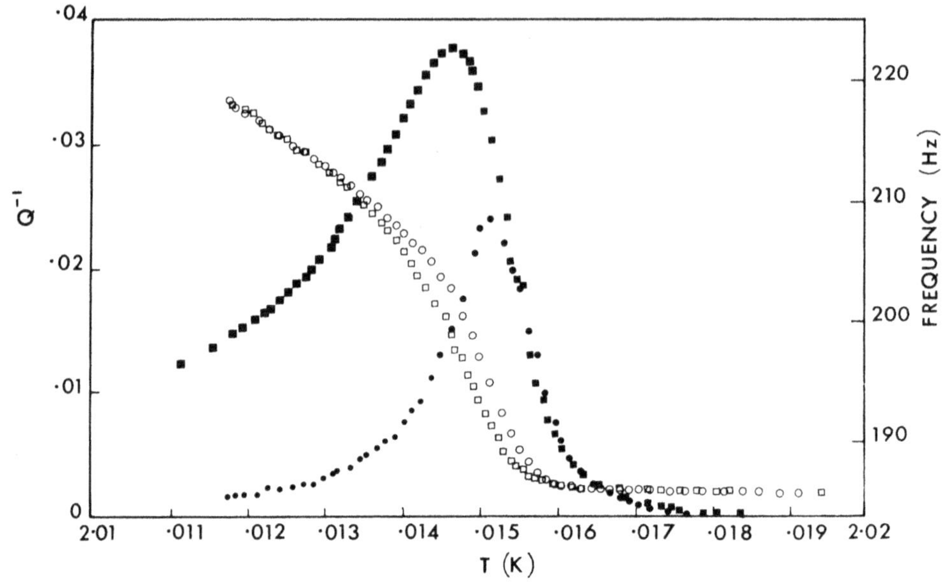

Figure 3. Resonance frequencies and dissipation at 72mV rms (circles) and 1062mV rms (squares). Open symbols are frequency and closed symbols are dissipation.

measurements the oscillator is excited by driving one chamber and measuring the absolute pressure swing in the other. Knowing the compressibility of the helium and the volume of the chambers, it is possible to calculate the total mass current flowing through the two channels. Since the hydrodynamics of the open channel are well understood, the flow through this channel can be calculated from the pressure gradient and subtracted from the total, thus yielding the supercurrent. Similarly, the phase slip rate, or acceleration of the superfluid can be calculated from the chemical potential gradient (related to the measured gradient). Assuming harmonic response (neglecting, for the moment, dissipative phase slip) we are able to calculate a superfluid velocity using the dimensions of the Vycor sample, its X factor, and porosity as determined from other measurements.

In Figure 4 we show dependence of the superfluid mass current as a function of the driving force. The current density, J_s, divided by the superfluid density is plotted against the superfluid velocity as calculated from the pressure gradient and the frequency. For low drive levels the currents are linear functions of the superfluid velocity. However, each current breaks away from this linear behaviour at some critical velocity indicated by arrows and if driven hard enough shows saturation effects. This behaviour of a saturating superfluid mass current with increasing pressure gradient is the typical critical velocity behaviour as seen in many previous experiments.

We are also able to measure the dissipation associated with the superflow. Here more subtle effects are seen. Figure 5 shows a plot of a representative total power dissipation, W, for a T/T_c of 0.9828, divided by the square of the flow velocity as a function of the superfluid velocity. The superfluid velocity is obtained from the current assuming a flow independent value of ρ_s. At low flow velocities the quantity W/v_s^2 is constant indicating a dissipation linear in

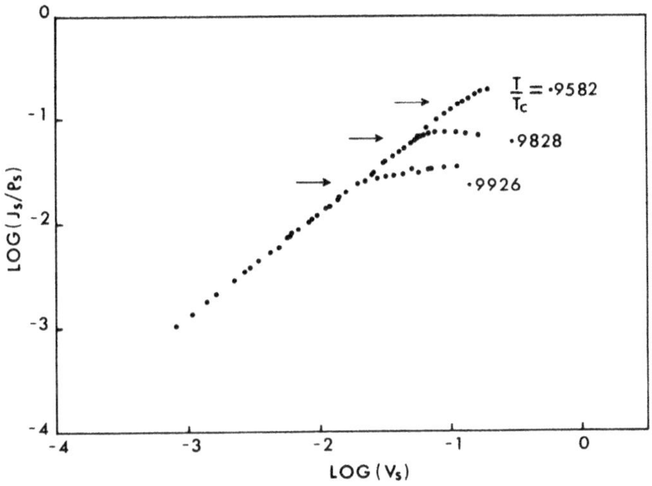

Figure 4

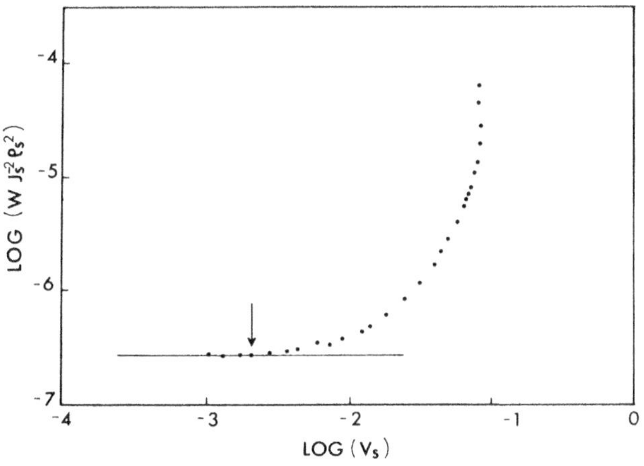

Figure 5

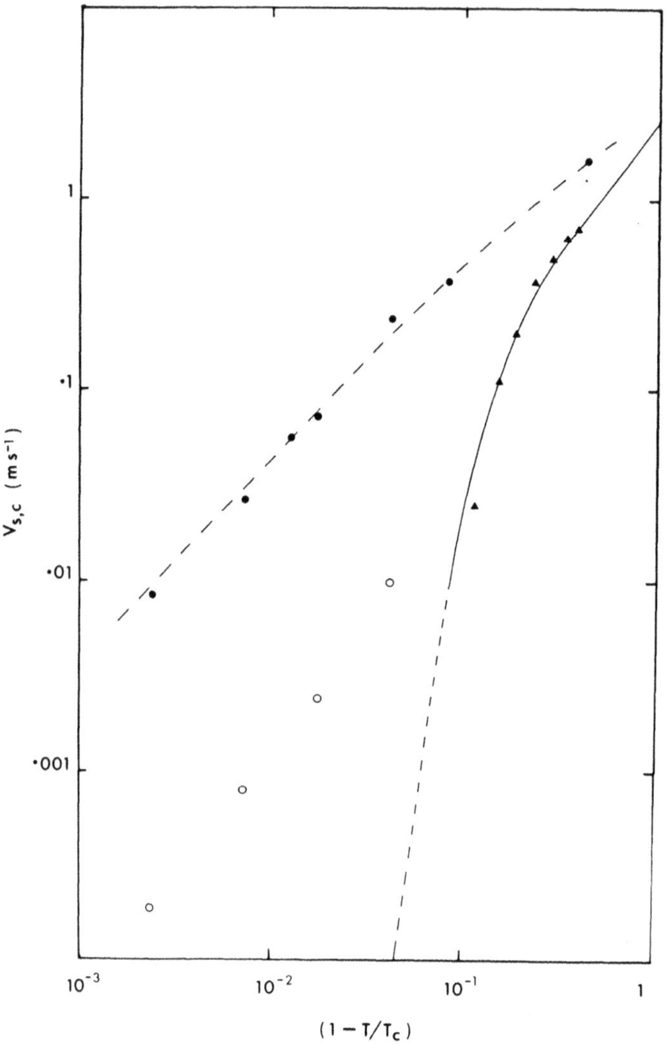

Figure 6. Critical velocities.

velocity squared. However at larger values the dissipation becomes non-linear. The steep up turn in the curve reflects the saturation of the supercurrent at high drive levels.

We summarize these findings in Figure 6 where the various critical velocities are plotted against the reduced temperature, $(1-T/T_c)$. The velocities where the saturation of the supercurrent first appears are plotted as solid circles while the values where the non-linear dissipation becomes evident are shown as open circles. The persistent current critical velocities of Kukich[2] are indicated as solid triangles. The line drawn through the Kukich data represents a critical velocity curve derived[2] from a simple thermal activated process over a barrier provided by a vortex-antivortex pair.

There are a number of comments that are suggested by the data shown in the final figure. At low temperatures there is a suggestion that the critical velocities determined from both the persistent current measurements and from supercurrent saturation behaviour in the present experiment appear to be heading for the same zero temperature value of a few metres per second. A zero temperature critical velocity of this magnitude is in accord with the value expected for a vortex pair separated by a few tens of nanometres. This is a distance on the order solid-solid correlation length as measured by low angle scattering[24] measurements and represents the minimum size for a non-simply connected path in the Vycor.

A second interesting feature is the trend of the saturation critical velocities (solid circles) as T_c is approached. We find that the saturation effect is strongly tied to the peak in Q^{-1} seen in temperature sweeps at constant drive level. It would appear that this peak in Q^{-1} can be made very narrow (close to T_c) provided the drive level is sufficiently small. This has implications for the dissipation peaks as seen in a number of torsional oscillator measurements.

A final point is to note that the velocities at which nonlinear dissipation sets in seem to follow rough power laws such that measurements with micron per second flow velocities should allow one to avoid nonlinear behaviour to within a reduced temperature 10^{-4} of T_c.

The data and analysis presented here is clearly of a preliminary nature. However the Helmholtz oscillator method appears promising and these measurements will be continued with a newly designed apparatus. In the future the emphasis will be on increasing the dynamic range and extending the measurements to temperatures below 1K.

ACKNOWLEDGEMENTS

J.D.R. would like to thank the Manchester Physics Department for its hospitality and the SERC (GR/F/45332) for its support to him as a Senior Research Fellow and also acknowledge the support of the National Science Foundation through grant NSF-DMR-8418605. Both authors wish to thank Stan Gillot and Gil West for their cheerful and encouraging technical support.

---

* Permanent address: Laboratory of Atomic and Solid State Physics, Clark Hall, Cornell University, Ithaca, N.Y., 14853-2501, U.S.A.

REFERENCES

1. R. H. Koch, V. Foglietti, W. J. Gallagher, G. Koren, A. Gupta, and M. P. A. Fisher, Phys. Rev. Lett. 63, 1511 (1989); for more recent work see V. M. Vinokur, M. V. Feigel'man, V. B. Geshkenbein, and A. I. Larkin, Phys. Rev. Lett. 65, 259 (1990).
2. G. Kukich, Ph.D. Thesis, Cornell University (1970), (Unpublished); J. S Langer and J. D. Reppy, Progress in Low Temperature Physics, Vol. 6, ed. C. J. Gorter (North-Holland, Amsterdam, 1970).
3. T. Minoguchi and Y. Nagaoka, Jpn. J. Appl. Phys. 26 Suppl. 26-3, 327 (1987). (Proc. of the 18th Int. Conf. on Low Temp. Physics). J. Machta and R. A. Guyer, Phys. Rev. Lett. 60, 2054 (1988).
4. D. F. Brewer, The Physics of Liquid and Solid Helium, Part II, Ch. 6, ed. K. H. Bennermann and J. B. Ketterson (John Wiley & Sons, New York, 1978).
5. A. J. Symonds, D. Phil. Thesis, University of Sussex (Unpublished). See also Ref. 4 p. 653.
6. J. G. Dash, The Physics of Liquid and Solid Helium, Part II, Ch. 5, ed. K. H. Bennemann and J. B. Ketterson (John Wiley & Sons, New York, 1978).
7. J. A. Herb and J. G. Dash, Phys. Rev. AI, 1427 (1973); Phys. Rev. Lett. 35, 171 (1975); M. W. Cole, J. G. Dash and J. A. Herb, Phys. Rev. B11, 163 (1974).
8. C. W. Kiewiet, H. E. Hall and J. D. Reppy, Phys. Rev. Lett. 35, 1286 (1975).
9. D. J. Bishop, J. E. Berthold, J. M. Parpia and J. D. Reppy, Phys. Rev. B24, 5047 (1981); B. C. Crooker, B. Hebral, E. N. Smith, Y. Takano and J. D. Reppy, Phys. Rev. Lett 51, 666 (1983).
10. P. C. Hohenberg, private communication (1978).
11. P. C. Hohenberg, A. Aharony, B. I. Halperin and E. D. Siggia, Phys. Rev. B 13, 2986 (1976).
12. D. Finotello, K. A. Gillis, A. Wong and M. H. W. Chan, Phys. Rev. Lett. 61, 1954 (1988). (1988).
13. M. W. Chan, Quantum Fluids and Solids - 1989, AIP Conf. Proc, 194, ed. G. Ihas and Y. Takano, (American Institute of Physics, N.Y., 1989).
14. M. H. W. Chan, K. I. Blum, S. Q. Murphy, G. K. S. Wong and J. D. Reppy, Phys. Rev. Lett. 61, 1950 (1988).
15. V. Kotsubo and G. A. Williams, Phys. Rev. B28, 440 (1983); Phys. Rev. Lett. 53, 691 (1984); Phys. Rev. B33, 6106 (1986). (1986).
16. C. Wang, and L. Yu, Phys. Rev. B33, 599 (1986).
17. V. Ambegaokar, B. I. Halperin, D. R. Nelson and E. D. Siggia, Phys. Rev. B21, 1806 (1980).
18. F. Gallet and G. A. Williams, Phys. Rev. B39, 4673 (1989).
19. N. Mulders and J. Beamish, Phys. Rev. Lett. 62, 438 (1989).
20. H. Cho, F. Gallet and G. A. Williams, Quantum Fluids and Solids - 1989, AIP Conf. Proc, 194, ed. G. Ihas and Y. Takano, (American Institute of Physics, N.Y., 1989).
21. R. P. Henkel, G. Kukich and J. D. Reppy, Low Temp. Phys. LT11, Vol. I, p. 178, (University of St. Andrews, 1968); H. W. Chan, A. W. Yanof, P. D. M. Pobell and J. D. Reppy, Low Temp. Phys. LT13, Vol. I, p. 229, (Plenum Press, N.Y., 1974).
22. A. Tyler and E. Vavasour, J.L.T.P. 23, 267 (1976).
23. J. S. Brooks, B. B. Sabo, P. C. Schubert and W. Zimmermann Jr., Phys. Rev. B19, 4524 (1979).9).
24. P. Wiltzius, F. S. Bates, S. B. Dierker and G. P. Wignall, Phys. Rev. 36, 291 (1987); D. W. Schaefer, B. C. Bunker and J. P. Wilcoxon, Phys. Rev. Lett. 58, 284 (1987).

LOW TEMPERATURE HEAT CAPACITY OF ^{4}He FILMS IN VYCOR

M. H. W. Chan, D. Finotello+, K. A. Gillis†, S. Mukherjee and P. Y. A. Wong

Department of Physics, The Pennsylvania State University
University Park, PA 16802

ABSTRACT

Precision heat capacity results of thin ^{4}He films adsorbed in Vycor are presented. We found a sharp heat capacity peak at the superfluid transition and evidence of monolayer completion. The temperature dependence of the heat capacity in various coverage regimes is analyzed in terms of one and two dimensional excitations.

INTRODUCTION

Although the study of the superfluid transition and other properties of ^{4}He film adsorbed in porous Vycor glass dates back to 1956[ref. 1,2], interesting physics is still being uncovered.

Vycor glass (Corning, type 7930) is an attractive substrate for the study of confined helium because the porous structure is highly interconnected and uniform. The distribution of the effective pore diameter is highly peaked at a value of about 70Å. The reason for the highly interconnected and yet uniform porous structure is related to the manufacturing process. Vycor is formed from a melt of borosilicate glass which is rapidly cooled below its phase separation temperature. After it is spinodally decomposed into bicontinuous SiO_2 rich and B_2O_3 rich phases, the latter, corresponding to 30% in volume fraction, is leached out with a suitable acid.[3] During the spinodal decomposition process, a specific Fourier component in the concentration fluctuation frequency spectrum shows dominant growth.[4] This is confirmed in a few recent small angle neutron and X-ray scattering measurements where the scattered intensity, I(q), is found to increase with decreasing wave vector, q, until $q = 0.03(\text{Å})^{-1}$, then drops with decreasing q[ref. 5]. Such a broad peak suggests the presence of a dominant length scale near $\lambda = 2\pi/q \sim 270$Å in the structure of Vycor glass. In other words, the porous structure in Vycor can be considered as three dimensionally homogeneous at length scales larger than 270Å. A rather revealing transmission electron micrograph of a thin Vycor glass slice obtained by P. Levitz, G. Ehret and J. M. Drakes can be found in the July 1989 issue of Physics Today.

SUPERFLUID TRANSITION

The superfluid transition of ^4He filled Vycor was examined quantitatively in a fourth sound study. The superfluid density was found to vanish at a transition temperature, T_c at 1.955K, according to a simple power law of the form,

$$\rho_s = \rho_{so} \left(\frac{T_c - T}{T_c} \right)^{\zeta} = \rho_{so} t^{\zeta} \tag{1}$$

This power law description is applicable between 5×10^{-3} and 1×10^{-1} in reduced temperature, t. The critical exponent, ζ, is found to be $0.65\pm.03$, in agreement with the bulk ^4He value of 0.674. Subsequent to the fourth sound study, the superfluid transition in the case when the Vycor pores were not filled was studied by means of the torsional oscillator method.[7] It was found in this and other experiments that a minimum coverage, $n_c \sim 28$ μmole/m^2 of ^4He is needed before superfluidity is possible even at T=0. For coverages n larger than n_c, the transition temperature T_c appears to scale linearly with $(n-n_c)$ up to $T_c \approx 1K$. Making use of this effect, film with T_c between 76 mK and 1K were studied. It was found that ρ_s in these films also vanishes with the same power law with the same bulk like exponent.[7] These results led to the suggestion that the superfluid transition of ^4He in Vycor in both the filled pore and thin film case exhibits three dimensional critical behavior belonging to the same universality class as that of the λ transition.[7]

If such an interpretation is correct, a sharp λ-like heat capacity peak at T_c is expected. Two attempts, one at Sussex,[8] and the other at SUNY, Buffalo,[9] were made to look for this peak in ^4He filled Vycor without success. Instead, a broad anomaly with a width of 0.3K centering at 2.1K was found. The Sussex study included several coverages where the Vycor pores were not completely filled. The most extensive heat capacity study of ^4He in Vycor was that of Tait and Reppy.[10] These authors found distinctly different behavior in the heat capacity C at different regimes of ^4He coverages, n. This study also found a hint of the superfluid transition in the form of a break in the slope of a plot of C/T vs T at temperatures and coverages that are in reasonable agreement with T_c as determined by ρ_s measurements.

HEAT CAPACITY MEASUREMENTS

Our heat capacity study is motivated by the ρ_s and the Tait and Reppy results. Our study confirms a number of the features reported in the earlier studies; in addition, due to the improved resolution, we have observed a number of new features including the sharp heat capacity peak related to the superfluid transition for films with $T_c<1K$. [ref. 11] No evidence of the sharp peak can be identified for $T_c \gtrsim 1K$. This is related to the fact that in addition to the sharp peak at T_c, a broad anomaly that increases rapidly with coverage is found. For $T \geq 1K$, including the filled pore case, the sharp peak is swamped by the broad feature. In the latest measurements,[12] we were able to show, through the notion of hyper-universality,[13] that the size and the shape of the sharp peak are consistent with the interpretation that the superfluid transition of ^4He confined in Vycor, in the thin film limit, is in the same universality class as the λ transition. The reason that the transition is three dimensional, despite the atomic size of the film thickness, is that as the reduced temperature $t \to 0$, the correlation length $\xi(t)$ exceeds the characteristic length of the porous structure and the film becomes three dimensionally connected. Away from the transition, or when the characteristic length of the experiment is short compared to ξ, the film exhibits two-dimensional characteristics. Some of our heat capacity results near T_c have already been published[11] and a recent review of the nature of the superfluid transition of ^{4}He in porous glass can be found in the conference proceedings of LT19[ref. 14].

In this paper, we shall discuss heat capacity results outside the superfluid transition region. An a.c. calorimetry technique is used in the measurement. In order to reduce the internal equilibration time, the sample used is a thin disc of Vycor 9mm in diameter and 0.6mm in thickness. The internal surface area is 5m^2 and the open pore volume is 0.0116 c.c. The Vycor disc is encapsulated in a thin layer of epoxy which also secured a thin (~0.1mm I.D.) capillary that connects the Vycor sample to a room temperature dosing unit. A heater and a carbon thermometer are glued onto each of the two sides of the disc. Heat capacity is measured by detecting the magnitude of the temperature oscillation due to a steady state sinusoidal heat input. The precision of a measurement is on the order of 0.2%. The background or addendum heat capacity of the sample, typically 1% of the total heat capacity, has already been subtracted in the data shown.

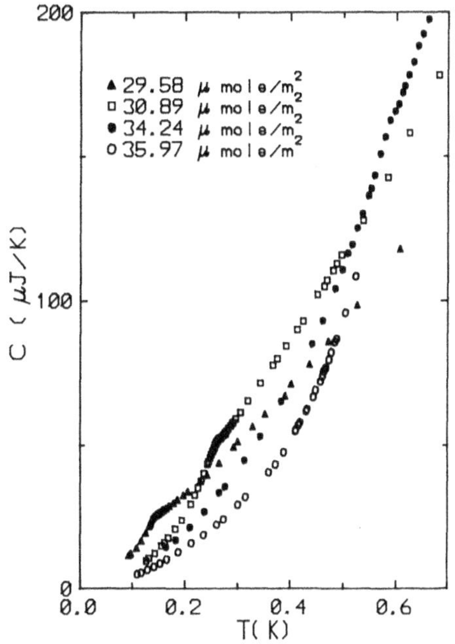

Figure 1. Heat capacity traces at four different coverages showing the sharp peak related to superfluid transition. The T_c's in ascending order of film coverages are 0.14, 0.26, 0.59 and 0.685K.

RESULTS AND DISCUSSIONS

Figure 1 shows heat capacity scans at four different coverages exhibiting the sharp peak at T_c. It should be noted that the heat capacity of the superfluid in the low temperature limit <u>decreases</u> with increasing coverage, in contrast to the behavior above T_c. The heat capacity value of a superfluid film at 36 μmole/m^2, in fact, drops below that of a film half its thickness. This interesting result is shown more clearly in Fig. 2 in a plot of the heat capacity as a function of surface coverage at three different temperatures, 0.1, 0.3 and 0.5K. This figure is interpolated from a large number of C vs T traces at closely spaced

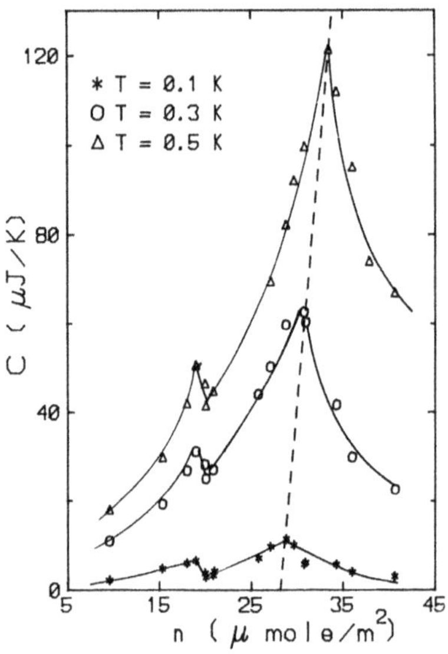

Figure 2. Heat capacity isotherm at various temperatures as a function of ^{4}He coverage. The first minima at 20 μmole/m^2 is a signature of monolayer completion. The second peak is related to superfluid transition.

coverages. Two peaks are found in these isotherms. The position or coverage of the first peak at 18 μmole/m^2 appears to be independent of temperature, while that of the second, or the more prominent peak, increases with temperature. This second peak is apparently the same one found in the heat capacity isotherm plot of Tait and Reppy,[10] separating the superfluid from the normal fluid region. This interpretation is consistent with Fig. 1, where C is found to increase with coverage for $T>T_c$ and decrease with coverage for $T<T_c$ in the low temperature limit. The peak position of the various isotherms, marked by the dashed line, traces out T_c as a function of coverage, This dashed line shows the

minimum critical coverage n_c for superfluidity to be 28 μmole/m^2, consistent with ρ_s measurements. The slope of T_c vs n in our study is also consistent with the torsional oscillator results.[7,11]

For $n<n_c$ the adsorbed film is often referred as an inert or localized layer. Fig. 2, however, shows a great deal of interesting thermodynamics. We interpret the minimum immediately after the first peak to be a signature of the completion of the first solid monolayer. Our estimate at 20 μmole/m^2 for the first layer is slightly higher than the 18.3 μmole/m^2 value based on ^{4}He vapor pressure isotherm measurements at higher temperatures.[15] We are relatively certain about the nature of the adsorbed film for $n<20$ μmole/m^2 and for $n>n_c$ at 28 μmole/m^2. For $n<20$ μmole/m^2, based on the effective areal density and the strength of the van der Waals' attractive potential, the adsorbed film is solid-like. For $n>28$ μmole/m^2, we know the film can become superfluid for some T above absolute zero. Based on Figure 2, we are not able to draw definite conclusions about the nature of the adsorbed film for coverages between 20 and 28 μmole/m^2. However, a close examination of the temperature dependence of C, to be shown below, indicates that the adsorbed film is probably also solid-like, similar to that with $n<20$ μmole/m^2.

We have examined the temperature dependence of the heat capacity at a variety of surface coverages to learn about the excitations of the adsorbed film. The results of the "best" fits are summarized in Table 1. It is remarkable that the measured heat capacity in the low temperature limit, irrespective of the different coverage regimes appears to favor a function that contains a linear T and a T^2 term. We expect the meaning

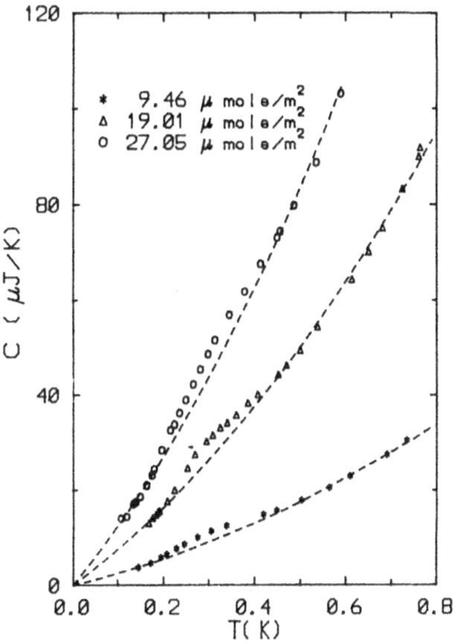

Figure 3. Heat capacity traces in the localized layer regime. Dashed lines represent fits of the "regular" heat capacity. Parameters of the fit are shown in Table 1.

and the origin of these two terms in the different coverage regime to be different. However, they appear to reflect the special geometry of the thin ^4He film inside the interconnected cylindrical pores of Vycor glass. In general, T and T^2 terms are associated with one and two dimensional excitations, respectively.

Figure 3 shows heat capacity as a function of temperature for three coverages in the localized layer. These traces are similar in that there is a broad anomaly centering near 0.3K that rides on top of a "regular" contribution. The height (53 erg/K) and the width (0.15K) of the broad anomaly appear to be coverage independent. The ($aT + bT^2$) functional form of the "regular" contribution suggests that the dominant excitations in the adsorbed film at these coverages resemble one and two dimensional phonons. To carry this interpretation further, we have calculated the effective one and two dimensional Debye temperatures,[2,8,10] which are also shown in Table 1. The fact that the Debye temperatures decrease with coverage may reflect the decreasing effective density of the adsorbed

TABLE 1. Temperature dependence of the heat capcity at various coverages. Heat capacity C is fitted as $C=aT+bT^2$; n, T_c, θ_{1D}, θ_{2D} are, respectively, the coverage, superfluid transition temperature, one and two dimensional Debye temperature. A broad anomaly in the localized region, n<28 μmole/m^2 have been subtracted.

n (μmole/m^2)	a (erg/K^2)	b (erg/K^3)	θ_{1D} (K)	θ_{2D} (K)	T_c (K)	Range of fit
9.46	231	243	53.9	21.4		(0.15K, 0.8K)
19.01	692	622	36.0	19.0		(0.15K, 0.8K)
27.05	1136	1076	31.2	17.3		(0.1K, 0.8K)
29.58	1440	813			0.140	(0.2K, 0.8K)
30.89	1515	1600			0.260	(0.25K, 0.8K)
33.54	195	2910			0.518	(0.1K, 0.35K)
35.97	156	2713			0.685	(0.1K, 0.4K)
40.56	178	2020			1.018	(0.1K, 0.4K)

film. The fact that the functional form for the 27 μmole/m² film is identical to that at lower coverages suggests the film remains solid-like. We have no explanation for the broad anomaly near 0.3K. The constancy of the size and location of the peak suggests that it is related to excitations along an interface, most likely the ^4He solid vapor interface. This broad anomaly vanishes once the total coverage is increased beyond n_c. A similar broad anomaly at coverages below n_c and a similar disappearance for coverages above n_c have been seen for ^4He film adsorbed on porous xerogel glass.[16]

Figure 4 shows a C/T vs. T plot of two films with superfluid transition temperatures of 0.26K and 0.14K. We have chosen to plot the data in this manner to show the temperature dependence of a normal fluid film. Although a functional form of $(aT + bT^2)$ is preferable, a functional form with a linear T plus T^3 terms also gives a reasonable fit. Because of the broad anomaly above T_c, mentioned earlier, it is not easy to apply the same analysis to a normal film with $T_c \lesssim 0.3K$.

Figure 5 shows the temperature dependence of the heat capacity of a superfluid film in the low temperature limit. In order to extend the analysis over a reasonable temperature range, only films with $T_c > 0.5K$ are included. The values of the dominant linear T terms are smaller than that found in normal fluid film and for most of the localized layer coverages. The magnitude of the linear T term is rather insensitive to coverage. A linear T term is expected if the normal fluid density, ρ_n, shows a T^2 dependence.[2] Such a dependence has, in fact, been reported in the

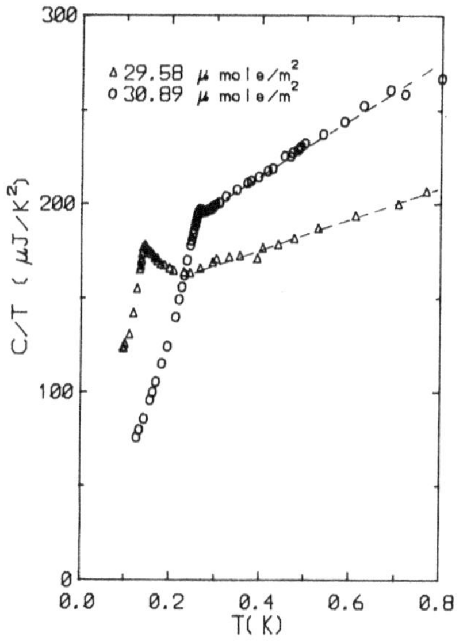

Figure 4. C/T vs. T plot highlighting the temperature dependence of two normal fluid films. Parameters of the fit are shown in Table 1.

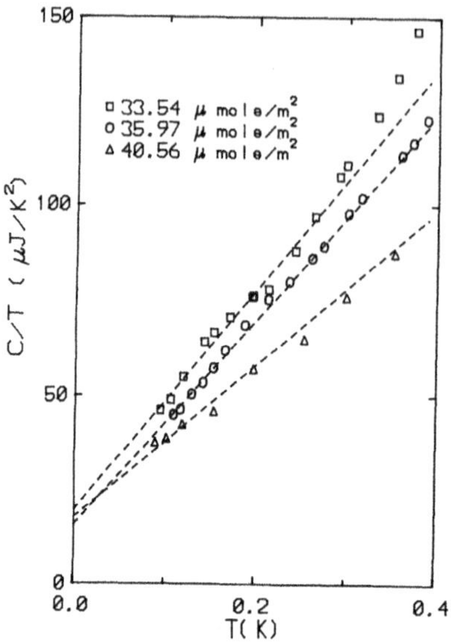

Figure 5. A plot of low temperature heat capacity of superfluid films showing the linear T plus T^2 dependence. Parameters of the fit are shown in Table 1.

torsional oscillator experiment.[7,17] However, the magnitude of the observed linear T term in C appears to be two orders of magnitude too large for ρ_n to be the sole contributor. As in the case for normal fluid film and films in the localized region, it is possible to have other excitations causing linear T heat capacity. As stated above, the T^2 term is probably related to two dimensional-like excitations including phonons, surface waves on the superfluid, as well as excitations along the solid-superfluid interface.

We have also extended the analysis of the heat capacity of the superfluid film to higher temperatures and we found the inclusion of a roton-like term, $e^{-\Delta/k_B T}$, is required.

We acknowledge useful conversations with S. Murphy. Y. Takano and J. D. Reppy. This work is supported by the National Science Foundation under grants DMR-8701651 and DMR-9008461.

REFERENCES

Present address: +Department of Physics, Kent State University Kent, Ohio.
†National Institute of Standards and Technology, Gaithersburg, MD

1. K. R. Atkins, H. Seki and E. U. Coudon, Phys. Rev. 102, 582 (1956).
2. For a comprehensive review of studies of the superfluid transition of ^4He in porous media up to 1975, see D. F. Brewer in The Physics of Liquid and Solid Helium, eds. K. H. Bennemann and J. B. Ketterson (John Wiley and Sons, New York, 1978)(, Ch. 6, p. 573.

3. M. E. Nordberg, J. Am. Ceram. Soc. 27, 299 (1944).
4. J. W. Cahn, J. Chem. Phys. 42, 93 (1965).
5. P. Wiltzius, F. S. Bates, S. B. Dierker and G. D. Wignall, Phys. Rev. 36, 2991 (1987); D. W. Schaefer, B. C. Bunker and J. P. Wilcoxon, Phys. Rev. Lett. 58, 284 (1987).
6. C. W. Kiewiet, H. E. Hall and J. D. Reppy, Phys. Rev. Lett. -35, 1286 (1975).
7. D. J. Bishop, J. E. Berthold, J. M. Parpia, J. D. Reppy, Phys. Rev. B24, 5047 (1981).
8. D. F. Brewer, J. Low Temp. Physics 3, 205 (1970).
9. R. A. Joseph and F. M. Gasparini, J. Phys. (Paris) Colloq. 39, C6-310 (1978).
10. R. H. Tait and J. D. Reppy, Phys. Rev. B-20, 997 (1979).
11. D. Finotello, K. A. Gillis, A. Wong and M. H. W. Chan, Phys. Rev. Lett. 61, 1954 (1988).
12. K. A. Gillis, S. Mukherjee, A. Wong and M. H. W. Chan, to be published.
13. D. Stauffer, M. Ferer and M. Wortis, Phys. Rev. Lett. 29, 345 (1972). P. C. Hohenberg, A. Aharony, B. I. Halperin and E. D. Siggia, Phys. Rev. B13, 2986 (1976).
14. M. H. W. Chan, Proceedings of 19th Intern. Conf. on Low Temp. Physics, Brighton, 1990.
15. D. F. Brewer, A. J. Symond and A. L. Thomson, Phys. Rev. Lett. 15, 182 (1965).
16. D. Finotello, A. Wong, K. A. Gillis, D. D. Awschalom and M. H. W. Chan, Proceed. 18th Int. Conf. on Low Temp. Physics, Jpn. J. of App. Phys., Vol. 26, Supp. 26-3, 283 (1987).
17. Y. Takano and J. D. Reppy, private communication.

VORTICES AND THE SUPERFLUID ^{4}He PHASE TRANSITION

IN TWO AND THREE DIMENSIONS*

Gary A. Williams

Physics Department
University of California
Los Angeles, CA 90024, U.S.A.

INTRODUCTION

The idea that thermally excited vortices might underlie the unusual properties of the superfluid ^{4}He λ-transition was first proposed by both Onsager[1] and Feynman[2] in their original papers on the quantization of circulation. In Feynman's words[2]: "If rotons are the smallest ring vortices... then there are states of higher energy corresponding to larger rings... As the temperature rises... then suddenly the rings of very largest length are of importance. ...The superfluid is pierced through and through with vortex line. We are describing the disorder of He I."

In spite of the simple and very physical picture of the superfluid transition that the vortex interpretation gives, this idea has never been accepted by the majority of theorists working in the field of phase transitions, and many of them have argued strongly against any role for vortices in three-dimensional (3D) systems.[3-5] It was only in two dimensions (2D) that Kosterlitz and Thouless[6] (KT) were able to show that vortices are the key excitations in the 2D superfluid transition, an initially controversial proposal that was subsequently confirmed by experimental measurements.[7] Nonetheless, there remains a widespread opinion that the vortex nature of the 2D transition is a curious anomaly unique to two dimensions, and that the 3D transition is somehow completely different in nature. We argue here that this is one of the most serious misconceptions in condensed matter physics, and that in fact the Feynman-Onsager-KT vortex mechanism provides a fundamental understanding of the transition in both 2D <u>and</u> 3D. The motivation for the recent extensions[8-10] of the KT ideas to 3D came from experimental studies[11] using ^{4}He films adsorbed in porous substrate materials. In this paper we first review the basic formulation of the 2D KT theory, discuss the experimental results, and then show how this can be simply generalized to describe the 3D λ-transition. We show that the 3D vortex theory is in good agreement with Monte Carlo simulations, and that this provides new support for the original Feynman-Onsager idea that roton excitations are vortex rings of the smallest size.

---

*Work supported in part by the US National Science Foundation, Grant DMR 89-12825.

2D KOSTERLITZ-THOULESS THEORY

In 2D Kosterlitz and Thouless[6] showed that pairs of vortices with opposite circulation were the important thermal excitations at the superfluid transition. The energy (normalized by $k_B T$) needed to excite a single pair of separation r is $U_o = 2\pi K_o \ell n(r-a_o)/a_o + 2E_c$, where $K_o = (\hbar^2/m^2)(\sigma_s{}^o/k_B T)$, a_o is the vortex core size, m is the helium atomic mass, E_c is the vortex core energy, and $\sigma_s{}^o$ is the "bare" areal superfluid density arising from excitations other than vortices. To analyze the difficult many-body problem of many pairs being excited near the critical temperature T_{KT}, there are two fundamental physical ideas that form the basis of the KT theory:

a) the vortex pairs are dipole structures that become oriented in an external flow field, in such a fashion that their net backflow cancels part of the external flow, reducing the superfluid density.

b) Dipole pairs of small separation are oriented by the flow field of larger pairs, partially canceling the flow and hence reducing the energy needed to excite the larger pairs.

The idea of part a) is implemented using dielectric linear response techniques. The reduction of the superfluid areal density σ_s from its bare value $\sigma_s{}^o$ is characterized by a dielectric constant $\epsilon = \sigma_s{}^o/\sigma_s = K_o/K_R$. The oriented pairs give rise to a susceptibility χ where $\epsilon = 1 + 4\pi\chi$. χ is calculated by multiplying the polarizability of a pair ($\sim K_o r^2$) and the number density of thermally excited pairs $[\sim 2\pi r \exp(-U(r))]$ and then integrating over all the pairs up to separation r, yielding for the renormalized superfluid density $K_R(r)$

$$\frac{1}{K_R(r)} = \frac{1}{K_o} + \frac{4\pi^3}{(a_o)^4} \int_{a_o}^{r} r^3 e^{-U(r)} dr \qquad (1)$$

The screened pair energy U(r) appearing in this equation is a consequence of part b) above, arising from the polarization of the smaller pairs in the flow field of the larger pairs. The "bare" energy U_o is reduced by the dielectric constant, but since ϵ is scale-dependent, this is written as an integral

$$U(r) = \int_{a_o}^{r} \frac{1}{\epsilon} \frac{\delta U_o}{\delta r} dr + 2E_c = \int_{a_o}^{r} \frac{2\pi K_R(r)}{r-a_o} dr + 2E_c \qquad (2)$$

These two equations can be solved by iteration starting from $r = a_o$ where $K_R = K_o$, and the renormalized macroscopic superfluid density is then $K_R(r \to \infty)$.

A well-known result of solving these equations is that as the temperature is raised (K_o decreased) there is an abrupt jump in K_R from the universal value $2/\pi$ to zero. This corresponds to a jump in superfluid density at the critical temperature T_{KT} related by

$$\frac{\sigma_s}{T_{KT}} = \frac{2}{\pi} \frac{m^2 k_B}{\hbar^2} = 3.5 \times 10^{-9} \text{ g/cm}^2\text{K} \qquad (3)$$

This jump to zero in σ_s is shown in the experimental data of Figs. 1a) and 2a). In Fig.1 the different curves correspond to ^{4}He films of different thickness d (since $\sigma_s{}^o$ is proportional to d) which is held constant as the temperature is raised. The jump to zero is well-determined by Eq.3, as shown by the solid line in the figure which has a slope corresponding to Eq.3. The same onset of superfluidity is shown in Fig.2a, which shows the third sound velocity ($c_3{}^2$ is proportional to σ_s) as a function of film thickness. In this experiment the temperature is held constant as the film thickness is changed, and at a particular critical thickness Eq.3 is satisfied and the third sound becomes completely overdamped (hatched lines).

^{4}HE FILMS ADSORBED IN POROUS MATERIALS

There are now extensive measurements of the superfluid transition of ^{4}He films adsorbed on porous substrates.[11-19] The average diameter, a, of the grains making up the porous material sets the scale for the effective dimensionality of the film. At length scales smaller than a, the film is 2D, and the KT screening arguments discussed in the previous section apply with little modification. At length scales greater than a, the 3D connectivity of the grains becomes important.

To investigate the variation of the superfluid transition with grain size, third sound measurements were undertaken[11-12] on a series of packed $A\ell_2O_3$ powders with nominal diameters ranging from 50 μm to 150Å, and the main results are shown in Fig. 2. In the largest 50μm powder the data is nearly identical to the third sound data on flat glass substrates, with the sharp loss of signal occuring at a critical KT film thickness (cross-hatched lines) that is the same on both the flat substrate and the 50μm

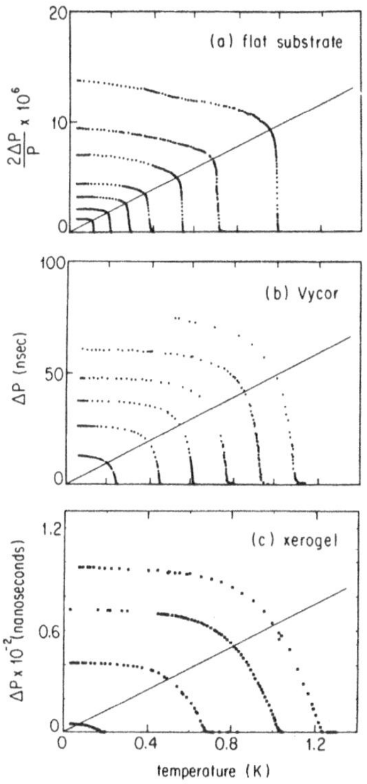

Fig. 1. Torsion oscillator data showing superfluid density versus temperature for several film thicknesses.
a) from Agnolet et al. (Ref. 7)
b) is from Ref. 13
c) is from Ref. 24.
The solid line in each graph is the universal KT line.[11,27]

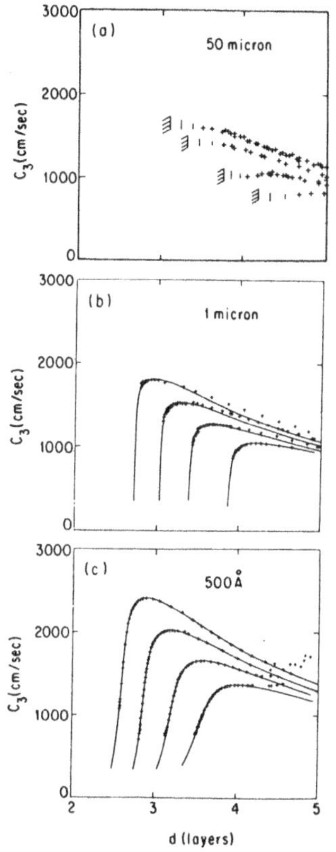

Fig. 2. Third sound velocity versus film thickness in atomic layers (1 layer = 3.6Å). The four data sets in each graph are taken at temperatures (top to bottom) 1.21K, 1.37K, 1.51K, and 1.62K.

powder. The transition in the powder is identical to the flat substrate KT transition because the vortex diffusion length in this experiment (~5μm) is considerably smaller than the 50 μm grain size, and only KT vortex pairs of separation smaller than the diffusion length are effective in reducing the superfluid density.[20]

As the powder grain size is reduced in Figs. 2b) and 2c), it is readily apparent that the transition still begins to occur at the critical KT thicknesses. Note that as the film is thinned the superfluid density begins to drop (i.e. a decreased third sound velocity) at precisely the same critical film thickness found in the 50 μm and flat-substrate data. The difference is that the decrease to zero of the superfluid density in the smaller powders is broadened and continuous, rather that the sharp jump found on the flat substrates. The smaller the powder diameter the more broadened the transition becomes, as can be seen most clearly in the 500A data of Fig. 2c). The third sound attenuation is also reduced as the powder size decreases, allowing data to be taken closer to the critical regime where $\sigma_s \to 0$. This data evolves continuously from the flat-substrate KT transition, and it is clear that the vortex pairs continue to play a key role in the porous materials.

The same characteristics have also been observed for films adsorbed in porous Vycor glass.[13,16] Fig. 1b) shows the original data[13] in Vycor with an average grain diameter ~ 150A, taken using a technique of measuring the period shift of a torsion oscillator containing the substrate and film. Comparing with the flat-substrate data of Fig. 1a), it is readily apparent that the transition still begins at the critical KT point (Eq. 3) for each film thickness, but that in the Vycor the KT jump is broadened, with a continuous decrease to zero at a temperature T_c that is 10-15% higher than T_{KT}. This broadening of the transition has recently been investigated as a function of the Vycor grain size by Shirahama et al.[16] They found the same results as discussed above for packed powders, that in the limit of very large grains the sharp KT jump is recovered, and that as the Vycor grain size is reduced the transition broadens continuously as a function of the grain size.

The broadening of the transition can be simply understood in terms of a finite-size limitation of the KT transition. The porous medium introduces a new length scale into the problem, the grain diameter a. On a flat substrate the KT recursion of Eqs. 1 and 2 is carried to $r \to \infty$, but for a finite system the recursion is stopped at the system size, and this leads to a finite-size broadening of the KT transition. By assuming that only vortex pairs of separation smaller than the grain diameter could be included in the recursion, Kotsubo and Williams found[11] a good fit to the broadening observed in the packed powders, shown as the solid lines in Figs. 2b) and 2c). A simple estimate for the upward shift in transition temperature $T_c - T_{KT}$ can be found by setting the Kosterlitz correlation length equal to the grain diameter, yielding[21]

$$T_c - T_{KT} \approx \left(\frac{b}{\ell n \frac{a}{a_0}} \right)^2 \tag{4}$$

where b is the nonuniversal constant of the correlation length. The Vycor results of Ref. 16 found quite good agreement with this functional form for the variation with grain size.

The finite-size characterization breaks down in the regime very close to T_c, where the coherence length becomes larger than the grain size and the 3D connectivity of the sample becomes important. In this regime there is a full 3D phase transition, as shown by the observation of power-law behavior of the superfluid density[13] and a sharp peak in the specific

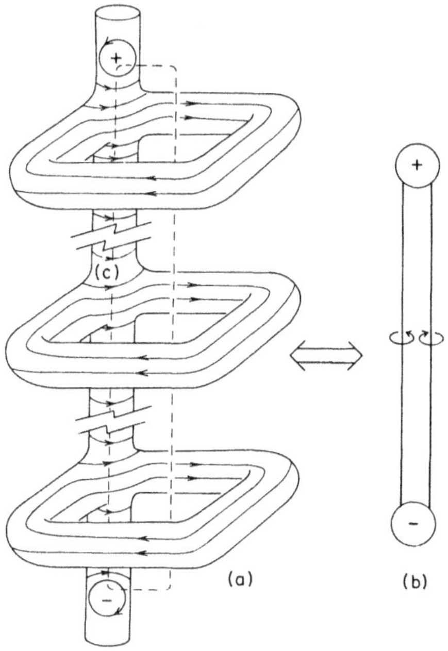

Fig.3. a) String structure on a cubic lattice of cylinders (only a portion of which is shown) illustrating the flow patterns in the adsorbed film due to the plus vortex at the top and the minus at the bottom. b) Equivalent flow pattern after coarse-graining: dipole field of two opposite circulation vortex lines spanning the plus-minus pair.

heat.[18] The key to explaining this crossover from 2D finite-size behavior to a 3D phase transition was first proposed by Minoguchi and Nagaoka[22] and independently by Guyer and Machta.[23] They postulated that the energy needed to excite a pair of vortices changes at the scale of the grain size, from logarithmic to linear dependence on the separation r between the plus and minus vortices,

$$U_0 = \begin{cases} 2\pi K_0 \, \ell n\left(\dfrac{r-a_0}{a_0}\right) & r < a \\ \pi\alpha K_0\left(\dfrac{r}{a}\right) & r > a \end{cases} \tag{5}$$

where α is a numerical constant. The linear energy results from the need to conserve the circulation of flow about a closed loop in the multiply-connected porous material; increasing the separation of the vortices linearly increases the number of such loops and hence also the energy. Fig. 3a) shows this "string" structure[8] on a cubic lattice of cylinders, with flow in one sense about a line of cylindrical trunks and in the opposite sense around the neighboring loops. This is a dipole structure, equivalent to two vortex lines of opposite circulation, and the lowest energy configuration is with the two lines parallel and separated by the lattice constant (Fig. 3b). Higher energy configurations are also possible where the two lines are separated by more than a lattice constant, and where the lines are not necessarily straight. The entropy of such strings is $S = k_B \, \ell n(c^r)$ where c is a constant (~5 for a cubic lattice), and hence there is a critical temperature T_c where the bare free energy $U_0 - TS$ is minimized for $r \to \infty$, corresponding to the possibility of an "unbinding" of the vortex pair.

The renormalized superfluid density from string excitations was calculated by Gallet and Williams[8], using the lowest energy straight strings, and also by Guyer and Machta[23] using a mapping to the Villain model. The results of Ref. 8 can be cast in the same form discussed above for the 2D case. The strings orient and reduce an applied flow, and short strings screen the energy of the longer strings. The string susceptibility is the product of the polarizability of a string ($\sim K_0 a r^2$) and the number density ($\sim 4\pi r^2 \exp(-U(r))$ integrated over all the strings of size less than r, giving a renormalized density

$$\frac{1}{K_R(r)} = \frac{1}{K_0} + \frac{\beta}{a^5}\int_a^r r^4 \exp(-U(r))\,dr \tag{6}$$

with β a constant, and a screened energy U(r) given by Eq. 2 with the linear U_0 of Eq. 5. The 2D Eqs. 1,2 are iterated from the scale a_0 up to the grain size a, and then the 3D string Eq. 6 is iterated to $r \to \infty$ to give the observable areal density σ_s, or equivalently the 3D volume density $\rho_s \sim \sigma_s/a$. The results are shown in Fig. 4 as a function of the grain size. The grain size dependence comes from an effective finite-size cutoff of the 2D recursion at the grain size a, because of the crossover of Eq. 5 to the thermodynamically unfavorable linear increase of the string energy. The difference $T_c - T_{KT}$ varies with grain size as given by Eq. 4, and as experimentally confirmed by the Japanese data in Vycor.[16] Close to T_c where $\sigma_s \to 0$, however, it finally becomes possible to excite 3D strings of length much larger than the grain size (since the string energy is proportional to σ_s), and this leads to a power-law dependence of the superfluid density, $\sigma_s \sim (T_c - T)^{\nu}$. The Gallet- Williams[8] calculation yields $\nu = 0.64$, while the Guyer-Machta approach[23] predicts $\nu = 0.67$, comparing well with the experimental result[13] $\nu = 0.63 \pm 0.03$. The correlation length of the transition, $\xi \sim a/K_R$, is essentially the mean string length being excited at a given $T - T_c$.

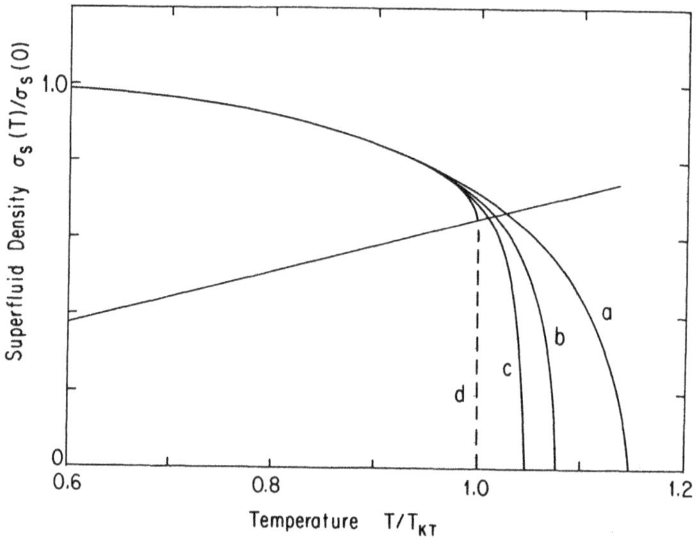

Fig. 4. Normalized superfluid density versus temperature from the vortex string calculation of Ref.8. The different curves a-d correspond to values a/a_0 = 6, 30, 150 and infinite, respectively. The solid line is the universal KT line.

The vortex interpretation also leads to an understanding of the specific heat and the dissipation at the transition. Since the model of Ref. 8 satisfies the Josephson scaling relation, there will be a critical specific heat peak at T_c due to the excitation of strings of infinite length. Superposed on this will also be a broad specific heat maximum somewhat above T_c, due to the dissociation of 2D vortex pairs of separation smaller than the grain size. Both of these features have been observed in experimental measurements on several different substrates.[18]

Finite-frequency dissipation peaks at T_c have been observed in Vycor over a wide range of frequencies[16,17], and a preliminary calculation[25] of the vortex dissipation is able to give a reasonable account of the main features of the data. A new length enters the problem at finite frequencies, the vortex diffusion length $r_D = \sqrt{D/\omega}$, where D is the diffusion constant describing the frictional drag forces on the vortex cores, and ω the angular frequency. At low frequencies where r_D is larger than the grain size there is no dissipation except in a narrow region about T_c where strings of length greater than r_D are excited. As the frequency is raised r_D decreases and hence strings of that length can be excited at larger values of $T-T_c$, broadening the width of the cusp-like dissipation peak. At the highest frequencies r_D becomes smaller than the grain size and the dissipation crosses over to the more usual 2D vortex pair dissipation[20], with a smooth, broad peak that extends well above T_c. This progression of the dissipation with frequency is shown in the graphs on the right side of Fig. 5, which compares the theory[25] with experiments on films in Vycor.[17] The graphs on the left show the broadening in ρ_s due to the finite-size cutoff[20] of the recursion relations at the scale of the diffusion length r_D.

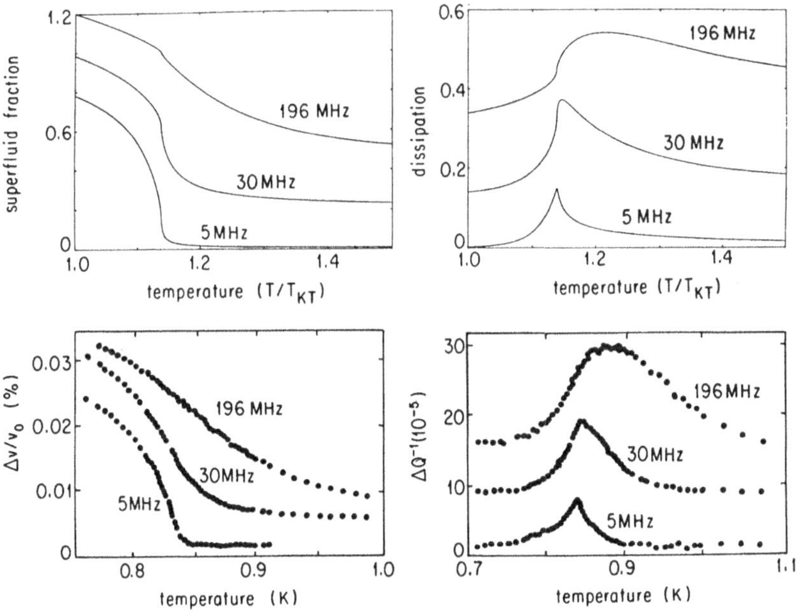

Fig.5. Finite-frequency dissipation (right graphs) and superfluid density (left graphs) versus temperature. The theory curves (top) are from Ref.25, compared with the Vycor data (bottom) of Ref. 17.

The vortex picture also provides a simple explanation of the increased broadening of the transition observed[14] in the very thinnest films having T_c values below ~ 0.1K. As seen from Fig.4, the transition will broaden for fixed grain size if the vortex core size a_o increases. Experimental evidence for such an increase in this regime has been observed by McQueeney[26], who finds a rapid rise in a_o as the film is thinned. This rise is in accord with expectations[41], since the core cannot be smaller than the mean interparticle spacing of the atoms making up the superfluid layer. It should be noted that an alternative interpretation of this broadening phenomenon as a crossover to an "ideal dilute Bose gas" has been proposed.[4,14] This interpretation, however, was based on data that was left uncalibrated, and the correlation with the universal KT line apparently was not noticed.[11,27]

There have recently been superfluid density measurements[19,24] on a new class of substrates, aerogel and xerogel, which find a considerably larger critical exponent, $\nu \approx 0.9$, than the $\nu = 0.63$ found in Vycor. The data[24] using xerogel is shown in Fig. 1(c). An interesting feature of these substrates is that at least some of them are known to be characterized by a fractal dimensionality D ranging between 1.8 - 2.4. The adsorbed superfluid will then also be distributed with the same fractal dimensionality. It is possible that the shifted exponent is a result of this change in dimensionality (and corresponding change in universality class), although further experiments with substrates of known D will be necessary to check this.[27]

VORTEX RINGS AND THE 3D λ-TRANSITION

As the above results indicate, vortex excitations can provide a consistent picture of superfluid transitions in three dimensions. There is a growing realization that the bulk λ-transition can also be described by such a mechanism[28], in agreement with the original speculation of Feynman and Onsager. A number of authors in the past have reiterated this idea (see the references in Refs. 8-11), but it is only recently that renormalization procedures using vortex ring excitations have allowed a quantitative calculation of the properties of the transition.[9,10] The energy to excite a circular ring of diameter r is, for large rings, $U_o = \pi^2 K_o (r/a_o)(\ell n(r/a_c)+C)$ where $K_o = 2(\hbar^2/m^2)(\rho_s^o\, a_o/k_BT)$ with ρ_s^o the bare superfluid density, a_c the core size, and C a constant related to the core energy. The basic physical ideas underlying the theories using these rings are:

a) vortex rings are dipole structures that become oriented in an external flow field, in such a fashion that their net backflow cancels part of the external flow, reducing the superfluid density.

b) Rings of small diameter are oriented by the flow field of larger rings, partially cancelling the flow and hence reducing the energy needed to excite the larger rings.

By comparing with the first section of this paper it can be seen that the 3D mechanism is identical to that of the 2D KT transition. The only difference is the dimensionality, which gives rise to power-law behavior of ρ_s in 3D, instead of the sudden jump found in 2D.

The implementation of part a) above again involves calculating a susceptibility by multiplying the ring polarizability (~ $K_o r^4$) times the number density of rings (~ $r^2 \exp(-U(r))$) and integrating over r, yielding

$$\frac{1}{K_R} = \frac{1}{K_o} + \frac{A_o}{(a_o)^7} \int_{a_o}^{r} r^6 \, e^{-U(r)} \, dr \qquad (7)$$

The screened energy is given as before by

$$U(r) = \int \frac{1}{\epsilon(r)} \frac{\delta U_o}{\delta r} dr + \pi^2 K_o C \qquad (8)$$

$$= \int_{a_o}^{r} \pi^2 K_R \left(\ell n \frac{r}{a_c'} + 1\right) \frac{dr}{a_o} + \pi^2 K_o C$$

where $a_c' = e^{-C} a_c$, and A_o is a numerical constant.

A difference between the 2D and 3D cases is that in 2D the vortex core size remains constant at a_o even near T_c, whereas in 3D it diverges at the transition. A particular model of the vortex core for purely circular rings was utilized in Ref. 9, and this resulted in a value for the superfluid critical exponent of $\nu = 0.5$, at considerable variance with the known value $\nu = 0.67$. A more realistic treatment of the effective core size has been suggested by Shenoy.[10] He takes into account the fact that the excited rings are not circular, but are distorted and wiggly, taking the form of a self-avoiding random walk. The screening effect of the wiggles on each other is similar to the case of polymer chains, where the Flory approximation is known to provide a good description.[29] Shenoy proposed that wiggly rings could be treated as circular rings using Eqs. 7 and 8, but with the screening effects giving rise to an effective core diameter a_c' such that $r/a_c' = (K_R r/a_o)^{-\theta}$, with $\theta \simeq 0.6$, the Flory exponent in 3D. Using this in Eq. 8 and iterating from $r = a_o$ to values past the correlation length ($\xi = a_o/K_R = m^2 k_B T/\hbar^2 \rho_s$) gives rise to a power-law phase transition $K_R \sim (K_o - K_{oc})^{\nu}$, where fits for ν give a value very close to the Shenoy prediction $\nu = 0.672$ for $\theta = 0.6$. The correlation length can be identified as the mean ring diameter being excited.[9]

Monte Carlo simulations of the XY model have been important in establishing the vortex nature of the phase transition.[30-33] Fig. 6 shows

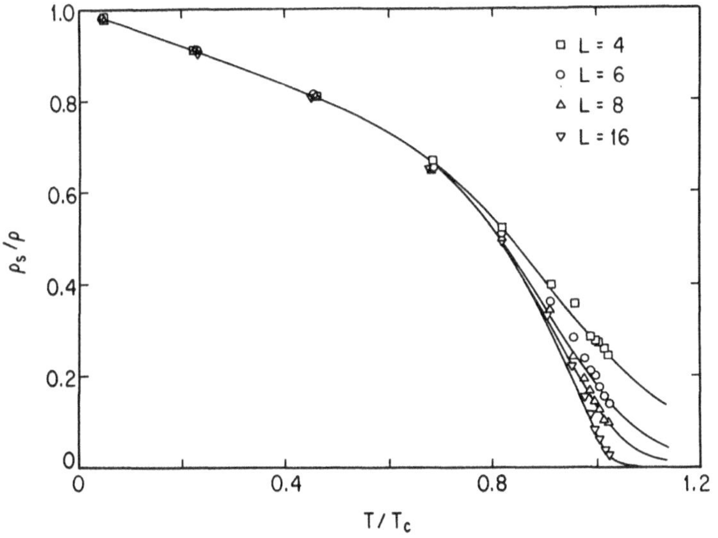

Fig. 6. Superfluid density of the XY model versus temperature, comparing the vortex theory (solid curves) with Monte Carlo simulations (data points from Ref. 30) on several different lattice sizes.

a comparison[34] of the superfluid density calculated from the above theory (solid lines) compared with Monte Carlo data[31] for the XY helicity modulus. At low temperatures vortices are not excited, and the initial linear decrease in ρ_s is due to excitation of the XY spin waves. At $T/T_c \sim 0.6$, however, the vortices begin to take over, and finally drive ρ_s to zero at T_c (for an infinite system). Because the Monte Carlo simulations are performed on a finite lattice size L, the iteration of the theory is truncated at a vortex diameter r_{max} that is on the order of the lattice size; taking $r_{max}/a_o = 0.43$ L is seen in Fig.6 to give a good description of the Monte Carlo data for four different lattice sizes. Near T_c the vortex theory is in detailed agreement[34] with the predictions of finite-size scaling, as might be expected since the coherence length of the theory is known to be correct.

Monte Carlo simulations also confirm other aspects of the vortex nature of the transition. The vortex density[30] shows a rapid increase with temperature to a value $n_v \sim 0.2$ per lattice site at T_c, and only a slow increase thereafter to $n_v = 0.33$ at $T = \infty$, the expected value when the spins are completely disordered. The vortex structures destroy the long-range coherence of the system. Such a density of vortex cores in liquid helium would certainly deplete any condensate fraction, as observed in neutron scattering experiments.[35] The Monte Carlo simulations are also able to directly visualize the vortex ring configurations.[32,33] The observation of large rings near T_c is graphic proof of the screening argument of Eq. 8. Without such a reduction of the screened energy it would not be possible for rings of such length to be thermally excited.

Fig. 6 has consequences for the interpretation of the superfluid density of liquid helium. It shows that the vortices have a dominant effect even at temperatures well below T_c, far from the critical regime. This contradicts the scenario proposed in Ref. 9, where both phonon and roton excitations were assumed to comprise the bare superfluid density ρ_s^o, and vortices were thought to be appreciable only within perhaps 50 mK of T_λ, the critical region. A much more likely scenario based on the XY model comparison is now suggested, that the bare density ρ_s^o has contributions only from the phonons (i.e. $\rho_s^o/\rho \approx 1$ since the T^4 phonon contribution is very tiny even near T_λ), and that the rotons instead are part of the vortex renormalization: they are the vortices of smallest size, the scale at which the recursion starts. Indeed, fits of the experimental superfluid density amplitude near T_λ to the above vortex model requires values of $\rho_s{}^o/\rho \approx 1$ for agreement, rather than the value $\rho_s{}^o/\rho \approx 0.01$ that might be expected from the scenario of Ref. 9.

Of course, the idea of rotons as a modified form of the smallest vortex rings has a long history, starting with Feynman and Onsager and extending to more recent investigations.[36] Although certainly the atomic nature of the liquid will change the energy and dispersion of the excitation from that of a macroscopic vortex ring, the main features needed for the renormalization are that the excitation have a dipolar structure and that it follow dielectric linear response to an applied flow field. Feynman[37] and others[38] have demonstrated that rotons have just these properties. The details of how the energy and dispersion change over to the vortex form at longer length scales is unfortunately not well known, although some progress in this direction has been made.[36]

The above considerations may also provide a basis for understanding the unusual scaling property of the superfluid density found by Maynard et al.[39] They showed that ρ_s/ρ at different pressures up to the melting curve collapses onto a single universal curve when plotted versus $T/T_\lambda(p)$, and that this holds over a wide temperature range, 1.2 - 2.17K. They also noted that the roton energy gap Δ_o scales in the same manner. The vortex theory curves shown in Fig.6 follow just this type of scaling, being only functions of T/T_c. The exact shape of the curves does depend on the initial vortex fugacity $y_o = \exp(-E_c/T)$, where E_c is the core energy of the smallest rings. E_c is a function of T/T_c in the XY model, but for the case of helium it could well be a separate function of pressure, thus spoiling the universal collapse. However, in the above scenario treating rotons as vortices, E_c is identified as the roton gap Δ_o, which is a function only of $T/T_\lambda(p)$. This would then lead to the universal scaling that is observed experimentally.

In summary, the vortex theory of the λ-transition developed in Refs. 9 and 10 has the following features: a) the critical exponents are correctly predicted[10], b) the Josephson scaling relation is satisfied[9], c) the coherence length is exactly correct[9], d) finite-size scaling is satisfied[34], e) the critical superfluid density amplitude is approximately correct, and (f) the high vortex density above T_λ provides a simple mechanism driving the condensate fraction to zero, since the liquid is filled with the normal cores. The main defect of the theory is that the Flory scaling proposal as yet has no rigorous justification, although it is interesting to note that similar techniques have independently been introduced to describe the vortex configurations in classical fluid turbulence.[40] The theory does provide a simple and very physical picture of the phase transition, in contrast to the very mathematical nature of perturbation theories such as high-temperature expansions and the $4-\epsilon$ expansion. Feynman and Onsager were probably the two most intuitive physicists of the twentieth century, and the current developments are further tribute to their astonishing insight.

REFERENCES

1. L. Onsager, Nuovo Cimento Suppl. 6:249 (1949).
2. R.P. Feynman, in "Progress in Low Temperature Physics," C.J. Gorter, ed., North-Holland, Amsterdam (1955), Vol.1, p.52.
3. P. Weichman, Phys. Rev. Lett. 61:2969 (1987).
4. P. Weichman and M.E. Fisher, Phys. Rev. 34:7652 (1986).
5. P.W. Anderson, "Basic Notions in Condensed Matter Physics," Benjamin/ Cummings, Menlo Park (1984).
6. J.M. Kosterlitz and D.J. Thouless, J. Phys. C 6:1181 (1973); see also A.P. Young, Phys. Rev. B 19: 1855 (1979).
7. D.J. Bishop and J.D. Reppy, Phys. Rev. B 22:5171 (1980); I. Rudnick, Phys. Rev. Lett. 40:1454 (1978); G. Agnolet and J.D. Reppy, Phys. Rev. B 39:8934 (1989).
8. F. Gallet and G.A. Williams, Phys. Rev. B 39:4673 (1989).
9. G.A. Williams, Phys. Rev. Lett. 59:1926 (1987).
10. S.R. Shenoy, Phys. Rev. B 40:5056 (1989).
11. V. Kotsubo and G.A. Williams, Phys. Rev. B 33:6106 (1986).
12. H. Cho, V. Kotsubo, and G.A. Williams, Can. J. Phys. 65:1532 (1987).
13. D.J. Bishop, J.E. Berthold, J.M. Parpia and J.D. Reppy, Phys. Rev. B. 24:5047 (1981).
14. B.C. Crooker, B. Hebral, E.N. Smith, Y. Takano, and J.D. Reppy, Phys. Rev. Lett. 51:666 (1983); J.D. Reppy, Physica B+C 126:335 (1984).
15. K. Shirahama, N. Wada, Y. Takano, T. Ito, and T. Watanabe, Jpn. J. Appl. Phys. 26, suppl. 26-3:293 (1987).
16. K. Shirahama, M. Kubota, S. Ogawa, N. Wada, and T. Watanabe, Phys. Rev. Lett. 64:1541 (1990).
17. N. Mulders and J.R. Beamish, Phys. Rev. Lett. 62:438 (1989).
18. D. Finotello, K.A. Gillis, A. Wong, and M.H.W. Chan, Phys. Rev. Lett. 61:1954 (1988).

19. M.H.W. Chan, K.I. Blum, S.Q. Murphy, G.K.S. Wong, and J.D. Reppy, Phys. Rev. Lett. 61:1950 (1988)
20. V. Ambegaokav, B.I. Halperin, D.R. Nelson, and E.D. Siggia, Phys. Rev. B 21:1806 (1980).
21. M. Barber, in "Phase Transitions and Critical Phenomena," C. Domb and J. Lebowitz, eds., Academic Press, London (1983), Vol.8.
22. T. Minoguchi and Y. Nagaoka, Prog. Theor. Phys. 80:397 (1988).
23. J. Machta and R.A. Guyer, Phys. Rev. Lett. 60:2054 (1988).
24. K. Blum, S. Murphy, M.H.W. Chan, D.D. Awschalom and J.D. Reppy, Jpn. J. Appl. Phys. 26, suppl. 26-3:275 (1987).
25. H. Cho, F. Gallet, and G.A. Williams, in "Quantum Fluids and Solids - 1989," G. Ihas and Y. Takano, eds., AIP, New York (1989) p.193.
26. D.F. McQueeny, Ph.D. thesis, Cornell University, 1988 (unpublished).
27. G.A. Williams, Phys. Rev. Lett. 64:978 (1990); M.H.W. Chan et al., ibid, p. 979.
28. H. Kleinert, "Gauge Fields in Condensed Matter," World Scientific, Singapore (1990) Vol.1.
29. S. Edwards, J. Chem. Soc. Faraday Trans. II, 75:1020 (1979).
30. G. Kohring, R. Shrock, and P. Wills, Phys. Rev. Lett. 57:1358 (1986).
31. Y. Li and S. Teitel, Phys. Rev. B 40:9122 (1989).
32. W. Janke, Ph.D. thesis, Freie Universität Berlin, 1985 (unpublished).
33. J. Epiney, Diploma thesis, ETH Zurich, 1990 (unpublished).
34. G.A. Williams, to appear in Proceedings of LT19, Physica B (1990).
35. E.C. Svensson, in "75th Jubilee Conference on Helium-4," J. Armitage, ed., World Scientific, Singapore (1983) p.10.
36. C.A. Jones and P.H. Roberts, J. Phys. A 15:2599 (1982).
37. R.P. Feynman, Phys. Rev. 94:262 (1954).
38. R.J. Donnelly and P.H. Roberts, Phys. Lett. 43A:199 (1973).
39. J. Maynard, J. Heiserman, and I. Rudnick, Phys. Lett. 58A:187 (1976).
40. A.J. Chorin, to appear in Commun. Math. Phys. (1990).
41. G.A. Williams, Jpn. J. Appl. Phys. 26, suppl. 26.3:305 (1987).

SCALING OF CONFINED ^{4}He AT THE SUPERFLUID TRANSITION

Francis M. Gasparini and Ilsu Rhee

Department of Physics and Astronomy
State University of New York at Buffalo
Buffalo, NY 14260 U.S.A.

INTRODUCTION

^{4}He at the superfluid transition has been well studied and has yielded valuable checks on aspects of critical behavior ranging from scaling, universality and explicit predictions of critical exponents and amplitude ratios. The most detailed results have been obtained in the thermodynamic limit, i.e., in situations where the smallest confining dimension for the helium sample is much larger than the magnitude of the critical correlation length, ξ. Since this length diverges at the transition, there are many experimental realizations, such as helium films or helium confined in pores, where the effect of confinement to dimensions comparable to ξ manifests itself as a modification of the properties realized in the thermodynamic limit, or as completely new behavior associated with a lower spacial dimension.

The modification of bulk behavior due to confinement to a small dimension d, i.e., a sample $\infty\times\infty\times d$, is described by theories of finite-size scaling,[1] see Barber[2] and Privman[3] for most recent reviews. According to this one may write for helium a free energy in the form[4]

$$F(t,d) = t^{2-\alpha} f(t^{-1/\theta}/d) + f_o(t), \tag{1}$$

where $t = |1-T/T_\lambda|$, α is the specific heat exponent, f_o is a smooth background function and θ is a critical exponent. A central ansatz of scaling theory is that $\theta=1/\nu$, where ν is the exponent of the three dimensional (3D) correlation length, $\xi = \xi_o t^{-\nu}$. That is the finite system manifests no new critical exponents, rather its behavior is governed by the bulk correlation length. There follows from Eq. 1 that the maximum, C_m, achieved by the specific heat, and its position, t_m, are given by

$$C_m \cong (A/\nu)\, \ell nd, \tag{2}$$

$$t_m \cong a_o d^{-1/\nu}, \tag{3}$$

where A is the amplitude of the nearly logarithmic divergence of the bulk system. Corrections of order α and confluent singularity corrections can also be incorporated in these equations.[5] From Eq. 1 there also follows that the specific heat difference between the bulk and confined system is given by

$$[C(T,\infty) - C(t,d)]t^{\alpha} = g(t^{-\nu}/d) - g(\infty). \tag{4}$$

Or, far enough from T_λ where $t^{-\nu}/d$ is small

$$[C(t,\infty) - C(t,d)]t^{\alpha} \cong g_o t^{-\nu}/d - g(\infty). \quad (5)$$

In the case of the superfluid fraction one may scale the confined system in an analogous way,[6]

$$\frac{\rho_s}{\rho} = \frac{\rho_{sb}}{\rho}(1 - \phi(t^{-\nu}/d)), \quad (6)$$

which to leading order, when $t^{-\nu}/d$ is small, may be written as

$$\frac{\rho_{sb} - \rho_s}{\rho_{sb}} \cong \phi_o\,(t^{-\nu}/d) + O(t^{-\nu}/d)^2. \quad (7)$$

One can obtain from Eq. 6 a shift equation for the vanishing of ρ_s by identifying a critical temperature T_c at which

$$\lim_{T \to T_c} \theta(t^{-\nu}/d) \cong 1. \quad (8)$$

For crossover into 2D one can refine Eq. 8 to allow for the expected universal jump in ρ_s at the Kosterlitz-Thouless transition.[7]

There are a number of experiments which have been done in an attempt to verify finite-size scaling in ^{4}He. These have been recently reviewed by Gasparini and Rhee.[8] What we will discuss here is a subset of these results, principally those from which one can scale the whole thermodynamic response as function of temperature and size. We will see that the overall results from existing experiments is that while one can scale the data with confinement size, the exponent of the temperature dependence is not ν as expected on the basis of bulk-correlation scaling.

THE SUPERFLUID FRACTION

In an experiment to verify size-scaling predictions, one has to exercise special care to insure uniformity in the confining dimension. In the experiment of Rhee et al.[7] this was achieved by using exceptionally flat silicon wafers which were chemically bonded at a separation dictated by a lithographically defined SiO_2 pattern. A schematic representation of this cell is shown in Fig. 1. The details of the fabrication of this cell have been discussed by Rhee et al.[9] The helium in the cell is confined between two parallel surfaces and its superfluid fraction is determined by measuring the mass loading on a torsional pendulum.

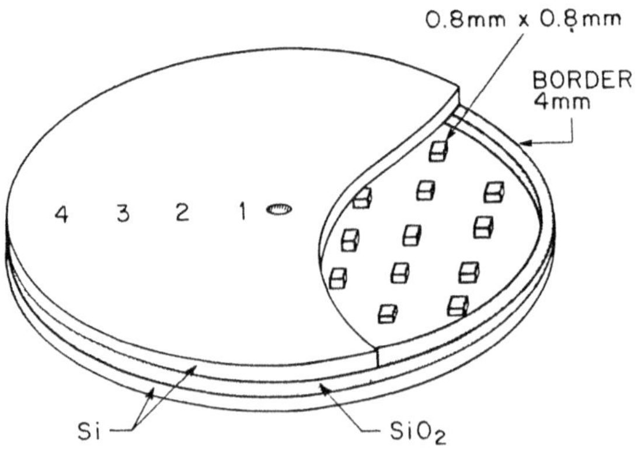

Fig. 1. Two Si wafers bonded at a separation determined by the SiO_2.

An important aspect of these cells is the ability to determine the spacing between the wafers after the cell is fabricated. This is done by looking at transmission interference fringes at wavelengths below the Si bandgap. This gap limits this procedure to spacings $\gtrsim 0.6\ \mu m$. Comparison of this type of measurement with the oxide thickness and with the average cell spacing as determined by helium mass loading give excellent agreement.[9]

The superfluid fraction of helium confined in a 0.519 μm cell is shown in Fig. 2.[7] The dashed line in this figure represents the bulk behavior. This line is determined by matching the data far from the transition with the known bulk behavior. The initial difference between the dashed line and the data (large values of t) is the result of finite-size 3D behavior. The more rapid variation for smaller t's is due to crossover into 2D behavior. This latter feature is emphasized in the inset of Fig. 2, where the dissipation of the oscillator is also plotted. This shows a maximum in the region where ρ_s varies most rapidly. This is the signature of Kosterlitz-Thouless behavior and comes from the mechanism of vortex unbinding.

The data in the finite-size region for four different cells, 3.9, 2.8, 0.519, 0.106 μm, were analyzed by Rhee et al. according to Eq. 6.[7] This is shown as the upper panel of Fig. 3. The variable $(t^{-\nu}d)^{-1/\nu}$ rather than $t^{-\nu}/d$ is used here. The quantities 0.672 and k are the exponent and amplitude of ρ_{sb}/ρ. If Eq. 6 were obeyed by the data, they should all collapse on a universal curve. This clearly does not happen, indicating a strong violation of scaling with the bulk correlation length exponent.

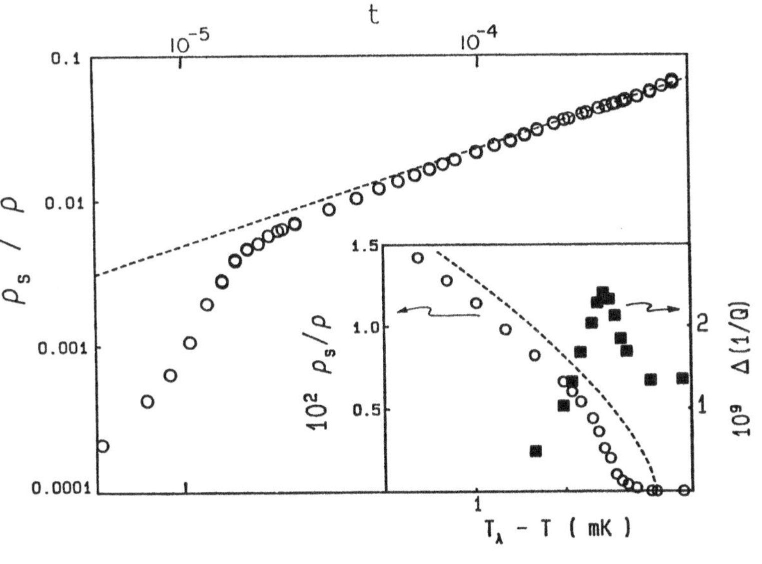

Fig. 2. Superfluid fraction vs. t for 0.519 μm cell. Inset: superfluid fraction and dissipation in 2D region. The dashed lines are bulk behavior.

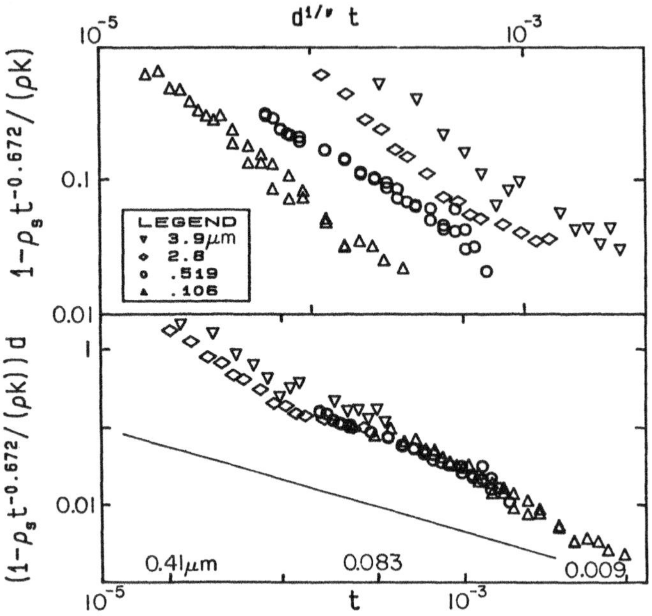

Fig. 3. Top: data plotted to test scaling with the exponent ν=0.672, Eq. 6. The data do not collapse as expected. Bottom: data collapse according to Eq. 7, but with an effective exponent of 1.14. d is in μm.

The bottom panel of Fig. 3 shows the same data scaled with d. This is a test of Eq. 7, according to which the leading term in the quantity $(1-\rho_s/\rho_{sb})d$ should be a simple power-law in t with exponent ν. Fig. 3 shows that these data do collapse when scaled with d, but on a universal curve of average slope -1.14. The expected slope of $-\nu$ is shown for comparison as a solid line. The bottom panel is thus a different way of showing the disagreement with correlation-length scaling. The top panel, however, makes a stronger statement because no assumptions on the form of the scaling function are made.

In an earlier determination of the superfluid fraction, Brooks et al.[10] and Schubert and Zimmermann[11] used the Helmholtz resonance between two chambers separated by a Nuclepore filter membrane. These filters have nearly cylindrical pores with relatively narrow size distribution for a given specified size. The superfluid fraction determined by Brooks et al.[10] was analyzed by Gasparini et al.[6] according to Eqs. 6,7. From this analysis it was found that the data did not collapse as a function of $t^{-\nu}/d$ and that good collapse could be achieved by scaling the data with d^2. Recently these data, as well as those of Schubert and Zimmermann, which are also for helium confined in Nuclepore, were reanalyzed by Gasparini and Rhee[8] by making use of bulk helium data which allowed a comparison with the confined helium measurements over a wider temperature range. This reanalysis is shown in Fig. 4 plotted in the same way as the top panel of Fig. 3. Again one finds that the data do not collapse. When scaled with d^2, in a plot analogous to the bottom panel of Fig. 3, these data for cylindrical confinement do collapse reasonably well. In particular, the data of Schubert and Zimmermann[11]

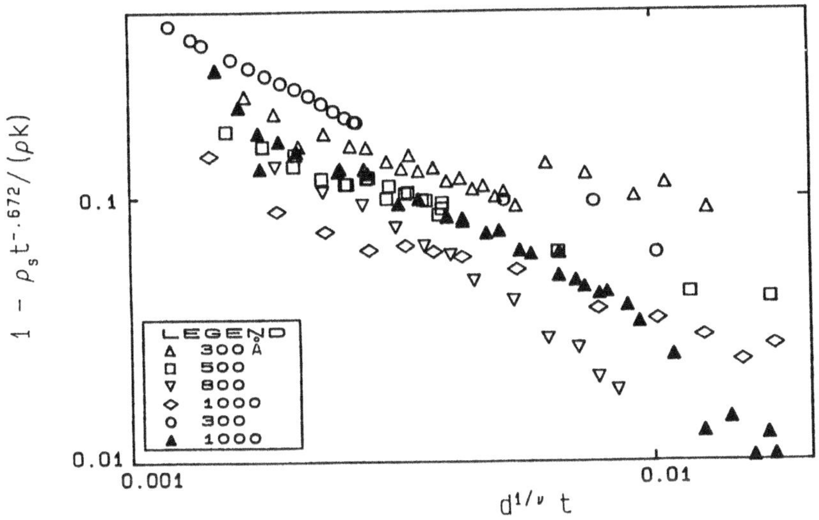

Fig. 4. Data from Ref. 10,11 plotted according to Eq. 6. d is in units of 1000Å. Compare with top panel of Fig. 3.

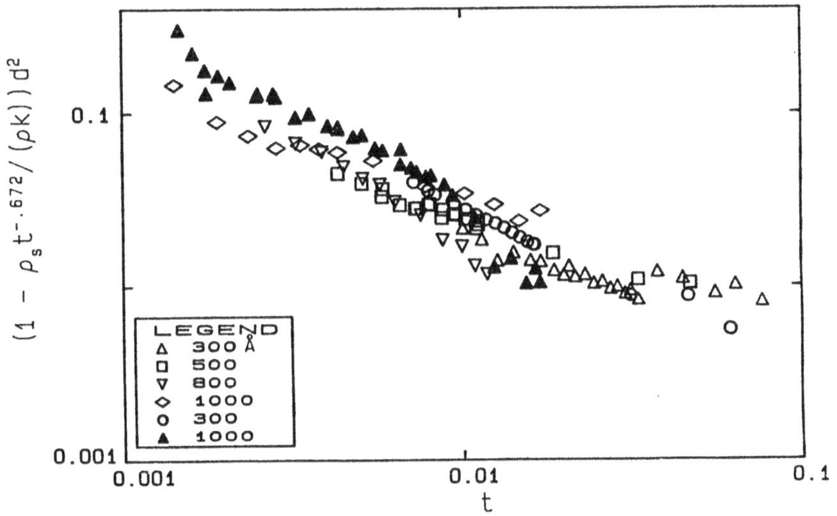

Fig. 5. Same data as in Fig. 4. Here the difference in superfluid fraction is scaled with d^2. Compare with bottom panel of Fig. 3.

(open circles, 300Å; and filled triangles, 1000Å; collapse particularly well. Note that the data in Fig. 5 fall a full decade farther from the transition than the data of Fig. 3. The temperature dependence defined by the data in Fig. 5 is very nearly a simple power law, especially if one ignores the data furthest into the 10^{-2} decade. These are outside a likely asymptotic dependence of the superfluid fraction difference. The exponent from these data is 1.0±0.1, where the error represents reasonable extrema lines. This exponent is rather close to that determined from the data in the planar confinement of the Si cells. The important and surprising difference is the d-scaling for the Si cells versus the d^2-scaling for the data in cylindrical pores. None of these results can be understood on the basis of Eqs. 6,7.

Along the abscissa of Fig. 3 there are numbers which indicate the value of the phase coherence length given by $\xi = 3.57t^{-0.6672}$Å. When one compares these values with the characteristic confinement dimensions one can see that deviations from bulk behavior are resolved experimentally when ξ is of the order of a few percent of the confinement. This initial deviation is thus predominantly a surface effect, and this suggests that even for confinement geometries characterized by a rather broad and irregular distribution of sizes one can obtain the exponent for the temperature dependence while forgoing the scaling with size. In other words, one may average the leading term of Eq. 7 over some unique distribution of sizes, but still retain an expected $t^{-\nu}$ dependence with a unique prefactor characteristic of a particular experiment. Closer to the transition this behavior is sharply modified as the helium in the smallest confinement becomes normal. In this limit this approach fails. This approach would also fail for certain geometries where there is a strong influence from correlation effects.

TABLE I. Scaling exponents for the superfluid density. References in brackets refer to the source of the data analysis. All errors are standard errors except when in parenthesis.

Reference	Confinement	Scaling Variable	Exponent
Brooks et al., Ref. 10 [Gasparini et al., Ref. 6]	Cylindrical, Nuclepore	t^{-s}/d^2	0.82±0.02
Rhee et al., Ref. 7	Planar, Si wafers	t^{-s}/d	1.14±0.02
Gasparini et al., Ref. 6	Planar, Mylar roll	t^{-s}	1.18±0.02
Henkel et al., Ref. 12 [Gasparini et al., Ref. 6]	Metricel	t^{-s}	0.95±0.14
Smith Ref. 13 [Gasparini et al., Ref. 6]	Planar, slit	t^{-s}	0.95±0.05
McQueeney, Ref. 14	Millipore	t^{-s}	1.035
Brooks et al., Ref. 10 and Schubert & Zimmermann, Ref. 11 [Gasparini & Rhee, Ref. 8]	Cylindrical, Nuclepore	t^{-s}/d^2	1.0±(0.1)

Results of analyses in the spirit discussed above, as well as for the data we have discussed explicitly, are summarized in Table I. It can be seen from this that from a variety of experiments one obtains an exponent for the temperature dependence which is close to unity. This is completely unexpected on the basis of correlation-length scaling.

THE HEAT CAPACITY

Measurements of the heat capacity for helium confined in Nuclepore filters of 300, 800, 1000 and 2000Å pore diameter were the first data to show disagreement with correlation-length scaling.[4] These are still the only data on heat capacity which can be tested against Eqs. 4 or 5. In particular there is a substantial region in t where the data are found to obey a simple power-law as suggested in Eq. 5. Results from this analysis yield $\alpha_s - \alpha = 0.472 \pm 0.012$ and $\alpha_s - \alpha = 0.464 \pm 0.016$ for $T<T_\lambda$ and $T>T_\lambda$, respectively[5]. Here α_s is the surface specific heat exponent, and one expects $\alpha_s - \alpha = \nu = 0.672$ as can be seen from Eq. 5. These data clearly do not support this. It is interesting that both branches of the heat capacity yield the same exponent, thus supporting a homogeneous value for α_s.

The heat capacity data have been analyzed in many ways, including the addition of correction-to-scaling terms to force agreement with the expected value for α_s.[5] This works at the cost of rather large correction terms. Recently, there have also been reports of renormalization-group calculations by Huhn and Dohm[15] which showed good agreement with the $T>T_\lambda$ branch of the heat capacity for 1000Å confinement. More recent calculations, however, of the whole scaling function shows that it is indeed a function of $d^{1/\nu}t$, thus in disagreement with experiment.[16]

If one looks to test Eqs. 2,3 for the specific heat maximum and the temperature at which it occurs, one can examine a variety of data for films. This has recently been reviewed by Gasparini and Rhee.[8] One finds that values of C_m do not reproduce well from experiment to experiment. On the other hand, values of t_m are somewhat better. This is to be expected, since for thick films corrections due to evaporation are very large, thus have to be done carefully. The position of the maximum, on the other hand, is not so sensitive. Eqs. 2,3 for films yield exponents of 0.466 ± 0.015 and 0.528 ± 0.043 for the data of Chen and Gasparini.[4] Other data are consistent with this value.[8] In the case of filled pores these equations yield exponents of 0.583 ± 0.046 and 0.502 ± 0.025. None of these results is in agreement with the exponent ν. Details of additional analysis with these data are discussed by Gasparini et al.[6] We point out that while Eqs. 2,3 focus on a readily identifiable feature of the heat capacity, the results from scaling all of the data according to Eqs. 4 or 5 give a much more stringent test of correlation-length scaling because of the much broader set of data both above and below T_λ which can be used in the analysis.

THE SHIFT IN T_c

In films the superfluid transition takes place at a temperature T_c below T_λ. This temperature can be identified in a number of experiments such as the point where the superfluid fraction vanishes or, in measurements above T_c, as the point at which the thermal conductance diverges. The details of the thermodynamic response near T_c is relevant to a phase transition characteristic of a 2D xy system. T_c itself, on the other hand, is expected to obey a shift equation which involves the exponent of the 3D correlation length

$$(1 - T_c/T_\lambda) = b d^{-1/\nu}. \tag{9}$$

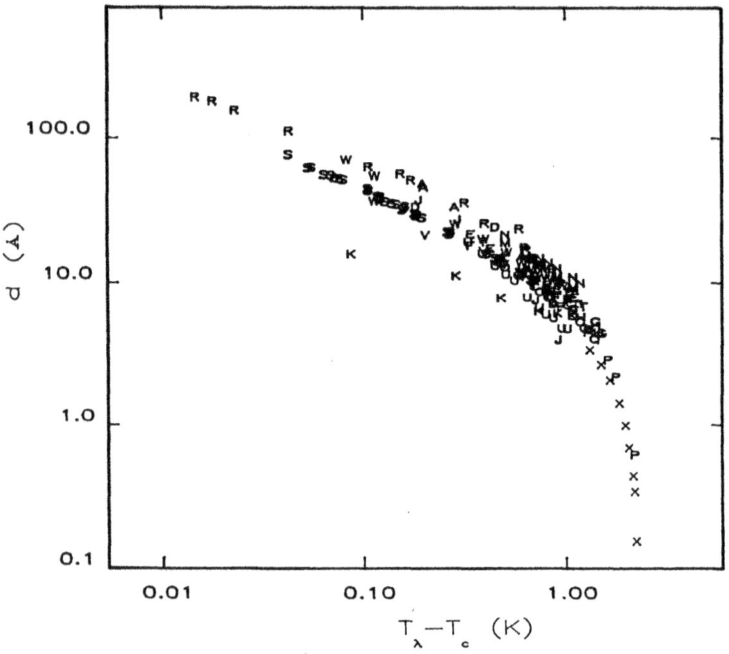

Fig. 6. Log-log plot of film thickness versus shift in transition temperature. d less than 3.6Å implies a submonolayer film. See Ref. 8 for details.

In fact, it is not clear if T_c should enter Eq. 9 or another characteristic temperature below T_c. Specifically, if one approaches of a thick film from below, the point at which a finite-size 3D description goes over into a 2D description is not at T_c, but rather a temperature below T_c. It is this crossover temperature which is more likely to obey Eq. 9, since it represents the point at which the 3D correlation length is some fraction of d. It may well be that the difference between these two temperatures is small on the scale of T_λ-T_c, hence irrelevant. However, this is not known since in experiments with films there is no signature to identify this crossover temperature. There are other issues regarding T_c which have to do with the topology of the films, i.e., strictly planar versus films in pores or on powders. These latter should show more complicated behavior associated with finite 2D behavior as well. There are also issues of how well T_c can in fact be extracted from a given measurement. These issues, and a broad range of data to test Eq. 9, have been reviewed by Gasparini and Rhee.[8] A plot of these data is shown in Fig. 6. The various symbols in this figure represent results from about twenty different experiments.[8] At best, one might say from this figure that a certain trend is established, but clearly not a unique functional form. Among these data three sets which follow T_c to the thick-film, asymptotic region where Eq. 9 should apply. These are the data of Sabisky and Anderson,[17] Maps and Hallock[18] and Yu et al.[19] These data yield exponents of 0.63, 0.79±0.05 and 0.52±0.01, respectively. We note that for the data of Yu et al. the exponent is in good agreement with that obtained from the heat capacity maximum of films, via Eq. 2, 0.528±0.043.

In the case of complete confinement it is best to look only at experiments in which the helium is confined in pores of nearly cylindrical geometry. In this way the constant b has the same geometrical significance. This would not be the case for powders or porous glasses. Data for a variety of experiments were analyzed recently by Gasparini and Rhee.[8] This is shown in Fig. 7. On this plot experiments which determine T_c via flow measurements are shown as squares. The circles represent values from the position of the heat capacity maximum. It can be seen that these two different features, t_c and t_m are coincident for cylindrical confinement. This is not the case for planar films.[19] The exponent which these data determine is 0.603±0.037 for all the data or 0.614±0.018 if one ignores the full squares as being anomalous. This exponent is less than ν=0.672 and is somewhat higher than the best exponent obtained for data on planar films which cover the widest range of thickness, up to 166Å, i.e., 0.52±0.01. Some of this difference is likely to be due to van der Waals effects.

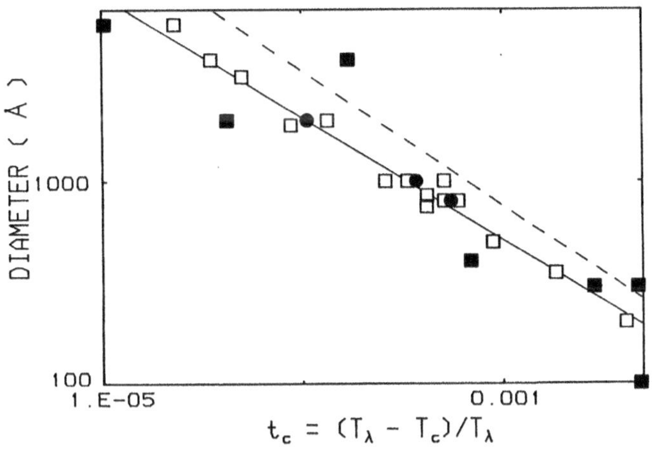

Fig. 7. Log-log plot of pore diameter versus shift in superfluid onset, squares. Filled circles represent specific heat maximum. See Ref. 8 for details.

RESULTS OF CALCULATIONS

Recent renormalization-group calculations of Huhn and Dohm[15] and Schmolke et al.[16] have yielded for the first time the scaling function for the specific heat. This is found to be in the same form as Eq. 4, thus in disagreement with the data which do not scale with the exponent ν. It is interesting, however, that the scaling function is very nearly a simple power law over a certain range of the scaling variable as suggested in Eq. 5 and as the data of Chen and Gasparini[4] first revealed.

For the superfluid density Schmolke et al.[16] have also calculated the scaling function. It is found to be in the form of Eq. 6 with a reasonable simple power-law representation in the ~10^{-4}-10^{-3} region of the variable $d^{1/\nu}t$, with d in μm. This simple power-law feature agrees with the data but, as we have seen, the data are not collapsed by the exponent ν.

Effects of the van der Waals field have been explored in the context of ψ-theory by Wang et al.[20] They find that the presence of this field destroys the expected scaling but, except for thin films, at a level typically too small to be seen experimentally. For the scaling of ρ_s the van der Waals effect goes in the opposite direction as far as reconciling the data with the exponent ν. For the shift of T_c in films, Wang et al.[20] and Finotello et al.[21] find that the presence of the van der Waals field may account for the shift exponent being as low as 0.52. Correcting for this would yield an exponent of ~0.57, closer to the value obtained for filled pores, where the van der Waals effect on T_c is small. This still leaves a disagreement with the expected exponent ν.

CONCLUSIONS, COMMENTS

From a variety of experiments, some involving only the determination of T_c or T_m as a function of confinement size, others involving the scaling of the complete thermodynamic response near T_λ, there are disagreements with the predicted scaling with the bulk correlation length. Given that this prediction is well supported by many model calculations, Monte Carlo work on finite systems and, in the case of helium, specific renormalization group calculations, it would seem that a resolution of this dilemma might come from the details of the helium-surface interaction. Specifically, there might be critical aspects of this interaction, beyond the van der Waals force, which affect the critical behavior in a way one had not suspected and not incorporated in any existing theory.

Another possibility is a modification of the scaling functions by the presence of correction terms of logarithmic form. This has recently been suggested by Privman,[22] and some aspects of his theory have already been tried by Rhee et al.[23] The deviations from scaling are not explained by these terms, however, results should be considered preliminary at this point. Privman remarks that the presence of these logarithmic terms emerges in the theory when one examines the scaling function as the dimensionality approaches 3. These terms would not be revealed in a calculation of the scaling function done in strictly three dimensions, such as in Ref. 16.[22]

ACKNOWLEDGEMENTS

We gratefully acknowledge the support of the National Science Foundation, Low Temperature Physics Program DMR8905771.

REFERENCES

1. M.E. Fisher, in "Critical Phenomena", Proceedings of the 51st Enrico Fermi Summer School, Varenna, Italy, M.S. Green, ed., Academic Press, New York (1971).
2. M.N. Barber, in "Phase Transitions and Critical Phenomena", Vol. 8, C. Domb and J.L. Lebowitz, eds., Academic Press, London, p. 145 (1983).
3. V. Privman, "Finite Size Scaling and Numerical Simulation of Statistical Systems," V. Privman, ed., World Scientific, Singapore (1990).
4. T.-P. Chen and F.M. Gasparini, Phys. Rev. Lett. 40, 331 (1978).
5. F.M. Gasparini, T.-P. Chen and B. Bhattacharyya, Phys. Rev. B23, 5797 (1981).
6. F.M. Gasparini, G. Agnolet and J.D. Reppy, Phys. Rev. B29, 138 (1984).
7. I. Rhee, F.M. Gasparini, and D.J. Bishop, Phys. Rev. Lett. 63, 410 (1989).

8. F.M. Gasparini and I. Rhee, to appear in "Progress in Low Temperature Physics," vol. XIII, D.F. Brewer, ed., North-Holland, Amsterdam.
9. I. Rhee, D.J. Bishop, A. Petrou and F.M. Gasparini, Rev. Sci. Inst. 61, 1528 (1990).
10. J.S. Brooks, B.B. Sabo, P.C. Schubert and W. Zimmermann, Jr., Phys. Rev. B9, 4524 (1979).
11. P.C. Schubert and W. Zimmermann, Jr., J. Low Temp. Phys. 44, 177 (1981).
12. R.P. Henkel, E.N. Smith and J.D. Reppy, Phys. Rev. Lett. 23, 1276 (1969).
13. E.N. Smith, Ph.D. thesis, Cornell University, unpublished (1971).
14. D.F. McQueeney, Ph.D. theses, Cornell University, unpublished (1988).
15. W. Huhn and V. Dohm, Phys. Rev. Lett. 61, 1368 (1988).
16. R. Schmolke, A. Wacker, V. Dohm and D. Frank, LT-19 (1990).
17. E.S. Sabisky and C.H. Anderson, Phys. Rev. Lett. 30, 1122 (1973).
18. J. Maps and R.B. Hallock, Phys. Rev. Lett. 47, 1533 (1981).
19. Y.Y. Yu, D. Finotello and F.M. Gasparini, Phys. Rev. B 39, 6519 (1989).
20. X.F. Wang, I. Rhee and F.M. Gasparini, LT-19 (1990).
21. D. Finotello, Y.Y. Yu, X.F. Wang and F.M. Gasparini, LT-19 (1990).
22. V. Privman, preprint (1990).
23. I Rhee, D.J. Bishop and F.M. Gasparini, LT-19 (1990).

DYNAMICS OF HELIUM CLUSTERS

Sandro Stringari

Dipartimento di Fisica, Università di Trento
38050 Povo,Italy

INTRODUCTION

In the last few years the properties of atomic and molecular clusters have attracted much experimental and theoretical work. In this context clusters of helium atoms are expected to exhibit rather unique properties because the atoms are light enough for quantum effects to be significant.

The experimental information on helium clusters is at present limited to thermodynamic measurements of small bubbles ($< 100 \AA$) of helium confined in metals[1] and to scattering of electrons or neutral atoms from free helium clusters generated by nozzle expansion[2,3]. By contrast in the last years there has been considerable progress in theoretical work. The theoretical study of helium clusters is stimulated by important conceptual questions that characterize these finite quantum systems. Does the behaviour of helium clusters differ from that of a piece of bulk matter? Is superfluidity a useful concept in a microscopic system? What are the main differences between ^{3}He and ^{4}He clusters? Are there important analogies between helium clusters (in particular ^{3}He) and atomic nuclei (magic numbers, shell effects ...)?

The purpose of this work is to discuss preliminary answers to the above questions. We mainly focus on the properties of excited states in ^{4}He clusters and discuss some implications on their superfluid behaviour. For a general discussion of the ground state properties of ^{4}He and ^{3}He clusters we refer to ref.4-7.

ELEMENTARY EXCITATIONS

The study of the elementary excitations of helium clusters has been the object of several theoretical calculations in the last few years. Most of these works are limited to ^{4}He clusters because of the simplifying consequences of Bose statistics.

A first classification of the elementary excitations in ^{4}He is provided by the liquid drop model (LDM)[8]. This model accounts for the macroscopic phenomena occurring in the hydrodynamic regime and is expected to provide the exact dispersion relations for the normal modes in the limit of very large clusters. According to the LDM one has both surface (ripplon type) and bulk (phonon type) excitations. The energies of surface modes are classified according to the law

$$\omega_\ell(surface) = \sqrt{\frac{4\pi}{3}\frac{\sigma}{m}\ell(\ell-1)(\ell+2)N^{-1/2}} \tag{1}$$

where ℓ is the multipolarity of the excitation. Note that the $\ell = 0$ and $\ell = 1$ surface modes cannot propagate, the $\ell = 0$ being by definition a compression mode, while the $\ell = 1$ surface state corresponds to a rigid translation of the cluster. The compression modes are characterized by the dispersion relation

$$\omega_\ell = ck_\ell \tag{2}$$

where c is the usual sound velocity and k_ℓ is fixed by the boundary condition for the ℓ-th order Bessel function

$$j_\ell(k_\ell R) = 0 \tag{3}$$

where $R = r_\circ N^{1/3}$ is the radius of the cluster. Equation (3) follows from the condition that the pressure should vanish at the surface. A major difference between surface and bulk excitations is connected with the N-dependence of their excitation energies ($N^{-1/3}$ for volume modes and $N^{-1/2}$ for surface modes) reflecting the different k dependence of the dispersion relation in the infinite medium. For the surface quadrupole ($\ell = 2$) and compression monopole ($\ell = 0$) modes, which are the lowest surface and compression modes respectively, the LDM predicts the following laws:

$$\omega_{\ell=2}(surface) = 10.4N^{-1/2}K$$

$$\omega_{\ell=0}(compression) = 25.6N^{-1/3}K \tag{4}$$

From eq.(4) it follows that the surface quadrupole mode is the lowest elementary excitation exhibited by helium clusters.

In light clusters one expects deviations from the prediction of the liquid drop model. In fact this model ignores finite size effects associated, for example, with the smoothing of the surface profile. Such a smoothing is expected to be particularly important in the case of compression modes because the compressibility of the system is considerably reduced at the surface. Such a behaviour is well known in nuclear physics where the energies of compression modes are known to deviate in a significant way from the LDM prediction.

The available theoretical calculations for the normal modes of helium clusters can be classified in the two following groups: a) microscopic calculations using a Feynman-type ansatz and correlated wave functions for the ground state[9-11]; b) calculations within mean field theories employing phenomenological effective interactions.

The calculations based on method a) assume that the wave function of the elementary excitations has the Feynman form

$$|\psi> = \sum_i f(\mathbf{r}_i)|0> \tag{5}$$

where $|0>$ is the wave function of the ground state and f is a local operator with a given multipolarity. The calculation of the excitation energy (we omit here for simplicity center of mass corrections)

$$\frac{<\psi|H|\psi>}{<\psi|\psi>} = \frac{\hbar^2}{2m} \frac{<0|\sum_i \nabla f^*(\mathbf{r}_i)\nabla f(\mathbf{r}_i)|0>}{<0|\sum_{i,j} f^*(\mathbf{r}_i)f(\mathbf{r}_j)|0>} \tag{6}$$

then requires i) the explicit knowledge of the excitation operator f, ii)the 2-body density matrix relative to the ground state (differently from the bulk where one can use the experimental value for the structure function, in the cluster the knowledge of the 2-body density matrix requires an explicit calculation). Several authors have recently worked in this direction[9-11]. The resulting values for the excitation energies are however still subject to uncertainties due to inacurracies in the choices of the operator f and of the 2-body density of the ground state. Interesting results using this method have been found for the roton structure of compressional excitations. According to the results of ref. 10 the roton minimum becomes less and less pronounced as the mass number of the cluster decreases. This is probably the consequence of the fact that light clusters are less dense compared to the bulk. One should finally keep in mind that the Feynman method provides only an upper bound for the excitation energy due to the intrinsic excitation of mutiphonon states contatined in eq.(5). The excitation of multiphonon states can be ignored in the study of macroscopic excitations (small q limit in infinite and semi infinite matter) where the Feynman ansatz is known to provide the exact solution. It is however very important in the roton region where the Feynman method is known to overestimate the experimental value of the excitation energy by a factor of 2.

The calculations based on method b) use the RPA equation for the density-density Green's function:

$$G^{-1}(\mathbf{r}_1,\mathbf{r}_2,\omega) = G_\circ^{-1}(\mathbf{r}_1,\mathbf{r}_2,\omega) - V_{p-h}(\mathbf{r}_1,\mathbf{r}_2) = 0 \tag{7}$$

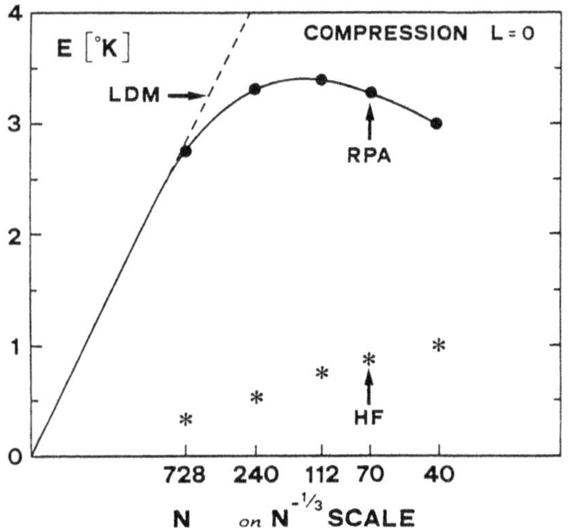

Fig.1 Excitation energies (in K) of the compression $\ell = 0$ mode. The dashed line is the LDM prediction (from ref.13).

where $G_\circ$ $(\mathbf{r}_1, \mathbf{r}_2, \omega)$ is the Green's function relative to the independent particle model fixed by the solution of the Hartree-Fock equations. Equation (7) correctly provides the phonon and ripplon dispersion relations when applied to the infinite and semi-infinite[12] problem respectively. In ref. 13 the RPA equations (7) have been solved employing an effective interaction V_{p-h} previously fixed to reproduce the bulk as well as the surface properties of liquid helium. The calculations of ref. 13 are similar to the ones currently carried out in nuclear physics where the rich phenomenology has permitted to check the quality of the RPA approach. In fig.1 we report the predictions of ref.13 for the monopole compression mode. The figure shows that light clusters are softer than an ideal sharp drop of uniform liquid (LDM prediction). This behaviour is

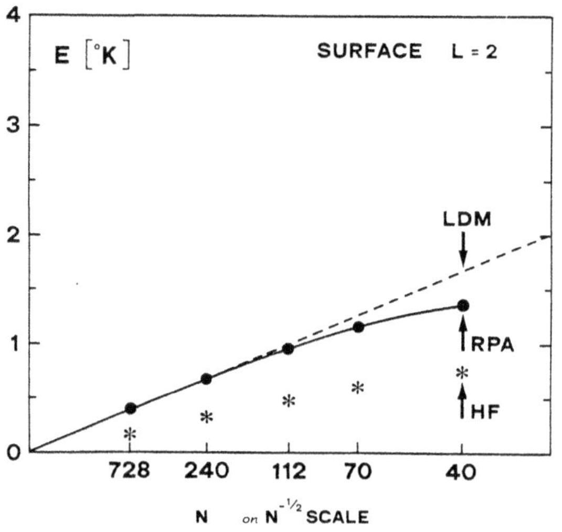

Fig.2 Excitation energies (in K) of the surface $\ell = 2$ mode. The dashed line is the LDM prediction (from ref.13).

very similar to the one exhibited by atomic nuclei and is confirmed by the calculations with method a). Equations (7) have been solved also for the quadrupole excitation (see fig.2). In this case the deviations from the LDM are found to be less important, significant effects being observable only for clusters contaning less than 100 atoms. The remaining uncertainties for the predictions of the excitation energies with method b) lie in the choice of the effective interaction. For example the interaction used in ref.13 has a very simple momentum dependence and does not account for the roton minimum in the excitation spectrum. Clearly a more detailed investigation of the q-dependence of the effective force is required in order to apply this method to the study of highly excited states.

STATISTICAL PROPERTIES AND SUPERFLUIDITY

The study of the elementary excitations in ^{4}He clusters reveals that the excitation spectrum at low energy is dominated by surface modes. For example using the LDM predictions we find that in a cluster with $N = 10^6$ atoms there are more than 300 surface states below the first monopole compression mode. As a consequence for a first estimate of the density of states and of the thermodynamic properties of helium clusters at low temperature it is sufficient to take into account only the surface modes. The restriction of the discussion to the low temperature regime is here suggested by the fact that free clusters cool down very quickly due to evaporation and that the final temperatures of the clusters produced via nozzle expansion in the experiments of ref.2,3 are expected to be of the order of $0.5K$[14]. In the low temperature regime very simple formulae can be derived. For example the density of states $\omega(E)$ in the microcanonical ensemble takes the following form[14]:

$$\omega(E) \simeq \frac{1}{E^{5/7}} exp[\alpha_s E^{\frac{4}{7}} N^{\frac{2}{7}}] \tag{8}$$

whith the coefficient α_s fixed by the surface tension of the liquid. The E and N dependence differs from the one given by the Bethe formula

$$\omega(E) \simeq \frac{1}{E} exp\sqrt{\frac{\pi^2}{\epsilon_F} EN} \tag{9}$$

holding for a degenerate Fermi system such as atomic nuclei or ^{3}He clusters (ϵ_F is the Fermi energy). From eq.(8) one gets the following formula for the specific heat of a cluster containing N atoms[14]

$$C(N) = \frac{28}{9}\Gamma(\frac{7}{3})\zeta(\frac{7}{3})\left(\frac{3}{4\pi}\frac{m}{\sigma}\right)^{\frac{2}{3}} T^{\frac{4}{3}} N^{\frac{2}{3}} \tag{10}$$

where Γ and ζ are the factorial and Reimann's functions, σ is the surface tension of liquid ^{4}He and N is the number of atoms in the cluster. Result (10), which is consistent with the well known Atkins'law[15] for the T-dependence of the surface tension, should be compared with the low T behavior of the specific heat in bulk liquid ^{4}He:

$$C_{bulk} = \frac{2\pi^2}{15c^3\rho} T^3 N \tag{11}$$

which is governed by the thermal excitation of phonons. Despite the weaker dependence on the mass number N exhibited by eq.(10), also for a cluster containing $N = 10^6$ atoms one finds that the ripplon contribution is dominant for temperatures of the order of $0.5K$.

The density of states $\omega(E)$ can be used to evaluate the rate of evaporation of helium clusters. To this purpose we make use of the Weisskopf formula

$$\Gamma = \frac{dN}{dt} = \frac{\Sigma m}{m^2}\int_0^{E-E_\circ} d\epsilon\epsilon \frac{\omega_{N-1}(E - E_\circ - \epsilon)}{\omega_N(E)} \tag{12}$$

In eq.(12) ϵ is the kinetic energy of the emitted atom, E is the excitation energy of the N-cluster, $E_\circ$ is the binding energy of the emitted particle and Σ is the cross section for the cluster formation (here assumed to be given by the geometrical approximation $\Sigma = \pi R^2$). In the limit $E_\circ, \epsilon \ll E$, eq.(12) can be rewritten in the following way

$$\Gamma = \frac{m\Sigma}{\pi^2} exp[\frac{\mu}{T}] \tag{13}$$

where $\mu \simeq E_\circ$ is the chemical potential and T is the temperature of the cluster defined in the microcanonical ensemble. Equation (13) can be used to estimate the cooling law of helium clusters due to evaporation. For this one needs the equation

$$\frac{dE}{dt} = -E_\circ \frac{dN}{dt} \tag{14}$$

together with the explicit form (10) for the specific heat. The resulting t-dependence for the temperature of the cluster is reported in fig.3[14]. It is worth noting that within the approximations here employed this law turns out to be independent of the mass number N. Of course in the macroscopic limit this is no longer true because the specific heat of the system is no longer given by eq.(10).

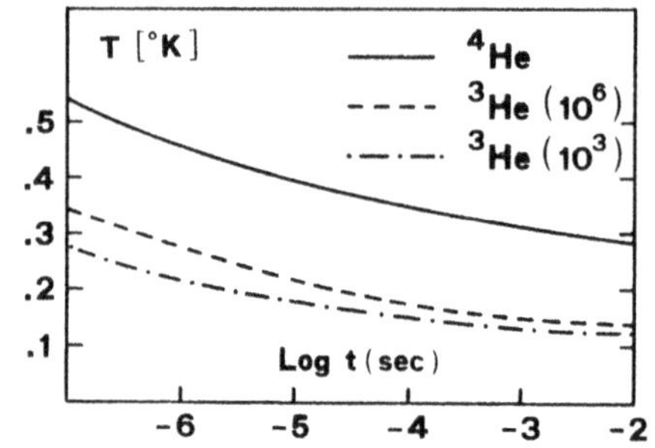

Fig.3. Cooling of helium clusters as a consequence of evaporation. (from ref.14).

The thermal excitation of ripplons is expected to influence in a significant way the superfluid behaviour of helium clusters. For a finite system a natural definition of superfluidity can be obtained in terms of the departure of the moment of inertia Θ from the rigid value Θ_{rig}. It is well known that the hamiltonian H' of a body in a coordinate system rotating with the body is related to the hamiltonian H in the laboratory system by the equation

$$H' = H - \omega L_z \tag{15}$$

where ω is the angular frequency (which has been taken along the z-axis) and $L_z = \sum_i l_z^i$ is the z-component of the angular momentum of the system. Equation (15) shows that the angular momentum of the rotating body can be calculated as the response of the system to the conjugate "field" ω. The ratio

$$\frac{\Theta}{\Theta_{rig}} = \frac{1}{\Theta_{rig}} \frac{\partial}{\partial \omega} < \omega \mid L_z \mid \omega > \tag{16}$$

between the moment of inertia $\frac{\partial}{\partial \omega} < \omega \mid L_z \mid \omega >$ of the system and the rigid value $\Theta_{rig} = \frac{2}{3} m N < r^2 >$ will be taken in the following as our definition of the normal (non superfluid) fraction of the system. In eq.(16) $\mid \omega >$ is the statistical state associated with the hamiltonian H' at a given temperature T. Eqs. (15-16) provide a natural generalization of the concept of superfluidity introduced by Landau in bulk liquid and have been extensively employed in problems with cylindrical symmetry.

In ref.16 the normal component of the moment of inertia has been calculated via a path integral Monte Carlo simulation as a function of the temperature. The results of this calculation reveal the normal fraction is enhanced with respect to the bulk case at low temperature with a consequent lowering of the domain of temperature where the system behaves as a normal fluid. It is interesting to discuss the predictions for the normal fraction due to the thermal excitation of ripplons. The explicit calculation yields the following result[17]:

$$\frac{\Theta}{\Theta_{rig}} = \frac{50}{27}\frac{1}{mr_\circ^2}\left(\frac{3m}{4\pi\sigma}\right)^{\frac{4}{3}}\Gamma(\frac{5}{3})\zeta(\frac{5}{3})T^{5/3}N^{-\frac{1}{3}} \tag{17}$$

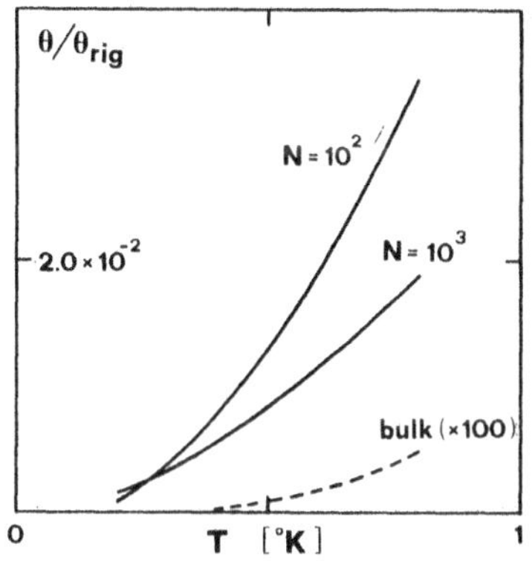

Fig.4. Normal fraction as a function of T for two different clusters (from ref.17).

Equation (17) should be compared with the expression for the normal density holding in the bulk at low T:

$$\frac{\rho_n}{\rho} = \frac{2}{45}\frac{\pi^2}{\rho mc^5}T^4 \tag{18}$$

In fig.4 we report the predictions for the moment of inertia of two different clusters. The results reported in the figure differ from the predictions of eq.(17) in the low T region where they exhibit the typical $e^{-\beta\omega_2}$ behaviour (ω_2 is the excitation energy of the lowest surface mode) ignored by eq.(17). Figure 4 clearly shows that the thermal excitation of the normal component results to be much more active in the cluster than in the bulk. Of course at higher temperatures the role of compression modes has to be taken into account. In particular it would be interesting to study the consequences of the results of ref.10 for the roton structure of light clusters on the T-dependence of the normal fraction in the range $T = 1 - 2K$.

REFERENCES

1. E.G. Syskakis, F. Pobell and H. Ullmaier, Phys. Rev. Lett. **55**,2964 (1985);
2. J. Gspann, in *Physics of Electrons and Atomic Collisions*, S. Datz ed. (North Holland, Amsterdam 1982);
3. H. Buchenau, E.L. Knuth, J. Northby, J.P. Toennies and C. Winkler, J. Chem. Phys. , **92**, 6875 (1990);
4. R. Melzer and J.G. Zabolitzky, J. Phys. A **17**, L565 (1984);
5. V.R. Pandharipande, S.C. Pieper and R.B. Wiringa, Phys. Rev. B **34**, 4571 (1986);
6. S. Stringari and J. Treiner, J. Chem. Phys., **87**, 5021 (1987);
7. S. Stringari, Int. School E. Fermi, CVII Course, Varenna 1988 (North Holland, Amsterdam in press);
8. A.Bohr and B.R. Mottelson, *Nuclear Structure* (Benjamin, NY 1975) vol.II;
9. R.B. Wiringa, S.C. Pieper and V.R. Pandharipande, Argonne preprint, Phy-4676-Th-85;
10. M.V. Rama Krishna and K.B. Whaley, Phys. Rev. Lett. **64**, 1126 (1990);
11. S.A. Chin and E. Krotscheck, preprint CTP-TAMU 1/90;
12. E. Krotscheck, S. Stringari and J. Treiner, Phys. Rev. B **35**, 4754 (1987);
13. M. Casas and S. Stringari, J. Low Temp. Phys., **79**, 135 (1990);
14. D.M. Brink and S. Stringari, Z. Phys. D **15**, 257 (1990)
15. K.R. Atkins, Can. J. Phys. **31**,1165 (1953);
16. Ph. Sindzingre, M. L. Klein and D.M. Ceperley, Phys. Rev. Lett. **63**,1601 (1989);
17. L. Pitaevskii and S. Stringari, Z. Phys. D **16**, 299 (1990).

PHASE-SLIPPAGE STUDIES of the CRITICAL VELOCITY in HELIUM-4

E. Varoquaux *, W. Zimmermann, Jr. *‡• and O. Avenel •

* Laboratoire de Physique des Solides
Université Paris–Sud
91405 – Orsay – France

‡ Tate Laboratory of Physics
University of Minnesota
Minneapolis – MN – 55455 – USA

• Service de Physique du Solide et de Résonance Magnétique
CEN–Saclay
91191 – Gif–sur–Yvette – Cedex – France

1. The Critical Velocity Problem

The problem of the critical velocity for the flow of superfluid ^{4}He through small orifices has been long standing.[1-6] We reexamine it here in the light of the results obtained by phase slippage experiments. The occurence of phase slips [7-11] when the flow velocity v_s exceeds a certain threshold v_c marks the onset of dissipation in the superfluid. According to current views,[12] dissipative slips of the phase by 2π are caused in ^{4}He by vortex filaments crossing all the streamlines passing through the orifice. The initial stages of this process are, however, still unknown.

In this progress report, we first briefly review some of the relevant experimental data, obtained for the most recent part with the help of two-hole hydromechanical resonators,[7-11, 13-16] and in particular, the striking effect of temperature on v_c even at temperatures much below 1 K. In the most recent experiments, not only could individual 2π slips be resolved but also the quantum state of hydrodynamic circulation of the loop threading the two holes could be determined.[16] Another feature of the double-hole resonator is that an arbitrary 'bias' circulation can be trapped in a metastable way.[10-11] A summary of the salient properties of the 2π phase slips is given below, but the reader is referred to the original literature for details.

Two possible approaches to understanding the onset of phase slippage involve the depinning and repinning of remanent vorticity and the nucleation of new vorticity. Remanent vorticity is known to exist in superfluid ^{4}He,[17] vorticity which may, at high enough flow velocities, develop into a self-sustaining vortex tangle[18] and lead to turbulence. Depinning and repinning of the remanent vorticity has been suggested to provide a mechanism for dissipation of superflow at absolute zero[19] and also to provide an explanation for the discrete events that we refer to here as phase slips.[20] However, as discussed below in Section 5, this interpretation of the phase slip experiments is open to serious criticism.[9, 21]

Vortex nucleation, on the other hand, has been shown to take place in ion propagation[22, 23] according to a well-defined mechanism predicted by Muirhead *et*

Excitations in Two-Dimensional and Three-Dimensional Quantum Fluids
Edited by A.G.F. Wyatt and H.J. Lauter, Plenum Press, New York, 1991

al.[24] However, a similar process, which involves half-ring vortices, can not take place as such in orifice flow in part because critical velocities are much lower than for ions and in part because their temperature dependence is markedly different. We shall discuss this difficulty below in Section 6 and introduce a simple, but incomplete, model involving linear vortex filaments.

We thus attempt to evaluate the merits of the pinning model mentioned above versus those of nucleation: it is clear that the vortices involved in phase slips either have to preexist or else be created when required. We shall tentatively conclude that, in the present, fragmented state of our understanding, nucleation at the walls cannot be ruled out as a possible phase slip process in short narrow orifices while vortex depinning and repinning is unlikely.

2. Some Experimental Facts on Critical Velocities

As stated in the introduction above, the critical velocity is defined here by the velocity threshold for the onset of phase slips. In the work carried out at Orsay,[7-11,15-16] individual 2π slips were resolved. In other work,[14,25-31] velocity jumps due to slips of many times 2π were sometimes seen. We shall assume that, in the latter situation, the apparent critical velocity is not too severely depressed below the threshold for 2π slips. For older work, in which less precautions were taken to screen out external mechanical vibration, or in which some form of average of the flow was measured, such an assumption is not necessarily warranted.

Two critical velocity régimes have been observed,[28,30-31] as can be seen in fig.1, which is an updated version of Wilks' original graph[2] of critical velocities: 1) a low-velocity, temperature-independent régime and 2) a high-velocity, temperature-dependent one, which will be the object of our prime concern here. Occasional switching between the two régimes in the course of the same run has been observed.[30-31]

A Feynman-type formula for the critical velocity,

$$v_c \simeq \frac{\kappa_4}{2\pi d} \ln\left(\frac{d}{a_0}\right) , \tag{1}$$

is obeyed in the temperature-independent régime,[1-2] although other formulae have been offered.[3] In eq.(1), κ_4 is the quantum of circulation for ^{4}He, d is the channel dimension, and a_0 is the vortex core radius ($a_0 \sim 1.5$ Å).

For orifices and channels of width d a few micrometers or less (as well as in powders and channels, it seems), the high-velocity régime can prevail with values for v_c, summarized in Table 1, ranging from about 1 m/s to over 10 m/s* around 1 K. In recent work in Orsay and Minneapolis,[7-11,13-16] critical velocities for a given orifice were very reproducible during the same cool-down but exhibited variations (up to a factor of 2) after warming up to room temperature. They are also plotted in fig.1 but show no systematic size dependence, although they depend on samples and runs. These critical velocities follow very nearly a linear dependence on T, namely,

$$v_c = v_{c0}(1-T/T_0) , \tag{2}$$

from millikelvin temperatures (in ultra-pure ^{4}He) up to about 1.9 K with $T_0 \sim 2.4$ to 2.5 K at zero pressure. We stress that this value for T_0 is observed in a number of different experimental situations.[8,14,32-36] Most previous experiments which showed a temperature dependence were interpreted in terms of the Iordanski-Langer-Fisher[6] theory for homogeneous nucleation of vortices which involves the quantity $\rho_s/\rho T$, ρ_s being the superfluid density. Although this theory yields results which agree rather well with eq.(2) in the temperature range between 1.2 and 2 K, where $\rho_s/\rho T$ is approximately linear in T,[8] it fails to do so at temperatures below 1 K.

The average of the values of v_c in the temperature-dependent régime, given in Table 1 at $T \sim 1$ K, is 6.5 m/s. The zero-temperature critical velocity corresponding to this mean value, assuming eq.(2) to hold, is $v_{c0} \sim 11$ m/s. The highest recorded velocity is 21 m/s.[14] In the following discussion, we shall refer to the mean value

---

*The value of 0.55 m/s quoted in ref.[7] is now thought to be a lower bound for v_c. A higher bound is 5 m/s.[9]

Table 1

T–dependent			T–independent		
d [μm]	v_c at ~ 1 K [m/s]	ref.	d [μm]	v_c [m/s]	ref.
0.02	7.5	33	0.2	2	33
10	6	35	0.31	0.18	37
8	5	29	5	0.02	36
5	1.84	36	0.385	0.11	38
0.18	13	14	8	0.30	29
0.1	9	14	40	0.01	39
0.16	4	14	120	0.004	39
0.185	13	14	380	0.0023	39
0.24	10	14	510	0.0015	39
0.3	4	14			
0.375	2	14			
0.35	8	14			
0.3	1	7			

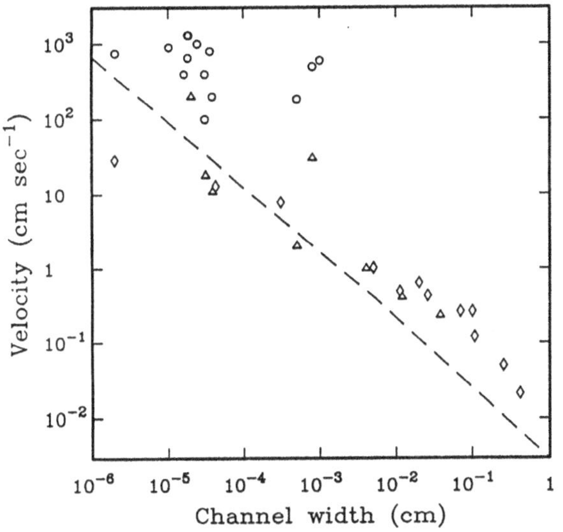

Fig.1. Critical velocity versus channel width (◇ – data collected by Wilks,[2] △ – temperature–independent data in Table 1, ○ – temperature–dependent data). The dashed curve is eq.(1).

rather than to the peak value as a conservative estimate of the zero–temperature value of the size–independent critical velocity, although the peak value might in fact be more significant.

Below about 120 mK, v_c turns out to be very sensitive to minute traces of ^{3}He, in the range of 10^{-8} or less,[8] as are ion propagation[40] and film flow.[41-42] We note, in view of the foregoing discussion, that, in both these processes, nucleation of small scale vortices is involved. The impurity effect is thought to come from a condensation of ^{3}He quasiparticles onto the vortex core,[43] changing the core energy and radius.

We have observed more recently[15] a pressure dependence of the quantities v_{c0} and T_0 which is shown in fig.2. The overall changes of T_0 and v_{c0} with increasing pressure are ~ ± 7 %, respectively, while that of ρ is 19 %. The product $v_{c0} \times T_0$ is approximately independent of pressure. This pressure dependence is markedly different from what is expected on the basis of simple hydrodynamical models,[1,18,20] namely, that v_{c0} is independent of pressure and that $T_0 \sim \rho_s$. This observation hints that some non–hydrodynamical quantity, like the roton parameters, or the vortex core radius, or the density very near the wall, is involved in the phase slip process (i.e., the bulk ρ is not a directly relevant parameter).

A last feature of temperature–dependent critical velocities is the statistical width of the critical transition.[16] This width increases with temperature. This effect is analyzed in detail in Section 4.

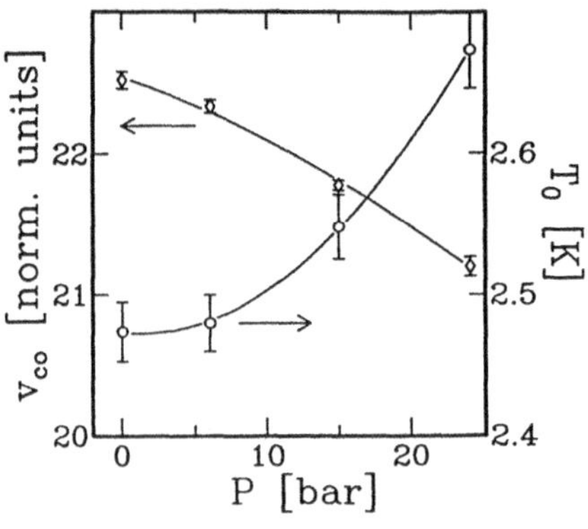

Fig.2. The zero temperature extrapolation of the critical velocity, in normalized units,[15] v_{co} (◇), to the left, and the characteristic temperature, T_0 (○), to the right, defined in eq.(2), versus pressure.

3. Experimental Features of the Phase Slips

Slips of the phase by 2π have been observed in two-hole resonators whose frequencies were governed either by the micro-orifice (weak-link dominated devices) or by a relatively open channel in parallel with the micro-orifice (*rf*-SQUID-like devices). In the former situation,[7-9] slips constitute the dominant feature. In the latter,[10,11] they are small features on top of large steps (the 'staircase' pattern). In both cases, the slips yield quite reproducible drops in the resonator amplitude and correspond to an energy loss given by the following relation:

$$\Delta E = \kappa_4 s_w \rho_s v_c \ , \qquad (3)$$

s_w being the cross-sectional area of the micro-orifice acting as a weak link. Eq.(3) is closely related to the *ac* Josephson relation.[7,9]

On occasions, giant slips or cascades which kill a sizable fraction of the resonator motion are observed. In some instances, virtually all resonator motion is

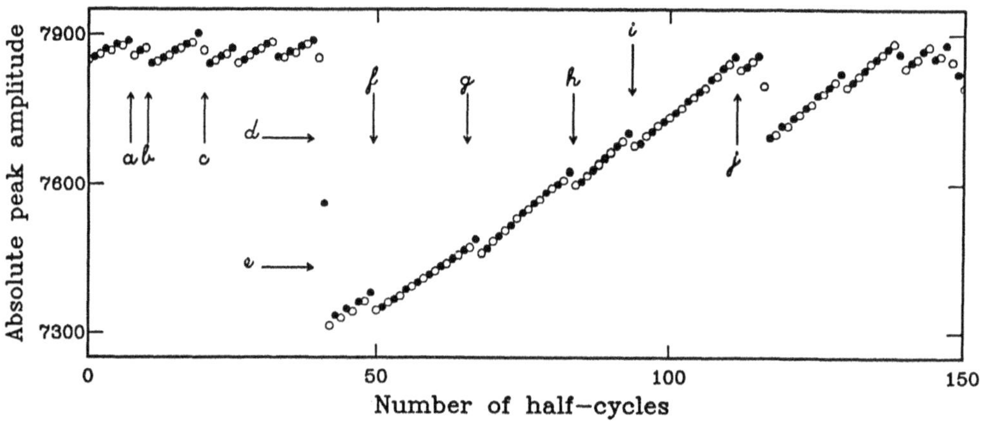

Fig.3. Absolute peak amplitudes of the membrane deflection at successive half-cycles of the resonator motion. Open and closed circles represent the peaks for opposite flow directions. Slips alternate nearly regularly in direction when no trapped vorticity is present as indicated by the labels *a*, *b* and *c*. A cascade occurs at *d* and trapped vorticity is first released in another cascade *e*, and then by single slips *f* to *j* occuring all in the same direction (opposite to that of cascade *d*). Two multiple slips take place after *j* and then, a normal course is resumed.

suppressed. These giant cascades, 'collapses', have also been seen by Sabo and Zimmermann,[44] by Hess,[30] and by Davies and Packard.[45] Giant slips in film flow have been seen by Adie and Armitage.[26] Cascades as well as individual 2π slips are shown in fig.3 where the absolute peak amplitude of the membrane deflection at each half cycle is plotted versus time. These events do not change the cell bias, but change the trapped circulation by an integral number of quanta and involve integral numbers of 2π slips. We have observed them to occur always in a given flow direction, the release of the trapped circulation taking place in the reverse flow direction, as indicated in the caption of fig.3. Cascades appear to be immediately preceded by a single slip in the reverse direction. The likelihood of cascades seems to increase with frequency and critical velocity, as they were prevailing at all temperatures down to 15 mK in one run at 27 Hz, and seen in large number at 9 Hz between 100 and 200 mK. They may also be triggered by mechanical vibration. However, their occurrence cannot be experimentally controlled, nor can the resonator frequency easily be changed, and thus these observations have an incidental character. Systematic studies have not been performed, but the very existence of collapses and cascades, concurrently with that of individual 2π slips, reveals the possibility of complex fluid behaviour in the vicinity of the orifice and, in particular, of asymmetry with respect to flow direction in the phase slip process. Also, when the large slips occur in rapid succession, the staircase pattern is observed to be suppressed. In addition, the average value of the critical flow is decreased.

No partial slip (i.e., non 2π) has ever been observed in the weak-link dominated resonator where the slip size can be accurately monitored.[7,9] Double 2π slips are not unusual even in situations where cascades are absent.

Bias currents of roughly any value (modulo 2π) can be trapped by severe overdriving.[10,11] We have found that these bias currents are extremely stable. They could arise from vortices created by these very strong excitations and remaining pinned somewhere in the resonator and would then be evidence for strongly pinned vorticity. Most of the time, ^{4}He falls, when cooled down through the λ transition, into a state of zero bias (or small bias compared to 2π). Thus, if remanent vorticity is present, it is either strongly pinned or else nearly free to move.

When using strong pulse excitations in the weak-link dominated resonator,[9] flow surges could be produced that would create up to 300 phase slips in rapid succession (in a total time of less than 4.5 ms) in the same direction of flow. An equal number of quanta of circulation were trapped in the loop when the diaphragm motion was brought to a halt at peak deflection by application of a suitable electrostatic potential. The trapped circulation could be released and quantified by slowly ramping up the resonator amplitude from a sub-critical level. Comparison[9] between the energy loss in the salvo of slips and that obtained from eq.(3) multiplied by the number of trapped quanta shows that the phase slip events overlapped sufficiently little so that the energy dissipated per slip was the same to better than 3%. That is, the successive events did not interfere with one another when separated by a 15 μs delay nor when as many as 300 followed one another in rapid succession in the same flow direction. It is likely that the critical velocity was not changed by more than a few % since $\Delta E \sim v_c$ by eq.(3), that emitted vortices cleared out of the orifice, and that any pinned vorticity, or vortex tangle, if disturbed, restored itself. This observation bears on the suggestion by Schwarz,[19,20] to be discussed below, that pinned vorticity might play a rôle in the phase-slip process: the favourable pinning sites would have had to be replenished in a time of the order of 15 μs or shorter.

4. Thermal Activation of the 2π Phase Slips

We shall take the functional dependence of eq.(2) and the value of T_0 to be universal features of the critical velocity régime under consideration. More extensive and systematic studies are, however, needed to establish this point firmly. The most straightforward interpretation of eq.(2) (and the only one so far to our knowledge) is that the instability of the superflow responsible for the zero temperature critical velocity v_{c0} can be thermally assisted with a rate given by

$$\Gamma = \Gamma_0 \exp(-E_a/k_B T) , \tag{4}$$

in which the activation energy E_a has the form

$$E_a = E_0(1-v_s/v_{co}) \; . \tag{5}$$

The characteristic energy E_0 is independent of the superflow velocity v_s.

As can be seen in fig.3 and in the inset of fig.4, slips trigger within narrow limits in velocity. However, there is a definite spread of the critical transition amplitudes. The rounding of the transition increases with temperature. Thus, there is a need for a more specific definition of v_c and a framework to present the data.

Let us start from the transition rate given by eq.(4). Consider now a half cycle of the resonator motion beginning at some time $t = 0$ and ending at t_f. The probability p that a slip takes place sometime during this half cycle is

$$p = 1 - \exp\left\{ -\int_0^{t_f} \Gamma_0 \, e^{-E_a(v_s)/k_B T} \, dt \right\} , \tag{6}$$

where E_a is given by eq.(5) and $v_s(t)$ can be taken with little error in general to be of the form $v_p \sin(\pi t/t_f)$, v_p being the peak velocity during the half cycle. The main contribution to the integral in eq.(6) comes from the region where v_s is close to its peak value. Performing an asymptotic evaluation of the integral, we find that

$$p(v_p) \simeq 1 - \exp\left\{ - \Gamma_0 t_f \sqrt{\frac{2k_B T \, v_{co}}{\pi \, E_0 \, v_p}} \; e^{-E_0(1-v_p/v_{co})/k_B T} \right\} . \tag{7}$$

Here the critical velocity is defined as the value v_c of the peak velocity for which $p(v_p) = 1/2$. This definition leads to

$$v_c = v_{co} \left\{ 1 - \frac{k_B T}{E_0} \ln\left[\frac{\Gamma_0 t_f}{\ln 2} \sqrt{\frac{2k_B T \, v_{co}}{\pi \, E_0 \, v_c}} \right] \right\} , \tag{8}$$

an expression which is equivalent to eq.(2) to the extent that the temperature and velocity variations in the logarithm can be neglected, i.e. when

$$\ln\left[\frac{\Gamma_0 t_f}{\ln 2} \sqrt{\frac{2k_B T \, v_{co}}{\pi \, E_0 \, v_c}} \right] \simeq \ln(\Gamma_0 t_f) = \gamma \; . \tag{9}$$

Such is the case over most of the temperature range except close to absolute zero or T_λ. We thus recover the linear temperature variation for v_c given by eq.(2) with

$$E_0 = \gamma k_B T_0 \; . \tag{10}$$

An effective width for the distribution p may be defined as

$$\Delta v_c = (dp/dv_p)^{-1}\big|_{v_c} , \tag{11}$$

Under the same approximation as for eq.(9), namely $T \ll T_0$, which implies $\Delta v_c \ll v_c$, we find that

$$\Delta v_c \simeq 2 v_{co} k_B T \, / \, (E_0 \ln 2) \; . \tag{12}$$

The width of the distribution is thus expected to increase linearly with T.

A slight complication arises in the analysis of the data from the fact that the quantity which is actually measured is the peak amplitude of the diaphragm motion. We have applied to this peak amplitude a small correction for the effect of the drive to obtain a quantity y which is directly proportional to v_p one-quarter cycle earlier. More troublesome is the need to take into account the state of trapped circulation in the loop threading the two holes. The trapped current contributes to the flow through the micro-aperture but not to the membrane deflection. This contribution must be determined by keeping track of all slips in each flow direction, starting from an initial state for which the circulation quantum number is known. An example of the distributions obtained for y is given in fig.4. Quantities proportional to v_c and Δv_c can be extracted from a fit of $p(y)$ to the data and are shown in fig.5.

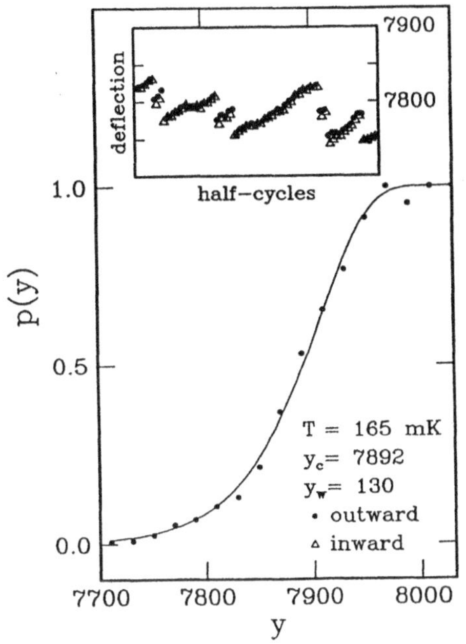

Fig.4. Statistical distribution p of the amplitude y related to the membrane peak deflection shown in the inset as explained in the text.

The drop in v_c at low temperature is due to ^{3}He impurities, as stated above.[8] For the three upper temperatures, the behaviours of v_c and Δv_c are consistent with eqs.(2) and (12) provided that we assume that Δv_c contains a constant additive contribution (due, perhaps at least in part, to residual mechanical vibration) not included in the model above. Taking the slope as indicated in fig.5 leads to $E_0 = 106$ K or 1.5×10^{-21} J. From the known value of T_0 at zero pressure,[8, 14-15] $T_0 = 2.47$ K, we obtain $\gamma = 43$, and with $t_f = 56$ ms, $\Gamma_0 = 7.75\ 10^{19}$ s^{-1}. These results[16] agree well with an earlier estimate[8] based on an upper bound for Γ_0.

The value of E_0 fixes the scale of the energy barrier for the phase-slip mechanism. It is large on an atomic scale but small on the scale of hydrodynamic processes: therefore, only a mesoscopic number of atoms is likely to be involved in the fluctuation assisting $v_c(T)$. It corresponds to the line energy of a vortex filament ~ 50 Å in length at an average distance of 10 Å from the wall, or else, of a bulk fluid vortex ring 19 Å in diameter (60 Å in circumference), and also is of the same order of magnitude as the energy barrier estimated by Schwarz[20] for the depinning of a vortex from a typical 50 Å asperity. Part of the critical velocity problem, as will be discussed below, is to find a justification for the appearance of such a length scale which would not be too dependent on the specific details of the wall geometry.[21]

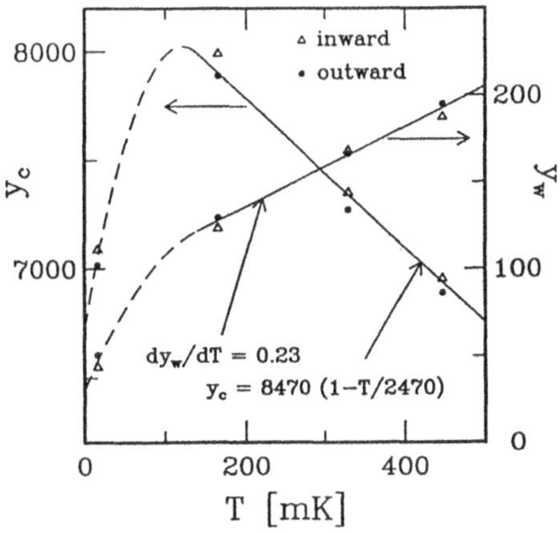

Fig.5. Critical amplitude y_c and width $y_w = y_c \Delta v_c / v_c$ versus temperature.

5. Depinning of Remanent Vorticity and Vortex Mills

The existence of remanent vorticity in the bulk superfluid has been established experimentally, notably by Awschalom and Schwarz.[17] An upper figure for the remanent line density in a slab geometry with thickness d is

$$L_r \lesssim 2 \ln(d/a_0) / d^2 . \tag{13}$$

Thus, there may exist in long narrow channels a large number of trapped vortices. If indeed also present in small orifices through very thin plates, trapped vortices may be expected to play a rôle in the phase-slip process. The effect of remanent vorticity on the critical velocity has been studied by Hulin *et al.*[29] using circular apertures with or without superleak guard filters. These authors have found a much stronger effect of the filter for a 20 μm orifice than for a 6 μm orifice. This tends to indicate that vorticity does not stay pinned in a tiny orifice where flows with high velocity take place.

In more direct studies, pinning forces have been found[46,47] to be either weak on the sites usually encountered, or very strong and of the order of the maximum vortex line tension on specially prepared pinning sites. We thus expect that pinned vorticity would either be washed away in the very first start–up of the resonator motion, or be more or less permanent.

We have evidence in our two-hole resonator experiments[10-11] for the latter possibility, coming from the observation of stable 'bias' currents, but not for the former. As already stated above, no early partial slips, as would in principle arise from vortex depinning, have been observed during the first amplitude sweep of the resonator immediately after initial cool-down (although no systematic search involving a large number of different cool-downs under varying conditions has been carried out).

We now turn to two specific proposals, made by Schwarz, to account for phase slips, the depinning and repinning process[19-20] and the helical vortex mill.[48]

In the first mechanism, vortices are assumed to hop from pinning sites to pinning sites, somehow loosing energy to the wall in the process. The velocity above which no pinned vortex state can exist in a channel of size d has been estimated by Schwarz[49] to be:

$$v_d = (\kappa_4/4\pi d) \ln(b/a_0) \quad , \tag{14}$$

where b is a characteristic size of the pinning site (e.g., the radius, for a hemispherical asperity).

For $d = 3000$ Å and $b = 50$ Å (see below), the depinning critical velocity is $v_d = 9.3$ cm/s. This value is too low by one to two orders of magnitude to account for phase slips. Also, it is channel-size dependent. Another difficulty with the depinning mechanism lies in the fact that the equivalent friction force exerted by the vortex feet on the wall necessary to account for the energy loss, eq.(3), turns out to be unreasonably high.[21] Vortices have to grow, spool out of the orifice and vanish. They cannot be used again in another phase slip.[19] The observation of salvoes of up to 300 phase slips[9] poses in an accute manner the problem of the replenishment of suitable pinning sites.

Depinning velocities are, according to eq.(14), purely hydrodynamical in nature (except for the fact that a_0 describes a quantum-mechanical vortex core) and independent of density. Thus, the pressure dependence shown in fig.2 is not explained in the depinning model.

Finally, the thermal activation of the depinning process leads to a linear temperature dependence, as in eq.(2), with T_0 given in terms of the pinning site size b by

$$T_0 = \frac{\rho_s f \kappa_4^2 b \ln(8b/a_0)}{2k_B \ln(\Gamma_0/\Gamma_{obs})} , \tag{15}$$

where f is a numerical factor of the order of 1/10.[20] The observed value of T_0 fixes the size of the pinning site to a well-defined value of ~ 50 Å. However, we would expect a distribution of pinning site sizes to exist that would vary from experiment to experiment. Hence, we expect a spread in T_0 from cell to cell and run to run, unless some unknown mechanism gives pinning sites that are uniform in size. We consider this to be rather unlikely.† Thus, the depinning mechanism[19-20] is most probably not involved in phase slips.[21]

Recently, Schwarz has proposed a new vortex mill[48] which is based on a well-studied instability of vortex filaments subjected to axial flow[54] and which, unlike the original suggestion by Glaberson and Donnelly,[55] has been demonstrated by numerical simulations to work. This helical vortex mill involves a rather special arrangement of a pinning site in a long channel. In its present form, the model would appear to yield a temperature and pressure-independent critical velocity having a $1/d$ dependence on channel size. It thus might be invoked to explain the temperature-independent critical velocity régime of eq.(1) in long channels.

It remains to be seen whether the helical vortex mill idea can be applied to a short orifice with some natural vortex configuration. It further remains to be seen whether the process could be thermally assisted with an energy barrier of the form $E_0(1 - v_s/v_{c0})$ with E_0 as low as $\sim$ 100 K and with E_0 and v_{c0} having the correct dependences on pressure. That the mill must continue to function under flow reversal places further constraints on such a process. Thus, there exists considerable uncertainty as to whether the helical vortex mill model could provide an account of the temperature-dependent régime of eq.(2).

6. Nucleation of New Vortices

As suggested by Iordanski[56] and by Langer and Fisher,[57] nucleation of small vortex rings can occur in the superfluid bulk at temperatures close to the lambda point, owing to the large value of the attempt frequency Γ_0. Although we lack fundamental understanding of this quantity, it is thought to reflect the overall number of possible states that the system can explore, and its magnitude is given by typical atomic frequencies multiplied by the relevant number of atoms. This homogeneous nucleation mechanism seems to be responsible for the decay of persistent currents observed first by Clow and Reppy[58] and has been used to interpret a number of experiments on channel flow.[32-33] This mechanism does not take into account the presence of walls. It fails to account for the temperature dependence of the critical velocity seen at low temperature. However, it becomes dominant when ρ_s is small and provides an important limiting case.

The nucleation of small half-rings, or horse-shoe vortices on the surface of ions has been put forward by Muirhead *et al.*[24] as a mechanism to explain the critical mobility of ions accelerated through superfluid ^{4}He. At 12 bar, the energy barrier is 3.1 K and the ion velocity is 59 m/s. These predictions have been verified experimentally at Lancaster[22-23] for negative ions (20 Å in diameter). Both the thermally activated and the quantum tunneling régimes have been observed. This beautiful work, also presented at this workshop, establishes clearly how vortex nucleation can take place at a boundary.

---

†This possibility may not be entirely discarded owing to the small number of independent experiments and orifice manufacturing techniques. For instance, it is possible that more or less identical 'cones' be grown during the vapor deposition (but they would be located outside of the orifice) or by carbon contamination during observation under the electron microscope. Or else, similar micro-crystallite edges may have been developed by etching and ion sputtering. Along another line of thinking, Meisel and Kumar[50] have suggested that the apparent universal behaviour of the pinning sites or nucleating asperities comes from a smoothing effect of the monolayer of helium atoms adsorbed on the wall. The smaller asperities, in the 10 Å range, which are known by electron microscopists[51] to be prevalent, would be 'snowed in' leaving only the larger size ones to stick out. We wish to point out that the orifice manufacturing techniques used in the two most significant sets of experiments[7-11, 13-16] are, in effect, quite different.[52-53] We consider as rather unlikely the possibility that the primordial asperities on different orifices, or even on the same orifice for different runs involving different contamination by water and air crystallites, be as identical to one another as the reported values for T_0, that is identical to within their quoted uncertainty ($\sim$ 2%).

For orifice flows in the temperature-dependent régime, the observed velocities are somewhat lower than in the ion case. If the thermal-activation analysis of Section 4 is thought to correspond to a nucleation process, the corresponding energy barrier E_a and attempt rate Γ_0 are significantly different from the ion case. Also, no transition toward a quantum tunneling process takes place. We can conclude that direct nucleation of small (a few Å in radius) half-ring vortices, as considered in particular by Volovik,[59] is unlikely to be taking place at the orifice walls and we have to seek some other process.

Let us then consider the formation of a rectilinear vortex filament of length l lying parallel to a plane wall at a distance $z \ll l$. If z is large compared to the core radius a_0, the energy and impulse of the filament and its image in the wall are half those of a vortex pair:[5-6]

$$E(z) = \frac{1}{4\pi} \rho_s \kappa_4{}^2 l \ln \frac{2z}{a_0} ,$$

$$p(z) = \rho_s \kappa_4 l z .$$

If the fluid moves by the wall with velocity u_s normal to the axis of the filament, the vortex filament 'free' energy[5] becomes

$$E(z) - p(z) u_s = \rho_s \kappa_4 l \left[\frac{\kappa_4}{4\pi} \ln \frac{2z}{a_0} - z u_s\right]$$

and reaches an extremum as z is varied for

$$z = z_m = \frac{\kappa_4}{4\pi u_s} . \tag{14}$$

Substituting this value of z into the 'free' energy yields the energy barrier for this process:

$$E_a = \frac{1}{4\pi} \rho_s \kappa_4{}^2 l \left[\ln \frac{\kappa_4}{2\pi u_s a_0} - 1\right] . \tag{15}$$

Following ref.[6], we introduce a critical observation rate Γ_{obs} and define the critical velocity by the equality

$$\Gamma_0 \exp\left[-E_a(u_s)/k_B T\right] = \Gamma_{obs} .$$

Writing $\bar{\gamma} = \ln(\Gamma_0/\Gamma_{obs})$ – a quantity slightly different from γ of eq.(9) – we find that the critical velocity u_c for vortex filament formation is given by:

$$u_c = \frac{\kappa_4}{2\pi e a_0} \exp\left[-\frac{4\pi \bar{\gamma} k_B T}{\rho_s l \kappa_4{}^2}\right] . \tag{16}$$

As T goes to zero, u_c tends to u_{co}, which is expressed by:

$$u_{co} = \kappa_4 / 2\pi e a_0 , \tag{17}$$

The zero temperature limit of the critical velocity is independent of the length of the nucleated filament but inversely dependent on the core parameter a_0 and, hence, directly affected by pressure and ^3He impurity content. Above u_{co}, no energy barrier impedes the formation of vortex filaments over a smooth wall. Indeed, for $u_s \simeq u_{co}$, E_a given by eq.(15) can be written to first order as

$$E_a = E_0 \left(1 - \frac{u_s}{u_{co}}\right) , \quad \text{with } E_0 = \rho_s l \frac{\kappa_4{}^2}{2\pi} . \tag{18}$$

At absolute zero, and for $u_s = u_{co}$, vortex filaments can form at a distance z_c from the wall given by eq.(14) with $u_s = u_{co}$: $z_c = e a_0/2$. At that distance, they stagnate in metastable equilibrium in the flowing superfluid. It is known that such vortex pairs develop instabilities[60-61] along their length with a characteristic wavelength which is about 8 times their spacing, in this case $8 e a_0$. We speculate that this 'trailing vortex' instability might fix the scale for the characteristic length l so that the vortex configuration actually nucleated is a segment of this length somehow bending down to meet the wall at each end. If so, this simple approach would lead to the following estimate for E_0:

$$E_0 = \rho_s \frac{\kappa_4{}^3}{\pi^2 u_{c0}} \quad . \tag{19}$$

Taking, as a conservative estimate, $a_0 = 1.5$ Å, we find that $u_{c0} = 39$ m/s, $z_c = 2$ Å, $l = 33$ Å and $E_0 = 27$ K. If, on the other hand, we rely on the experimental estimate for v_{c0}, ~ 11 m/s, we find $a_0 \sim 5$ Å, $z_c = 7$ Å, $l = 108$ Å and $E_0 = 100$ K. Thus we find estimates for E_0 which are in order of magnitude agreement with the value determined from experiment. The latter value of a_0 corresponds to the superfluid healing length derived by Sobyanin and Stratonnikov[62] from surface tension measurements. The possible relevance to the present work of this type of measurements was noted by Meisel and Kumar.[50]

The fact that the stagnation distance z_c is of the order of the healing length, a distance at which we do not expect the classical hydrodynamical expressions for the line energy and impulse of a vortex to hold, greatly invalidates any detailed quantitative results. Also, we do not know how this fact really affects the vortex pair instability.‡ However, some qualitative features of this rudimentary model are of value. The temperature dependence of u_c given by eq.(16) is found, when that of ρ_s is taken into account, to be very nearly linear from 1.9 K down to absolute zero. The zero temperature limit u_{c0} is governed by a_0, a quantity which might possibly lead to the observed pressure and impurity concentration dependences. Also, E_0, and therefore T_0, are universal quantities, and at least of the right order of magnitude. The product $v_{c0} \times T_0$ might turn out to be nearly pressure independent if the superfluid density in eq.(19) is considered as being dominated by wall attraction and only weakly dependent on hydrostatic pressure.

Once nucleated, vortex filaments presumably evolve toward horse-shoe vortices with diameter ~ l. Whether these vortices can grow or not is governed by the flow profile close to the wall. If the geometry allows it, they expand to produce a phase slip. If the flow velocity is under-critical, they shrink back to the wall. This *growth-at-the-wall* problem must now be tackled.

Vortex trajectories far from walls (i.e., a few healing lengths away) are governed at low temperatures by the equations for ideal fluid motion, as stressed by Schwarz, and as proved true even rather close to boundaries by the success of the Muirhead-Vinen-Donnelly calculation. The growth of the vortex away from the wall must be studied in this framework.

Velocities as large as 40 m/s may be needed for spontaneous growth on smooth surfaces as seen above. If half-rings of radii ~ 15 Å are formed, local velocities of ~ 23 m/s are required to sustain their growth. These velocities are significantly larger than the measured average flow velocities. Local velocities u_s past walls are not straightforwardly related to mainstream average velocities and depend strongly on both the shape of the orifice and the detailed structure of its walls.

This problem was already considered by Vinen in his classic paper[4] on critical velocities in ^{4}He for somewhat idealized asperity geometries. His conclusion was that large enhancement factors could be expected past sharp protuberances. The case of sharp edges, which give rise to divergences in potential flow, has been considered in more detail by Brown and Hess.[35] The flow enhancement factor due to more realistic semi-elliptical surface defects protruding a height a over walls and having an extension b in the direction of the flow has been shown by Krasnov[64] to be

$$u_s/v_s = 1 + a/b \ . \tag{20}$$

A hemicylindrical defect with $a = b$ gives a flow enhancement factor of 2. If we assume that there exists a stack of these defects on top of one another with different length scale from 1000 Å down to atomic distances, the combined enhancement factor might be 4, or 8, or possibly more. Also, besides geometrical defects of atomic dimensions, wall impurities may also play a rôle, as suggested by Shevchenko.[65] Thus, unless special precautions are taken to eliminate all surface imperfections, local velocities are enhanced by a factor which can be estimated to be not less than 5, as discussed, in particular, by Hess.[25] This leads to experimental

‡Pumir and Siggia[63] have found, by numerical simulations, that vortex pair instabilities are strongly dependent on the structure of the vortex core.

local flow velocities of up to 50 m/s and above. These velocities are high enough to warrant vortex nucleation and growth.

7. Conclusion

There appear to be two rather distinct régimes for critical velocities in superfluid ^{4}He, a low-velocity, size-dependent, and temperature-independent régime and a high-velocity, size-independent, and temperature-dependent one. The mechanisms responsible for each of these régimes are likely to be different. Several authors[4,29] have reached the conclusion that a plausible explanation of the size-dependent régime lies in the existence of trapped vorticity. Vortex mills can then be at work.[48] None of the recent observations based on phase slippage experiments reported here contradicts such a conclusion. As a matter of fact, the observation of stable bias currents brings additional evidence for the existence of strongly pinned vorticity.

The high-velocity régime, in which phase slips are observed to occur, poses a different problem. It is apparently not directly related to the geometrical size of the orifice. The fact that it is affected by temperature down to millikelvin temperatures, and also by minute concentrations of ^{3}He impurities, shows that very small energies are involved in the process. The pressure dependence that we have reported is different from what is expected for purely hydrodynamical processes involving the bulk fluid density.

In their present forms, neither the vortex depinning mechanism nor the helical vortex mill proposed by Schwarz[19-20,48] seem to account for all the features of the high-velocity régime and of phase slips. We have chosen to favour vortex nucleation within the orifice because the supply of suitably pinned vortices cannot be inexhaustible and because depinning forces are so weak. Thermal nucleation accounts in a natural way for one of the most remarkable features of the high-velocity régime, namely, its temperature dependence. It also accounts for the statistical rounding of the critical transition. We have attempted to construct a phenomenological model which would yield the observed energy barrier. The outcome of this model is eq.(19), which bears a close resemblance to the energy barrier proposed by Chester[66] for homogeneous nucleation, apart from the fact that u_{co} contains the influence of the walls. This rudimentary model contains *ad hoc* assumptions to make this influence universal. It can only serve an illustrative purpose here, as these assumptions cannot at present be justified further. The problem of the dynamical behaviour of nascent vortices in the immediate vicinity of a wall remains untackled.

The velocities necessary both for vortex nucleation and for sustaining the growth of the nucleated vortex away from the nucleation sites appear to be higher than the observed critical velocities. To cope with this difficulty, we have followed the line of reasoning of other authors[4,25] and call for an enhancement factor due to wall defects of ~ 5. Our conclusion that this factor might be sufficient to warrant vortex nucleation and growth differs from that of Vinen,[4] and to a lesser extent, from that of Hess,[25] for the same problem. The main reason for this divergence is that critical velocities in the high-velocity régime observed more recently are now consistently and significantly higher than the older data in the size-dependent régime. But it is quite clear that this provisional conclusion calls for more extended experimental confirmation.

Acknowledgements

We gratefully acknowledge very useful conversations with Dominique D'Humières, Jean-Pierre Hulin, Pradeep Kumar, Alain Pumir, Riitta Salmelin, and Klaus Schwarz.

References

1. R.P. Feynman, in Prog. Low Temp. Phys., vol.I , ed. C.J. Gorter (North-Holland, Amsterdam, 1955) p.17.

2. J. Wilks, *The Properties of Liquid and Solid Helium* (Clarendon Press,Oxford, 1967).
3. W.E. Keller, *Helium-3 and Helium-4* (Plenum, New York, 1969);
R. de Bryun Ouboter, K.W. Taconis, W.M. van Alphen, Prog. Low Temp. Phys. vol.V, ed. C.J. Gorter (North-Holland, Amsterdam, 1969).
4. W.F. Vinen, in *Liquid Helium,* Proc. Int. School of Physics Enrico Fermi, Course XXI (Academic Press, New York, 1963) p.336.
5. A.L. Fetter, in *The Physics of Liquid and Solid Helium*, part 1, ed. K.H. Benneman, J.B. Ketterson (Wiley–Interscience, New York, 1976) p.238.
6. J.S. Langer, J.D. Reppy, Prog. Low Temp. Phys., Vol. VI, ed. C.J. Gorter (North-Holland, Amsterdam, 1970) Ch.1.
7. O. Avenel, E. Varoquaux, Phys. Rev. Lett. **55**, 2704 (1985).
8. E. Varoquaux, M.W. Meisel, O. Avenel, Phys. Rev. Lett. **57**, 2291 (1986).
9. E. Varoquaux, O. Avenel, M. Meisel, Can. J. Phys. **65**, 1377 (1987).
10. O. Avenel, E. Varoquaux, Phys. Rev. Lett. **59**, 1168 (1987).
11. O. Avenel, E. Varoquaux, Jpn. J. Appl. Phys. **26–3** 1798 (1987).
12. P.W. Anderson, Rev. Mod. Phys. **38**, 298 (1966).
13. B.P. Beecken, W. Zimmermann, Jr., Phys. Rev. **B 35**, 74 (1987).
14. B.P. Beecken, W. Zimmermann, Jr., Phys. Rev. **B 35**, 1630 (1987).
15. O. Avenel, E. Varoquaux, W. Zimmermann, Jr., Proc. LT19 (1990).
16. W. Zimmermann, Jr., O. Avenel, E. Varoquaux, Proc. LT19 (1990).
17. D.D. Awschalom, K.W. Schwarz, Phys. Rev. Lett. **52**, 49 (1984).
18. K.W. Schwarz, Phys. Rev. Lett. **50**, 364 (1983).
19. K.W. Schwarz, Phys. Rev. Lett. **57**, 1448 (1986).
20. K.W. Schwarz, Phys. Rev. Lett. **59**, 1167 (1987).
21. O. Avenel, E. Varoquaux, Phys. Rev. Lett. **64**, 416 (1988).
22. R.M. Bowley, in *Quantum Fluids and Solids–1989*, eds. G.G. Ihas, Y. Takano (AIP, New York, 1989) p.149.
23. P.C. Hendry, N.S. Lawson, P.V.E. McClintock, C.D.H. Williams, R.M.Bowley, Phys. Rev. Lett. **60**, 604 (1988).
24. C.M. Muirhead, W.F. Vinen, R.J. Donnelly, Phil. Trans. R. Soc. London **A 311**, 433 (1984).
25. G.B. Hess, in *Low Temperature Physics LT–13*, vol.1, ed. K.D. Timmerhaus, W.J. O'Sullivan, E.F. Hammel (Plenum Press, New York, 1974) p.302.
26. C.J. Adie, J.G.M. Armitage, *75th Jubilee Conference on Helium-4* (World Scientific, Singapore, 1983) p.174.
27. B.C. Crooker, Ph.D. thesis, Cornell University (1984), unpublished.
28. J.P. Hulin, C. Laroche, A. Libchaber, B. Perrin, Phys. Rev. **A 5**, 1830 (1972).
29. J.P. Hulin, D. D'Humières, B. Perrin, A. Libchaber, Phys. Rev. **A 9**, 885 (1974);
J-P. Hulin, Ph.D. Thesis, (Université de Paris, 1973), unpublished and private communication.
30. G.B. Hess, Phys. Rev. **B 15**, 5204 (1977).
31. W. Zimmermann, Jr., unpublished.
32. H.A. Notarys, Phys. Rev. Lett. **22**, 1240 (1969).
33. S.J. Harrison, K. Mendelssohn in *Low Temperature Physics LT–13*, vol.1, ed. K.D. Timmerhaus, W.J. O'Sullivan, E.F. Hammel (Plenum Press, New York, 1974) p.298.
34. E.Y. Andrei, W.I. Glaberson, Phys. Lett. **79A**, 431 (1980).
35. Y.J. Brown, G.B. Hess, J. Low Temp. Phys. **49**, 267 (1982).
36. M.E. Banton, J. Low Temp. Phys. **74**, 211 (1974).
37. W.E. Keller, E.F. Hammel, Physics **2**, 221 (1966).
38. E.N. Smith, Ph.D. Thesis (Cornell University, 1971), unpublished.
39. H.E. Corke, A.F. Hildebrandt, Phys. Rev. **A 2**, 1492 (1970).
40. G.G. Nancolas, R.M. Bowley, P.V.E. McClintock, Phil. Trans. R. Soc. London **A 313**, 537 (1985).
41. D.B. Crum, D.O. Edwards, R.E. Sarwinski, Phys. Rev. **9**, 1312 (1974).
42. M.W. Toft, J.M.G. Armitage, J. Low Temp. Phys. **52**, 342 (1983).

43. L.S. Rent, I.Z. Fisher, Sov. Phys. JETP **28**, 375 (1969).
44. B.B. Sabo, Ph.D. Thesis (University of Minnesota, 1973), unpublished.
45. J. Davis, A. Amar, R. Packard, Poster at QFS89, Gainesville, FL (1989).
46. S.G. Hedge, W.I. Glaberson, Phys. Rev. Lett. **45**, 190 (1980).
47. P.W. Adams, M. Cieplak, W.I. Glaberson, Phys. Rev. **B 32**, 171 (1985).
48. K.W. Schwarz, Phys. Rev. Lett. **64**, 1130 (1990).
49. K.W. Schwarz, Phys. Rev. Lett. **47**, 251 (1981); Phys. Rev. **B 31**, 5782 (1985).
50. M.W. Meisel, P. Kumar, Proc. LT19 (1990).
51. C. Colliex, private communication (1990).
52. P. Sudraud, P. Ballongue, E. Varoquaux, O. Avenel, J. Appl. Phys. **62**, 2163 (1987).
53. B.P. Beecken, W. Zimmermann, Jr., J. Vac. Tech. **A 3**, 1839 (1985).
54. D.W. Moore, P.G. Saffman, Phil. Trans. R. Soc. London **272**, 403 (1972).
55. W.I. Glaberson, R.J. Donnelly, Phys. Rev. **141**, 208 (1966).
56. S.V. Iordanski, Sov. Phys. JETP **21**, 467 (1965).
57. J.S. Langer, M.E. Fisher, Phys. Rev. Lett. **19**, 560 (1967).
58. J.R. Clow, J.D. Reppy, Phys. Rev. Lett. **9**, 291 (1967).
59. G.E. Volovik, JETP Lett. **15**, 81 (1972).
60. S.C. Crow, AIAA Journal **8**, 2172 (1970).
61. S.E. Widnall, Ann. Rev. Fluid Mech. **7**, 141 (1975).
62. A.A. Sobyanin, A.A. Stratonnikov, Sov. Phys. JETP, **45**, 613 (1987).
63. A. Pumir, E.D. Siggia, Phys. Fluids **30**, 1606 (1987).
64. Yu. Krasnov, Sov. J. Low Temp. Phys. **5**, 519 (1979); **6**, 598 (1980).
65. S.I. Shevchenko, Sov. J. Low Temp. Phys. **9**, 69 (1983).
66. G.V. Chester, in *The Helium Liquids*, ed. J.G.M. Armitage and I.E. Farquhar (Academic Press, London, 1975) p.35.

DISCUSSION ON RESTRICTED GEOMETRY

Chairmen: M. H. Chan, R. Guyer

Vortices : The paper of Williams and related ideas; e.g. the papers by Griffin and Glyde.

Discussion involving Griffin, Platzman and Williams explored a number of related questions. Platzman argued that a Kosterlitz-Thouless (KT) transition, able to be studied by examining the free energy, necessarily involves the balance of the amplitude of two terms in the free energy, the strength of the interaction and kT. This balance made sense only in D=2. "So what is this about a KT transition in D=3?" Williams stated that he used the phase "KT mechanism" to refer not to this free energy balance but rather to a scenario in which a rudimentary excitation, e.g. a string pair, is bootstraped into prominence. The screening of the flow field of the larger excitation by that of the smaller excitation was responsible for this effect. Thus in D=3 a "KT mechanism" works to bring the vortex rings into prominence.

Since Williams related the vortex rings to the rotons, Griffin, calling attention to the "no flow" excitation spectrum (phonons, maxons and rotons) and the possibility that flow might bring other excitations into the picture, asked for clarification of this connection. Williams replied that the connection was speculative. However, support for it may be gleaned from study of XY model. There, one notices that vortices play an important role over a wide temperature range. Griffin asked why the vortices, not just the small vortices associated with the roton, were not seen in neutron scattering. "They are too non-local to be seen by the neutrons". In any case it was agreed that the description Williams was making of the D=3 transition did not depend on the vortex ring-roton identification.

Platzman averred that what Williams was trying to do was to show that what Reppy and co-workers see in restricted geometry is a KT transition smeared by finite size effects?

Williams : Yes.

Platzman to Williams: Then, why is there a universal exponent.

Williams : There is, close enough to the phase transition.

Gasparini to Williams : It seems that the model you use has the right symmetry, irrespective of the language and the excitations that you put into it, will give you the right results, i.e. the exponents. So why are we focusing on the excitations, rotons, etc? The symmetry will give the right exponents!

Williams : You want to understand the physical mechanism underlying the lambda transition.

Gasparini : Why?

Putterman : (Asserting that he was extraordinarily qualified to comment, symmetry vs mechanism?, in that he neither worked in critical phenomena nor sought funding to do so). It is a sorry day when we say that the physics doesn't matter, that it is just a question of the number of degrees of freedom of the order parameter. While how you deal with the phase transition may be a matter of taste, away from equilibrium, when you are doing dynamics, it is crucial that you know what the key excitations are.

Chan turned the discussion back to the nature of the transition in restricted geometry and appealed to the appearance of the specific heat data in support of identification of the phase transition as a lambda transition. Finally Ferrell reminded everyone that the second sound mode was the soft mode at the lambda transition "Vortices or no, you don't want to forget second sound".

^3He-^4He Films : A number of people commented on aspects of the physical picture of films. The emphasis of much of what was said was oriented toward clarifying what was meant by a layer.

In response to a question from Dahm, Krotscheck stated that, although his calculations were as yet incomplete, he felt that the excitations he sees correspond to motions in the film that can be primarily identified with a single layer. In addition Krotscheck argued that, to the degree that one can ignore the first two (solid) layers, his picture is in accord with that of Guyer. Bozler asked about the length scale associated with oscillations in the behaviour of ^3He films, e.g. oscillations in T_1, specific heat, etc. Guyer replied, while the oscillations are due to evolution of the occupation of energy levels in the film, the essential length scale is approximately on "layer"; i.e the number of atoms in a thin piece of bulk material, a piece of bulk material of size 1cm*1cm*3.6 angstroms. The structure of the film, the bumps in the density, are a manifestation of the structure in the wavefunctions associated with the energy levels.

The discussion then turned to assessment of the possibility of "seeing" excitation among the energy levels available to particles in a film. Platzman remarked on the analogy to heterostructures.

Platzman : By doing neutron scattering, or using some other probe, could you see transitions from the filled states to the empty states?

Guyer : Both NMR measurements, T1, and third sound measurements show evidence for thermal activation into the empty states.

Platzman wondered about the relationship of the single particle excitations of the film to the model states used in the calculation.

Platzman : Would any important features of the model state spectrum survive in the interacting system?

Guyer : Experiment shows that they do. A correct theory of the type you are alluding to has yet to be worked out.

Hallock re-capitulated the experimental evidence, for small ^{3}He concentratrions on ^{4}He films, for single particle excitation levels.

In the context of this discussion it is useful to recall the energy level spectroscopy of Gasparini and co-workers (from specific heat data) and the elaborate comupational scheme being implemented by Krotscheck and co-workers.

There was considerable discussion on the Penn State heat capacity result of ^{4}He in Vycor away from the transition region.

Wyatt : Can one explain the linear T term in the heat capacity in terms of two level systems?

Chan : The adsorbed films, particularly the first couple of layers are very likely to be glass-like and the linear T term may be related to two level systems. However, we have not made a model to compare the magnitude of the linear T term.

Harrison : Some years ago we found that the heat capacity of the filled pore Vycor could be explained by a model of a set of Helmholtz oscillators with frequencies distributed with constant density of states. That is,the excitation of the ^{4}He were treated as fractions, or the localized models below the crossover of a fractal system.

Vilches : In interpreting the capacity vs, coverage isotherms, it should be noted that the heat capacity reflects not only the phase transition but also a conversion term from one phase to the other if there is a phase coexistence.

THE INERTIA OF THE ^{4}He CRYSTAL SURFACE

AND THE CALCULATION OF THE PHONON TRANSMISSION

D. O. Edwards, M. S. Pettersen and H. Baddar

Physics Department
The Ohio State University
Columbus, Ohio 43210 USA

The temperature dependence of the surface inertia of atomically rough ^{4}He crystals is discussed in terms of quantum kinks moving at velocities near the speed of sound in the liquid. The Wolf-Nozieres boundary conditions, which determine the transmission of high frequency phonons through the surface are derived. Neglecting the dependence on the azimuthal angle, we fit some of the published crystal shape measurements to find the surface tension, and the stiffness as a function of the angle from the c-axis. The surface stress, estimated from the temperature dependence of the surface stiffness along the melting curve, is about three times larger than the surface tension. The boundary conditions involve the surface elastic coefficients, which may be determined from the phonon transmission.

I. INTRODUCTION

In this paper we discuss two topics connected with the inertia of the surface of atomically rough ^{4}He crystals. The concept of the surface inertia ν_s was introduced by Kosevich and Kosevich[1] and Puech and Castaing[2]. According to fits[2,3,4,5] to the Kapitza conductance[6,7,8] of the surface, which is given by a thermal average of the phonon transmission coefficient, ν_s plays the dominant role in the transmission of high frequency (collisionless) sound through the rough crystal surface.

The first of our two topics is the following. If the crystal grows by the motion of quantum kinks[7] on steps, then the surface inertia can be expressed in terms of the inertial mass of the kinks, as explained in Mukherjee et al.[4] (referred to as MEP). In section II of the present paper we discuss the temperature dependence of ν_s used in MEP. In particular we give the explicit form for the energy spectrum of the kinks.

Although the surface inertia apparently provides the dominant term in the phonon transmission probability, there are other contributions which must be included before the surface inertia can be obtained from the Kapitza conductance. These contributions come from the surface tension and its dependence on orientation and surface strain. Recently the dynamics and thermodynamics of the crystal surface have been revised by Andreev and Kosevich[10] and Nozieres and Wolf[11]. Using this new analysis, Wolf and

Nozieres[12] (WN) have derived the static boundary conditions at the surface in the 2D case where, if $\hat{z}$ is perpendicular to the surface, there are no dependences on the y-coordinate. One of their boundary conditions differs from the version used in previous calculations. The new version clears up a difficulty noticed by Maris and Huber[3]: when applied to the transmission of sound, the earlier equations did not conserve energy.

The second topic and main objective of the present paper is to present, in section V, the 3D version of the WN boundary conditions, which is necessary to calculate the phonon transmission coefficient. We also briefly consider, in section VI, the curvature dependence of the surface tension which WN neglect. Our equations contain many new terms, including several involving the "surface elastic coefficients", which are the 2D surface equivalents of the bulk elastic coefficients (bulk and shear modulus etc.). The ^{4}He crystal provides a unique opportunity to study these coefficients. They may be measured by studying the dependence of the transmission coefficients into the three crystal phonon modes, as a function of the direction of the incident thermal phonon in the liquid.

In section III we review the definition of the surface properties, including the relationship between the surface inertia ν_s and Gibbs' surface mass M_s. In section IV we fit the published experimental data to find the orientation and strain dependence of the surface tension. This gives the first estimate of the surface stress, which is found to be approximately three times the surface tension.

II. QUANTUM KINK CONTRIBUTION TO THE SURFACE INERTIA

In this section we review the temperature dependence of the quantum kink contribution to the surface inertia used in MEP. The surface inertia ν_s is related to the number of steps per unit length N_s, the average number of kinks of a given kind (+ or -) per unit length of a step $N_k = \langle N_+ + N_- \rangle$, and the kink mass m_k according to equation (5) of MEP:

$$\nu_s = m_k N_k / (\sigma_C N_s \zeta) \qquad (1)$$

Here ζ is a dimensionless roughness factor $\zeta \equiv \sigma_C \langle (N_+ - N_-)^2 \rangle$ where $\langle\ \rangle$ denotes an average over the surface of the crystal. The cross-section σ_C is defined by $\sigma_C \ell = v_C$ where v_C is the atomic volume in the crystal and ℓ is the spacing between atoms on a step. Equation (1) is obtained by assuming that, due to kink-kink interactions, the positive and negative kinks move together with the same drift velocity. The crystal is able to grow or shrink because the surface roughness, caused by thermal fluctuations, produces imbalances between N^+ and N^-. The formula is derived by equating the kinetic energy of the kinks, due to their drift velocity, to the kinetic energy associated with the surface inertia.

The dependence of ν_s on N_s and N_k in (1) is different from the formulas derived by Kosevich and Kosevich[1]. They assume that kinks of opposite signs move independently of each other.

In equation (1) the quantities ζ, N_k and N_s are, in principle, temperature dependent. However, the data on the Kapitza conductance and the roton growth resistance may be well fitted[4,5] assuming that they are independent of temperature. Presumably this is because the surface is very rough and these quantities have reached their maximum values. On the other hand, the energy spectrum of a quantum kink, which is not like that of a classical particle, does give rise to a temperature dependence of ν_s which seems to agree with the experimental data. (However this comparison with the Kapitza conductance used the old, incorrect boundary conditions, not those described in section V, below).

The way in which the quantum kink energy spectrum was obtained in MEP is as follows: In the model proposed by Kosevich and Kosevich[1] the effective mass at low velocities m_k^0 is due to the kinetic energy of the superfluid liquid as it moves towards the moving kink. As the kink moves the crystal grows or melts and the kink acts as a sink or source for the superfluid. The kink is approximated as a hemispherical sink of radius r_0 moving on the surface of the crystal. If the liquid is treated as incompressible (infinite velocity of sound), the result for the kink mass is

$$m_k^0 = \frac{m_4}{2\pi} \left(\frac{\sigma_C}{\ell r_0} \right) \frac{(v_L - v_C)^2}{v_L v_C}, \tag{2}$$

where v_L and v_C are the atomic volumes of liquid and crystal. Assuming that $\sigma_C/\ell r_0 \approx 1$, this equation gives $m_k^0 \approx 0.002\ m_4$.

With such a small mass the thermal velocity of the kinks would exceed the speed of sound in the liquid c_L and the calculation must be modified to take this into account. This is done by solving the hydrodynamic equations for a moving source in an ideal compressible liquid. In this case the energy E of the moving liquid contains both kinetic and elastic (compressibility) terms. Treating the deviations from the equilibrium density as small one finds,

$$E(v) = \frac{1}{2} m_k^0 v^2 \frac{\gamma^2}{2} \left[\left(1 + \frac{2}{\gamma^2}\right) \frac{\gamma}{2\beta} \operatorname{atn}(\beta\gamma) + \frac{1}{2} \right]. \tag{3}$$

Here v is the kink velocity, which never exceeds c_L, and $\beta \equiv v/c_L$, $\gamma \equiv 1/\sqrt{(1-\beta^2)}$.

The Lagrangian L may be found by integrating $E = H = v dL/dv - L$. This gives for the momentum

$$q = dL/dv = E/v + \int (E/v^2) dv. \tag{4}$$

The result of numerically integrating (4) with (3) is shown in Fig. 1 as a graph of E/qc_L versus $q/m_k^0 c_L$. The changes in the density of the liquid near the kink begin to be large at $q/m_k^0 c_L \approx 2$. However the spectrum must tend to the asymptotic form $dE/dq \to c_L$ as $v \to c_L$, and we are already fairly close to this when the small amplitude approximation breaks down.

The figure also shows the kinetic energy of a "relativistic" particle of the same rest mass, with c_L playing the part of the speed of light. In view of the approximations in the hemispherical sink model, the difference between the numerical results and relativistic formula is unimportant.

The temperature-dependent effective mass of the kinks $m_k(T)$, which is used in (1), was calculated in MEP from the mean momentum of a slowly drifting 1D Bose distribution of "relativistic" kinks:

$$\bar{q} \equiv m_k(T) u = \int_{-\infty}^{\infty} \delta n_q q\, dq \Big/ \int_{-\infty}^{\infty} n_q^0 dq \tag{5}$$

where $n_q^0 + \delta n_q$ and n_q^0 are the drifting and equilibrium Bose occupation numbers. The mass $m_k(T)$ is the same as the "rest mass" m_k^0 at low temperatures, where only the "non-relativistic" part of the spectrum is sampled. As shown in Fig.2 of MEP, the thermally averaged mass ratio $\gamma(T) \equiv m_k(T)/m_k$ varies by about a factor of 3 between 0 and 1 K.

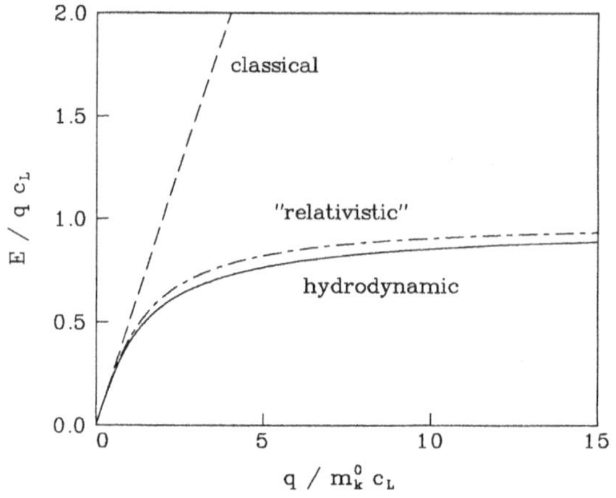

Fig.1. The kinetic energy of a kink E in terms of its momentum q. E/qc_L is plotted versus $q/m_k^0 c_L$, where c_L is the speed of sound in the liquid and m_k^0 is the kink effective mass at low velocity. The graph shows a comparison between the energy derived from hydrodynamics, using the model suggested by Kosevich and Kosevich (Ref.1), with the kinetic energy of a classical or a "relativistic" particle.

Since the periodicity of the atoms along each step on the surface is ℓ, the kink states must form a band with maximum wave vector π/ℓ and maximum energy $\pi\hbar c_L/\ell$. This corresponds to a band width of ~ 30 K (not 60 K as wrongly stated in MEP).

III. THE DEFINITION OF THE SURFACE PROPERTIES: SURFACE STRESS AND STRAIN

We first consider a simplified microscopic picture of a plane equilibrium crystal surface (Fig. 2) to better understand the definition of the surface properties. The figure shows the density ρ and the stress σ_{xx} as a function of the z coordinate which is assumed to be perpendicular to the surface. The pressure p in the bulk liquid has been added to σ_{xx}: $\sigma_{xx} + p$ is asymptotically equal to zero in the liquid but not in the crystal if it is subject to non-hydrostatic stress.

Figure 2 does not show any oscillations in the density of the crystal due to the localization of the atoms - one is either considering a continuum approximation to the crystal or the orientation of the surface is of low crystal symmetry. In the latter case the oscillations have almost disappeared due to averaging in the x and y directions.

To define surface properties, one conventionally chooses a dividing surface located at $z = z_s$ (see for instance Ref.13). Then for example the surface mass per unit area M_s is defined by

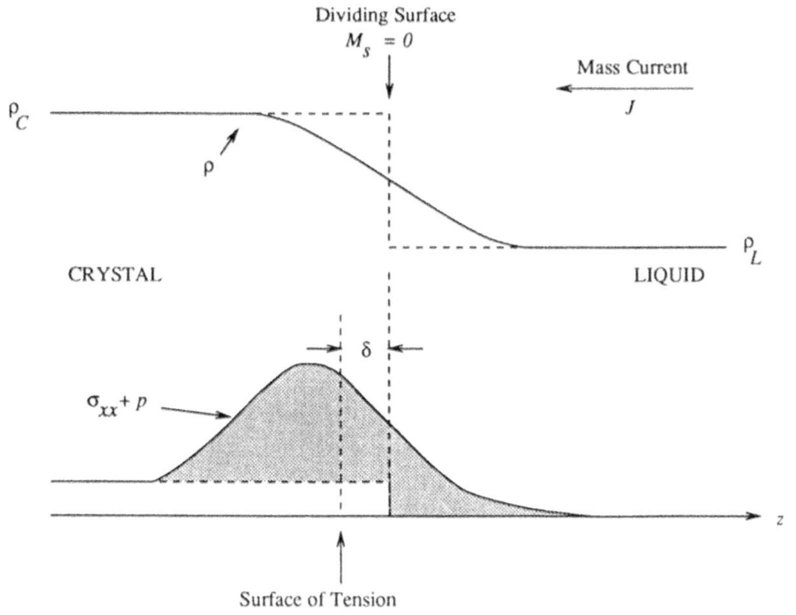

Fig.2. The definition of the properties of the crystal surface (schematic). The dividing surface which makes the surface mass $M_s = 0$ and the surface of tension, separated by distance δ, are both illustrated. The shaded area between the curve of σ_{xx}+p and the "ideal" behavior (the step function) is the surface stress β_{xx}.

$$M_s = \int_{-\infty}^{\infty} \rho(z)dz - \int_{-\infty}^{z_s} \rho_C dz - \int_{z_s}^{\infty} \rho_L dz \tag{6}$$

where ρ_C and ρ_L are the asymptotic densities deep in the crystal and liquid. Similarly,

$$\beta_{xx} = \int_{-\infty}^{\infty} \sigma_{xx}(z)dz - \int_{-\infty}^{z_s} (\sigma_{xx})_C dz + \int_{z_s}^{\infty} p dz \tag{7}$$

is one component of the two-dimensional surface stress tensor β_{ij}. The other components β_{yy} and $\beta_{xy} = \beta_{yx}$ are defined in an analogous way. The surface stress has not yet been measured in helium, but recently it was measured[14] for the reconstructed (111) surface of Si. Note that σ_{zz}, σ_{zx}, σ_{zy}, do not depend on z in equilibrium and therefore there are no surface properties corresponding to these quantities.

The surface tension $\gamma = F_s - M_s\mu$ is the surface contribution, per unit area, to the grand free energy of liquid plus crystal. F_s is the surface

Helmholtz free energy, defined from the local density of the Helmholtz free energy using an equation analogous to (4), μ is the chemical potential per unit mass in the liquid.

One should note that, in general, these surface properties depend on the choice of dividing surface. For example, the surface stress depends on z_s when the crystal is not hydrostatically stressed, e.g. when $\sigma_{xx}+p \neq 0$ deep in the crystal. On the other hand one can form invariant quantities such as the surface tension or the quantity $\beta_{xx} - (\sigma_{xx}+p)M_s\Delta\rho$, where $\Delta\rho \equiv \rho_C - \rho_L$, which do not depend on z_s.

If the crystal surface is in steady motion, due to freezing or melting, one can regard Fig.2 as a view in the frame of reference of the surface, with a mass current J perpendicular to the surface. The kinetic energy density associated with the motion is $J^2/2\rho_C$ and $J^2/2\rho_L$ deep in the crystal and liquid, and the kinetic energy of the surface is

$$T_s = \int_{-\infty}^{\infty} t(z)dz - \int_{-\infty}^{z_s} J^2/2\rho_C dz - \int_{z_s}^{\infty} J^2/2\rho_L dz$$
$$\equiv (M_s+\nu_s)J^2/2\rho_L\rho_C \tag{8}$$

where ν_s is the (invariant) surface inertia. The kinetic energy density t(z) due to freezing contains contributions from the motions of quantum kinks on the surface.

In another frame of reference in which the interface has velocity V_z, the local current density is $J'(z) = J - \rho(z)V_z$ and

$$T'_s = T_s + \tfrac{1}{2}M_s V_z^2. \tag{9}$$

In the same way, since the total z-momentum of the system per unit area is $P_z = \int J dz$ and J is independent of z, one finds

$$P_{zs} = 0, \qquad P'_{zs} = M_s V_z. \tag{10}$$

The surface inertia ν_s appears in the kinetic energy T_s but not in the momentum.

For an interface between two fluids, Gibbs suggested two special choices for the dividing surface[13]. The first is chosen so that $M_s = 0$. The second, called the "surface of tension" is chosen so that the first moment of the stress about z_s is zero. (Note that for two fluids the stress tensor deep inside either phase is just $-p\delta_{ij}$.) This means that the surface stress, which in this case is just $\gamma\delta_{ij}$, effectively acts in the surface of tension. For a curved surface the radii of curvature in the Laplace equation are those of the surface of tension. Using the surface of tension, the surface mass $M_s = M_s^0$ is non-zero and is referred to as the Gibbs adsorption.

Typically the distance δ between these two special surfaces, which is given by $M_s^0 = \delta\Delta\rho$, where $\Delta\rho$ is the difference in density between the two phases, is of atomic size. For instance, δ has been estimated as -0.3 Å for the liquid ^4He to vapor interface[15]. Nevertheless, for short-wavelength perturbations of the surface δ must be taken into account. It has an important effect on the spectrum of high frequency ripplons at the surface of liquid ^4He[15].

For the interface between two fluids the Gibbs adsorption is given by[13] $M_s^0 = -(\partial\gamma/\partial\mu)_T$ where the chemical potential μ may be varied independently of the temperature T only by varying the curvature of the surface. In the case of two fluids in contact the Gibbs adsorption M_s^0 or the length δ specifies both the curvature dependence of the surface tension and its dependence on the stress within either phase.

For the equilibrium between a crystal and its melt, although ways to generalize Gibbs' second choice of dividing surface are available[16], the choice $M_s = 0$ used by Wolf and Nozieres[12] (WN) is clearly simpler. In this case, WN show that γ depends on the 2D surface strain u_{ij} in the crystal. Rather than the actual microscopic value of the strain at the surface, u_{ij} with $i,j = x,y$ is the strain field in the interior of the crystal extrapolated to the dividing surface.

If $\vec{w}(\vec{r})$ is the elastic displacement vector at the (undisplaced) position $\vec{r}$, then the strain field in the interior of the crystal is given in terms of the unsymmetrized "distortion" tensor[17] $w_{ij} = \partial w_i/\partial r_j$ by[17]

$$u_{ij} = (w_{ij} + w_{ji} + w_{ki}w_{kj})/2 \tag{11}$$

using the usual summation convention.

An incremental strain u_{ij} changes the area of part of the surface dS in the ratio

$$dS'/dS = 1 + \mathrm{tr}(u) \tag{12}$$

to first order in the strain. With $M_s = 0$, $\gamma dS'$ is the Helmholtz free energy of this part of the surface, so the surface stress is given by

$$\beta_{ij} = \partial(\gamma dS'/dS)/\partial u_{ij}. \tag{13}$$

This gives, to first order in the strain,

$$\beta_{ij} = \gamma\delta_{ij} + \partial\gamma/\partial u_{ij}, \tag{14}$$

an equation given by Shuttleworth[18]. Note that the symmetry of β_{ij} is ensured since, according to (11) above, u_{ij} is symmetric.

An alternative convention[10], similar to that used in elasticity[17], is to define the surface tension as the free energy per unit area of the unstrained surface. The appearance of many of the equations is then considerably changed: for example the first term on the right in (14) disappears.

Calculation of the Surface Strain

If the original position of the dividing surface is at $z = z_s$, during the passage of a sound wave, the new position of a point on the surface of the crystal, originally at $\vec{r}_0 = (x_0,y_0,z_s)$, becomes $\vec{r}' = \vec{r}_0 + \hat{n}_0 h(x_0,y_0) + \vec{\xi}(x_0,y_0)$. Here $\hat{n}_0$ is the outward normal to the original unperturbed surface, h is the change in position due to freezing (or melting) at constant strain and $\vec{\xi}(x_0,y_0)$ is the elastic displacement of the surface due to the sound wave.

To find the boundary conditions governing h and $\vec{\xi}$, WN minimize the Helmholtz free energy of the system (surface plus bulk phases), expressed as an integral over x_0 and y_0, with respect to variations of $h(x_0,y_0)$ and $\vec{\xi}(x_0,y_0)$. Although $\vec{\xi}$ is defined only on the surface, this is sufficient to determine the surface strain which enters the expression for the surface tension. After the boundary conditions are found, $\vec{\xi}$ is identified as the 3D strain field evaluated at the surface.

To find the boundary conditions to first order, the free energy must expressed to second order in h and $\vec{\xi}$ and their derivatives with respect to x_0 and y_0. Variations in h and $\vec{\xi}$ produce changes in the surface area, the surface orientation and the surface strain.

The displacement $h(x_0,y_0)$ produces a change in the orientation of the surface with respect to the crystal axes, but the displacement $\vec{\xi}$ does not. We define the orthogonal unit vectors $\hat{x}$, $\hat{y}$ and $\hat{n}$ to define the orientation of the surface. The outward normal to the surface is $\hat{n}$, and $\hat{x}$ lies in the plane containing the c-axis and $\hat{n}$. The normal to the perturbed orientation of the surface may be specified by the components n_x, n_y in the directions of the original unperturbed $\hat{x}_0$, $\hat{y}_0$ unit vectors. If $\delta\hat{n}$ is the difference between $\hat{n}$ and the unperturbed $\hat{n}_0$, for small angles, $\delta n_x = n_x = -\partial h/\partial x \equiv -h^x$ and $\delta n_y = -h^y$. (We use superscript x and y to mean differentiation with respect to x_0 and y_0.) To first order, in powers of h

$$\delta\hat{n} = (-h^x, -h^y, 0). \tag{15}$$

Similarly

$$\delta\hat{x} = (0, h^y\cot\theta, h^x), \qquad \delta\hat{y} = (-h^y\cot\theta, 0, h^y), \tag{16}$$

where θ is the angle between $\hat{c}$ and $\hat{n}$. Equation (16) is easily derived from the fact that $\hat{y}$ is parallel to $\hat{n}\times\hat{c}$ and $\hat{x}$ is parallel to $\hat{y}\times\hat{n}$, where $\hat{c} = (\sin\theta, 0, \cos\theta)$.

The distortion w_{xx} in the plane of the surface due to $\vec{\xi}$ is found by differentiating $\vec{\xi}$ in the direction $\hat{x}$, and then taking the component in the direction $\hat{x}$:

$$w_{xx} = \hat{x}.(\hat{x}.\vec{\nabla})\vec{\xi}(x_0,y_0) \tag{17}$$

with $\vec{\nabla} = \hat{x}_0\partial/\partial x_0 + \hat{y}_0\partial/\partial y_0 + \hat{n}_0\partial/\partial z_0$. Since $\vec{\xi}$ does not depend on z_0, the last term in $\vec{\nabla}$ may be ignored. To second order in the small quantities h and $\vec{\xi}$ this gives

$$w_{xx} = \xi_x^x + h^x\xi_z^x + h^y\cot\theta(\xi_x^y + \xi_y^x) \tag{18}$$

where $\xi_x^y \equiv \partial\xi_x/\partial y_0$ etc. The analogous expression for w_{yy} is

$$w_{yy} = \xi_y^y + h^y\xi_z^y - h^y\cot\theta(\xi_x^y + \xi_y^x). \tag{19}$$

The ratio of the strained area compared to the unperturbed area is, to second order,

$$dS'/dS_0 = 1 + \xi_x^x + \xi_y^y + \xi_x^x\xi_y^y - \xi_x^y\xi_y^x + \tfrac{1}{2}(h^x+\xi_z^x)^2 + \tfrac{1}{2}(h^y+\xi_z^y)^2. \tag{20}$$

We require w_{xy} or w_{yx} only to first order in the h and $\vec{\xi}$. To this accuracy

$$w_{xy} = \xi_x^y, \qquad w_{yx} = \xi_y^x. \tag{21}$$

Apart from the terms in y in the preceding equations, the result for u_{xx} differs slightly from the 2D equation used by WN. They did not include the second order term $w_{xx}^2/2$ in (11), the definition of u_{ij}, although the term $w_{zx}^2/2$ was included.

IV. THE STRAIN AND ORIENTATION DEPENDENCE OF THE SURFACE TENSION OF ROUGH ^{4}HE CRYSTALS

Before writing down the boundary conditions at the surface, we review the orientation and strain dependence of the surface tension and the extent to which they can be determined from the available experimental data. We first consider the orientation dependence.

Measurements of the shape of the crystal give the surface stiffness tensor α_{ij} where

$$\alpha_{xx} = \gamma + \partial^2\gamma/\partial n_x{}^2, \quad \alpha_{yy} = \gamma + \partial^2\gamma/\partial n_y{}^2, \quad \alpha_{xy} = \alpha_{yx} = \partial^2\gamma/\partial n_x \partial n_y. \tag{22}$$

The data[19,20,21,22] are mostly measurements of α_{xx} as a function of θ and ϕ. Here θ and ϕ are the spherical coordinate angles for the surface normal $\hat{n}$ measured with respect to the crystal axes. The angle θ is between $\hat{n}$ and $\hat{c}$, while ϕ is between $\hat{x}$ and the (10$\bar{1}$0) axis. From the hexagonal symmetry of the crystal the relevant ranges of these angles are $\theta = 0$ to $90°$ and $\phi = 0$ to $30°$. As shown in Fig.3, there are measurements[22] of α_{xx} at two different ϕ which suggest that the dependence of γ on ϕ is fairly small and may reasonably be neglected. We shall make this assumption throughout the rest of this paper. It means that α_{ij} has its principal axes along $\hat{x}$ and $\hat{y}$, so that $\alpha_{xy} = \alpha_{yx} = 0$.

Since γ does not depend on ϕ, if $\delta\gamma$ is a small change in the surface tension due to a change in the orientation of the surface $\delta\hat{n}$, we may write

$$\delta\gamma = \gamma'\delta m + \tfrac{1}{2}\gamma''(\delta m)^2 + \ldots \tag{23}$$

where $m = \cos\theta$ and $\gamma' = d\gamma/dm$, $\gamma'' = d^2\gamma/dm^2$. Since $\delta m = \delta(\hat{n}.\hat{c})$ and $\delta\hat{n} = (\delta n_x, \delta n_y, -\tfrac{1}{2}(\delta n_x^2 + \delta n_y^2))$ to second order, we find

$$\begin{aligned} &\partial\gamma/\partial n_x = \gamma'\sin\theta, && \partial\gamma/\partial n_y = 0, \\ &\partial^2\gamma/\partial n_x^2 = \gamma''\sin^2\theta - \gamma'\cos\theta, && \partial^2\gamma/\partial n_y^2 = -\gamma'\cos\theta, \\ &\partial^2\gamma/\partial n_x\partial n_y = 0. && \end{aligned} \tag{24}$$

To fit the data we have used the polynomial

$$\gamma = \sum_{\ell=0}^{6} a_\ell m^{2\ell}. \tag{25}$$

The result is shown in Fig.3. The fit ignores the critical behavior of γ at the c, a and s facetting transitions. Our assumptions about the high mobility of the interface are not valid unless we are far away from a facetting direction at the given temperature, or far above the facetting temperature at the given direction.

Figure 3 shows that a rather small θ dependence in γ gives rise to rather strong dependence in α_{xx} and α_{yy}.

The Surface Elastic Constants

We next consider the strain dependence of γ. To obtain the strain dependence of β_{ij} to first order, we need γ to second order in the change of surface strain u_{ij}. Taking into account the symmetry of the crystal, six coefficients are necessary[23]:

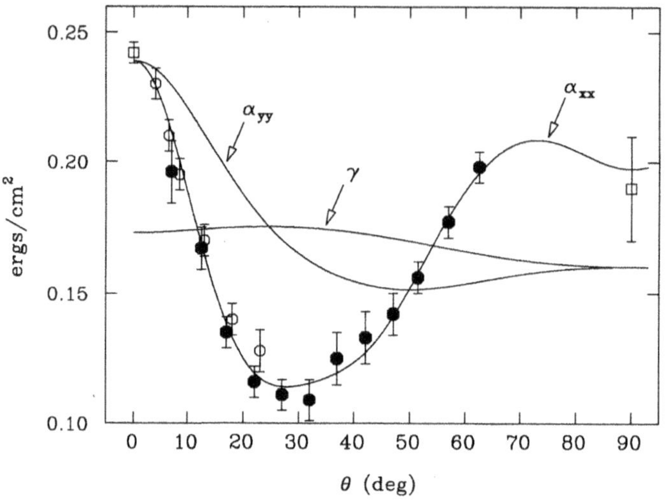

Fig.3. The surface tension γ, and the surface stiffnesses α_{xx} and α_{yy} of the rough ^4He crystal, derived from equation (25), as a function of θ, the angle between the normal to the surface and the (0001) or c-axis. The equation has been fitted to the data points for α_{xx} (Ref. 22) which are for $\phi = 13°$ (closed circles) and 5° (open circles), where ϕ is the angle between $\hat{x}$ and the $(10\bar{1}0)$ axis. The squares are from Ref.21. In addition, values of $(\partial^2\alpha_{xx}/\partial n_x{}^2)/\alpha_{xx}$ from Ref.21 (not shown) have been fitted at $\theta = 0°$ and 90°.

$$\gamma = \gamma_0 + b_{xx}u_{xx} + b_{yy}u_{yy} + \tfrac{1}{2}b_{xxxx}u_{xx}^2 + \tfrac{1}{2}b_{yyyy}u_{yy}^2 + b_{xxyy}u_{xx}u_{yy}$$
$$+ 2b_{xyxy}u_{xy}{}^2. \qquad (26)$$

In writing this equation we have assumed the principal axes of the stress tensor (14) to be along $\hat{x}$ and $\hat{y}$. This is consistent with our assumption that γ does not depend on ϕ.

The coefficients b_{xx} and b_{yy} are, from (14), contributions to the surface stress. The four other coefficients are the surface elastic constants, since from (9) and (21) they give the rate of increase in surface stress with surface strain. All six coefficients may be expected to depend on the orientation of the surface.

If the dependence of γ on u_{ij} were isotropic, the number of coefficients would be reduced to three, with $b_{xx} = b_{yy}$, $b_{xxxx} = b_{yyyy}$, and $b_{xxyy} = b_{xxxx} - 2b_{xyxy}$.

To determine a value of $(b_{xx}+b_{yy})$ we make use of the fact that the quantum kink model of the rough surface, described in MEP, indicates that the temperature dependence of the surface entropy $S_s = -(\partial\gamma/\partial T)_u$ is very small. This is calculated from the entropy of the 1D Bose gases of kinks on each step. In addition, the temperature independent part of S_s, associated with the presence of the steps on the rough surface, must also be small, since the stiffness of the rough surface below 1 K, where the melting curve is comparatively flat, is almost independent of temperature. Consequently, we attribute the temperature dependence of the surface stiffness, observed by Gallet et al.[24] above 1 K, to the fact that the pressure, and consequently the crystal strain, changes along the melting curve.

We have already pointed out, in footnote 16 of MEP, that if this dependence is attributed to the change of chemical potential μ along the melting curve, i.e., to the Gibbs adsorption, then $\delta \approx -1.0$ Å. Translating this result into a value of $b_{xx}+b_{yy}$, we find

$$b_{xx} + b_{yy} = -3K\delta\Delta\rho/\rho_C = +0.8 \text{ erg/cm}^2, \qquad (27)$$

where $K = 3 \times 10^8$ erg/cm³ is the bulk modulus of the crystal. This indicates that the mean surface stress $\beta_{ii} \approx \gamma_0+(b_{xx}+b_{yy})/2 = 0.6$ erg/cm² is approximately three times as large as the surface tension $\gamma_0 \approx 0.2$ erg/cm².

At the present time we have no data to determine b_{xx} and b_{yy} separately. The best way to obtain the surface stress and surface elastic constants may be to measure the transmission of thermal phonons as a function of the direction of incidence. As shown in the next section, the surface stress and surface elastic constants contribute to the transmission of sound through the surface with the same frequency dependence as the surface stiffness and surface inertia.

V. THE WOLF-NOZIERES BOUNDARY CONDITIONS

Wolf and Nozieres derive the boundary conditions at the surface by writing an expression for the total free energy of the system F, including bulk and surface terms. They then use the condition that, in static equilibrium, F must be a minimum with respect to variations of the displacements ξ_x, ξ_z and h. A similar procedure is used in Ref.16. The resulting Euler-Lagrange equations give the boundary conditions for σ_{xz}, σ_{zz} and the pressure in the liquid p. (All y-dependences are omitted.) The first two equations represent the conditions for mechanical equilibrium involving the transfer of momentum across the surface. The final equation represents the condition for chemical equilibrium involving the transfer of matter and energy across the surface. After generalizing their procedure to include the y direction, we find a number of new terms. The changes from the unperturbed values of σ_{xz}, σ_{yz}, σ_{zz} and p, are

$$\Delta\sigma_{xz} = -\partial\beta_{xx}/\partial n_x h^{xx} + (b_{xx}-b_{yy})h^{yy}\cot\theta + (3b_{xx}+b_{xxxx})\xi_x^{xx}$$
$$+ (b_{yy}+b_{xyxy})\xi_x^{yy} + (b_{xx}+b_{yy}+b_{xyxy}+b_{xxyy})\xi_y^{yx}, \qquad (28)$$

$$\Delta\sigma_{yz} = -\partial\beta_{yy}/\partial n_x h^{yx} + (b_{xx}-b_{yy})h^{yx}\cot\theta + (3b_{yy}+b_{yyyy})\xi_y^{yy}$$
$$+ (b_{xx}+b_{xyxy})\xi_y^{xx} + (b_{xx}+b_{yy}+b_{xyxy}+b_{xxyy})\xi_x^{xy}, \qquad (29)$$

$$\Delta p+\Delta\sigma_{zz} = \beta_{xx}(h^{xx} + \xi_z^{xx}) + \beta_{yy}(h^{yy} + \xi_z^{yy}), \qquad (30)$$

$$\Delta p(1-\rho_C/\rho_L) = \alpha_{xx}h^{xx} + \alpha_{yy}h^{yy} + \beta_{xx}\xi_z^{xx} + \beta_{yy}\xi_z^{yy}$$

$$+ (b_{xx}-b_{yy})(\cot\theta\xi_x^{yy} + \xi_y^{yx})$$

$$- \partial\beta_{xx}/\partial n_x\xi_x^{xx} - \partial\beta_{yy}/\partial n_x\xi_y^{yx} + (\nu_s/\rho_L)\partial J/\partial t. \qquad (31)$$

In (31) we have included the contribution from the surface inertia ν_s. Using the same procedure as WN it is easy to show that these equations conserve energy.

We note that the precise form of these boundary conditions depends on our definition of the strain (11). An alternative definition, omitting the quadratic terms in (11), would give the same physical results but would change the meaning of the quantities b_{xxxx} etc. in (26), and would change the terms involving these quantities in the boundary conditions.

For a sound wave in the liquid incident on the surface with constant amplitude, the terms on the right-hand side of these equations are all proportional to the square of the frequency ω. Therefore they all contribute to the low frequency transmission probability which is proportional to ω^2.

In a later publication we intend to present numerical calculations of the phonon transmission coefficient based on these boundary conditions. Here we shall just discuss the order of magnitude of the various kinds of terms.

Assuming a phonon incident from the liquid with wave vector $\vec{k}$, to find the transmission probability one must write the displacement $\vec{\xi} = (\xi_x,\xi_y,\xi_z)$ in terms of the amplitudes of the three transmitted phonon modes in the crystal, taking into account their directions and polarizations. The displacement h is related to ξ_z and the vertical displacement of the liquid ξ_{Lz} by $h = (\xi_z-\xi_{Lz})\rho_L/\Delta\rho$. Therefore, other things being equal, the terms in h contribute to the transmission amplitude by a factor $\rho_L/\Delta\rho \approx 10$ times more than those in ξ.

Similarly, $(\nu_s/\rho_L)\partial J/\partial t$ is proportional to $\nu_s\omega^2 h$. If we compare this with a term such as $\alpha_{xx}h_{xx}$, which is proportional to $\alpha_{xx}k_x^2 h$, the ratio is $\nu_s c_L^2/(\alpha_{xx}\sin^2\theta_i)$ where θ_i is the angle of incidence and c_L is the velocity of sound in the liquid. Using the fitted value[4] of $\nu_s = 2.1 \times 10^{-10}$ gm/cm^2 at low temperature, this ratio is $\sim 1.4/\sin^2\theta_i$. This accounts for the dominant role of the inertia in the transmission.

Our numerical calculation[5], based on the old boundary conditions, gives roughly a factor of 2 between the transmission amplitude with and without ν_s. On the other hand, the surface tension terms proportional to h should be easily distinguishable from the effect of ν_s by measuring the phonon transmission as a function of θ_i.

We now show that the magnitude of the surface elastic constants are probably large enough to make them measurable at large θ_i. A reasonable guess for the magnitude of b_{xxxx} for instance is $\Delta K a$, where $\Delta K = K_C - K_L = 0.7 \times 10^8$ erg/cm^3 is the difference in bulk modulus between crystal and liquid, and a is an atomic length, perhaps 1 Å. This gives $b_{xxxx} \approx 0.7$ erg/cm^2. The quantity $3b_{xx} + b_{xxxx} \approx 2$ erg/cm^2 is approximately a factor of 10 larger than γ or α_{xx}, compensating for the ratio $\rho_L/\Delta\rho$ between terms in h and terms in ξ_x, ξ_y, ξ_z.

VI. THE EFFECT OF CURVATURE

For a crystal surface in equilibrium with its melt, the 2D curvature tensor of the surface c_{ij} can be varied independently of the surface strain u_{ij}. The strain can be controlled by using side walls to distort the crystal. Ref.25, for instance, suggests that the effect of non-hydrostatic strain on the equilibrium with the melt be studied by using a parallel plate arrangement where the plates are piezoelectric transducers. At the same time the distance between these side walls would control the curvature of the surface. The chemical potential μ and pressure p in the liquid are then determined by the temperature, curvature and surface strain. The surface tension is also a function of the same variables.

The displacements h, due to freezing, and ξ, due to strain, both contribute to the curvature. For instance, to first order $c_{xx} = h^{xx}+\xi_z^{xx}$. Neither term is already included in the dependence of γ on the strain, given by (26). The effect of h^{xx} on γ must have a different physical origin from the effect of ξ_z^{xx}. The latter represents bending of the atomic planes of the crystal, whereas h^{xx} does not. Consequently the dependence of γ on curvature has ξ_z^{xx} as an independent thermodynamic variable from h^{xx}.

It is clear that, even with the constraint of thermodynamic equilibrium, that there must be a multitude of new coefficients describing the dependence of γ on h^{xx}, ξ_z^{xx}, ξ_x^{xx}, etc. Fortunately, in the equations governing the transmission of a sound wave, the fact that each of these quantities is a second space derivative means that the curvature terms all have at least one additional factor of $k_\parallel$, the horizontal wave vector, compared to the terms previously considered.

It follows that the curvature dependence of γ does not contribute to the first order terms in the transmission amplitudes which are proportional to the angular frequency ω (corresponding to a transmission probability) but only to the second order terms proportional to ω^2. When the amplitude is squared, the curvature terms will appear in products proportional to ω^3. The curvature contribution will be smaller than the first order terms by a factor of $k_\parallel a$ where a is some atomic length. At ~0.5 K, assuming a ~ 1 Å, $k_\parallel a$ is of the order of ~0.05. In any case, the transmission probability is nearing unity at such temperatures and the effect of any of the terms will be quite small. Therefore the curvature terms probably may be neglected in the phonon transmission problem.

Acknowledgements

This work was supported by the Low Temperature Physics Program, Division of Materials Research of the US National Science Foundation, Grant number DMR 8905385. We would like to acknowledge useful conversations with S. Balibar and W. F. Saam, and help with a literature search from T. M. Levin.

REFERENCES

1. A. M. Kosevich and Yu. A. Kosevich, Fiz. Nizk. Temp. 7, 809 (1981); [Sov. J. Low Temp. Phys. 7, 394 (1981)].
2. L. Puech and B. Castaing, J. Physique Lett. 43, L601 (1982).
3. H. J. Maris and T. E. Huber, J. Low Temp. Phys. 48, 99 (1982).
4. D. O. Edwards, S. Mukherjee and M. S. Pettersen, Phys. Rev. Lett. 64, 902 (1990).

5. M. S. Pettersen and D. O. Edwards, submitted for publication in the Proceedings of LT19, March 1990.
6. T. H. Huber and H. J. Maris, Phys. Rev. Lett. 47, 1907 (1981) and J. Low Temp. Phys. 48, 463 (1982).
7. B. Castaing, G. Bonfait and D. Thoulouze, Physica B109-110, 2093 (1982); L. Puech, B. Hebral, D. Thoulouze and B. Castaing, J. Physique Lettres 43, L.809 (1982).
8. P. E. Wolf, D. O. Edwards and S. Balibar, J. Low Temp. Phys. 51, 489 (1983).
9. A. F. Andreev and A. Ya Parshin: Zh. Eksp. Teor. Fiz. 75, 1511 (1978); [Sov. Phys. JETP 48, 763 (1978)].
10. A. F. Andreev and Yu. A. Kosevich, Zh. Eksp. Teor. Fiz. 81, 1435 (1981); [Sov. Phys. JETP 54, 761 (1981)]. See also Yu. A. Kosevich, Zh. Eksp. Teor. Fiz. 90, 481 (1986); [Sov. Phys. JETP 63, 278 (1986)] and M. Yu. Kagan and Yu. A. Kosevich, Zh. Eksp. Teor. Fiz. 91, 826 (1986); [Sov. Phys. JETP 64, 487 (1986)].
11. P. Nozieres and D. E. Wolf, Z. Phys. B 70, 399 (1988).
12. D. E. Wolf and P. Nozieres, Z. Phys. B 70, 507 (1988).
13. See, for instance, R. B. Griffiths, in: "Phase transitions in Surface Films", ed. by J. G. Dash and J. Ruvalds, Plenum Press, New York (1980).
14. R. E. Martinez, W. M. Augustyniak and J. A. Golovchenko, Phys. Rev. Lett. 64, 1035 (1990).
15. D. O. Edwards and W. F. Saam, in: "Progress in Low Temperature Physics", Vol. VIIA, ed. by D. F. Brewer, North Holland, Amsterdam (1978), p. 183.
16. J. I. D. Alexander and W. C. Johnson, J. Appl. Phys. 58, 816 (1985).
17. L. D. Landau and I. M. Lifshitz, "Theory of Elasticity", tr. by J. B. Sykes and W. H. Reid, Pergamon Press, London (1981).
18. R. Shuttleworth, Proc. Phys. Soc. London A63, 444 (1950).
19. A. V. Babkin, K. O. Keshishev, D. B. Kopeliovich and A. Ya. Parshin, Pis'ma Zh, Eksp. Teor. Fiz. 39, 519 (1984) [JETP Lett. 39, 633 (1984)].
20. A. V. Babkin, D. B. Kopeliovich and A. Ya. Parshin, Zh. Eksp. Teor. Fiz. 89, 2288 (1985); [Sov. Phys. JETP 62, 1322 (1985)].
21. P. E. Wolf, F. Gallet, S. Balibar, E. Rolley and P. Nozieres, J. de Physique 46, 1987 (1985); F. Gallet, S. Balibar and E. Rolley, J. de Physique 48, 369 (1987).
22. O. A. Andreeva and K. O. Keshishev, Pis'ma Zh. Eksp. Teor. Fiz. 46, 160 (1987) [JETP Lett. 46, 200 (1987)]. A later paper, O. A. Andreeva, K. O. Keshishev and S. Yu. Osip'yan, Pis'ma Zh. Eksp. Teor. Fiz. 49, 661 (1989) [JETP Lett. 49, 759 (1989)], which was not available to us at the time the fit shown in Fig.3 was made, confirms the dependence of α_{xx} on θ, but shows that α_{yy} is a strong function of ϕ at $\theta = 90°$. We hope to present a more accurate version of the present work, including the dependence on ϕ, in a later publication.
23. The calculation is analogous to that for 3D orthorhombic crystals in Ref.17. In the case of the surface however we may not neglect the linear term.
24. F. Gallet, P. E. Wolf and S. Balibar, Phys. Rev. Lett. 52, 2253 (1984).
25. S. Balibar, D. O. Edwards and W. F. Saam, submitted to J. Low Temp. Phys., March 1990.

THERMAL EXCITATIONS AND HELIUM 3 CRYSTALLIZATION

S. Balibar*, D. O. Edwards**, F. Graner* and E. Rolley*

*Laboratoire de Physique Statistique de l'ENS, associé au CNRS et à l'Université P.M. Curie, 24 rue Lhomond, 75231 Paris Cedex 05, France

**Department of Physics, The Ohio State University, Columbus, OH 43210, USA

ABSTRACT

The main reason for studying the crystallization of helium is that, at low temperature, its two isotopes can be made very pure and diffusion problems in bulk phases can be neglected. Then, the dissipation which limits the growth rate takes place at the moving liquid-solid interface, and some information on the crystallization processes can be obtained. The one fact that thermal excitations are different in helium 4 and in helium 3 makes the crystallization of these two quantum systems rather different from one another. Helium 4 was studied some years ago. In helium 3, recent experiments and theoretical articles show that the growth rate is maximum at $T_m = 0.32$ K where the latent heat vanishes and the growth is isothermal. Below T_m, a temperature difference usually develops between the liquid and the crystal, and the growth rate is much smaller, as a consequence of the necessary flow of a large latent heat through the Kapitza resistance of the liquid-solid interface. This Kapitza resistance is smaller than previously thought, because of the good coupling between transverse modes in the (Fermi) liquid and transverse phonons in the solid. The latent heat is released on the liquid side of the interface, and the intrinsic (or isothermal) growth rate is found to be limited by the momentum exchanges during the reflection of (Fermi) quasiparticles by the crystal lattice.

INTRODUCTION

At first sight, helium 3 crystals[1] behave at 0.32K like helium 4 crystals at 1.3K[2,3]. By this we mean that, at these particular temperatures, both crystals relax towards their equilibrium shape in a fraction of a minute. Since no facets are then present on their surface, (both are completely rounded, i.e. rough[3,4]), they actually look like honey drops. However, contrary to such viscous fluid drops, where a mass flow occurs inside the moving surface, the shape relaxation of helium crystals proceeds through local growth or melting at different parts of their interface with the surrounding liquid. The relaxation time is thus determined by the growth properties of the crystals.

The first interesting property of helium isotopes is that they can be made pure enough that during crystallization from the liquid phase, mass diffusion in the bulk phases can be neglected: the growth rate is only determined by the flow of mass and heat through the interface. Moreover, it is well known that helium 4 and helium 3, being two very different quantum systems, have low temperature phases with very different excitations. Since it is these excitations whose reflection and transmission at the interface control the growth rate[5,6,7,8,9], the similarity between helium 3 at 0.32K and helium 4 at 1.3K is quite fortuitous. Away from these temperatures, the two crystals may grow at rates which differ by several orders of magnitudes. As an example, at 0.1K, rough crystal surfaces in helium 4 relax within less than a millisecond[10-11] ,while in helium 3 several days are needed[12,13].

The purpose of this review article is to describe experiments[1] which were recently done in helium 3, then to explain why the observed phenomena are so different from what has been seen [2,3,10,11] in helium 4. As we shall see, helium 4 crystals usually relax isothermally, which means that the temperature difference between the liquid and the solid is negligible (except if a heat current is forced, on purpose, through their interface). Such an isothermal situation also occurs in helium 3 at 0.32K because the latent heat of crystallization is then zero. However, away from 0.32K, a significant heat has to cross the helium 3 crystal surface, which gives rise to a temperature difference between the two phases and to the large dissipation controlling the growth. Eventually, we shall see that in helium 3, the Kapitza resistance between the liquid and the solid phase is surprisingly small[1,14] around 0.3K, and recall that in helium 4 it was anomalously large [15,16,17].

THE EXPERIMENT BY GRANER ET AL.

Graner et al.[1] crystallized helium 3 in an optical cell, which was observed through windows and a superfluid helium 4 thermostat. A particular procedure enabled them to prepare crystals with a shape which was out of equilibrium and which then relaxed under the effect of gravity and surface tension (fig. 1). From the knowledge of these two driving forces, and the easy measurement of the local growth (or melting) velocities, they analyzed the growth dynamics at 0.32 ± 0.03 K where the latent heat is small or zero. As we shall see, this simple experiment allowed them to measure two of the three dissipation coefficients of the interface, namely the isothermal growth rate k and the Kapitza resistance R_k.

However, two experimental problems had to be solved, namely the minimization of impurity and temperature gradients. When using ordinary helium 3, containing 200 ppm

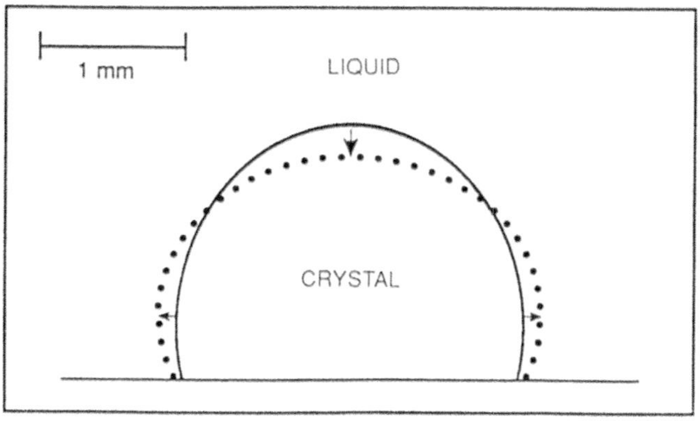

Figure 1. In the experiment by Graner et al.[1], the relaxation of the shape of a helium 3 crystal is observed close to 0.32K where the latent heat vanishes. It proceeds at constant solid mass and by melting at the top and crystallization on the sides of the crystal. The dotted line represents the equilibrium shape.

Figure 2. The equilibrium shape of a helium 3 single crystal, as observed by Rolley et al.[4]. The crystal width is about 3 mm, larger than the capillary length, so that the shape is flattened by gravity. From an analysis of such shapes, Rolley et al. deduced that the surface stiffness of helium 3 crystals is 0.60 ± 0.11 erg/cm^2 between 0.15 and 0.40 K.

helium 4, the liquid+solid system never reaches equilibrium, because concentration gradients never relax in the crystals. Consequently, Graner et al.[1] used purified helium 3 with 5 ppm helium 4 only. This residual amount was completely adsorbed as a fraction of a monolayer on the two sintered copper pieces which where fixed inside their experimental cell. This sintered copper also provided a good thermalization without which big temperature differences would have dominated the tiny effects of gravity and surface tension on the crystal shapes. They estimated that, despite the necessary illumination through windows and infrared filters, temperature gradients in their cell were smaller than 0.5 μK/mm, and consequently negligible in their experimental conditions.

As analyzed by Graner et al.[1], the force which drives the growth or the melting of crystals is the mechanical part of the difference in chemical potential $\Delta\mu = \mu_L - \mu_C$ between the liquid and the solid. This is given by

$$\Delta\mu_m = - \{(\rho_C - \rho_L)\gamma(z - z_0) + \gamma C\}/\rho_C \tag{1}$$

where ρ_C (resp. ρ_L) is the crystal (resp. liquid) density, z is the height of the interface, z_0 the equilibrium height of a flat interface, $C = 1/R_1 + 1/R_2$ the curvature of the interface and γ its surface stiffness.

The interface moves at a velocity v, so that they defined an effective growth rate k_{eff} such that

$$v = k_{eff}\Delta\mu_m. \tag{2}$$

As explained in reference 1, the effect of possible drifts in the cell pressure was eliminated by measuring the difference in growth velocity between two different points as a function of the difference in $\Delta\mu$ between the same two points (generally, the crystal top and a point on the crystal side where the interface happened to be immobile). The main difficulty was to measure the crystal curvature C, since only one of the curvature radii was visible, but simple hypotheses such as revolution symetry around the crystal top led them to a reasonable estimate of C. Finally, the surface stiffness γ was obtained from a previous analysis of equilibrium shapes[4] (fig.2). It should be noted that, in the case of helium 3 crystals, as for all bcc crystals, the surface tension α was shown to be isotropic within about 1%, so that the two quantities α and $\gamma = \alpha + \partial^2\alpha/\partial\phi^2$ are not very different.

As shown on fig.3, Graner et al. found that the growth (or melting) velocity is proportional to the applied force $\Delta\mu_m$, as expected for a rough surface. They also showed that the effective growth resistance 1/keff has a minimum value at 0.32 ± 0.01 K (fig.4)

INTERPRETATION

In order to understand this experiment, one may have the following and simple idea: at 0.32K, no latent heat is released and the crystal shape relaxes according to the value of an intrinsic growth rate. Away from this temperature, heat has to be conducted and thermal effects slow down the growth dynamics. This is correct but needs to be rigorously demonstrated.

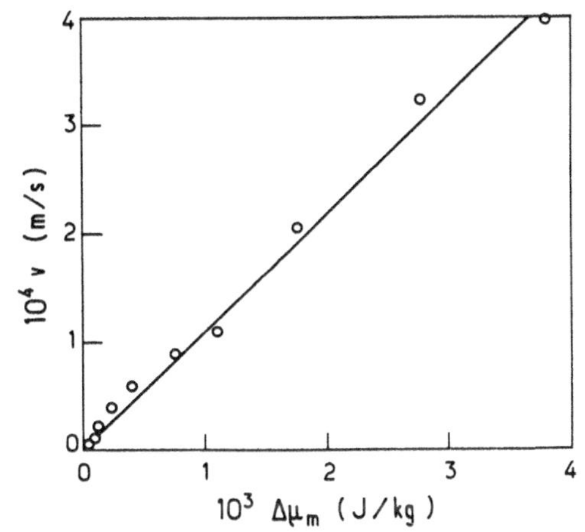

Figure 3. As expected for a rough crystal surface, Graner et al.[1] found a growth velocity v proportional to the driving force ($\Delta\mu m$ is the mechanical part of the difference in chemical potential between the liquid and the solid).

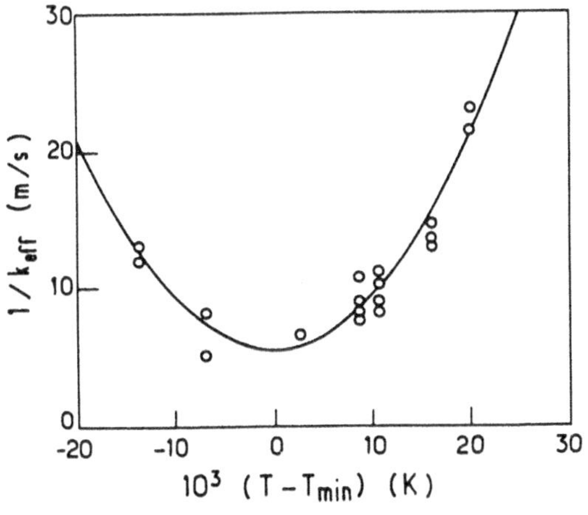

Figure 4. The effective growth resistance of helium 3 crystals, as a function of temperature, around 0.32K. The resistance is minimum when the latent heat is zero.

As first introduced by Castaing and Nozières[6], then futher developped by Bowley and Edwards[7] and Balibar, Edwards and Saam[12], the dynamics of rough crystal surfaces, close to equilibrium, may be described by the two linear equations

$$\Delta\mu/T = aJ + bJ_E \tag{3}$$
$$\Delta T/T^2 = bJ + cJ_E. \tag{4}$$

Here $J = \rho_C v$ and J_E/T are the currents of mass and entropy crossing the interface from liquid to crystal (fig.5). We shall follow here the recent presentation by Balibar Edwards and Saam[12] of the non-equilibrium thermodynamics involved in this problem.

As a consequence of Onsager's reciprocity relation, the same coefficient b appears in eqs. (3) and (4). As we shall see, the quantities a, b, c are related to the growth resistance of the surface, to its Kapitza resistance, and to the way how, during growth, the liquid entropy is released and the solid entropy absorbed in the liquid and in the solid. The Kapitza resistance is defined by $R_k = \Delta T/J_E$ with $J = 0$. It follows from (4) that

$$R_k = cT^2. \tag{5}$$

In a steady state of growth, heat flows from the walls of the container into the crystal, and into the walls out of the liquid, at rates[7]

$$Q_C = JTS_C - J_E, \qquad Q_L = JTS_L - J_E \tag{6}$$

per unit area of the interface. Thus the entropy of the newly formed solid is supplied by the heat from the walls Q_C plus the heat J_E arriving through the interface. Similarly the

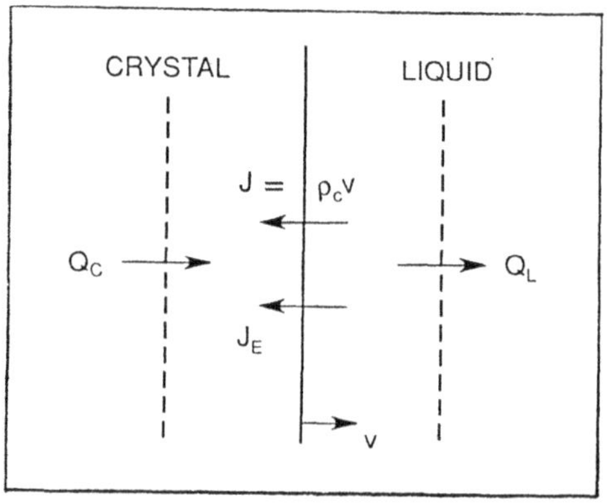

Figure 5. The moving interface seen from the frame of reference of the crystal, from reference 7. The crystal grows at velocity v, and a mass current density $J = \rho_C v$ crosses the interface. The flux of entropy in the interface frame of reference is J_E/T. Q_L and Q_C are the heat current densities through stationary planes in the crystal and in the liquid.

entropy of the liquid being frozen is removed by J_E and Q_L. Note that, as expected, $JT(S_L - S_C) = (Q_C - Q_L)$ is the total flow of heat from unit area of the interface to remove the latent heat $L = T(S_L - S_C)$.

A way to clarify the physical meaning of the cross coefficient b is to rewrite (4) as

$$J_E = \Delta T/R_k - Jb/c, \tag{7}$$

and substitute in (6):

$$Q_C = J(TS_C + b/c) - \Delta T/R_k, \qquad Q_L = J(TS_L + b/c) - \Delta T/R_k. \tag{8}$$

From (8), it appears that the quantity $-b/cT$ is the entropy per unit mass which is convected through the interface by the helium passing from the liquid to the solid phase. The heat conducted through the interface is $\Delta T/R_k = J_E + Jb/c$. The value of b/c determines which part of the latent heat has to flow to or from the interface, on each side. Of course, the heat currents Q_C and Q_L generally also depend on the thermal impedances of the liquid phase and of the solid phase.

Furthermore, from (3) and (4), the growth velocity v can be expressed as

$$v = k\{\Delta\mu - (b/c)(\Delta T/T)\} \tag{9}$$

where

$$k = \{(\rho_C aT)(1 - b^2/ac)\}^{-1}. \tag{10}$$

The quantity k, being the ratio $v/\Delta\mu$ when ΔT is zero, is the isothermal growth rate. The apparently simpler quantity $k_E = (\rho_C aT)^{-1}$ is the adiabatic growth rate, but experiments usually give an approximation of the isothermal growth rate k and it is difficult to imagine a situation where the conduction current $\Delta T/R_k$ exactly cancels the convection current $-Jb/c$ to obtain the adiabatic condition $J_E = 0$.

The Kapitza resistance is determined by τ, the mean thermal phonon transmission coefficient[15] through the interface from the liquid to the crystal:

$$R_k^{-1} = (3/4)(\rho_L S_{\Phi L} c_L \tau). \tag{11}$$

Here, c_L is the sound velocity in the liquid, and $S_{\Phi L}$ is the phonon contribution to the liquid entropy. R_k could similarly be expressed in terms of the phonon entropy and velocity in the solid, but a summation over the three modes would be needed. In helium 4, it was shown by the theory of Bowley and Edwards[7], and by the experiments of Wolf et al.[17] that the quantity $-b/cT$ is positive and small compared to S_C or S_L at all temperatures between 0 and the λ transition.

On the contrary, in helium 3, it appears[9,12,14] that

$$-b/cT = S_C = (k_B \log 2)/m, \tag{12}$$

the spin entropy of the solid per unit mass. This means that when an atom sticks to the growing crystal, nearly all the entropy exchange takes place on the liquid side, which seems reasonable if one realizes that quasiparticles are much better coupled to each other than with long wavelength phonons in the low temperature crystal. Graner et al. supposed that b/c is given by eq. (12), and their results actually support this assumption, but they do not justify it.

The entropy production at the interface can be written as

$$\partial S/\partial t = aJ^2 + 2bJJ_E + cJ_E^2 \tag{13}$$

so that the quantity b^2 - ac is necessarily positive. It can also be written as

$$\partial S/\partial t = J^2/(k\rho_C T) + (\Delta T)^2/(R_k T^2) = J^2/(k\rho_C T) + R_k(J_E + Jb/c)^2/T^2. \tag{14}$$

One can define an effective growth resistance k_{eff}^{-1} by

$$k_{eff}^{-1} = \rho_C T(\partial S/\partial t)/J^2 = k^{-1} + (\rho_C R_k/T)(J_E/J + b/c)^2 \tag{15}$$

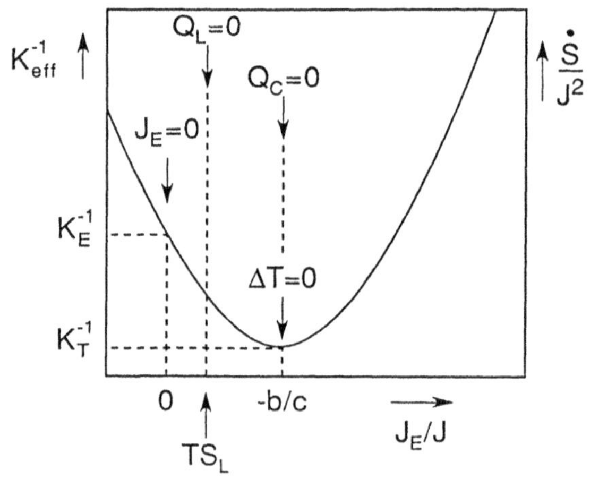

Figure 6. The effective growth resistance of helium crystals as a function of the thermal conditions in the experiment, as presented in ref. 12. In the case of helium 4, the value of b/c is close to zero, and there is not much difference between isothermal ($\Delta T = 0$) and adiabatic ($J_E = 0$) growth. On the contrary, in helium 3 , as indicated on this figure, there are large difference between the various types of growth conditions.

This equation is illustrated in fig. 6 which shows how the apparent growth resistance depends on the thermal conditions of the experiment i.e. on the ratio J_E/J. Isothermal conditions ($\Delta T = 0$) correspond to $J_E/J = -b/c$ and to the minimal entropy production or growth resistance. No heat current through the solid, i.e. $Q_C = 0$, corresponds to $J_E/J = TS_C$, and $Q_L = 0$ to $J_E/J = TS_L$. The adiabatic situation corresponds to $J_E = 0$.

With these equations in mind, let us now consider the relaxation of helium 3 crystals. Around 0.32K, the phonon mean free path in solid helium 3 is larger than 1 mm so that the thermal impedance Z_C of Graner's crystals is negligible. On the contrary, the thermal impedance of the liquid Z_L is very large. One has

$$\Delta T = Q_L Z_L = R_k(J_E + Jb/c) \qquad (16)$$

and from (6)

$$J_E + Jb/c = J(TS_L + b/c)/(1 + R_k/Z_L). \qquad (17)$$

The total entropy production per unit area has a surface term given by (14) and a small volume term $Q_L \Delta T/T^2$. We now suppose that

$$-b/c = TS_C(1 - \eta(T)) \qquad (18)$$

where $\eta(T)$ is small, as explained above. From eqs (15) to (18), one sees that the effective growth resistance which was measured by Graner et al. is

$$k_{eff}^{-1} = k^{-1} + (R_k \rho_C/T)(L + \eta(T)TS_C)^2/(1 + R_k/Z_L). \qquad (19)$$

As explained by Balibar, Edwards and Saam[12], this equation is more precise than eq.9 in ref.1 where η and R_k/Z_L were neglected. It shows that the existence of a liquid channel for heat conduction, in addition to the (interface+solid) channel, slightly reduces the effective growth resistance. Also, if $\eta(T)$ is small and slowly varying with temperature, the effective growth resistance versus temperature should have a minimum value close to 0.32K where L=0. The resistance should vary approximately as the square of the temperature difference from 0.32K. This is precisely what was observed by Graner et al. whose results can be represented as

$$k_{eff}^{-1} = 550 + (3.9\mathrm{x}10^6)(T - 0.32)^2 \text{ cm/s}.$$

Since the latent heat L is about $9\mathrm{x}10^6(T-0.32)$ erg/g, this experiment gives

$$k^{-1} = 550 \text{ cm/s}, \qquad R_k = 1.3 \text{ cm}^2\text{K/W}$$

at T = 0.32K. The value of R_k is indeed small compared to the thermal resistance of a 1 mm thick liquid layer ($Z_L > 2000$ cm^{2}K/W according to Anderson et al.[20]). Graner's experiment also shows that η is less than 1.4×10^{-2} at 0.32K.

HELIUM 3 VERSUS HELIUM 4

As explained by Graner et al.[1], the value of the growth rate k is consistent with a theory by Puech et al.[9] who predicted

$$k = (3/4)(\rho_C\rho_L/\Delta\rho^2)(m/p_F)[t/(1-t/2)], \qquad (20)$$

where t is the sticking probability of helium 3 quasiparticles on the crystal surface, and p_F their Fermi momentum; m is the mass of helium 3 atoms and $\Delta\rho = \rho_C - \rho_L$. This theory supposes that the intrinsic growth resistance k^{-1} results from momentum exchanges at the interface during the reflection or sticking of quasiparticles on the crystal lattice which is at rest in the reference frame of the laboratory. It differs by a large squared density ratio from an earlier proposal by Andreev and Parshin[5] who supposed that the quasiparticles were directly interacting with the interface moving at the velocity v. Graner's results are not consistent with this early proposal but consistent with eq.20 if the sticking probability t is about 10%, a reasonable value.

As for the value of the Kapitza resistance, it first appeared surprisingly small. Indeed, a simple calculation of the transmission probability of longitudinal phonons from the liquid into the solid leads to a value which is larger by a factor 6. As explained by Graner, Bowley and Nozières[14], the reason for this apparent discrepancy lies in the existence of transverse modes in Fermi liquids. At low enough temperature or high enough frequency, one knows[21] that transverse zero sound modes propagate in helium 3 at a small velocity (about 20 m/s). It is shown in ref.14 that , although liquid helium 3 is not yet a degenerate Fermi liquid at 0.32K, and transverse sound at thermal frequencies is intermediate between the ballistic (zero sound) regime and the hydrodynamic, damped, regime, a definite viscoelastic behavior exists, which significantly increases the transmission of transverse phonons from the solid into the liquid. Graner, Bowley and Nozières showed that the experimental value of R_k is quite consistent with a reasonable estimate of this transverse impedance at 0.32K. Eventually, Graner's experimental value, which seemed also contradictory to an earlier experiment described in ref.9, has recently been very well confirmed by a new experiment performed by Bossy and Castaing[22].

From these results and their analysis, one may draw the following picture of the process of shape relaxation of a helium 3 crystal around 0.32K. At precisely 0.32K, the crystal melts at the top and grows on its sides (fig.1). This induces a flow of mass since there is a small difference in density between the two phases. Most often, the quasiparticles are reflected when hitting the crystal lattice, giving rise to a certain amount of momentum exchange and dissipation. The crystal shape relaxes in less than a minute.

As soon as the latent heat is non zero, and whatever its sign, it is released in the liquid via a direct energy exchange among the quasiparticles. In the particular geometry of Graner's experiment, this heat has to be transfered from the top to the sides. The most efficient channel is a transmission to the transverse phonons in the solid through the Kapitza resistance of the interface. As a result, an additional source of surface dissipation appears, which controls the relaxation of the crystal shape, and it is already 4 times slower at 0.30 or 0.34K than at 0.32K. Around 0.1K, the relaxation would take about 3 days since the latent heat is large and the Kapitza resistance approximately varies[14] as T^{-3}. The transverse modes in liquid helium 3 play a significant role in the heat transmission through the solid-liquid interface.

Going down to lower temperatures, the thermal conductivity of liquid helium 3 increases as $1/T$, so that the latent heat should start being conducted through the liquid around 50 mK, and the relaxation time should decrease. Of course, in the sub-millikelvin region, where the liquid is superfluid and the solid antiferromagnetic, rough surfaces in helium 3 may become highly mobile like in helium 4, with spin waves possibly playing a role similar to that of phonons in helium 4.

Let us recall what was found in helium 4. It is well known that the excitations in superfluid helium 4 are phonons and rotons, and that transverse phonons in solid helium 4 have a smaller velocity than longitudinal phonons in the liquid. As explained by Bowley and Edwards[7], all dissipation coefficients are controlled by the reflection and transmission of these thermal excitations. When the temperature goes down from about 1.4K, the growth rate increases first exponentially, due to the disappearance of rotons, then as T^{-4} due to the one of phonons[2,7,10] (at low temperature, the main dissipation results from the reflection of transverse phonons on the solid side). The growth rate of rough surfaces in helium 4 is so high below 0.6K that deformations of the surface propagate as crystallization waves[10]: when shaking a crystal with a large rough horizontal surface, capillary waves appear as on the surface of water. This very high mobility allows the crystal surface to follow the oscillating pressure of a sound wave (even at thermal frequencies!) and the phonon transmission probability τ decreases as T^2, so that the Kapitza resistance varies as T^{-5}, an anomalous and remarquable property[15,16,17]. Finally, the cross term b being small, and solid helium 4 an even better thermal conductor than superfluid helium 4, the relaxation of rough crystal surfaces in helium 4 is fast and rarely far from isothermal[7,12].

We shall conclude with a last comparison between helium 3 and helium 4. The quantity $1 - b2/ac$ is close to 1 in helium 4, so that isothermal growth proceeds approximately at the same velocity as would adiabatic growth (remember that $k_E/k = 1 - b^2/ac$). We think that this is a consequence of the large similarity between excitations in solid and liquid helium 4. On the contrary, we have seen that the large difference between excitations in solid and liquid helium 3 induces a very different behavior of crystals. The quantity $1 - b^2/ac$ is now very small (3×10^{-4} at 0.32K) and the isothermal relaxation of crystals at 0.32K proceeds far more rapidly than at 0.1K where the situation is close to adiabatic. It would be, of course, very interesting to see if the recovery of a symetrical situation in helium 3 below 1 mK actually leads to growth properties similar to what was found in helium 4.

S.Balibar thanks W.F. Saam for many fruitful discussions and for his efficient role in the organisation of a very stimulating visit at The Ohio State University (sept.89 - march 90). D.O. Edwards wishes to acknowledge support from the National Science Foundation, Division of Material Research, under Low Temperature Physics Program grant DMR 8905385.

REFERENCES

1. F.Graner, S.Balibar and E.Rolley, J. Low Temp. Phys. 75, 69 (1989).
2. J.Bodensohn, K.Nicolai and P.Leiderer, Z. Phys. B64, 55 (1986).
3. F.Gallet, S.Balibar and E.Rolley, J. de Physique 48, 369 (1987).
4. E.Rolley, S.Balibar, F.Gallet, F.Graner and C.Guthmann, Europhys. Lett. 8, 523 (1989).
5. A.F.Andreev and A.Ya.Parshin, Zh. Eksp. Teor.Fiz. 75, 1511 (1978) (Sov. Phys. JETP 48, 763 (1978)).
6. B.Castaing and P.Nozières, J. de Physique 41, 701 (1980).
7. R.M.Bowley and D.O.Edwards, J. de Physique 44, 723 (1983).
8. A.F.Andreev and V.G.Khnizhnik, Zh. Eksp. Teor. Fiz. 83, 416 (1982) (Sov. Phys. JETP 56, 226 (1982)).
9. L.Puech, G.Bonfait and B.Castaing, J. Low Temp. Phys. 62, 315 (1986).
10. K.O.Kehishev, A.Ya.Parshin and A.B. Babkin, Zh. Eksp. Teor. Fiz. 80, 716 (1981) (Sov. Phys. JETP 53, 362 (1981)).
11. G.Agnolet, Jpn. J. Appl. Phys. 26, A79 (1987).
12. S.Balibar, D.O.Edwards and W.F.Saam, submitted to J. Low Temp. Phys. (march 1990).
13. E.R.Grilly, J. Low Temp. Phys. 4, 615 (1971).
14. F.Graner, R.M.Bowley and P.Nozières, To appear in J. Low Temp. Phys. (1990).
15. H.J.Maris and T.E.Huber, J. Low Temp. Phys. 48, 99 (1982).
16. L.Puech, B.Hebral, D.Thoulouze and B.Castaing, J. Physique Lettres 43, L.809 (1982).
17. P.E.Wolf, D.O.Edwards and S.Balibar, J. Low Temp. Phys. 51, 489 (1983).
18. D.O.Edwards, S.Balibar and P.E.Wolf, 75th Jubilee Conf. on Helium 4, Ed. J.G.M.Armitage, World Scientific (1983) p.70.
19. P.Nozières and M.Uwaha, J. de Physique 48, 389 (1987).
20. A.C.Anderson, J.I.Connolly, O.E.Vilches and J.C.Wheatley, Phys. Rev. 147, 90 (1966).
21. P.R.Roach and J.B.Ketterson, Phys. Rev. Lett. 36, 736 (1976).
    F.Dalfovo and S.Stringari, Il Nuovo Cimento 6D, 445 (1985).
22. J.Amrit and J.Bossy, Proc. of LT19, Physica B 165&166, 529 (1990).

ANISOTROPY OF SURFACE STIFFNESS, GROWTH KINETICS AND ROUGHENING TRANSITION IN ^{4}He

Konstantin Keshishev and Olga Andreeva

Institute for Physical Problems
USSR Academy of Sciences
Moscow, USSR

INTRODUCTION

The solid-superfluid interface in ^{4}He has attracted the attention of theoreticians and experimentalists during the last decade. Many essential problems concerning the thermodynamics and growth kinetics of the interface are still unsolved. The surface free energy α and the surface stiffness $\tilde{\alpha}_{ij} \equiv \alpha + +\partial^2\alpha/\partial\theta_i\partial\theta_j$ (i=1,2; θ_i are the angles which determine the crystallographic orientation of the surface) determine the equilibrium properties of the interface. The growth dynamics are characterized by the kinetic growth coefficient $K \equiv m\, v/\delta\mu$ (v is the growth rate along the normal to the surface $\vec{n}$, m is the ^{4}He atomic mass, and $\delta\mu$ is the difference between the chemical potentials of the two phases) which is also a function of the surface orientation θ_i.

Below we describe the direct measurement of the angular dependence of both quantities $\tilde{\alpha}$ and K. Our experimental method is based on the technique developed for the study of the spectrum and damping of crystallization waves[1]. In accordance with the theory[2] the spectrum is described by the following expression:

$$\omega^2 - \frac{\rho_1 |\vec{k}|}{(\rho_s - \rho_1)^2}\tilde{\alpha}_{ij}k_i k_j - \frac{\rho_1}{\rho_s - \rho_1} g\,|\vec{k}| + \frac{\rho_s \rho_1}{(\rho_s - \rho_1)^2}\frac{i\omega|\vec{k}|}{K} = 0\ , \qquad (1)$$

where ρ_s and ρ_1 are the densities of the solid and liquid phases respectively, ω and $\vec{k}$ are the frequency and the wave vector, g is gravity constant. The dispersion relation looks more simple in the plane of section determined by the vectors $\vec{n}$ and $\vec{k}$. In this case, using only the angle θ which fixes the orientation of $\vec{n}$ in the section and ignoring gravitation and damping, we have the following expression for an experimental determination of the surface stiffness:

$$\tilde{\alpha} \equiv \alpha + \partial^2\alpha/\partial\theta^2 = \frac{(\rho_s - \rho_1)^2}{\rho_1}\frac{\omega^2}{k^3}\ . \qquad (2)$$

The damping of the waves results primarily from the finite value of K. For the case of small damping ($\kappa \ll k$, where κ^{-1} is the damping length) the growth coefficient is given by:

$$K = \frac{1}{3}\frac{\rho_s\, \rho_1^{1/3}\omega^{1/3}}{(\rho_s - \rho_1)^{2/3}\tilde{\alpha}^{2/3}\kappa}\ . \qquad (3)$$

Excitations in Two-Dimensional and Three-Dimensional Quantum Fluids
Edited by A.G.F. Wyatt and H.J. Lauter, Plenum Press, New York, 1991

Using a method of exciting and detecting crystallization waves[1], one can record the configuration of the plane wave at a certain instant and thereby determine the wavelength and damping. Above 0.5 - 0.6 K the method cannot be used owing to the high roton damping.

EXPERIMENTAL METHOD

We used a ^3He-refrigerator with base temperature 0.32 K designed for optical measurements. In comparison with previous experiments[1] the essential changes were made in the design of the optical pressure cell. In the present case it consists of a cylindrical metal container (23 mm inner diameter, 29 mm long) and two optical windows in the ends (Fig.1). The axis of the cylinder runs horizontally and coincides with optical axis of the cryostat and of the overall apparatus. The flat copper base is in thermal contact with refrigerator through a cold finger, indium plated flange and a flexible thermal link (a stack of 15 μm copper foil stripes). The flexible helical capillary (0.2 mm o.d., 0.1 mm i.d.) connects the cell with the outer high pressure system. On the inner wall of the cell, placed symmetrically on the left and right are two wire capacitors for exciting the crystallization waves. The capacitor consists of two 30 μm diameter insulated wires bifilarly wound on a cylindrical fiberglass plate. To symmetrize the experimental conditions with respect to rotation of the cell around its axis, both capacitors are made in the form of cylindrical surfaces bounded by a sector of 120°. One more wire capacitor (a cylinder 2 mm diameter, 4 mm long) at the top of the cell is used as the source of a strong electric field to obtain oriented samples[1,3]. The cell is inserted in a mobile yoke mounted on a ball bearing. It can be rotated, along with the sample inside, through an angle of 60° in either direction from the horizontal position. The rotation is achieved by moving a kevlar thread by means of a small electromotor thermally anchored to the 1.2 K ^4He bath. Two reference marks (5 mm long pieces of a 20 μm diameter wire) are mounted on the outer surface of the front window. The marks are placed diametrically opposite and parallel to the bottom. They are intended for optical measuring of the cell turning angle. The angle was measured to an accuracy 3×10^{-3} rad by means of a telescope supplied with a goniometer measuring angles in the frontal plane.

In accordance with the design of the apparatus it is convenient to choose a Cartesian coordinate system so that the Z-axis runs vertically, the Y-axis coincides with the rotation axis and the origin is at the center of the cell. The X-axis then becomes horizontal.

During an experiment the crystalline sample usually fills the entire lower part of the cell. The longitudinal size of the equilibrium interface is 29 mm, while the transverse size is 20 mm. Both values considerably exceed the capillary length, 1.4 mm. Therefore the curvature of the meniscus except for 5 mm margins is rather small. The maximum deviation from horizontal for the central region is less than 3×10^{-2} rad. A dc voltage applied to the right and left capacitors reduces the contact angle. Thus the inclination of the surface in the XZ plane can be reduced by an order of magnitude. We assume that the interface is horizontal to this accuracy. An ac voltage of frequency 300-3000 Hz applied to one of the side capacitors excites a plane wave with vector $\vec{k}$ parallel to the X-axis. Equation (2) corresponds exactly to this case if the cell turning angle is taken as θ. When the cell is turned through an angle θ the interface relaxes to the horizontal corresponding to the minimum energy in the gravitational field in several seconds. This spontaneous process is the result of melting of certain parts of the crystal and of growing of other parts, while the orientation of the crystal lattice with respect to the cell remains completely fixed.

In the preliminary stage of every experimental run a number of small faceted crystals were dropped to the flat bottom and then were melted until a

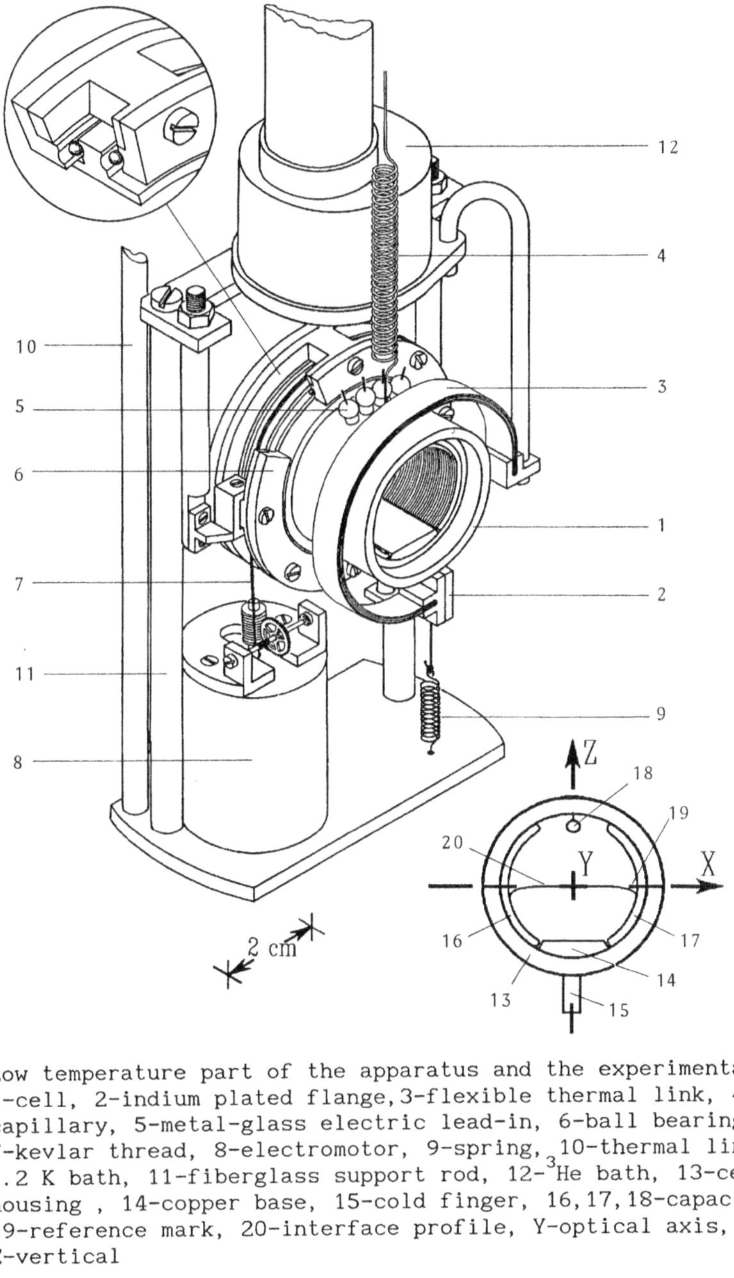

Fig. 1. Low temperature part of the apparatus and the experimental cell. 1-cell, 2-indium plated flange, 3-flexible thermal link, 4-filling capillary, 5-metal-glass electric lead-in, 6-ball bearing, 7-kevlar thread, 8-electromotor, 9-spring, 10-thermal link to 1.2 K bath, 11-fiberglass support rod, 12-^{3}He bath, 13-cell housing , 14-copper base, 15-cold finger, 16, 17, 18-capacitors, 19-reference mark, 20-interface profile, Y-optical axis, Z-vertical

sample of the desired orientation was grown. The small inclination of the normal to the facet with respect to the vertical in the XZ-plane was measured optically. A laser beam falling at glancing angle to the facet was simultaneously reflected from the facet and from a horizontal mirror. The angle between the two reflected beams was measured with accuracy 5×10^{-4}rad. The orientation of samples in the horizontal plane was estimated rather approximately. The error was 1°-5° from sample to sample depending on sizes and orientation of the lateral facets of the seed lying on the bottom. The method described was employed for the study of surface stiffness and growth coefficient anisotropy[4,5].

DETERMINATION OF SINGULAR ORIENTATIONS

In hcp ^{4}He all orientations except three are rough. This fact is confirmed by existence of crystallization waves down to 30 mK[(6)]. Besides that, three roughening transitions were observed[1,7,8] at temperature $T_{R1} = 1.28$ K, $T_{R2} = 1$ K and $T_{R3} = 0.36$ K. While cooling, a facet of new crystallographic orientation appears at the corresponding temperature. Comparing the hcp structure symmetry with peculiarities of the growing crystals one can uniquely identify the basis of a hexagonal prism ($T_{R1} = 1.28$ K) as the (0001) plane. In the case of the lateral facet ($T_{R2} = 1$ K) one of two possibilities (either $(10\bar{1}0)$ or $(11\bar{2}0)$) is realized. The third facet ($T_{R3} = 0.36$ K) can be identified as either the $(10\bar{1}0)$ or $(11\bar{2}0)$ plane, depending on whether the lateral facet is $(10\bar{1}0)$ or $(11\bar{2}0)$. In ideal hcp structure the angle between the [0001] and $[10\bar{1}0]$ directions is equal to $\mathrm{arctg}(4\sqrt{2/3}) \cong 62,1^\circ$, while it is $\mathrm{arctg}(2\sqrt{2/3}) \cong 58.5^\circ$ for the [0001] and $[11\bar{2}0]$ pair. According to Wolf et al.[8] the angle between the (0001) plane and the third facet is $60^\circ\pm3^\circ$, so the accuracy is insufficient for the unique identification of the second and third transitions. Two years later Wolf et al.[9] using indirect arguments drew a conclusion in favor of the $(10\bar{1}0)$ and $(10\bar{1}1)$ planes.

We have taken the opportunity to determine directly the Miller indices, corresponding to the second and third roughening transitions. One of the samples was fortunately dropped so that the deviations of the normals $\vec{n}_1, \vec{n}_2, \vec{n}_3$ to the first, second and third facet from the XZ plane were $+6\times10^{-3}$rad, -5×10^{-3}rad and -2×10^{-3}rad respectively. The measurement was made at a temperature of 0.33 K as stated above. It was noticed[8] that the size of the third facet in growth, and moreover in equilibrium, is very small. That is actually so if the facet is oriented arbitrarily. Growing the crystal so that it occupied the lower half of the cell and using rotation, we oriented the facet horizontally in the XZ plane. In this case in the regime of slow growth the facet spreads over the whole interface, and its orientation with respect to the cell and gravity was fixed. The same procedure was followed with the (0001) facet. Unfortunately the minimum inclination of $\vec{n}_2$ from the vertical in XZ plane, corresponding to the upper limit of the cell turn, was 12.6°. Therefore the above value -5×10^{-3}rad is the result of calculation. For the angle of interest between the (0001) facet and the third one we obtained a value of $62.0\pm0.2^\circ$. Thus we determined the Miller indices corresponding to the second and third roughening transitions. They are in fact $(10\bar{1}0)$ and $(10\bar{1}1)$ in accordance with that declared by Wolf et al[9].

SURFACE STIFFNESS ANISOTROPY

Using the same favorable sample we measured the angular dependence of the surface stiffness $\tilde{\alpha}(\theta)$ in the $(11\bar{2}0)$ section at a temperature of 0.33 K. These data are shown in Fig.2. The angle θ is taken from the [0001] direction and the value $\theta = 62.1^\circ$ corresponds to the $[10\bar{1}1]$ direction. In Fig.2 the smoothed data for two other sections passing through the c-axis are also shown. One of the sections (dashed line) coincides within 1° with the $(10\bar{1}0)$ plane, the other (dotted line) crosses the $[10\bar{1}0]$ direction at $\varphi \approx 10^\circ$(see insert in Fig.2). These data were obtained at 0.4 K on two other crystals.

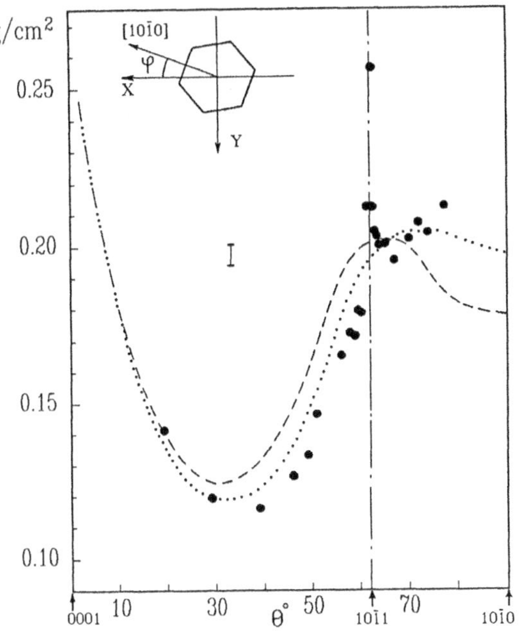

Fig.2. Angular dependence of surface stiffness in different sections. θ-angle between surface normal and [0001] direction, φ (insert)-angle between section plane XZ and [10$\bar{1}$0] direction; solid circles-φ=0, dotted line-φ=10°, dashed line-φ=30°, arrow under abscissa, marking [10$\bar{1}$1] direction, relates to φ=0 section only.

Shown in the center, the error bar gives the scatter in the experimental points. So far as no azimuth anisotropy near c-axis is observed, we have measured the stiffness in the (11$\bar{2}$0) section starting from θ = 19.3°. The limiting factor at θ = 77.4° was discussed above. In Fig.2 besides the significant general anisotropy of surface stiffness (which is very similar to the behaviour of $\tilde{\alpha}(\theta)$ in other sections passing through the c-axis), a singularity at θ = 62° is observed. According to previous data[3,4,5], a sharp increase of $\tilde{\alpha}$ in the vicinity of the (10$\bar{1}$1) facet is the characteristic feature of the atomically smooth state. In comparison with the (0001) and (10$\bar{1}$0) facets[5], the anomalous behaviour is concentrated in a significantly narrower angular range of about ±5°. This remark also relates to the growth coefficient angular dependence (see next section). For this reason we could approach to within 0.2° of the [10$\bar{1}$1] direction, much closer than to the [0001] and [10$\bar{1}$0] directions. All data shown in Fig.2 (see also Fig.1[(5)]) correspond to samples, whose c-axis (or another singular direction) deviate from the examined section within several milliradians. In all these cases we neglect the slight deviation and, using equation (2), we find a diagonal component of the $\tilde{\alpha}_{ij}$ tensor. Moreover, in the case when the section coincides with a crystallographic symmetry plane we measure one of two principal components of the tensor.

GROWTH COEFFICIENT ANISOTROPY

Recently we reported data[5] concerning the angular dependence of the growth coefficient for the sections passing through the [0001] and [10$\bar{1}$0] axes. In the case discussed in the previous section we used the opportunity to examine the (11$\bar{2}$0) section. The dependence K(θ) for this section at 0.33K is shown in Fig.3 (solid circles). For the sake of completeness we also present here the data[5] concerning the vicinity of the [0001] and [10$\bar{1}$0]

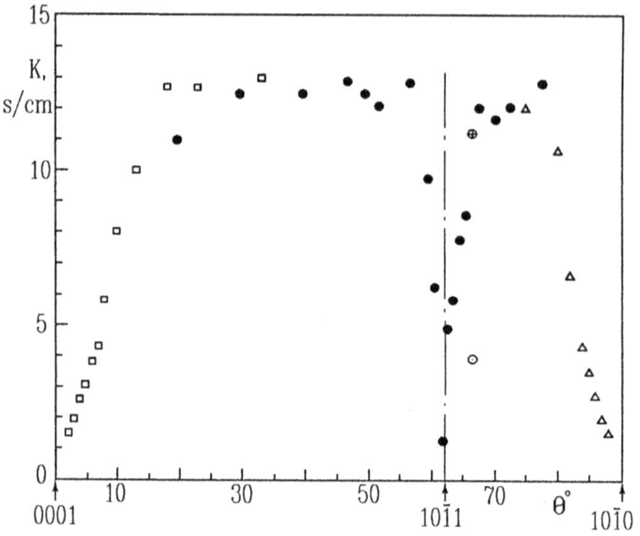

Fig.3. Angular dependence of growth coefficient in different sections. Section (11$\bar{2}$0): solid circles-T=0.33K, open circle-T=0.43K, crossed circle-K(T=0.43K)×(0.43/0.33)^4. Another section running through: [0001]-squares, [10$\bar{1}$0]-triangles.

directions in other sections. In all the cases the general feature is a linear decrease of K as θ tends to a singular orientation. The only quantitative difference is the narrowness of this region around the [10$\bar{1}$1] direction.

Assuming that the roughening temperature of the (10$\bar{1}$1) facet is equal to 0.365 K[8], we tried to follow the change of K(θ) caused by heating and by transformation of the facet to the rough state. We supposed that the growth coefficient, being equal to zero at the smooth facet, should increase sharply at the transition point up to a specific rough value. The necessity of such disappearance of the K(θ) singularity is rather obvious and has been proved[9] by measuring the (0001) facet mobility near T_{R1}. For this reason we have measured the K(T) dependence in the temperature interval 0.33K<T<0.43K at two fixed orientations: θ = 62.4° and θ = 66.4°, corresponding to the beginning and to the end of the K(θ) linear behaviour. In spite of our expectations no indication of transition was observed up to 0.43 K. At the θ = 62.4° orientation after a slight heating the growth coefficient fell preventing measurement and remained zero within our sensitivity up to 0.43 K. At the same time, while heating, the K(θ=66.4°) value monotonically decreased in accordance with the known temperature dependence $K \propto T^{-4}$ [1,2]. Two points in Fig.3 illustrate this result: the lower point (open circle) is the K value measured at 0.43 K, the upper (crossed circle) corresponds to the normalized value K(T=0.43K)×(0.43/0.33)^4. Unfortunately we were limited to 0.43 K because of the large damping. Thus our measurements show that the smooth state of the (10$\bar{1}$1) orientation lasts at least up to 0.43 K.

The growth coefficient of the vicinal surfaces as a function of temperature and of angle θ was considered by Nozieres and Uwaha[10]. Our data concerning K(θ) dependence have a similarity to their prediction. We examined also the temperature dependence of the growth coefficient at the angles corresponding to the $K \propto \theta$ regime. Two cases were studied: (i) θ = 5.2° near [10$\bar{1}$0] in the interval 0.34K<T<0.44K and (ii) θ = 4.7° near [0001], 0.33K<T<0.43K. In both cases a $K \propto T^{-4}$ dependence was observed. According to the theory[10] the growth coefficient should be proportional to either T^{-3} or T^{-4}, depending on the relation between the parameters of surface structure and of the liquid excitations. However, one cannot insist on the agreement of the theory and experiment. The fact is that the $K \propto \theta$ regime is observed in the angular region where, judging from $\tilde{\alpha}(\theta)$ (Fig.1), steps interaction is strong, not weak as assumed by the theory.

DISCUSSION

Most of modern theoretical studies of the roughening transition are based both on exactly solvable models and on renormalization group analysis, that give a number of universal predictions, interpreting the roughening transition in the frame of Kosterlitz-Thouless theory. One of these predictions is the universal relation[11] between the surface stiffness $\tilde{\alpha}_{ij}$, roughening temperature T_R and the periodicity of the crystalline planes along the normal to the surface d:

$$k_B T_R = \frac{2}{\pi} (\det \tilde{\alpha}_{ij})^{1/2} d. \tag{4}$$

Hitherto it was assumed[9,12,13] that this relation had been proved for all three roughening transitions in ^{4}He. The widely developed renormalization group theory[9,14] predicts, in particular, a linear angular dependence of surface stiffness in the vicinity of singular directions. However, the monotonically increasing $\tilde{\alpha}(\theta)$ when approaching the [0001] direction (Fig.1) or the [10$\bar{1}$0] direction[5] to within an angle of $\approx 3\times10^{-2}$rad at 0.4 K (low enough in comparison with T_{R1} or T_{R2}) is itself a strong argument against the theory. Since no temperature dependence of the stiffness is observed[1,3,5,9], the previous statement is valid for the [10$\bar{1}$1] direction (Fig.1), but the minimum angle is 6×10^{-3}rad in this case. The assumption that the $\alpha \propto \theta$ law should follow to smaller angles, meets a serious objection, because in this case the ratio d/ξ (ξ is correlation length), which determines the critical region, appears to be anomalously small even at $T \ll T_R$.

Nevertheless, let us accept this version and turn to the experiment. In Fig.4 the smoothed angular dependences of the diagonal components of the $\tilde{\alpha}_{ij}$ tensor are represented in isometric projection. We used here our previous[5] and new (Fig.1) data and that of Parshin and co-workers[3] obtained by quite another method at T = 1.22 K. The orientation of the surface normal $\vec{n}$ is determined now by the polar angle θ and by the longitude φ in a coordinate system related to crystallographic axes (insert in Fig.3). In spite of the incompleteness this plot clearly demonstrates the general features of the stiffness anisotropy. Close to the [0001] direction, where $\tilde{\alpha}_{ij}$ is independent of φ, $\tilde{\alpha}_{11}$ being equal to $\tilde{\alpha}_{22}$ rises as high as 0.33÷0.36 erg/cm^2 at $\theta = 5\times10^{-3}$rad. In the framework of our assumption this value should obviously be considered as the lower bound of the corresponding component value of $\tilde{\alpha}_{ij}$ (θ=0, T=T_R). Similar estimations of $\tilde{\alpha}_{ij}$ (θ_i=0, T=T_R) should be accepted for all singular facets. The interplane distance corresponding to the molar volume 20.983 cm^3/mole [15] is d = 2.99 Å. Thus, using relation (4), one gets an estimate $T_{R1} \geq 1.36$K. In the vicinity of the second facet only $\tilde{\alpha}_{22}$(θ = = 90°, φ = 1.5°) = 0.36 erg/cm^2 is measured close enough to the [10$\bar{1}$0] direction. The nearest experimental value of the second component is $\tilde{\alpha}_{ij}$ (θ=77.4°, φ=0) = 0.21 erg/cm^2. Taking into account the behaviour of both $\tilde{\alpha}_{11}$ and $\tilde{\alpha}_{22}$ in the vicinity of every smooth orientation, it is natural to extrapolate the $\tilde{\alpha}_{11}$ (θ, φ=0) dependence, say, up to θ = 88° (dashed line in Fig.4). It gives $\tilde{\alpha}_{11}$(θ = 88°, φ=0) = 0.25 erg/cm^2. The translation period along the [10$\bar{1}$0] direction is d = 3.18 Å, and in accordance with (4) we get the following estimate: $T_{R2} \geq$ 1.40 K (whereas the latest experimental data prove to be 0.95÷1.05K for T_{R2}). Near the [10$\bar{1}$1] orientation only one of the two principal components $\tilde{\alpha}_{11}$(θ=62.1° ± 0.3°, φ=0) = 0.22 erg/cm^2 is measured. However one can use rather a moderate and realistic estimate: $\tilde{\alpha}_{22}$ (θ=62.1°, $\varphi \to 0$) ≅ ≅ 0.2 erg/cm^2. The translation period along the [10$\bar{1}$1] direction is d=2.80Å. The respective theoretical value T_{R3} = 0.76 K looks also rather pessimistic even if the new information about the (10$\bar{1}$1) transition, discussed above is taken into account.

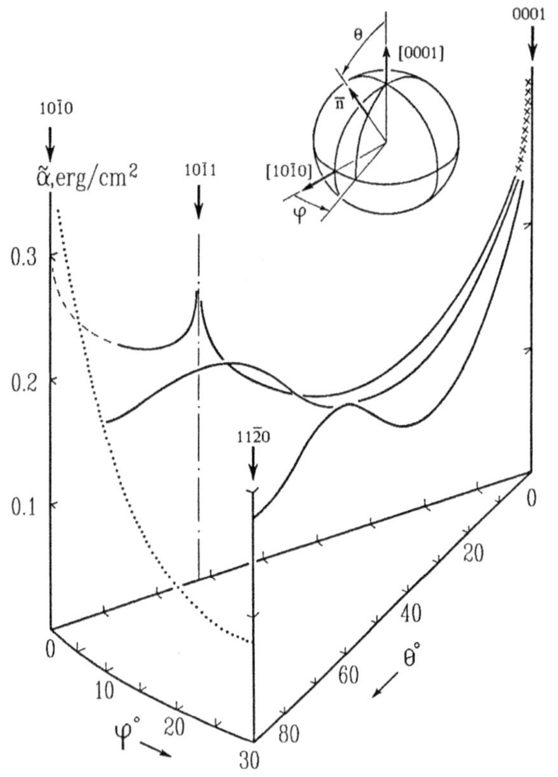

Fig. 4. Diagonal components of surface stiffness tensor as a function of surface orientation θ and φ (see insert). Solid lines-$\tilde{\alpha}_{11}(\theta, \varphi = \mathrm{const})$ in sections $\varphi = 0, 10^\circ, 30^\circ$; dashed line-extrapolation $\tilde{\alpha}_{11}$ ($\varphi = 0$, $\theta \to 90^\circ$); crosses-$\tilde{\alpha}_{11}$ (θ, φ = const) at T=1.22K[3]; dotted line-$\tilde{\alpha}_{22}$ (φ, $\theta = 90^\circ$)[5].

Thus there is a strong contradiction between equation (4) and the experimental data. According to this equation the next "high temperature" candidate for the roughening transition is the (11$\bar{2}$0) plane having d = 1.83 Å. Measured at 0.4 K the both principal components $\tilde{\alpha}_{11} = 0.18$ erg/cm^2 and $\tilde{\alpha}_{22} = 0.08$ erg/cm^2 are measured exactly at that orientation. The roughening temperature determined by relation (4) is 0.18 K. A direct verification of this prediction would finally solve the problem. In conclusion one should emphasize that the experimental situation with roughening transition in ^{4}He that looked very clear just a year ago, looks now rather controversial.

REFERENCES

1. K. O. Keshishev, A. Ya. Parshin, and A. V. Babkin, Sov.Phys.JETP 53:362(1981)
2. A. F. Andreev, and A. Ya. Parshin, Sov.Phys.JETP 48:763(1978)
3. A. V. Babkin, D. B. Kopeliovich, and A. Ya. Parshin, Sov.Phys.JETP 62:1322(1985)
4. O. A. Andreeva and K. O. Keshishev, JETP Lett.46:200(1987)
5. O. A. Andreeva, K. O. Keshishev, and S. Yu. Osip'yan, JETP Lett.49:756(1989)

6. G. Agnolet, D. D. Osheroff, and W. O. Sprenger, Jpn.J.Appl.Phys. 26:355(1987)
7. J. E. Avron, L. S. Balfour, C. G. Kuper, J. Landau, S. G. Lipson, and L. S. Shulman, Phys.Rev.Lett 45:814(1980)
8. P. E. Wolf, S. Balibar, and F. Gallet, Phys.Rev.Lett. 51:1366(1983)
9. P. E. Wolf, F. Gallet, S. Balibar, E. Rolley, and P. Nozieres, J.Physique 46:1987(1985)
10. P. Nozieres and M. Uwaha, J.Physique 48:389(1987)
11. D. S. Fisher and J. D. Weeks, Phys.Rev.Lett.50:1077(1985)
12. S. Balibar and B. Castaing, Surf.Sci.Reports 5:87(1985)
13. S. Balibar, F. Gallet, and E. Rolley, Physica Scripta T29:265(1989)
14. P. Nozieres and F. Gallet, J.Physique 48:353(1987)
15. E. R. Grilly, J.Low Temp.Phys.11:33(1973)

LIQUID-SOLID ^{4}He INTERFACE: EQUILIBRIUM SHAPE AND EXITATIONS

Alexander Andreev

Institute for Physical Problems

USSR Academy of Sciences

Moscow, USSR

INTRODUCTION

In the past decade, extensive studies, both theoretical and experimental, are made on the phenomena occurring at the phase boundary between solid and liquid ^{4}He. The neighbourhood of a quantum crystal with a superfluid quantum liquid makes the properties of the interface very unusual.

It was argued[1] that in view of the large amplitude of quantum fluctuations, nearly all the faces of a ^{4}He crystal must be in the rough state, even at zero temperature. Such a quantum-rough state of the interface is characterized by a non-dissipative processes of crystallization and melting. The most interesting consequence of the non-dissipative crystallization is the existence of weakly damped crystallization waves[1] which have been discovered experimentally in ^{4}He by Keshishev et al.[2] The experimentally observed spectrum of the crystallization waves and the temperature dependence of their attenuation agree[2] with that predicted [1,3,4]. Another consequence of the non-dissipative crystallization is the anomalous temperature dependence of the Kapitza resistance at the liquid-solid ^{4}He interface predicted by Marchenko and Parshin[5]. The anomalous Kapitza resistance has been also observed experimentally[6-9].

However, Fisher and Weeks[10] argued, quite generally, that all faces must be in the smooth state at sufficiently low temperature, whether the crystal is quantum or classical (see also[11]). This means each type of face has its

Excitations in Two-Dimensional and Three-Dimensional Quantum Fluids
Edited by A.G.F. Wyatt and H.J. Lauter, Plenum Press, New York, 1991

own roughening transition from a low temperature smooth state to a high temperature rough one at a critical temperature T_R . Thus the theory[10,11] predicts an infinite sequence of roughening temperatures. So far, three transitions have been observed experimentally on the surfaces of a ^{4}He crystal, taking place at temperatures of 1.2, 0.9 and 0.35 K [14-16,2]. No new transitions are observed on lowering the temperature further to 0.07 K[16].

Both renormalization group analyses and results for exactly solvable models show that the roughening transition is of Kosterlitz-Thouless type, i.e. it is an infinite-order transition with a very weak singularity in the free energy (see the review of van Beijeren and Nolden[17]). An important result of the theory of the roughening transition is a simple relation[10,18]

$$T_R = (2/\pi)a^2\gamma^{1/2} \qquad (1)$$

between the roughening temperature of a crystal face and the surface stiffness tensor γ_{ik} , $i,k = 1,2$, of a rough state of the surface right at the transition temperature. In the above, a is the distance between equivalent lattice planes parallel to the surface, and $\gamma = det(\gamma_{ik})$. Experiments performed to clarify the nature of the roughening transition were both for[19] and against[20,21] the Kosterlitz-Thouless picture. Recent experiments by Keshishev[22] on the anisotropy of both the surface stiffness tensor and the growth coefficient in ^{4}He crystals, certainly contradict the Kosterlitz-Thouless picture and, in particular, the relation (1).

The aim of this paper is to review the present contradictory state of our understanding of the phenomena occuring at liquid-solid ^{4}He interfaces and to suggest a point of view to clarify it.

QUANTUM ROUGHNESS AND RESONATING-VALENCE-BOND STATE

The above mentioned problem of quantum roughness of a liquid-solid ^{4}He interface, is a subject of great interest. Recently Levitov[23] showed that the interface roughening problem for a quantum crystal can be mapped exactly into the problem of the short-range resonating-valence-bond (RVB) system as realized by a quantum hard-core dimer gas on a two-dimentional lattice.

Anderson[24] proposed that La_2CuO_4 is in a RVB state, and that the high-T_c superconductivity observed in the material when doped with Ba or Sr must be understood in terms of this unusual insulating state. "Short-range" RVB state

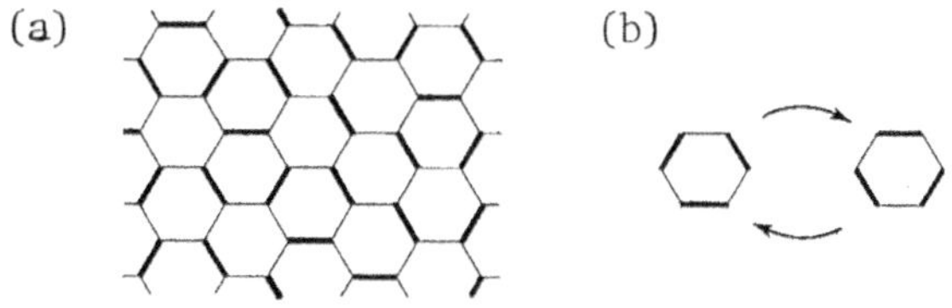

Fig. 1. (a) The resonating valence-bond (RVB) state is a coherent superposition of states like the one shown here. Lines denote strong valence bonds, across which the electrons form singlet pairs; each site is strongly bonded to precisely one neighbour. (b) An elementary resonance (elementary process of tunneling motion) between two valence-bond configurations.

was considered by Kivelson et al.[25]. A prototype for the short-range RVB state is a quantum hard-core gas of dimers depicted in Fig. 1 for a hexagonal lattice. Heuristically, valence bonds can be viewed as real-space Cooper pairs which repel one another, a joint effect of the Pauly principle and the Coulomb interaction.

Fig.2 shows the mapping of dimers on a hexagonal lattice to a surface of a cubic crystal. It is clear from Fig.2 that the elementary process of tunneling motion shown in Fig. 1b is identified in the constructed mapping with a process of a cube sticking to (or removing from) the surface, i.e. with an elementary process of crystallization (or melting).

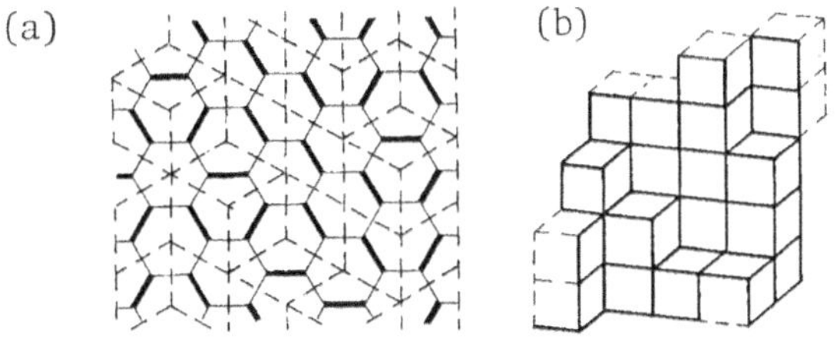

Fig. 2. Mapping of dimers to a surface of a cubic crystal. (a)Covering dimers by 60° rhombi one get a picture of a surface of a crystal shown separetely in (b). The broken lines in (b) show faces of cubes which are either invisible or absent in the tiling (a).

The quantum dimer gas was discussed[25] as a candidate for a system on a lattice having a quantum-liquid ground state[24]. Two main questions to be answered are the following: (a) Is the ground state ordered (crystalline) or disordered (liquid)? (b) What is the nature of low-energy excitations? In the constructed mapping the crystalline and the liquid states of the dimer gas are identified with the smooth and the rough states of the corresponding crystal surface. Low-energy excitations of the dimer gas ("resonons", see[25]) are identified with the crystallization waves. It is clear that the presence of holes (vertices not covered by dimers) in doped states of the dimer gas, stabilizes a liquid state even if it is unstable for undoped (close-packed) system. Levitov showed[23] that holes in the 3D crystal picture correspond to screw dislocations which pierce the surface. Thus in the presence of a sufficiently large amount of dislocations the surface can favor a rough state. The importance of dislocations in the roughening problem was already discussed[17].

FLUCTUATIONS OF THE INTERFACE

The question about thermal and quantum fluctuations of the interface plays the key role in the roughening problem[10,11]. Consider the height-height correlation function

$$K(\underline{R}) = \langle [z(\underline{R})-z(0)]^2 \rangle \tag{2}$$

where the interface is described in cartesian coordinates, $z=z(\underline{r})$, $\underline{r}=(x,y)$. $K\ \underline{R})$ describes the mean square height difference between two points on the interface separated by the distance $R=|\underline{R}|$, or it describes the mean square thickness of an interface of characteristic size R. The interface of any rational slope is smooth or rough depending on whether $K(\infty)$ is finite or infinite. Any rational interface of finite thickness will certainly be localized by the lattice periodic potential. Irrational interfaces are always rough.

Let us repeat well-known arguments to calculate the thickness of a rough interface at nonzero temperature due to fluctuative crystallization-melting processes. Consider first a planar interface $z(\underline{r})=const$. Due to thermal fluctuations it is displaced, by an amount $\delta z(\underline{r})$. The corresponding energy cost is

$$U = (1/2)\int dxdy\ \gamma_{ik}(\partial z/\partial x_i)(\partial z/\partial x_k). \tag{3}$$

Expanding $\delta z(\underline{r})$ in a Fourier series, substituting it in (3), and assigning an energy $T/2$ to each mode, we find

$$\langle(\delta z)^2\rangle = (T/4\pi^2)\int d^2k(\gamma_{ik}k_ik_k)^{-1} \tag{4}$$

The integral is divergent both at large and small k. We cut it off at $k_{min} \simeq R^{-1}$ where R is the size of the interface, and at $k_{max} \simeq 1/a$. We have

$$K(R) \simeq \langle(\delta z)^2\rangle \simeq (T/\gamma)\, \log(R/a). \tag{5}$$

The logarithmic divergence of the correlation function $K(R)$ at $R \to \infty$ is the basis for the Kosterlitz-Thouless picture of the roughening transition. Similar calculations of the thickness of a presumed rough interface at $T=0$ due to zero-point crystallization waves give finite result at $R \to \infty$[10]. This means that a rough state of a rational interface is impossible at $T=0$. The quantum roughness is realized at $T=0$ for irrational interfaces.

However, we argue below that there is a different mechanism of the interface fluctuations which is characterized at nonzero temperature by a correlation function $K(R)$ proportional to R (crystallization-melting fluctuations give such result for 1D interfaces). Due to such fluctuations the interface is rough at any nonzero temperature. This rules out the Kosterlitz-Thouless picture of the roughening transition. It is important to note that, generally, the displacement δz of the interface is a sum

$$\delta z = \zeta + u \tag{6}$$

of a "growth" displacement ζ and a "strain" displacement u. To raise the interface one can either grow some more crystal at constant strain, or strain the crystal. In the roughening the strain displacement u is irrelevent since there is no displacement of the interface relative to the lattice periodic potential. Consider fluctuations of the interface due to thermal defects of the type shown in Fig.3. The defect is similar to the Frank-Read source known in the crystal growth theory[11]. It consists of a step of a finite length l on the crystal surface and a semicircle dislocation in the bulk. The Burgers vector of the dislocation is normal to the surface, so the dislocation is screw near its ends. Along the line AB with A and B far from the defect, the change of the growth displacement is $\Delta\zeta=-a$, and the strain displacement is $\Delta u=+a$. The change of the total displacement (6) is zero.

Let O and $\underline{R}$ be two points at the interface separated by a macroscopi-

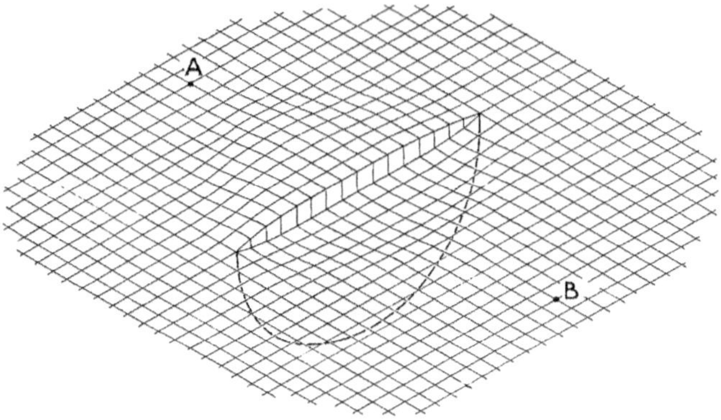

Fig. 3. The basic defect on the crystal surface consisting of a step of a finite length and a semicircle dislocation.

cally large distance R. Consider intersections of a line connected our points with defects of the type described above. The growth height difference $\zeta(\underline{R})-\zeta(0)$ is $a(N_+-N_-)$ where N_+ and N_- are the numbers of "positive" and "negative" intersections. Since each defect has a finite energy ε and a step length l, the average numbers of the intersections are of order

$$\langle N_+\rangle = \langle N_-\rangle = N \simeq (R/l)\ exp(-\varepsilon/T). \tag{7}$$

The height-height correlation function is

$$\langle[\zeta(\underline{R})-\zeta(0)]^2\rangle \simeq a^2\langle(\delta N)^2\rangle \simeq a^2 N \simeq R\ (a^2/l)\ exp(-\varepsilon/T). \tag{8}$$

It is proportional to R. Such fluctuations destroy the smooth state. All crystal surfaces must be rough at any nonzero temperature.

NATURE OF THE SURFACE TRANSITIONS

The preceding result shows that three surface transitions observed experimentally in ^{4}He at temperatures of 1.2, 0.9 and 0.35 K, are not roughening transitions of the Kosterlitz-Thouless type. Any of three interfaces is rough both above and below its transition temperature. The equilibrium faceting do not arise at all at any nonzero temperature. Below the transition temperature both the size of the equilibrium facet and the magnitude of the cusp in the angular dependence of the surface energy are zero. This is in accordance with the experiments[20-22]. All attempts to measure the facet size[20,21] or the magnitude of the surface energy cusp[22] failed. Only an upper limit of the cusp magnitude was established which is very small[22].

It was shown experimentally[22] that all three transitions are of the same character. What is the nature of transitions? The following picture seems to be natural. Below a transition temperature, a singular orientation of the interface appears. Namely, the angular dependence of the surface stiffness $\gamma(\theta)$ for vicinal surfaces $(\theta \to 0)$ is[22] either of a power-law type:

$$\gamma(\theta) = A\, \theta^{-\alpha} \tag{9}$$

with a small exponent $\alpha \simeq 0.2$, or logarithmic:

$$\gamma(\theta) = B\, \log(\theta_0/\theta) \tag{10}$$

where A, B and θ_0 are positive constants. Above the transition temperature the singularity in $\gamma(\theta)$ disappears.

At the transition temperature, there occurs a change in the growth characteristics $v(\Delta\mu)$ where v is the interface growth velocity and $\Delta\mu$ is the difference in chemical potential across the interface. The growth characteristics was measured by Balibar and his group[19]. Above the transition the growth velocity is proportional to $\Delta\mu$ at small $\Delta\mu$:

$$v = K\, \Delta\mu \tag{11}$$

where K is a growth coefficient. Below the transition the growth coefficient of the singular surface is zero. The angular dependence of the growth coefficient for vicinal surfaces is linear[22]:

$$K(\theta) = K_0\, |\theta| \tag{12}$$

where K_0 is a constant. The growth characteristics of a presumed smooth interface would be exponential at small $\Delta\mu$[19]:

$$v \propto \exp[-C(\beta^2/T\Delta\mu)] \tag{13}$$

where C is a positive constant and β is the step energy. Since $\beta=0$ for rough interfaces, it seems natural to suggest that the growth characteristics of

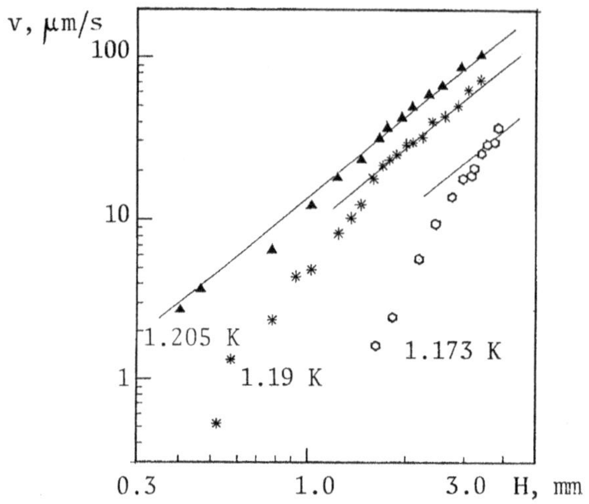

Fig. 4. Logarithmic plot of the growth characteristics of the (0001) surface at temperatures T = 1.205, 1.19, 1.173 K below the transition temperature, measured in Ref. 19. Here $H=[\rho_s/(\rho_s-\rho_l)g]\Delta\mu$ is the level difference and ρ_l, ρ_s are the liquid and solid densities.

the singular surface is of a power-law type:

$$v = L\ (\Delta\mu)^{\delta} \tag{14}$$

where L is a constant and $\delta>1$.

Fig. 4 shows logarithmic plot of experimental results[19] on the growth velocity of the (0001) surface below the transition temperature. Solid lines in Fig. 4 correspond to eq. (14) with $\delta\simeq5/3$ which is temperature independent. At very small $\Delta\mu$, the growth characteristics is of an exponential type. One could explain that by a pinning of the interface by defects of the crystal lattice. The pinning should be important at small enough $\Delta\mu$. It is well known that the kinetic faceting results from the anisotropy of the growth velocity of eq. (12). According to our picture, it was the kinetic faceting which manifested itself in all experiments. The relaxation time of the kinetic faceting can be very long due to the pinning mentioned above. Note that such a possibility was mentioned by Keshishev et al.[20] to explain the failure of their measurements of the equilibrium facet size.

Thus a problem arizes to explain the critical behaviour determined by eqs. (9), (10), (12) and (14) theoretically.

REFERENCES

1. A. F. Andreev and A. Ya. Parshin, Sov.Phys.JETP 48:763 (1978).
2. K. O. Keshishev, A. Ya. Parshin, and A. V. Babkin, JETP Lett. 30:56 (1979), and Sov.Phys.JETP 53:362 (1981).
3. A. F. Andreev and V. G. Knizhnik, Sov.Phys.JETP 56:226 (1980).
4. R. M. Bowley and D. O. Edwards, J.Phys.(Paris) 44:723 (1983).
5. V. I. Marchenko and A. Ya. Parshin, JETP Lett. 31:724 (1980).
6. T. E. Huber and H. J. Maris, Phys.Rev.Lett. 47:1907 (1981).
7. B. Castaing, G. Bonfait, and D. Thoulouze, Physica 109&110 B+C: 2093 (1982).
8. L. Puech, B. Hebral, D. Thoulouze, and B. Castaing, J.Phys. Lett.(Paris) 43:809 (1982).
9. P. E. Wolf, D. O. Edwards, and S. Balibar, J.Low Temp.Phys. 51: 489 (1983).
10. D. S. Fisher and J. D. Weeks, Phys.Rev.Lett. 50:1077 (1983).
11. P. Nozieres, Shape and Growth of Crystals, Lectures at the Beg-Rohu Summer School, August 1989.
12. E. Fradkin, Phys.Rev. B28:5338 (1983).
13. S. V. Iordanskii and S. E. Korshunov, JETP Lett. 39:128 (1983), and Sov.Phys.JETP 60:528 (1984).
14. J. E. Avron, L. S. Balfour, C. G. Kuper, J. Landau, S. G. Lipson, and L. S. Schulman, Phys.Rev.Lett. 45:814 (1980).
15. S. Balibar and B. Castaing, J.Phys.Lett.(Paris) 41:329 (1980).
16. P. E. Wolf, S. Balibar, and F. Gallet, Phys.Rev.Lett. 51:1366 (1983).
17. H. van Beijeren and I. Nolden, Roughening Transition, Preprint, Institut fuer Theoretische Physik, Rheinisch-Westfaelische Technische Hochschule, Templergraben 55, 5100 Aachen-FRG, (1989).
18. C. Jayaprakash, W. F. Saam, and S. Teitel, Phys.Rev.Lett. 50: 2017 (1983).
19. P. E. Wolf, F. Gallet, S. Balibar, E. Rolley, and P. Nozieres, J.Phys.(Paris) 46:1987 (1985).
20. K. O. Keshishev, A. Ya. Parshin, and A. I. Shal'nikov, Soviet Scientific Review, Section A: Physics Reviews, I. M. Khalatnikov, ed., 4:155 (1982).

21. A. V. Babkin, K. O. Keshishev, D. B. Kopeliovich, and A. Ya. Parshin, JETP Lett. 39:633 (1984).
22. K. Keshishev and O. Andreeva, This workshop.
23. L. S. Levitov, Phys.Rev.Lett. 64:92 (1990).
24. P. W. Anderson, Science, 235:1196 (1987).
25. S. A. Kivelson, D. S. Rokhsar, and J. P. Sethna, Phys.Rev. B35: 8865 (1987); S. A. Kivelson, Phys.Rev. B36:7237 (1987); B39: 259 (1989).

CHARGE MOTION IN SOLID HELIUM

Leonid P. Mezhov-Deglin, and Andrey I. Golov

Institute of Solid State Physics, Ac. Scie. USSR
142432, Chernogolovka, Moscow distr., USSR

1. INTRODUCTION

This paper is dedicated to the memory of the famous soviet physicist Alexander I.Shal'nikov. The first successful experiments with charges in solid helium fulfilled in his laboratory at 60th years were the part of wide program of investigations of properties of solid helium. One of the main purpose of the program was the search of quantum tunneling effects in helium crystals (later on named by quantum crystals).

In series of the experiments[1] the steady-state properties of the diode with radioactive charge source in solid ^{4}He and ^{3}He were investigated (current dependences on temperature $I(T)$ or applied voltage $I(U)$). Charges (microscopic charged defects) were created in thin layer near the radioactive β-target (source). The sign of the charges moving across a sample from the source to the metal collector was governed by the polarity of the applied voltage U. Nearly exponential temperature dependences of the current $I(T)$ at low fields were observed which permitted to estimate the effective activation energies of charge diffusion processes ε in solid helium. This investigations were followed[2] by the direct time-of-flight measurements of the charge velocity in a triode (a control grid was inserted between the electrodes). Later on the detailed investigations of the behavior of the velocity V of charges in the range of thermally activated motion were carried by Keshishev[3].

Since the first successful experiments a number of experimental and theoretical investigations were performed (experiment: Shal'nikov, Mezhov-Deglin, Keshishev and Tsymbalenko, Marty and Williams, Dahm et al., Mezhov-Deglin, Efimov,Golov; theory: Shikin, Andreev et al., and others, see the reviews[4,5]).

The most informative review of the results published up to 1985 was made by Dahm[6]. The author[6] came to a conclusion that despite on extensive investigations the structure of positive and negative charges in solid helium and mechanisms of their motion had not been established.

The main purpose of this paper is to represent and discuss our recent results[7,8] obtained in ranges not investigated thoroughly earlier: i.e. at temperatures below 1K and high electrical fields up to $1.3 \cdot 10^5$ V/cm, in hcp and bcc crystals from pure ^{4}He and ^{4}He containing up to 5% ^{3}He impurity atoms.

Excitations in Two-Dimensional and Three-Dimensional Quantum Fluids
Edited by A.G.F. Wyatt and H.J. Lauter, Plenum Press, New York, 1991

2. EXPERIMENTAL METHOD

The experimental cell was consisting of β-active source (titanium-tritium target) and metal collector frozen into solid helium as in first Shal'nikov's experiments. The distance between the diode electrodes were L = 0.3 mm in the most of experiments. Helium crystals were grown at constant pressure. The method of determining the velocity of injected charges V from transient of collector current on step-like application of the external voltage had been known too[6].

The transients $I(t)$ or steady state currents in diode for space charge limited currents can be calculated by solving Poisson's equation with charge density $\rho = j/V(E)$ and boundary condition $E(x) = 0$ at x = 0. Assuming a power-low dependence of the charge velocity from electrical field $V(E) \propto E^n$ the transit time τ deduced from the position of the maxima of the $I(t)$ curves (the time of arrival of a front of charged particles) can be related to the time of flight τ^* by a single charge the same distance L in homogeneous electrical field $E = U/L$ by relation $\tau(U) = g\tau^*(E)$, or the average front velocity $V(U) \equiv L/\tau = gL/\tau^*$, where the parameter g changes monotonically from 1.27 to 1.52 with increasing n from 1 to 10. In the first approximation $V(U) = (1.4 \pm 0.1)V(E)$. Naturally, the reconstruction the form of $V(E)$ curve from the "experimental" nonmonotonic $V(U)$ dependence (which was true for positive charges in hcp ^{4}He crystals in strong fields) demands the special numerical calculations.

As was shown by experiments and subsequent numerical calculations the trapping of mobile charges did not change significantly the transit time τ, but lowered many times the steady state current at low fields. It permitted us to control the quality of the specimen and to study the charge - defect interactions. The strong differences in charge mobilities observed in some samples , grown at the same pressure (en example is shown for $V_+(T)$ on Fig.3),or, for eminent, in the upper and lower parts of the sample in diode, having two separated collectors, may be explained by the strong anisotropy of mobility.

3. THE RESULTS OF THE MEASUREMENTS

3.1. Weak fields

In weak fields $E \leq 10^4$ V/cm the deviations of the field dependence of velocity $V(U)$ from linear low was less than 20% so it may be written that $V = \mu E$, where μ is the charge mobility.

3.1.1. HCP samples

Temperature dependences of the velocity of positive V_+ and negative V_- charges in different samples are shown in Fig.1. The dashed curves join the experimental points,the continuous lines correspond to exponential dependence $V(T) \sim \exp(-\varepsilon/T)$. For positive charges three regions could be distinguished in the $V_+(T)$ curves: high temperatures $T \geq$ 1K usually described by an exponential dependence, a transition region 0.9K > T > 0.7K, and a low temperature region T < 0.7 K, where the velocity again fell strongly with lowering the temperature. The shape of the $V_+(T)$ curves in the transition region depended strongly on the pressure: at $p \geq$ 32 atm in all the samples maxima were observed, but at lower pressures $p \leq$ 30 atm there were only changes of the slope of $V_+(T)$ curves. On the right scale here is shown the

1)

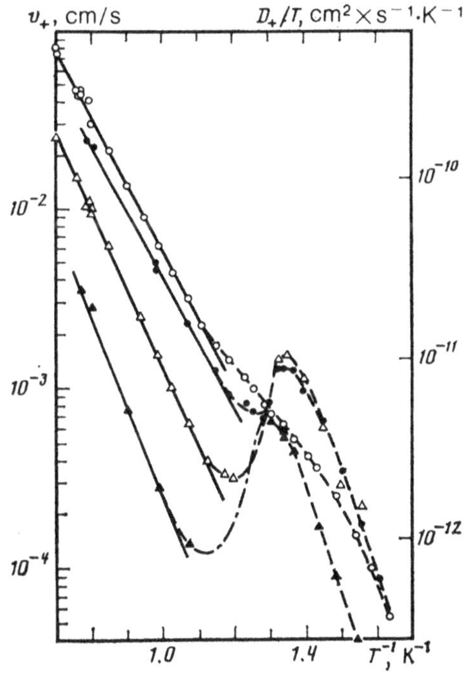

2)

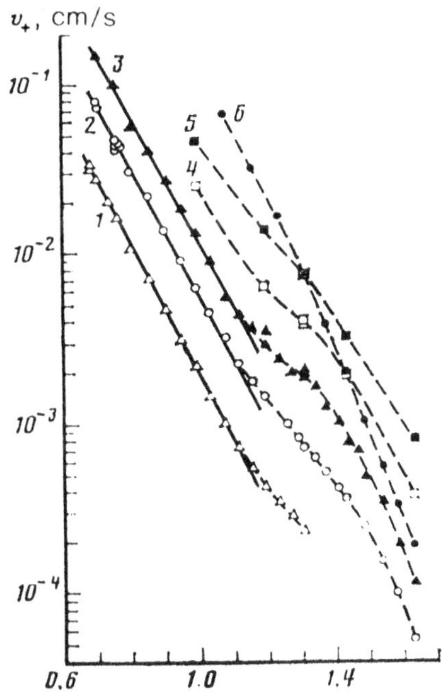

FIG. 1. Typical temperature dependence of the velocity $V_+(T)$ at U = 300V obtained for hcp ^{4}He samples grown at different pressures: o - p = 28.8atm (159), o - 31.3 (184), Δ - 33.0 (186), Δ - 35.9 (172).

FIG. 2. Evolution of the shape of $V_+(T)$ curves with increasing the voltage: 1 - U = 100V, 2 - 300, 3 - 600, 4 - 1000, 5 - 1600, 6 - 2800. Sample 159, p = 28.8atm, hcp ^{4}He.

ratio of the diffusion coefficient D_+ to temperature. The values of D_+ was deduced from charge mobility μ_+ using the Einstein relation $D = \mu kT/e$, or for our diode and U = 300V D/T = $0.7 \cdot 10^{-8} V(300V)$. Clear, that the activation energies, estimated from slopes of $D(T)$ lines, are 1 - 2 K higher than from $V(T)$ dependences at $T \geq 1K$.

Evolution of the shape of $V_+(T)$ curves with increasing the voltage is shown at Fig.2.

In the range of typical scattering of the data from sample to sample in anisotropic hcp ^4He an addition of 0.16% impurity ^3He atoms in ^4He gas did not change particularly the temperature dependences of charge velocities in low fields. Increasing the ^3He content up to 5% results in increasing of the ε_+ and smoothing the low temperature dependence (Fig.4).

In contrast to $V_+(T)$ the temperature dependences of $V_-(T)$ in the working range $V = 10^{-1} - 5 \cdot 10^{-5}$ cm/sec were close to exponential in all the hcp ^4He samples (Fig.3).

3.1.2. BCC samples

In pure ^4He the bcc phase exists in narrow ranges of temperatures and pressures only. Addition of 5% ^3He atoms permitted to investigate the temperature and field dependences of charge velocities in bcc ^4He in wide ranges. Temperature dependences of mobilities of positive (black symbols) and negative (open symbols) charges in two bcc samples (^4He + 5% ^3He) are shown in Fig.5 It ought to be mentioned that in all the bcc samples in the working range of velocities or mobilities (down to $5 \cdot 10^{-9}$ cm^2/Vsec) the temperature dependences of $V(T)$ were close to exponential. More over comparison of the charge mobilities in bcc ^3He and ^4He samples (our measurements and the literature data from[6] has shown that at nearly equal molar volume their values are very close and differ significantly from the mobilities in hcp ^4He samples.

3.2. Strong electrical fields

3.2.1. Positive charges in hcp ^4He samples

Keshishev had observed[3] the faster than linear field dependences $V_+(E)$ in samples grown at 32.4 and 40.5 atm and slower than linear in a sample grown at 25.5 atm. A considerable widening of field and temperature working ranges showed that below 0.9K field dependences $V_+(U)$ were more complex and sometimes nonmonotonic. One might distinguish two types of field dependences $V_+(U)$ (Fig.6): curves with a maximum and curves in which the derivative $\delta V_+/\delta U$ only decreased in a field above a certain threshold. The first type dependences were observed for all samples grown at pressures higher than 30 atm and some samples at p < 30 atm. The second one for the part of samples grown at low pressures (p < 30 atm) only. Evolution of the $V_+(U)$ dependences in strong fields with lowering temperature (numbers along the curves) in samples of both types is shown in Fig.7,a,b.

It is seen that the second type field dependence was obtained for samples with low mobility in weak fields. The shape of $V_+(U)$ curves in this samples was nearly independent from temperature in contrast to the first-type dependences and the threshold voltages were considerably lower (Fig.7,b). All of this might be ascribed to high anisotropy of hcp crystals.

Introduction of ^3He impurity atoms did not change qualitatively the $V_+(U)$ dependences in strong fields.

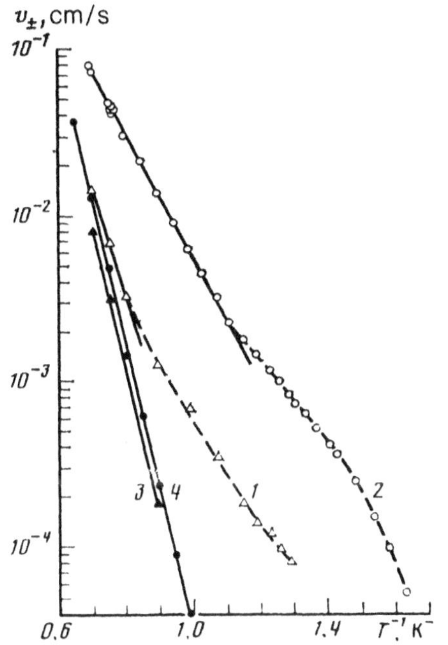

Fig. 3. Temperature dependences of the velocity of positive and negative charges at $U = 300V$ in two samples of hcp ^4He. Sample 157 (28.8atm): 1 - $V_+(T)$, 3 - $V_-(T)$, Sample 159 (28.5atm): 2 - $V_+(T)$, 4 - $V_-(T)$.

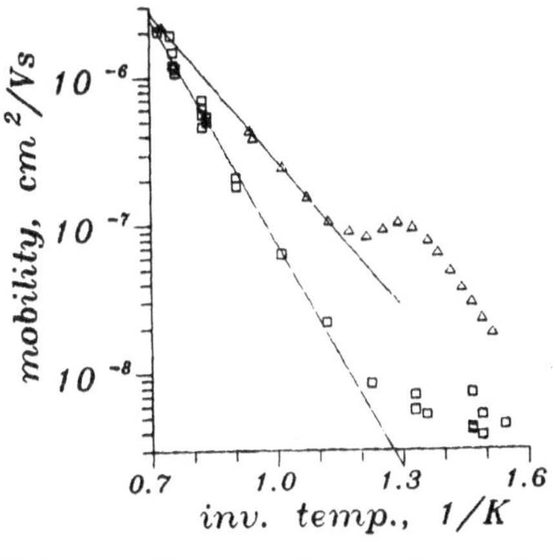

FIG. 4. Temperature dependences of mobilities of positive charges in hcp samples:
Δ - pure ^4He, $V_m = 20.55\ cm^3/mole$,
□ - $^4He + 5\%\ ^3He$, $V_m = 20.5 cm^3/mole$.

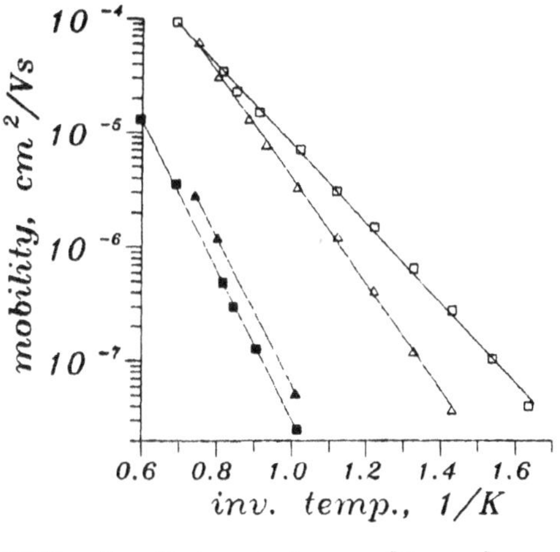

FIG. 5. Temperature dependences of mobilities of positive (black symbols) and negative (open ones) charges in bcc $^4He + 5\%\ ^3He$ samples: □ - $V_m = 21.2\ cm^3 mole$, Δ - $V_m = 21.3\ cm^3 mole$.

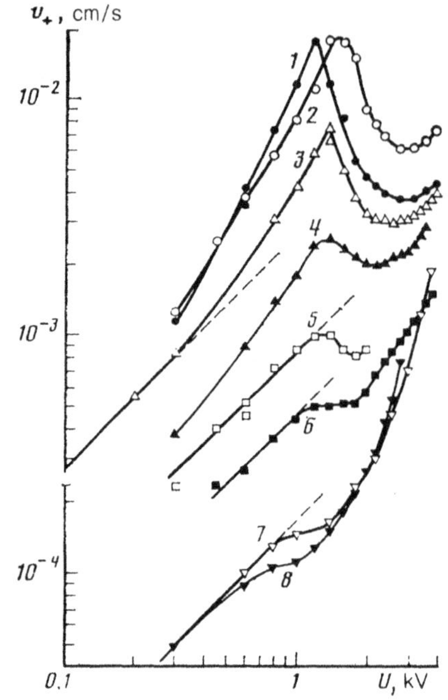

FIG. 6. Field dependences of the positive charge velocity $V_+(U)$ at $T = 0.72 \pm 0.01$K obtained for seven hcp ^{4}He samples grown at different pressures: 1 - sample 189 (34.0atm), 2 - 184 (31.3), 3 - 107 (30.8), 4 - 156A (29.1), 5 and 6 - two parts of sample 156 (29.1atm), 7 - 157 (28.5), 8 - 109 (26.2). The dashed lines rspresent the linear dependence.

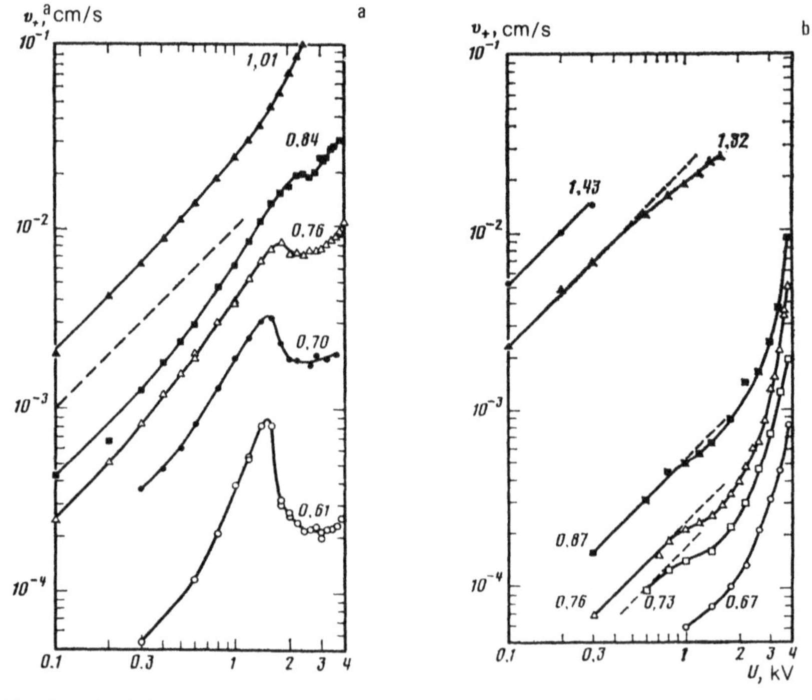

FIG. 7. Evolution of the $V_+(U)$ curves on increase in temperature (the numbers alongside the curves give the temperatures T in K): a - sample 159 (28.8atm), b - 157 (28.5). Hcp ^{4}He.

3.2.2. Negative charges in strong fields

The velocity of negative charges V_- in strong fields rose monotonically in all the samples. Typical example of $V_-(U)$ dependence in hcp ^4He sample is shown in Fig.8, (temperatures are shown by numbers along the solid curves, the dashed lines represent the power laws $V \propto U^n$, where n = 1,3,5). Obviously that at T = 0.85K $V_-(U)$ dependence in fields $U/L \geq 9 \cdot 10^4$ V/cm is closer to $V_- \propto U^5$.

The similar field dependences observed earlier[8] in bcc ^3He samples for charges of both signs could be approximated by the relationships of the type:

$$V(T,U) = A\exp(-\varepsilon/T)\sinh(\gamma U/T), \qquad (3.1)$$

or, assuming $D_0 = Ab$, $b = \gamma L/e$

$$V(T,E) = (D/b)\sinh(eEb/kT), \text{ and } D = D_0\exp(-\varepsilon/T) \qquad (3.2)$$

where D_0, ε, and b are fitting parameters. The values of parameters obtained in this way for negative charges in hcp ^4He were approximately in the range from $2 \cdot 10^{-6}$ cm^2/s, 15K, and 0.35 Å to $2 \cdot 10^{-4}$ cm^2/s, 21K, and 0.8Å.

4. DISCUSSION

4.1. Let us summarize the main recently known experimental data:

1) In the region of thermally activated diffusion the mobilities of positive and negative charges and appropriate activation energies depend strongly on the structure of crystal lattice of the sample and its molar volume (and not on the isotopic mass and statistic, crystal quality and small impurity concentration).

At low pressures in bcc samples

$$\mu_- > \mu_+, \quad \varepsilon_+ > \varepsilon_- \approx \varepsilon_0,$$

and in hcp samples

$$\mu_+ > \mu_-, \quad \varepsilon_- > \varepsilon_+ \approx \varepsilon_0$$

Here ε_0 is the vacancy creation energy from independent measurements.

With sufficient increasing the pressure the difference between ε_+ and ε_- diminishes (more over at certain pressures ε_- and ε_+ may exchange their positions at the same crystal structure), and both they exceed ε_0.

2) Temperature dependences of mobility of positive charges in hcp ^4He and ^4He + 5% ^3He samples grown at pressures $p \leq$ 40atm at low temperatures T < 1K differ significantly from the exponential law. Depending on the pressure and crystal properties (orientation, apparently) maxima or inflections on $V_+(T)$ curves are observed in the region T = 0.9 - 0.7 K (Fig.1).

3) In the same hcp ^4He samples maxima or inflections are observed on $V_+(U)$ curves in fields about certain thresholds. There exist some correlations between the shape of $V_+(U)$ curves in strong fields and the values of low temperature mobilities μ_+ in weak fields (and on crystal molar volume and orientation too). The fact of charge velocity drop in strong fields was recently confirmed by Tsymbalenko measurements in triode[9].

8)

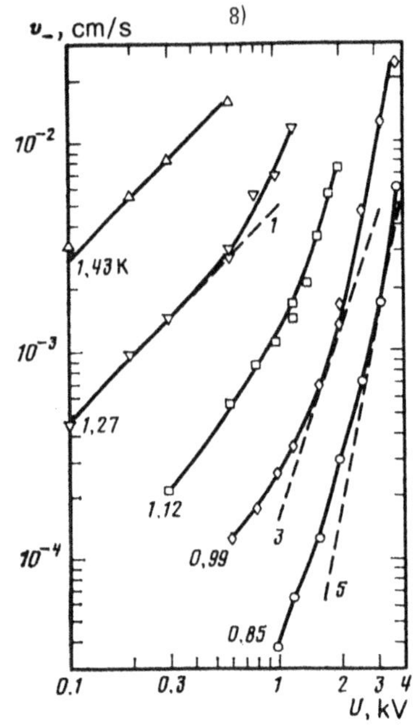

9)

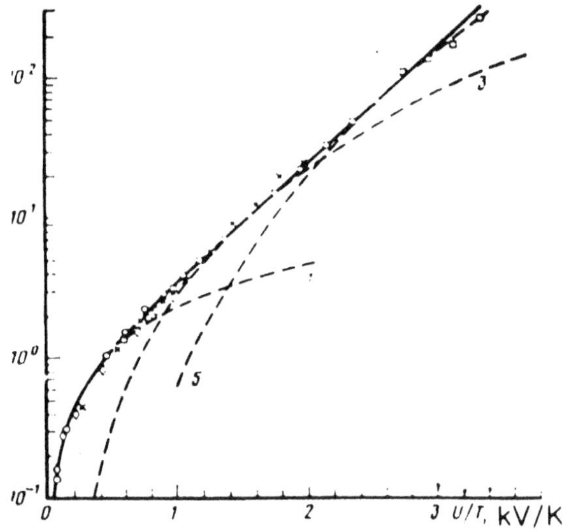

FIG. 8. Field dependence of the velocity of negative charges $V_-(U)$ obtained at different temperatures. The dashed lines represent the power law $V \propto U^n$, where n = 1, 3, or 5. Sample 107 (30.8atm), hcp ^{4}He.

Fig. 9. Dependence of the ratio $V_-/A\exp(-\varepsilon_-/T)$ on the reduced voltage U/T in hcp ^{4}He sample 157 (28.5atm) at different temperatures (K): v - 1.43, o - 1.33, x - 1.11, □ - 1.02. The solid curve represents $\sinh(\gamma U/T)$ and the dashed curves represent $V \propto U^n$ where n =1, 3, or 5.

4) In contrast to $V_+(U)$ dependences the velocity of negative charges $V_-(U)$ in hcp ^{4}He samples rises monotonically with voltage applied. At $T < 1$K the field dependence $V_-(U)$ in high fields is stronger then cubic. In bcc ^{3}He the analogous field dependences were obtained for charges of both signs earlier. In wide range of fields and temperatures these charges velocities may be fitted with satisfactory accuracy by the formula (3.2):

$$V(T,E) = (D(T)/b)\sinh(eEb/kT)$$

Here $D(T)$ is the diffusion coefficient of Arrhenius type and in the frames of simple thermoactivated hopping model b is half the projection of a displacement vector on the direction of applied electric field. The values of parameter $2b$ in different samples were two - five times less than interatomic distance.

The agreement between the activation energies and diffusion coefficients of positive charges and ^3He impurity atoms in hcp ^4He and between the same parameters for negative charges motion and self-diffusion of atoms in bcc ^3He at low pressures proved that the vacancy assisted motion might be the main mechanism of charge transport. But most of the recent experimental results obtained in strong fields and low temperatures were not predicte by the earliest theories of viscous flow of negative charge cavity (Shikin[5]) or inelastic scattering of delocalized vacancies by point charged defect (Andreev, Meyerovich[4]).

4.2. Charges in solid helium. What are they?

It is generally accepted[6] that the negative charge in solid helium at small pressure is an electron localized in a cavity of radius nearly 10Å (that slightly decreases with increasing the pressure). From the microscopic point of view such a small cavity in discrete crystal lattice is a cluster of N vacancies, $N(p)$ varies approximately from 100 to 50 over the pressure range 30 - 130 atm. Because of the desrete nature of $N(p)$, the pressure scale breaks up into intervals, whithin each of which the particular value of N remains constant[10].

The positive charge consists of bare molecular ion (He_2^+ or He_3^+) surrounding by the shell of deformed helium crystal due to electrostriction[6]. Polarization forces rapidly drop as the distance from the ion arises, but the atoms from the first coordination sphere of radius ~ 2.5 Å have the binding energy of the order 200K. This value is an order of magnitude larger than binding energy in solid helium at low pressures. So it is worthwhile to assume that in solid helium at low pressures, where the interatomic distance is 3.6 Å, positive charge is the large molecular defect consisting of cluster ion He_n^+ ($n > 10$), surrounded by deformed crystalline lattice.

The estimation of an additional electrostrictive pressure at 10Å from a center of charge is $\Delta p \approx 6$atm. This may be one of of the reasons of existence of an additional potential barrier for charge - vacancy interaction[5] or of small effective cross-section in a case of inelastic scattering of vacancy waves on the charge[4].

We can assume that the charge transport is a consequence of its transitions between nonequal states with slightly different energy. In the case of negative charge they may be the transitions, governed by absorbtion and emission of vacancies, between the states (cavities) with different N. As was proposed by one of the authors[10] at different pressures the energy gap from the "ground state" N to the nearest excited state N^* may differ from zero to a certain value. It might be an another reason for the existence of the additional potential barrier for vacancy-assisted transport of negative charges observed at some pressures in hcp and bcc solid helium.

4.3. Theory of Andreev and Savishchev

A new symmetry approach to the problem of charge structure and charge motion was proposed by Andreev and Savishchev[11]. The charge motion in helium crystals involves the displacement of the localized charge complex to one of the neighboring translationally nonequvalent localized state with equal energy. The most important feature of the model is that the symmetry of the set of corresponding displacement vectors may differ considerably from the symmetry group of the surrounding crystal lattice and the length of a displacement vector of the charge complex may be less then the interatomic distance.

Displacement of the charge complex consisting of a number of atoms (or vacancies for negative charge cavity) is a result of manyparticle tunneling processes in which a trajectory of vacancy goes through a lattice sites occupied by charged complex. There are two assumptions at the next consideration. The first is that vacancies in pure ^{4}He are delocalized quasiparticles and displacement processes can be considered as the quantum-mechanical inelastic scattering of vacancies by localized charge complexes (conception developed by Andreev and Meyerovich earlier). The second is that the value of inelastic cross-section is a hundred times less than the area of the crystal elementary cell, which may be connected to the complication of manyparticle tunneling transition and exponential character of its probability.

Taking into account the finite value of the vacancy bandwidth, the temperature and the energy gain on charge displacement to a neighboring state in an electrical field, which must be absorbed by the vacancy in the case of inelastic scattering leads to the prediction of nonmonotonic field dependences of the charge drift velocity in strong electrical field (for example, to decreasing like $V(E) \propto (\Delta - eEu)^2$ for $\Delta - eEu \ll T$, where Δ and u are the width of the vacancy energy band and the displacement vector). The temperature dependence of the charge mobility consisting of two different exponential asymptotes and the possibility to determine the set of displacement vectors (i.e. the charge complex symmetry) from the drift velocity anisotropy measurements are predicted too.

So the new approach explains qualitatively a number of peculiarities in the temperature and field dependences which were observed in investigations of the positive charge motion in hcp ^{4}He at low pressure. But there are a number of other results in hcp ^{4}He (particularly at larger pressures) that can not be explained now: 1) fast nonlinear increasing and subsequent drop of the velocity $V_+(U)$ in strong fields; 2) high maxima on $\mu_+(T)$ dependences at transition temperature; 3) sufficient deviation of charge activation energy from vacancy one ε_0, and nearly exponential field dependence for negative charges in hcp ^{4}He (and for both signs in bcc ^{3}He, though the possibility of an application of this model to charge motion in paramagnetic ^{3}He is under question).

4.4. Charge-dislocation complexes and other hypothesis

The most of alternative possible explanations of observed peculiarities in $V_+(T)$ and $V_+(U)$ in 1 ^{4}He are discussed in papers[6,7]. They are suppose possible quantum self-diffusion of positive charge without any vacancies, the possibility of phase transition in charge structure at low temperature, etc. To our opinion, one of the most interesting is the next explanation[7] of the fall of the positive charge velocity in hcp ^{4}He:

In electrical fields above a certain threshold the charges may create dislocation loops which are capable to capture charges and move together with them across a crystal (in the same way quantum vortex rings appear around charges in superfluid helium). Like in a liquid, the decrease of the velocity with increasing field may be explained by an increase in the loop radius. Freshly formed dislocations are high mobile, their mobility depends exponentially on temperature with activation energy close to the energy of vacancies in hcp ^{4}He. This is in agreement with $V_+(T)$ dependence observed at voltages larger than the threshold one, for example at U = 2.800V, that corresponds to the minimum of the $V_+(U)$ curve (Fig. 2.).

5. CONCLUSION

Extending of the range of measurements of the charge motion velocity in hcp and bcc ^{4}He and ^{3}He crystals and comparison of the results with recent theoretical model have shown that even for qualitative explanation of the results we need to take into account the microscopic structure of charges in different crystal lattices. To clear out the structure of charges a number of experiments may be proposed: measurements of an anisotropy of mobility in oriented crystals, spectroscopy investigations, etc.

REFERENCES

1. A. I. Shal'nikov, About charge motion in liquid and solid helium, Zh. Eksp. Teor.Fiz., 41: 1059 (1961) [Sov. Phys. JETP, 14: 755 (1962)].
   A. I. Shal'nikov, About charge motion in solid helium,, Zh. Eksp. Teor.Fiz., 47: 1727 (1964) [Sov. Phys. JETP, 20: 1161 (1965)].
2. K. O. Keshishev, L. P. Mezhov-Deglin, and A. I. Shal'nikov, Mobility of charges carriers in ^{4}He crystals, Pis'ma Zh. Eksp. Teor.Fiz., 12: 234 (1970) [Sov. Phys. JETP Lett., 12: 160 (1970)].
3. K. O. Keshishev, Charge motion in crystalline helium, Zh. Eksp. Teor.Fiz., 72: 521 (1977) [Sov. Phys. JETP, 45: 273 (1977)].
4. A. F. Andreev, Diffusion in quantum crystals, Usp. Fiz. Nauk., 118: 251 (1976) [Sov. Phys. Usp. 19: 137 (1976).
   A. F. Andreev and A. E. Meierovich, Theory of transport of charges and impurity atoms in quantum crystals, Zh. Eksp. Teor.Fiz., 67: 1559 (1974) [Sov. Phys. JETP, 40: 776 (1975)].
5. V. B. Shikin, Charge mobility in liquid, solid and dense gaseous helium, Usp. Fiz. Nauk., 121: 457 (1977) [Sov. Phys. Usp. 20: 226 (1977).
6. A. J. Dahm, Charge motion in solid helium, in: "Progress in Low Temperature Physics", Vol. 9, ed. D. F. Brewer, North-Holland, Amsterdam (1985).
7. A. I. Golov, V. B. Efimov, and L. P. Mezhov-Deglin, Motion of injected charges in hcp ^{4}He crystals, Zh. Eksp. Teor.Fiz., 94: 198 (1988) [Sov. Phys. JETP, 67: 325 (1988)].
8. A. I. Golov and L. P. Mezhov-Deglin, Motion of injected charges in solid ^{4}He containing up to 5% ^{3}He, to be published in Proceedings of LT-19, Brighton (1990).
9. V. L. Tsymbalenko, Studies of "dark" currentin crystalline ^{4}He Fiz. Nizk. Temp., 15: 115 (1989).
10. A. I. Golov, Discrete states of an electron bubble in solid helium, Pis'ma Zh. Eksp. Teor.Fiz., 49: 346 (1989) [Sov. Phys. JETP Lett., 49: 396 (1989)].
11. A. F. Andreev and A. D. Savishchev, Charge motion in solid helium, Zh. Eksp. Teor.Fiz., 96: 1109 (1989).

^{4}He FILMS ON GRAPHITE STUDIED BY NEUTRON SCATTERING

H.J. Lauter[1], H. Godfrin[1], V.L.P. Frank[1,2] and P. Leiderer[2]

1 Institute Laue-Langevin, BP 156X, F-38042 Grenoble, France
2 University of Konstanz, D-7750 Konstanz, West Germany

INTRODUCTION

A ^{4}He film shows particular properties. For example the superfluid onset temperature is shifted to lower temperatures with decreasing film thickness and a reduced superfluid fraction is detected [1-3]. This may be caused by a different excitation spectrum with respect to the bulk [e.g. 3]. In addition a ^{4}He film has two interfaces the gas-liquid boundary and the liquid-solid boundary. At the gas-liquid boundary quantized capillary waves (ripplons) [4-7] can be excited. Concerning the liquid-solid boundary it was measured that if the film is deposited on a graphite substrate, that the two layers adjacent to the graphite are solid [8]. The interesting interface is thus the liquid-solid ^{4}He interface. Here the freezing-melting wave could be excited [9]. In addition this interface could play a role in the anomalous small Kapitza-resistance. In this report we will give some new information about excitations in a ^{4}He film and at its boundaries.

EXPERIMENT

In contrast to previous neutron scattering experiments of ^{4}He films [10,11], which where carried out using graphite powder as substrate, we used this time Papyex [12] as substrate. Papyex has besides a powder contribution oriented graphite crystallites which show a 30º (FWHM) mosaic distribution with respect to the c-axis. A densest first layer (which under consideration of the pressure of the second layer has a density of 0.115 atoms/Å^2 [8]) consisted of 312 cc (STP) ^{4}He. The subsequent layers are taken to have densities of 0.95 atoms/Å^2 [8] , 0.078 atoms/Å^2, 0.078.atoms/Å^2...

The inelastic neutron scattering experiments were performed on the time-of-flight spectrometer IN6 [13] at the ILL with a chosen wavelength of 5.12Å. The detectors are located in an angular range of 11.9º - 113º ; this corresponds, for elastically scattered neutrons, to momentum transfers (Q) between 0.254 - 2.046Å^{-1}. The energy resolution is only slightly dependent on Q and is about 0.055 μeV for elastically scattered neutrons. The sample was mounted with the c-axis perpendicular to the scattering plane. The data obtained from the sample cell before any ^{4}He was adsorbed were used as background and subtracted from subsequent measurements with adsorbed ^{4}He.

EXCITATIONS AT THE LIQUID-SOLID BOUNDARY

In previous experiments it was shown that excitations can be measured at the liquid-solid ^{4}He interface, which have a constant dispersion ("flat modes") indicating a localized

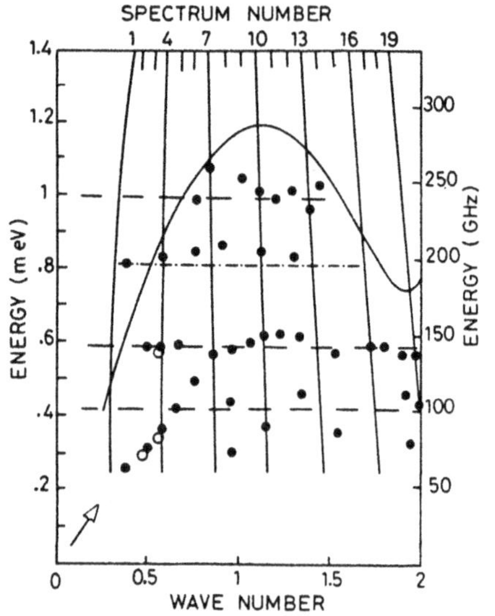

Fig.1. Position of the signals in the energy-wavenumber plane from 6.4 ^{4}He layers on top of the Ne coated graphite powder at T=0.5 K [10]. The spectra have been measured on a time-of-flight spectrometer, therefore the spectrum number indicates the path along which the spectra are taken with constant scattering angle. The solid line represents the bulk ^{4}He signals. The dashed-dotted line shows the peak position of the multiple scattering (diffraction of neutrons by the (002) graphite reflection and creation of a roton). Accidently, this signal is superimposed on a localized mode which are otherwise marked by dashed lines. The arrow points to the first indication of the ripplon, which has been measured in 4.8 ^{4}He layers on Papyex (o) [10].This mode does not depend on the amount of adsorbed ^{4}He beyond a coverage of 4 layers [11]. Now we changed from the powder substrate to Papyex and the flat mode lost about 30% in intensity with respect to the roton intensity. Thus a more homogeneous substrate and/or the preferential orientation may contribute to the decrease of the signal.

excitation [10,11]. This is shown in figure 1, where for a 6.4 ^{4}He layer film on Ne coated graphite powder in addition to the bulk excitations these flat modes show up. Also in the case of Ne coated graphite a solid ^{4}He layer is next to the Ne layers. The proof that these flat modes are localized at the liquid-solid ^{4}He boundary was given by the fact that this mode still persists in a completely with ^{4}He filled sample cell and that the intensity of this mode does not depend on the amount of adsorbed ^{4}He beyond a total coverage of 4 layers [11]. Now we changed from the powder substrate to Papyex and the flat modes lost about 30% in intensity with respect to the roton intensity. Thus a more homogeneous substrate and/or the preferential orientation may contribute to the decrease of this signal.

The main origin of the flat modes was suggested to be an exchange of atoms between the second solid ^{4}He layer (pure graphite substrate) and the liquid ^{4}He. In order to test this idea we preplated the graphite with two layers of H_2. The ^{4}He isotherm taken on the preplated graphite [14] shows that the density of the first ^{4}He layer of 0.072 atoms/Å^2 is not high enough to be a solid at least at the temperature of T=1.96 K. From Ref.15 it can be deduced that about a density of 0.08 atoms/Å^2 is needed to solidify an incommensurate monolayer of ^{4}He.

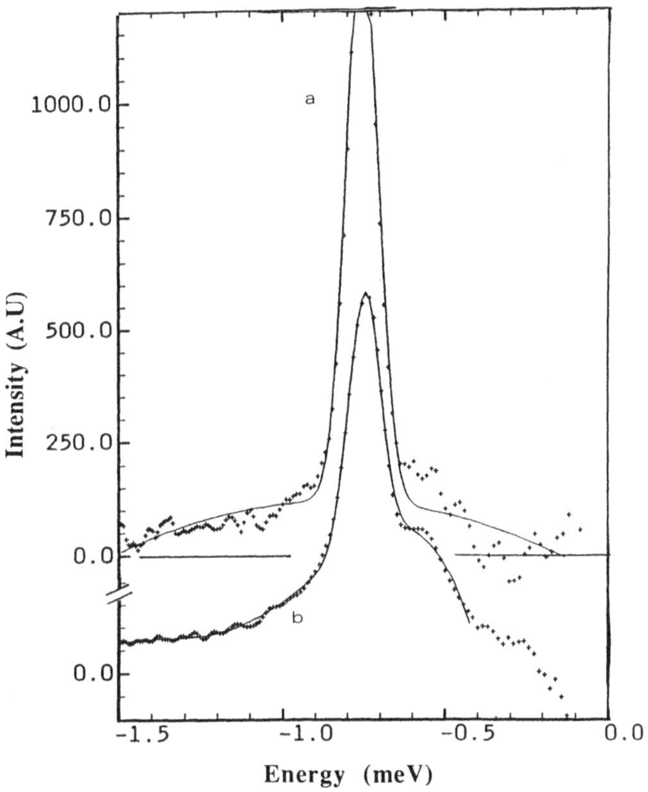

Fig.2. Inelastic neutron scattering scans along the path of spectrum number 18 in fig.1 with an elastic Q = 2.00 Å^{-1}. In a) (sample 1) the graphite was preplated by two layers of H_2 and subsequently 4.6 layers of ^{4}He are adsorbed. In b) (sample 2) a pure graphite substrate was used with 5.06 adsorbed layers of ^{4}He in total. In both cases the roton shows up at an energy transfer of 0.755 meV. The flat mode intensity is at about 0.6 meV. The fit is a very rough one and focuses mainly on the roton. The temperature is 0.7 K. The ordinate is displaced by 250 units between 2a) and 2b). The energy scale is negative due to a convention on the IN6 spectrometer for the energy transfer. The absolute value is the energy transfered to the excitation.

The comparison of a spectrum taken with and without preplating is shown in fig.2. The spectra have been taken at a constant angle, which corresponds to a Q of 2.00 $Å^{-1}$ on the elastic line. From figure 1 it is seen that this corresponds roughly to spectrum number 18 and this scan crosses the bulk dispersion at a Q of 1.92 $Å^{-1}$ and an energy of 0.755 meV. In both cases the roton excitation can be clearly seen. At a lower energy of 0.6 meV the intensity of a flat mode shows up. It is evident that this intensity is decreased by a factor of two in the case of the preplated sample with respect to the intensity of the not preplated sample. This already shows that the suppression of the solid ^4He layers decreases the intensity of the flat modes by an appreciable amount, so that the conclusion that the solid layers are involved in the local modes is correct. Remember that the intensity of the flat modes does not depend on the ^4He layer thickness in the range of thickness displayed in fig.2 [11]. These solid layers are present in most examples of Kapitza-resistance measurements and should be taken into consideration, which means that the excitation of a localized mode can render possible a transmission of a phonon from a solid through an interface to liquid ^4He.

Another comparison has to compare the intensity of the bulk roton signal of the film on the different substrates. The intensity increases in the case of the preplated sample (called sample 1 with a filling of 4.6 layers) by a factor of 2.2 (same linewidth) with respect to the not preplated sample (sample 2 with a filling of 5.06 layers including the solid ones). This is mainly due to the replacement of the solid ^4He layers by liquid layers. A more careful consideration gives the following result. The filling was 5.06 layers of ^4He for sample 2. It turns out (see paragraph: Phonons and Rotons in a ^4He-film) that 3.5 layers do not contribute to the intensity of phonons and rotons in the case of sample 2, thus a layerthickness of 1.56 does only contribute to the signal intensity. An intensity gain of 2.2 was measured in switching to sample 1 (see fig.2, this factor is also confirmed at other Q's). This gives 3.43 layers which do contribute to the signal intensity for sample 1. The filling of sample 1 was 4.6 layers. Thus 1.2 layers do not contribute to the intensity of phonon and roton excitations. This is about the same number as on the pure graphite substrate, where 3.5 layers are inert, but two of them are solid so 1.5 inert liquid layers are left. The information which can be taken from the phonon and roton signal intensity for the two different substrates is twofold. It firstly confirms that no solid ^4He layers are present in the case of the preplated sample and secondly it demonstrates a liquid ^4He layer (1.2 to 1.5 layers) which does not contribute to the intensity of the phonons and rotons.in a ^4He film on different substrates.

PHONONS AND ROTONS IN A ^4He-FILM

A long outstanding question is how the dispersion curve of bulk liquid ^4He is changed if the ^4He is confined in a film. A very simple approach was given in Ref.3, where a change in the static structure factor was calculated in going from 3-dimensional ^4He to 2-dimensional ^4He. This shift results in a lowering of the roton minimum and a shift towards lower energies. Our scans hardly approached the roton minimum in Q and on the other hand this theory was not made for the phonons and maxons. Thus an other theory was considered [5,16]. Here a clear increase of the energy in the maxon region was predicted due to the higher density regions of the liquid layers near the solid. However our fits to the experimental data, exhibited in fig.3, show that a lowering of the dispersion curve takes place in the maxon region if the film thickness is reduced. The lowering of the dispersion curve is continued towards the roton minimum. However the roton minimum was not covered by the scans. The phonon branch is not influenced. But phonons were only visible for our thickest film. Our data from the completely with ^4He filled cell agree with the bulk Dispersion curve [17]. But we have no explanation for the softening of the maxon and the roton branch, where the rotons have with negative group velocity .

The intensity of the phonons and rotons is plotted in fig.4. A calibration to the bulk data was done for all film thicknesses at a Q of 1.92 Å-1. The data taken with the

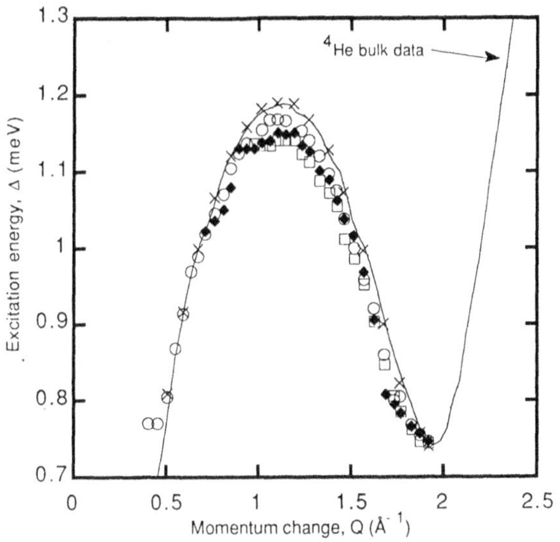

Fig.3. Variation of the dispersion relation of the phonons and rotons as a function of film thickness at a temperature of 0.7 K. The total ^{4}He layer thickness is 5.06 layers (○), 4.16 layers (◆) and 3.85 layers (❑). (×) marks data from the completely with he filled sample cell.

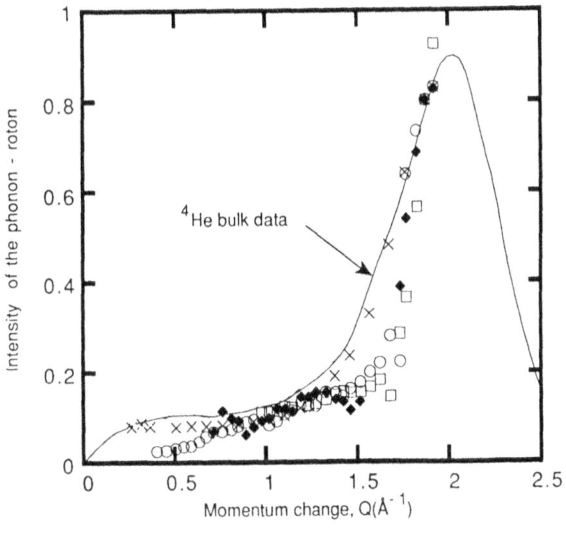

Fig.4. Intensity of the phonons and rotons as a function of Q at a temperature of 0.7K.The total ^{4}He layer thickness is 5.06 layers (○), 4.16 layers (◆) and 3.85 layers (❑). (×) marks data from the completely with he filled sample cell. The intensities of the different coverages are calibrated to 0.8 , the value of the bulk data at 1.92 Å^{-1}

completely ^{4}He filled cell show towards lower Q a small deviation to lower intensities with respect to the bulk data. We do not know yet whether this is significant. Important deviations appear however for the ^{4}He films. The thick film (5.06 layers) looses much intensity below a Q of 0.7 $Å^{-1}$. This Q corresponds just to the liquid filmthickness. Thus the loss of the third dimension reduces the intensity of the phonon. The thinner films show at low Q no more intensity. Thus a loss in intensity is detected if the filmthickness correlates with the momentum transfer of the excitation and the intensity disappears even at still lower Q.

Another remarkable feature in figure 4 is the loss of intensity with respect to the bulk data around 1.7 $Å^{-1}$. This is again the region, where rotons have negative group velocity and where a shift in the dispersion relation was measured. We can only suggest that this effect may again result from the reduced dimensionality.

The behavior of the intensity of the phonon is plotted in fig.5 against the filmthickness for a Q of 1.2 $Å^{-1}$. The intensity extrapolates to zero intensity at a total coverage of 3.5 layers. This coincides with measurements in Ref.18, where a loss of superfluidity is detected at about 3 adsorbed layers.

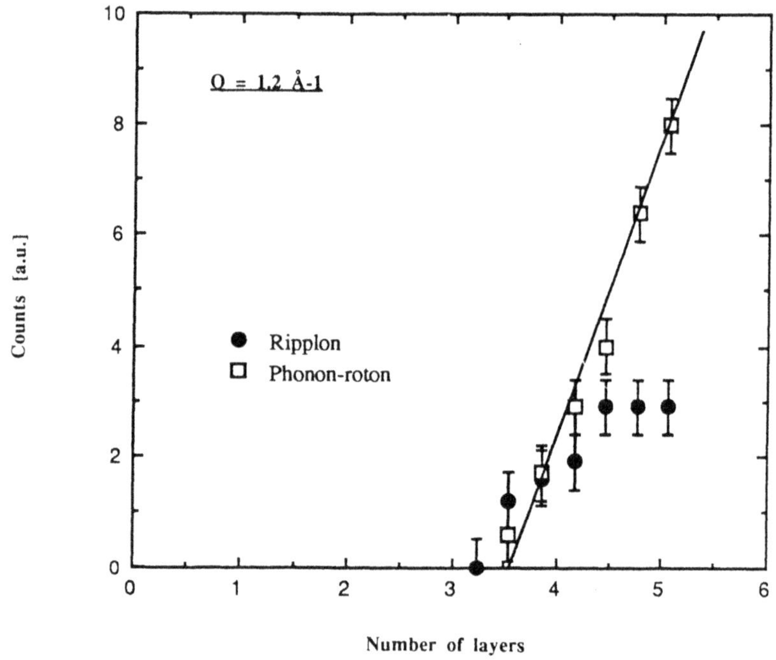

Fig.5. Intensities of the phonon (❑) and ripplon (❍) as a function of coverage at T=0.7K.

RIPPLON

Quantized capillary waves (ripplons) are the elementary excitations of a free liquid surface. Their existence at the bulk ^{4}He surface and in films has been predicted by theory and indirectly confirmed by experiment [4,19]. At long wavelengths the ripplon dispersion relation is easily evaluated using hydrodynamic relations for an incompressible fluid:

$$\omega^2 = (\alpha_0/\rho_0)\, k^3 \qquad \{1\}$$

where α_0 is the zero temperature surface tension, ρ_0 the ^{4}He density at zero pressure and k the wavevector. The temperature dependence of the surface tension ($\alpha(T)$) at very low temperatures can be deduced from the ripplon dispersion relation. Detailed measurements

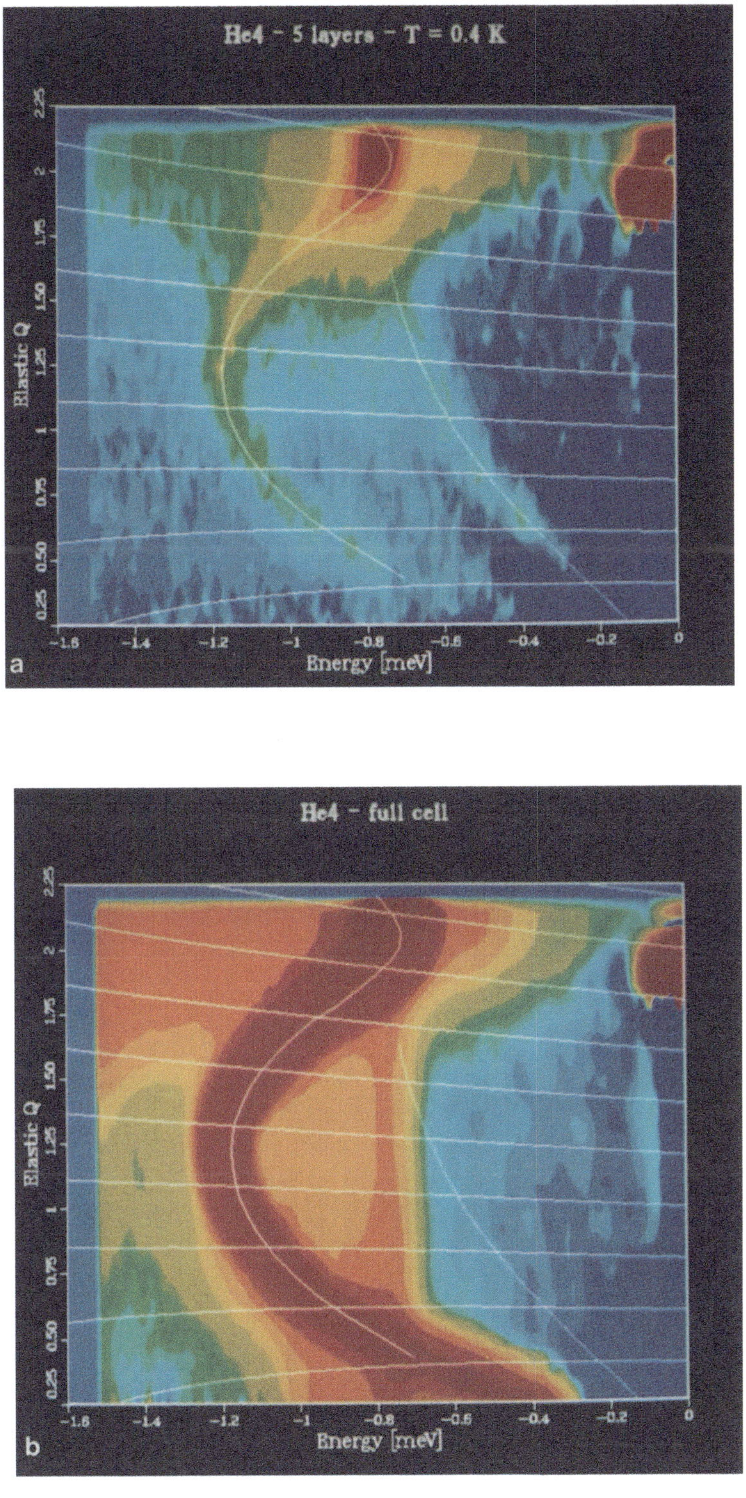

Fig.6. Intensity on the phonon-roton curve and on the ripplon curve in the energy-Q plane. Fig.6a shows the signal from 5.06 adsorbed layers on graphite and fig.6b the signal from the completely filled sample cell (including the graphite).

[20] of $\alpha(T)$ revealed a much larger temperature dependence than expected from formula {1}. Several modified dispersion curves have been proposed which differ mainly for wavevectors above 0.5Å^{-1}. The idea of a 'surface roton', with a minimum at ~2K, was introduced by Reut and Fisher [21] improving the agreement with the available thermodynamic data. Edwards et. al. [4,20], taking into account the curvature dependence of α, were able to fit the experimental data on the excess surface entropy. Their model involves two parameters: a length δ=d(ln α_0)/dK where K=(r_1^{-1} + r_2^{-1}) is the curvature of the surface, and an area a = dδ/dK. Within the precision of the entropy data, several sets of parameters have been used (a=+1.5Å^2, δ=0 [20] and a=+1.0Å^2, δ=-0.336Å [4]), the latter giving a better agreement. Such a large variation in the parameters corresponds to very different ripplon dispersion curves at wavevectors ~1Å^{-1}, with a common trend indicating the presence of a downward curvature. Little direct experimental evidence was available [10], however, on the ripplon dispersion curve at these wavevectors. Such a study requires a microscopic probe like inelastic neutron scattering (INS), but due to the low neutron cross section of ^{4}He the measurement has to be performed on samples with a large surface to volume ratio. The success of a neutron total reflection experiment is however not yet excluded [22].

A first outline of the experiment of which some results are depicted in fig.6 was already given in Ref.23. In figure 6 the different colors indicate the behavior of the intensity as a function of energy and momentum transfer. It is clearly seen in fig.6a that besides the intensity on the phonon-roton curve there is intensity on an energetically lower lying branch. Evidence of the existence of this branch has been given previously on measurements [10,11]. This branch coincides with the calculated dispersion of the ripplon using the parameter set in Ref.4. The agreement is very good, it seems to be even up to 1.5A-1. At still higher Q the roton intensity combined with the one of the flat modes becomes to high to distinguish the ripplon signal. This good agreement allows to say that the temperature dependence of the surface tension is really based on an experimentally verified dispersion relation. It still remains to prove the modified parameter set [4] by a theory.

In fig.6b the result of the completely filled sample cell is shown. In the region where the ripplon should show up, the colors are the same as in fig.6a. Only near the phonon-roton intensity and the flat bar of the multiple scattering [10,11] the attribution of the colors to intensity has been modified. Thus this figure shows that no signal of the ripplon intensity is visible in fig.6b the filled cell, although below 0.7 Å^{-1} it would have been distinguishable from the overwhelming quasi bulk phonon roton intensity. This disappearance proves that the ripplon signal is really bound to the gas-liquid interface.

A more detailed description of the ripplon dispersion is seen in fig.7. The measured dispersion agrees indeed very well with the calculation in Ref.4 up to Q = 1Å^{-1}. But between 0.6 to 1 Å^{-1} a small but detectable increasing slope of the dispersion curve with decreasing film thickness is detectable. The behavior at higher Q lets assume a minimum in the dispersion curve. However this comes about due to crossing the flat mode at 0.6 meV and each negative slope is due to the fit-routine, which clamps with increasing Q to the higher intensity of the flat mode. An interaction between the ripplon and the flat mode in a thin film is not excluded [11] but needs further consideration. We can confirm a positive slope of the dispersion curve in agreement with Ref.4 up to 1.3 Å^{-1} for certain coverages. At higher Q's the ripplon intensity gets lost.

The intensity of the ripplon is plotted in fig.8 as a function of Q for the different coverages. The ripplon is most intense at low Q loosing its intensity rapidly up to 0.6 Å^{-1}. At higher Q's again the interference with the flat modes makes a correct data analysis impossible in particular to the half automatic fit-routine. Perhaps the experiments on a H_2 preplated graphite surface will render the analysis more effective.

From preferential Q's, which are free from the flat mode intensity, ripplon intensities can be taken by careful "by hand" assisted fitting as a function of coverage. For a Q of 1.2 Å^{-1} this has been done and is shown in fig.5. The ripplon intensity seems to saturate

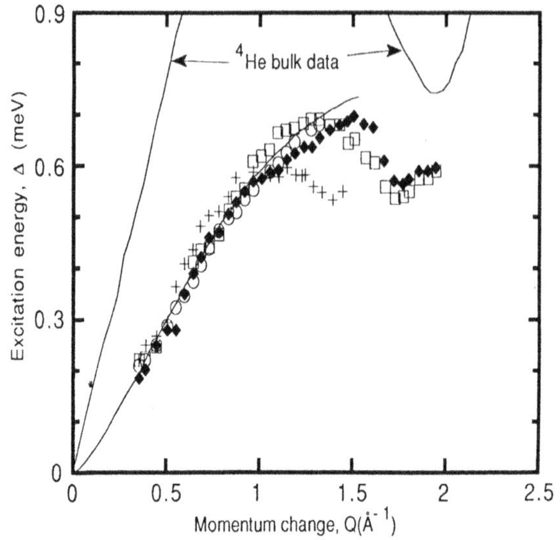

Fig.7. The ripplon dispersion for various coverages.The total He layer thickness is 5.06 layers (○), 4.16 layers (◆), 3.85 layers (❑)and 3.54 layers (+).

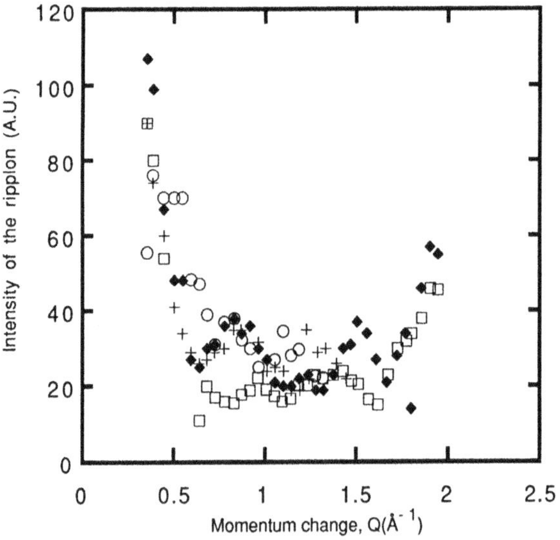

Fig.8. The intensity of the ripplon as a function of Q for various coverages. The total ^{4}He layer thickness is 5.06 layers (○), 4.16 layers (◆), 3.85 layers (❑)and 3.54 layers (+).

beyond a coverage of 5 layers. This is in agreement with the picture of a surface wave, whose intensity does not depend on the amount of bulk liquid below the surface. On the other hand the ripplon intensity disappears near 3 layers. This is in agreement with a penetration depth of about an atomic layer.[4,6] However this intensity disappears definitely at a lower coverage than the phonon-ripplon intensity. This phenomena is not yet understood.

Other modes are predicted for a thin film between the ripplon and the continuum (the phonon-roton dispersion curve) [5,6,16] but up to now no other modes could be detected.

CONCLUSION

In summary it turns out that a liquid helium film exhibits a lot of interesting features in the excitation spectrum. The excitation spectrum of the bulk phonon-roton curve is modified in a helium film as well as the dependence of the intensity on the momentum transfer. There are in addition excitations which have no dispersion (localized modes) at the solid helium-liquid-helium interface. Finally the dispersion curve of the ripplon at the gas-liquid boundary could be measured for the first time to relatively high momentum transfers.

ACKNOWLEDGEMENTS

This work has been partially supported by the West Germany Federal Ministry of Research and Technique (BMFT).

REFERENCES

[1] M.Chester, Comments Solid State Phys.10, 91 (1981)
[2] Saam W.F. and Cole M.W., Phys.Rev.B11, 1086 (1975)
[3] Götze W. and Lücke M., J.Low Temp.Phys.25, 671 (1976)
[4] Edwards D.O. and Saam W.F., Prog.in Low Temp.Phys., Ed. D.F.Brewer, Vol. VII A, 283 (1978), North Holland
[5] Krotscheck E., Stringari S. and Treiner J., Phys.Rev.B, 35, 4754 (1987)
Epstein J.L. and Krotscheck E., Phys.Rev.B37, 1666 (1988)
[6] Ji G. and Wortis M., Phys.Rev.B34, 7704 (1986)
[7] Vinen W.F., Springer Ser.in Sol.St.Sci., Ed. K.Ohbayashi & M.Watabe, Vol.79, 189 (1989), Springer Verlag.
[8] Lauter H.J., Schildberg H.P., Godfrin H., Wiechert H. and Haensel R., Can.J.Phys., 65, 1435 (1987) and references therein.
[9] Keshishev K.O., Parshin A.Ya. and Babkin A.B., Sov.Phys.JETP53, 362 (1981)
[10] Lauter H.J., Godfrin H and Wiechert H,: "Phonon Physics", Kollar J. et a. edts., (World Scientific, Singapore 1985) p.842
[11] Lauter H.J., Frank V.L.P., Godfrin H. and Leiderer P., Springer Ser.in Sol.St.Sci., Ed. K.Ohbayashi & M.Watabe, Vol.79, 99 (1989), Springer Verlag.
[12]Papyex is a trademark of Carbon Lorraine, France
[13]"Guide to Neutron Research Facilities" from the ILL
[14] measurements of Moller T. at the ILL
[15] Motteler F.C. and Dash J.G. private communication ; Motteler Ph.D.thesis, Univ. of Washington, 1985
[16] Krotscheck E. this workshop
[17] Woods A.D.B. and Cowley R.A., Rep.Prog.Phys.36, 1135 (1972)
[18] Chan M.H.W. this workshop
[19] Vinen W.F.,Springer Ser.in Sol.St.Sci., Ed. K.Ohbayashi & M.Watabe, Vol.79, 189 (1989), Springer Verlag.
[20] Edwards D.O., Eckardt J.R. and Gasparini F.M., Phys.Rev.A 9, 2070 (1974)
[21] Reut S and Fisher I.Z., Sov.Phys.JETP, 33, 981 (1971)
[22] Saam W.F., Phys.Rev.A8, 1048 (1973).
[23] Godfrin H., Frank V.L.P., Lauter H.J. and Leiderer P., in "Phonons 89",Hunklinger S., Ludwig W. and Weiss G. edts. (World Scientific, Singapore 1989) p.904

INTERFACE BETWEEN TWO THIN ^{3}He-^{4}He LIQUID MIXTURES FILMS

Jean Pierre Laheurte, Jean Claude Noiray and Jean Pierre Romagnan

Laboratoire de Physique de la Matière Condensée (U.A. 190)
Université de Nice-Sophia Antipolis,
Parc Valrose, F-06034 Nice Cedex

The reduction of one dimension size in liquids is of course a quite natural way to enhance the influence of surface effects with respect to bulk contribution. The very first experiments on thin ^{4}He films started in 1949. It was quickly understood that the study of such films, which thicknesses being as low as one or a few atomic layers, opened a new and fascinating research field : the two dimensional (2D) world and also the transition from this 2D to a three dimensional (3D) behaviour. Among the most spectacular results or discoveries on these films study, there is obviously the description of the superfluid transition in films as belonging to an universal class of transitions characteristic of the 2D world. This Kosterlitz Thouless transition describes the order disappearance by the thermal activation of a free vortex in the fluid[1]. Another quite interesting concept can be analysed in its limits with the help of helium mixtures films. It is the notion and existence of interface when it should appear between very thin films. Obviously the spatial extension and the localisation of an interface is directly related to the density profile in fluid systems. The region where the profile density sharply changes from one constant value corresponding to the bulk value of the first phase to another constant value associated to the bulk value of the second phase is the interface region. The spatial extension of the interface can reach a few atomic layers. So what is the meaning of an interface, when the sizes of the two phases become also of the order of a few atomic layers ? Do we expect to observe a one-phase system with a rather continuous density profile or is it possible that the two-phases system will still exist and that the spatial extension of the interface could be reduced to one or less than one atomic layer ? To answer this question we can use ^{3}He-^{4}He mixtures films as a very appropriate experimental system.

The isotopic helium mixtures do not remain homogeneous below 0.872 K. There is a concentration and temperature domain in which is observed a separation between 2 phases[2], with localisation of the dense phase (rich in ^{4}He) at the bottom of the mixture container. This phase separation is connected to the mass and statistics difference between the isotopes. The ^{4}He atoms looking after their superfluid fundamental state are of course disturbed by the presence of too much ^{3}He atoms in the surrounding. So a segregation pulling out ^{3}He atoms from ^{4}He rich region will help to get the ^{4}He superfluid order. On the other hand the mass difference for the two isotopes introduces, in this temperature range, where quantum effects play a major role, a difference for the volume occupied by each isotope atom. So in bulk systems below 0.872K we observe two phases systems where the ^{4}He rich phase is superfluid.

Already in bulk systems the importance of "surface fields" coming from the presence of a substrate and of the free surface has been noticed and discussed. It was explained[3], already in 1967, that the existence of walls induces a ^{4}He preferential adsorption. The reason is quite simple. In the vicinity of the walls, the van der Waals attractive interaction potential between He atoms and the substrate is quite large. In a first approximation it is $V(z) \approx -\theta/z^3$, where z is the atom distance to the walls, while θ is a constant depending on the substrate nature. Typical values for θ are between 30 $(K/(a.l)^3)$ to 50 $(K/(a.l)^3)$ at a distance corresponding to one atomic layer (a = 3.6 Å). So there is, close from the walls, a kind of "potential well". It is easy to understand that the fluid energy will be lower when accumulating the largest number of atoms in this well. As ^{4}He atoms are smaller they will preferably concentrate close to the walls. This argument is not by itself sufficent, as the free energy of the mixture also contains an entropy term. However this gives the physical reason for this ^{4}He preferential adsorption near walls.

On the other side the "surface field" coming from the free surface is increasing locally the ^{3}He concentration. The reason is still coming from a volume size effect, which origin is of course the mass of the isotope. At the free surface the atoms will only have half of the surrounding space filled with a dense liquid phase. So the bonds of these atoms are reduced with respect of those inside the liquid and this increases their kinetic energy. In order to minimize the total energy of the fluid it is clear that we must put at the free surface the smallest possible number of atoms, which means the ^{3}He atoms of larger size.

In some sense these two surface effects working in opposite direction have a tendancy to make a separation more favourable and so trigger such a segregation. This has been observed experimentally[4].

Now when we look at thin mixtures films we could at first sight assume that these surface effects will become more and more efficient than in bulk systems. And so we could expect a phase separation developing before the one detected in bulk systems. However this simple consideration forgets that in the same time the size reduction makes more difficult to get the superfluid order. This is the reason for the decrease of the superfluid transition temperature in ^{4}He films when the thickness is reduced.

Clearly ^{3}He-^{4}He mixtures films are a very rich and complex system in which we can expect to follow the spatial extension of an interface and eventually its disappearance down to films thicknesses of the order of a few atomic layers. This is also probably one of the unique system in which we can "see" a two phase separation with the help of superfluid probes.

Various tentatives to theoretically describe a structural transition in ultra thin binary mixture films were proposed[5]. In 1988 Desideri and Sornette used a mean field model to get concentration profiles[6]. The main idea was to assume that the natural separation in helium mixtures observed in bulk systems and enhanced by the existence of two surface fields (substrate and free surface) was frustrated by the creation of spatial variations in concentrations and superfluid order parameter in the direction perpendicular to the substrate. This description use the same ingredients than those when studying wetting, surface transitions, proximity effects in supraconductors and also the healing length origin for a superfluid film. So this theoretical description is based on a Landau-Ginzburg model with an additional constraint which is global conservation of ^{3}He and ^{4}He atoms in the film. This treatment neglects of course, and it can be a severe source of disturbance, the two dimensional nature of the film. For instance by the finite size effects correction to the three dimensional behaviour of the mixtures but also by not including specifically the two dimensional nature of the "Kosterlitz-Thouless" superfluidity.

Nevertheless the free energy in such a model reads

$$F\{\phi(z),\psi(z)\} = \int d^2r \int_0^d dz\, f(\phi(z), \psi(z)) \qquad (1)$$

with $\phi(z) = X(z) - X_3$ is the difference between the local concentration at a distance z from the walls X(z) and the homogeneous concentration X_3, and $\psi = A\, e^{i\varphi}$ is the superfluid order parameter at z.

The free energy density being

$$f(\phi,\psi) = G(\phi,\psi) + \frac{1}{2}\xi_c^2\left(\frac{d\phi}{dz}\right)^2 + G'_S(\phi) + G''(\phi) + \frac{h^2}{8\pi^2 m^{x2}}(|\nabla\psi|^2) + \gamma\phi.|\psi|^2 \qquad (2)$$

The first term $G(\phi,\psi)$ is the bulk free energy density. The second term accounts for the non uniform ^{3}He concentration and introduces a typical bulk concentration correlation lenght ξ_C; the $G_S(\phi)$ terms describe the existence of two surface fields and their contributions in the film energy. The term proportional to $|\nabla\psi^2|$ is the usual expansion of the thermodynamic superfluid potential and is at the origin of the healing length for superfluids in contact with an interface. Finally the last term represents the coupling between concentration and superfluid orders. To simplify the free energy evaluation drastic approximations were made. The first was to replace the two order parameters problem by a single order parameter ϕ. This is the limit of strong coupling between superfluidity and ^{4}He concentration so that

$$1 - \phi = |\psi|^2 \qquad (3)$$

Then the following expressions were proposed to numerically solve the problem

$$G(\phi) \approx \alpha\,(\phi - \phi_S)^2\,(\phi - \phi_N)^2 \qquad (4)$$

where $\phi_S = X_S(T) - X_3$ and $\phi_N = X_N(T) - X_3$, the X_S and X_N values being the bulk equilibrium concentration of the superfluid and normal phase.

The "α" factor being a slowly varying function of T was taken as constant. The surface contributions were oversimplified by choosing

$$G'_S(\phi) = h_S\phi(o)\,\delta\,(z)$$

and

$$G''_S(\phi) = 0$$

G'_S embodies the influence of van der Waals interactions between liquid and substrate. The choice of a positive surface field value h_S implies to get $\phi(0) < 0$ in order to reduce the energy. This is well the expected effect of ^{4}He preferential adsorption. This mathematical expression of G'_S is however a very crude approximation as it neglects the spatial extension of the van der Waals potential ($\alpha\, z^{-3}$). The other surface contribution G''_S was neglected in view of the large difference between the typical strength of the He-He interaction (~ 3K) compared to the He-substrate interaction (30-50 K). This simplification of course means that the Andreev states for ^{3}He atom in the film are neglected or may be represented by an appropriate scaling of the other surface field h_S.

Finally it appears a renormalisation of the ξ_C correlation length replaced by some ξ length which includes, because of the coupling between ψ and ϕ, all contributions due to spatial variations of the order parameters. Then the free energy density is approximated by the three first terms on the right hand side of equ (2).

Such a problem without constraint was solved by Lipkowsky and Gompper[7]. It predicts abrupt phase separation displaced to lower temperature with respect to bulk system. In the phase diagram at constant temperature the concentration displacement with respect to bulk is in the films scaled by d^{-1}. Furthermore below a critical thickness $d/\xi \approx 4$ it is found that no phase separation occurs anymore. In the Nice group they imposed a constraint coming from the atoms conservation in the films.

Their main conclusion, within the approximation made to numerically solve the problem, was that the profile density does not show the existence of an interface for thin films but rather smooth concentration profile. Even for films of thicknesses reaching 10 atomic layers, the profile was such that the spatial extension of the interface was quite large.

The analysis in terms of surface concentration at the walls or at the free surface gives however some interesting informations. The coexistence domain smoothly decreases with the total film thickness d. It also disappears when $d/\xi \leq 4$ which is reminiscent of the "phase separation" disappearance observed in a film in contact with a reservoir. Obviously the constraint of atoms conservation is rounding out the transition. The evolutions of substrate and liquid-gas interfaces concentrations with ^3He addition to a ^4He film indicate a regime change which could be analyzed in term of "stratification transition". For instance when 4 atomic layers of ^3He are added to 6.2 atomic layers of ^4He. Also the temperature influence on these two extreme concentrations show for a film containing the equivalent of 6.4 atomic layers of ^4He and 3.6 atomic layers of ^3He that a regime change towards "stratification" below 0.4 K could be expected.

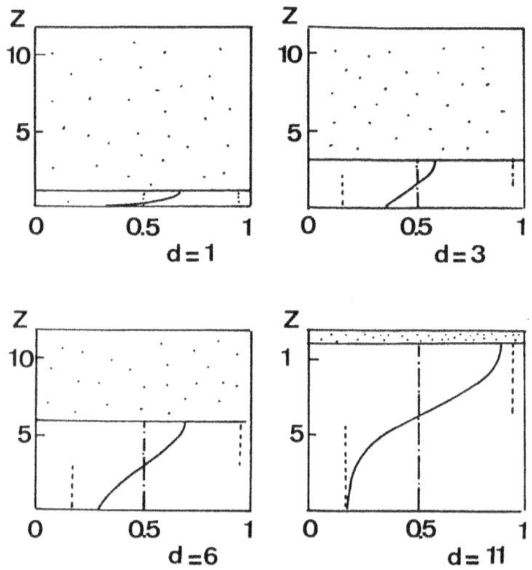

Fig. 1. Evolution, at T = 0.4 K, of the concentration profile vs. z the distance to the walls for several total thicknesses d, with an average concentration X = .5 (d and z are expressed in units of ξ, ξ = 1 a.l.). The dashed line indicate $X_S(T)$ and $X_N(T)$ bulk values of equilibrium phases[6].

This theoretical approach can be compared to recent predictions concerning the influence of wetting on phase equilibria, for a fluid confined between two parallel walls exerting surface fields. In the mean field analysis Fisher and Nakanishi[9] predict a shift of the critical point for a fluid of thickness d with respect to the bulk value. The critical temperature variation scales as : $\Delta T_c = T_{c,\infty} - T_{c,d} \propto d^{-2}$, when the two surface fields are identical. However in 1990, Parry and Evans[9] found that if the two walls exert different and competitive fields, then the phase equilibria is strikingly different. For a fluid wetting one wall and driyng the other one above a critical wetting temperature T_W, then the mean field analysis forbids a two phase system existence above T_W for finite wall separation. It is clear that mean fields descriptions always predict strong constraints which almost exclude the observation of two phase systems in ultra thin films of Helium mixtures. Recent calculations of J. Treiner et al[10] emphasise the fact that all these mean field descriptions put a too strong weight on the cost of the interface creation. In ultra thin films the bending of some order parameter over a distance < 1 atomic layer is energetically prohibitive. Then they are not appropriate to describe systems developing layered structure. Different theoretical approaches [11], specially in the case of submonolayer coverage of ^3He, rely on the filling of localized energy levels for ^3He atoms. They usually find that the fundamental state is localized at the free surface confirming the ^3He preferential adsorption. Recently it was assumed that ^3He atoms could also fill localised states inside the film and closer to the walls. The physical idea being that for thin films a ^3He atom is strongly clamped by the van der Waals potential with the substrate even if it sits at the free surface. So the energy loss by introducing this atom inside the film can be compensated by its kinetic energy reduction, because of the effective mass increase with pressure. In such a case thermal excitation could help the filling of such energy levels and so will be a kind of transition from a pure towards an impure superfluid phase. Simultaneously we expect that the thickness of the interface will increase of even that the interface will disappear.

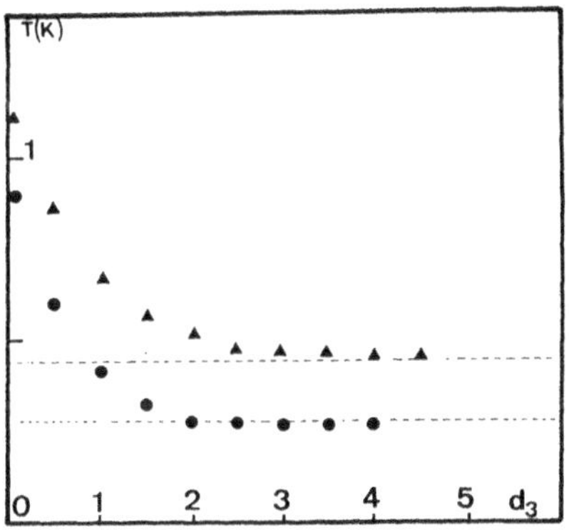

Fig. 2. Superfluid onset temperatures T_o in mixture films for various ^3He thicknesses d_3. Triangles $d_4 = 3$ a.l.; cercles $d_4 = 2.6$ a.l. T_o goes asymptotically towards a limiting value when increasing d_3 (dashed lines). This asymptotical value is the one measured[14] for $d_3 = 12$ a.l [15].

Experimental detection of the existence of a two phase system in thin ^3He-^4He mixture film is of course a challenge, but it is probably one of the rare if not unique fluid in which such observation is possible for very small thicknesses (≈ few atomic layers). The superfluid probe gives us a chance to look at the existence of an interface in ultra thin films and so confirm or not the descriptions related to mean field analysis. The experimental works on mixture films are quite important using various techniques[12,13].

In this work we will use experimental studies of σ_S, the superfluid mass per unit area obtained by Mac Queeney et al [14] with torsional oscillator techniques associated to transition temperatures and third sound velocity measurements from the Nice group [5,15], to try to clarify the film structures. The superfluid transition temperatures T_o of mixture films when adding ^3He are presented fig. 2 for systems with respectively $d_4 = 2.6$ atomic layers and $d_4 = 3$ atomic layers; d_4 represents, as equivalent thickness, the amount of ^4He atoms adsorbed, while d_3 is the equivalent thickness of ^3He atoms adsorbed in the film.

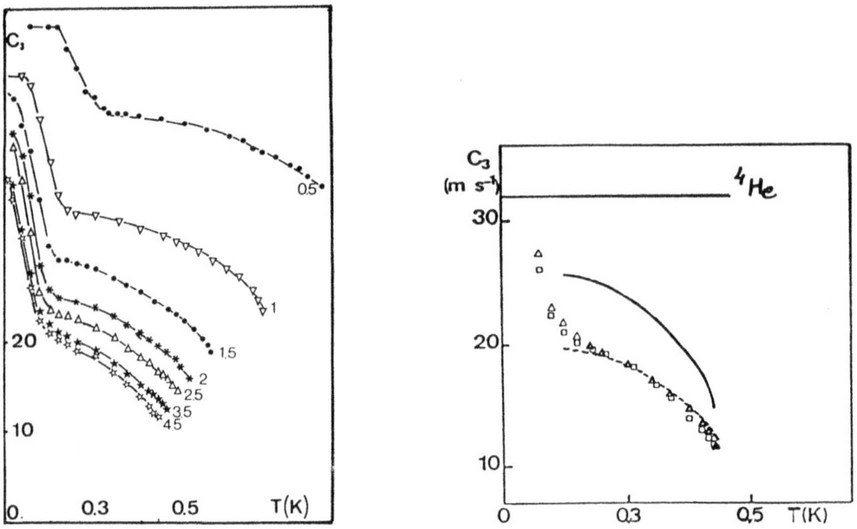

a b

Fig. 3. a) Third sound velocity[15] C_3 (m/s) vs. T in mixtures films with constant ^{4}He coverage $d_4 = 3$ a.l. The ^{3}He coverage is given on each curve in (a.l.). For $d_3 > 4$ a.l. we observe coalescence of the curves.
b) Theoretical analysis (equ. 5) for T > 0.2 K and $d_3 = 4.5$: full line with $d_s = d_4$, dashed line across the experimental points with $d_s = d_4 + 0.4$.

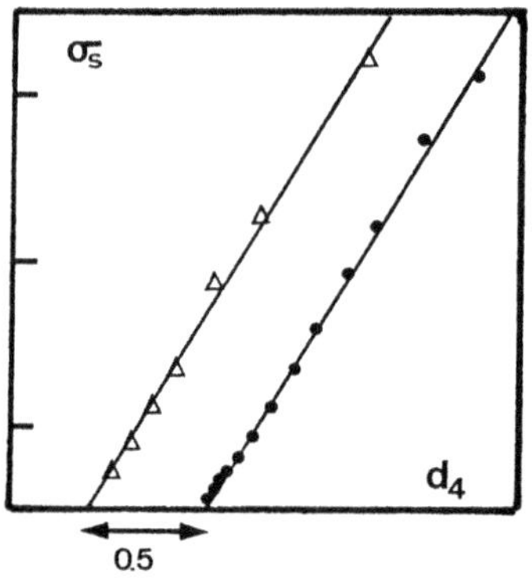

Fig. 4. Evolution of the superfluid density σ_s with the ^{4}He thickness d_4 in the T = 0 limit[14]. Triangles for a pure ^{4}He film, cercles when the ^{4}He film is loaded by $d_3 = 12$a.l. of ^{3}He. In both cases σ_s is linear with d_4 and with the same slope. Normal phase proximity reduces σ_s, as if there was 1/2 extra dead layer.

Clearly T_o decreases as d_3 increases and T_o reaches asymptotically a constant value. This behavior means that for a given amount of ^{4}He, T_o, and so from the Kosterlitz-Thouless theory σ_S which is directly connected to T_o, remains constant as more ^{3}He is added beyond a critical coverage $d_{3c}(d_4)$. This critical coverage is around 4 atomic layers for $d_4 = 3$ and about 2 atomic layers for $d_4 = 2.6$. It must be pointed out that the asymptotic values for T_o are identical to the onset superfluid temperatures measured by McQueeney et al[14] for the same amounts of ^{4}He but loaded by 12 atomic layers of ^{3}He. This result means that above d_{3c} the addition of more ^{3}He does not change the quantitative superfluid nature of the film as σ_S is unchanged. Of course this is a strong evidence that above d_{3c} we have a stratified structure with two phases respectively rich in ^{4}He and ^{3}He. This evidence is confirmed by the measurements of third sound velocity (see fig. 3). The third sound velocity which depends both on σ_S and the superfluid phase thickness d_S is unchanged even in a large range of temperature when $d_3 > d_{3c}$. This shows that both σ_S and the superfluid film thickness are not affected by ^{3}He addition above d_{3c}. This result is a strong confirmation of the existence of a stratified structure in these mixture films even for films of total thickness around 7 atomic layers.

The natural and following question is to find some information about the interfacial region. This is possible by analysing third sound velocity measurements. This velocity C_3 is obtained from : $C^2{}_3 = (\sigma_S/d_S{}^4)\ f(d_n/d_S)$ where the corrective factor $f(d_n/d_S)$ proposed by Ellis et al(16) accounts for the loading by a normal phase of thickness d_n of the superfluid phase of thickness d_S.

Comparing to the unloaded pure ^{4}He film with velocity C_{30} and superfluid mass $\sigma_4(T)$, we find :

$$\frac{C_3^2(T)}{C_{30}^2(T)} = \frac{\sigma_S(T)}{\sigma_4(T)}\ f\left(\frac{d_n}{d_S}\right) \tag{5}$$

The ratio $\sigma_S(T) / \sigma_4(T)$ is given by the experimental work of McQueeney et al [14]. Then the results analysis for a sample in a stratified structure $d_4 = 3$ and $d_3 = 4$ or 4.5 atomic layers fig. (3) show a qualitative agreement for the temperature dependence only. In order to get a quantitative agreement we have to assume that the restoring force is weaker in the mixture film than in the ^{4}He pure film. When extending the superfluid film thickness to $d_S = d_4 + 0.4 = 3.4$, we find a remarkable agreement between the experimental data and this analysis above $T = 0.2$ K. The behavior below $T = 0.2$ K or for lower ^{3}He coverages will not be discussed here, as we intend to focus on systems where a stratified structure is clearly identified. The extension of the superfluid thickness above d_4 is quite reasonable as we expect that the ^{3}He atoms belonging to the superfluid phase and which weaken σ_S will also increase the superfluid film thickness. The influence of these ^{3}He atoms is however quite probably connected to the interfacial zone. Effectively torsional oscillator measurements[14] have shown that at very low temperature the loading of 12 atomic layers of ^{3}He on ^{4}He films was described as if its effect was only to require half atomic layer of ^{4}He in addition before superfluidity occurs with respect to pure ^{4}He (fig. 4).
The superfluid density being identical in pure ^{4}He films and mixture film. So a reasonable picture is a pure ^{4}He phase close to the walls with above the thick ^{3}He film. The ^{3}He presence disturbing the superfluid properties in the very vicinity of the ^{3}He-^{4}He interface. This result is in good agreement with the previous observation that the superfluid phase was extended by 0.4 atomic layers thickness when analysing third sound velocity measurements. It means that probably in the interfacial zone there is a mixture formation which reduces of course the σ_S value by proximity effect of the ^{3}He atoms. However it also increases the effective thickness of the superfluid phase as observed by the weakeness of the restoring force. We notice that in the case of a very sharp interface without ^{4}He-^{3}He exchange, we will observe a σ_S decrease but not a reduction of the restoring force, as d_S will remain at least constant or will decrease.

The spatial extension of the interface seems to be of the order of this one layer mixture zone. Obviously this assumption is in agreement with the observation that σ_S remains contant for $d_3 > d_{3c}$ (see fig. 2). The concentration profile should not be affected by ^{3}He addition from 4 up to 12 layers. Also the coalescence of third sound velocity curves (fig. 3) above d_{3c} confirms again that neither d_S of σ_S change for larger ^{3}He load. This suggests that all interfacial effects are built with already $d_3 = d_{3c}$.

Another confirmation of the interface localisation of ^{3}He penetrating in the superfluid phase comes from the temperature dependence of σ_S[14]. The Cornell group observed that in their thin superfluid film ($d_4 < 3$) loaded by a thick ^{3}He film ($d_3 = 12$), the temperature evolution of σ_S could be described by two mechanisms connected to ^{3}He-^{4}He phase separation. The evaporation of ^{4}He atoms from the superfluid phase to the thick (almost bulk ?) normal phase following the expected $\alpha\ T^{3/2}\ e^{-C/T}$ dependence. The other mechanism being the dissolution of ^{3}He in the thin superfluid phase also following the expected $\alpha\ T^2$ dependence. The fact that this last contribution is independent in their analysis of the ^{4}He thickness shows that the ^{3}He dissolution does not affect the all superfluid film but is rather concentrated in the interfacial layer between the two liquid films.

Of course we expect that the increase of ^{4}He thickness pulling the interface away from the substrate, will strongly reduce the interfacial effects in a stratified structure on the third sound velocity. This is clearly the case in the analysis of the third sound measurements fig. 5 of the Nice group on a mixture film with 6.2 atomic layers of ^{4}He loaded by ^{3}He [5].

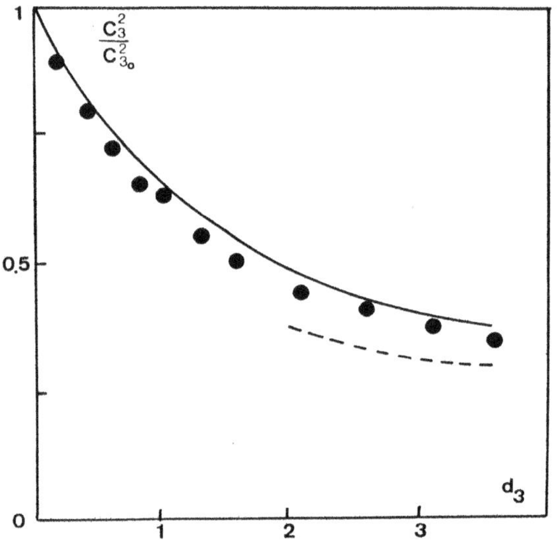

Fig. 5. Square of relative third sound velocity $C_3\ (d_4,d_3) / C_3\ (d_{4,0})$ is presented with respect to ^{3}He coverage d_3 [5], for a constant ^{4}He coverage $d_4 = 6.2$, at $T = 0.4$ K. The full line which describes a complete separation between pure ^{3}He and ^{4}He films fit correctly the data. A displacement of the interface line ≈ 0.4 layer (as in the case of the previous analysis $d_4 = 3$), dashed line, does not work. This strongly suggests a thinning of the interfacial zone as d_4 or the distance to the walls increase.

The description of two pure phases ^4He and ^3He with a sharp interface is quite in agreement with experimental data. This is also a model which fits third sound data of Ellis et al[12] with $d_4 > 5.6$. These results not only confirms that in a stratified structure the interfacial effects resume over a small distance < 1 atomic layer, but they also suggest that the interface becomes sharper as the ^4He thickness increases. Putting that differently it suggests that the interface is more well defined and the ^4He-^3He atoms exchange is reduced when this interface is put far away from the substrate.

In conclusion the analysis of third sound velocity as well as superfluid mass measurements demonstrate the existence of an interface between a superfluid thin film and a normal film even for systems as thin as 5 atomic layers. For stratified structures this interface becomes more and more well defined as the superfluid film thickness increases, or may be what is more pertinent when the interface location is far from the wall. All interfacial effects seem to resume over a very short distance probably in one atomic layer separation between phases. In this interfacial zone we observe ^4He and ^3He atoms exchange between phases which lead to modification and extension of the theoretical boundary line ~ 0.5 atomic layer in models describing σ_S or C_3. This experimental analysis shows that mean field models are not appropriate to describe the layered structure.

References

1. J.M. Kosterlitz and D.J. Thouless, J. Phys. C6, 1181 (1973)
   D.R. Nelson and J.M. Kosterlitz, Phys. Rev. Lett., 39, 1201 (1977)
2. G.E. Watson, J.D. Reppy and R.C. Richardson, Phys. Rev. 188, 384 (1969).
   E.H. Graf, D.M. Lee and J.D. Reppy, Phys. Rev. Lett., 19, 417 (1967).
3. J.R.G. Keyston et J.P. Laheurte, Phys. Lett, 24A, 132 (1967)
4. J.P. Romagnan, J.P. Laheurte, J.C. Noiray and W.F. Saam, J. Low Temp. Phys., 30, 425 (1978)
5. J.P. Laheurte, J.C. Noiray, J.P. Romagnan and D. Sornette, J. Phys. France, 47, 39 (1986); Phys. Rev. Lett. 53, 2421 (1984).
6. J.P. Desideri and D. Sornette, J. Phys. France, 49, 1411 (1988).
7. R. Lipowsky and Gompper, Phys. Rev., 29, 5213 (1984).
8. M.E. Fisher and H. Nakanishi, J. Chem. Phys., 75, 5857 (1981);
   H. Nakanishi and M.E. Fisher, J. Chem. PHys., 78, 3279 (1983).
9. A.O. Parry and R. Evans, Phys. Rev. Lett., 64, 439 (1990).
10. J. Treiner, Private communication
11. D.S. Sherrill and D.O. Edwards, Phys. Rev. B, 31, 1338 (1985);
    F.M. Gasparini, B. Bhattacharyya, and M.J. DiPirro, Phys. Rev. B, 29, 4921 (1984);
12. F.M. Ellis, R.B. Hallock, M.D. Miller and R.A. Guyer, Phys. Rev. Lett., 46, 1461 (1981) ;
    F.M. Ellis and R.B. Hallock, Phys. Rev. B., 29, 497 (1984).
13. B. Bhattacharyya and F.M. Gasparini, Phys. Rev. Lett., 49, 919 (1982);
    X.W. Wang and F.M. Gasparini, Phys. Rev. B, 34, 4916 (1986).
14. D. McQueeney, G. Agnolet and J.D. Reppy, Phys. Rev. Lett., 52, 1325 (1984).
15. J.P. Romagnan, J.P. Laheurte, J.C. Noiray et M. Papoular, Phys. Rev. B, 37, 7 (1988).
16. F.M. Ellis, R.B. Hallock, M.D. Miller and R.A. Guyer, Phys. Rev. Lett., 46, 1461 (1981).

DISCUSSION ON INTERFACES

D. O. Edwards and J. P. Laheurte

After the session there was a lively discussion. In some cases the answers to the questions can be found in the printed reports of the lectures. The following is a partial summary of the rest of the discussion supplied by the participants after the session. It is divided under two headings.

1. The Liquid-solid interface and the roughening transition

Balibar : Comment on the talks by Keshishev and Andreev. Contrar to what was said by Kostia Keshishev, I do not think that there exists any experimental data contradicting the universal relation of roughening

$$k_B T_R = \frac{2}{\pi} \gamma_R a^2$$

The problem when checking this relation is that γ_R is the value of the surface stiffness at T_R and in the precise direction ($\phi = 0$) which undergoes the roughening transition.

T_R is not easy to know with great precision and the crystals have to be very carefully aligned since γ is predicted to show a singularity near $\phi = 0$. However, for the first transition in helium 4 ((0001) surfaces at 1.30K), lots of data exist and I tried to compare our (indirect) check of the universal relation with the results of the various experiments done in Moscow (Babkin 1985, et al Andreeva et al. 1987-89). The agreement is surprisingly good as can be seen in my contribution to the LT19 meeting in Brighton. It confirms that the universal value is 0.315 erg/cm^2; also that the coupling parameter is about 0.58 from the magnitude of the renormalizatim effect; and that the cut-off wavevector for short wavelength fluctuations, which determines the angular width of the renormalization effect, is of order 0.3/a. All these values were justified when fitting Nozieres theory on our experimental measurements (average curvature of crystals around T_R, exponential decay of the step energy, and the critical variation of the growth rate at various driving forces around T_R), and we now see that they agree also very well with the Russian results, especially with Babkin's results which I now consider as the best direct check of the universal relation of roughening. To my opinion, all other possible disagreements with this relation originate in the use of γ values which are not taken at the right temperature or angle.

My second comment concerns Sacha Andreev's talk. I can certainly admit that if the crystal was really invaded by a high density of dislocations, this would affect the roughening transition because the basic ingredient, which is the lattice periodicity in the bulk, would be destroyed. But some surface disorder is not enough. Also I am rather surprised that he tries to reinterpret the results of Wolf et al.(1985) with power laws which have no justification at all and don't go very well through the data points. Wolf's results are the best experimental evidence I ever saw of the growth of facetted crystals via the 2-D nucleation of terraces. Our fits with well established exponential relations are very good over four decades in growth velocity. They are part of our analysis of the roughening transition which shows very good agreement with Nozieres' theory.

Bozler to Balibar : The differences in the theory for surface roughening seem clear. But, it also seems that there are differences in the interpretation of experiments. Are roughening transitions in fact observed? are there any "make or break" experiments that can resolve this?

Balibar : There is no doubt that facets appear or disappear around certain temperatures and that the growth changes character there. The problem is to know whether the measurements are well described by the various possible theories or not. Nozieres' theory, according to which the roughening transition is of Kosterlitz-Thouless type, was shown by us to fit very well three types of measurements (average curvature around T_R, step energy vanishing at $T \leq T_R$, growth rate around T_R) with a unique choice of parameters. It is only this simultaneous agreement with 3 sets of measurements which makes us confident in the validity of this theory. The fact that we now also find very nice agreement with the angular variation of the surface stiffness, as measured by Babkin et al. (198?) and Andreeva et al. (1987-89) is an obvious additional support to our point of view. Unfortunately, the roughening transition is probably too gradual or continuous for a single experiment to be decisive. But no other theory has yet been shown to fit existing data.

Andreev : Direct experimental evidence for the roughening transition could be measurements of the equilibrium facet size or the magnitude of the surface energy cusp near the transition point. All attempts to measure these quantities failed. To explain the failure of their measurements of the equilibrium facet size, Keshisheve et al (1982) noticed that it is probably kinetic faceting which is manifesting itself in the experiment. The experiments of Balibar et al (1985) are indirect and they can be interpreted in different ways. As to the theory, it is probably important to emphasize that the surface excitations, consisting of steps of finite length and semi-circle dislocations as well as the dislocation loops in the bulk of the crystal, destroy the Kosterlitz-Thouless picture of theroughening transition.

Keshishev : I would like to note that the fitting shown by S Balibar, looks no better than a straight line in log-log scale with $\tilde{\alpha} \sim \Theta^{-1/5} \div \Theta^{-1/6}$ passing through our experimental points and through the data of Babkin et al. It is also important to emphasize that according to all experiments there is no temperature dependence of the surface stiffness. It seems also rather difficult to explain in the manner proposed by S Balibar, the too large discrepancy between experimental and theoretical values T_R for the second transition.

Gasparini : In the case of the 4He superfluid transition in films a characteristic of the K-T transition is the change of a hydrodynamic response from being linear as a function of drive to being non-linear. Is there an analgous response in the case of the roughening transition which can be measured?

Balibar : Yes, it is growth velocity v as a function of the corresponding driving force which is the difference in chemical potential $\Delta\mu = \mu_L - \mu_c$ between the liquid and the solid. It was shown beautifully by my student Franzois Gallet (J.Physique 48, 369 1987) that there is a continuous change from non-linear behaviour of $v(\Delta\mu)$ below T_R to linear behaviour $v = k\Delta\mu$ above the roughening temperature T_R. Our precise checks of Nozieres' theory were based on the analysis of this change in response function.

Lauter to Edwards : In the dispersion curve the parameters $a = 1Å^2$ and $\delta = -0.336Å$ have been used (Edwards & Saam). (The parameters take into account to curvature dependence of the surface tension) How would the dispersion curve change if the set from Ebna and Saam($a = 2.83Å2$ $\delta = 0.336Å$) were to be used?

Edwards : The calculation in the review by Edwards & Saam is now obsolete because the surface tension at OK is now known to be 0.356 erg cm^{-2} (through the work of Ikushima and his group) rather than 0.378.erg cm^{-2} The latter value was extrapolated from the old measurements of Atkins and Norahara using the temperature dependence measured at Ohio State. The experiments of Ikushima et al confirm the temperature dependence very precisely. It is not known whether the a and δ fitted using 0.356 erg cm^{-2} would improve the agreement with Ebner and Saam.

2. Helium Films

Harrison : Lauter made a bold statement that the localized modes created by the solid/liquid 4He interface account for the anomalous Kapitza resistance. He finds that two layers of H_2 remove the localized layers. Is there any evidence that two layers of H_2 also changes the Kapitsa resistance up to the Khalatnikov value?

Lauter : The idea is that the localized modes at the liquid-solid 4He interface allow heat to by pass the mismatch of the transmission of phonons across the solid-liquid interface (the Kapitza resistence). A phonon experiment has not been done to prove this idea.

Wyatt to Lauter : How important is the substrate for the thin film effects that have been discussed. The experiments have only been done on a very few materials i.e graphite, nuclepore and vycor.

Lauter : As soon as effects from the holding potential die out with increasing coverage the substrate nature becomes unimportant (eg. Ripplon on a thick film). However, first and second layer structure and excitations (eg. "flat modes") depend heavily on the substrate.

Guyer to Reppy : Is there evidence for a contribution to ρ_n, as seen in torsional on oscillator experiments, that behaves like the rotons of Lauter.

Reppy : For thicker films, with transition temperatures above 1K, the answer is yes.

Guyer to Laheurte : Doesn't the result of Krotscheck, the evolution his ^3He wavefunction with ^4He thickness support your idea about the sharpening of the interface.

Laheurte : Yes, certainly. The work of J Theiner also suggests the main contribution for such behaviour. It comes from the balance between Van de Waals and kinetic energy of a ^3He atom when it is put right at the free surface or just inside the film For low ^4He coverage, the pressure dependence of the ^3He effective mass may be favourable for localization of the ^3He inside the film. For higher ^4He coverage (pressure reduction) the ^3He atoms will sit at the free surface. This will sharpen the interface.

Edwards to Laheurte : Do I understand that the shift of 0.4 atomic layers (a l) in the position of the interface that you have found is actually in the opposite direction to the 0.5 a l shift found by Reppy and his collaborators at Cornell?

Laheurte : No, it comes from the same thing. The shift of superfluid interface of 0.4 a l is obtained by putting some ^3He atoms below the interface and so the surrounding of ^4He by ^3He will reduce superfluid mass in the interfactual zone with respect to pure ^4He. This is the reason for the reduction of σ_s in the Cornell experiment which is analysed as being an addition of 0.5 a l layers of ^4He before you observe superfluid behaviour.

Hallock to Laheurte : The notion of ^3He-^4He mixture interface is attractive in that it appears to explain some of the third sound velocity data you presented. But, our third sound resonance mixture data from 1983-85 and some of your data show that below T~0,15 - 0.2K the third sound velocity has no dramatic temperature dependence. This seems to suggest that the interface disappears and this is rather puzzling. For lower temperatures our the third sound damping studies show structure which is apparently related to the structure we have recently seen by NMR. The recent ideas of Pooloff and Treiner may be relevant. To explain the structure in our third sound frequency data near 0.2K. I speculated at the Banff conference that above 0.2K some of the ^3He might jump into the ^4He film to some unspecified location. It is possible that this location might be a state close to the substrate surface as recently suggested by Pooloff and Treiner. This might serve to explain some of the observations.

ADSORBED QUANTUM GASES

H. Godfrin

Institut Laue-Langevin
156X 38042 Grenoble Cedex, France

PROBLEMS AND RESULTS

There is considerable current interest in the behavior of thin films of quantum gases (H_2, D_2, HD, ^4He, and ^3He) adsorbed on solid substrates. Present experimental and theoretical studies may be divided in two groups. The first one corresponds to phenomena related to the general aspects of structural phase diagrams, the existence of commensurate phases and domain walls, lattice dynamics, and other properties also found in adsorbed classical gases. ^3He, due to its very low mass, is the system with the largest quantum corrections, the "experimental quantum limit". All these properties are therefore very similar for the different quantum gases, as seen on the general aspect of their phase diagrams.

The second group corresponds to intrinsic quantum properties, like superfluidity, nuclear magnetism, degeneracy and quantum exchange. These properties display substantial differences depending on the statistics, the presence of nuclear spin, the symmetry of the molecule, etc., and each gas must be discussed separately.

The choice of the substrate influences both types of properties: for example, the existence of commensurate phases depends on substrate and adsorbate characteristics. Although the present trend is to concentrate the research effort on a few substrates like Grafoil or MgO, technical reasons often dictate other choices. The combination of several substrates and adsorbates yields a much richer behavior than that of three dimensional systems. The presence of the substrate, however, also introduces severe constraints on the experiments: for example, a large background signal in heat capacity measurements, very large neutron diffraction peaks which hide those from the adsorbates, a substantial broadening of NMR linewidths, etc..

A further difficulty results from the small number of atoms involved in a surface physics experiment. The sensitivity of the experimental techniques is often a serious problem, which is usually solved by using substrates with a high specific area (10 to 1000 m2/g). The heterogeneity effects in these substrates are difficult to evaluate. Measurements on a series of substrates of different coherence length have been performed at Seattle (heat capacity) and at Grenoble (neutron scattering), which show that the structure and the phase transitions can be substantially affected by heterogeneity.

As a consequence of these difficulties associated to surface physics experiments, it is often necessary to combine several techniques to achieve a reasonable knowledge of 2D systems, even in the simple cases. In 3D samples, structure determinations can usually be performed directly by diffraction methods. In 2D, results of thermal and dynamical studies had

Excitations in Two-Dimensional and Three-Dimensional Quantum Fluids
Edited by A.G.F. Wyatt and H.J. Lauter, Plenum Press, New York, 1991

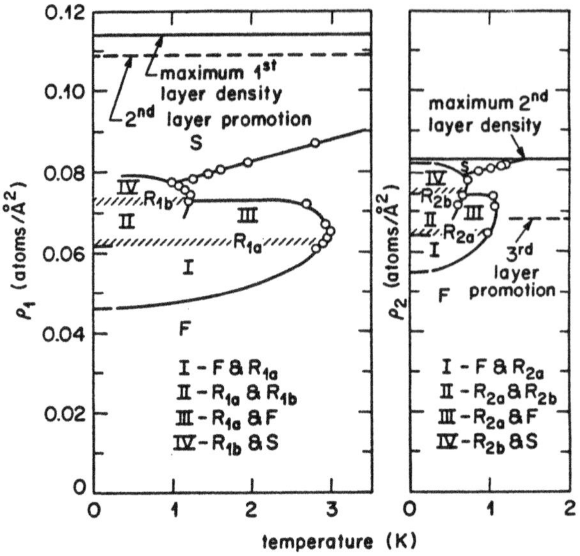

Figure 1. Phase diagram of the first and second atomic layers of ^{3}He adsorbed on graphite. F, R, S denote fluid, registered and solid phases respectively. (ref. 2 and 5).

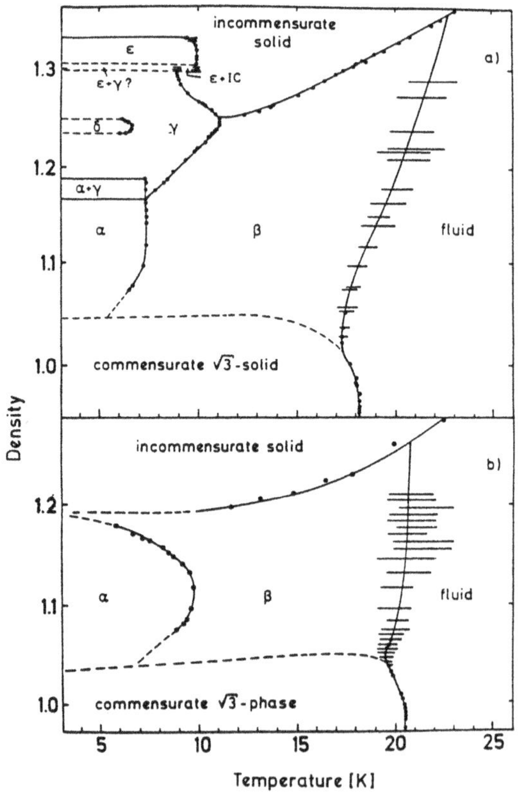

Figure 2. Phase diagrams for D_2 (a) and H_2 (b) adsorbed on graphite, from ref. 15.

to be used in addition to deduce the nature of the structural phases. For instance, it is possible to distinguish a degenerate Fermi fluid (liquid ^3He) from a localized paramagnet (solid ^3He) using NMR or heat capacity techniques.

The early results in this field were obtained on heterogeneous substrates like jeweler's rouge, small carbon particles, alumina powder, mylar, Vycor, etc.. Although many of the ideas used today to describe two-dimensional systems were already developed at this stage, the results were mainly qualitative, due to the poor characterization of the substrates and often to the strong three-dimensional nature of the investigated films. This situation has changed substantially since the introduction of exfoliated graphite substrates by the Nancy group and the extensive measurements performed on adsorbed films by the Seattle group (1,2), which opened a new field of research in surface physics. Besides graphite, another system has also been used which can be considered as a good substrate: preplated solid surfaces. Although these systems are somewhat more difficult to characterize, they have also been used with good results.

The development of the experimental techniques during the last years has resulted in high quality data, obtained by different methods, on well characterized systems. It is possible to compare now the results of different techniques: the experimental data base covers now most of the phase diagram of Helium and Hydrogen films adsorbed on graphite and MgO. Solid, liquid, commensurate phases and coexistence regions have different signatures in NMR, neutron scattering, heat capacity or adsorption isotherm measurements; furthermore, the thermal variation of these properties often provides a good physical understanding of the nature of the two-dimensional phases.

The physics of adsorbed quantum gases has become quantitative. This quality has introduced a new difficulty: it is now necessary to establish the absolute value of the coverages (areal density of the adsorbate) with higher precision; that is, experiments performed on the same system in different laboratories use coverage scales which must be compared within 1% at least. A typical sample presents a surface area of tens of square meters and the amount of gas adsorbed onto the substrate is therefore on the order of millimoles, which can be easily measured. Adsorption isotherms were used initially to determine the "monolayer coverage", but this provided only a semi-quantitative estimate. The heat capacity melting peak of the commensurate phase was then used (2) to provide an absolute coverage scale: the areal density obtained from the graphite lattice parameter is 0.0636 atoms/$Å^2$. As already noted by the Seattle group, the number of atoms needed to obtain the commensurate phase depends slightly on the adsorbed gas, an effect which may be due to zero point motion. Also, at the high temperatures where the "best commensurable phase" melts the number of thermal vacancies and insterstitials is large. The real commensurable coverage is not properly defined by this measurement, if an accuracy better than 5% is required.

The "best commensurable phase" can be investigated by neutron scattering at low temperatures (3). The coverage providing the tallest diffraction peak may be considered to correspond to 0.0636 atoms/$Å^2$. However, this does not agree with the lattice parameters determined in the first layer incommensurate phase: in this large range of the phase diagram an homogeneous triangular phase is assumed, and the measured lattice parameter allows a direct determination of the areal density (3). A 5% discrepancy is observed with respect to the neutron commensurable phase determination. The additional amount of gas needed may be due to the heterogeneity of the substrate; this implies that the substrate is not characterized by a unique value of its surface area, beyond a 5% precision: this value will depend on the coverage region investigated, and the effects of heterogeneity should be carefully estimated for the different types of measurements.

The ^3He NMR results from Grenoble (4), for example, are expressed in a coverage scale based on the neutron diffraction values in the incommensurate phase. To compare coverages, one scales along the coverage axis the ^3He adsorption isotherms obtained in the neutron scattering and that from the NMR experiments. Given the excellent accuracy of the adsorption isotherms, the procedure is in fact quite reliable and does not introduce relevant errors. The neutron scattering coverage scale is believed to be the correct one for coverages near and above monolayer completion, where the defects in the graphite corrugation play a minor role. Recent ^3He heat capacity measurements (5), on the other hand, are based on the coverage scale of the nitrogen adsorption isotherm at 74 Kelvin. Although the areal densities of this experiment seem to correspond well with those obtained at Grenoble, this may be due to the combination of several factors.

The discrepancy between the coverage scales is presently on the order of 5%, and even larger figures are commonly found. Considering that this is typically the difference in density between solid and fluid phases, it is not surprising to find modifications of the phase diagram within such narrow coverage ranges.

Coverages have been often characterized in the literature by the determination of "layer promotion" coverages; this concept is ill-defined, because it depends on the sensitivity of the technique used to observe atoms promoted to the successive layers. Combining different techniques there is presently a more precise knowledge of the areal density of each layer as a function of the total coverage. Neutron scattering, for instance, provides good values for the evolution of the density of solid layers as a function of the total coverage, whereas heat capacity is able to detect accurately even small amounts of atoms in fluid phases. More work on the coverage scale and its limitations is needed.

The experimental progress has started to motivate a substantial theoretical development; in fact, a renewed interest, since much of the experimental interest on the adsorbed gases was due to the theoretical work performed more than one decade ago.

One of the most difficult problems is the correct evaluation of the interaction potentials. Our knowledge of the adsorption potentials is still limited, and the interaction of gases with solid surfaces is an important subject of research (6). Recently, infrared absorption techniques have been used to study induced dipolar interactions (7). The lateral variation of the potential, or corrugation, is the object of discussions (8), even for classical gases adsorbed on graphite.

The interatomic potentials, on the other hand, are very well known. Several numerical simulations and calculations of the structure and the dynamics of rare gases on graphite have been performed recently (9-13). In some cases, a detailed comparison is possible between experiments and theoretical predictions; a typical example is the zone center phonon gap of the commensurate phase: the dynamics constitute an excellent tool to investigate the accuracy of the potentials (8,9). The excitations of the incommensurate phases, on the other hand, have been studied directly only very recently (8): the sensitivity of inelastic neutron studies allows now preliminary determinations of the phonon density of states; most of the information on the dynamics has been derived indirectly from heat capacity measurements. (2,5). Numerical calculations exist mainly for the commensurate phases, and results for the incommensurate phases are beginning to be available (14).

The main features of the phase diagrams of adsorbed quantum gases are well known now; this includes the $\sqrt{3}$ commensurate phases, and the incommensurate phases found in the first and second atomic layers. Recent studies, however, have revealed the existence of more complex commensurate phases . The submonolayer phase diagram of the hydrogen isotopes is particularly interesting (15). In ^{3}He, it has been proposed (see discussion in ref. 5) that the second layer forms a commensurate phase with respect to the first layer. Also, the first layer may become commensurate with the graphite substrate at high coverages (3). Theoretical results are scarce on this topic, although numerical simulations seem to be particularly suited for the determination of these structural aspects.

A novel research domain is constituted by the domain-wall phases, which exist as solids and even as liquids. Several aspects of the commensurate-incommensurate phase transitions are of present theoretical interest (16), and the possibility to vary continuously the atomic density makes the adsorbed quantum gases a particularly attractive system.

The investigation of melting in two dimensions, which was one of the original motivations of the investigations in films, is also an active field of research (17). The situation is still controversial; the order and the universal character of the transition, the influence of heterogeneity, the number of defects at melting, the possibility of a Kosterlitz-Thouless phase have been discussed in detail.

Wetting of films on solid substrates is another controversial subject; classical and quantum gases are presently being investigated by adsorption isotherms and quarz-crystal microbalance techniques (18).

Multilayer growth of hydrogen films has been studied by adsorption isotherms and heat capacity techniques on MgO and graphite substrates (15,19). The multilayer phase diagrams display a very interesting periodic structure in coverage, and layering critical points are observed, establishing a connection with the roughening transition of 3D crystals. This effect and the possibility to observe supercooled liquid phases will be discussed at this workshop. The latter point is related to the search for a possible superfluid phase of hydrogen.

^{3}He and ^{4}He films have been used to study Fermi and Bose gases and liquids in two dimensions. Early studies were limited to the high temperature regime, involving essentially virial expansions. Present investigations cover the temperature regime where those systems display substantial degeneracy effects (5,20). ^{3}He, in particular, has been studied by NMR and heat capacity techniques, and several new results will be discussed in this workshop (21). Multilayer ^{3}He films display a thickness dependent magnetization (22) which has been attributed to the particular level distribution of non-interacting Fermi particles in a finite size box (23). The substantial degree of layering found in numerical calculations of the density profile of liquid helium films (11) makes the previously mentioned interpretation somewhat uncertain; this subject will be discussed during this workshop.

There is presently a renewed interest in the excitations in 4He films; electrons, neutrons and hydrogen atoms constitute efficient probes to test the inelastic response of this system. The ripplon dispersion relation has been investigated for wavevectors up to $2Å^{-1}$ (24,25). Theoretical calculations have been developed for this system.

^{3}He solid films at very low temperatures also allow to investigate the properties of 2D magnets. Ferromagnetic and antiferromagnetic exchange interactions have been observed at Grenoble by NMR techniques in submonolayer and multilayer films (4). Recent heat capacity (5) and NMR (21,26) data provide a rather complete experimental picture, which has motivated a substantial theoretical activity. Ruderman-Kittel (27) and multiple spin exchange (28) interactions are believed to originate the large nuclear magnetic effects observed experimentally. At high coverages the second layer magnetism can be described with the 2D ferromagnetic Heisenberg model (29); this magnetism persists in the full pores regime, as shown by Bozler and coworkers (30); nuclear ordering is thought to be due to a breaking of the Heisenberg symmetry due to dipolar and size effects.

The 2D antiferromagnet is of course of present interest, due to the research on high temperature superconductivity. Research on the second layer of ^{3}He in the antiferromagnetic regime is in progress.

CONCLUSIONS

There is presently a large spectrum of investigations involving adsorbed films, requiring complementary techniques applied to well chosen substrate-adsorbate combinations. A characteristic of the recent results is certainly their quantitative aspect. A coherent picture of the structure and excitations of quantum gases adsorbed on "good" substrates has gradually been constructed, and investigation on heterogeneous or "bad" substrates is in progress: surfaces, interfaces and porous media, which constitute an important part of our physical environment, are now much better understood.

REFERENCES

1. J. G. Dash, Films on Solid Surfaces, Academic Press, N.Y. (1975).

2. M. Bretz, J. G. Dash, D.C. Hickernell, E.O. McLean and O.E. Vilches, Phys. Rev. A8 (1973) 1589 and A9 (1974) 2814.

3. H.J. Lauter, H. Godfrin, V.L.P. Frank and H.P. Schildberg, Proc. 19th. Conf. on Low Temp. Phys., (1990) to be published; H.J. Lauter, H.P. Schildberg, H. Godfrin, H. Wiechert and R. Haensel, Can. J. Phys. 65 (1987) 1435; S. K. Sinha, Methods of Experimental Physics (1987) Academic Press, Vol. 23 Part B, p. 1.

4. H. Franco, R.E. Rapp and H. Godfrin, Phys. Rev. Lett. 57 (1986) 1161.

5. D.S. Greywall, Phys. Rev. B41 (1990) 1842; D.S. Greywall and P.A. Busch, Phys. Rev. Lett. 62 (1989) 1868 and Phys. Rev. Lett. 65 (1990) 64 and to be published.

6. M.W. Cole, D.R. Frankl, and D.L. Goodstein, Rev. Mod. Phys. 53 (1981) 199.

7. T.E. Huber and C.A. Huber, Phys. Rev. B 40 (1989) 12527.

8. V.L.P. Frank,H.J. Lauter, H. Godfrin and P. Leiderer, this conference.

9. J.M. Gottlieb and L.W. Bruch, Phys. Rev. B40 (1989) 148.

10. J.P. Bouchaud and C. Lhuillier, Z. Phys. B75 (1989) 283; B. Bernu, C. Lhuillier and D.M. Ceperley, to be published.

11. E. Krotscheck, Phys. Rev. B32 (1977) 5713.

12. F.F. Abraham and J.Q. Broughton, Phys. Rev. Lett. 59 (1987) 64.

13. P.A. Withlock, G.V. Chester and M.H. Kalos, Phys. Rev. B38 (1988) 2418.

14. L.W. Bruch and J.M. Gottlieb, Phys. Rev. B37 (1988) 4920.

15. J. Cui, S.C. Fain, H. Freimuth, H. Wiechert, H.P. Schildberg and H.J. Lauter, Phys. Rev. Lett. 60 (1988) 1848; H. Freimuth, H. Wiechert, H.P. Schildberg and H.J. Lauter, Phys. Rev. B (1990) in print.

16. M. den Nijs, in Phase Transitions, Academic Press, 1988, Vol. 12, p. 219.

17. K. Strandburg, Rev. Mod. Phys. 60 (1988) 161; M.S. Pettersen, M.J. Lysek and D.L. Goodstein, Phys. Rev. B 40 (1989) 4938.

18. A.D. Migone, A. Hofmann, J.G. Dash and O.E. Vilches, Phys. Rev. B 37 (1988) 5440.

19. J. Ma, D.L. Kingsbury, F. Liu and O.E. Vilches, Phys. Rev. Lett. 61 (1988) 2348.

20. T.W. Kenny and P.L. Richards, Phys. Rev. Lett. 64 (1990) 2386.

21. J. Saunders, C.P. Lusher and B.P. Cowan, and C.P. Lusher, J. Saunders and B.P. Cowan, these proceedings.

22. R. Higley, D. Sprague, R. Hallock, Phys. Rev. Lett. 63 (1989) 2570
J. M. Valles, R. H. Higley, B.R. Johnson and R. Hallock, Phys. Rev. Lett. 60 (1988) 428.

23. R.A. Guyer, Phys. Rev. Lett. 39 (1977) 1091.
R.A. Guyer, K. McCall and D. Sprague, Phys. Rev. B40 (1989) 7417.

24. J.J. Berckhout, O.J. Luiten,, I.D. Setija, T.W. Hijmans, T. Mizusaki and J.T.M. Walraven, Phys. Rev. Lett. 63 (1989) 1689.

25. H.J. Lauter, H. Godfrin, V.L.P. Frank and P. Leiderer, this conference.

26. H. Godfrin, R.E. Rapp and H.J. Lauter, Proc. 19th. Conf. on Low Temp. Phys., (1990) to be published.

27. S. Tasaki, Prog. Theor. Phys. 79 (1988) 1311.

28. M. Roger, Phys. Rev. Lett. 64 (1990) 297 and these proceedings.

29. H. Godfrin, R.R. Ruel and D.D. Osheroff, Phys. Rev. Lett. 60 (1988) 305 and J.Phys.(Paris) C8-49 (1988) 2045.

30. L.J. Friedman, A.L. Thompson, C.M. Gould and H.M. Bozler, Phys. Rev. Lett. 62 (1989) 1635.

NUCLEAR MAGNETISM OF ^{3}HE FILMS ADSORBED ON THE SURFACE OF GRAPHITE

J. Saunders, C.P. Lusher, B.P. Cowan

Millikelvin Laboratory
Royal Holloway and Bedford New College
University of London, Egham, Surrey, TW20 0EX, U.K.

1. INTRODUCTION

The properties of ^{3}He adsorbed on the surface of graphite exhibit a rich variety of phenomena as the areal density is changed. The system permits the study of two dimensional quantum liquids and solids, as well as 2D solids with an overlayer of fluid. Also because of the existence of registered solids, certainly in the first layer and possibly in the second layer, one can study quantum exchange in systems for which there is no 3D equivalent. In addition the multilayer fluid film shows evidence of quantum size effects.

Here we describe measurements of the nuclear magnetic susceptibility of these systems. At low temperatures, where effects of quantum degeneracy are important, the susceptibility clearly distinguishes between localised and delocalised phases. In solid phases deviations from Curie law at the lowest temperatures provide a measure of the atomic exchange. In the fluid phases the temperature independent susceptibility observed at low temperatures provides a measure of quasiparticle interactions in the film. In regions for which it appears the film consists of both solid and fluid phases it is possible to determine the number of spins in each component and in some cases their density.

This article will discuss: (1) exchange in two dimensional solid ^{3}He, its dependence on areal density and the relative importance of the possible mechanisms of atomic exchange; (2) interactions in two dimensional Fermi liquids; (3) evidence for quantum size effects in multilayer films; (4) phase transitions in the submonolayer film; (5) evidence for the delocalisation of ^{3}He atoms and a structural phase transition in dilute solid isotopic mixture films. All the nuclear magnetic susceptibility measurements reported here were made by continuous wave NMR at 1.08MHz; the substrate used was exfoliated graphite (grafoil); temperatures were measured by a ^{3}He melting curve thermometer. These measurements are complemented by studies of the heat capacity and where necessary we refer to recent measurements at low millikelvin temperatures.

2. EXCHANGE IN TWO DIMENSIONAL SOLID ^{3}HE

There are a number of forms of two dimensional solid provided by helium adsorbed on graphite. By studying the first and second layer solids in both the incommensurate and registered phases it should be possible to understand the importance of their structure and the influence of the surface binding potential on atomic exchange processes. The registered phases also serve to localise the ^{3}He atoms at significantly larger interatomic spacing than in bulk. In addition it is possible to investigate the influence of fluid overlayers, which provide an indirect channel for exchange.

The submonolayer solid, above an areal density of approximately 0.08Å^{-2}, is known from neutron scattering studies[1] to form an incommensurate solid with a triangular lattice. At 0.064Å^{-2} the helium is in perfect 1/3 registry with the grafoil substrate and at intermediate coverages a mixed solid phase is found, whose structure is not firmly established[2]. As the density is increased through the incommensurate solid the atomic exchange frequency determined from NMR spin relaxation measurements[3] decreases as expected with areal density ($J \sim \rho^{-26}$). The maximum exchange energy observable at 0.08Å^{-2}, is around 1mK. However measurements of the susceptibility[4,5] find no departures from Curie law at low temperatures and suggest J< 50μK. On the other hand heat capacity measurements[6] do find a spin contribution with a T^{-2} dependence, the coefficient of which (e_2) is consistent with an exchange energy of 0.30mK. These discrepancies have yet to be understood and may have their origin in multiple spin exchange[7,8]. A competition exists between the various ring exchanges which are antiferromagnetic or ferromagnetic for cycles involving even or odd numbers of atoms respectively. This cancellation affects most strongly the calculated Weiss constant; the heat capacity (coefficient e_2) is not as sensitive to the sign of the exchange constants. It has been suggested, in the context of bulk solid, that while the picture of multiple spin exchange provides a useful semiempirical description and is supported by Monte Carlo calculations, it may not be the physically most appropriate model. However the problem of exchange in the two dimensional solid is further elucidated by experiments on the second layer, where the atomic exchange interaction appears to be considerably stronger than in the submonolayer film.

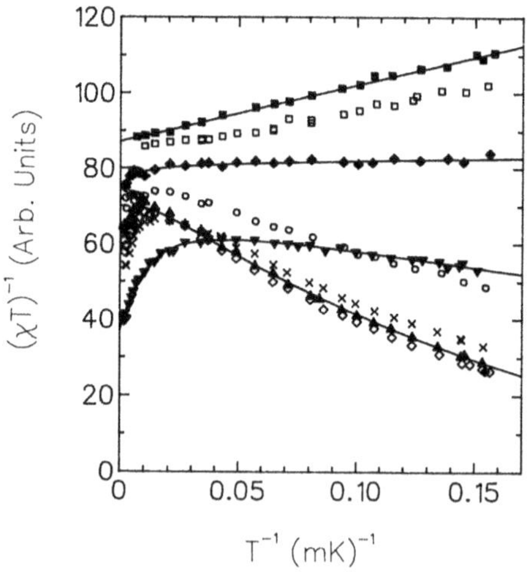

Fig. 1. Temperature dependence of the susceptibility at surface densities $\rho_T=\rho_3+0.114$ Å^{-2}; 0.180(■),0.182(□),0.190(◆) 0.198(○),0.210(▲),0.220(◇),0.230(×),0.270(▼). Solid curves are fits to the data.

Promotion to the second layer commences near $0.108Å^{-2}$, with the second layer atoms first forming a two dimensional fluid. A cusp-like feature in the NMR linewidth[10] provides a precise indicator of promotion and was used to determine all the areal densities in this work. In these experiments, in order to study the second layer, we coated the grafoil with a monolayer of ^4He, which is preferentially adsorbed at the surface. The ^4He is nonmagnetic and also forms a dense relatively incompressible solid. (As will be seen the magnetic anomalies observed here are consistent with earlier work on pure films[11,12], confirming the view that the compressed first layer is inert). For second layer areal densities (ρ_2) less than $0.055Å^{-2}$ the nuclear magnetic susceptibility is consistent with Fermi liquid behaviour (section 3). Above this coverage solidification of the second ^3He layer is clearly observed in the nuclear magnetic susceptibility, confirming the conclusion of heat capacity measurements[13]. We find that the susceptibility first follows a Curie Weiss law at $\rho_2=0.066Å^{-2}$ (fig.1), and take this to indicate the completion of solidification. The susceptibility is consistent with that of a solid with antiferromagnetic exchange. In the case of the pure ^3He film it has been proposed that at $0.064Å^{-2}$ the second layer atoms are localised into a structure in 4/7 registry with the first helium layer[14].

At a second layer density of $0.069Å^{-2}$ promotion to the third fluid layer occurs. This is clearly seen from the high temperature behaviour of the susceptibility (fig.1) and from the heat capacity isotherms[6]. As the coverage is increased atoms go both into the fluid third layer and compress the second layer solid. Now in order to understand the coverage dependence of the exchange and the possible structure of the second layer solid it is important to know its areal density. This can be obtained, quite directly, from the susceptibility measurements. The measured susceptibility is fit to the sum of two terms, a solid and a fluid contribution. The solid term is taken to be the ten term high temperature series expansion for a Heisenberg nearest neighbour ferromagnet on a 2D triangular lattice[15]. The fluid is modelled as a 2D Fermi liquid, using the exact temperature dependence of a 2D gas, $\chi(T)=(C/T_F)(1-\exp(-T_F/T))$. The effective Fermi temperature is left as an adjustable parameter to account for interactions. Thus in fitting the data it is possible to determine the effective exchange interaction J, the fluid magnetic degeneracy temperature T_F and the ratio of numbers of fluid and solid atoms N_f/N_s. From N_f/N_s ,knowing the total areal density and attributing the solid signal to the second layer, the surface density of the second solid layer can be determined.

The dependence of these parameters on coverage (shown as the equivalent total surface density for a pure film $\rho_T=0.114Å^{-2}+\rho_3$, where ρ_3 is the areal density of ^3He in this work) are shown in Fig.2 . The peak in the exchange frequency corresponds to that found earlier in the magnetisation isotherms[11]. Remarkably, the dependence of the second layer density on coverage is not smooth but distinctly shows evidence of what we refer to as stepwise compression. There are two pieces of experimental evidence which corroborate this data. (i) At a coverage $0.293Å^{-2}$ neutron scattering studies[16] find a triangular lattice of density $0.084Å^{-2}$. (ii) The density dependence of the melting temperature T_m of the second layer solid[17] shows similar structure to ρ_2 (within a small scaling of coverages). This behaviour is consistent with[18] $dT_m/d\rho \approx 63KÅ^2$.

The first plateau at $\rho_2 \approx 0.075Å^{-2}$ occurs in the vicinity of the exchange peak. It is rather suggestive that the density $0.076Å^{-2}$ is in 2/3 registry with the helium first layer[19]. This density is also in 2/5 registry with the graphite substrate, but the periodicity of the substrate potential is negligible in the second layer[14]. Thus one interpretation is that this 2/3 structure is stable over the coverage range $0.21<\rho_T<0.24Å^{-2}$ and resists distortion; atoms rather go into the fluid third layer. This should be contrasted with the behaviour on first solidification; before and after third layer promotion ρ_2 increases until the plateau at 0.075Å^{-2} is reached. It is plausible that in the coverage range $0.18<\rho_T<0.21$Å^{-2} the second layer solid consists of a coexistence of 4/7 and 2/3 registered phase[6] or possibly passes through a sequence of registered phases[20]. At the second plateau the second layer consists of incommensurate solid.

It is instructive to look at the dependence of measured second layer properties on the areal density of the second layer, using the coverage dependence of the second layer density determined here. We show in fig.3 the effective exchange energy inferred from our nuclear magnetic susceptibility data and isotherms of the spin heat capacity inferred from measurements of the total heat capacity of the multilayer pure ^{3}He film[6]. Both quantities show a remarkable cusp-like peak at 0.075Å^{-2}. It is tempting to conclude that this strong exchange anomaly must be associated with the particular (probably registered) structure at this density. We should point out that the original analysis of the spin heat capacity[6] used an indirectly inferred second layer density and consequently found the peak to occur at 0.08Å^{-2} where a coexistence of 2/3 phase and incommensurate phase is likely.

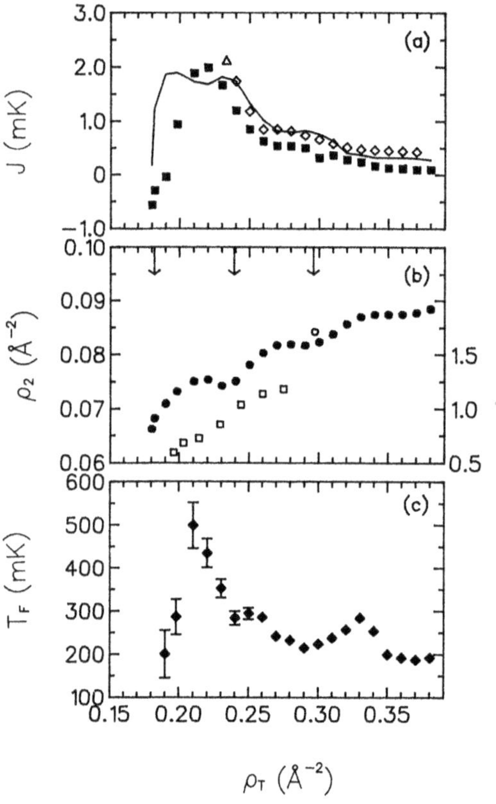

Fig. 2. Dependence on total surface density ρ_T of
a) exchange constant J; these results (■), ref.12 (△), ref.6 (◇). Solid line is a result of multiple spin exchange calculation using these ρ_2.
b) second layer solid areal density ρ_2; these results (●), ref.16 (○). Arrows indicate "layer promotions" as identified in ref.6 . Also shown is melting temperature of second layer solid, T_m (□).
c) fluid magnetic degeneracy temperature T_F.

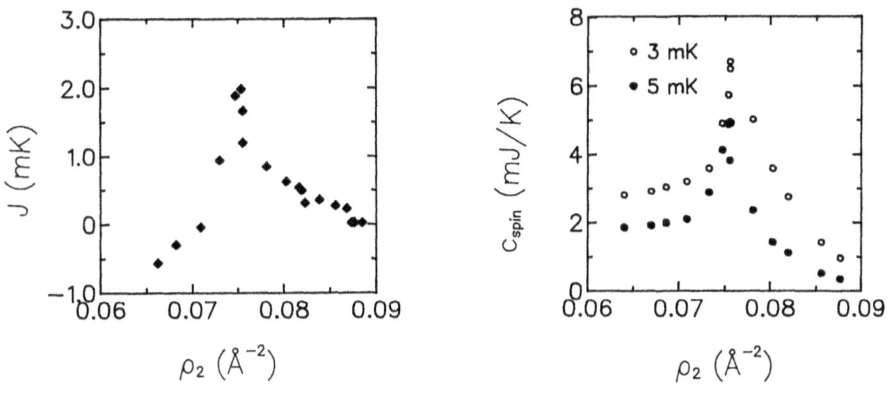

Fig. 3. Dependence of exchange constant and spin heat capacity on areal density of second layer solid.

In the submonolayer film, where the solid surface density is known unambiguously, recent heat capacity measurements[21] find a spin contribution whose isotherms are very similar in form to that of the second layer isotherms as reanalysed here. In particular a sharp peak is found at 0.076Å^{-2}. This density would correspond to a 2/5 registered phase with respect to the substrate. Thus the anomalies in both the first and second layers appear to be related to registered structures, unless there is some alternative reason for a "magic" ^{3}He density. If the first layer anomaly is attributable to exchange it is not known whether this is antiferromagnetic or ferromagnetic since, as already mentioned, no evidence of exchange effects has yet been detected in nuclear susceptibility measurements at submonolayer coverages.

Other features in the second layer isotherms (fig.3) which can be identified more tentatively are found in the vicinity of promotion to the third layer and around 0.08Å^{-2} at the formation of the incommensurate solid. Some of the variation in the exchange at the "cusp" may be attributable to the influence of the third fluid overlayer.

The theoretical discussion of exchange in this system in terms of the multiple spin exchange model[8] does not include the influence of the second layer structure. In these calculations the second layer solid has been assumed to be a triangular lattice. The gross features of the density dependence of the exchange are reproduced by this model when our measured second layer areal densities are used. The model can account for the crossover from antiferromagnetic to ferromagnetic exchange through the competition between the antiferromagnetic and ferromagnetic ring exchanges. An effective exchange constant $J_{eff}= \theta/3$ is given by $-(J_p +3K +5S/8)$, where $J_p=J_{nn}-2J_T$ is the effective pair exchange constant and J_{nn}, J_T, K, S are the exchange constants for two, three, four and six particle exchange. A further simplifying assumption is to take $K \approx S$ and write $K \approx S = rJ_p$; r is then a frustration parameter giving the ratio in strength of the antiferromagnetic to ferromagnetic exchange cycles. Then $\theta=-3J_p(1+3.625r)$. In the second layer antiferromagnetic solid, prior to third layer promotion, if the parameter r is close to the value for exact frustration ($r=-0.275$) then $\theta \ll |3J_p|$ and the inferred entropy of $0.25k_B \ln 2$ at the heat capacity peak is accounted for.

From fits of the susceptibility to $\chi=C/(T-\theta+B/T)$ we determine the Curie Weiss constant to be $\theta=-1.7$mK and -0.9mK at second layer densities 0.066 and 0.068 respectively. In this model[22] $B= J_p^2(3+3r/2-963r^2/64)$, and at the higher density it is possible to infer separately $r \approx -0.3$. There is therefore consistency between both sets of measurements and the model, since the heat capacity data at $\rho_T=0.184Å^{-2}$ are best fit by $J_p=-4.8mK$ and $r=-0.25$.

An estimate of the density dependence of exchange from WKB calculations[7] is given by $J_n/k_B=73C_n s_n(\sigma/a)^{19/2}\exp[-S_n(\sigma/a)^5]$ (K), where a is the interatomic spacing, σ is a pair potential parameter, s_n is a symmetry factor, C_n is a prefactor of order unity, and n denotes the number of participating atoms. The crossover to ferromagnetic exchange is a consequence of the hierarchy of dimensionless actions $S_3<S_4\sim S_6<S_2$. It has been argued that atomic exchange involving motion in the third dimension would further favour three particle exchange. Indeed the inclusion in the description of exchange of the z component of the atomic wavefunction, which reflects the atomic zero point motion in the surface binding potential, may well account for the relative strength of exchange in the second layer as compared to the first layer. There is a recognised need for Monte Carlo calculations of such effects. The influence of the fluid overlayer, which would tend to introduce a steric hindrance of out of plane motion, and contribute to the coverage dependence of the exchange also requires consideration. The solid curve[23] in fig.2a. gives the calculated effective exchange constant $\theta/3$. It gives a reasonable account of the density dependence of the exchange except in the vicinity of the peak. As we have seen the discrepancy here may be attributable to special features of the exchange in a registered solid.

There are however other mechanisms which can operate in the second layer. Firstly there is a form of indirect exchange[24] in which an atom is promoted into a virtual third layer fluid state, the vacancy tunnels to a neighbouring site and the atom returns to the vacant site. This results in antiferromagnetic nearest neighbour exchange, which is suppressed as the fluid overlayer forms. In the presence of fluid overlayers indirect exchange of the RKKY type can occur, which is ferromagnetic[25]. Thus competition between these two mechanisms will also lead to a crossover from antiferromagnetic to ferromagnetic behaviour. Calculations of the latter process rely on a model of the multilayer film, and show an oscillatory component to the dependence of the exchange on surface density associated with the process of layer by layer growth[26,27]. As discussed in section (4) our results for the fluid susceptibility are consistent with "layering" and in fig.2 we show the oscillations in the fluid magnetic degeneracy temperature. While the observed structure in J does not provide conclusive evidence for the presence of indirect exchange, there is however a pronounced dip in J that occurs at $0.30Å^{-2}$. This coincides as predicted with promotion to the fifth layer.

In all cases the nuclear magnetic susceptibility is well described by the high temperature series expansion using a single effective exchange constant. This is not the case for the second layer spin heat capacity for $\rho_T<0.26Å^{-2}$ or the submonolayer solid if not incommensurate. In the most extreme case the heat capacity is found to vary as T^{-1} rather than T^{-2}. According to our density measurements at a coverage of $\rho_T= 0.26Å^{-2}$ the second layer solid density is $0.08Å^{-2}$ corresponding to the border of the incommensurate solid region (as in the first layer). This supports the suggestion[6] that the anomalous temperature dependence is a result of the presence of registered solid. Such a tendency is found both in the first and second layers. It would therefore seem possible that the treatment of atomic motion in terms of exchange in a low density quantum solid, stabilised by an underlying lattice, in terms of few particle multiple exchange in a two dimensional crystal is inadequate. A description of the quantum excitations and quantum fluctuations of this system has yet to be developed. Perhaps the relevant objects are domain walls[28].

3.TWO DIMENSIONAL FERMI LIQUID

One of the interesting features of two dimensional Fermi liquids is the possibility they offer to vary the areal density from the weakly interacting gas limit (where first principles calculations of the influence of interactions on thermodynamic properties should be feasible) to strongly interacting fluid. The first layer of ^3He on grafoil is generally believed to be a Fermi liquid at low surface densities. This results from the atomically flat nature of the substrate on a mesoscopic (~100Å) length scale. However the maximum density is limited by a transition to a solid-fluid coexistence region at 0.035$Å^{-2}$ and nuclear magnetic susceptibility measurements are complicated by the large signal from ^3He atoms localised by the weak substrate heterogeneity[4] (as discussed further in section (5)). In the second layer the ^3He atoms are still bound with an energy ~25K to the first layer. Replacing the first layer with ^4He atoms, which are preferentially bound to the surface, eliminates the large Curie law contribution to the susceptibility. It is expected that the interactions between ^3He quasiparticles will not be influenced by the substrate, because of the high compressibility of the dense first layer.

The fluid susceptibility is shown in fig.4, normalised by the zero temperature ideal gas susceptibility $\chi_0(0)$ which in two dimensions is independent of areal density. The susceptibility is temperature independent at the lowest temperatures as expected for a degenerate Fermi liquid and exhibits large susceptibility enhancements above the ideal gas value. At the highest fluid densities the raw data show a slight increase at the lowest temperatures due to a very small density of localised spins ($<5.10^{-4}$ $Å^{-2}$). This density is determined from the intercept of $\chi T(T\rightarrow 0)$ and is attributable to atoms entering the first layer as it is slightly compressed with increasing coverage. The Curie constant was determined from the limiting value of χT at high temperatures and was found to be proportional to areal density, as expected, and in agreement with the more precise determination in the second layer solid.

The susceptibility enhancements, shown as a function of areal density in fig.5, are greater than observed in ^3He-^4He mixture films on a nuclepore substrate[29]. The difference between the systems is presumably that ^3He adsorbed on an incompressible solid monolayer on graphite is more two dimensional than the other system where a thicker ^4He film has been used. The mixture data extrapolate to a hydrodynamic effective mass which

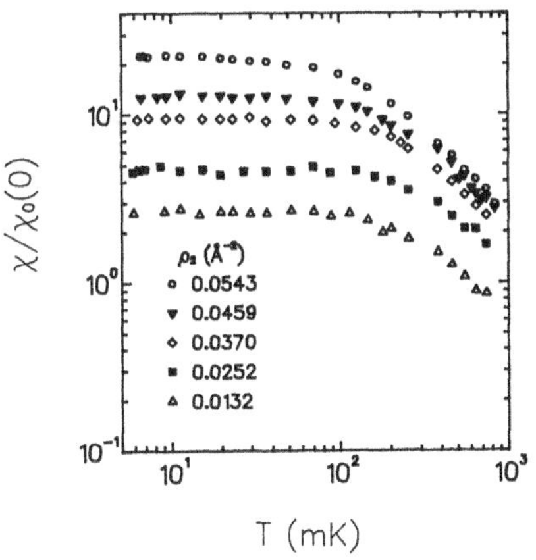

Fig. 4. Normalised susceptibility of second layer fluid.

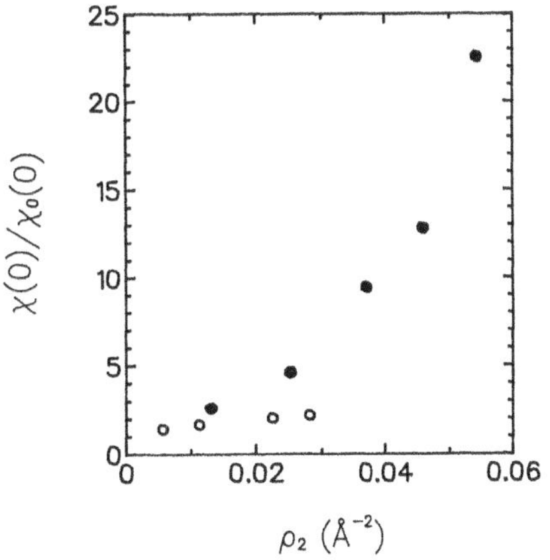

Fig. 5. Susceptibility enhancement as a function of areal density. (●) this work, (o) ref.29.

depends on ^4He film thickness. It might be expected that in our case the susceptibility enhancement would tend to unity in the limit of low areal densities. It would be of considerable interest to compare our measured susceptibility enhancements approaching this limit with the microscopic calculations of such non-uniform systems[30]. And studies in which the ^4He film is varied could show up variations in the hydrodynamic mass and the interactions which are predicted to show weak oscillations with layer by layer growth[30].

Comparing the susceptibility enhancements in 2D, as observed here, with the results for bulk ^3He shows that interactions are more important in 2D. In fig.6. the enhancements are plotted against an effective nearest neighbour separation, assuming an fcc structure in bulk and a triangular lattice in 2D. This plot emphasises the wider range of a_{nn} accessible in two dimensions.

Since the susceptibility enhancement is given by Landau Fermi liquid theory to be $(m^*/m)(1+F_0^a)^{-1}$, it is possible to infer F_0^a from these results given values of m^*/m obtained from the heat capacity[6]. We note that the heat capacity does exhibit a linear temperature dependence but that an anomaly has been detected[31] around 3mK. We have determined F_0^a using the effective mass inferred from the linear term, assuming that the second layer Landau parameters do not depend on the isotopic composition of the first layer. In fig.7 the relative dependence of these Landau parameters is compared with the results of the almost localised Fermion model[32]. This provides a microscopic description of ^3He in terms of a small number of parameters and therefore can predict the relative dependence of the Landau parameters. The model is a lattice gas description in which the atoms are distributed in a regular array of cells. The energy for double occupancy of a cell (creation of an interstitial) is U. This system corresponds to the Hubbard model with a half-filled band and is solved in the Gutzwiller approximation[33]. Taking a symmetric density of states there is a further parameter $p \approx 1$ which depends on the precise form of the density of states. In terms of the normalised interaction $I=U/U_c$ it is found that $I=[1-m/m^*]^{1/2}$ and $F_0^a=-p[1-1/(1+I)^2]$. Thus the theoretical curves in fig.7 are determined by the single parameter p. Agreement is good both with the bulk and 2D results. This reinforces the view of liquid ^3He as a nearly localised rather than a nearly ferromagnetic system.

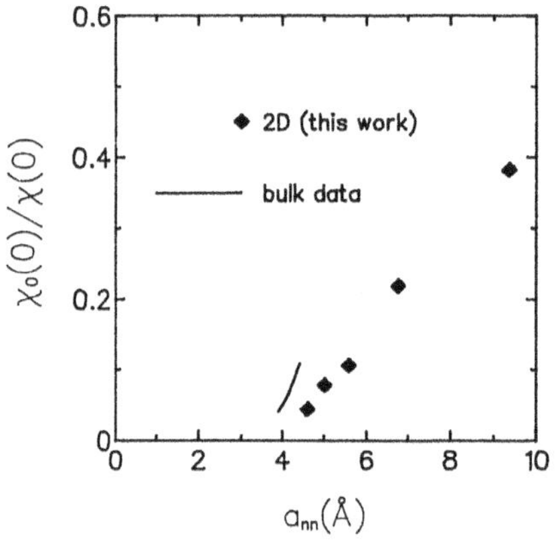

Fig. 6. Inverse of the susceptibility enhancement at T=0 vs. nearest neighbour spacing in 2D and 3D.

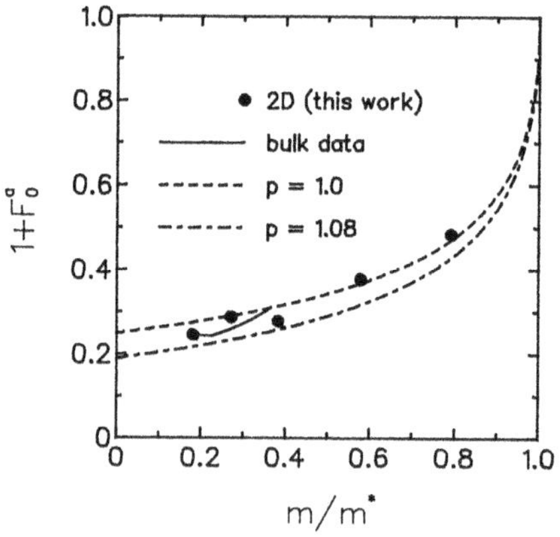

Fig. 7. Relative dependence of the Landau parameters $1+F_0^a$ and m^*/m. Theoretical curves are the prediction of the almost localised Fermion model.

It is of interest to compare the full temperature dependence of the susceptibility with theory through $T \sim T_F$. In this respect a semi-phenomenological extension of the lattice gas model to finite temperatures in 2D would be of value[34]. One approximate expression valid at all temperatures in the dilute limit[35], $\chi(T)=\chi_0(T)/(1-F\chi_0(T))$ where $\chi_0(T)$ is the ideal gas susceptibility, is not a good fit above 0.01Å^{-2}. As found for bulk ^{3}He the phenomenological replacement of the Fermi temperature by an effective magnetic degeneracy temperature[36] provides a reasonable fit over the whole temperature range. The two dimensional results offer the prospect of testing theories of interacting Fermi systems outside the realm of validity of Landau Fermi liquid theory. In order to establish conclusively that this 2D system is indeed a Fermi liquid measurement of a transport property such as the spin diffusion coefficient is desirable.

4.QUANTUM SIZE EFFECTS IN THIN MULTILAYER FILMS

Recent microscopic calculations[37] of the density profile of thin quantum fluid films adsorbed on a solid surface support the idea of a layered structure. The calculated density shows oscillations which are more pronounced and smear out more slowly with increasing coverage the stronger the binding potential. Measurements on the magnetisation of mixture films[38] show structure consistent with two liquid layers. The subsequent evolution of the magnetisation is similar to that expected for bulk liquid. On the other hand in our experiments with grafoil as an adsorbate, structure is observed out to five liquid layers. This may be due to the better homogeneity of the substrate.

In our experiments the first two layers are solid, the first layer being ^{4}He. The results are fit as described in section (2). For these purposes the procedure allows the determination of the low temperature fluid susceptibility $\chi_L=C/T_F$ and the areal density of fluid ρ_L. These results are shown in fig.8 ,where the susceptibility has been normalised by the susceptibility of an ideal 2D gas at T=0, $\chi_0(0)$, which is independent of surface density. There is evidence for a sequence of steps and plateaux in the dependence of the susceptibility on areal density. The arrows indicate the number of liquid "layers" at intervals of 0.065Å^{-2}.

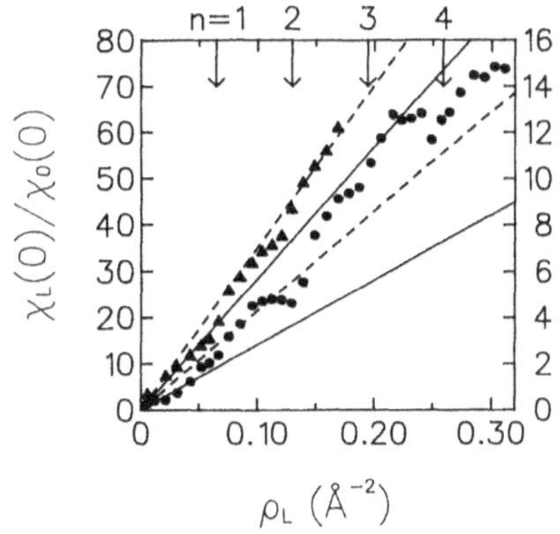

Fig. 8. Coverage dependence of low temperature fluid susceptibility (●) and heat capacity, ref.6. (▲). Straight lines indicate the bulk liquid behaviour for γ/γ_0 (dashed) and χ/χ_0 (solid). In each case upper (lower) line uses melting pressure (zero pressure) parameters.

Correlated with this behaviour there is less distinct structure in the heat capacity. We show values of γ, the coefficient of the linear term of the fluid heat capacity, normalised by γ_0, the density independent value appropriate to an ideal Fermi gas.

A recent model[39] shows how such structure can arise in a self bound film of uniform density, as a result of quantum size effects due to the finite film thickness. Neglecting the substrate potential the states are taken as $\psi \sim \sin(\pi mz/d)\exp(i\underline{k}.\underline{r})$, where the z part of the wavefunctions are the particle in a box states of energy $\hbar^2\pi^2m^2/2m_3d^2$. The film thickness d is determined by the number of atoms and the atomic interactions. As these available states are occupied, with increasing coverage, a series of 2D Fermi fluids is formed with quantum numbers m=1,2,3,...The highest occupied m state increases steplike as a function of coverage. The Fermi level is found to be approximately constant due to the decrease in the energy of the m states as the film thickens. Since the total susceptibility is the sum over the m occupied 2D Fermi systems then the susceptibility of an ideal film evolves in a sequence of steps of equal magnitude.

Questions which arise are: how to handle interaction effects in the film; and how the crossover to 3D behaviour occurs. There may be some clue in the general trend of both the heat capacity and susceptibility data, which is to increase in proportion to the total number of fluid atoms as expected for the growth of bulk liquid at constant density. The bulk liquid line is shown in fig.8 , where the behaviour both at zero pressure and on the melting curve are shown (the slopes of the lines depend on the values of the Landau parameters m^*/m and F_0^a). Although the film density under the influence of the substrate potential is close to that of low pressure liquid[40] we note that the trend of the data is more in accord with the melting curve parameters.

Since we have seen in section (3) that the Landau parameters of a 2D Fermi liquid at an areal density of 0.055Å^{-2} are approximately the same as for bulk liquid at the melting pressure, we rather propose that the Fermi liquid parameters of each occupied m state are those appropriate for a 2D Fermi fluid. The implication is that the interactions between quasiparticles in the different m states are small. Within the model the orthogonality of the particle in a box wave functions and the relatively large energy difference between the different m states contribute to this. Here the substrate potential must be taken into account. Experimentally it is found that above 100mK the steps are far less distinct. If the simple model of the film is correct it is necessary to ask what determines the effective mass which appears in the energy associated with motion perpendicular to the plane, since this will determine the characteristic temperature at which the structure becomes smeared. A detailed interpretation of the inferred low temperature behaviour remains to be done. Paramagnon theories of Fermi liquid interactions that suggest a further substrate induced enhancement of the Stoner factor towards a ferromagnetic instability[41] are not supported by these data.

Our conclusion is that the data for ^{3}He on grafoil support the existence of quantum size effects in multilayer fluid films. A qualitative account of this system is provided by a recent theoretical model[39]. It is worthwhile noting that this model for the fluid overlayer leads to a prediction of an RKKY indirect exchange interaction for the second solid layer, which oscillates as a function of coverage with a period of one liquid layer[27]. As discussed in section (2) this has so far not conclusively been observed.

5. SUBMONOLAYER FILMS

The established features of the phase diagram of submonolayer films are well known: a fluid phase at low areal densities, a 1/3 registered phase at $0.064Å^{-2}$, and an incommensurate solid phase above $\sim 0.08Å^{-2}$. Recent measurements of the nuclear magnetic susceptibility to low millikelvin temperatures[4] find substantial reductions in the susceptibility below Curie law at coverages below registry, attributable to quantum degeneracy in the film. The systematics of this data are examined by fitting the results to a sum of a fluid and a solid term, $\chi T = x_s C_0 + x_f C_0 (T/T_F)[1-\exp(-T_F/T)]$, where C_0 is the Curie constant for fractional monolayer coverage $x=1$ ($0.108Å^{-2}$), and x_s, x_f are the coverages of solid and fluid. Below $x \sim 0.3$ there is a solid signal equivalent to a monolayer fraction ~ 0.02. This is consistent with the expected number of atoms localised by substrate heterogeneity. It was established that these sites could be preferentially occupied by ^4He atoms (introduced at 4K followed by an anneal at 20K) resulting in a dramatic reduction in signal size at low temperatures. It is found that above $x \sim 0.3$ the fluid phase becomes unstable with respect to a coexistence of fluid and solid. More tentatively a further transition is identified at $x \sim 0.45$ and is suggested to correspond to the formation of a registered solid plus defects phase.

There are several outstanding issues: (1) to explain the instability of the uniform fluid phase; (2) to understand the nature of the coexistence region; since coexistence must obtain on each homogeneous platelet the solid component may well consist of mesoscopic clusters for which quantun fluctuations will be important. There is no reported heat capacity signature associated with crossing a phase boundary into the coexistence region. (3) If the registered phase plus defects indeed exists it should constitute an interesting form of quantum glass that should be described by a distribution of effective exchange energies.

6.SUBMONOLAYER ISOTOPIC SOLID MIXTURE FILMS

Finally we report some very recent measurements on isotopic mixture films intended to elucidate the nature of atomic tunnelling in the submonolayer solid. These experiments were motivated by the following simple idea. The most widely used theory in the description of exchange in quantum solids is the multiple spin exchange theory. The susceptibility can provide a measure of exchange through the deviations from Curie law, however as discussed already there is a cancellation between different ring processes. On the other hand in a dilute 2D solid solution all such processes will contribute additively to the tunneling rate of a ^3He atom. This proposal must be tempered with the following concerns; at what concentration does the elastic interaction of the ^3He impurities localise the "impuritons"[42], does phase separation occur and what is the role of solid defects?

We note that it is known that both ^3He and ^4He have very similar submonolayer phase diagrams[18]. This is because the van der Waals interactions both between atoms and with the substrate are the same for both isotopes. The areal density at zero temperature at which the transition from a "mixed solid" phase to the incommensurate solid has not been established but is expected to be around $0.08Å^{-2}$. We further note that the calculated position of the maximum in the single atom probability density above the surface is virtually identical. However we observe strong quantum degeneracy effects when the total areal density is in the range $0.08 < \rho < 0.09$ $Å^{-2}$, significantly well below the densities for promotion to the second layer ($\sim 0.11Å^{-2}$) found for isotopically pure films.

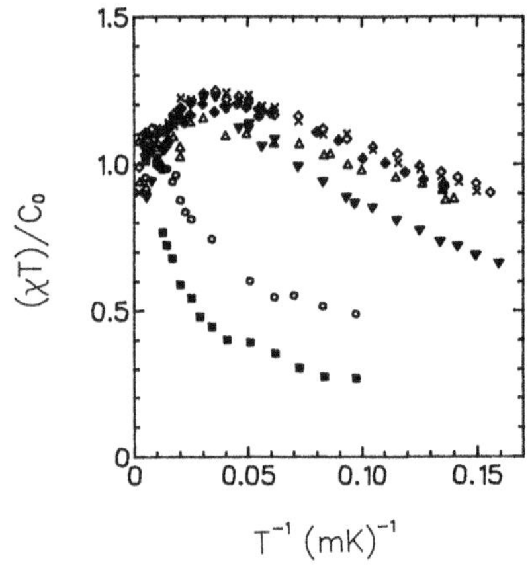

Fig. 9. Temperature dependence of χT for solid mixture at 0.064(△), 0.068(◇), 0.072(◆), 0.076(X), 0.080(▼), 0.082(○), 0.086(□) $Å^{-2}$.

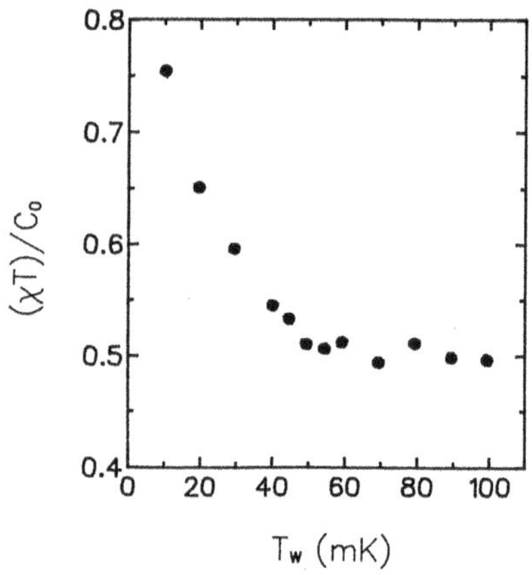

Fig. 10. Determination of structural phase transition temperature at coverage 0.082 $Å^{-2}$.

These measurements (fig.9) were made with a ^3He surface density of 0.013$Å^{-2}$; the total coverage was increased by adding ^4He. At registry a small but significant decrease in χT becomes apparent below 25mK. In the pure film there are no observable deviations from Curie law. At areal densities 0.068, 0.072 and 0.076 $Å^{-2}$ while the low temperature behaviour is very similar there is an initial increase in χT as the sample cools to 25mK, similar to that observed in the pure film just above registry[4]. Then at 0.08$Å^{-2}$ there is a dramatic decrease in the low temperature susceptibility, while at 0.082$Å^{-2}$ the susceptibility at 10mK is a factor two smaller than the Curie value. It is natural to conclude that these quantum degeneracy effects in a 2D solid are large in the incommensurate solid and inhibited by the disorder of the mixed solid phase. Although it is tempting to ascribe this to the tunneling of impuritons we note that the degree of degeneracy increases when the coverage is taken to 0.086$Å^{-2}$, while the dependence of the exchange energies on interatomic spacing would suggest the opposite (given that the solid is free of disorder for $\rho > 0.082Å^{-2}$).

Moreover this mobile phase is unstable with respect to a structure in which the ^3He are more localised. It has proved possible to observe this phase transition and to locate the critical temperature simply by measuring the nuclear magnetic susceptibility. On cooling from 200mK to 10mK, for example, the susceptibility is observed to increase with a time constant of approximately five hours. At an areal density of 0.082$Å^{-2}$ the critical temperature was determined by warming to 200mK for two hours to ensure homogeneity, cooling to some intermediate temperature T_w, waiting for 15 hours and subsequently cooling to 10mK to detect whether the transition had occurred at T_w. By plotting (χT) against T_w (fig.10) it was possible to establish a critical temperature of 50 ± 5 mK.

Above an areal density of 0.086$Å^{-2}$, the susceptibility at low temperatures decreases with increasing coverage and at 0.095$Å^{-2}$ is approximately Curie law. For submonolayer densities above 0.09$Å^{-2}$, the linewidth is coverage independent and temperature dependent; the cusp in the linewidth seen for pure ^3He at second layer promotion is not observed.

These results lead us to offer the following preliminary speculations.

The structural phase transition we observe is phase separation. At $T>T_c$ (and $T<T_c$ before "phase separation" has occurred) there is an isotopic stratification in the film which controls the "impuriton" mobility. Such an effect would disappear on phase separation. This suggests that there is a delicate balance between quantum effects and the equilibrium structure of the film (including how two dimensional it is). This may well be of relevance when considering exchange in the various putative registered phases discussed in section (2).

ACKNOWLEDGEMENTS

This work was supported by SERC GR/E/34582. We thank D.C.Smith, A.K.Betts, M.Kelly and F.Greenough for technical support. The contributions of V.A.Mikheev to this work and the support of the Royal Society are also gratefully acknowledged. We have benefitted from correspondence and discussions with M.Roger, R.A.Guyer and M.G.Richards.

REFERENCES

1. M.Nielson, J.P.McTague, W.Ellenson, J.Phys.(Paris),32,C4-10 (1977)
2. for a review see J.G.Dash and M.Schick in The Physics of Liquid and Solid Helium, Part II, ed. K.H.Bennemann & J.B.Ketterson, (Wiley 1978)
3. B.Cowan, L.A. El-Nasr, M.Fardis, A.Hussain, Phys.Rev.Lett. 58, 2308 (1987)
4. J.Saunders, C.P.Lusher, B.P.Cowan, Phys.Rev.Lett. 64, 2523 (1990)
5. H.Godfrin, unpublished.
6. D.S.Greywall, Phys.Rev. B41, 1842 (1990)
7. M.Roger, Phys.Rev. B30, 6432 (1984)
8. M.Roger, Phys.Rev.Lett. 64, 297 (1990)
9. M.C.Cross, Jap.Journ.App.Phys. 26, Suppl.26-3, 1855 (1987)
10. M.G.Richards, J.Phys.(Paris),39,C6-1342 (1978)
11. H.Franco, R.Rapp, H.Godfrin, Phys.Rev.Lett. 57, 1161 (1986)
12. H.Godfrin, R.Ruel, D.Osheroff, Phys.Rev.Lett. 60, 305 (1988)
13. D.S.Greywall and P.A.Busch, Phys.Rev.Lett. 62, 1868 (1989)
14. V.Elser, Phys.Rev.Lett. 62, 2405 (1989)
F.Abraham, J.Q.Broughton, P.W.Leung, V.Elser, Europhys.Lett. 12, 107 (1990)
15. G.A.Baker, H.E.Gilbert, J.Eve, G.S.Rushbrooke, Phys.Lett. 25A, 207 (1967)
16. H.Lauter et.al. Can.J.Phys. 65, 1435 (1987)
17. S.van Sciver, Phys.Rev. B18, 277 (1978) ; heat capacity peaks observed in this work interpreted as second layer melting in ref.6.
18. In the first layer $dT_m/d\rho \approx 185$ KÅ^2; S.V.Hering, S.van Sciver, O.E.Vilches, J.Low Temp.Phys. 25, 793 (1976)
19. The possibility of this phase has been suggested in ref.6.
20. K.Machida, M.Fujita Phys.Rev.B to be published
21. D.S.Greywall to be published
22. M.Roger private communication
23. The parameters of this curve are; $S_3=48.6, S_4=S_6=58, S_2=67.3$; all $C_n=1.2$, $\sigma=2.65$Å.
24. M.Heritier, J.de Phys.Lett. 40,451 (1979)
25. H.Jichu and Y.Kuroda, Prog.Theor.Phys. 67, 715 (1982)
26. S.Tasaki, Prog.Theor.Phys. 79, 1311 (1988)
27. R.A.Guyer, Phys.Rev.Lett. 64, 1919 (1990)
28. A.Luther, J.Timonen, V.Pokrovsky, Phase Transitions in Surface Films, ed. J.G.Dash and J.Ruvalds, (Plenum 1979)
29. J.M.Valles, R.Higley, B.R.Johnson, R.B.Hallock, Phys.Rev.Lett. 60, 428 (1988). Data shown are for 9.5 ^{4}He "layers".
30. E.Krotscheck, M.Saarela, J.L.Epstein, Phys.Rev.Lett. 61, 1728 (1988)
J.L.Epstein, E.Krotscheck, M.Saarela, Phys.Rev.Lett. 64, 427 (1990)
31. D.S.Greywall and P.A.Busch, Phys.Rev.Lett. 65, 64 (1990)

32. reviewed in P.Wolfle and D.Vollhardt, Superfluid ^{3}He (Taylor and Francis 1990)
33. D.Vollhardt, Rev.Mod.Phys. 56, 99 (1984)
34. K.Seiler,C.Gros,T.M.Rice,K.Ueda and D.Vollhardt, J.Low Temp.Phys. 64, 195 (1986) have performed such a calculation in 3D.
35. S.Havens-Sacco and A.Widom, J.Low Temp.Phys. 40, 357 (1980)
36. H.Ramm,P.Pedroni,J.R.Thompson and H.Meyer, J.Low Temp.Phys. 2, 539 (1970)
37. E.Krotscheck, Phys.Rev. B32, 5713 (1985)
38. R.Higley, D.Sprague and R.B.Hallock, Phys.Rev.Lett. 63, 2570 (1989)
39. R.Guyer, K.McCall,and D.Sprague, Phys.Rev. B40, 7417 (1989)
40. discussed in D.F.Brewer, Physics of Liquid and Solid Helium, ed. K.H.Bennemann and J.B.Ketterson, (Wiley 1978)
41. D.Spanjaard, D.Mills, M.Beal-Monod, J.Low Temp.Phys. 34, 307 (1979)
42. A.Andreev, Progress in Low Temperature Physics vol.8, ed.D.F.Brewer (North Holland 1982) reviews bulk solid isotopic dilute mixtures.

LOW-FIELD NMR STUDIES OF THE BOUNDARY LAYERS OF ^{3}He ON GRAFOIL

Hans M. Bozler

Physics Department
University of Southern California
Los Angeles CA 90089-0484 USA

ABSTRACT

The boundary layer of ^{3}He on graphite has shown properties of a two-dimensional ferromagnet. Both, in the cases of ^{3}He filling pores of Grafoil and with a few monolayers, magnetization far in excess of the Curie value are observed at millidegree temperatures. Our experiments on the low-field NMR spectra of the ^{3}He boundary layers in filled pores show collective modes for $T < 1$ mK. One of these modes extrapolates to nonzero frequency and amplitude as the field is dropped to zero. The Heisenberg symmetry is broken by the weak dipolar energy. This dipolar energy and finite size effects allow for order in zero field and finite temperature.

INTRODUCTION

The layer of ^{3}He at the boundary between the liquid and virtually any container exhibits properties that are dramatically different from either the liquid state and also bulk solid. The magnetism of the boundary layer can be easily distinguished from bulk liquid by a temperature dependant magnetization. In most cases this surface magnetization appears to have a ferromagnetic tendency.[1,2,3,4]

Of the many possible substrates upon which we could study surface magnetism, Grafoil (exfoliated graphite) provides the best opportunity to observe the properties of a two dimensional magnetic system. This is because Grafoil has relatively large atomically flat surfaces which can be oriented. Although Grafoil is the best candidate for studies of two dimensional magnetism, it is by no means ideal. Finite size, magnetic anisotropy, and impurities ultimately play some role in the results.[5,6]

In earlier work on Grafoil,[7] we found that we could observe a frequency shift in the NMR line due to the polarization of the surface spins as shown in Figure 1. This frequency shift had the angular dependence on magnetic field consistent with a two dimensional magnetic sheet.

The demagnetizing field due to the ^{3}He nuclear dipole moments, λM_0, gives a negative/positive frequency shift when the applied field was perpendicular/parallel to the Grafoil planes. (For a fully polarized

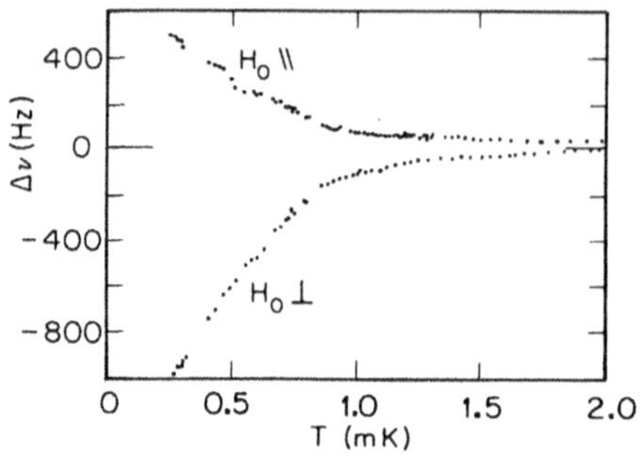

Fig. 1. The measured frequency shift vs temperature. Identical samples are placed parallel and perpendicular to H_0 (16.9 mT).

triangular lattice sheet in the first layer with a = 3.2 Å , λM_0 is 4.78 gauss. In the case of the second layer, where a = 3.78 Å, λM_0 = 3.09 gauss) In large applied fields ($H_0 \gg \lambda M_0$) the applied field holds the static magnetization parallel to the field, while in small applied fields, the demagnetizing field forces the spins into the planes.

We can calculate[8,9] the expected frequencies in the presence of the dipolar field by first calculating the equilibrium spin angle. If the applied static field $\vec{H}_0$ is at an angle θ_h with respect to the normal, the static magnetization $\vec{M}_0$ will be at an angle θ_m given by

$$h_0 \sin(\theta_m - \theta_h) = \tfrac{1}{2}\sin(2\theta_m) \tag{1}$$

where $h_0 = H_0/\lambda M_0$. Following a small angle tipping pulse, the magnetization precesses around its equilibrium position at a frequency $\omega\gamma\lambda M_0$ with the dimensionless frequency ω given by

$$\omega^2 = (h_0 \cos\theta_h - \cos\theta_m)^2 + h_0 \sin\theta_h (h_0 \sin\theta_h + \sin\theta_m). \tag{2}$$

In the high field limit Eqns. 1 and 2 predict the angular dependence observed in Fig. 1. At low fields, the magnetization tips away from the field and the frequency shift becomes sensitive to the magnitude as well as the direction of the field.

The observation that ^{3}He on Grafoil provides a laboratory for studying two dimensional magnetism has stimulated a number of experiments. Recent experiments by Godfrin and coworkers[10,11,12] have measured the magnetization of several monolayer films as a function of coverage and temperature. They have discovered that the exchange energy is very sensitive to coverage with a large enhancement near 2.5 layers. Greywall has also made detailed heat capacity throughout this regime.[13,14]

The recent efforts at U.S.C. have been oriented towards observing the magnetism in very low applied fields.[8,9,15,16] The motivation behind the this work has been to perform measurements in fields so low that the spin behavior would be dominated by the local dipole fields.

APPARATUS

The apparatus for the U.S.C. low field experiments has been described extensively in reference 17 as well as 8,9,15,16. The most important features of our experiments are as follows. The NMR is performed with a pulsed very low field SQUID apparatus which has a linear, frequency independent (0 - 50 kHz) response. This allows us to use applied fields more than two orders of magnitude below that required in previous experiments. Since the SQUID pick up is nonresonant and linear, we can directly compare the magnetizations observed at various applied fields. The Grafoil is placed so that the planes are either perpendicular or parallel to the applied field. In all cases, our ^{3}He sample completely fills the Grafoil.

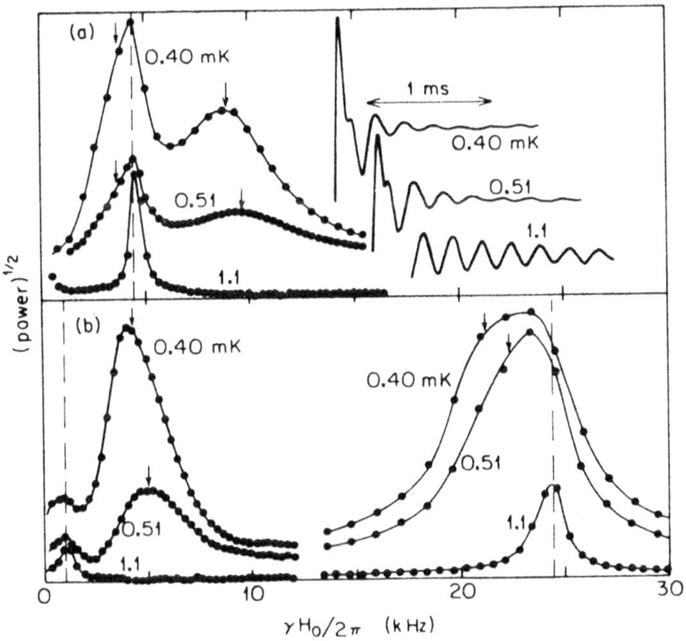

Fig. 2. (a)Fourier-transform power spectra and associated free-induction decays with an applied field of 1.3 G. Arrows point to the center frequencies of the low-temperature modes as fitted by a Gaussian line. Vertical dashed lines indicate the high-temperature (15mK) Larmor frequency. (b)Two additional Fourier-transform families which show displaced modes growing above (for H_0=0.3 G) and below (H_0=7.5 G) the Larmor frequency.

RESULTS

H_0 Perpendicular to Grafoil

Our first set of measurements of the surface magnetism of ^{3}He in low fields provided rather surprising results.[8,9] Figure 2 shows representative sequences of NMR power spectra and, in one case, the signals from which they are derived. The signals appearing in the Figure 2 are almost entirely due to the ^{3}He surface spins, since the surface signal is so large as to make the bulk liquid contribution negligible. Here, the applied field was perpendicular to the Grafoil. Above 3 gauss the frequency shifted negatively as expected. At such fields the boundary layer spins should be aligned with the static H_0 field, pointing perpendicular to the Grafoil sheets. The 2.5 kHz negative shift implies a 15% polarization of regions in the ^3He surface layer.

Below 3 gauss (where the static H_0 field is comparable to the expected local fields for polarized ^3He domains), the NMR line splits into two modes as the temperature is lowered below 1 mK. In all of our data the higher frequency mode appears to grow at a displaced frequency rather than continuously shifting away from the Larmor line. The amplitude at the Larmor frequency diminishes with temperature while at the same time the new modes are growing in amplitude. Figure 3 shows the frequencies where we observed resonances at a low temperature (0.42 mK)as a function of field. The circles and crosses are meant only to group the modes according to higher and lower frequencies. The solid line in Figure 3 shows the prediction from Eqns. 1 and 2 assuming that the Grafoil has a $\pm 15°$ spread of angles about the nominal (perpendicular) direction.

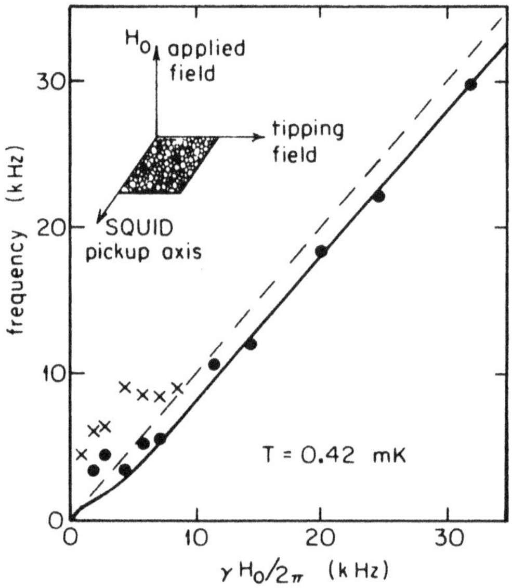

Fig. 3. NMR frequencies vs applied field at 0.42 mK. The dashed line is the Larmor frequency. Additional modes, grouped in higher and lower frequencies, are shown as crosses and filled circles. The solid line is the predicted spectrum for a 2D array with an anisotropy energy due only to the dipole fields of the aligned spins. The inset depicts the cell geometry.

In addition to the anomalous frequencies at low fields, we also observed that the amplitude rapidly decreased below 3 gauss. This can be easily explained since the nuclear dipole energy of a polarized sheet is minimized when the polarization lies in the sheet (refer to Eqn. 1). Thus when the field is sufficiently small, the spins precess in a direction that is orthogonal to the SQUID pick-up coil. The displaced mode which appears above the Larmor frequency at the lowest fields must then be explained through the dynamics of the spins lying flat in the Grafoil planes.

We interpret the extra peaks in the NMR spectra as arising from the collective modes within the ^{3}He surface layers ordered over large domains by the exchange interaction. The shift of the frequency from the Larmor frequency is a measure of the anisotropy energy for reorienting the magnetization of these domains.

H_0 Parallel to Grafoil

While the cross-over between positively and negatively shifted NMR lines was a good indicator of the dipolar energy of the surface spins, this spin flop transition made the study of the limiting amplitude and frequencies at low fields impractical. We therefore proceeded with taking our remaining data with H_0 parallel to the planes of the Grafoil.[15,16] In this configuration we observed that we could observe two distinct modes. The first shifted mode had its frequency proportional to the applied field. The second displaced mode had both its frequency and amplitude offset so that they were finite even at zero field.

The frequency shifts and amplitudes of these modes are shown in Figures 4 and 5. The finite amplitude and frequency of the displaced mode, as the field approaches zero, provides compelling evidence that some order exists in zero field and finite temperature. This may appear surprising in view of the general result by Mermin and Wagner[18] that in 2d, long wave length spin waves destroy ferromagnetic order at any finite temperature for systems with Heisenberg symmetry. On closer examination this places only a rather weak limitation on real systems since finite size along with

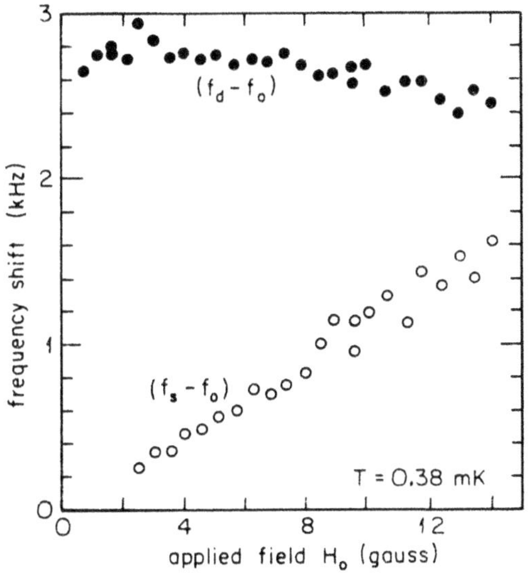

Fig. 4. Field dependence for the frequency shifts of the two modes relative to the Larmor frequency. The open circles record the shifted mode difference frequency (f_s-f_0) which is linear in the applied field. The closed circles record the displaced mode difference (f_d-f_0) that shows an almost constant displace-ment of 2.7 kHz above the Larmor frequency.

perturbations that modify the long wave length spin wave spectrum cause the Mermin Wagner result to no longer apply.

The origin of these modes is not yet clear. At low fields, the majority of the signal comes from the displaced mode at zero field. One possibility is that the modes are intrinsic collective modes of an ordered state. Alternatively entire domains (~100Å) could be precessing as a rigid moment locked to some magnetic impurity.

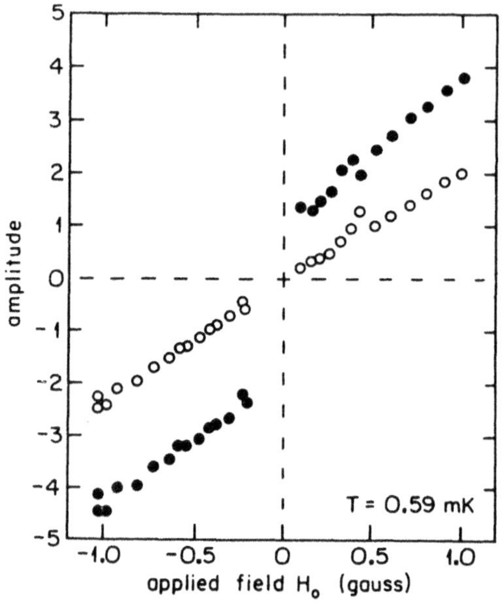

Fig. 5. The amplitude vs field dependence for the two modes contributing to the surface signal at T = 0.59 mK. The shifted mode (open circles) extrapolates to zero amplitude for zero field while the displaced mode (closed circles) shows an offset as the field passes through zero.

DISCUSSION

Physical interchanges between the adsorbed atoms are most likely responsible for the ferromagnetic tendency.[19] The film experiments[12] have shown that the magnetization of one layer (the second) can be described by a two-dimensional Heisenberg model. In this system we also have long-range dipole interactions between the ^{3}He nuclei. The most obvious effect of the dipole interactions is to cause the spin-flop transition discussed above. A second and more subtle effect is to allow long-range order at finite temperatures. Yafet et. al[20] have shown that the dipole energy alters the spin-wave stiffness at long-wavelength so that the spectrum goes as $E \propto k$ instead of $E \propto k^2$. This change in the long-wavelength spin-wave energy spectrum is sufficient to allow order.

The most surprising feature of this result is that a very small perturbation has a large effect on the ordering. In this case, the exchange rate J is about 1 mK while the dipole coupling δ is of the order of 0.1 μK. But, low temperatures and zero field the pure Heisenberg correlation length, ξ, diverges exponentially in J/k_BT. The dipolar energy, $\delta\xi^2$, is associated with a cluster of aligned spins of size ξ. We can estimate T_c by the solution to $k_BT_c \sim \delta\xi^2$, which yields

$$k_BT_c \approx \frac{2c_1 J}{\ln(c_2 J/\delta)} \tag{3}$$

(here c_1 and c_2 are constants of order unity). The transition temperature therefore vanishes very slowly with δ, and is determined primarily by J.

The rapid growth of ξ at temperatures below J/k means that there is a competition between the growth of the dipolar energy and the finite structural coherence of the Grafoil. Estimates of this coherence are around 100 Å and we cannot discriminate between effects due to the finite size and dipole driven ferromagnetic order. A more uniform substrate like ZYX should provide an unambiguous result; however, because of the much more limited surface area, this remains a challenging experiment.

Our experiments at low applied fields were designed to study the growth of the magnetization of the boundary layer of ^{3}He when the system is not substantially polarized by the applied field. At much higher applied fields, and at a selected coverage (where the effective exchange interaction J is a maximum) Godfrin et. al[12] have been able to make precise measurements of the susceptibility and compare it with a pure Heisenberg exchange model. In such a comparison, the effective exchange constant is determined from the high temperature behavior ($T \gtrsim J$).

Since, in our experiment, the SQUID signal is directly proportional to the magnetization, no renormalization of the data is necessary to compare the various magnetic fields. The upper five curves in Figure 6 shows our the temperature dependence the entire amplitude of the magnetization at several applied fields. Taking the entire surface signal, we find a value for J = 0.8 mK. The data in these upper curves in Figure 6 do not include a subtraction of a monolayer signal as was done by Godfrin et. al.[12] If we subtract a portion due to one solid monolayer following a Curie Law, the estimated value of J will increase to 0.9 mK.

For comparison, I have indicated the trend in the data from reference 7 taken at much higher fields. The normalization between this data and the low field data is approximate and is intended only to show the relative temperature dependence.

We see from Figure 6, that there is a departure in the low temperature behavior of M/H_0 from the high temperature exponential behavior expected for a Heisenberg model even in the lowest applied fields. This picture is qualitatively similar to the data of Godfrin et. al, however progressive saturation appears at much lower fields and values of J/T than one might expect. We see that this saturation continues down to the lowest fields we used (0.18 gauss). If we choose a fixed temperature and find the relation between M and H below 1mK, we find very good agreement with

$$M \propto H^{1/\delta}$$

with $1/\delta$ about 0.5 at 0.47 mK and slightly temperature dependent. This analysis ignores the fact that the signal has broken into two modes.

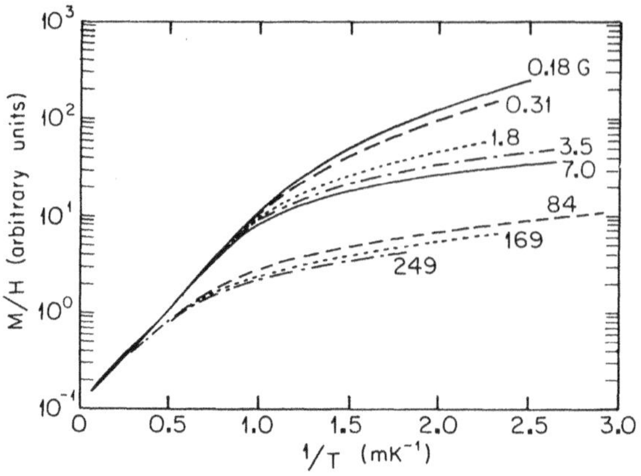

Fig. 6. Magnetization vs inverse temperature. The curves represent the total magnetization divided by the applied field. Thus, the high temperature limit is field independent. The upper five curves are the total signal from references 15,16 and other than the field, there is no scaling. The lower three curves are from reference 7 and have been adjusted to agree approximately at high temperatures.

The maximum polarizations observed at these low fields by measuring M are much lower (less than 1%) than would be inferred by the frequency shifts. The observed frequency shifts can be used to estimate polarization by using Equation 3 for the case where $\theta_h = \theta_m = \pi/2$. For full polarization of a triangular (second layer of ^{3}He on graphite) the computed λ implies a shift of 4.77 kHz. Our kHz shifts imply polarizations on the order of 30% or greater. This may indicate that the large frequency shifts involve only a fraction of those spins which reside in the boundary layer. However, these polarizations are far in excess of the free spin values. If we treat our data by subtracting the amplitude corresponding to the spins in the first layer, we infer significantly higher polarizations. However, these are still well below the polarization reported by Godfrin et. al at a similar J/T. Of course, we also have a much smaller applied field. It is not yet possible to entirely reconcile the magnetic behavior of the boundary layer with filled pores at very low fields with the higher field film experiments. It is possible that differences exist beyond simply rescaling J. The recent studies of heat capacity by Greywall[14] indicate that at the 2.5 layer coverage where the exchange appears to be a maximum, there is promotion to the fourth layer.

Recently, Tasaki[21] has developed a model for surface magnetism calculating the indirect exchange between the localized spins and the adjacent liquid like layer. He finds that at high densities (corresponding to the case where there are a number of liquid layers above the boundary layer), there is a crossover between ferromagnetic exchange and a weakly antiferromagnetic behavior. He thus explains the experimental results of Friedman et. al[8,9,15,16] in terms of ordering into an antiferromagnetic V_2 phase where in zero field three atom clusters have radially outward pointing spins. Tasaki then explains the spin dynamics by using the formalism developed for the antiferromagnetic phase of solid ^{3}He.[22] This picture has the appeal that it appears to account for the frequencies observed. Also, the reduced susceptibility (relative to the frequency shifts) could be more easily explained. However, it is not clear how this model can explain the high temperature behavior.

The origin and nature of the boundary magnetism of ^{3}He remains with many open questions. Much of the experimental picture is clearer now then it was a few years ago. The work of Godfrin and coworkers along with the detailed heat capacity measurements by Greywall have shown that the enhanced magnetism most likely stems from exchange processes at the solid liquid interface. Our work shows that there exists a rich set of phenomena in the dynamics of this 2D spin system.

ACKNOWLEDGEMENTS

This work has been a a collaboration with several coworkers. In particular, the first studies of surface magnetism on grafoil were instigated by A.L. Thomson who has remained as part of this collaboration. The more recent experimental work was performed primarily by L.J. Friedman. The work has benefitted by the insights of C.M. Gould and the theoretical interpretation by M.C. Cross, and P.B. Weichman. This work has been supported by the National Science Foundation through grant DMR88-00291.

REFERENCES

1. A. I. Ahonen, T. Kodama, M. Krusius, M. A. Paalanen, R. C. Richardson, W. Schoepe, Y. Takano, J. Phys. C 9:1665 (1976); A.I. Ahonen, T.A. Alvesalo, T. Haavasoja, M.C. Veuro, Phys. Rev Lett. 41:494 (1978).
2. D. F. Brewer and J. S. Rolt, Phys. Rev. Lett. 29:1485 (1972).
3. H. M. Bozler, T. Bartolac, K. Luey, A. L. Thomson, Phys. Rev. Lett. 41:490 (1978).
4. H. Godfrin, G. Frosatti, D. Thoulouze, M. Chapellier, W. G. Clark, J. Physics (Paris), Colloq. 39:C-287 (1978).
5. J. K. Kjems, L. Passell, H. Taub, J. G. Dash, A. D. Novaco, Phys. Rev. B 13:1446 (1976).
6. C. Bouldin, E. Stern, Phys. Rev. B 25:3462 (1982); P. W. Stephens, P. A. Heiney, R. J. Birgenau, P. M. Horn, D. E. Moncton, G. S. Brown, Phys. Rev. B 29:3512 (1984).
7. H. M. Bozler, D. M. Bates, A. L. Thomson, Phys. Rev. B 27:6992 (1983).
8. L. J. Friedman, S. N. Ytterboe, H. M. Bozler, A. L. Thomson, and M. C. Cross, Phys. Rev. Lett. 57:2943 (1987).
9. L. J. Friedman, S. N. Ytterboe, H. M. Bozler, A. L. Thomson, and M. C. Cross, Proceedings of the Banff Conference on Quantum Fluids and Solids (Oct. 1986), Canadian Journal of Physics.
10. H. Franco, H. Godfrin, D. Thoulouze, Phys Rev. B 31:1699 (1985).
11. H. Franco, R. E. Rapp, H. Godfrin, Phys. Rev. Lett. 57:1161 (1986).
12. H. Godfrin, R. R. Ruel and D. D. Osheroff, Phys. Rev. Lett. 60:305 (1988).
13. D. S. Greywall, Phys. Rev. B. 41:1842 (1990).
14. D. S. Greywall, Quantum Fluid and Solids-1989, ed. by G. G. Ihas, Y. Takano, AIP Conference Proceedings 194, (Gainsville, 1989) p. 213.
15. L. J. Friedman, A. L. Thomson, C. M. Gould, H. M. Bozler, P. B. Weichman, and M. C. Cross, Phys. Rev. Lett. 62:1635 (1989).
16. L. J. Friedman, A. L. Thomson, C. M. Gould, H. M. Bozler, P. B. Weichman, and M. C. Cross, Quantum Fluid and Solids-1989, ed. by G. G. Ihas, Y. Takano, AIP Conference Proceedings 194, (Gainsville, 1989) p. 201.
17. L. J. Friedman, A. K. Wennberg, S. N. Ytterboe, H. M. Bozler, Rev. Sci. Instrum. 57:410 (1986).
18. N. D. Mermin, H. Wagner, Phys. Rev. Lett. 17:1133 (1966).

19. J. M. Delrieu, M. Roger, J. H. Hetherington, J. Low Temp. Phys. 40:71 (1980); M. Roger, Phys. Rev. B 30:6432 (1984); M. Roger, J. M. Delrieu, Jap. Jour. Appl. Phys. 26, Suppl. 26-3, 267 (1987); H. Jichu, Y. Kuroda, Prog. Theor. Phys. 67:715 (1982); H. Jichu, Y. Kuroda, Prog. Theor. Phys. 69:1358 (1983).
20. Y. Yafet, J. Kwo, E. M. Gyorgy, Phys. Rev. B 33:6519 (1986).
21. S. Tasaki, Prog. Theor. Phys. 79:922 (1988); S. Tasaki, Prog. Theor. Phys. 80:946E (1988); S. Tasaki, Prog. Theor. Phys. 81:946 (1989); S. Tasaki, Prog. Theor. Phys. 82:1032 (1989).
22. D. D. Osheroff, M. C. Cross, D. S. Fisher, Phys. Rev. Lett. 44:792 (1980).

SEARCH FOR A FLUID PHASE IN FILMS OF MOLECULAR HYDROGEN ISOTOPES ADSORBED ON MgO*

O. E. Vilches[†], F. C. Liu[†], D. L. Kingsbury[†], J. Ma[†], M. Bienfait[‡], J. Suzanne[‡], J. M. Gay[‡], M. Maruyama[‡], P. Zeppenfeld[‡], D. Degenhardt[⋆], H. J. Lauter[⋆], F. Rieutord[⋆], G. Coddens[||]

† Dep. of Physics
U. of Washington
Seattle, WA 98195, U. S. A.

‡CRMC2 and F. des Sciences
Campus de Luminy
Case 913
13288 Marseille Cedex 09, France

⋆Institut Laue Langevin
156X
F-38042 Grenoble, France

||Laboratoire Léon Brillouin
CEN de Saclay
91191 Gif-sur-Yvette Cedex, France

INTRODUCTION

Films of molecular hydrogen in its various isotopic combinations physisorbed on solid substrates are a fertile ground for studying quantum and dimensionality effects in the substances with the lowest triple point temperatures. The first monolayer of H_2, of HD, and of D_2 adsorbed on exfoliated graphites have been studied in great detail[1-7]. The very rich phase diagrams that have emerged from these studies are dominated by two types of solids: one commensurate with the substrate triangular lattice of adsorption sites, the other one (of higher density) incommensurate with the substrate. At lower densities than the one required for approximately perfect commensuration a two-phase coexistence region between a commensurate solid (with vacancies) and its vapor preempts condensation into a liquid, while at densities in between the commensurate and incommensurate solids several types of "domain wall

* Work supported by NSF (DMR Low Temp. Prog. 8611466 and 8913454 and INT 8612187) and CNRS

solids" have been proposed, and many of those are observed. Only recently the second and higher atomic layer phases of these molecules have been studied in some detail[8-10].

The objectives of the multilayer studies may be summarized as follows:

a) Is it possible to find a particular atomic layer with a phase diagram similar to the one in bulk hydrogens (3D), but with the triple and critical points shifted to lower temperatures in agreement with expectations from the quantum theorem of corresponding states (QTCS) in two dimensions (2D)?

b) Is it possible that in multilayer films a surface melted layer exists to rather low temperatures?

These objectives are interesting since molecules of H_2 and D_2 should obey Bose-Einstein statistics and may be candidates, in the fluid phase, for some form of superfluid condensation in 3D or 2D, in analogy with ^4He thick and thin films. Estimates based on the 3D ideal Bose gas condensation temperature and the density of liquid H_2 at its triple point (similar to estimates on ^4He) would place this transition below 6 K. This requires supercooling droplets or thick films of H_2 from 13.8 K, the 3D triple point. On the other hand, dimensionality, the QTCS, and calculations of the limiting value at $T = 0$ K of the minimum quantum parameter value needed for solidification[11] predict that the 2D triple point should be somewhere between 5 and 8 K. Of course estimates of the superfluid onset temperature (pressumably a Kosterlitz-Thouless transition) point to a temperature below 2 K for a monolayer film.

In this paper we report on measurements we have been carrying on for some time on films of H_2, HD, and D_2 adsorbed on MgO using a variety of techniques: adsorption isotherms, heat capacity, and elastic (ENS) and quasi elastic (QENS) neutron scattering. Results for multilayer films adsorbed on graphite are reported in other papers in this volume[12].

EXPERIMENTAL SURFACES

All the experiments that we have performed require a large adsorption area in order to enhance the signal to background ratio. Considerable effort has been devoted to the fabrication of MgO powder substrates that are composed of microcrystals with a very large fraction of (100) planes exposed for adsorption. The most uniform powders that we have fabricated come from burning pure Mg ribbons in either air or a mixture of O_2 and Ar. The "smoke" so produced, when examined under an electron microscope, is made up of essentially perfect cubes of sizes ranging from 700 Å edge up to 2 μm, yielding an area to mass ratio of about 8 m^2/gram. This powder has to be collected very quickly when produced in air in order to prevent degradation, and is baked in high vacuum above 900 C for about 24 hours. Details of its fabrication and characterization by Kr-adsorption isotherms are given by Coulomb et al[13,14]. These powders are extremely good for adsorption isotherm work, reaching uniformity comparable to the one of the best exfoliated graphites, and have also been used in scattering studies from D_2/MgO and rare gases/MgO[10,15,16].

Less uniform powders may be prepared by starting from a slurry of $Mg(OH)_2$ with 1 to 5% $MgCl_2.6H_2O$. The slurry is then dried at 150 C, calcined in air at 700 C, compacted in steel dies at 15 MPa, compressed isostatically at 210 MPa, and finally sintered in flowing dry Ar at 1300 C. The pellets so produced are composed of MgO sintered "cubes" of fairly uniform size, with an adsorption area of about 2 m^2/gram. The adsorption area and densification of the pellets is very dependent on the sintering temperature; sintering at 900 C produces pills with about 8 m^2/gram at 65% of bulk density, while the 1300 C pills are compacted to over 75% of bulk density. Details of the effect of various parameters have been reported by Mikijelj and Whittemore[17]. For our purposes, these pellets are then baked in high vacuum at about 900 C for several hours and checked for uniformity by Kr-adsorption isotherms at liquid N_2 temperatures before transfering to the experimental cells. While the quality of the surface is not as good as the one of MgO smoke, the pellets have much better thermal conductivity and a much larger surface to sample volume ratio. We have used them for both heat capacity measurements of H_2 bilayer films, and QENS studies of thick and thin films of HD.

Other experimental details are described with results for the different molecules. Where possible, we have combined our results with those of Ref. 15 since those results provide an excellent calibration for all measurements.

RESULTS

Substrate characterization. Our initial characterization tool is measurement of a Kr adsorption isotherm at liquid N_2 temperature. A substrate with excellent surface homogeneity will produce an isotherm with a large, essentially constant, pressure step at 0.135 Torr corresponding to 2-D liquid-vapor coexistence, and a small, almost constant, pressure step at 0.56 Torr corresponding to crossing the solidification line. The amount of Kr adsorbed at the small step may be used to correlate all the different MgO substrates used, and thus to correlate all the hydrogen coverages. The actual area per molecule of Kr at this transition is not exactly known, although from x-ray scattering measurements at lower densities it is possible to estimate that it should be between 15.4 and 14.3 Å^2/atom[16,18].

Thermodynamic measurements. Isotherm measurements over the first 7 layers of H_2, the first 4 layers of HD, and the first 5 layers of D_2 have been performed. Fig. 1a shows a single isotherm for each of the gases starting at the completion of the first layer. The H_2 measurements were made on two samples of MgO smoke different than the one used for the HD and D_2 measurements. Inserts 1b and 1c show in an expanded vertical scale a portion of the HD and D_2 isotherms at the completion of the first layer. The H_2 isotherms have been previously reported by Ma et al[9], with Fig. 1d of that reference showing a similar inset to the ones shown here. Degenhardt[10,15] has reported similar measurements for D_2/MgO for the region between the completion of the first and the third layers. Large vertical steps in Fig. 1a indicate layer by layer growth, with possible two-phase coexistence regions. Feature A is present for the three molcules studied, as well as in Degenhardt's sample used in his ENS study. It reflects, in D_2/MgO, a coexistence between commensurate (2×4) and (2×6) surface structures on the first layer. It occurs at the same coverage (within 2%) for HD and D_2, and for the same scaled coverage for H_2 in our samples and D_2 in Degenhardt's sample. The scaling is simple: the ratio between the hydrogen coverages at the middle of step A and the Kr coverage at the middle of the small melting step at 77 K is 1.37 ±0.01 for all the MgO samples used. In the study of Refs. 10 and 15 further addition of D_2 compresses the monolayer into a triangular lattice of density 0.0982 Å^{-2} oriented in such a way that there is always commensuration every two rows of Mg atoms. We expect that a similar effect occurs in HD and D_2.

The small step called B in Fig. 1a at the completion of the 2nd layer exists only in the H_2 and HD isotherms. The D_2 isotherms appear to have a small inflexion point at comparable coverages. Step B merges with the large vertical step of the second layer at lower temperatures than the ones shown in Fig 1a. Heat capacity measurements done on H_2 adsorbed on a sintered MgO substrate show that merging occurs at 7.20±0.05 K, while the combination of isotherms and heat capacity shows that the larger second layer step goes away at 10.05±0.05 K. We interpret these temperatures as those corresponding to the solid-liquid-vapor triple and liquid-vapor critical points of a single layer of H_2 on top of the incommensurate solid first layer of H_2 (see Ref. 9 for details). The similar features on the HD bilayer film give $T_t \approx 9.9 \pm 0.1$ K and $T_c \approx 10.5 \pm 0.3$ K, deduced from isotherms. If the D_2 inflexion point corresponds to the same feature, then $T_t \approx 12.1 \pm 0.2$ K and $T_c \approx 12.4 \pm 0.4$ K. This last critical point occurs at the same temperature found by Degenhardt[15].

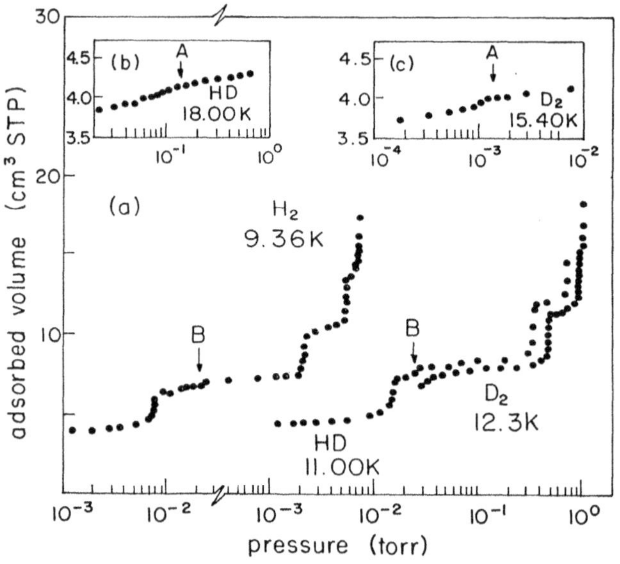

Fig. 1. H_2, HD, and D_2 isotherms on MgO smoke at the temperatures shown. Arrow A corresponds to the (2×4) to the (2×6) transition observed in D_2 by ENS[10]. Arrow B marks the fluid-solid coexistence line of the second layer.

All layers above the second one do not appear to have "feature B". We searched the region between the third and fourth layers of H_2 and HD with a very high density of points but found neither a step nor an inflexion point. Thus only two phases seem to coexist in all the layers above the second one for the three molecules adsorbed on this substrate. Triple and critical points for each one of the layers are indicated in Table I. Figs. 2, 3, and 4 show P-T phase diagrams for the three systems studied. For

Table I. Critical and triple points (in K) of e-H_2, HD, and e-D_2 in 3-D and in the various layers physisorbed on MgO. The dimensionless quantum parameter η is defined in text.

		e-H_2	HD	e-D_2
Bulk	T_c	33.0	35.9	38.3
	T_t	13.81	16.61	18.69
	η	0.0764	0.0509	0.0382
2nd layer	T_c	10.05±0.05	10.5±0.3	12.4±0.4[a]
	T_t	7.20±0.05	9.9±0.1	12.1±0.2
3rd layer	T_c	10.3±0.3	12.1±0.3	13±1[a]
4th layer	T_c	10.4±0.2	12.4±0.4	12.9±0.4
5th layer	T_c	10.1±0.3		13.0±0.5
6th layer	T_c	10.3±0.9		

[a] Degenhardt, Ref. 15

clarity, many multilayer points are not shown but may be inserted in the figures using Table I values. All pressures have been corrected for thermomolecular effects.[19]

Elastic Neutron Scattering. Degenhardt et al[10] studied several years ago the region between the first and second layers of D_2 adsorbed on MgO smoke using ENS. Deuterium atoms are good coherent scatterers and the signal due to the physisorbed film can be easily detected. In our interest to find if a fluid phase existed to very low temperatures, we studied with ENS whether or not a surface melted layer of D_2 would appear in thick films at temperatures well below the bulk triple point. Elastic neutrons can detect both the solid component as somewhat narrow diffraction peaks, and the liquid component as a broad spectrum. The experiments were done at ILL-Grenoble on the D1B line using a wavelength λ=2.518 Å. This line has a 400 channels linear detector that collected data over a 2θ range of 80 degrees. The MgO substrate was the same one used in Ref. 10. This sample was baked and pumped before the experiment, and a Kr isotherm at liquid N_2 temperature was run to verify its quality. Guided by the results of the H_2 isotherms that at 10K it was possible to build 6 or more layers without capillary condensation, we dosed the equivalent of about 15 layers at low temperature, closed the fill line, and warmed the sample up to 16.58 K where we observed its annealing by monitoring the neutron spectra. When the solid signal was stable the sample was warmed up to 16.90 K, and from there on we measured 18 spectra at temperatures somewhat evenly spaced in log|t|, where $t = (T - T_t)/T_t$ is the reduced temperature difference to the bulk triple point. Each spectrum took approximately 8 hours of counting time.

A preliminary analysis of the results is shown in Fig. 5. In Fig. 5(a) one can see a common problem to all surface melting studies: while the temperature rises the total thickness of the adsorbed film decreases. This information is deduced from measurements of the dead volumes of the cell and fill line while cold, and the vapor pressure of D_2.

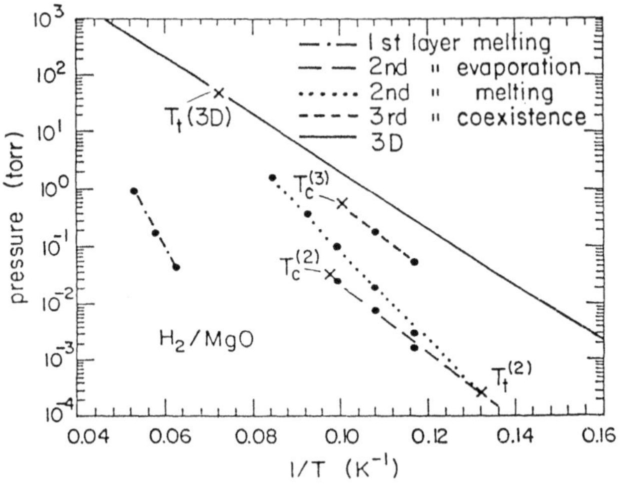

Fig. 2. Vapor pressure of adsorbed layers of H_2 on MgO vs $1/T$. Only coexistence lines are shown. Coexistence lines of the 4th, 5th, and 6th layers are not shown.

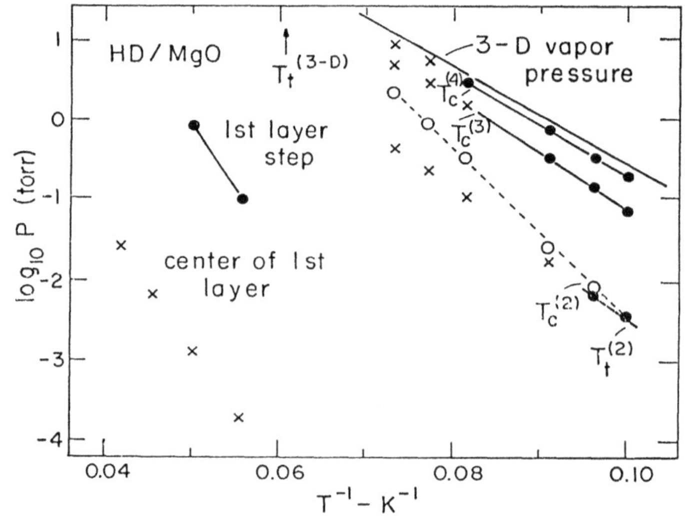

Fig. 3. Vapor pressure of adsorbed layers of HD on MgO vs $1/T$. Phase coexistence shown with solid and dash lines.

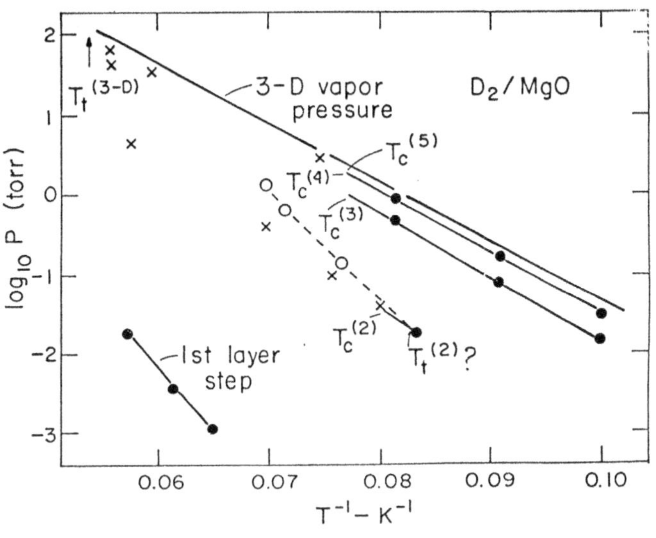

Fig. 4. Vapor pressure of adsorbed layers of D_2 on MgO vs 1/T. Phase coexistence shown with solid lines. Dash line shows inflexion point of isotherms between 2nd and 3rd layers, see text.

The first interesting observation is that the D_2 solid film is growing in the fcc structure (rather than hcp[20]), with a conventional cubic cell edge of 5.118 Å. This leads to a nearest neighbour distance in the close packed planes of 3.619 Å, and a surface density for these planes of 0.0882 $Å^{-2}$. We observe the (111), (200), and (311) reflections within our detector range. The (220) peak is almost exactly superimposed with a reflection from the stainless steel cell container and can not be measured accurately from the background substracted spectra.

At the 3D triple point there are still two solid layers next to the substrate. Although when we performed the experiment we expected no liquid at the temperature at which we started the scans, more recent QENS measurements with HD/MgO show appreciable mobility down to 11.9 K (see below and Zeppenfeld et al[21]). The two possible solid fractions shown have been obtained from the intensity of the ENS spectrum solid lines under different assumptions. The solid symbols in Fig. 5a assume 2 layers of solid film at T_t and 100% solid at 16.90 K, while the $\times$'s have been adjusted to 2 layers of solid film at T_t and 1.5 layers of liquid at 16.9 K (about 10% in reduced temperature, the same liquid thickness of the QENS experiment at the same reduced temperature). The two normalizations give the two sets of results for the liquid thickness shown in Fig. 5b, where symbols corresponding to Fig. 5a have been used. It is expected that for long range forces (like Van der Waals' forces) the liquid thickness should grow as $|t|^{-1/3}$; for reference, a straight line with a 1/3 slope has been drawn above the points[22]. It is safe thus to conclude that for these thick films a liquid fraction is observed well below T_t of the bulk system. How mobile these liquid fractions are may be determined by QENS measurements.

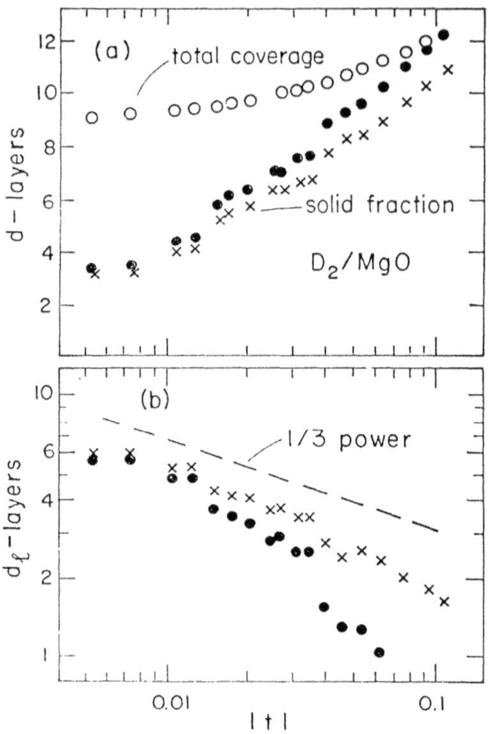

Fig. 5. Surface melting of a D_2/MgO thick film. (a) Total film thickness in cell, and solid film thickness under two different assumptions, versus reduced temperature. (b) Liquid thickness vs reduced temperature deduced for the two assumptions of top figure. Surface melted layers of Van der Waals films should grow with a 1/3 power dependence on reduced temperature.

Quasi Elastic Neutron Scattering. QENS is an excellent tool to study mobility in 3D and 2D for systems formed by molecules that contain atomic hydrogen. It has been used successfully in the study of melting of single layers and multilayers of CH_4 on graphite and MgO[23]. Mobile molecules change the energy of scattered thermal neutrons by small amounts (in the μev range) producing an energy broadening about the elastically scattered peak. Atomic hydrogen has a very large incoherent neutron scattering cross section (about 80 barns), but while ortho-H_2 also has a large σ_{inc}, para-H_2 (the equilibrium form at low temperature) has a relatively small σ_{inc} ($\approx$4 barns) at the temperatures of interest. Conversion of ortho- to para-H_2 during the experiment would make measurements unreliable. We thus chose HD as the system to study. The loss of sensitivity (σ_{inc} $\approx$50 barns) is certainly compensated by its stability.[24]

We have performed two experiments, one with thick films to study the mobility of a possible surface melted layer, the second one to study the mobility of the sec-

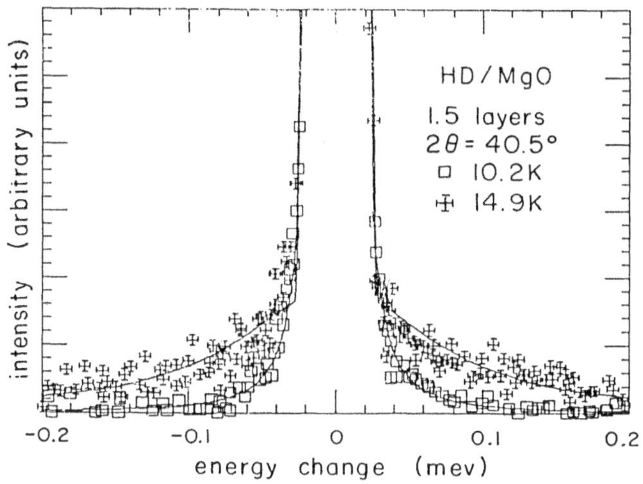

Fig. 6. Quasielastic neutron scattering spectrum of 1.5 layers of HD/MgO at the scattering parameters shown in figure.

ond, third, and fourth layers of HD/MgO. The first experiment was performed at the Mibemol instrument of the Laboratoire Léon Brillouin (LLB) in Saclay using 1.28 meV neutrons, and a 200 m^2 area pill of sintered MgO as the substrate. A full report on this experiment has been published.[21] The main result is that for a film that had 8 layers at very low temperatures the first broadening due to liquid scattering is observed at 11.89 K (when the film is still about 8 layers thick), with the amount of liquid being equivalent to half a layer. The fitted translational diffusion coefficient for this film is either 4 or 1 ($\times\ 10^{-5}\ cm^2 s^{-1}$) depending on whether 2D or 3D diffusion equations are used to fit the width of the lorentzian liquid lineshapes as a function of wavevector. By the triple point, all of the film minus 2 layers is fluid, and the measured diffusion coefficient is the one of the 3D liquid.

The second experiment, just concluded, was performed at the IN5 line of the Institute Laue-Langevin (ILL) at Grenoble[25]. The same MgO pill used in the thick films studies was used. The experiment attempted to determine if at the middle of the second layer step observed in the HD/MgO experiments the film was indeed liquid. Measurements at 7 K were below the resolution limit, thus showing only the elastic scattered peak broadened by the instrumental resolution. Measurements at 10.5 K though show a small broadening. We collected data at several scattering vectors, the data taken at 2θ=40.5 degrees with 8 Åwavelength neutrons being shown in Fig. 6 for two temperatures. The data analysis is similar to the one used for the thick film experiment, except that in this case we used a 2D hexagonal lattice model[26,27] to fit the linewidths and obtain (at low Q) the diffusion coefficient. The fraction of mobile film x is obtained by fitting the recorded spectra with a fraction x of a lorentzian plus a fraction $1 - x$ of perfectly scattered neutrons from the solid, convoluted with the instrument resolution, all multiplied by a function of scattering vector Q and the radius of gyration, to take into account that the HD molecule is a free rotor at the temperature of the experiment. The data of Fig. 6 gives x=0.33, thus 0.5 layers (with 20% undertainty) are responsible for the broadening. The translational diffusion

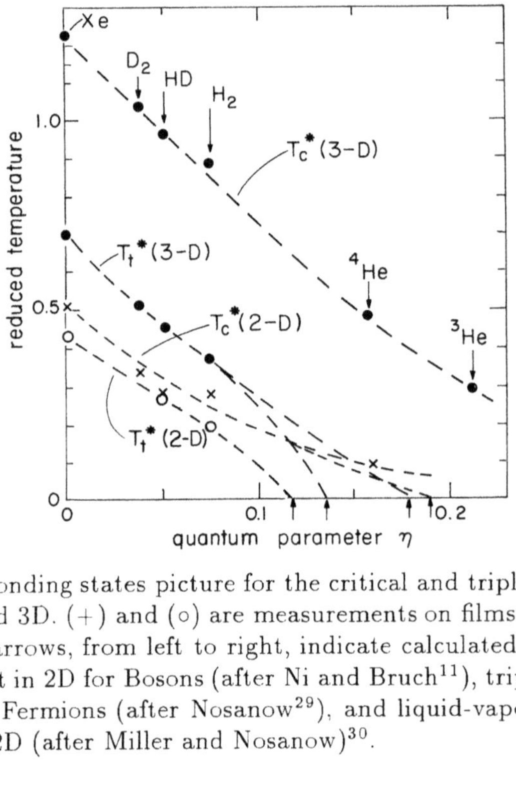

Fig. 7. Corresponding states picture for the critical and triple point temperatures in 2D and 3D. (+) and (o) are measurements on films physisorbed on MgO. Vertical arrows, from left to right, indicate calculated limiting values of η: triple point in 2D for Bosons (after Ni and Bruch[11]), triple points in 3D for Bosons and Fermions (after Nosanow[29]), and liquid-vapor critical point of Fermions in 2D (after Miller and Nosanow)[30].

coefficient is $0.8\pm0.2\times10^{-5}$ cm^2/s at 10.5 K, and about doubles at 14.9 K. The data is not of enough quality to determine the liquid-vapor boundary.

CONCLUSIONS AND SUMMARY

Adsorption isotherms combined with heat capacity, QENS, and ENS measurements provide a lot of information about the thermodynamic state, structure, and mobility of the physisorbed molecules of the various forms of hydrogen. The isotherms show distinctly that when adsorbed on MgO crystallites, the films grow in a layer by layer mode, with the second layer of H_2 and HD having phase diagrams similar to the ones in 3D, but with the critical and triple point temperatures shifted to lower temperatures in agreement with expectations from the QTCS[28] . We show in Fig. 7 a "corresponding states picture" similar to the ones prepared by Nosanow[29] for 3D systems. Our picture emphasizes the so-called quantum regime of high quantum parameter $\eta=(h/2\pi)^2/m\sigma^2\epsilon$, with h being Planck's constant, and σ and ϵ being the two parameters in a Lennard-Jones or fitted interparticle potential. The values we found in our measurements for the reduced triple and critical points (k_BT_t/ϵ and k_BT_c/ϵ) are not the lowest ones found for physisorbed films (see Ref. 12 for the values of the critical and triple point temperatures of bilayer H_2 and D_2 films adsorbed on graphite). Wiechert's measurements certainly emphasize the importance of the substrate in the fluid properties of these films. Careful choice of new substrates may produce an effective interaction that further lowers the triple point to such a temperature that we may start looking for a superfluid molecular hydrogen film.

REFERENCES

1. M. Bretz and T. T. Chung, J. Low Temp. Phys. 17, 479 (1974)
2. M. Nielsen, J. P. McTague, and W. Ellenson, J. Phys. (Paris) Colloque 38, C4-10 (1977)
3. H. Freimuth and H. Wiechert, Surf. Sci. 162, 432 (1985); ibid 178), 716 (1986)
4. F. C. Motteler and J. G. Dash, Phys. Rev. B31, 346 (1985)
5. F. A. B. Chaves, M. E. B. P. Cortez, R. E. Rapp, and E. Lerner, Surf. Sci. 150, 80 (1985)
6. J. Cui and S. C. Fain, Phys. Rev. B39, 8628 (1989)
7. J. Cui, S. C. Fain, H. Freimuth, H. Wiechert, H. P. Schildberg, and H. J. Lauter, Phys. Rev. Letters 60, 1848 (1988); ibid 60, 2704 (1988)
8. J. L. Tell and H. J. Maris, Phys. Rev. B28, 5122 (1983), H. J. Maris, G. M. Seidel, and T. E. Huber, J. Low Temp. Phys. 51, 471 (1983), R. H. Torii, H. J. Maris, and G. M. Seidel, Bull. Am. Phys. Soc. 32, 395 (1987)
9. J. Ma, D. L. Kingsbury, F. C. Liu, and O. E. Vilches, Phys. Rev. Letters 61, 2348 (1988), and Jian Ma, Ph.D. dissertation, U. of Washington (1989) (unpublished)
10. D. Degenhardt, H. J. Lauter, and R. Haensel, Jpn. J. Appl. Phys. 26, Suppl 26-3, 341 (1987)
11. X. Z. Ni and L. W. Bruch, Phys. Rev. B33, 4584 (1986)
12. H. Wiechert, see article in this volume on multilayer films of H_2 and D_2 adsorbed on graphite
13. J. P. Coulomb and O. E. Vilches, J. Phys. (Paris) 45, 1381 (1984)
14. J. P. Coulomb, T. S. Sullivan, and O. E. Vilches, Phys. Rev. B30, 4753 (1984)
15. D. Degenhardt, Doctoral Dissertation, U. of Kiel (1988) (unpublished)
16. D. Degenhardt, H. J. Lauter, and R. Frahm, Surf. Sci. 215, 535 (1989)
17. B. Mikijelj and O. J. Whittemore, Am. Cer. Soc. Bull. 66, 809 (1987)

18. J. L. Jordan, J. P. McTague, J. B. Hastings, and L. Passell, Surf. Sci. 150, L82 (1985), and J. L. Jordan, Doctoral Dissertation, UCLA (1983), Brookhaven National Laboratory publication No. 33760
19. A. Van Itterbeck and E. de Grande, Physica 13, 289 (1947)
20. I. F. Silvera, Rev. Mod. Phys. 52, 393 (1980)
21. P. Zeppenfeld, M. Bienfait, F. C. Liu, O. E. Vilches, and G. Coddens, J. Phys. (Paris), 51 (in press, September 1, 1990)
22. J. G. Dash, in "Proc. Solvay Conference on Surface Science", F. de Wette, ed, (Springer-Verlag, 1988); Contemp. Phys. 30, 89 (1989); Science 246, 1591 (1989)
23. M. Bienfait, in "Dynamics of Molecular Crystals", J. Lascombe, ed (Elsevier Sci. Pub., Amsterdam, 1987) p. 353
24. W. D. Seiffert, B. Weckermann, and R. Misenta, Z. Naturforsch. 25, 967 (1970)
25. F. C. Liu, O. E. Vilches, M. Bienfait, M. Maruyama, P. Zeppenfeld, and F. Rieutord, to be published
26. C. T. Chudley and R. J. Elliott, Proc. Phys. Soc. (London), 77, 353 (1961)
27. C. Riekel, A. Heidemann, B. E. F. Fender, and G. C. Stirling, J. Chem. Phys. 71, 530 (1979)
28. J. de Boer, in "Molecular Theory of Gases and Liquids", J. O. Hirschfelder, C. F. Curtiss, and R. B. Bird, eds. (J. Wiley & Sons, New York, 1954) p. 424
29. L. H. Nosanow, J. Phys. (Paris) Colloque, 41, C7-1 (1980)
30. M. D. Miller and L. H. Nosanow, J. Low Temp. Phys. 32, 145 (1978)

LATTICE DYNAMICS OF QUANTUM GASES ADSORBED ON GRAPHITE INVESTIGATED BY INELASTIC NEUTRON SCATTERING

V.L.P. Frank[§,*], H.J. Lauter[§], H. Godfrin[§] and P. Leiderer[*]

[§] Institute Laue-Langevin, BP 156X, F-38042 Grenoble, France
[*] University of Konstanz, D-7750 Konstanz 1, West Germany

INTRODUCTION

During the last decade, substantial progress in the knowledge of the properties of two-dimensional (2D) matter has been achieved thanks to studies realized on monolayers of gases adsorbed on well characterized substrates like graphite. The existence of two relevant length scales in this problem, the adsorbate-adsorbate equilibrium distance and the separation between adsorption sites, leads to a rich phase diagram, where commensurate (C) and incommensurate (IC) structures can be present[1,2]. In a commensurate phase, the lack of translational invariance produces an energy gap (Δ) at the zone center in the acoustic branch of the phonon dispersion relation. The magnitude of this gap at low temperatures is related to the lateral variation of the adsorption potential, the so called corrugation. In an incommensurate phase, on the other hand, the translational invariance has been recovered and the phonon branches go to zero frequency at the zone center.

Changes in temperature and/or coverage induce phase transitions that are the result of a delicate balance between the interaction of the adsorbed gas and the substrate, and between the adsorbed gas molecules themselves. The interaction potential between the gas molecules is known in general in great detail. However, the knowledge of the details of the gas-adsorbate interaction potential is still sparse, since not many experimental techniques are available and it is very difficult to model it theoretically. The adsorption potential itself has been determined mainly with molecular beam scattering, but the magnitude of the in-plane corrugation is very difficult to obtain in this way. Recently, inelastic neutron scattering measurements of the corrugation have been performed for the hydrogen[3,4] and helium[5] isotopes, krypton[6], methane[7] and nitrogen[8] adsorbed on graphite. These results have provided quantitative values against which the theoretical models of the adsorbate-substrate interaction can be compared.

The case of the quantum gases (the helium and hydrogen isotopes) is particularly interesting due to the similarity of their phase diagrams[9-15]. They all present a commensurate $(\sqrt{3} \times \sqrt{3})R30^{\circ}$ phase at low coverages and temperatures. At higher coverages, they present an incommensurate phase which is the stable one up to monolayer completion. The C-IC transition goes through a domain wall phase (DW) at low temperatures and at higher temperatures through a domain wall liquid (DWL). The details of this C-IC transition varies slightly for the different quantum gases and the richest phase diagram is obtained for D_2. All these phases have been studied in detail by adsorption isotherms, specific heat and neutron diffraction measurements. The aim of neutron inelastic measurements is to determine the excitation spectrum of the different phases and to deduce the interaction potentials involved.

The detailed knowledge of the adsorption potential is not only relevant for adsorbed

Excitations in Two-Dimensional and Three-Dimensional Quantum Fluids
Edited by A.G.F. Wyatt and H.J. Lauter, Plenum Press, New York, 1991

monolayers, but is necessary to understand the gradual change of bulk properties that occur near an interface.

EXPERIMENTAL RESULTS

The measurements were performed on the three axis spectrometer IN14 and IN3 at the high flux reactor of the Institute Laue-Langevin. The spectrometer were used in the fixed final wavevector (K_F) mode with a Be filter after the sample and a horizontally curved analyzer. The sample consisted of 31.70g of Papyex[16] sheets located in an aluminum container, oriented with their c–axis normal to the scattering plane. The total surface area was determined by adsorption isotherms and neutron diffraction.

Usually difference spectra are presented, where the scan taken with the graphite sample before adsorbing the gas is subtracted as background. For ^{3}He, however, a corrected background must be used, since the large neutron absorption of ^{3}He modifies drastically the signal coming from the graphite substrate. Therefore, in this case, the background scan has been multiplied by a transmission factor, T, to take this effect into account. This factor is a function of the sample geometry, the neutron energy-momentum transfer and the ^{3}He coverage. The transmission per unit height has been calculated for a cylindrical sample integrating numerically the following formula:

$$T_{(R,\theta,\rho)} = \iint r\ dr\ d\varphi\ \exp\left[-\left(\Sigma_{(K_I)}\, l_{I(r,\theta,\varphi)} + \Sigma_{(K_F)}\, l_{F(r,\theta,\varphi)}\right)\right]$$

where R is the radius of the sample, θ the scattering angle, ρ the ^{3}He density, l_I and l_F the lengths of the initial and final path within the sample and Σ the corresponding inverse absorption length (coverage dependent). The transmission of the sample increases monotonically by about 30% between 0 and 1THz (1THz = 48.28K) for our experimental setup. An alternative method to determine the background is to use data obtained at the same coverage, but higher temperatures where the layer melts (above 3K for ^{3}He). The melting, however, produces a broadening of the elastic peak due to quasielastic scattering related to the higher diffusion coefficient of the fluid. At still higher temperatures desorption occurs and again the signal from the substrate is modified. Thus, only the corrected background method can be used for this particular system. It must be pointed out that the elastic and inelastic signals are well separated, that corrections due to background subtraction are small and that the background does not show any peak in the inelastic region of interest; therefore, this procedure does not introduce significant systematic errors in the evaluation of the spectra within our present error bars.

The graphite sample, and hence the adsorbed layer, is essentially a 2D powder: the in-plane crystallographic directions are not oriented. Thus, the measured signal consists of a directional average over all phonons for which scattering is allowed (see insert in fig. 1). Scans were taken at a momentum transfer $Q = 1.70Å^{-1}$. This corresponds to the distance between zone centers in reciprocal space for the C phase. Due to the powder average and the polarization factor, these scans focus on the phonons at the zone center (i.e., the gap region) and on the transverse phonons at the zone boundary. In scans performed at $Q = 0.85Å^{-1}$, on the other hand, one works within the first Brillouin zone, and thus only longitudinal phonons can be excited[3,17,18].

The Hydrogen isotopes

In the first inelastic neutron scattering experiments[19] only an "Einstein mode" could be detected in the commensurate phase and no phonon dispersion measured. More recent work, performed with higher resolution[3], could assign the zone center phonon gap and the transverse and longitudinal zone boundary phonons. In figure 1 are shown the results for D_2, HD and H_2. Deuterium, which has a coherent neutron cross section, allows a separate measurement of the transverse and longitudinal phonons using the technique described in the paragraph above (see insert in fig. 1). HD and H2, on the other hand, possess a large

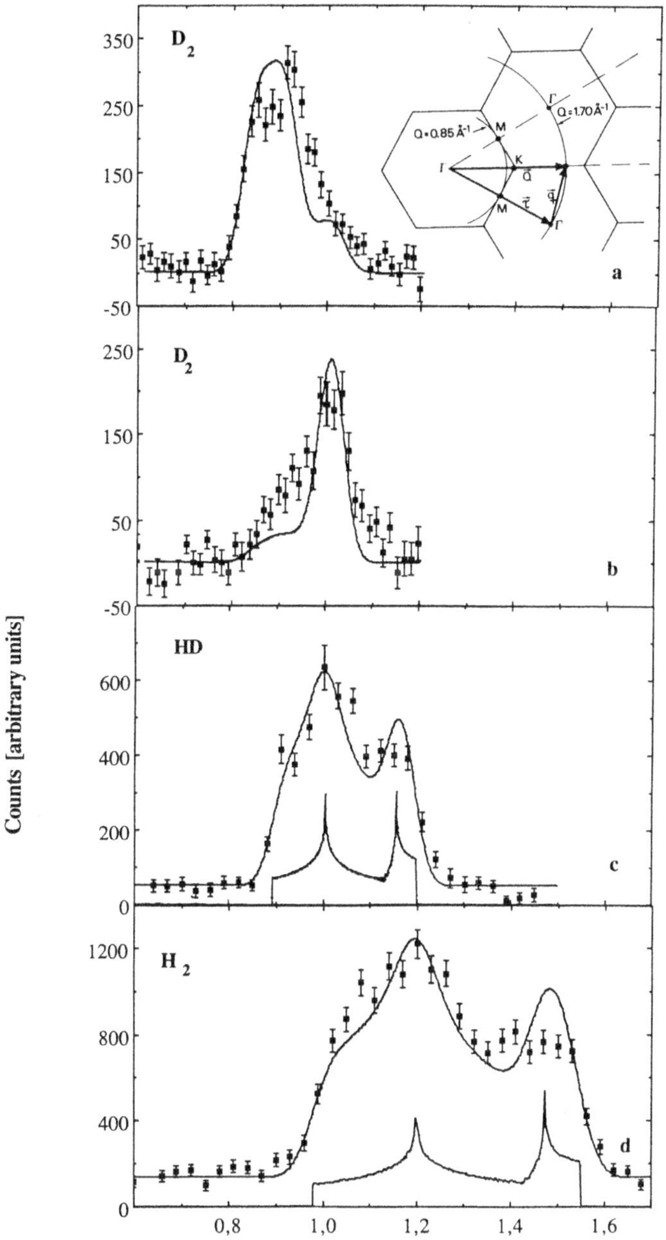

Figure 1. Neutron inelastic data of the hydrogen isotopes adsorbed on graphite in the commensurate phase (the background has already been subtracted). At the top are the deuterium (D_2) spectra for Q=1.70Å^{-1} (a) and Q=0.85Å^{-1} (b), the full line is the fit to the data as explained in Refs. 3 and 4. The insert shows the region of reciprocal space covered by these two scans. One example of a scattering triangle is sketched: it corresponds to the excitation of a transverse zone boundary phonon (**Q** is the total momentum transfer, τ is a reciprocal lattice vector and **q** is the phonon momentum). The deuterated hydrogen (HD) and the hydrogen (H_2) spectra taken at Q=1.70Å^{-1} are shown in parts (c) and (d), respectively. The curve through the experimental points is the fit with the phonon density of states convoluted with the energy resolution of the spectrometer. The unconvoluted density of states is drawn beneath it.

incoherent neutron cross section and thus, only the density of states can be measured. Nevertheless, the main features of the phonon spectrum could be identified and they are listed in Table I together with the theoretical results.

The zone center frequencies, which are a measure of the hydrogen-graphite corrugation, are in reasonable agreement with the calculations, but the width of the DOS is overestimated by theory. The interaction between the adsorbed molecules is weaker than expected and more theoretical work would be welcome. The mass dependence of the phonon spectrum cannot be explained with an isotopic effect only. This aspect will be discussed after the experimental results for the He isotopes have been presented.

Table I

Parameters characterizing the phonon density of states for the in-plane modes of the commensurate phase of the hydrogen isotopes adsorbed on graphite (the uncertainty of the experimental values is estimated to be ≈2 K).

	H_2	HD	D_2
Zone center gap[4] (neutron scattering)	47.3 K	43.2 K	40.0 K
calculations from reference 20	39.1 K	-	31.2 K
calculations from reference 21	46.6 K	-	36.9 K
DOS width[4] (neutron scattering)	27.5 K	14.7 K	9.5 K
calculations from reference 21	42.1 K	-	14.8 K

At higher coverages (around ρ=1.15) a striped domain wall phase could be identified using neutron diffraction and LEED[22]. Inelastic neutron spectra performed on D_2/graphite, show the appearance of a new excitation with an energy ≈29 K, lower than the phonon gap of the C phase[23]. The fact that the energy of this excitation does not change when modifying the coverage (in the limits of the DW phase) suggests that it corresponds to the excitation of a single domain wall. If it were due to the interaction of domain walls it should shift in energy as the distance between domain walls is changed. The peak corresponding to the commensurate structure remains, but is reduced in intensity and begins to broaden due to lifetime effects. A very interesting aspect of these excitations is that they exist with a well defined shape in the domain wall liquid phase, where the diffraction data only show the signal of a liquid[23].

The inelastic spectra obtained for the IC phase of D_2 (with $\rho > 1.3$) show that the frequency of the modes increases linearly with coverage up to monolayer completion, indicating a gradual stiffening of the lattice. They can be fitted assuming a homogeneous triangular lattice, where, in first approximation, the effect of the substrate has been ignored. Such a fit is shown in figure 2 for ρ=1.48. The sound velocities obtained from this model are c_L=2200 m/s for the longitudinal modes and c_T=1280 m/s for the transverse ones. The nearest neighbor distance at this coverage is 3.50Å. This is comparable to solid hcp D_2 at 279 bar[24] (a=3.526 Å, c=5.750 Å), where the sound velocities in the hexagonal planes are c_L=2160 m/s and c_T=1130 m/s.

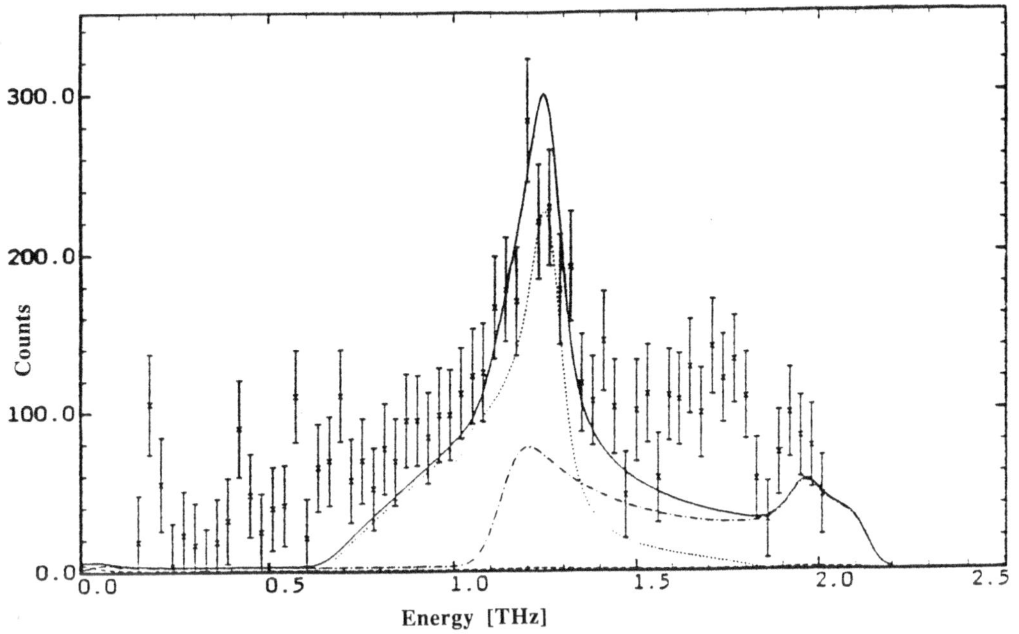

Figure 2. Inelastic spectra taken at Q=1.70Å^{-1} of D_2 adsorbed on graphite in the incommensurate phase (ρ=1.48, T=4 K). The full line is the fit to the data. The model considers a triangular lattice interacting through central harmonic forces, taking into account the mosaicity of the sample, the powder average and the resolution of the spectrometer as described in ref.3. The dotted line and the dashed-dotted line are the contribution of the transverse and longitudinal phonon branches, respectively. The fitted force constant between adjacent adsorbed molecules is 0.2 N/m, the corresponding longitudinal and transverse sound velocities are 2200 m/s and 1280 m/s, respectively.

The Helium isotopes

Only recently the zone center phonon gap of the commensurate phase of the He isotopes has been measured[5, 25]. The large neutron absorption cross section of ^{3}He makes it extremely difficult to obtain reliable data. As discussed above, a corrected background has to be subtracted. ^{4}He, on the other hand, has a very small coherent neutron cross section which makes its measurement very difficult.

In figure 3a are shown the INS data obtained for a ρ=0.9 ^{3}He coverage at T=0.85 K. A clear peak can be seen at 0.27 THz which has been identified as the phonon gap. The dashed line is the background corrected for the large neutron absorption cross section of ^{3}He. Figure 3b depicts the same data with the background subtracted. The full line is a fit with a model that considers a triangular array of interacting helium atoms. Each atom also interacts with the graphite through a restoring force to its equilibrium position, the center of the graphite hexagon. All the forces are assumed to be central harmonic forces. The mosaicity and powder averages were performed as described in ref. 3.

The origin of the extra intensity at frequencies up to 0.7 THz is not yet clear. It can be ascribed to two possible origins: it is either the signal of the transverse zone boundary phonon or it is related to the out-of-plane bouncing mode. This out-of-plane mode has been found at 1.5 THz for a single He atom[27]. It seems unlikely that the presence of adjacent He atoms

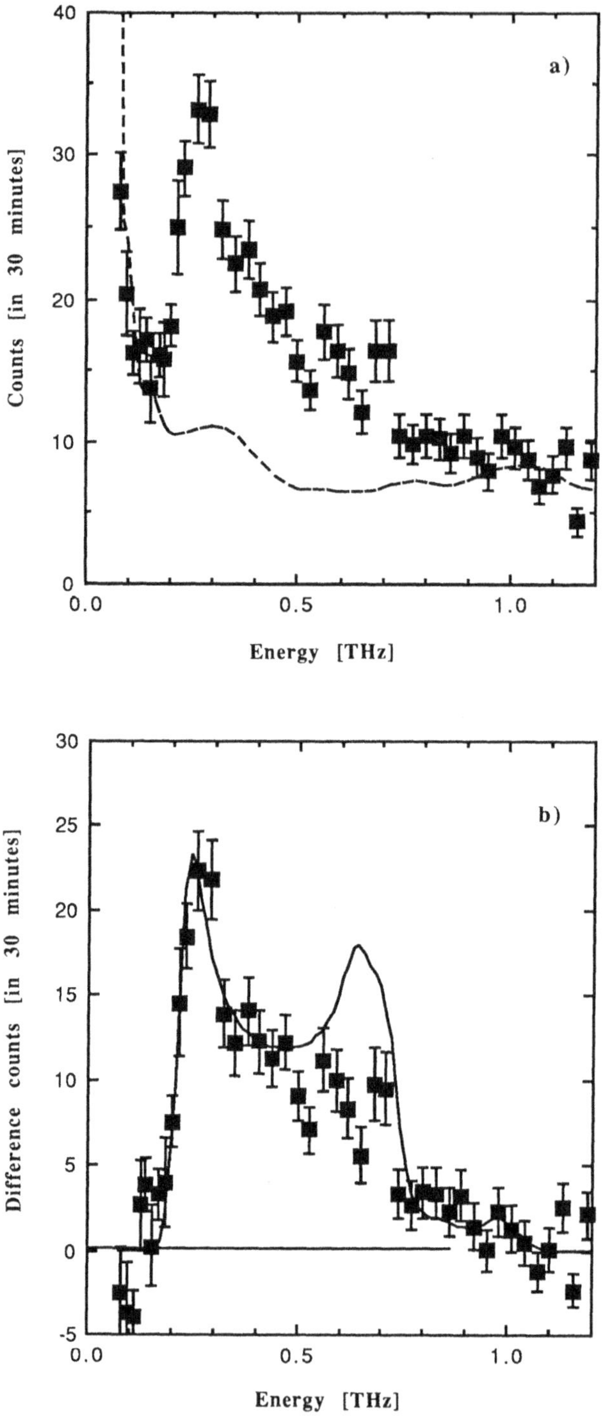

Figure 3. a) Neutron inelastic spectra for ^3He ($Q=1.70 Å^{-1}$, $\rho=0.9$ and $T=0.85K$). The dashed line is the background corrected for absorption as described in the text. b) The same data with the background subtracted. The full line is the fit with a model that considers a triangular array of ^3He atoms and takes account of the mosaicity and powder averages of the sample, as well as the spectrometer resolution.

should modify substantially this value. On the other hand, if the transverse phonon is at ≈0.65 THz, as suggested by the data, then the phonon dispersion relation of ^3He would be the largest measured for the quantum gases. This would imply that the effective interaction between the ^3He atoms is very strong, much stronger indeed than for the hydrogen isotopes. If this feature is confirmed by further experiments, then the picture of the domain wall phase for ^3He will be completely different from that of hydrogen. The sharp superheavy domain walls found in the DW phase of the hydrogen isotopes is a result of the weak adatom-adatom interaction potential compared with the adatom-substrate one. In the case of ^3He the DW would consist of very broad domain walls (a rough calculation suggest that they would be ≈7 commensurate lines wide), since it would be energetically more favorable for the system to loose commensurability keeping the adatom-adatom separation as large as possible.

Table II

Parameters characterizing the phonon density of states for the in-plane modes of the commensurate phase of the helium isotopes adsorbed on graphite (the uncertainty of the experimental values is estimated to be ≈2 K).

	^3He	^4He
Zone center gap[5, 25] (neutron scattering)	11 K	11 K
calculations from reference 26	35 K	28 K
calculations from reference 20	16 K	16 K
DOS width[5] (neutron scattering)	(?)38 K	-

DISCUSSION

Usually, when modelling the adsorbate-substrate potential Steele´s[28] Fourier decomposition has been used

$$V(\bar{\mathbf{r}},z) = V_0(z) + \sum_g V_g(z) \exp(i\bar{\mathbf{g}}.\bar{\mathbf{r}})$$

where z is measured perpendicular to the substrate surface, **r** is a vector in the plane of the surface (its origin for graphite is taken at the center of the honeycomb cell), and the vectors **g** are reciprocal lattice vectors of the substrate surface lattice. This sum is truncated at the first shell of reciprocal lattice vectors $|g| = g_o$ and the functions $V_o(z)$ and $V_{go}(z)$ are treated near the minimum ($z=z_o$) of V. The leading amplitude of the corrugation V_{go} is negative for the quantum gases adsorbed on graphite and the minimum adsorption energy is found[29] at **r**=0.

For the $(\sqrt{3} * \sqrt{3})R30^{\circ}$ commensurate lattice, ignoring a possible hybridization of the adlayer and substrate dynamics[30], the frequency for the in-plane zone center phonon gap is given for an adsorbate of mass M by

$$\Delta = g_o [- 3 V_{go} / M]^{0.5}$$

in an harmonic approximation. For the quantum gases, large deviations from this simple formula is expected due to their large zero point motion and the strong anharmonicity of the in-plane part of the adsorption potential.

In theoretical calculations of the properties of these systems, a form for the adsorbate-adsorbate and for the adsorbate-substrate interaction potential has to be assumed. For the interaction between rare gas atoms a Lennard-Jones (12,6) form:

$$\varphi(r) = 4 \varepsilon [(\sigma/r)^{12} - (\sigma/r)^6]$$

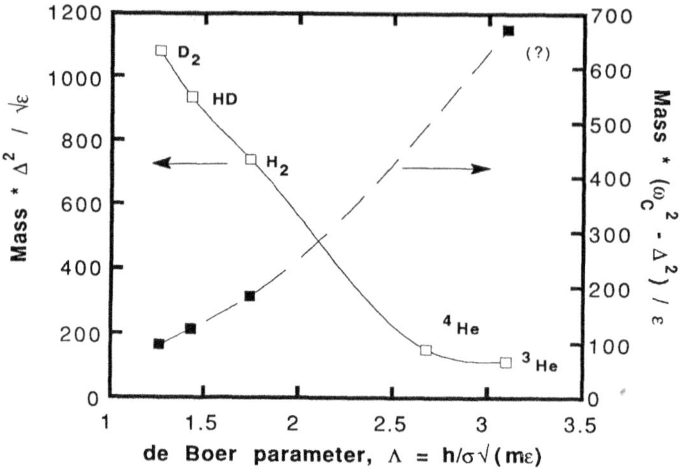

Figure 4. Gap values and DOS width plotted vs. the de Boer parameter Λ for the quantum gases (Λ is the ratio between the de Broglie wavelength of the relative motion of two atoms with energy ε and the minimum in the LJ potential). The values plotted are reduced quantities and are explained in the text.

seems quite reasonable, but also more accurate potentials have been proposed[31]. For the hydrogen isotopes calculations Novaco[21] used the Silvera-Goldmann[32] potential.

For the interaction with the substrate the Lorentz-Berthelot rules for unlike interactions have been generally accepted:

$$\sigma_{AB} = (\sigma_{AA} + \sigma_{BB}) / 2$$

$$\varepsilon_{AB} = (\varepsilon_{AA}\ \varepsilon_{BB})^{0.5}$$

combining the Lennard-Jones parameter of both systems. This practice is questionable, since it is known that the geometric mean often overestimates ε_{AB} and other combining rules have been suggested[33,34]. To circumvent this problem, a set of Lennard-Jones parameters have been constructed that match the values of the bound state energies as determined by selective adsorption experiments[21,35,36]. The results of these calculations, as shown in Tables I and II, are quite encouraging and reproduce very well the experimental gap values. An excellent review on the substrate corrugation problem for rare gases and also more complicated molecules has been presented recently by Bruch[37].

Figure 4 presents the gap and DOS width values for the quantum gases. Assuming the Lorentz-Berthelot combination rule, V_{go} is proportional to $(\varepsilon_{AA}\ \varepsilon_{BB})^{0.5}$, where ε_{AA} corresponds to the adsorbate and ε_{BB} to graphite. The quantity $M\ \Delta^2 / \varepsilon_{AA}{}^{0.5}$ should always be a constant. There is still a dependence on σ that is left, but it should not modify dramatically the result, since it changes only slightly between He and H_2. The strong departure of a constant seen in figure 4, is an evidence of the importance of the quantum character of these molecules. The large zero point motion, increasing with the de Boer parameter, averages the surface corrugation producing a lower effective corrugation.

The DOS width is a measure of the interaction between adsorbed molecules. All the quantum gases adsorbed in the C phase are diluted respect to their bulk value. A larger zero point motion increases the overlap of the wave functions and produces a stronger effective interaction. The quantity $M\ (\omega_C{}^2 - \Delta^2) / \varepsilon_{AA}$, where ω_C is the upper cutoff frequency of the DOS, should also be constant. A clear increasing tendency is displayed by the hydrogen isotopes[4], unfortunately the ^{3}He value is still only a guess.

CONCLUSIONS

The agreement between theoretical calculations for the commensurate phase and experimental values for the phonon gap is reasonable, but the interaction between adatoms is still overestimated. The measurements performed in the C and IC phases allow the determination of elastic constants which are comparable to bulk systems under large positive and negative pressures. An adequate analysis of these data permit the determination of the interaction potential at distances that are hard to reach in 3D matter. A precise knowledge of the graphite surface has an intrinsic interest that goes beyond the adsorbate monolayer regime.

ACKNOWLEDGEMENTS

This work has been partially supported by the Federal Ministry of Research and Technology (BMFT) of the Federal Republic of Germany.

REFERENCES

1. N.D.Shrimpton, B.Joos and B.Bergersen, Phys. Rev. B 38, 2124 (1988)
2. J.Villain and M.B.Gordon, Surf. Sci. 125, 1 (1983)
3. V.L.P.Frank, H.J.Lauter and P.Leiderer, Phys.Rev.Lett. 61, 436 (1988)
4. H.J.Lauter, V.L.P.Frank, P.Leiderer and H.Wiechert, Physica B. 156&157, 280 (1989)
5. V.L.P.Frank, H.J.Lauter, H.Godfrin and P.Leiderer, "Phonons 89", pp. 1001, Eds. S.Hunklinger, W. Ludwig & G.Weiss, World Scientific, Singapore, 1990. The value for the gap quoted in this work was 13 K. This corresponds to the peak value of the spectrum. Taking into account the convolution with the resolution function of the spectrometer shifts it down to 11 K.
6. V.L.P.Frank, H.J.Lauter and H.Taub, unpublished data.
7. T.Moeller, H.J.Lauter, V.L.P.Frank and P.Leiderer, in "Phonons 89", pp. 919, Eds. S.Hunklinger, W. Ludwig & G.Weiss, World Scientific, Singapore, 1990
8. F.Y.Hansen, V.L.P.Frank, H.Taub, L.W.Bruch, H.J.Lauter and J.R.Dennison, Phys.Rev.Lett. 64, 764 (1990)
9. M.Bretz, J.G.Dash, D.C.Hickernell, E.O.McLean and O.E.Vilches, Phys.Rev. A 8, 1589 (1973)
10. Hering S., PhD. thesis, University of Washington, 1974
11. F.C.Motteler and J.G.Dash, Phys. Rev. B 31, 346 (1985)
12. F.A.B.Chaves, M.E.B.P.Cortez, R.E.Rapp and E.Lerner, Surf. Sci. 150, 80 (1985)
13. H.Freimuth and H.Wiechert, Surf. Sci. 162, 432 (1985)
14. H.Freimuth and H.Wiechert, Surf. Sci. 178, 716 (1986)
15. H.Freimuth and H.Wiechert, Surf. Sci. 189, 548 (1987)
16. Papyex is produced by Carbone Lorraine, 45 Rue des Acacias, F-75821 Paris Cedex 17
17. S.W. Lovesey, Theory of neutron scattering from condensed matter, Oxford Science Publications, Oxford (1986)
18. Taub H., Carneiro K., Kjems J.K. and Passell L., Phys.Rev B 16, 4551 (1977)
19. M.Nielsen, J.P.McTague and W.Ellenson, J.Phys.(Paris), Colloq. 38, C4-10 (1977)
20. J.M.Gottlieb and L.W.Bruch, Phys. Rev. B 40, 148 (1989); J.M.Gottlieb and L.W.Bruch, Phs. Rev. B 41, 7195 (1990)
21. A.D.Novaco, Phys. Rev. Lett. 60, 2058 (1988)
22. H.J.Lauter, H.P.Schildberg, H.Godfrin, H.Wiechert and R.Haensel, Canad. J. Phys. 65, 1435 (1987) and references therein; J.Cui and S.C.Fain, J. of Vacuum Sci. & Technology A5, 710 (1987); J.Cui, S.C.Fain, H.Freimuth, H.Wiechert, H.P.Schildberg and H.J.Lauter, Phys. Rev. Lett. 60, 1848 (1988)
23. V.L.P.Frank, H.J.Lauter and P.Leiderer, Jpn. J. Appl. Phys. Suppl. 26, 347 (1987)
24. M.Nielsen, Phys. Rev. B 7,1626 (1973)
25. V.L.P.Frank, H.J.Lauter, H.Godfrin and P.Leiderer, preliminary data obtained in a test measurement. The value of 11 K quoted for the gap takes into account the spectrometer resolution.
26. X.-Z. Ni and L.W.Bruch, Phys. Rev. B 33, 4584 (1986)

27. M.W.Cole, D.R.Frankl and D.L.Goodstein, Rev. Mod. Phys. 53, 199 (1981) and references contained therein.
28. W.A.Steele, "The Interaction of Gases with Solid Surfaces", Pergamon, New York, 1974.
29. W.A.Steele, Surf. Sci. 36, 317 (1973)
30. E. de Rouffignac, G.P.Alldredge and F.W. de Wette, Phys. Rev. B 24, 6050 (1981); F.W. de Wette, B.Firey and E. de Rouffignac, Phys. Rev. B 28, 4744 (1983)
31. R.A.Aziz and H.H.Chen, J. Chem. Phys. 67, 5179 (1977); R.A.Aziz, V.P.S.Nain, J.S.Carley, W.L.Taylor and G.T. McConville, J. Chem. Phys. 70, 4330 (1979)
32. I.F.Silvera and V.V.Goldmann, J. Chem. Phys. 69, 4209 (1978)
33. M.Rigby, E.B. Smith, W.A.Wakeham and G.C.Maitland, "The Forces between Molecules", Clarendon, Oxford, 1986.
34. K.T.Tang and J.P.Toennies, Z. Phys. D 1, 91 (1986)
35. A.D.Novaco and J.P.Wroblewski, Phys. Rev. B 39, 11364 (1989)
36. L.Mattera, F.Rosatelli, C.Salvo, F.Tommasini and U.Valbusa, Surf. Sci. 93, 515 (1980); E.Ghio, L.Mattera, C.Salvo, F.Tommasini and U.Valbusa,J. Chem. Phys. 73, 556 (1980)
37. L.W.Bruch, in "Phase Transitions in Surface Films", NATO Advanced Study Institute, Erice, Italy, 19-30 June 1990

THERMODYNAMIC AND NEUTRON-DIFFRACTION STUDIES OF H_2 AND D_2 MULTILAYERS PHYSISORBED ON GRAPHITE

H. Wiechert

Institut für Physik
Johannes Gutenberg-Universität
D-6500 Mainz, F.R.G.

1. INTRODUCTION

The evolution of physisorbed films from two-dimensional (2D) to bulk behavior is a topic of high current interest.[1] In this context the behavior of the quantum systems H_2, HD and D_2 is of particular relevance because of the following reasons:

1. Due to the strong influence of the quantum zero-point energy the hydrogen isotopes are highly compressible. This leads to strongly compressed monolayers[2,3] before further layer condensation occurs. This property makes these systems significant for the exploration of conditions of multilayer growth. It is generally believed,[1,4-6] that the incompatibility between the adsorbate and bulk lattice structures gives rise to lateral strains between the overlayers, which may cause reentrant incomplete wetting phenomena[7] in the weakly physisorbed quantum systems. Thus the behavior of the first few layers next to the substrate is of crucial importance for the character of the multilayer growth.

2. The phase diagrams and structures of hydrogen isotope multilayers adsorbed on graphite are unknown so far. Recently, detailed specific-heat studies[8-12] mapped out the phase diagrams of the monolayers, and with neutron diffraction[2,3,12-17] and LEED[17,18] the structures of the observed phases were identified. At low coverages these systems exhibit a commensurate $(\sqrt{3}\times\sqrt{3})$ R30° phase[2,3] due to the strong influence of the substrate corrugation potential, and undergo a commensurate-incommensurate (C-IC) transition via a sequence of different domain-wall phases[8-18] to an equilaterally spaced triangular compressed incommensurate phase at higher coverages. First exploratory neutron-diffraction measurements of D_2 multilayers[19] found common oblique bi- and trilayer structures. A recent study of the multilayer growth of H_2 adsorbed on MgO[20] discovered the occurrence of a conventional van der Waals phase diagram including triple and critical points in the second layer and a succession of layering transitions at higher coverages.

3. The weakly bound states of hydrogen multilayers on substrates may be close to the limiting case where a quantum system at low temperatures condenses as a liquid rather than as a solid.[21] H_2 is a Bose particle and is expected to undergo a Bose-Einstein condensation to a superfluid phase provided the solidification can be suppressed to temperatures below 6.6 K[22] and probably even lower in a real system.[23] For the second layer of H_2 on MgO a triple line at 7.2 K was found,[20] which is still too high for a superfluid transition. The question remains: Do the molecules in the second layer of H_2 on graphite condense into localized states?

Excitations in Two-Dimensional and Three-Dimensional Quantum Fluids
Edited by A.G.F. Wyatt and H.J. Lauter, Plenum Press, New York, 1991

In order to provide a contribution to these challenging questions, in this paper a short overview will be presented on recent heat-capacity and neutron-diffraction experiments of H_2 and D_2 multilayers adsorbed on graphite.

2. EXPERIMENTAL DETAILS

For the heat-capacity measurements we employed a conventional quasi-adiabatic technique (Nernst calorimeter).[9,11] The calorimeter was a cylindrical copper cell (diameter 30 mm, height 60 mm) filled with a piece of graphite foam (trademark of Union Carbide). This substrate consists of a uniform powder with single-crystal domain size of about 900 Å in diameter.[24]

The neutron diffraction experiments were performed on the two-axes powder diffractometers D1B and D16 at the Institute Laue-Langevin in Grenoble. The sample consisted of a stack of exfoliated graphite sheets, which were oriented parallel to the neutron-scattering plane. Data were taken as well on Papyex as on ZYX graphite (trademarks of Carbone Lorraine and Union Carbide) with crystallite sizes of about 300 and 2000 Å in diameter, respectively. Peak positions and relative intensities were identical within experimental error for these two substrates. To analyze the data, powder-averaged Lorentzian-squared lineshapes[25] were fitted to the data.

Natural H_2 and D_2 gases were used, which adsorbed on graphite convert to their ground states: para-hydrogen and ortho-deuterium. The density ρ (normalized to $\rho = 1$ for the ideal commensurate $\sqrt{3}$-phase) was determined from the known structure of the incommensurate (IC) 2D solid phase below monolayer completion. To agree with this scale, the specific-heat coverages had to be adjusted by 4 % for D_2 and 2 % for H_2. Separate vapor pressure measurements were done to correct the densities for desorption, which for the weakly bound hydrogen isotope multilayers becomes appreciable above about 12 K.

3. HEAT-CAPACITY RESULTS

A. Hydrogen

Fig. 1 shows a series of representative heat-capacity scans for second layer H_2 adsorbed on graphite foam at densities between $\rho = 1.656$ and 2.462. The background due to the calorimeter has been subtracted from the data and the total heat capacity of the film is plotted versus temperature. The bottom trace indicates for comparison the contributions of the approximately completed monolayer ($\rho = 1.44$, dotted line) and of desorption (solid line) to the total heat capacity at $\rho = 2.134$. A first broad anomaly appears at $\rho = 1.656$, grows with density and shifts to a maximum temperature of 9.31 K at $\rho \approx 2.03$. At $\rho = 1.751$ a second small heat-capacity signal could be detected at T = 5.96 K. It increases in height with density, but always stays at the same temperature. Beyond $\rho \approx 2.2$ the high-temperature anomaly diminishes again and converges to the low-temperature peak. Finally, above $\rho \approx 2.5$ only one sharp peak remains, which shifts rapidly to higher temperatures with growing density (not shown in Fig. 1, but see Fig. 2) indicating the melting transition of the completed second layer.

The constancy of the peak temperature at $T_t = 5.96$ K suggests the existence of a triple line. The maximum specific heat is $C/N_2k_B \approx 2.9$, where N_2 is the number of H_2 molecules in the second layer participating in the melting process and k_B is the Boltzmann constant. The heat of fusion obtained by integrating C vs. T is 0.93 cal/mole, only about 1/30 of the 3D heat of fusion. Both values are roughly a factor of 3 smaller than those observed for second-layer H_2 adsorbed on MgO,[20] and appreciably smaller than those found, e.g., for monolayer melting of Ne/graphite.[26,27] This indicates that the first-order character of the melting transition from the second-layer solid to the liquid phase is very weak. The heat-capacity anomalies signalizing the liquid-vapor coexistence line are very well pronounced cusp-shaped peaks exhibiting their similarity to those of 2D Ising transitions.[28] Contrary to that for H_2/MgO[20] only

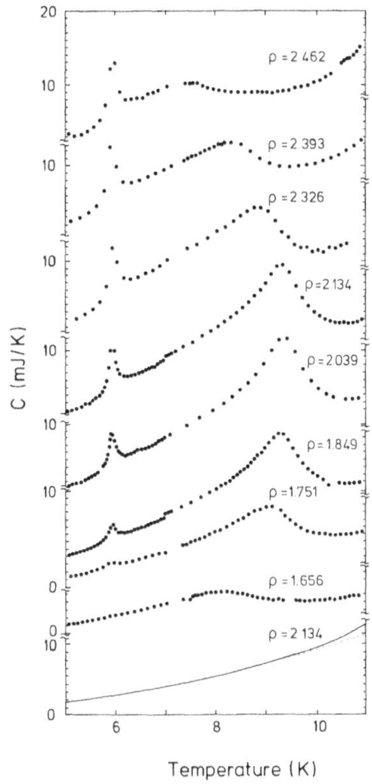

Fig. 1. Heat-capacity data of second-layer H_2 on graphite versus temperature at eight densities between ρ = 1.656 and 2.462. The bottom trace shows for comparison the contribution of the first full monolayer (dotted line) and of desorption (solid line) to the total heat capacity at ρ = 2.134.

very small "bumps" could be detected. The critical point of the second-layer liquid-vapor coexistence region was determined to be at T_c = 9.31 K and ρ_c = 2.03. An analysis - to be presented in a separate paper - shows evidence that this critical point belongs to the 2D Ising universality class with critical exponents β = 0.127 ± 0.015 and $\alpha \approx 0$.

In Fig. 2 the locations of all heat-capacity peaks are plotted in a phase diagram (open triangles). In addition to the second layer, we also studied the third layer up to densities of $\rho \approx 4.2$. The specific-heat signals of the third layer are somewhat more rounded than those of the second layer, but give clear indications of the phase boundaries. A third layer triple line was found at T_t = 9.35 K and a third layer critical point at T_c = 10.08 K. The diagram provides clear evidence of solid-vapor (S + V) and liquid-vapor (L + V) coexistence regions. This is supported by heat-capacity isotherms, which revealed a linear variation with density in these regions, which is a necessary condition for two-phase equilibrium. End points of coexistence regions obtained from heat-capacity isotherms are included in the diagram (crosses with error bars). The temperature range of the third layer liquid-vapor coexistence region is much smaller (1/4 to 1/5) than that of the second layer, which may point to a gradual suppression of the triple line with growing layer number. For comparison, for H_2 on MgO[20] no sign of a triple line in the third and further layers was found. The critical points seem to converge to a common temperature ($T_r \approx$ 10.1 K), which may be the roughening temperature.[29,30] This temperature is about 73 % of the 3D triple point temperature (T_p(3D) = 13.81 K), slightly lower than that of H_2 on MgO[20] (75 %) or that of Ne and Ar on graphite (80 %).[27] The melting lines of the completed individual layers denoted by numbers (1), (2) and (3) in Fig. 2 shift to higher temperatures with increasing density due to a combination of compression by the topmost layer and

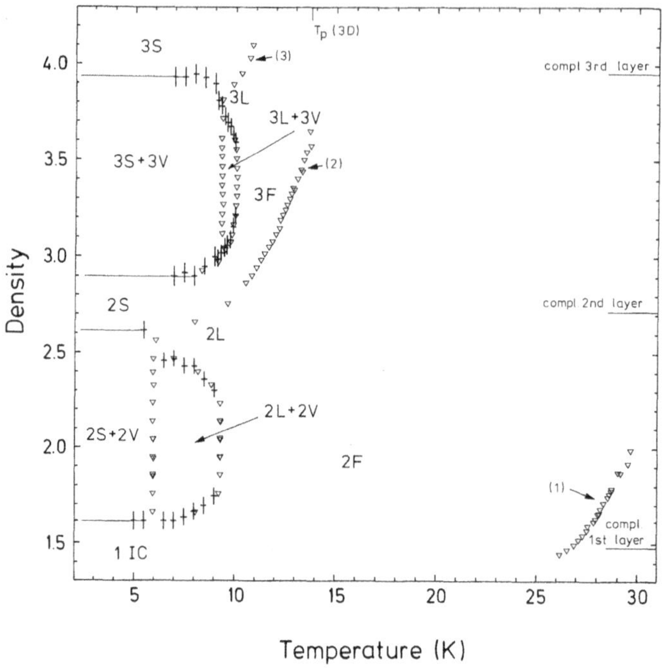

Fig. 2. Phase diagrams of the second and the third layer of H_2 adsorbed on graphite. The open triangles represent the locations of specific-heat anomalies, the crosses with error bars are from heat-capacity isotherms. The lines labeled (1), (2) and (3) mark the melting transitions of the first, second and third layer, respectively. All data are corrected for desorption. The symbols IC, S. V, L and F are, respectively, incommensurate solid, solid, vapor and hypercritical fluid; 1, 2 and 3 stand for first, second and third layer. The solid lines indicate the boundaries of solid-vapor coexistence regions in each layer. T_p(3D) marks the 3D triple point at 13.81 K, and the lines on the right ordinate mark the densities of the completed first, second and third layers.

substrate attraction. The first layer remains solid far above the 3D triple point and probably also the second layer. This phenomenon is symmetric with surface melting[1] and is designated as substrate freezing,[27,31] the tendency of the adsorbed film to stabilize a dense solid layer next to the substrate.

The phase diagrams of H_2 multilayers on graphite provide an indication for the existence of 2D liquid phases in addition to solid and vapor phases in the second and the third layer. Thus one may infer that lattice-gas[1,29] or solid-on-solid models,[1,30] which most often contain only two phases (solid and vapor), are unrealistic for the description of H_2 multilayers. The triple point of the 2nd layer occurs at T_t = 5.96 K. This is the lowest triple point found in nature so far. For H_2 on MgO a triple point at T_t = 7.2 K was observed.[20] The temperature of 5.96 K is lower than 6.6 K, the anticipated temperature, where an ideal 3D Bose condensation into a superfluid state should occur according to theoretical predictions.[22] The phase diagram provides no support for the existence of such a condensation, even if the freezing of the low-temperature solid phase already appears to be rather smooth. Thus one may conclude that a transition to a superfluid 2D phase - if it exists at all - should only occur on substrates which stabilize the 2D liquid phase to substantially lower temperatures.

B. Deuterium

In case of D_2 adsorbed on graphite foam, the density range between ρ = 1.5 and 6.5 (second to fourth layer) has been explored by specific-heat measurements. From

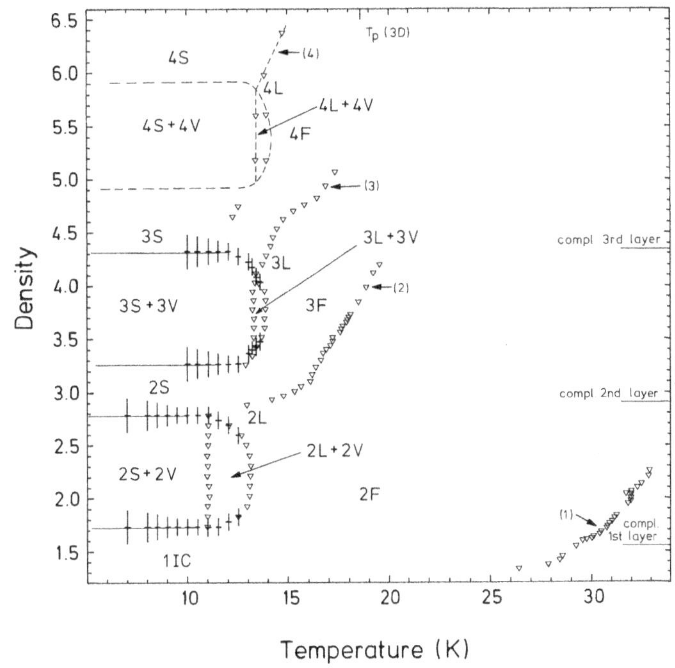

Fig. 3. Phase diagrams of the second, third and fourth layer of D_2 adsorbed on graphite. All data are corrected for desorption. The nomenclature is the same as that explained in the caption of Fig. 2. The solid lines indicate experimentally determined phase boundaries, the dashed lines are speculative. Four distinct layers with solid-vapor (S + V) and liquid-vapor (L + V) coexistence regions are clearly discernible, likewise the melting lines of the completed layers (1)-(4). T_p(3D) marks the 3D triple point at 18.69 K, and the lines on the right ordinate mark the densities of the completed first, second and third layers.

heat-capacity isotherms two-phase equilibrium regions were deduced. The locations as well of all specific-heat peaks (open triangles) as of end points of linear portions in heat-capacity isotherms (crosses with error bars) are summarized in the phase diagram of Fig. 3. As for H_2 on graphite van der Waals-like phase diagrams result with solid-vapor (S + V) and liquid-vapor (L + V) coexistence regions, triple and critical points for layers (2) - (4). Thus the D_2 multilayers also provide evidence for the existence of three separate phases, i.e., solid, liquid and vapor, in each layer. Due to the reduced influence of the quantum zero-point energy, the triple and critical temperatures of the layers are higher than those for H_2. For the second layer the following values were determined: T_t = 11.04 K and T_c = 13.10 K, and for the third layer: T_t = 13.20 K and T_c = 13.86 K. The data for the second layer scale approximately as expected from the quantum theory of corresponding states applied to 2D systems.[32] The heat of fusion at the triple line of second-layer D_2 is 1.62 cal/mole. The temperature difference $\Delta T = T_c - T_t$ for the second layer is reduced compared to that for H_2, whereas the ΔT-values for the third layers of both systems are approximately equal. This may indicate that the second layer of H_2 is more strongly decoupled from the first layer owing to the enhanced zero-point motions of the molecules than in case of D_2. The critical points of the individual layers seem to tend towards a common temperature of about 14 K (roughening temperature). The heat-capacity measurements reveal that the first and second layer definitely remain solid above the 3D triple point (T_p(3D) = 18.69 K), which again points to the phenomenon of substrate freezing.

These experiments have shown, that heat-capacity measurements are a powerful method to study phase diagrams of multilayers. To obtain, however, microscopic informations on the nature of the observed phases, one has to apply diffraction techniques.

4. NEUTRON-DIFFRACTION RESULTS

A. Deuterium

Let us first report on neutron-diffraction results obtained for ortho-D_2 physisorbed on graphite (Papyex), because this system has a neutron-scattering cross-section by a factor of about three larger than that of para-H_2 and thus should yield a clearer picture of the phenomena occurring in these quantum systems. The evolution of neutron-diffraction patterns as one passes through layers two to four with growing film thickness at a fixed temperature of T = 4 K is illustrated in Fig. 4. The data displayed have been background-subtracted, leaving only the signal from the adsorbed layer. They are plotted versus the wave-vector Q. Unfortunately, the intense (002)-graphite background reflection at Q = 1.876 $\mathring{A}^{-1}$ prevents the recording of reliable data in the range $1.8 < Q < 1.92$ $\mathring{A}^{-1}$ as indicated by the double-arrow in Fig. 4. Data points in this range have been normally omitted for clarity. The first diffraction profile at $\rho = 2.07$ shows the (10) Bragg reflection of the incommensurate (IC) equilateral triangular structure of the compressed first layer. The first sign of a solid second layer appears in the next pattern at $\rho = 2.36$. In addition to the main peak at Q = 2.15 $\mathring{A}^{-1}$, a new peak emerges at Q = 1.93 $\mathring{A}^{-1}$ and a small shoulder at the leading edge of the main peak at Q = 2.09 $\mathring{A}^{-1}$, which is hardly to be seen in this

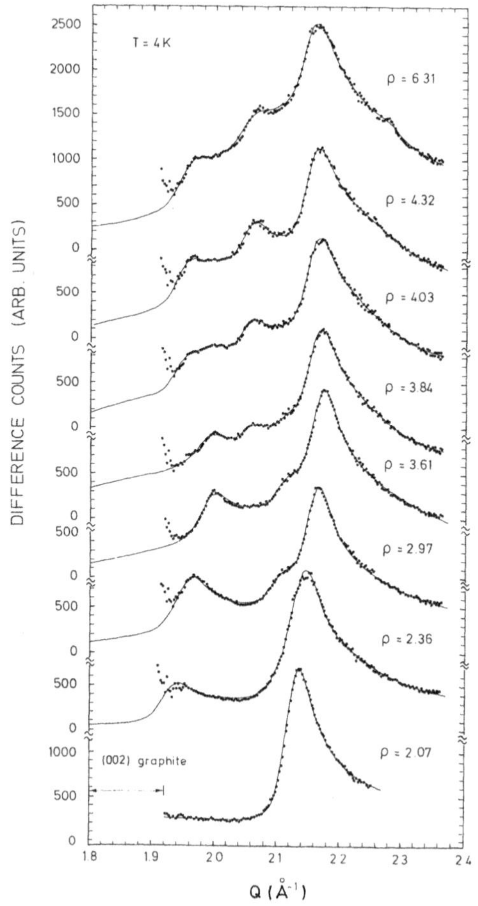

Fig. 4. Evolution of neutron-diffraction patterns with density at T = 4 K for D_2 on graphite (Papyex). The double-arrow indicates the range of severe interference with the (002)-graphite reflection. The empty cell background has been subtracted from the data. The solid lines are fits of powder-averaged Lorentzian-squared line shapes on top of a broad Gaussian background to the data.

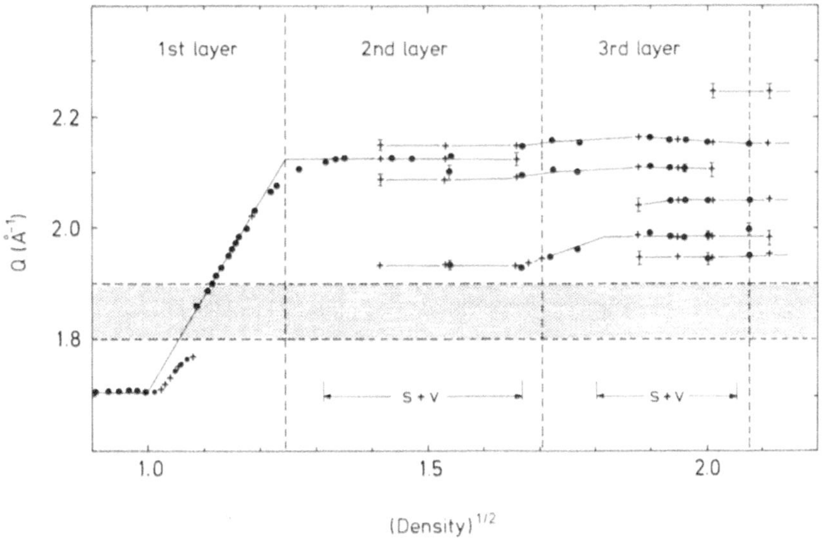

Fig. 5. The neutron-diffraction peak position Q vs. the square root of density ρ for D_2 on Papyex (solid circles) and D_2 on ZYX graphite[19] (crosses) at T = 4 K. The vertical dashed lines indicate the densities of the complete first, second and third layer. Values for second and third layer completion were calculated by assuming common oblique unit cells (see text). Below monolayer completion only the positions of the main peaks are presented.[16,17] The solid lines serve to guide the eye. The dotted range shows the Q-region of strong interference with the graphite (002)-reflection. At the bottom the two-phase coexistence regions (S + V) from heat-capacity measurements are displayed.

spectrum, but becomes clearly discernible in the next two spectra at ρ = 2.97 and 3.61. The low-Q peak shifts to higher Q-values with increasing density signalizing a compression of the second layer. Entering the third layer region above ρ = 3.6, results in the appearance of new peaks. The shoulder at the leading edge of the most intense peak now vanishes and two new reflections start to arise at Q = 2.05 $Å^{-1}$ and Q = 1.95 $Å^{-1}$ for ρ = 4.32, being first in coexistence with the second-layer reflections at ρ = 4.03. Finally, at the highest density (ρ = 6.31) a new peak at Q = 2.276 $Å^{-1}$ forms, which may be attributed to the (101) reflection of bulk D_2 in hcp structure.[33] The solid lines in Fig. 4 represent powder-averaged Lorentzian-squared intrinsic line shapes[25] with a coherence length of 300 Å, which empirically gave the best fits to the data. From the fits the exact peak positions and intensities were extracted. It is evident from the figure that the Bragg reflections are residing on a broad Gaussian-shaped background, which grows with increasing film thickness and may be caused by defects or disordered ranges in the overlayers.

The complex behavior of the diffraction patterns with growing density becomes clearer in a plot of the peak positions as a function of the square root of density. Fig. 5 shows data obtained for D_2 adsorbed as well on Papyex (solid dots) as on ZYX[19] (crosses) at T = 4 K. Both sets of data are in excellent agreement with each other. The data of the first layer have already been presented previously[13-17] and will not be discussed here and also the ZYX results.[19] In the density range ($1.32 \leq \rho \leq 1.551$) below monolayer completion the system forms a homogeneous equilateral triangular IC phase, which covers the total substrate surface and is uniformly compressed as more molecules are incorporated in the layer. This fact was used to calibrate the density scale of the present work. The upper knee of the line at $\sqrt{\rho}$ = 1.245 indicates, where the formation of the second layer starts. In the density range of the second layer indicated by the dashed vertical lines, a coexistence between a peak attributed to the densest first layer and three peaks of a new bilayer structure is visible. This means that the system adopts a bilayer structure on some crystallites,

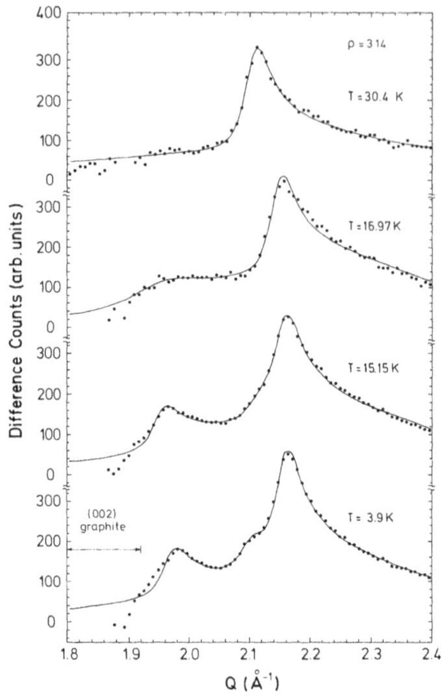

Fig. 6 Neutron-diffraction scans at ρ = 3.14 (low-temperature value) showing the melting of the bilayer of D_2 on Papyex. The solid lines are best fits using a composite line shape of a broad Gaussian background and powder-averaged Lorentzian-squared line shapes. The reduced quality of the fits at low Q-values results from the disturbing overlap with the (002) graphite reflection.

whereas others are still covered by the monolayer. Increasing the density results in a growing number of bilayer crystallites until the total surface is covered by the bilayer. An analogous effect occurs in the third layer. For comparison, the solid-vapor coexistence ranges (S + V) found from heat-capacity measurements (cf. Fig. 3) are indicated at the bottom of Fig. 5. It is evident that the peak positions approximately stay constant in the coexistence ranges, but begin to shift (and certain peaks vanish) as soon as the boundaries of these ranges are traversed. Thus there is good internal consistency between the results gained by both methods.

This is also valid for the melting transition. An example is shown in Fig. 6 at a density of ρ = 3.14, which is the low-temperature value of the density before desorption effects set in. At T = 3.9 K the characteristic spectrum of the completed bilayer is observed. As the temperature is raised the diffraction pattern remains practically unchanged until at T > 15 K the intensities of the low-Q peak and the shoulder-like feature centered at Q = 1.95 and 2.11 $\AA^{-1}$, respectively, drop and disappear at T $\approx$ 16 K. This melting temperature is in agreement with the heat-capacity result (cf. Fig. 3). At T = 16.97 K only a broad anomaly of a liquid second layer remains besides the first-layer reflection. The spectrum taken at T = 30.4 K demonstrates that the first layer is solid far beyond the 3D triple point, which provides direct evidence for substrate induced freezing.

To analyze the results several models were tested. The three-peak structure of the solid bilayer suggests that the system is subject to considerable strains between the layers, otherwise the shoulder-like feature would not appear. Assuming a uniaxially compressed first layer and a lower density equilaterally spaced triangular structure of the second layer, explains the peak positions, but not the intensity ratios. A model that fits to the observed features is that of a common oblique bilayer, which was originally proposed for helium on graphite.[34] As already published previously

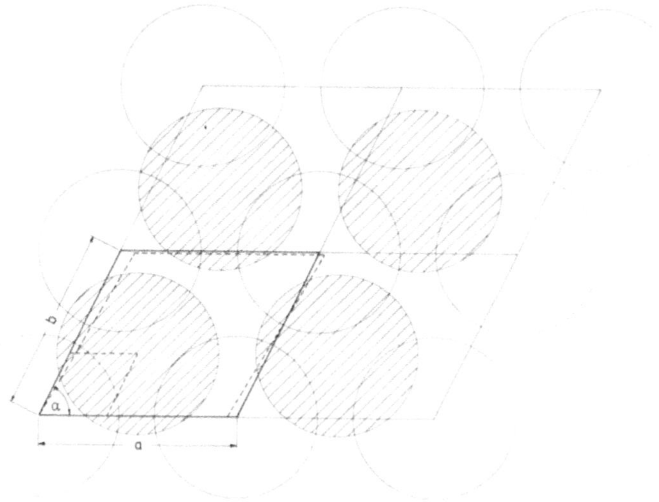

Fig. 7. The proposed common oblique lattice bilayer structure of D_2 on graphite.[19] The open and hatched circles represent molecules in the first and second layer, respectively. The solid lines indicate oblique unit cells and the dashed lines the hexagonal undistorted unit cell of the densest monolayer IC structure.

for D_2/ZYX graphite,[19] this model describes as well peak positions as intensities correctly. It is schematically illustrated in Fig. 7. The first-layer D_2 molecules are represented by open circles and the second-layer molecules by hatched circles. The solid lines indicate the oblique unit cells and, for comparison, the dashed large parallelogram the hexagonal undistorted unit cell of the densest monolayer IC structure. The lattice parameters a, b and α were determined from fits to the experimental data. The best consistency with the observed intensities results from a common unit cell, in which the first-layer molecule is located at (0, 0) and the second-layer molecule at (0.349, 0.381) in reduced coordinates. The lattice constants of the complete bilayer are listed in Table 1. It is remarkable that the lattice constant a approximately amounts to that of bulk hcp D_2,[3] whereas b is still compressed. Obviously, the system is affected by strong strains due to competing adhesive and cohesive forces, which distort the bilayer from hexagonal symmetry ($\alpha = 64.2^\circ$). From the structure the density ρ and the particle density per layer n was calculated. It is striking that the value of n for the bilayer ($n = 0.0932$ Å^{-2}) is reduced compared to that of the first layer ($n = 0.099$ Å^{-2}). Approaching the bulk solid, the system seems to prefer a more loosely packed structure, which is reasonable. During formation of the bilayer the first layer releases part of its lateral stress by promoting up to 5.8 % of the molecules to the second layer. An analogous effect was recently found in computer simulations of methane multilayer growth.[35] Table 1 also contains data for a common trilayer structure deduced from the diffraction spectra. It is evident that the tendency to looser packing arrangements and thus to the bulk phase is continued. The enhanced strain in the trilayer may be reflected by the larger value of the angle α. Wether the strain finally hinders the growth of the film into an arbitrarily thick layer and thus leads to incomplete wetting is unknown so far. The appearance of the (101) reflection at the highest investigated densities (cf. Figs. 4, 5) may be a first sign of bulk crystallisation, but it may also result from the approach of the interlayer distance to its bulk value. Even if the discussed models describe the observed features of the diffraction data, it should be emphasized that this interpretation is not unambiguous. Other models may apply as well, but none has been found so far that fits to the data.

B. Hydrogen

For H_2 multilayers adsorbed on ZYX graphite in general similar, but in detail slightly different features have been observed as for D_2. For brevity we will present

Table 1. Total densities ρ (in units of the complete $\sqrt{3}$-phase), particle densities per layer n and lattice parameters (a, b, α) of completed multilayers of D_2 and H_2 adsorbed on graphite. Data for the bulk solids are from Ref. 3.

Element	Layer	ρ	n (Å^{-2})	a (Å)	b (Å)	α
D_2	1	1.551	0.099	3.415	3.415	60°
	2	2.921	0.0932	3.584	3.325	64.2°
	3	4.317	0.0918	3.554	3.380	65.04°
3D solid (hcp)			0.09065	3.569	3.569	60°
H_2	1	1.473	0.094	3.505	3.505	60°
	2	2.71 ± 0.03	0.0865 ± 0.001	3.76 ± 0.05	3.37 ± 0.02	65.3° ± 0.6°
	3	3.95 ± 0.03	0.084 ± 0.001	3.74 ± 0.04	3.47 ± 0.02	66.8° ± 0.5°
3D solid (hcp)			0.08233	3.745	3.745	60°

here only the main results. In Fig. 8 a series of diffraction patterns is shown at T = 2 K and various densities. The first spectrum at ρ = 1.51 displays the Bragg peak of the slightly overfilled first layer. Increasing the density leaves the position and shape of the peak initially nearly unchanged until at ρ = 3.41 about half of the third layer is adsorbed. No sign of a shoulder-like feature in the leading edge of the main peak as for D_2/graphite can be seen for the second layer of H_2. If one assumes that the two layers adopt uncorrelated IC equilaterally spaced triangular structures of different densities, one would expect the occurrence of two reflections in the Q-range studied: a peak from the densest first layer and a signal from the less closely packed solid second layer. The position of the second-layer peak can be estimated from the upper boundary of the solid-vapor coexistence range obtained from the heat-capacity measurements. This yields a value of Q ≈ 1.82 Å^{-1} just in the region of strong overlap with the (002)-graphite reflection marked by the double-arrow in Fig. 8. Thus the presence of this peak is completely obscured by the strong interference with the substrate.To give a hint to its possible position it is indicated by the dashed line. At low temperatures there was sometimes a slight tendency to a somewhat steeper slope of the data at the leading edge of the supposed second-layer peak, but that was hard to prove. Due to the lack of any clearly identifiable feature, it is difficult to make any definite statement on the nature of the second layer. We believe that it can be regarded as a nearly perfect 2D IC solid, which is "freely floating" on the first layer. The phase diagram of the second layer determined by heat-capacity measurements also supports this interpretation. Obviously, in contrast to D_2 the larger contribution of the zero-point energy for H_2 causes the stabilisation of this solid phase before the third layer is formed. At ρ = 3.41 the shoulder-like feature at the leading edge of the first-layer reflection already known from the D_2 results appears. We take it as a sign for the formation of a common oblique bilayer next to the substrate due to the pressure exerted on the system by part of the third layer. As the film thickness is raised beyond the third layer (see diffraction patterns at ρ = 4.18 and 5.88), the system reorganizes probably to a common oblique trilayer. The total densities, particle densities per layer and the lattice constants, which are derived from the data under the assumption of common bi- and trilayers, are summarized in Table 1. As the observation of a third peak was prevented by interference with the substrate, the data are less certain than those for D_2 indicated by the estimated errors. It is reassuring that they exhibit the same tendencies, i.e., the existence of lateral strains in the layers and the approach to more loosely packed layers with increasing layer number.

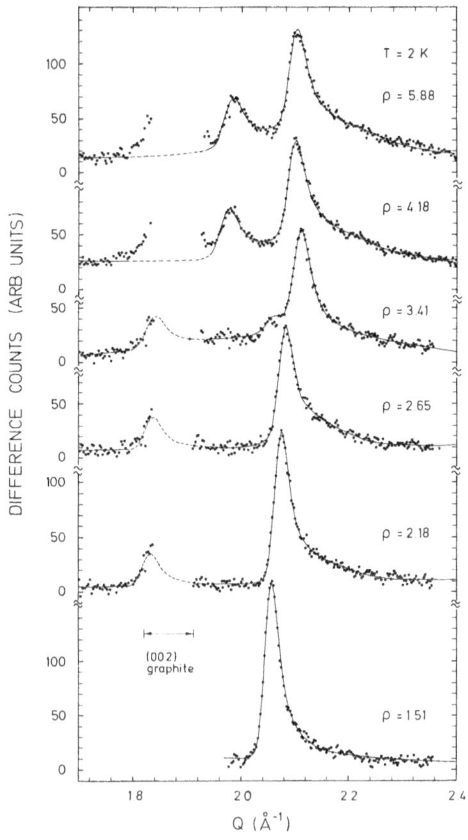

Fig. 8. Diffraction profiles from H_2 films on ZYX graphite at T = 2 K and various densities indicated in the figure. The solid lines are fits of powder-averaged Lorentzian-squared line shapes with a coherence length of 2000 Å to the data. The dashed peaks show suggested Bragg reflections, which probably occur in the Q-range of strong disturbance from the (002) graphite substrate reflection. For densities ρ = 4.18 and 5.88 they may also exist, but are not displayed.

5. CONCLUSION

The study of the phase diagrams and structures of H_2 and D_2 multilayers on graphite gives an impression of the richness of phenomena, which are encountered in physisorbed quantum systems. The present results have shown that the topologies of the multilayer phase diagrams are similar to those of 2D van der Waals systems including triple and critical points, which suggest the existence of three phases, i.e., solid, liquid and vapor, in each layer. Due to the strong influence of quantum effects the second layer of H_2 seems to form a nearly freely floating IC solid phase on top of the first layer. The triple point of the second layer is 5.96 K, the lowest triple point temperature found in nature to date. As further layers are adsorbed considerable strains arise, which may lead to the formation of common oblique bi- and trilayer structures. For D_2 this behavior occurs already in the second and third layer. Due to elastic strains within the multilayers, the layers are not independent of each other demonstrated by the shift of the triple and critical points to higher temperatures with growing film thickness and by reduced temperature widths of the liquid-vapor coexistence regions. Consequently it appears that the approach to bulk crystallisation in these quantum systems is a fascinating, but rather complicated process.

ACKNOWLEDGMENTS

The experimental work reported has been carried out in collaboration with St.-A. Arlt, H. Freimuth and G. Schauß concerning the heat-capacity measurements and with H. J. Lauter concerning the neutron-diffraction measurements. I would like to express my gratitude to them and also to J. G. Dash, S. C. Fain and O. E. Vilches for stimulating discussions. Research was supported by the SFB 262 (Mainz), the Materialwissenschaftlichen Forschungszentrum (Mainz) and the German Federal Minister of Research and Technology (BMFT) under Contract No. 05 453 AA I8.

REFERENCES

1. For recent reviews, see, e.g.: J. G. Dash, in: "Solvay Conference on Surface Science", F. W. de Wette, ed., Springer Verlag, Berlin (1988), p. 142; S. Dietrich, in: "Phase Transitions and Critical Phenomena", Vol. 12, C. Domb and J. L. Lebowitz, eds., Academic Press, London (1988), p. 1; and references cited therein.
2. M. Nielsen. J. P. McTague. and W. Ellenson, J. Phys. (Paris) Colloq. 38: C 4-10 (1977).
3. M. Nielsen, J. P. McTague, and L. Passell, in: "Phase Transitions in Surface Films", J. G. Dash and J. Ruvalds, eds., Plenum, New York (1980), p. 127.
4. R. J. Muirhead, J. G. Dash, and J. Krim, Phys. Rev. B 29: 5074 (1984).
5. D. A. Huse, Phys. Rev. B 29: 6985 (1984).
6. F. T. Gittes and M. Schick, Phys. Rev. B 30: 209 (1984).
7. M. Bienfait, J. L. Seguin, J. Suzanne, E. Lerner, J. Krim, and J. G. Dash, Phys. Rev. B 29, 983 (1984).
8. H. Wiechert and H. Freimuth, in: " Proc. 17th Intern. Conf. on Low Temp. Phys. LT-17", U. Eckern et al., eds., North-Holland, Amsterdam (1984), p. 1015.
9. H. Freimuth and H. Wiechert, Surf. Sci. 162: 432 (1985).
10. F. C. Motteler and J. G. Dash, Phys. Rev. B 31: 346 (1985); F. C. Motteler, Ph. D. dissertation, University of Washington (1986).
11. H. Freimuth and H. Wiechert, Surf. Sci. 178: 716 (1986).
12. H. Freimuth, H. Wiechert, and H. J. Lauter, Surf. Sci. 189/190: 548 (1987).
13. H. Wiechert, H. Freimuth, H. P. Schildberg, and H. J. Lauter, Japn. J. Appl. Phys. 26, Suppl. 26-3: 351 (1987).
14. H. P. Schildberg, H. J. Lauter, H. Freimuth, H. Wiechert, and R. Haensel, Japn. J. Appl. Phys. 26, Suppl. 26-3: 345 (1987).
15. H. J. Lauter, in: "Phonons 89", S. Hunklinger, W. Ludwig, and G. Weiss, eds., World Scientific, Singapore (1990), p. 871.
16. H. Freimuth, H. Wiechert, H. P. Schildberg, and H. J. Lauter, Phys. Rev. B 42: 587 (1990).
17. J. Cui, S. C. Fain, H. Freimuth, H. Wiechert, H. P. Schildberg, and H. J. Lauter, Phys. Rev. Lett. 60: 1848 (1988); 2704 (1988).
18. J. Cui and S. C. Fain, Phys. Rev. B 39: 8628 (1989).
19. H. P. Schildberg, H. J. Lauter, H. Freimuth, H. Wiechert, and R. Haensel, Japn. J. Appl. Phys. 26, Suppl. 26-3: 343 (1987).
20. J. Ma, D. L. Kingsbury, F. Liu, and O. E. Vilches, Phys. Rev. Lett. 61: 2348 (1988).
21. X. Z. Ni and L. W. Bruch, Phys. Rev. B 33: 4584 (1986).
22. V. L. Ginzburg and A. A. Sobyanin, JETP Lett. 15: 242 (1972).
23. H. J. Maris, G. M. Seidel, and T. E. Huber, J. Low Temp. Phys. 51: 471 (1983).
24. R. J. Birgeneau, P. A. Heiney, and J. P. Pelz, Physica 109/110 B: 1785 (1982).
25. H. P. Schildberg and H. J. Lauter, Surf. Sci. 208: 507 (1989).
26. D. M. Zhu, D. Pengra, and J. G. Dash, Phys. Rev. B 37: 5586 (1988).
27. D. M. Zhu and J. G. Dash, Phys. Rev. B 38: 11673 (1988).
28. H. K. Kim and M. H. W. Chan, Phys. Rev. Lett. 53: 170 (1984).
29. M. J. de Oliveira and R. B. Griffiths, Surf. Sci. 71: 687 (1978).
30. M. P. Nightingale, W. F. Saam, and M. Schick, Phys. Rev. B 30: 3830 (1984).
31. M. S. Pettersen, M. J. Lysek, and D. L. Goodstein, Phys. Rev. B 40: 4938 (1989).
32. J. Ma, T. S. Sullivan, and O. E. Vilches, Japn. J. Appl. Phys. 26, Suppl. 26-3: 263 (1987).
33. M. Nielsen, Phys. Rev. B 7: 1626 (1973).
34. K. Carneiro, L. Passell, W. Thomlinson, and H. Taub, Phys. Rev. B 24: 1170 (1981).
35. J. M. Phillips and C. D. Hruska, Phys. Rev. B 39: 5425 (1989).

SCATTERING OF ATOMIC HYDROGEN FROM THE SURFACE OF LIQUID HELIUM FILMS

J.T.M. Walraven

Van der Waals laboratorium, Universiteit van Amsterdam
Valckenierstraat 65/67, 1018 XE Amsterdam, The Netherlands

1. Introduction

Liquid helium covered surfaces are indispensable in experiments with atomic hydrogen gas (H) at low temperature ($T \lesssim 1$ K). Helium is used because H does not dissolve in this liquid, it is chemically inert and offers the weakest concievable atom-surface interation to the H atoms. These properties enable confinement of H gas while surface adsorption is unimportant except at the lowest temperatures. Therefore helium films are very effective in supressing massive surface catalized recombination in which solid molecular hydrogen is formed.

Information on the atom-surface interplay has been obtained from a variety of experiments, as many experiments turn out to be very sensitive for surface adsorption. At sufficiently low temperatures destabilization processes in the adsorbed gas are rate limiting in the chain of events which lead to the molecular state H_2. Such processes provide information on static properties of the helium surface such as the adsorption energies of the hydrogen isotopes. These properties are discussed in the literature for coverages of ^{4}He and ^{3}He as well as for ^{3}He-^{4}He mixtures. General reviews of the field are published by Greytak and Kleppner[1] and by Silvera and Walraven.[2]

For the purpose of the present conference dynamical aspects of the atom-surface interaction are of most interest. As H gas does not dissolve in liquid helium, H atoms are in particular sensitive to excitations of the liquid surface although it should be pointed out that at low temperatures the surface is in fact rather difficult to excite and that most atoms scatter off elastically. The dominant inelastic scattering process is the sticking channel in which H atoms adsorb onto the surface under excitation of a single ripplon. The name sticking is somewhat misleading as the adsorbed atoms can move freely over the surface as in a two-dimensional gas and are desorbed after a short residency time. The surface may also be excited in a direct inelastic scattering process, without adsorption. Experimentally the sticking phenomenon is studied by measuring transport properties such as the thermal boundary resistance across the gas-liquid interface or capillary flow in the Knudsen regime. The quantities to determine are the sticking coefficient s or the thermal accommodation coefficient α. These quantities are of crucial importance for the interpretation and succes of experiments in which H

Excitations in Two-Dimensional and Three-Dimensional Quantum Fluids
Edited by A.G.F. Wyatt and H.J. Lauter, Plenum Press, New York, 1991

is compressed to high density in a small volume in attempts to observe Bose-Einstein Condensation (BEC).[1,2] In such experiments heating of the gas due to recombination tends to drive the sample out of thermal equilibrium with the walls. Both the sticking process and direct inelastic scattering contribute to the heat tranfer from gas to surface. However the direct inelastic process turns out to be quasi-elastic and its contribution to the accommodation coefficient may be neglected at temperatures below 0.5 K. Beam experiments with an angular resolution of approximately 1° are required to discriminate between purely specular reflection and the quasi-elastic channel. The aivalable information on sticking and thermal accomodation was summarized up to 1988 by Berkhout and Walraven.[3]

The paper is organized as follows. First, in sections 2,3 and ,4 the theory is briefly introduced. Then experiments on sticking, thermal accommodation and near elastic scattering are discussed.

2. Theoretical background

The theory of inelastic scattering of H from the surface of ^{4}He has been developed in several articles in the literature.[4-9] The common approach is to treat the liquid as an incompressible non-viscous fluid continuum with a step-function-like density profile to describe the surface. Such an approach is justified since the surface will be collectively excited as may be anticipated from the long thermal wavelength of the incident atoms and the weak atom surface interaction. For a gas of H atoms at temperature T_g= 200 mK the thermal wavelength $\Lambda \equiv (2\pi\hbar^2/mk_BT_g)^{1/2} \simeq 39$ Å, much larger than the mean separation between the atoms in the liquid. As very little is known of excitations of the surface of ^{3}He, all papers written on the subject deal with ^{4}He. For the present discussion it is useful to consider a liquid layer of thickness d covering a horizontal planar substrate. For d being small ($d \lesssim 300$ Å) one deals with properties of a helium film. In the limit $d \to \infty$ the layer behaves as bulk liquid and the properties of the free surface should be eminent. The elementary excitations of the helium surface are the ripplons. Using a normal mode decomposition, fluctuations $h(\boldsymbol{r})$ in the thickness of the liquid layer may be related to ripplon creation and annihilation operators through

$$h(\boldsymbol{r}) = A^{-1/2} \sum_{\boldsymbol{q} \lesssim \boldsymbol{q}_{max}} h_{\boldsymbol{q}} e^{i\boldsymbol{q}\cdot\boldsymbol{r}} , \tag{1}$$

where $\boldsymbol{r}$ is a position vector in the plane of the undisturbed surface and $h_{\boldsymbol{q}}$ is the amplitude operator for ripplons with wavevector $\boldsymbol{q}$, defined in terms of ripplon creation ($r_{\boldsymbol{q}}^+$) and annihilation ($r_{\boldsymbol{q}}$) operators by

$$h_{\boldsymbol{q}} = (\frac{\hbar q \tanh(qd)}{2\rho_0\omega_q})^{1/2}(r_{\boldsymbol{q}}^+ + r_{-\boldsymbol{q}}) , \tag{2}$$

where ρ_0 is the mass-density of ^{4}He in its ground-state and ω_q is related to q by the dispersion relation

$$\omega_q^2 = [g_e q + (\frac{\gamma}{\rho_0})q^3] \tanh(qd) . \tag{3}$$

Here g_e represents an effective gravity-like acceleration and γ is the surface tension. For ^{4}He the surface tension is $\gamma = 3.78 \times 10^{-4}$ Jm^{-2}. The sum in eqn. (1) is restricted to wavenumbers $\boldsymbol{q} \lesssim \boldsymbol{q}_{max}$, as for wavelengths smaller than the interatomic spacing of the liquid the continuum picture breaks down.

Several characteristic ripplon wavelengths are relevant to the analysis. First note that the ripplon dispersion is composed of two contributions. The term linear in q results from a gravity-like restoring force $g_e \equiv g + 3\alpha_3/d^4$, where $3\alpha_3/d^4$ is the acceleration on a ^{4}He atom at a distance d above the substrate due to the Van der Waals interaction with that substrate. For a typical helium film this Van der Waals contribution is 10^7 times larger than that of gravitaion proper (g), so that gravity effects may be neglected eccept for liquid layers of thickness $d \gtrsim 1\mu$m. The q^3 term describes the dispersion due to capillarity. Eqn. (3) implies that for ripplon wavelength λ larger than a cross-over-wavelength λ_0 the gravity-like term is dominant. This λ_0 depends quadratically on the film thickness. For $d = 100$ Å one calculates $\lambda_0 \simeq 3600$ Å. The hyperbolic tangent in eqn. (3) describes the change in dispersion from a wave in a thick layer of liquid, $d \gg \lambda$, to a wave in a thin layer of liquid, $d \ll \lambda$. A final characteristic wavelength is the thermal wavelength λ_T of the ripplons, defined by $\hbar\omega_{q_T} = k_B T_g$. For $T_g = 200$ mK one calculates $\lambda_T \simeq 100$ Å. This means that ripplons with a wavelength of less than 100 Å are not thermally excited under typical experimental conditions. Consequently dynamical surface roughness on this length scale finds its origin in the zero-point fluctuations of the surface. The rms amplitude of these fluctuations is calculated to be approximately 2 Å.[10]

3. Surface adsorption - The sticking channel

The sticking probability, defined as the probability per collision for an atom to scatter into a surface bound state was first calculated by Zimmerman and Berlinsky[4] and by Kagan and Shlyapnikov.[5] Goldman studied the sensitivity of the sticking probability for the shape of the adsorption potential.[7] In the sticking process the atoms are scattered into a surface bound state. The physical nature of the bound state and the adsorption potential are described by Mantz and Edwards.[11] The adsorption potential only supports a single bound state $\phi(Z)$ normal to the surface. The wavefunction of the adsorbed atoms is free-particle-like and may therefore be written as

$$\psi_{\boldsymbol{k}}(\boldsymbol{R}, Z) = A^{-1/2} \exp^{i\boldsymbol{k}\cdot\boldsymbol{R}} \phi_k(Z), \tag{4}$$

where $\boldsymbol{k}$ is the wavevector of the atom parallel to the surface, $\boldsymbol{R}$ is the position vector in the plane of the surface and Z is the distance above the undisturbed surface. The adsorption process may be mediated both by ripplons or bulk phonons. Kagan and Shlyapnikov have shown that the ripplon channel dominates over the phonon channel under typical experimental conditions.[5] The adsorbed atoms rapidly thermalize. Zimmerman and Berlinsky showed that the typical thermalization time due to atom-ripplon interaction is much shorter than the residency time of the atoms on the surface, which ranges from 1×10^{-8} s at 400 mK to 3×10^{-4} s at 100 mK for H on ^{4}He.[13] As the atom-ripplon coupling is very weak the adsorption rate may be calculated in linear response with first order time-dependent perturation theory:[4]

$$\Gamma = \frac{2\pi}{\hbar} \sum_{\boldsymbol{q}} \frac{\hbar q}{2\rho_0\omega_q} [1 + n(\omega_q)][1 + N(\boldsymbol{k} - \boldsymbol{q})] | < B | \frac{\partial U_q}{\partial z} | k_\perp > |^2 \delta(E_i - E_f), \tag{5}$$

where $|B>$ is the surface bound state and $k_\perp$ is the normal component of the wavevector of the incident atoms. Apart from the overlap matrix element which contains the normal derivative of the (q-dependent) adsorption potential and the δ-function included for energy conservation, eqn. (5) contains the atom-ripplon coupling strength which depends on the thermal occupation $n(\omega_q)$ of the ripplons. $N(\boldsymbol{k})$ is the occupation of adsorbed atomic states. This term is negligible except near the saturation limit of the adsorbed H. For incident energies much smaller than the adsorption energy ϵ_a the low temperature limit is reached where the kinetic energy of the incident atoms is negligible and only the relatively large adsorption energy has to be accommodated. Momentum conservation then yields that adsorbed atom and excited ripplon should have equal and opposite momentum. The resulting ripplon is calculated to have a wavelength of 48 Å and only the $\boldsymbol{q}$ vectors close to this value enter into the sum over final states in eqn. (5). As these ripplons have a wavelength clearly shorter than the wavelength of the thermal ripplons, their thermal occupation will be negligible ($n(\omega_q) \approx 0$) and the adsorption is purely induced by the zero-point fluctuations of the liquid surface. Further, in this limit $\tanh qd \simeq 1$. One easily verifies that the only temperature dependence that enters eqn. (5) arises from the temperature dependence of the average normal velocity of the atoms. As a result the thermally averaged adsorption rate should fall as $\Gamma \sim T_g$ for $T_g \to 0$. Dividing by the wall collision rate shows that the sticking probability should vanish as $T_g^{1/2}$ for $T_g \to 0$. Note that the surface temperature does not appear. The importance of two-ripplon emission was calculated by Statt to be of order 20 % in comparison with the single ripplon process.[6] Goldman found that for temperatures below 100 mK s is very sensitive for the long range part of the interaction potential and that for some potential shapes the low temperature behavior is only reached for temperatures below 1μK.[7] Thus far, the experimental temperatures have not been sufficiently low to observe the $T_g^{1/2}$ behavior.

4. Direct scattering - The quasi-elastic channel

The importance of direct scattering over small angles was analyzed theoretically by Hijmans and Shlyapnikov[8] and by Tiesinga, Stoof and Verhaar[9] in relation to the mirror experiment described in section 7. The direct channel involves scattering from continuum state to continuum state. In linear response this rate is written to first order in perturbation theory as

$$\Gamma = \frac{2\pi}{\hbar} \sum_{k'_\perp, \boldsymbol{q}} \frac{\hbar q \tanh(qd)}{2\rho_0 \omega_q} [1 + n(\omega_q)] | < k'_\perp | \frac{\partial U_q}{\partial z} | k_\perp > |^2 \delta(E_i - E_f), \tag{6}$$

where $k_\perp$ and $k'_\perp$ are the normal components of the wavevectors of the incident and scattered particle respectively and $\boldsymbol{q}$ is as before the momentum of the ripplon excited. In this case the factor $\tanh qd$ apears explicitely. Long wavelength ripplons are dominant due to the fact that the adsorption energy no longer has to be accommodated. For small angle scattering ($q \ll k, \omega_q \ll \epsilon_k$) eqn. (6) may be simplified using $n(q) + 1 \approx n(q) \approx k_B T_s / \hbar\omega_q, dU_q/dz \approx dU_0/dz$. In this limit the thermally activated ripplons dominate the scattering and not only the gas temperature but also the surface temperature T_s is important. Note that thermal activation favors the low frequency ripplons according to $n(q) \sim 1/\omega_q$. Hijmans and Shlyapnikov show that in the above approximation the matrix element can be evaluated analytically and turns out to depend only on the adsorption energy and not on the detailed shape of the potential. This property was pointed out in a more general context by Tiesinga *et al.*.[9] Evaluating eqn. (6) the thermally averaged scattering probability due to the direct process scales like $W_d \sim T_g T_s$.

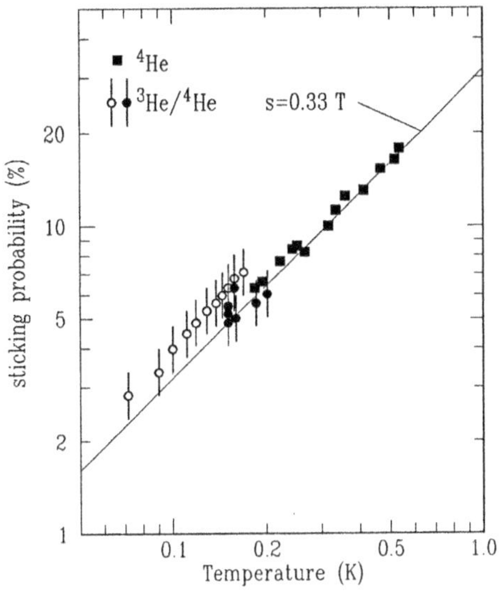

Figure 1. The sticking probability as determined by the capillary method.

5. Capillary flow experiment

As an example how the sticking probability may be determined experimentally, in this section the capillary flow experiment of Berkhout *et al.*[12] will be discussed. Other methods are summarized in ref..[3] In the experiment of Berkhout *et al.* a buffer volume V is filled with H to a density n_g. The gas was then allowed to flow through a capillary to a volume kept free of H atoms. Under these conditions $T_g = T_s$. The flow Φ leaving the capillary depends on the H flux entering the capillary and the transmission coefficient (Clausing factor) K:

$$\Phi = \frac{1}{4} n_g \bar{v}_g A K \equiv N/\tau \tag{7}$$

where n_g and $\bar{v}_g \equiv (8k_B\, T_g/\pi m)^{1/2}$ are the density and thermal velocity of the H atoms in the gas repectively, N the number of atoms in the buffer volume, and A the entrance orifice of the capillary. By measuring τ the Clausing factor follows from $K = \tau_0/\tau$, where $\tau_0^{-1} \equiv \frac{1}{4}\bar{v}_g A/V$. To determine s the relation between K and s must be known. In the Knudsen flow regime and for $s = 1$ the Clausing factor assumes its Knudsen value K_K, purely determined by geometrical factors and thus temperature independent. For capillaries with large ratio of length ℓ over radius a, $K_K = 8a/3\ell$. In the geometry of Berkhout *et al.* $K_K = 0.0284$.[12] For $s = 0$ the Clausing factor is 1, that is, every particle entering the capillary will scatter specularly until it leaves at the other end. The relation between s and K for several values of interest of s was determined by a one speed Monte Carlo computer simulation. It would be interesting to see how sensitive this result is for thermal averaging. The capillary flow results are given in fig. (1).

The sticking coefficient turns out to be proportional to T and may be given within 10% experimental accuracy as $sT^{-1} = 0.33\,\mathrm{K}^{-1}$ in the range 73 mK$< T <$526 mK. It is remarkable that within experimental error s is the same for ^{4}He surfaces and for surfaces of ^{3}He-^{4}He mixtures.

6. Thermal accommodation

The accommodation coefficient is defined as

$$\alpha = \lim_{T_g \to T_s} (\dot{Q}_g - \dot{Q}_o)/(\dot{Q}_g - \dot{Q}_s) , \tag{8}$$

where $\dot{Q}_g$ is the average kinetic energy flux from the gas at temperature T_g to the surface at temperature T_s. $\dot{Q}_o$ is the average outgoing kinetic energy flux of the atoms leaving the surface and $\dot{Q}_s$ the same quantity assuming all incident atoms to fully thermalize to the surface temperature. For $T_g \rightarrow T_s$ the net heat flux from gas to surface reduces to

$$\dot{Q} = \alpha\Phi \; 2k_B(T_g - T_s) , \tag{9}$$

where $\Phi = \frac{1}{4} n_g \bar{v}_g A$ is the wall collision rate, with n_g and $\bar{v}_g$ the density and thermal velocity of the H atoms and A the surface area under consideration. The thermal accommodation coefficient is calculated by Kagan, Shlyapnikov and Glukhov,[17] Statt[6] and Goldman.[7] For temperatures below 500 mK where direct inelastic scattering is negligible Goldman gives a simple and very useful equation to relate α and s:

$$\alpha(T) = s(T)\{1 + \frac{1}{2}\beta^2 \frac{\partial^2 \log s(T)}{\partial \beta^2}\} \tag{10}$$

where $\beta = (k_B \, T)^{-1}$. This expression enables a direct comparison of the experimental results for α and s. For a linear temperature dependence it implies $\alpha = 3/2s$.

The most complete set of data on the accommodation coefficient was obtained by Helfrich *et al.*[14] who measured the heat loss of an electrically heated bolometer suspended in the H gas by fine wires. The results of these authors is in very nice agreement with the results of Berkhout *et al.* if one accounts for the factor 3/2 discussed above.

7. Mirror experiment

To demonstrate the existance of a strong elastic peak Berkhout *et al.*[15] measured the specular reflectivity of a helium surface at normal incidence with helium coated hemispheric mirror. This experiment is quite similar in concept to the capillary flow experiment described before. A buffer volume V_B is filled with hydrogen to a density of 10^{14} atoms/cm^3. The gas expands through an orifice and is detected on a pumping plate H-flux detector[16] which also keeps the secondary volume free of hydrogen. The exponential decay time of the flux escaping the buffer volume is measured. If specular reflection is dominant this quantity should depend strongly on the position of the hemispheric mirror as ilustrated in fig. (2). If the center of the mirror does not coincide with the exit orifice of the buffer volume V_B, as sketched in fig. (2a), the buffer vulume will be emptied as in the absence of the mirror. However, if mirror-center and exit-orifice coincide, as shown in fig. (2b), the beam of atoms expanding through the diaphragm is reflected back into the buffer volume, causing the exponential decay time to be much longer. Under these conditions, the loss from the storage volume depends on the reflectivity of the mirror which in turn depends on the inelastic scattering phenomena at the helium surface (see fig. (2a)). If the dimensions and temperature of the cell are accurately known one may calculate the loss factor χ representing the probability that an atom is *not* scattered back into the buffer volume

$$\chi \equiv \frac{\tau_0}{\tau} \approx -\frac{\tau_0}{N}\frac{dN}{dt} . \tag{11}$$

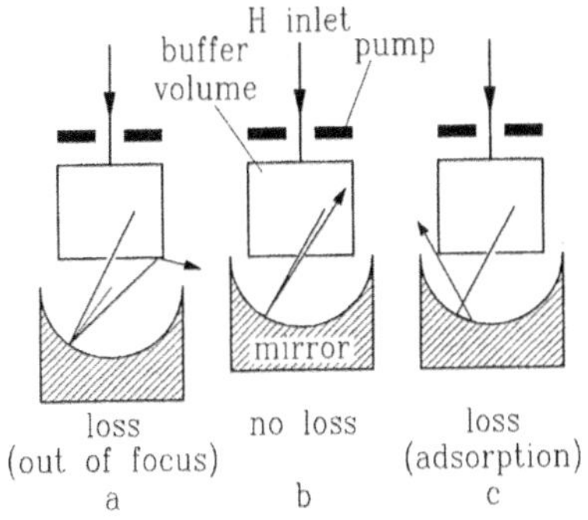

Figure 2. The priciple of the mirror experiment. For a discussion see the text.

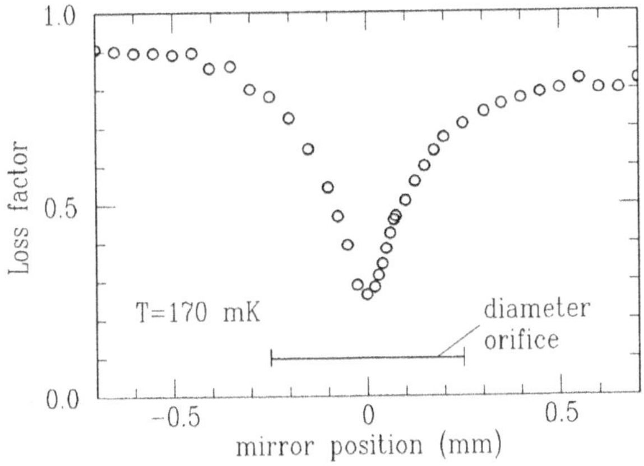

Figure 3. The loss factor as a function of the vertical mirror position as observed by Berkhout *et al.*.

Here N is the total number of atoms in the buffer volume and $\tau_0 \equiv 4V_B/\bar{v}A$ the first-order decay time in the absence of the mirror, $\bar{v} = (8k_BT/\pi m)^{1/2}$ is the average thermal velocity of the atoms and A is the cross-sectional area of the orifice. In fig. (3) the loss factor is shown as a function of the position of the mirror. With the mirror far from focussing conditions the loss is found to approach unity as in the absence of the mirror. As seen from the figure the loss from the buffer volume may be reduced by a factor 4-5 by proper adjustment of the mirror. This probably represents the first demonstration ever of focussing of an atomic beam by means of a mirror. In the absence of lateral misalignment and spherical aberration the loss at the minimum of the position scan corresponds to the deviation from perfect reflectivity of the mirror.

Plotting the loss factor at the minimum of the scan curve as a function of temperature one obtains the results shown in fig. (4). The main results are the solid triangles, obtained with pure ^{4}He. This is best described by the solid curve which results from four independent loss factors, i=1–4, accumulated according to[15]

$$\chi = 1 - \prod_{i=1}^{i=4}(1 - \chi_i) \tag{12}$$

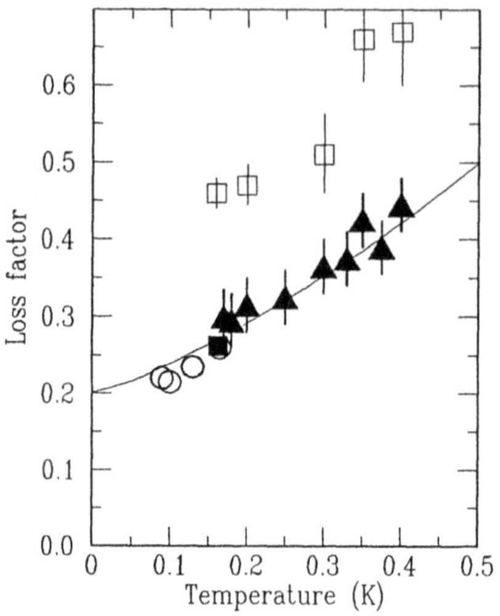

Figure 4. Measured loss factors at the minimum of the scan curves as a function of temperature.

The loss contributions are adsorption at normal incidence ($\chi_1 T^{-1} = 0.5\,\mathrm{K}^{-1}$), direct inelastic scattering ($\chi_2 T^{-2} = 0.5(1)\,\mathrm{K}^{-2}$), lateral misalignment ($\chi_3 = 0.2$) and other geometrical factors ($\chi_4 = 0.042$). No significant effect was observed upon adding up to 0.1% ^{3}He (solid square and open circles). This quantity should be sufficient to assure full monolayer coverage. This is remarkable as such a layer changes the adsorption energy by a factor 2.5 and the surface tension by a factor 2. A similar lack of effect was observed by ref..[12] If the helium film thickness is reduced as much as possible the effective reflectivity decreases, see open squares in fig. (4). This may be explained by the presence of static roughness due to the substrate.

8. Acknowledgements

This work is part of the research program of NWO and of the Stichting voor Fundamenteel Onderzoek der Materie (FOM), which is financially supported by the Nederlandse Organisatie voor wetenschappelijk onderzoek (NWO).

References

1. T.J. Greytak and D. Kleppner, in: NEW TRENDS IN ATOMIC PHYSICS, Vol.II G. Grynberg and R. STORA (Eds.), Elsevier Sci. Publ., Amsterdam (1984) 1125-1230.

2. I.F. Silvera and J.T.M. Walraven, Prog. Low Temp. Phys. D.F. Brewer (Ed.), Elsevier Sci. Publ. **X** (1986) 139-370.

3. J.J. Berkhout and J.T.M. Walraven, in: SPIN-POLARIZED QUANTUM SYSTEMS, S. Stringari (Ed.), World Scientific Publ., Singapore (1989) 201-208.

4. D.S. Zimmerman and A.J. Berlinsky, Can. J. Phys. **61** (1983) 508-513.

5. Yu. Kagan and G.V. Shlyapnikov, Phys.Lett. **95A** (1983) 309-312.

6. B.W. Statt, Phys.Rev.B. **32** (1985) 7160-7164.

7. V.V. Goldman, Phys.Rev.Lett. **56** (1986) 612-615.

8. T.W. Hijmans and G.V. Shlyapnikov, Phys. Lett. **142** (1989) 45-48.

9. E. Tiesinga, H.T.C. Stoof and B.J. Verhaar, Phys. Rev. (1990).

10. M.W. Cole, Phys. Rev. A **1** (1970) 1838-1840.

11. I.B. Mantz and D.O. Edwards, Phys. Rev. B **20** (1979) 4518-4526.

12. J.J. Berkhout, E.J. Wolters, R. van Roijen and J.T.M. Walraven, Phys.Rev.Lett. **57** (1986) 2387-2390.

13. D.S. Zimmerman and A.J. Berlinsky, Can. J. Phys. **62** (1984) 590-596.

14. J. Helffrich, M. Maley, M. Krusius and J.C. Wheatley, Phys. Rev. B **34** (1986) 65506-553.

15. J.J. Berkhout, O.J. Luiten, I.D. Setija, T.W. Hijmans, T. Mizusaki and J.T.M. Walraven, Phys.Rev.Lett. **63** (1989) 1689-1692.

16. J.J. Berkhout, O.H. Höpfner, E.J. Wolters and J.T.M. Walraven, Jpn. J. Appl. Phys. **26** (1987) 231-232 LT-18.

17. Yu. Kagan, G.V. Shlyapnikov and N.A. Glukhov, JETP Lett. **40** (1984) 1072-1076.

DISCUSSION ON ADSORBATES

Chairmen: H.Bozler and O.Vilches

Krotscheck to Walraven : Normally one sees interesting effects when the energy of the probe agrees with the energy of one of the collective modes. In helium films, one would expect such a level crossing at about 1 K energy of the H-atom with the ripplon, and at larger or equal to 10 K with the phonon/roton. Is it possible to make experiments in that regime of energies of the H-atoms?

Walraven : To observe the rotons one has to go to temperatures where our current detection method does not work due to the vapor pressure of He interfering with the flow of H atoms. Possibly the bulk excitations may be probed with atomic deuterium that can be made to dissolve in liquid ^{4}He.

Mezhov-Deglin to Walraven : What is the gas density used in the (spin polarized hydrogen optics) experiments?

Walraven : 10^{14} to 10^{15} atoms per cm^3.

Mezhov-Deglin to Walraven : Is it possible to study excitations in an adsorbed layer of H atoms on Liquid ^{4}He? You mentioned that the binding energy is about 1 K, thus could one detect surface waves at $T < 0.8$ K like one does with ^{3}He films?

Walraven : It should be possible to meet the conditions where the adsorbed gas is dominating the adsorption kinetics. This region has been studied by Yu, Kagan and co-workers. One expects a Kosterlitz-Thouless type transition. In this regime it should be possible also to address the excitations in the layer. Maybe one can measure sound propagation.

Leiderer to Vilches : In your HD-surface diffusion measurements, how low in temperature did you measure?

Vilches : Due to lack of time (we suffer from the ILL syndrome) we measured 10.3 K and 7.5 K. At 7.5K the signal is all solid, or within the instrument resolution. At 10.3 K the data is poor but we can see the liquid signal.

Leiderer to Vilches : Have you seen surface activated diffusion in the solid?

Vilches : It is difficult to decide. The instrument resolution is poor and MgO is a randomly oriented powder so one has to average over all orientations.

Lauter to Vilches : What is the reason that the 2nd layer triple line of H_2 on MgO is higher than the one on graphite? Does it have to do with the first layer structure?

Vilches : One has to look at the "effective interaction" between H_2 atoms being influenced strongly by the substrate. The 1st layer of H_2/MgO. It is possible that molecules in the 2nd layer of H_2/MgO interact from closer sites to the normal equilibrium distance in bulk than molecules on the 2nd layer of H_2/graphite. This would make the effective interaction weaker and lower T_t.

Mezhov-Deglin to Vilches : Did you estimate the diffusion coefficient of molecules in the liquid layer on the surface of the bulk H_2? Can you see the changing in the form of the crystals at temperatures below the roughening temperature?

Vilches : Yes, we fit the translational diffusion coefficient to the quasielastic scattered neutron peak. It is of the order of magnitude of the bulk liquid diffusion coefficient, but at half an atomic layer it is about a fifth of the bulk D. We can not see changes in the form of the crystal since our "crystals" are only 1 to 10 atomic layers thick.

Chan to Wiechert : Did you look for superfluidity in the 2nd layer of the hydrogen gas?

Wiechert : No, we only looked at specific heat measurements. There are only three phases: solid, liquid and gas. I wouldn't expect it.

Platzman to Frank : I wonder if you have ever considered that in these very quantum systems there is no such a thing as a zone boundary phonon. There was a discussion a long time ago by Werthamer and Fredkin about how you should think about lattice dynamics in quantum systems.

Frank : In a very dilute system (4.3Å or so) the adatom-adatom interaction is of the same order of magnitude as the substrate corrugation. Also the substrate potential and corrugation is anharmonic, but contains only even terms.

Other comments : This particle-particle potential and the corrugation potential are of the same magnitude,

Donnelly to Lauter or Frank : What happens to the roton spectrum as you make the films thinner and thinner? What is the present state of the "surface roton" once proposed as being connected with rotons near a surface?

Lauter : They cannot see any levels. They measured with several resolutions and found no effects.

Donnelly : The data discussed by Lauter does indeed have dispersionless energy levels which go below the roton minimum. They are, however, connected with the solid-liquid interface and not connected with true roton excitations. Thus this old idea does not seem to be confirmed.

Wyatt to Chan : I thought that the data which you showed yesterday supported the idea of a two dimensional surface roton.

Chan and Reppy : In heat capacity and ρ_n measurements we found the need to put in a term of the form $\exp(-\Delta/kT)$. We can't tell whether it is "roton-like" or not. For He in Vycor, delta is on the order of 5 K for a film with $T_c \sim 1K$ and on the order of 2.8K for $T_c \sim 0.5K$.

Griffin : Lauter et al find a localized excitation (no dispersion) related to liquid-solid interface. It appears to have the right energy to explain the above heat capacity and ρ_n measurements. However, this excitation is in no way a 2D roton, of the kind suggested many years ago.

Bozler showed a transparency of Grafoil and generating a discussion of the actual size of the uniform (or coherent) regions and their orientation. Answers centred on neutron scattering measurements on Papyex that indicate that 50% of the surface is oriented randomly, neutrons are not capable of distinguishing "positive curvature" from "negative curvature" regions, coherence length may be 150Å. Worries come as to what is the role of the 50% misoriented surface in the magnetic properties. Bozler also pointed that the size of coherent regions becomes important at ~ 0.5 mK the magnetic coherence length becomes comparable to the structural coherence length of the sample.

Gasparini to Saunders : Is there any evidence in your measurements, when compared to Greywall's heat capacity measurements, to indicate what is the origin of his extra heat capacity in the linear region? When Greywall's measurements are extrapolated to zero temperature the heat capacity does not go through zero.

Saunders : It is difficult to say what the contribution of the first layer is to be susceptibility measurements. There is an uncertainty due to the density of the first layer.

Lauter : (He showed a transparency of the phase diagram of ^3He/graphite). In the first monolayer of ^3He on graphite, the phase between the commensurate phase and the incommensurate phase is a domain wall phase. In the region where you suppose the second layer to become commensurate with the first one another event takes place, the first layer becomes commensurate with the substrate. This has to be taken into account (see contribution to LT-19, H J Lauter et al).

Godfrin to Saunders : When the film is ferromagnetic, we can normally fit the susceptibility quite well with a Curie-Weiss law. But, in the coverage range where your data indicate a continuous transition from $J<0$ to $J>0$, do you obtain good Curie-Weiss fits?

In Grenoble data (H Franco, Ph.D. thesis) there are strong deviations, even a strong ferromagnetic divergence at low T for films which appear to be antiferromagnetic at higher temperatures.

Saunders : At second layer of 0.066 density the deviations from simple Curie-Weiss are very small, they get the same parameter (-0.3) that is obtained from the specific heat data.

Godfrin : We should compare carefully our results, our larger deviations appear at low temperatures, outside your measuring range.

Godfrin to Bozler : The $H^{1/2}$ dependence of the magnetization curve you have observed has also been seen in 3D solid ^3He (Osheroff, Godfrin and Ruel), but a different fit was preferred using the multiple spin exchange (MSE) mean field expression $H = aM + bM^3 + cM^5 +$..related to MSE coefficients. Your data have the same very characteristic form, this may be an evidence for the existence of MSE. One can also fit the solid ^3He susceptibility data using Ceperly's parameters without having to fit anything.

Bozler : The power law that fits gradually decreased with decreasing temperature. Any power law fit is somewhat arbitrary. Fits to square root behaviour mainly illustrate that such a fit is possible. Trying to show fractional power law behaviour near phase transitions is notoriously difficult in many systems.

Reviewers Comments from Discussion Section on Adsorbates

This session dealt with adsorbed films. The interesting physics comes from the two-dimensional phenomena exhibited by these films, from the substrate-adsorbate interactions, and from the cross-over from two to three dimensions. These quantum films studied include the heliums (3-He, 4-He, and mixtures), the hydrogens (including deuterium and spin polarized atomic hydrogen). The hydrogen films discussed in this session were adsorbed on films of superfluid 4-He. In all cases, the films studied were grown on crystalline substrates. These were graphite in the form of Grafoil, Papyex, etc. and MgO.

The discussion topics centered on the following subjects:

1. Adsorbed Atomic Hydrogen

Surface densities achieved in the experiments described by Walraven are high enough to be near or above the required density for a Kosterlitz-Thouless type transition. Calculations are needed the effect of this transition on the chemical potential of the 3-D vapor in equilibrium with the physisorbed film, and on the three body recombination rate (expected to decrease by a factor of 6). Observing this transition is made difficult because of the low number of atoms involved.

Optics with polarized cold H atoms is also possible (as discussed by Walraven). In the discussion, spectroscopic experiments were suggested to probe excitations on the surface of the underlying helium film taking advantage of level crossings with the ripplon (at about 1 K energy) or phonon/rotons (at about 10 K). These experiments will have to be done at temperatures lower that 0.8 K due to the vapor pressure of the 4He scattering the H atoms. Additional experiments may be possible using polarized atomic D. It is possible to dissolve deuterium atoms in liquid ^{4}He, thus a new dilute solution quantum system may be created and the properties of the D quasiparticles remain to be measured.

2. H_2, HD, and D_2

The 2nd layer triple point of H_2 on MgO (7.2 K) lies higher than the one on graphite (5.96 K). Effective interactions seem to play a role here, but these interactions have not been calculated. These low triple points are excellent manifestations of quantum corresponding states in 2-D, and have brought the fluid state to fairly low temperatures.

An interesting possibility is whether by suitable choice of the proper substrate this triple point could be brought down to even lower temperature so the condensation into a Kosterlitz-Thouless superfluid state might be observable.

3. ^{4}He Films

The quality of measurements of the excitation spectrum for thin and thick films of ^{4}He on graphite using inelastic neutron scattering has improved dramatically (as seen in the contributions by V. Frank and H. Lauter). Questions were raised as to the possibility that one could observe the once proposed "surface roton". Although the data shows dispersionless energy levels which go below the roton minimum, these levels are connected with the solid-liquid interface in the films and are not with true roton excitations. These levels (localized excitations) seem to be of the right energy to explain the heat capacity and normal fluid density measurements of both Chan and Reppy (this work was presented in the session on ^{4}He in Restricted Geometry). However, these measurements require the addition of a term $\exp[-\Delta/kT]$ in addition to linear and quadratic temperature dependent terms.

An additional comment was made that for these quantum systems one should think about the possibility that there are no zone boundary phonons. All that is needed for this to happen is that the particle-particle potential be very anharmonic. In the fairly dilute quantum systems realizable by physisorption this may occur. For example, He atoms interact with each other at distances well above the minimum in the potential well. This case was treated many years ago by Fredkin and Werthamer. A possible complication in physisorbed systems it the role of substrate corrugations.

4. ^{3}He Films

Measurements of the nmr and specific heat of ^{3}He monolayer and multilayer films (see contributions by H. Godfrin and J.Saunders), and of a solid film in equilibrium with bulk liquid (H. Bozler) have revealed a very rich magnetic phase diagram, with paramagnetic, ferromagnetic, and antiferromagnetic regions. Most of these experiments have used exfoliated graphite as a substrate and have not shown significant effects due to the complications of the disorder in the substrate. About 40% to 50%

of the surface of Grafoil or Papyex is randomly oriented, but most of that is still crystalline.

There are differences in the way coverage is reported. Also, background effects are accounted in different ways. However, there is a fair amount of agreement between the data for ^{3}He films grown directly on graphite, and ^{3}He films grown on ^{4}He.

In the discussion, the question of fitting susceptibility to the Curie-Weiss Law or to series expansions (assuming a magnetic structure) were raised. At sufficiently high temperatures (refer to the contribution by Saunders), the differences between these fits is small, but may result in substantially different conclusions. Also, the low temperature magnetization which can be fit by a fractional power law (at fixed T as discussed by Bozler) is not unique. A possibility of fitting the surface magnetization with multiple spin exchange coefficients was suggested.

Part of the picture for these two dimensional systems is complicated by the delicate behavior of the layers closer to the substrate. Neutron scattering measurements indicate that at temperatures where particular magnetic regimes start there are also rearrangements of the underlayers. Models that treat the first (or second) physisorbed layers as rigid and ascribe all the magnetic changes to either the uppermost layer or the layer in equilibrium with the liquid are then not valid.

ELECTRONS AT THE SURFACE OF QUANTUM SYSTEMS

P. Leiderer

Fakultät für Physik, Universität Konstanz
7750 Konstanz, Federal Republic of Germany

INTRODUCTION

The two lightest elements hydrogen and helium provide unique systems to study surfaces and interfaces of condensed matter in the presence of large quantum fluctuations. Surface phenomena to be mentioned here are, e.g., surface melting, roughening and layering.[1] Since in the quantum systems the gas density above the surface is relatively high in the regions of interest, investigations using conventional surface techniques like LEED and EELS are of limited applicability, because they usually require high vacuum conditions. A technique wich has been applied successfully is neutron scattering, which - although not being particularly sensitive to surfaces - can be used if the quantum systems are adsorbed on substrates with a large intrinsic surface area, like exfoliated graphite.[2] Information about quantum surfaces has been obtained also from measurements of specific heat, adsorption isotherms, third sound propagation[19-22] and, as a somewhat peculiar method, from the investigation of surface state electrons (SSE). It is the latter which is the topic of this paper.

Electrons near the surface of a dielectric are attracted due to the image potential $V(z)=(1/z)e^2(\epsilon-1)/4(\epsilon+1)$. (Here e is the elementary charge, ϵ is the dielectric constant of the material and z is the distance of the electron from the surface). For the condensed phases of hydrogen and helium this attraction is rather weak because of their low polarizability. In addition, at atomic distances the negative electron affinity of H_2 and He leads to a potential barrier, which is $\approx 10^4$K for He [3] and $\approx 3\times10^4$K for H_2 [4]. As a result the electrons are trapped in a potential well perpendicular to the surface, forming states similar to the electronic radial distribution in a hydrogen atom, however with a much smaller binding energy in the ground state (8K for He, 160K for H_2) and a much larger effective Bohr radius (a_B = 76Å for He, 17Å for H_2).

In contrast to the quantization perpendicular to the surface, the electrons can move parallel to the surface as nearly free particles, provided the substrate is smooth enough. An example where this condition is fulfilled in an almost ideal way is the surface of liquid helium (see Fig.1), where at low temperatures the motion of individual electrons is only limited by scattering from quantized surface waves ("ripplons"). The 2-dimensional electron systems which can thus be generated on liquid helium are a classical counterpart to the degenerate 2-dimensional electron system on MOSFET- and related semiconductor structures.

Excitations in Two-Dimensional and Three-Dimensional Quantum Fluids
Edited by A.G.F. Wyatt and H.J. Lauter, Plenum Press, New York, 1991

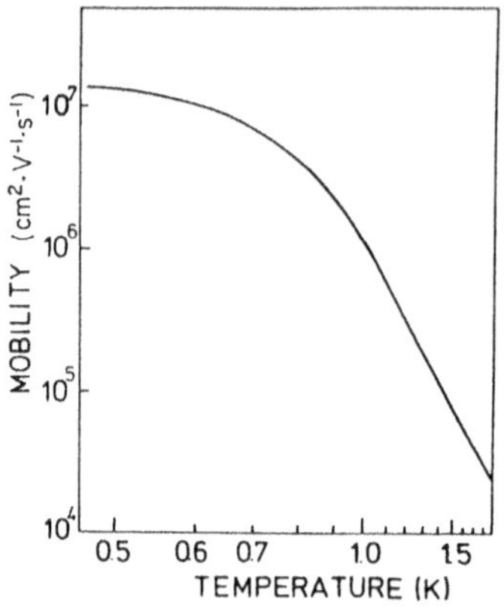

Fig.1. Electron mobility on bulk ^4He. The dominant electron scattering mechanisms are scattering from ripplons and gas atoms at low and high temperature, respectively (after ref.3)

The reason for classical statistics to apply for the surface state electrons on liquid helium is that the electron density obtainable is limited to $n<n_{crit}=2.4\times10^9$ cm^{-2} (see below). This corresponds to a Fermi temperature of only 50 mK, much lower than the actual temperature in most of the experiments up to now.

A number of review articles has been devoted to the system of electrons on liquid helium,[3] and some of the most important results are: i) the excited electronic states are hydrogen-like, and spectroscopic measurements in the range of a hundred GHz agree well with calculations; ii) the mass of the SSE is close to the free electron mass, as measured, for example, by cyclotron resonance; iii) Coulomb interaction in the 2-dimensional layer of electrons leads to collective excitations, i.e. 2D longitudinal plasmons; iv) one of the most intriguing effects in this system is the formation of an electron solid, the so-called Wigner crystal, which appears when the plasma parameter (the ratio of Coulomb and thermal energy) rises above about 130.[5]

It would be extremely interesting to investigate the Wigner crystal also at densities higher than the above-mentioned limit n_{crit} and trace out the complete phase diagram, in particular the melting of the crystal towards the degenerate Fermi liquid at low T and the vicinity of the critical point which is expected above 10 K. Unfortunately, these regions are not accessible with liquid helium as a substrate, because there an electrohydrodynamic (ehd) instability develops at n_{crit}:[6,7] The surface deforms spontaneously, and consequently also the initial homogeneity of the electron density at the surface is destroyed. The electrons then accumulate in macroscopic dimples, each holding on the order of 10^5 electrons, and due to Coulomb repulsion between the dimples a regular lattice is formed which can be considered as a macroscopic Coulomb crystal. Although quite interesting in itself, this lattice is even more classical than the (microscopic) Wigner crystal formed by single electrons.

Two routes have been pursued to circumvent the ehd instability. If the electrons are localized on a thin helium film rather than on bulk liquid, a density of up to 10^{11} electrons/cm^2 could be obtained due to stabilization of the film by van der Waals forces. A second possibility is to use a solid substrate like H_2, and also in this case the experimental maximum electron density could be increased by more than an order of magnitude. As a disadvantage of these substrates, the mobility of the SSE was reduced considerably compared to the high value on bulk helium due to interaction with surface imperfections. In the course of the experiments it turned out, however, that this originally undesirable effect can be used as a very sensitive method to study surfaces by means of SSE.

Electrons can also be applied in another form for investigating quantum surfaces and interfaces: If they have high enough energy, they can overcome the potential barrier at the surface and penetrate into the condensed phase, where they rapidly relax into electron bubbles with a diameter of some twenty Å (depending on pressure) and consequently an effective mass six orders of magnitude above the mass of free electrons. At low enough temperatures these "negative ions" can be trapped at the liquid helium surface from underneath, again forming a classical 2D Coulomb system. In addition, the electron bubbles can be trapped at the interface between liquid and solid helium, where they can be used to study melting and crystallization under very clean conditions.[8]

In the following, I shall first discuss SSE on bare hydrogen and subsequently electrons on helium films supported by solid hydrogen as well as by other solids. Electron bubbles will be treated by Vinen.[9]

ELECTRONS ON SOLID H_2

Experiments with electrons on bulk solid H_2 were first performed by Troyanovskii et al.,[10,11] who measured the ac conductivity of the SSE at frequencies in the MHz range. The crystal was grown in this case from the liquid phase at the triple point. A similar preparation procedure was used by Edel'man and Faley[12] for cyclotron resonance studies of the SSE at 20 GHz, and by Cieslikowski et al.[13,14] for mobility measurements in the kHz

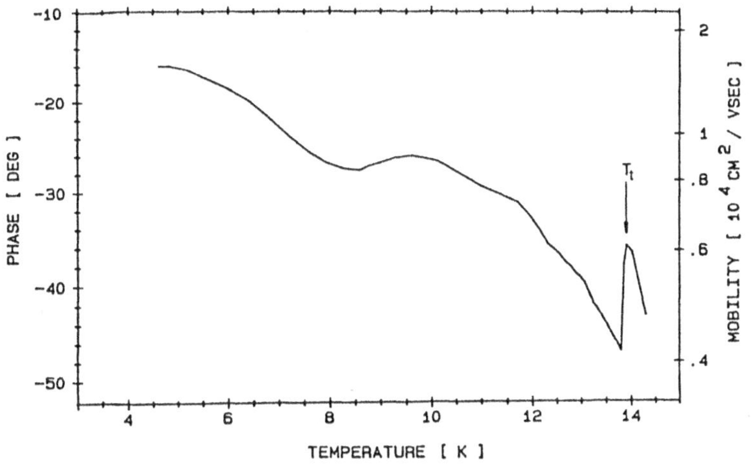

Fig.2. Electron mobility on bulk hydrogen. The curve was taken during cooling. The solidification at the triple temperature T_t is reflected in a sharp drop due to electron scattering from surface irregularities in the solid phase. The left hand scale refers to the phase shift of the electron motion with respect to the driving ac field (after ref.13)

range. In all these experiments a characteristic temperature dependence was obtained for the electron mobility, as shown in Fig.2.

At temperatures above about 10 K one important mechanism for energy loss of the electrons is gas atom scattering, which in dense gas (i.e. at the highest temperatures) is further increased by the tendency for the formation of electron bubbles.[15] A second contribution arises from scattering by surface roughness, which appears as soon as the solid is formed, showing up in Fig. 2 as a sharp step at the triple temperature T_t = 13.9K. The height of this step varies with the parameters of the growth process and apparently depends on the surface quality of the crystal.

At T ≤ 10 K gas atom scattering becomes extremely small, and hence the principal scattering mechanism is due to surface defects. (The contribution of scattering from surface excitations on the solid (Rayleigh waves), as calculated by Troyanovskii et al.,[11] is more than two orders of magnitude below the experimental data and hence is negligible here. Likewise, the system of electrons is not noticeably influenced by the ortho-para conversion of hydrogen.) From an analysis of their mobility data between 4 and 10 K Troyanovskii et al. conclude that the main surface imperfections responsible for the scattering are defects with a dimension on the order of the lattice constant and a density of 2×10^{12} cm^{-2}. Another model for the surface defects was considered by Edel'man and Faley,[12] who supposed that their solid hydrogen surfaces had a terrace structure with flat sections about 10^{-5} cm in size and steps much larger than an atomic distance. In confirmation of the estimate of rather high steps results for the same H_2 surfaces covered with a superfluid ^4He film of 200 Å thickness were presented. It was argued that if the scattering centres had atomic distances such a film should yield a substantial increase of the electron mobility up to the value known from bulk helium. Since this was not observed, it was inferred that the surface imperfections had at least a height comparable to the film thickness.

As reported by Edel'man and Faley and also by Cieslikowski et al.,[13] irregular variations of the electron mobility were observed quite frequently for temperatures around 8 K. A slight indication of this effects is also visible in Fig.2. It was suggested by Edelman that this might be a signature of a roughening transition,[12] although bulk effects like cracks resulting from differences in the thermal expansion of H_2 crystal and sample cell also appear possible.

Whereas all the H_2 crystal surfaces discussed so far had been grown from the liquid phase (with a crystal thickness d ≥ 1mm), more recently also thin H_2 films (thickness ~ 1μm), quench-condensed at T =1.5K << T_t, were investigated with SSE by Kono et al..[16] Such films might be an interesting substrate for 2-dimensional electron systems, since they allow to reach high electron densities already at low holding voltages.

In the experiments of Kono et al. it was observed that SSE on the freshly prepared films displayed an extremely low conductivity σ, apparently due to a very rough surface. However, σ could be increased by orders of magnitude by thermal annealing of the films. As an example, Fig.3 shows data for a H_2 film treated in subsequent annealing steps. It was found that σ could be described approximately by an Arrhenius law* $\sigma = \sigma_0 \exp(-E/kT)$, with an activation energy E which decreased from 23 to 10K in the course of

* For the data in Fig.3 this applies only to temperatures T ≥ 2K, because in this experiment at smaller T the H_2 surface was covered with a helium film.

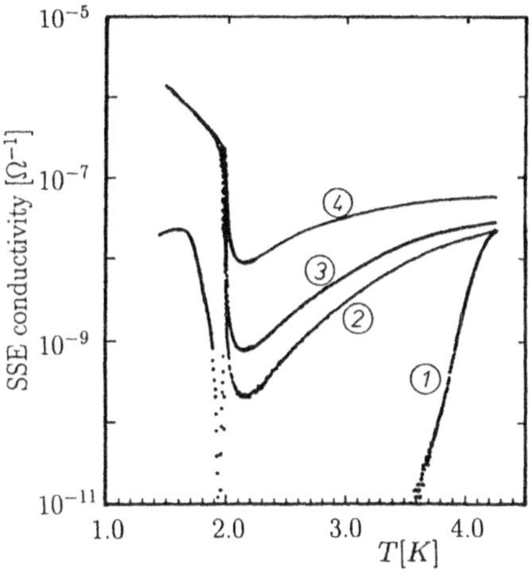

Fig.3. Electron conductivity on a quench-condensed hydrogen film for various annealing steps: 1-as prepared at 1.5K; 2-briefly annealed at 4.2K; 3-annealed for 12 h at 4.2K; 4-briefly annealed at 8K. In addition to the hydrogen a small amount of helium was admitted to the cell, which formed a saturated film of about 300 Å thickness below T ≈ 1.9K, giving rise to a steep increase in the conductivity. The electron density was $4 \times 10^8 cm^{-2}$ in this case (ref. 16)

the annealing procedure. The temperature dependence of σ was interpreted as thermal activation of electrons trapped by surface irregularities; the decreasing activation energy then is an indication that upon annealing the traps become shallower and that therefore the degree of roughness of the surface decreases. Although a quantitative analysis is still lacking, the measurement of σ appears as a promising tool to study surface conditions. It should be added that the films were completely smooth as judged by optical inspection, i.e. roughness on the scale of the wavelength of light could not be detected. Additional information was obtained from measurements with the H_2 surface being covered with an additional film of superfluid ^4He, similar to the experiments of Edel'man and Faley on bulk H_2.[12] In contrast to these earlier results, however, the He film was found to have a drastic influence on the conductivity (see Fig.3, T ≤ 1.9K). In fact, for saturated He films the value for σ coincided with the conductivity on bulk He (exept for the as-quenched films, where it was lower by about a factor of 10), indicating that the amplitude of the surface roughness of the H_2 films was distinctly smaller than the He film thickness of 300 Å.

ELECTRONS ON HELIUM FILMS

Electrons on saturated helium films with a typical thickness of 200-400 Å have already been mentioned briefly in the previous section. Now I discuss this system in more detail, first very thin He films in the monolayer range, then the transition region with a thickness of some ten monolayers and finally the saturated films of about a hundred monolayers. The isotope investigated was nearly exclusively ^4He. The substrate for the helium was in general again solid H_2, except for the films with several hundred Å thickness, where also results for more conventional dielectrics like glass or polymers exist. Neon substrates, which are in many respects similar to hydrogen, have been reviewed by Kajita.[17]

Thin He Films

One motivation for coating a hydrogen surface with helium was the idea that a He film on the order of a monolayer might heal the defects on the hydrogen. SSE on such films were investigated by Paalanen and Iye[18] from 1.25 to 4.2K, and down to lower temperatures by Cieslikowski et al..[13,14] These measurements showed that a He film did not improve the electron mobility μ, as one might have expected on the basis of the above argument, but that on the contrary μ was reduced in the presence of helium. Furthermore, as the He film thickness was increased, the mobility did not drop monotonously but showed a pronounced oscillatory behavior. Mobility maxima occurred whenever the coverage reached an integral number of monolayers, whereas minima were observed at approximately half-filled topmost layers. This is illustrated in Fig.4, taken from an experiment where the film thickness was varied by changing the temperature at a fixed total amount of helium (gas phase + adsorbed film) in the cell.

For coverages in the submonolayer regime the increase in electron scattering was interpreted by Paalanen and Iye as being due to density fluctuations in the 2-dimensional system of adsorbed helium atoms, which are proportional to the compressibility of this system. For a densely packed, complete monolayer of He on H_2 the compressibility becomes small and hence the mobility reached nearly the value on the bare H_2 substrate. For thicker films, qualitatively the same picture should apply for the topmost layer, as long as well-defined layers exist. The mobility signal therefore provides a measure of the extent of layering of the helium film. From the fact that 9 mobility maxima could be resolved in the experiment by Cieslikowski et al. at temperatures below 0.8K one can conclude that the layering effects in liquid helium * near a solid wall extend to at least 9 interatomic distances. At higher T they become more and more washed out.

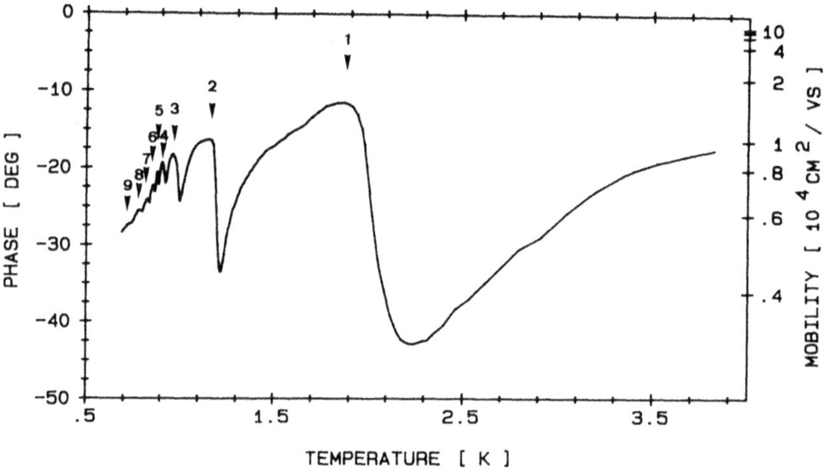

Fig.4. Electron mobility on bulk solid hydrogen, covered with a ^{4}He film whose thickness increases as the temperature is lowered. The figures indicate the number of monolayers completed at the respective temperatures. (after ref.13)

* It is known from other substrates that helium usually forms one or two solid layers next to a wall, caused by the van der Waals pressure. For H_2, due to its small interaction with He, it is not clear yet whether the first layer of He solidifies.

Pronounced layering of He films, although not over quite as large a distance, has been observed also in several other experiments using specific heat, third sound, and adsorption isotherms.[19-22] An unexpected result was reported recently by Shirron et al.[21] who observed modulations in the third sound velocity with a periodicity distinctly smaller than a monolayer. However, in another third sound experiment Mistura et al.[22] found good agreement between the location of maxima in the third sound velocity and the completion of monolayers as determined from adsorption isotherms. A formation of layers was also obtained in numerical simulations by Krotscheck.[23]

He Films at Medium Thickness

In the intermediate region between monolayers and thick saturated films the mobility eventually starts to rise, because the image potential gradually decreases and approaches the value of bulk helium. Consequently also the interaction of the electrons with defects and surface excitations is reduced. This behavior, already shown in Fig.3, is in some more detail illustrated in Fig.5, where the transition region is plotted on an expanded scale. Strikingly the transition is not smooth, but a sharp dip is observed around a film thickness of 10 monolayers with a half width of only ~2 monolayers *. This dip turns out to be a rather general phenomenon, because not only has it been observed for all the quench-condensed H_2 substrates, but also for He films on solid H_2 grown from the bulk liquid,[13,18] He films on solid Ne,[17] and even for He films on sapphire[24] (although in the latter case the dip occurred at a much larger film thickness).

As to now a conclusive interpretation for the mobility dip apparently does not yet exist. Andrei[24] and Paalanen and Iye[18] have ascribed it to electron ripplon scattering, which is small for very thin and very thick films and has a maximum at a film thickness near the value where the dip is

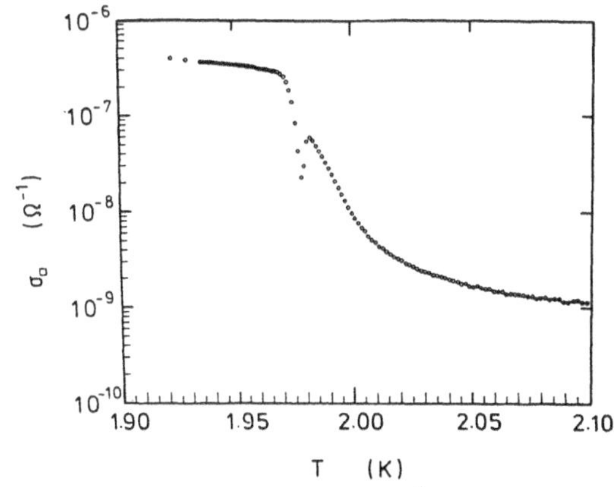

Fig.5. Conductivity of SSE on a ^{4}He film in the transition region from a monolayer to a thick saturated film. The dip occurs at a film thickness of 35 Å, as estimated from measurements with a quartz balance (ref.16)

* The structure due to layering as discussed in the previous section does not appear in this figure because of the relatively high temperature and possibly also because of the roughness of the substrate.

observed. However, at a temperature around 2K this contribution ought to be rather weak, and it is not clear why it should lead to such a sharp structure. A resonance enhancement can also be excluded, because in the investigations of Kono et al.[16] the dip was found to be not affected at all when the frequency of the ac conductivity measurement was varied between 1 and 100 kHz. As another source for the dip one might consider the formation of a polaronic state, which has been discussed quite extensively in the literature:[25-29] Under the influence of the image and external electric fields an electron exerts a pressure upon the film surface, and thus can produce a dimple in the He film underneath it. The mobility of this polaron (= electron + dimple) will be drastically reduced compared to the quasifree electrons. Estimates show that for He films on H_2 the energy gain on forming a polaron is indeed largest for a film thickness of about 10 monolayers.[16] Although this energy is smaller than k_BT for the experiments discussed here, the tendency to form a dimple could give rise to precursor effects similar to the soft mode observed near the macroscopic dimple instability.[7]

It should be added that for He films on classical substrates a sharp drop with decreasing film thickness, qualitatively similar to the one in Fig.5, has been observed, which has been interpreted as a transition from a delocalized to a polaron state.[24] Present theories, on the other hand, predict that polarons on a perfectly smooth substrate should develop only at temperatures substantially lower than the range of the experiments. Apparently pinning by surface roughness plays an important additional role for the localization of the electrons.[30]

Saturated He Films

Whereas the formation of one-electron dimples is still not fully understood, the overall effect of the electron pressure on the film surface is well established. The influence of the electron layer becomes particularly large for thick films, because the restoring van der Waals force decreases as d^{-4} (or even as d^{-5} for large thickness d, when retardation has to be taken into account). Charging the film therefore decreases its average thickness, as first observed by Etz et al. using ellipsometry.[31] As a result the film stabilizes itself as the surface charge is increased, so that for instance on He films supported by a mylar or a glass substrate electron densities up to 10^{11} cm^{-2}, two orders of magnitude higher than on bulk helium, could be achieved. At such high densities the film thickness is reduced to 70 Å by the electrostatic pressure and is then essentially independent of the initial thickness of the saturated film. These results were at variance with calculations by Tatarskii[32] which suggested that films become unstable for densities greater than $2x10^{10}cm^{-2}$. It was shown by Hu and Dahm,[33] however, that if the separation of the electrons from the film due to quantum motion is taken into account, the charge-saturated film is stable against surface perturbations for all densities, until eventually tunneling to the substrate sets in.

The mobility of SSE on saturated He films is in general extremely low (except, as discussed, for H_2 and Ne) due to pinning at surface irregularities. It could be shown by Yiang and Dahm[34] that the degree of pinning was substantially reduced and the mobility therefore increased when using a very smooth surface, like float glass, which in addition was carefully cleaned. With such a substrate these authors studied the sliding of the SSE at high density, where a 2-dimensional electron solid is formed, and found features in the conductivity and noise characteristics similar to sliding charge density waves.

Electronic transport studies on saturated films yielded also another interesting result, shown in Fig. 6: Lehndorff and Dransfeld[35] measured the microwave absorption of SSE at a frequency of 9 GHz and a temperature T≥1K

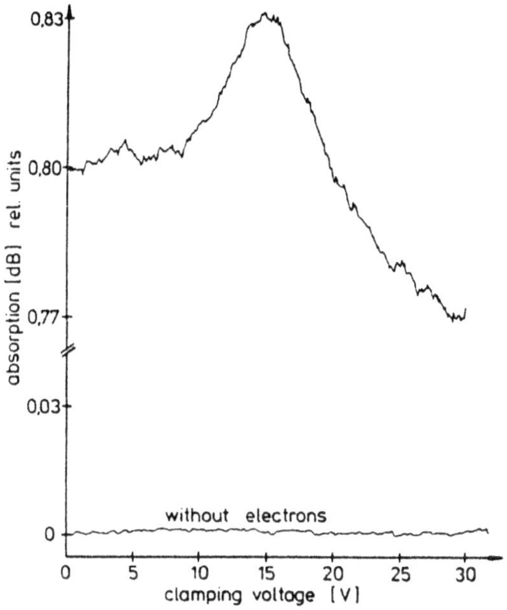

Fig.6. The microwave absorption of electrons on a helium film at a freqency of 9 GHz and a temperature of 1.12K as a function of the applied clamping voltage. A polymer foil, 6 µm thick, of Hostaphan (Mylar) served as a substrate for the helium film (thickness d ≈ 150Å). The electron density was about $2x10^{10}cm^{-2}$ (top curve). A run without electrons (bottom curve) is shown for comparision (taken from ref.35)

and found a broad maximum as a function of the external holding field. It was suggested that with the mylar substrate used in this experiment polarons were formed on the saturated He film, and that the absorption maximum reflected the lateral resonance vibration of the electrons localized in their dimples. Further investigations of this phenomenon for variable film thickness, magnetic field and lower temperature should give valuable information about the polaron state.

CONCLUSIONS

The various examples show that surface state electrons on hydrogen and helium are a very sensitive probe in order to study surface properties. The interaction of the electrons with these surfaces is on one hand sufficiently weak to provide a conductivity signal which can conveniently be measured, on the other hand the coupling is strong enough to give rise to large effects. The discussion has also shown that still considerable experimental and theoretical work has to be done towards a full understanding of the observed effects. Prospects for obtaining more quantitative results about quantum surfaces, possibly by a combination of SSE with other techniques, appear promising.

As pointed out in the beginning, this survey has focussed on effects related to the interaction of SSE with surface structures and excitations. Two-dimensional electron systems are of course of much more general interest. Recently the value of He and H_2 substrates for investigations of phenomena like magnetotransport and localization in 2D has been demonstrated,[36-39] and many more suggestions for experiments with surface state electrons exist.[40]

ACKNOWLEDGEMENTS

In the course of this work I have greatly benefitted from a close cooperation with K.Kono und U.Albrecht, and from numerous discussions with D.Cieslikowski, A.J.Dahm and V.B.Shikin. Support by the Deutsche Forschungsgemeinschaft, SFB 306, is also gratefully acknowledged.

REFERENCES

1. see, e.g., J.G. Dash, Phase transitions at interfaces: Roughening, surface melting and triple point wetting, Solvay Conference on Surface Science (University of Texas, 1987)
2. H.J. Lauter, Inelastic neutron scattering from adsorbates, Phonons 89, eds. S.Hunklinger, W. Ludwig and G.Weiss (World scientific, 1990), p.871, and contribution at this conference
3. M.W. Cole, Rev. Mod. Phys. 46, 451 (1974); C.C. Grimes, Surf. Sci. 73, 379 (1978); V.S. Edel'man, Sov. Phys. Usp.23, 227 (1980); F.I.B. Williams, Surf. Sci. 113, 371 (1982); Yu.P. Monarkha and V.B. Shikin, Sov. J. Low Temp. Phys. 8, 279 (1982); N. Studart and O. Hipolito, Revesta Brasileira de Fisica 16, 194 (1986); V.B. Shikin and Yu.P. Monarkha, "Two-dimensional charged systems in liquid helium", Editorial board for physical and mathematical literature (Moscow, 1989)
4. V.V. Zav'yalov and I.I. Smol'yaninov, Experimental observation of photoresonance in electrons above a solid hydrogen surface, Sov. Phys. JETTP 65, 194 (1987)
5. C.C. Grimes and G. Adams, Evidence for a liquid-to-crystal phase transition in a classical, two-dimensional sheet of electrons , Phys. Rev. Lett. 42, 795 (1979)
6. P. Leiderer and M. Wanner, The dimple crystal - a charge-induced instability of the helium surface, Surf. Sci. 98, 68 (1980)
7. P. Leiderer, Electrons at helium interfaces, Physica 126B, 92 (1984)
8 J. Bodensohn, K. Nicolai and P. Leiderer, The growth of atomicaly rough ^{4}He crystals, Z.Phys. B64, 55 (1986)
9. W.F. Vinen, Ions trapped below the surface of superfluid helium, presented at this conference
10. A.M. Troyanovskii, A.P. Volodin, and M.S. Khaikin, Electron localization over the surface of crystalline hydrogen and neon, JETP Lett. 29, 382 (1979)
11. A.M. Troyanovskii and M.S. Khaikin, Electron mobility in a two-dimensional layers over the surface of solid hydrogen, Sov.Phys.JETP 54, 214 (1981)
12. V.S. Edel'man and M.I. Faley, Investigation of electrons localized above solid hydrogen by means of the cyclotron resonance method, J. Low Temp. Phys. 52, 301 (1983)
13. D. Cieslikowski, P. Leiderer, and A.J. Dahm, Mobility oscillation of electrons on thin He films, Can. J. Phys. 65, 1525 (1987)
14. D. Cieslikowski, A.J. Dahm, and P. Leiderer, Investigation of thin helium films with surface-bound electrons, Phys. Rev. 58, 1751 (1987)
15. J.L. Levine and T.M. Sanders, Mobility of electrons in low-temperature helium gas, Phys. Rev. 154, 138 (1967)
16. K. Kono, U. Albrecht and P. Leiderer, Surface state electrons on a hydrogen film, to be published; U. Albrecht, H. Dilger, P. Leiderer, and K. Kono, Surface state electrons on solid hydrogen films, Proceedings of the 19th International Conference on Low Temp. Physics (Brighton 1990)
17. K. Kajita, Electronic properties of two-dimensional electrons formed on liquid helium and solid neon, Jap. J. of Appl. Phys. 26-3, 1943 (1987)
18. M.A. Paalanen and Y. Iye, Electron mobility on thin He films, Phys. Rev. Lett. 55, 1761 (1985)

19. S.E. Polanco, J.H. Quateman, and M. Bretz, Layering and thermal conductivity of ^{4}He films on uniform graphite, J. Phys. (Paris), Colloq.39, C6-344 (1978)
20. J.D. Maynard and M.H.W. Chan, Third sound on HOPG surface, Physica 109&110B, 2090 (1982)
21. P.A. Shirron, K.A. Gillis, and J.M. Mochel, Structure in thin superfluid helium films, J. Low Temp. Phys. 78, 157 (1990)
22. G. Mistura, G. Zimmerli, and M.H.W. Chan, Study of a layered superfluid film, poster presented at the Nato ASI "Phase transitions in surface films" (Erice 1990)
23. E. Krotscheck, Liquid helium on a surface: Ground state, excitations, condensate fraction, and impurity potential, Phys. Rev. B32, 5713 (1985)
24. E.Y. Andrei, Observation of the polaronic transition in a two-dimensional electron system, Phys. Rev. Lett. 52, 1449 (1984)
25. L.M. Sander, Electron localization on a liquid-helium surface, Phys. Rev. B11, 4350 (1975)
26. M. Saitoh, Theory of strongly coupled ripplonic polarons at finite temperature, J. Phys. C: Solid State 16, 6995 (1983)
27. F.M. Peeters and S.A. Jackson, Frequency-dependent response of an electron on a liquid helium film, Phys. Rev. B31, 7098 (1985)
28. M.H. Degani and O. Hipolito, Polaronic behavior of electrons on a liquid-helium film, Phys. Rev. B32, 3300 (1985)
29. F.M. Peeters and S.A. Jackson, Temperature dependence of the dynamical response of an electron on a thin liquid-helium film, Phys. Rev. B34, 1539 (1986)
30. Yu.Z. Kovdrya, F.F. Mende, and V.A. Nikolaenko, High-frequency conductivity of electrons above a helium film in the regime of localisation on substrate potential inhomogeneities, Sov. J. Low Temp. Phys. 10, 589 (1984)
31. H. Etz,W. Gombert, W. Idstein, and P. Leiderer, Stability of charged ^{4}He films Phys. Rev. Lett. 53, 2567 (1984)
32. V.V. Tatarskii, On stability of helium charged film, Sov. J. Low Temp. Phys. 12, 255 (1986)
33. X.L. Hu and A.J. Dahm, Comments on the stability of charged thin helium films, to be published
34. H.W. Jiang and A.J. Dahm, Conduction properties of a new two-dimensional charge-density wave, Phys. Rev. Lett. 62, 1396 (1989)
35. B. Lehndorff and K. Dransfeld, Microwave absorption of electrons bound to films of helium, J. Phys. France 50, 2579 (1989)
36. P.W. Adams and M. Paalanen, Localization in a nondegenerate two-dimensional electron gas, Phys.Rev.Lett. 58, 2106 (1987)
37. P.W. Adams and M. Paalanen, Anderson localization of electrons in dense ^{4}He gas, Phys. Rev. Lett. 61, 451 (1988)
38. P.W. Adams and M.A. Paalanen, High-field magnetoconductivity of electrons on hydrogen, Phys. Rev. B37, 3805 (1988)
39. M.J. Lea, A.O. Stone, and P. Fozooni, The Hall effect in 2D electrons on liquid helium, Europhys. Lett. 7, 641 (1988), and contribution at this conference
40. A.J. Dahm, Electrons on thin helium films, presented at this conference

ELECTRONS ON HELIUM FILMS*

A.J. Dahm

Physics Department
Case Western Reserve University
Cleveland, Ohio 44106, USA

INTRODUCTION

The properties of two-dimensional electron fluids and crystals bound to a bulk helium surface have been studied extensively for the last decade[1]. These systems are limited to the classical regime due to the instability of a highly charged surface to fluctuations. The maximum density on a bulk helium surface is 2.4×10^{9} cm^{-2}. At this density the melting temperature $\propto \sqrt{n}$ is ~ 1 K compared to a Fermi energy $\propto n$ ~ 50 mK. Here n is the electron areal density. The high melting temperature is a result of the large potential energy reduction for the Coulomb interaction in the lattice phase where small electron separations are excluded.

Studies of a two-dimensional quantum electron fluid are of greater intrinsic interest. Peeters and Platzman[2] pointed out that the quantum regime can be attained with electrons on a thin helium film supported by a dielectric or metallic substrate. The van der Waals attraction of the helium atoms by the substrate stabilizes the helium surface and allows for high electron densities, thus increasing the Fermi energy relative to the melting temperature. Furthermore, the dielectric response of the substrate screens the electron-electron interaction and lowers the melting transition. The possibility then exists of creating a degenerate electron fluid on the surface of a liquid helium film. I will discuss attempts to approach the quantum regime and possible experiments which might be undertaken on this ideal, degenerate two-dimensional fluid. I will also discuss phenomena which occur in this system which are analogous to properties of other condensed matter systems.

LIMITATIONS ON THE ELECTRON DENSITY

Electron densities on the order of 10^{11} cm^{-2} have been measured[3] on dielectric and metallic substrates although the charges were immobile presumably due to the roughness of the substrates used. For electrons on thin films, the image charge in the dielectric beneath the film is the dominate source of the pressure on the film which might lead to an instability of the surface. The separation of an electron from its image charge is less than the inter-electron spacing, and the continuous surface

*Supported in part by National Science Foundation Grant No. DMR-87-13922

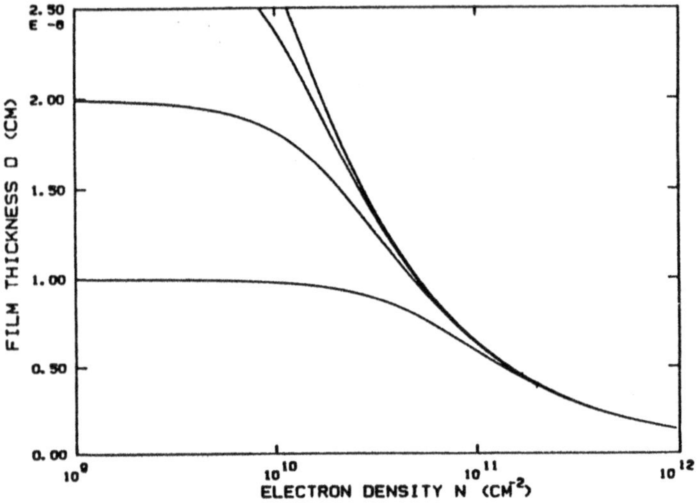

Fig. 1. Helium film thickness versus density for a glass substrate with van der Waals coefficient 4×10^{-15} ergs. Initial film thicknesses are 100, 200, 300 and 400 Å.

charge distribution cannot be assumed in calculating the stability limit of the films. The problem of the stability limit for discrete charges was formulated by Vil'k and Monarkha[4] and examined by Tartarskii[5].

Although the image field increases the local pressure on the film and contributes to the surface instability, the average electronic pressure suppresses the film thickness[3,5] and enhances the van der Waals restoring pressure. The result is that the film is marginally stable against surface fluctuations for all saturated electron densities[6]. The maximum electron density is limited to $\sim 2\times10^{11}$ cm^{-2} by electron tunnelling to the substrate as the film thins at large electronic pressures. A graph of film thickness versus saturated electron density is shown in Fig. 1. Film thicknesses are nearly the same for glass and metal substrates.

The influence of the electron separation from the helium surface on the stability is interesting. An electron is in the ground state for motion perpendicular to the film with a time dependent coordinate $z(t)$ relative to the film surface. For a film of thickness d the image charge is located a distance $2(d+z)$ below the electron. The average separation of the electron from the film, $\bar{z}$, is tens of Angstroms. This separation reduces the image field sufficiently to stabilize the charged film[6]. If the electron were in the plane of the surface[5], the critical density would be limited by surface instabilities to $\sim 2\times10^{10}$ cm^{-2}.

QUANTUM REGIME

Peeters and Platzman[2] estimated the melting temperature in a one component model to show that the electron fluid can exist below the Fermi temperature on thin films. Specifically they show that the fluid state is stable at T = 0 for low densities!

Consider an electron on a film supported by a substrate of dielectric constant ε. Substrate screening can be represented by the image charge of magnitude $\delta e = e(\varepsilon-1)/(\varepsilon+1)$. The average interaction between electrons separated by a distance r is

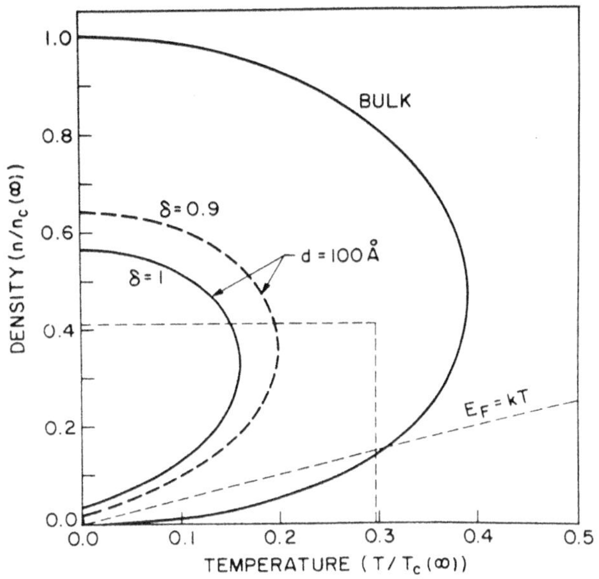

Fig. 2. Phase diagrams for bulk helium and 100 Å helium films: solid curve - metallic substrate; dashed curve - sapphire substrate. $n_c = 2.4\times10^{12}\ \mathrm{cm}^{-2}$ and $T_c = 33$ K.

$$V(r) = -e^2/r + \delta e^2/[r^2 + (2\bar{d})^2]^{1/2}, \tag{1}$$

where $\bar{d} = d + \bar{z}$. For thin films, $d \ll r$, the interaction can be written as the sum of a Coulomb and a dipole term as

$$V(r) \to -(1 - \delta)e^2/r + (\delta/2)(2e\bar{d})^2/r^3. \tag{2}$$

Let a be the lattice spacing. One can estimate the maximum value of $\bar{d}/a$ from Fig. 1 as ~ 0.3. Thus, the Coulomb term in Eq. (2) dominates for most dielectric substrates ($\varepsilon \lesssim 12$).

For metallic substrates, $\delta = 1$, the interaction is dipolar for thin films. The potential energy of the system is then

$$V(r) = -\Sigma_{i<j}\ p^2/2r_{ij}^3 = -(p^2/2a^3)\Sigma_{i<j}\ (a/r_{ij})^3, \tag{3}$$

where $p = 2ed$. The summation in Eq. (3) is smaller for the lattice phase. At low temperatures the kinetic energy is dominated by the zero-point energy, which is proportional to $n\hbar^2/m$ for both phases. Since the electrons are more localized in the lattice phase, the zero-point energy increases upon crystallization. The free energy difference between the solid and fluid phases at T = 0 is

$$\Delta F = F_s - F_f = -\alpha p^2/2a^3 + \beta\hbar^2/2ma^2. \tag{4}$$

where α and β are constants. At sufficiently low densities, large a, the fluid phase is stable at T = 0. Peeters and Platzman estimate the fluid to exist at T = 0 for $n < 10^{11}\mathrm{cm}^{-2}$ for a 100 Å film on a metal substrate. Their estimated phase diagram is shown in Fig. 2. Film suppression (see

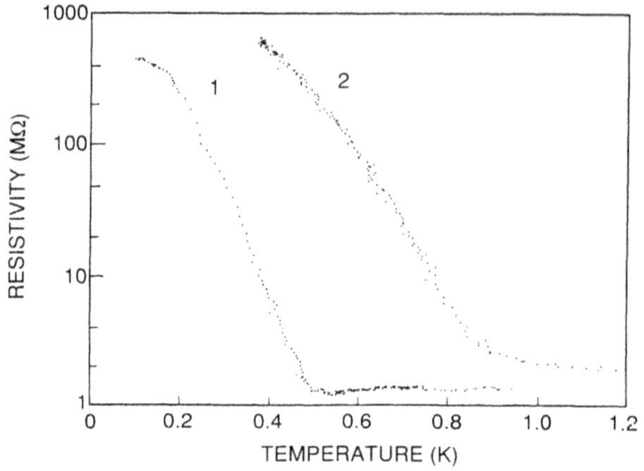

Fig. 3. Resistivity versus temperature for two samples: (1) $n = 2\times10^{9}$ cm^{-2}, d = 380 Å; (2) $n = 4\times10^{9}$ cm^{-2}, d = 340 Å.

Fig. 1) was not taken into account in this figure. The free energy difference vanishes again at high densities, $a \ll \bar{d}$, where the Coulomb interaction dominates. This occurs at T = 0 when

$$\Delta F = -\gamma e^2/a + \beta\hbar^2/2ma^2 = 0. \tag{5}$$

leading to a reentrant fluid phase. The limit $a \lesssim \bar{d}$ is not attainable experimentally.

PHASE DIAGRAM

Saitoh[7] calculated the melting curve under the assumption that melting occurs by the dissociation of dislocation pairs[8]. In this theory the melting curve is given by $T = a^2\mu/4\pi k$, where μ is the shear modulus of the lattice and k is Boltzmann's constant. He obtained an analytical expression for the shear modulus in the presence of the screened interaction. Jiang and Dahm[9,10] measured the melting point for various densities and film thicknesses on a glass substrate. They identified the melting point by a break in the resistivity versus temperature plot. An example of their data is shown in Fig. 3. They interpret the rapid rise in resistivity below the break as due to pinning of the lattice by variations in the substrate image potential on a non-ideal substrate. Their results are in excellent agreement with the theoretical predictions of Saitoh as shown in Fig. 4. For these crystals the melting point is reduced by a factor as large as three by the screening. However, the crystals were still classical.

LIMITATIONS DUE TO SCATTERING

Some experiments on the quantum fluid, such as the 2D Hall effect, require a large scattering time τ. The cyclotron frequency at one Tesla is $\sim 2\times10^{11}$ sec^{-1}. This corresponds to a filling factor of unity for a density of 5×10^{10} cm^{-2}. Substrate roughness and ripplon scattering put severe constraints on τ. The ripplon scattering rate for the classical electron gas is temperature independent and varies inversely with holding field E, which is proportional to the saturated density. The theoretical

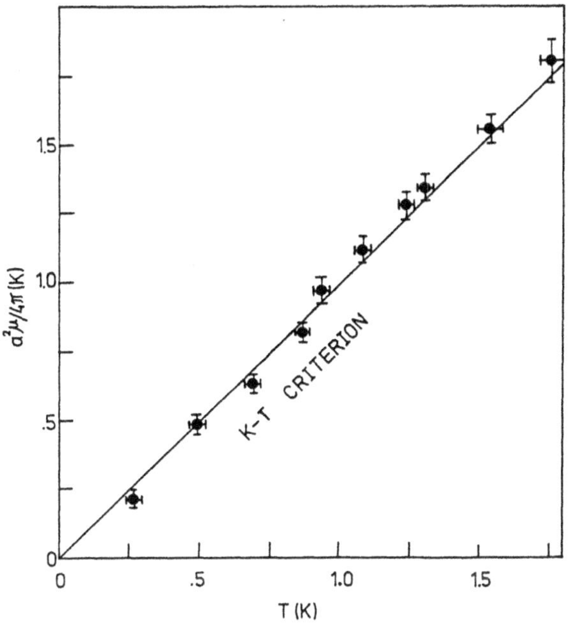

Fig. 4. Normalized shear modulus versus resistive transition temperature. The density and film thickness range from $n = 6\times10^{8}$ cm^{-2}, $d = 480$ Å, $T = 0.27$ K to $n = 2\times10^{10}$ cm^{-2}, $d = 150$ Å, $T = 1.75$ K.

scattering time for non-interacting electrons at a density of $10^{10}cm^{-2}$ is ~ 3 ps. Ripplon scattering is stronger by a factor of order two[11,12] for the highly correlated fluid. Experimental values[10,13] of the scattering time on glass substrates at this density and d ~ 250 Å are ≤ 1 ps. These measurements may include some substrate roughness scattering.

The ripplon rate is reduced for the degenerate electron fluid. Ripplons which dominate the scattering have a wave vector of 2p/ħ, where p is the electron momentum, and a characteristic temperature of 10 to 30 mK. For the non-interacting electron gas the collision frequency is given by[14]

$$\nu(p) = \nu_0 2mkT/p^2 \ ; \qquad \nu_0 = (eE)^2/4\sigma\hbar, \tag{6}$$

for ripplon frequencies ≤ T. Scattering is dominated by large values of the momentum. Then the scattering rate is temperature independent for the classical fluid and is reduced by a factor of T/T^F for the quantum fluid. Here T^F is the Fermi energy. The scattering decreases more rapidly when thermal ripplons have insufficient momentum to scatter electrons across the Fermi circle. The theoretical scattering time for the non-interacting degenerate fluid at a density of 5×10^{10} cm^{-2} is ~ 5 ps at 10 mK. The condition that an electron complete a cyclotron orbit without scattering at unity filling factor is marginal and in our crude estimate is independent of density at a fixed temperature. A more exact calculation of the scattering time in a magnetic field is needed.

The most severe challenge for films of thickness on the order of 100 Å is to find a sufficiently smooth substrate. A plot of the mobility as a

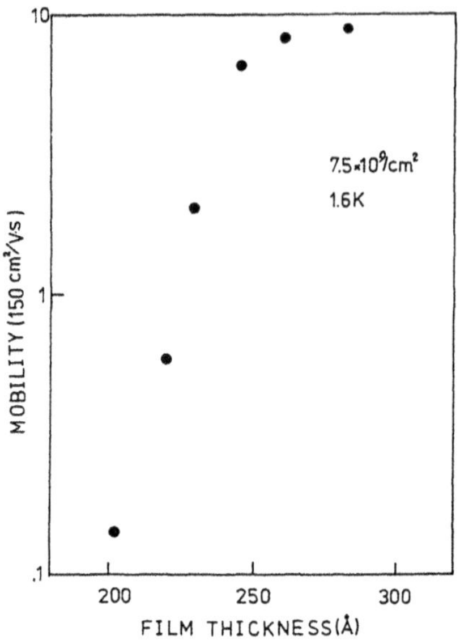

Fig. 5. Fluid mobility versus film thickness.

function of film thickness is shown in Fig. 5 for a glass substrate. The scattering time is sharply reduced below 250 Å for this substrate. The helium film can be reduced to ~ 150 Å on some glass substrates before the scattering is appreciably enhanced. We find commercial GaAs wafers to be smoother than glass, and coating a substrate with neon or hydrogen reduces substrate scattering. A possible smooth material is cleaved mica, but the larger dielectric constant of GaAs is preferred for screening. At this time we have not been successful in obtaining mobile electrons on 100 Å films.

EXPERIMENTS ON THE QUANTUM FLUID

Melting and the Phase Diagram

An extremely interesting topic is melting of the quantum fluid. Measurements of the shear modulus[15] and specific heat[16] in the vicinity of the melting point as well as the data presented in Fig. 4 lend support to the model of melting by the dissociation of dislocation pairs[8] for the classical electron lattice. This is a classical melting theory, and dislocation pairs are thermal equilibrium defects in the crystal. It is straight forward to vary the density and cross the phase boundary at $T = 0$ where no thermal defects exist. Chui and Esfarjani[17] have recently predicted that quantum melting for a Coulomb lattice will occur via the spontaneous generation of dislocations. They calculated the density dependence of the quantum mechanical self-energy of a dislocation and predict dislocations to be generated spontaneously at a particular density. This mechanism might also apply to a dipole lattice which has smaller dislocation energies.

A measurement of the effect of a magnetic field on the quantum phase boundary is of strong interest. Application of a magnetic field does not affect the melting point in the extreme classical limit. However, a magnetic field alters the energetics of the quantum fluid phase and limits the zero point motion of the electrons in the solid phase to a cyclotron

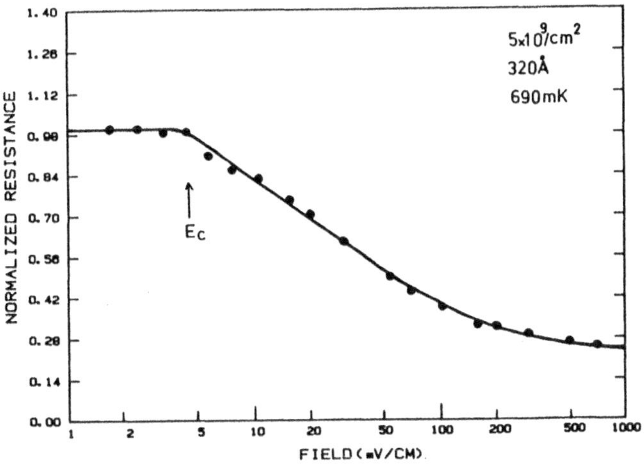

Fig. 6. Resistivity versus driving field.

radius. One might expect to observe reentrant behavior as a function of magnetic field near the fractional filling levels as has been suggested to explain data on two-dimensional electrons arrays in heterojunctions[18]. In general, the entire phase diagram for various degrees of screening and different magnetic fields is of interest.

2D Hall Effect

A measurement of the 2D Hall effect and magnetoresistance minima at the Hall plateaus in this idealized system is of obvious interest. The observation of a Hall plateau requires "impurity" levels at the Fermi level. These would have to be provided by variations in the substrate potential. It may be difficult to have these trapping levels present and still limit the substrate scattering to allow cyclotron orbits to be completed. Ripplon scattering can probably be suppressed at very low temperatures. Magnetoresistance measurements[19] on the classical electron gas show interesting behavior at large fields which has been attributed to scattering with broadened Landau levels.

2D Superconductivity

A small wake in the helium surface occurs in the path of an electron moving in the presence of a holding field pressing the electron toward the surface. In the presence of the holding field this "trough" provides an attractive potential for another electron. Thus, there exists an attractive electron-electron interaction via this electron-ripplon coupling which is analogous to the electron-phonon coupling in a metal. This coupling is expected to lead to a superfliud state[2]. There is no theoretical estimate of the superconducting transition temperature. Strong scattering by ripplons may suppress superconductivity.

Ripplon Scattering from the Quantum Fluid

In general, the problem of ripplon scattering from this highly correlated quantum fluid of electrons is of both theoretical and experimental interest. The low temperature regime where scattering across the Fermi circle cannot occur is analogous to the Bloch T^5 law in metals. The temperature dependence in this regime differs for capillary waves and third sound, i.e., it depends on the thickness of the helium film.

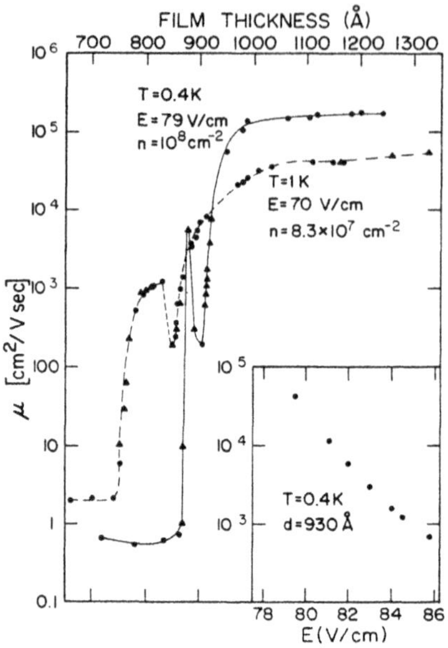

Fig. 7. Mobility versus film thickness at two temperatures. Insert: Mobility versus field at fixed thickness.

ANALOGS OF OTHER CONDENSED MATTER SYSTEMS

Sliding Charge-Density Waves

The crystal phase of electrons on a thin helium film is pinned by variations of the image potential in the rough underlying substrate. The conduction properties of this weakly pinned, periodic array of charges[9,20,21] are analogous to those of sliding charge-density waves in other systems[22]. The rapid rise in resistivity with decreasing temperature due to pinning is shown in Fig. 3. The nonlinear resistivity is shown in Fig. 6. At a critical field Ec the resistivity begins to decrease as regions of the lattice become temporarily depinned and move forward in an incoherent manner leading to a large 1/f noise. At larger fields, approximately 100 mV/cm for this crystal, the entire lattice becomes depinned and "jerks" across the random potentials. When this occurs there is an abrupt increase in the broadband noise spectrum. The resistivity at large fields is limited by ripplon scattering. This system also exhibits a frequency dependent conductivity with both real and imaginary parts varying approximately as $\omega^{-1/2}$.

This two-dimensional, weakly pinned system is ideal for the study of the properties of sliding charge-density waves. Unlike other systems, the pinning potential and inter-particle potential can be changed in situ by varying the film thickness and density, respectively. Finally, pinning in stronger potentials, obtained merely by reducing the film thickness, should lead to the formation of an electron glass.

Polarons

A single electron at rest on a liquid helium surface in the presence of a holding field will create a depression in the surface with lateral dimensions on the order of the capillary length. At low temperatures and large holding fields an electron will become self-trapped in such a depression. This state is labeled a ripplonic polaron in analogy with electrons self-trapped by ionic displacements in polar crystals.

The strongly coupled polaron is particularly interesting because it represents a localization in an ideal system and corresponds closely from a theoretical point of view to a continuum electron-phonon model. The ripplonic polaron is valuable in testing predictions of the model because the coupling strength can be varied, in situ, from weak to strong coupling by varying the holding field and helium film thickness. The polaron can also be probed by varying the temperature and magnetic field.

Andrei[23] reported the detection of the polaronic state. She measured the mobility and effective mass of electrons as a function of film thickness at fixed holding field. A transition to a low mobility state with a large effective mass occurred at a temperature dependent film thickness ~ 900 Å or as a function of field at a fixed film thickness. Her mobility data are shown in Fig. 7. Peeters and Jackson[24] reproduced the sharp mobility dip theoretically as a function of coupling strength which is varied experimentally by varying the holding field or the film thickness dependent capillary length. The interpretation of the dip in the mobility is interesting. Ripplons which scatter most efficiently from a dimple are those with a wavelength λ comparable to its size. As the dimple grows its thermal velocity decreases. A relative scattering maxima occurs when the ripplon phase velocity $\propto \lambda^{-1}$ matches the thermal velocity of the dimple.

There are two discrepancies between the interpretation of this experiment and theoretical predictions. A single electron dimple is predicted to have dimensions of a capillary length which is an order of magnitude greater than the inter-electron spacing. Secondly, the holding field was an order of magnitude less than theoretical estimates. Degani and Hipolito[25] suggest that the features observed in Andrei's experiment might be a result of the formation of multi-electron dimples. A detailed understanding of these data has not been given.

Recently, microwave absorption from electrons on a film supported by a plastic substrate has been attributed to polaron formation[26]. Plastic substrates are less smooth than glass, and one must distinguish between microwave absorption from electrons trapped in surface potentials and from polarons. Surface defects could assist polaron formation. A systematic study of ripplonic polarons on smooth substrates is needed to understand this very unique state.

Localization

We attribute the decrease in mobility with decreasing film thickness shown in Fig. 5 as resulting from electron localization in the potential variations of the image potential caused by substrate roughness. The drop in the mobility occurs at different film thicknesses for different glass substrates and cannot be attributed to mobile polarons. This represents another example of the versatility of this system to investigate phenomena which have been investigated in other condensed matter systems, but with the added advantage of being able to vary the coupling parameter over a wide range, in situ. Adams and Paalanen[27] investigated the localization of classical electrons on a hydrogen substrate. It may be possible to study the localization in the quantum limit on a helium film. This system would be a valuable tool for testing two-dimensional localization theories.

SUMMARY

Preliminary work on electrons supported by a helium film suggests that the quantum regime of this fluid is attainable. Screening of the Coulomb interaction by the substrate reduces the melting temperature, and a quantum fluid at T = 0 has been predicted. Scattering of electrons by ripplons and substrate roughness remain a problem. If a sufficiently smooth substrate can be found, then electrons on a helium film will form an ideal system for the study of the properties of a two-dimensional quantum fluid. It is also a testing ground for theories of other condensed matter properties such as sliding charge-density waves, polarons and localization.

REFERENCES

1. A. J. Dahm and W. F. Vinen, Physics Today 40, 43 (1987).
2. F. M. Peeters and P. M. Platzman, Phys. Rev. Lett. 50. 2021 (1983); P. M. Peeters, Phys. Rev. B 30, 159 (1984).
3. H. Etz, W. Gombert, W. Idstein, and P. Leiderer, Phys. Rev. Lett. 53, 2567 (1984).
4. Yu. M. Vil'k and Yu. P. Monarkha, Fiz. Nizk. Temp. 10, 901 (1984) [Sov. J. Low Temp. Phys. 10, 469 (1984)].
5. V. V. Tartarskii, Fiz. Nizk. Temp. 12, 451 (1986) [Sov. J. Low Temp. Phys. 12, 255 (1986)].
6 X. L. Hu and A. J. Dahm, Phys. Rev. B 42, 2010 (1990).
7. M. Saitoh, Phys. Rev. B 40, 810 (1989); Surf. Sci. 229, 356 (1990).
8. For a brief review of this theory, see K. J. Strandburg, Rev. Mod. Phys. 60, 161 (1988) and references therein.
9. H. W. Jiang and A. J. Dahm, Surf. Sci. 229, 352 (1990).
10. H. W. Jiang, M. A. Stan, and A. J. Dahm, Surf. Sci. 196, 1 (1988).
11. M. A. Stan and A. J. Dahm, Phys. Rev. B 40, 8995 (1989).
12. V. A. Buntar, Y. Z. Kovdrya, V. N. Grigor'ev, Yu. P. Monarkha, and S. S. Sokolov, Fiz. Nizk. Temp. 13, 789 (1987) [Sov. J. Low Temp. Phys. 12, 451 (1987)].
13. H. W. Jiang and A. J. Dahm, Jap. J. Appl. Phys. 26, 751 (1987).
14. V. B. Shikin and Yu. P. Monarkha, J. Low Temp. Phys. 16, 193 (1974).
15. G. Deville, A. Valdes, E. Y. Andrei, and F. I. B. Williams, Phys. Rev. Lett. 53, 588 (1984).
16. C. D. Glattli, E. Y. Andrei, and F. I. B. Williams, Phys. Rev. Lett. 60, 420 (1988).
17. S. T. Chui and K. Esfarjani, preprint.
18. H. W. Jiang, H. L. Stromer, D. C. Tsui, L. N. Pfieffer, and K. W. West, Phys. Rev. Lett. 65, 633 (1990).
19. R. W. van der Heijden, M. C. M. van de Sanden, J. H. G. Surewaard, A. T. A. M. de Waele, H. M. Gijsman and F. M. Peeters, Europhys. Lett. 6, 75 (1988).
20. H. W. Jiang and A. J. Dahm, Phys. Rev. Lett. 62, 1396 (1989).
21. K. Kajita, Jpn. J. Appl. Phys. 25, Suppl. 3, 1943 (1987);Surf. Sci. 196, 29 (1988).
22. For a review, see G. Gruner, Rev. Mod. Phys. 60, 1129 1988).
23. E. Y. Andrei, Phys. Rev. Lett. 52, 1449 (1984).
24. F. M. Peeters and S. A. Jackson, Phys. Rev. B 34, 1539 (1986).
25. M. H. Degani and O. Hipolito, Phys. Rev. B 32, 3300 (1985).
26. B. Lehndorff and K. Dransfeld, J. Phys. (France) 50, 2579 (1982).
27. P. W. Adams and M. A. Paalanen, Phys. Rev. Lett. 58, 2106 (1987).

TUNNELING AND EVAPORATION FROM THE ELECTRON LIQUID TRAPPED AT A HELIUM SURFACE

P M Platzmann

AT & T Bell Laboratories
600 Mountain Avenue
Murray Hill, New Jersey 07974

Tunneling and evaporation from a well characterized many body quantum mechanical system is an important and intriguing experimental and theoretical problem. 2-D electrons trapped at the bulk helium vacuum interface is such a system. There are strong coulomb correlations among the electrons whose density may be easily varied over a wide range. The strong correlations are manifested by a crystalization at sufficiently high densities and low temperatures.[1] Escape from such a system can be initiated by the application of external fields which are roughly a hundred volts per cm. In addition, the system is extraordinarily clean and homogeneous, i.e. free of defects, local field hot spots, impurities, etc.

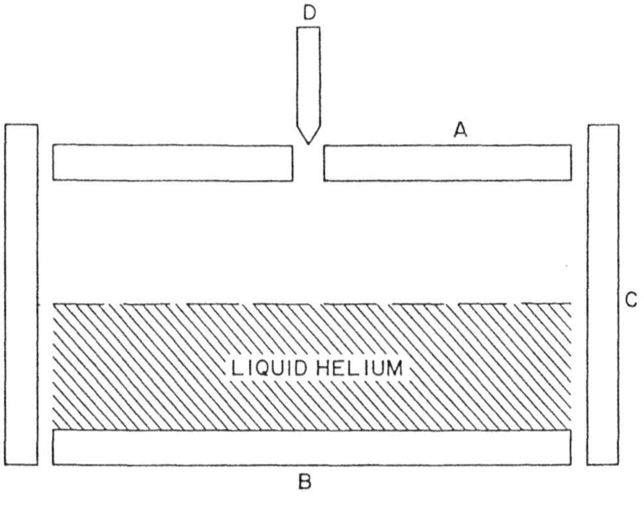

Fig. 1 Diagram (cross section) of the cylindrical experimental sample container. A is the upper capacitor plate, B is lower plate, C is the ring plate that provides the radial containing field, and D is the point of discharge for generating free electrons.

The deceptively simple experimental cell is shown in Fig. 1.[2] It has two horizontal parallel capacitor plates 2 cm in diameter, separated by a distance of 1 cm. The upper plate has a small hole through its center. A point electrode located just above this hole is charged to about -400 V to initiate a discharge. The discharge is performed at a temperature of about 0.98 K where the gas pressure is sufficient to sustain the discharge. The bottom plate is held at dc ground potential. During the discharge the (negative) potential of the upper plate, V_p, is gradually decreased so as to provide the pressing field that forces the electrons onto the surface. Values of V_p ranging between -15 and -300 V have been used.

The space between the plates is filled approximately half way with liquid helium. A cylinder surrounding the plates, but insulated from them, is held at a negative potential, V_r, so as to provide a radial field to prevent the electrons from reaching the walls of the container. In this geometry the charge forms a circular disc with maximum density, n_0, at its center. The diameter of the disc depends on V_p, V_r, and the total charge, Q, as described later. The electrons move in the plane of the liquid surface in response to the horizontal components of the ac field used to measure the capacitance. If the diameter of the charge disc is close to that of the retaining cylinder, the capacitance between the plates is substantially reduced by the shielding effect of the electron layer. As the diameter of the disc decreases, its shielding effect decreases so that the capacitance between plates increases. Thus the capacitance can be used as a measure of the disc diameter and, consequently, of the total charge on the surface. A typical calculated density profile is shown in Fig. 2. The main point of this figure is to show that the density is quite uniform at the center and to remark that escape from the edges is exponentially small relative to the central flat region because of the simple energetics of the problem which we will now discuss.

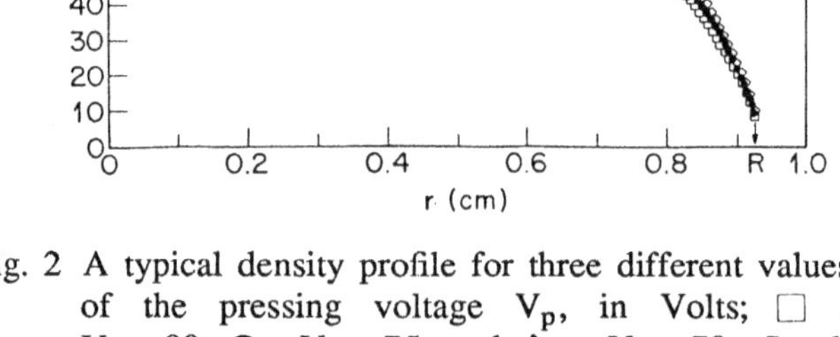

Fig. 2 A typical density profile for three different values of the pressing voltage V_p, in Volts; □ , V_p=–80; ● , V_p=–75; and ◇ , V_p=–70. In all cases the radial potential V_r=–97.

The motion of an electron normal to the surface is governed by the image potential in the Helium.

$$V_1(z) = \frac{-\Lambda e^2}{z}, \; z > O \tag{1}$$

$$=V_0(=1\ eV),\ 1\ z < 0$$

with $\Lambda = 0.007$ and an external pressing field E_p which is required to maintain a finite density of electrons, which has a potential $eE_p z$. In addition we must include the coulombic potential U_c of the other 2D electrons. In these highly correlated systems the neighbors nearest to the electron which is escaping are at a distance of approximately $n^{-1/2}$ so that the coulomb effects will only be important for $z \leq n^{-1/2}$ i.e. approximately 10^3–10^4Å for the system under consideration. For $z < n^{-1/2}$ we may write,

$$U_c(z) \cong -4\ e^2 n^{3/2} z^2 / \text{v} \tag{2}$$

This formula for U_c may be seen as the expansion of the dimensionless self-consistent $U_c / e^2 n^{1/2}$, which has a maximum at $z = 0$, with respect to the small dimensionless parameter $zn^{1/2}$. At larger distances $z > n^{-1/2}$ the effect of the other electrons is that of a uniform electric field $E_c = 2\pi ne$. If $\delta \equiv E_p / E_c < 1$ then tunneling and evaporation are allowed.

The quantity v expresses our ignorance of the many body aspects of the problem. Their are two extreme values of v which characterize or more precisely bracket the physics of the situation. The "adiabatic" value of v is $\text{v}_A = .9$. This value is obtained(3),(4) by moving an escaping electron some distance z above the surface, allowing the other electrons to relax to the new position and computing the energy. The other is obtained by taking the frozen configuration at zero temperature i.e. moving an electron out from the triangular lattice. This value is simply determined to be $\text{v}_F = 1.2$.(4)

Neglecting very weak coupling to ripplons we can in this one body approximation write the total effective one body potential U for $0 < z < n^{-1/2}$ as,

$$U = -\Lambda e^2 / z + eE_p\ z - 4\ e^2 n^{3/2} z^2 / \text{v} \tag{3}$$

When $z < 0$, since V_0 in Eq. (1) is so large, we may assume $U = \infty$. The first two terms in Eq. (3) describe a Stark shifted 1D hydrogen atom with the Bohr radius $a_B \cong \hbar^2 / m\Lambda e^2 \cong 76.2$ Å and the Rydberg energy,

$$E_R(E_p) \cong m\Lambda^2 e^4 / 2\hbar^2 - 1.5eE_p a_B. \tag{4}$$

The last term in Eq. (4) changes the Rydberg energy by $12\text{v}^{-1} e^2 n^{3/2} a_B^2$, which is too small to be significant. The height of the barrier $U_{\max}$ is determined ($z \gg a_B$) by the last two terms in Eq. (3) i.e.,

$$U_{\max} = \left(\frac{\pi}{2}\right)^2 e^2 n^{1/2} \text{v} \delta^2. \tag{5}$$

Since we know that tunneling rates are ten to fifteen orders of magnitude less than the attempt frequency and since the important potentials are slowly varying on a scale of a_B, we want to use what amounts to a WKB approximation to compute tunneling rates.[5] If we think of a single electron moving in then 1-D one particle potential (Eq. (3)) then we can write down the answer to the tunneling rate in the conventional WKB approximation as,

$$W^T \cong E_R/\hbar \exp -2\left[\int_0^{z_{max}} \sqrt{2m(eE_p z - 4e^2 n^{3/2} z^2/v + E_R)}\, dz\right] \tag{6}$$

when $\delta \leq .5$, $z_{max} \cong \left[\dfrac{\Lambda v}{8n^{3/2} a_B}\right]^{1/2} < (\pi n)^{-1/2}$. In Eq. (6) we have neglected the image potential which is the reason the lower limit on the integral is zero. The evaporation rate in this same one particle approximation assuming thermal equilibrium with the bath is given by,[6]

$$W^E = \frac{T}{\hbar} S_T \exp\left[-\frac{E_R + U_{max}}{k_B T}\right]. \tag{7}$$

when S_T is a sticking coefficient of order unity. The value of density and temperature where $W^T \cong W^E$ clearly depends on E_p. When $E_R \cong U_{max}$ then the two rates are equal when,

$$T = .06\,(n)^{3/4} \tag{8}$$

Here T is in degrees Kelvin and n is in $10^8 cm^{-2}$. At 1 °K evaporation dominates. True tunneling should be observed for high density samples in the .1 °K range.

What does the experiment tell us. In Fig. 3 we show a representative set of data.[2] The escape rate (per electron) is shown as a function of temperature. At the

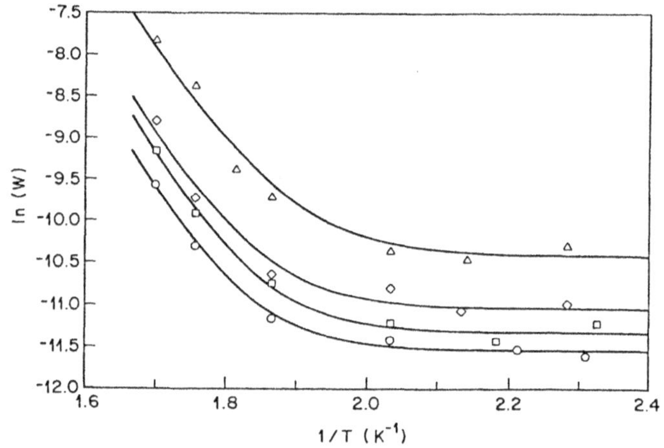

Fig. 3 The log of the escape rate vs 1/T at approximately constant charge. The curves are for four different values of the pressing voltage V_p in Volts and density n in $10^8 cm^{-2}$, □ V_p=−38, n=1.49; ○ V_p=−37, n=1.44; ◇ V_p=−34, n=1.42; △ V_p=−29.5, n=1.37;

lowest temperatures around .4 °K, there is little or no temperature dependence. The authors originally interpreted this to be the quantum tunneling regime. However, we have seen that we *expect* tunneling to dominate for such systems at *much* lower temperatures. In addition the absolute value of the tunnel rate obtained from Eq (6) is many (ten) orders of magnitude smaller than the observed rate. At higher temperatures the escape rate approaches an exponential dependence characteristic of activation over a barrier. Use of Eq's (5) and (7) show that both the activation energy and absolute rate is in good quantitative agreement with the data.

The question we would now like to address for the remainder of the talk is how can we understand the temperature independent regime. Is it some type of many body tunneling which we have badly underestimated by our one particle treatment or is it some type of many body evaporation.

In order to determine the true many body tunneling rate, we need to solve the many body Schroedinger equation for N *distinguishable particles* i.e.,

$$\sum_{i=1}^{n} \frac{\partial^2 \psi(\mathbf{r}_1 \cdots \mathbf{r}_n)}{\partial^2 \mathbf{r}_i} + \frac{2m}{\hbar^2}(E-U) = 0. \tag{9}$$

The Pauli principle plays no role since $E_F \equiv \hbar^2 \pi n/m \ll k_B T \ll e^2 n^{1/2}$. We want solutions which correspond to one particle (for example particle number one) tunneling i.e. all $z_i \cong 0$ for $i \neq 1$, and $z_1 \equiv z$ arbitrarily large. In a recent publication[7] it was shown that a many particle WKB approximation was very accurate approximation ($a_B n^{1/2} \ll 1$) for this problem. This approximation assumes a quasi classical wave function of the form

$$\psi \cong e^{-S/\hbar} \tag{10}$$

which in turn leads naturally to a set of pseudo classical Newtons equations

$$m\mathbf{r}_i = \partial U/\partial \mathbf{r}_i \tag{11}$$

The solution of these equations (with the wrong sign of the force) $z = z(t)$, $\rho_i = \rho_i(t) = \rho_i(z)$ $(i \neq 1)$ minimizes the action

$$S = -i\int \mathbf{p}\cdot d\mathbf{r} = \int_{z_1}^{z} (2m(U-E)(1+\rho_i'(z)^2)^{1/2} \tag{12}$$

While it is not possible to analytically find $\rho_i(z)$ the form of Eq. (12) shows unambiguously that

$$\int \sqrt{2m(U_A-E)}\, dz \le S \le \int \sqrt{2m(U_F-E)}\, dz. \tag{13}$$

which in turn implies that the observed temperature independent tunneling regime around .5 °K can *not* be tunneling.

Many Body Evaporation is a whole subject in itself. Recently[8] Azbel has conjectured that such evaporation could possibly lead to a temperature independent regime and even a regime where the evaporation rate increases as the temperature is lowered. We[9] have made the first attempt to calculate such Many Body Evaporation by rewriting Eq. (7) and Eq. (5) as,

$$W^E = \frac{T}{\hbar} e^{-E_R/k_{BT}} \int d\rho P_\Gamma(\rho) e^{-(\frac{\pi}{2})^2 e^2 n^{1/2} \delta^2 v(\rho)} \tag{14}$$

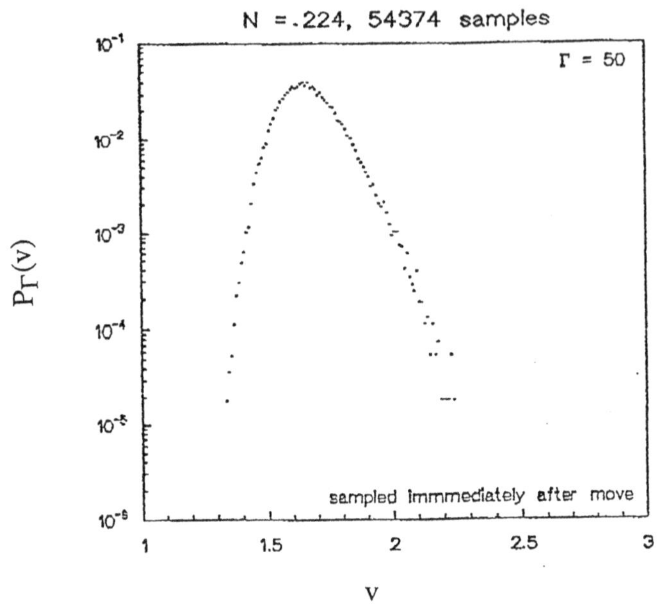

Fig. 4 The probability of finding a given v for Γ=50. The x axis is linear in v but the units used in the numerical computation indicate the peak value is v≅1.

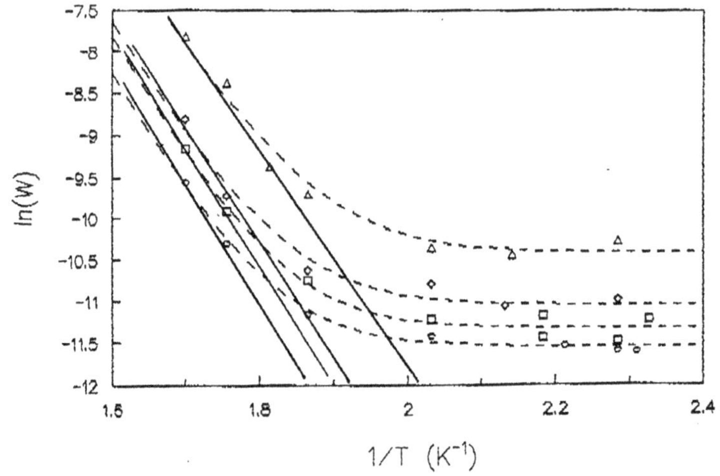

Fig. 5 The resultsos using equation (14) with probability functions generated by a Monte Carlo procedure using 224 particles with periodic boundry conditions compared with the data in ref (2).

Here the function $P_\Gamma(\rho) \equiv P_\Gamma(\nu(\rho))$ is the probability of finding, in equilibrium, the configuration ρ. We calculate a v(ρ) for each configuration as if the configuration is frozen. As the temperature $\Gamma \equiv e^2 n^{1/2}/k_B T$ is changed relative to the melting temperature of the electron liquid the frozen configuration[6] which are important changes. The function $P_\Gamma(\rho)$ for Γ=50 is shown in Fig. (4). Such a probability distribution leads to a slightly modified evaporation rate. In Fig. (5) we show the comparison of one set of experimental data with Eq. (14). This description is clearly unable to explain the data.

In summary then, we are left with a real puzzle. We do not understand even qualitatively the temperature dependence of the escape rate in this very simple strongly correlated 2-D electron liquid.

References

1. C. C. Grimes and G. Adams, Phys. Rev. Lett. 42:795 (1979); and D. S. Fisher, B. I. Halperin and P. M. Platzman, Phys. Rev. Lett. 42:738 (1979).
2. J. K. Goodkind, G. Saville, A. Ruckenstein and P. M. Platzman, Phys. Rev. B 38:8778 (1988).
3. Y. Iye, K. Kono, K. Kajita and W. Sasaki, J. Low Temp. Phys. 38:293 (1980).
4. R. Price (Private Communication).
5. L. Landau, E. M. Lifshitz, "Quantum Mechanics," Pergamon Press, New York (1981).
6. J. Iche and P. Nozieres, J. Phys. (Paris) 37:1313 (1976).
7. M. Ya Azbel and P. M. Platzman, Phys. Rev. Lett. September 11 (1990).
8. M. Ya Azbel, Phys. Rev. Lett. 64:1553 (1990).
9. R. Price and P. M. Platzman (to be published).

ELECTRONS ON HELIUM IN A QUANTIZING MAGNETIC FIELD

M. J. Lea

Department of Physics
Royal Holloway & Bedford New College
University of London,
Egham, Surrey,TW10 0EX, U.K.

INTRODUCTION

Electrons above the surface of superfluid helium are held in a vertical potential well formed by the repulsion of the helium atoms and the attraction of the image or polarization charges induced in the liquid, which may be augmented by a vertical electric holding field F_z. Below 1K all the electrons can be in the quantum ground state as far as vertical motion is concerned. However, the electrons are free to move parallel to the helium surface (the x–y plane) and form a non-degenerate two-dimensional electron gas (2DEG), density n, with a Maxwell-Boltzmann distribution of thermal velocities. Their mobility μ in a horizonatal electric field is limited by collisions with He atoms from the vapour and with the surface wave excitations, or ripplons, on the liquid helium. Electron-electron correlations are also very important and lead to a phase transition to a classical electron solid at low temperatures. This solid phase prevents the Fermi degenerate gas regime from being reached for electrons on bulk liquid. This coupled 2-D system of electrons and helium is of great interest both for the properties of the electrons and for the information which can be obtained about the helium itself.

The application of a vertical magnetic field B causes the electrons to move in circular orbits at the cyclotron frequency ω_c Because of the high electron mobility these orbits are quantized into Landau levels, which have a profound effect on the electrical transport properties of the 2DEG. The strong temperature dependence of the helium excitations and the variation of the coupling to the electrons with F_z enable these quantum effects to be studied over a wide range of experimental conditions. This paper will concentrate on the properties of a non-degenerate 2DEG in the quantum limit, $\hbar\omega_c/kT >> 1$. This system forms an interesting contrast to the 2DEG systems in semiconductors where the Fermi degenerate region is easily reached.

BASIC IDEAS

This section presents some basic ideas as a framework for subsequent discussion. The zero field conductivity σ_0 and resistivity ρ_0 is given by the Drude model as

$$\sigma_0 = ne\mu_0 = ne^2\tau_0/m = 1/\rho_0 \tag{1}$$

Measurements of the zero field mobility[1] in the atom scattering regime are in good agreement with calculations by Saitoh[2]. For ripplon scattering, the independent electron theories[2,3,4] give a reasonable account of the data at low electron densities[4]. However, as the density increases electron correlations become important. First, the "complete control" region is reached, as explained by Buntar' et al.[5,6], where the electron system reaches internal equilibrium between ripplon scattering events. This modifies the transport relaxation time and the mobility as the electrons drift coherently. Secondly, a density dependent decrease in mobility is observed[4], which has been explained by Mehrotra[7] as due to extra scattering from viscoelastic modes in the correlated electron fluid.

In a field σ and ρ become tensors. The classical Drude model gives

$$\sigma_{xx} = \sigma_0/(1+\mu^2B^2) \quad ; \quad \sigma_{yx} = \mu B \sigma_{xx} \qquad (2)$$
$$\rho_{xx} = \rho_0 \quad ; \quad \rho_{xy} = B/ne$$

but ignores the quantization of electron orbits[8] into Landau levels, for $\omega_c\tau = \mu B > 1$, with energies

$$E_N = (N + 1/2)\,\hbar\omega_c \pm 1/2\hbar\omega_c \qquad (3)$$

including spin splitting. Figure 1(a) shows the quantizing field $B = 1/\mu$, based on the zero field mobility for independent electrons[2], and the quantum limit, $\hbar\omega_c/kT = 1$. Note that below 1K the holding field F_z has a strong effect on the mobility. The Landau levels will be broadened, due to collisions or other mechanisms, to a width W and a density of states $g(E)$ as shown schematically in Figure 1(b), compared to the constant density of states in zero field. In the extreme quantum limit, $\hbar\omega_c >> kT$, only the lowest, $N = 0$, spin-up level will be occupied and the electron wave function will be a flat disk of the form $\psi = A\exp(-r^2/2l^2)$ where the characteristic orbit radius is the magnetic length $l = (\hbar/eB)^{1/2}$ which is only 12.8 nm for $B = 4$T.

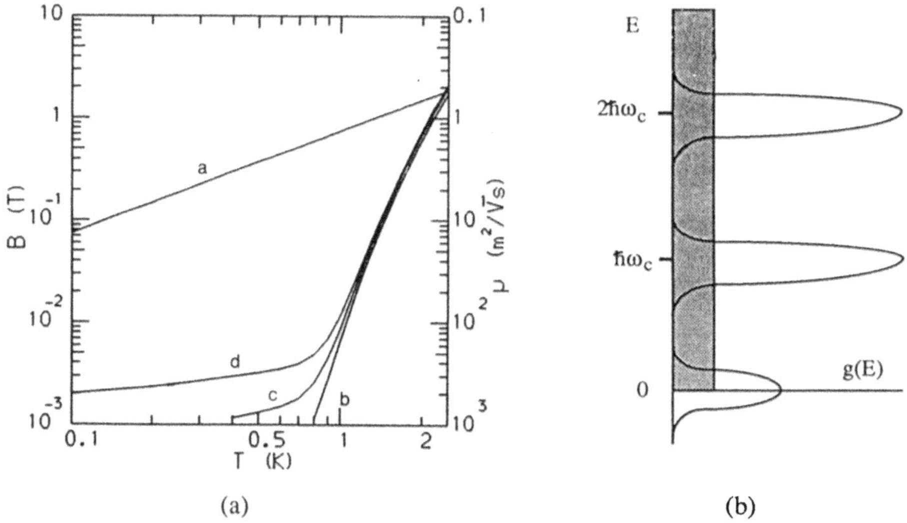

(a) (b)

Figure 1. Quantized energy levels in a magnetic field.
(a) Line **a**: the magnetic field for $\hbar\omega_c = kT$ for free electrons.
Lines **b, c, d**: the quantizing field $B = 1/\mu$ for independent electrons[2] for $F_z = 0, 2\times 10^4$ and 4×10^4 V m^{-1} respectively.
(b) Schematic diagram of the quantized Landau levels. The lowest level is spin polarized.

In the collisionless limit, $\omega_c\tau >> 1$, an electric field $\mathcal{E}$ in the x-y plane will cause the orbits to drift at right angles to both $\mathcal{E}$ and B with a velocity $v_d = \mathcal{E}/B$ giving $\sigma_{xx} = 0$ and $\sigma_{yx} = ne/B$. Collisions cause diffusion of the orbits down the potential gradient produced by $\mathcal{E}$. The Einstein relation for a non-degenerate gas then gives

$$\sigma_{xx} = ne^2L^2 / kT\tau \tag{4}$$

where L is the diffusion length and τ is a scattering time. Saitoh[9] and Dykman and Khazan[10] have calculated σ_{xx} directly, but eq.(4) is very useful conceptually. (For classical orbits $L^2 = mkT/e^2B^2$ and eq(2) is recovered). In the extreme quantum limit, $L^2 = l^2/2 = \hbar/2eB$ and hence

$$\sigma_{xx} = \frac{ne}{B}\frac{W}{2kT} \quad ; \quad \rho_{xx} = \frac{B}{ne}\frac{W}{2kT} \tag{5}$$

where $W = \hbar/\tau$ is the collision width of the Landau level. Ando et al.[8] considered the effects of diffusion on the Hall effect and found

$$\sigma_{yx} = \frac{ne}{B} - \Delta\sigma_{yx} \quad ; \quad \rho_{xy} = \frac{B}{ne} \tag{6}$$

where $\Delta\sigma_{yx} \sim \sigma_{xx}/\omega_c\tau$. Note that eq.(5) is quite different from the Drude model and that the specific field dependence of the magnetic length alone gives rise to a linear magnetoresistance. However, a more dominant effect is the field dependence of the scattering time due to the changed density of states as discussed below.

EXPERIMENTAL METHODS

Experimentally, the Sommer-Tanner[11] technique is used to measure the conductivity and resistivity of electrons on helium. An array of electrodes is placed parallel to the electron sheet, using a Corbino disk[1,12,13] or rectangular geometry[14,15] as shown in Figure 2. The complex AC impedance between 2 electrodes, due to the electrons, is then measured using a capacitance bridge or else one electrode is driven with a small AC voltage (angular frequency ω) and the voltages or currents at other electrodes are measured with a lock-in amplifier. Some circuit analysis is then performed to obtain the resistance of the 2-D electrons. The Corbino disk geometry, Figure 2(a), is the simplest and enables σ_{xx} to be found both in zero and finite fields[16], though the Hall voltages are shorted out and σ_{yx} cannot be obtained. Complications arise for the rectangular geometry Figure 2(c) in a field[14,15,17] due to the Hall effect. If the electron sheet is fully screened by the electron-electrode capacitance C' F/m then, in zero field, a heavily damped voltage wave will propagate on the 2-D transmission line formed by the electron and electrodes, with a 2-D 'skin depth' $\delta = (2/\omega C'\rho_0)^{1/2}$ (assuming $\omega\tau << 1$). In a field, the electron currents flow round the edge of the electron sheet with a skin depth $\delta_{\parallel} = (2/\omega C'\rho_{xx})^{1/2}$ while the skin depth perpendicular to the edge $\delta_{\perp} = (\rho_{xx}/\rho_{xy})\,\delta_{\parallel}$. For high electron mobilities, particularly below 1K, $\delta_{\parallel} >>$ the sample size, while $\delta_{\perp}$ can be very small, confining the currents to the edges. (At higher frequencies,

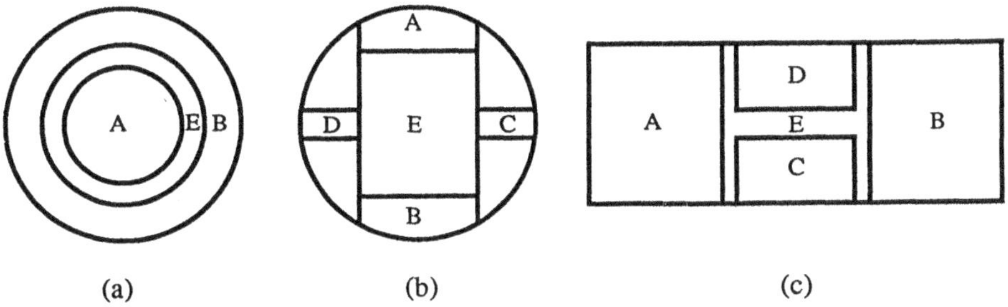

Figure 2. Some electrode geometries used for transport measurements of a 2DEG on helium.

$\omega\tau >> 1$, these edge modes become edge magnetoplasmons[18]). Simple, lumped impedance, analysis becomes increasingly difficult below 1K and data analysis in a magnetic field proceeds via measurements of the 2-D skin depth[19]. In principle, both ρ_{xx} and ρ_{xy} can be found, though spurious edge effects can restrict measurements of ρ_{xy} when $\rho_{xy} >> \rho_{xx}$.

VAPOUR ATOM SCATTERING

Above 1.2K electrons are scattered predominantly by ^4He atoms from the vapour. Because of the large mass ratio and the low thermal velocities of the atoms the scattering can be regarded as quasi-elastic. The mobility in zero field is inversely proportional to the vapour density and depends exponentially on the temperature, as shown in Figure 1. Experiment[1] and theory[2] are in good agreement. Iye[1] used the field dependence of σ_{xx}, as given by the Drude model, eq.(2), to obtain μ_0. However, the fields were small, $\leq 0.2T$. The magnetoconductivity σ_{xx} up to 4T has been measured using the Corbino geometry by van der Heijden et al.[16], as shown in Figure 3. Large deviations from the Drude model (dashed lines) occur at the higher fields. The same group[14] also measured ρ_{xx} and ρ_{xy}, using the electrode geometry of Figure 2(b), as shown in Figure 4. Strong positive magnetoresistance was found, particularly as the temperature decreased.

Ando, Fowler and Stern[8] have calculated the 2-D conductivity for elastic scattering from δ-function potentials in the extreme quantum limit. The basic idea is simple. The scattering is proportional to the local density of states. If the Landau level is collision broadened with a width $W = \hbar/\tau$ then the scattering rate in a field is enhanced over the zero field rate $1/\tau_0$ by a factor $\sim \hbar\omega_c/W$ as the electron states over an energy $\hbar\omega_c$ are concentrated into the width W.

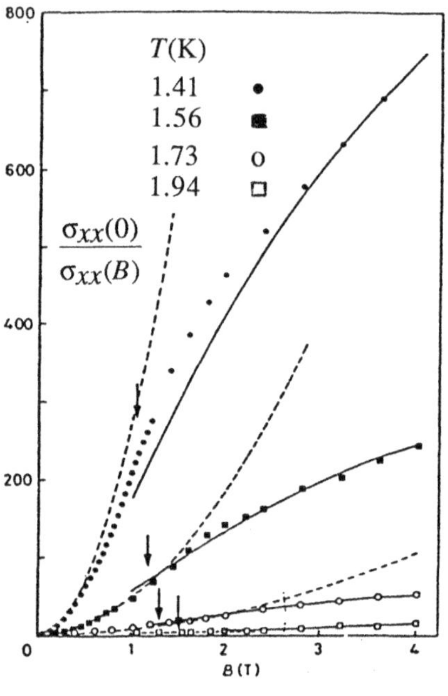

Figure 3

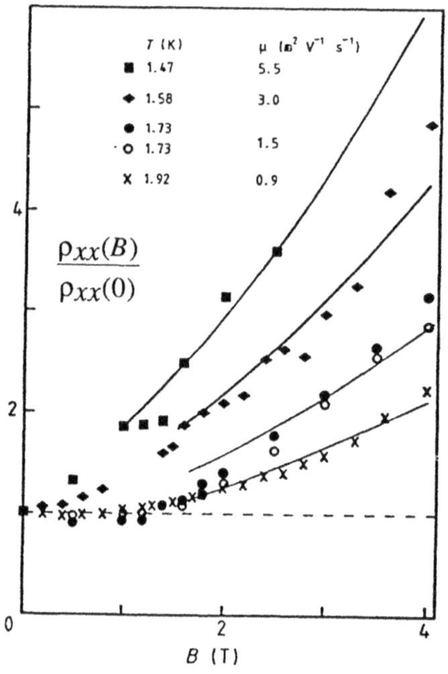

Figure 4

Figure 3. The magnetoconductivity σ_{xx} for electrons on helium, from van der Heijden et al.[16] The arrows show the field at which $\hbar\omega_c = kT$. The dashed lines show the Drude model, eq.(2) and the solid lines the quantum theory, eq.(8).
Figure 4. The magnetoresistivity ρ_{xx} for electrons on helium, from van der Heijden et al.[14] The dashed lines show the Drude model, eq.(2) and the solid lines the quantum theory, eq.(8). The mobility μ was used in the theoretical calculations.

For a semi-elliptic density of states this gives

$$\frac{1}{\tau} = \frac{1}{\tau_0} \frac{2}{\pi} \frac{\hbar\omega_c}{W} \quad ; \quad W = \frac{\hbar}{\tau_0} \sqrt{\frac{2}{\pi}\,\omega_c\tau_0} \tag{7}$$

as W is determined self-consistently. Van der Heijden et al[16] applied this theory to a non-degenerate gas and obtained

$$\sigma_{xx} = \frac{ne}{B} \frac{2}{\pi} \frac{\coth(\hbar\omega_c/kT)}{I_1(W/kT)} \left(\cosh\frac{W}{kT} - \frac{W}{kT}\sinh\frac{W}{kT}\right) \tag{8}$$

where I_1 is a modified Bessel function of order 1. For $\hbar\omega_c/kT >> 1$ and $W/kT << 1$ this reduces to

$$\sigma_{xx} = \frac{4}{3\pi} \frac{ne}{B} \frac{W}{kT} = 0.42 \frac{ne}{B} \frac{W}{kT} \tag{9}$$

which is close to the Einstein relation, eq(5).

Transforming to ρ_{xx}, assuming $\rho_{xy} = B/ne$, gives

$$\rho_{xx} = 0.34\, \rho_0 \frac{\hbar\omega_c}{kT} \sqrt{\omega_c\tau_0} \tag{10}$$

showing that $\rho_{xx} \propto B^{3/2} \rho_0^{1/2}$. The solid lines in Figures 3 and 4 show the calculations from eq.(8) which at high fields are close to the simple expressions in eqs.(9) and (10). Van der Heijden et al.[14] evaluated $\Delta\sigma_{yx}$ and found this to be small in the range of the experiments. Consequently the classical linear Hall effect is observed[14,15,20,21], as shown in Figure 5(a), for several electron densities[21]. As expected the Hall voltage is proportional to $1/n$, Figure 5(b). This linear region has been measured up to fields of over 3T[14,22], but at higher fields the AC measurements show large phase shifts due to the 2-D skin depth effects described above.

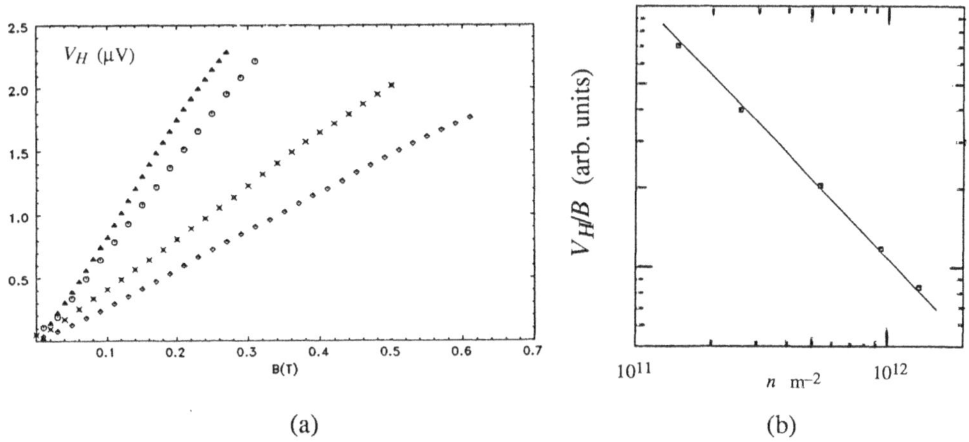

Figure 5. The Hall Effect in a 2DEG on helium.

(a) The linear Hall voltage V_H for electron densities n = 0.26 (Δ), 0.58 (o), 0.94 (×) and 1.34 (◊) × 10^{12} m^{-2}, from Stone[21].

(b) The gradient of the Hall voltage, dV_H/dB, versus the electron density n, on a log-log plot, from Stone[21].

RIPPLON SCATTERING: ELECTRON FLUID

Below 1K, the electron mobility is limited by ripplon scattering and can be very large. It can also be varied by a vertical electric holding field, as shown in Figure 1. Data for ρ_{xx} and ρ_{xy} in a field is somewhat limited, partly because of data analysis problems due to 2-D skin depth effects. Figure 6 shows ρ_{xx} at 0.5K for three values of the holding field[19], measured using the electrode geometry of Figure 2(c). Again, positive magnetoresistance is observed and the effect of the holding field is to increase the electron-ripplon coupling and ρ_{xx} at all fields. Saitoh[9] and Dykman[10,22] have analysed the extreme quantum limit theoretically using the interaction Hamiltonian given by Shikin and Monarka[3]. Because of the small cyclotron radius l the ripplons which produce scattering have wavevectors $q<1/l$ and energies $\hbar\omega_q<<kT$. The effective electron-ripplon scattering is thus also quasi-elastic in the quantum limit. Saitoh[9] calculated the scattering self-consistently, assuming that the Landau level width was due only to collision broadening, and has given analytic expressions for σ_{xx}, and hence ρ_{xx}.The theory gives similar magnetoresistance to the δ-function elastic scattering theory, eq(8), though the coupling has a complicated dependence on magnetic field. A comparison with the experiments is shown in Figure 6 where ρ_0 is in reasonable agreement with theory[2] but $\rho_{xx}(B)$ lies well below the theoretical predictions, particularly in the higher fields.

However, Dykman and Khazan[10] have shown that electron-electron correlations can have a large, though indirect, effect on the electron-ripplon scattering. The basic idea is that each electron sees a fluctuating electric field due to variations in the short-range order of neighbouring electrons. This leads to a fluctuating potential energy difference ~ $W_e = \hbar/\tau_e$ across an orbit radius l, where

$$\tau_e = 0.043\,(B/T)^{1/2}\,n^{-3/4} \tag{11}$$

and $$\sigma_{xx} = \frac{ne^2l^2\tau_e}{2\pi\hbar}\int_0^\infty l\,\frac{|V_q|^2}{\hbar\omega_q}\,q^2\exp(-q^2l^2/2)\,dq \tag{12}$$

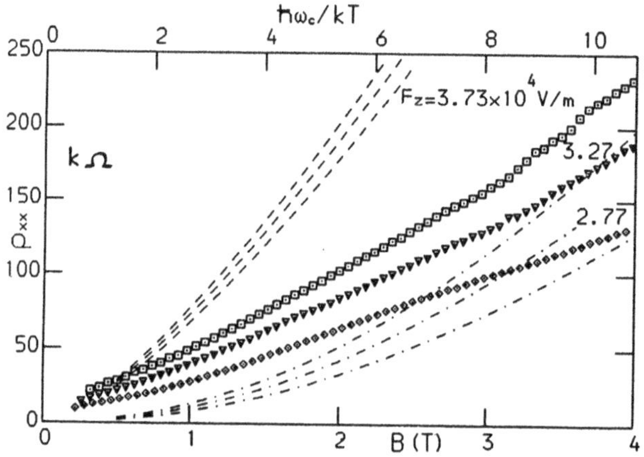

Figure 6. The magnetoresistivity ρ_{xx} for a 2DEG on helium at 0.5K for $F_z = 2.77$, 3.27 and 3.73×10^4 V m^{-1} respectively, from Stone et al.[19] The dot-dashed lines show the single-electron theory[9] while the dashed lines show the many-electron theory[10,22].

$$\text{This leads to} \quad \rho_{xx} = \frac{B}{ne} \frac{\tau_e (eF_z)^2 H}{4\hbar\alpha\sqrt{2\pi}} \tag{13}$$

where α is the surface tension and the factor $H(\sim 1)$ has to be evaluated numerically[22] though an analytic expression can be obtained under certain conditions[23]. Calculations of ρ_{xx} versus $\hbar\omega_c/kT$ using eq(13) are shown in Figure 6. The data lie between the two theories but tend towards the many-electron results at higher fields.

It is of interest here to consider the relation of this theory to the single particle theory and its validity. The many-electron fluctuations seem to be equivalent to a broadening of the Landau level to an effective width W_e ($>\hbar/\tau$, the scattering width) and hence a reduction in scattering due to the lower effective density of states. Following eq(7) an approximate, but useful, relation can be obtained between the scattering rate without many-electron effects, $W_1 = \hbar/\tau_1 \propto \hbar\omega_c/W_1 = \omega_c\tau_1$ as before, and with many electron effects $W_m = \hbar/\tau_m \propto \hbar\omega_c/W_e = \omega_c\tau_e$. Hence we expect $W_1 = A\sqrt{W_m W_e}$ where $A \sim 1$. This is borne out by a numerical comparison of the theories. Many-electron effects should be important if $W_e > W_1$ and hence $W_1 > W_m$ which is equivalent to comparing ρ_{xx} as calculated on the two theories, as shown on Figure 7 where $\rho_{xx}/\rho_{xy} = \rho_{xx}ne/B$ is plotted versus T for a field of 4T. From eq.(5) this is equivalent to plotting $W_1/2kT$ and $W_m/2kT$. In the ripplon scattering regime $W_1 > W_m$ and many electron effects are important. However, the theory given by Dykman also assumes $W_e/2kT \ll 1$ and this is also shown on Figure 7. At 4T this condition is satisfied but $W_e/2kT$ increases as the field decreases and this probably explains the cross-over between the two theories seen experimentally in Figure 6.

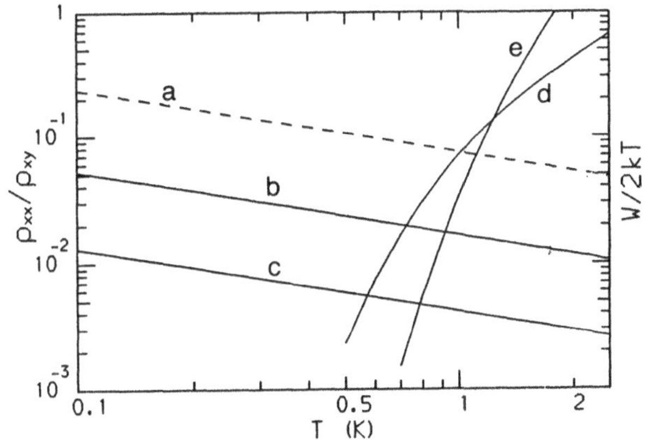

Figure 7. Calculations of ρ_{xx}/ρ_{xy} using the single electron and many electron theories, for $n = 2 \times 10^{12}$ m^{-2}, B =4T and $F_z = 0$.

Line **a**: The energy uncertainty due to many electron effects, $W_e/2kT$ versus T.
Lines **b**, **c**: ρ_{xx}/ρ_{xy} for ripplon scattering for (**b**) the single electron theory[9], $= W_1/2kT$ and (**c**) the many electron theory[10,22], $= W_m/2kT$. The *lower* value of ρ_{xx} should be observed experimentally.
Lines **d**, **e**: ρ_{xx}/ρ_{xy} for vapour atom scattering for (**d**) the single electron theory, $= W_1/2kT$ and (**e**) the many electron theory, $= W_m/2kT$. The *lower* value of ρ_{xx} should be observed experimentally.

Above 1K, eq.(14) can be used to estimate the many electron effects for ^{4}He atom scattering, as shown in Figure 7. In this region, $W_1 > W_e$ and hence many electron effects should not be significant, and ρ_{xx} should be given by eq.(8) as found experimentally[14].

Below 1K, ρ_{xy} should be given by the classical result B/ne to a good approximation. A linear Hall voltage is found experimentally[20,24], but skin depth effects become increasingly important as the mobility increases and the electron currents become confined to the edges in the Hall geometry.

RIPPLON SCATTERING: ELECTRON SOLID

As the temperature is reduced the 2DEG undergoes a phase transition into a classical 2-D electron solid[25], at $T_m = 0.216\, n^{1/2}$ K. The melting temperature is independent of magnetic field up to 7T[26]. In zero field the resistivity shows a loss peak[27] near the transition and decreases by a factor of ~ 2. Stan and Dahm[27] have suggested that this may be due to ripplon scattering from dislocations or from soft shear modes in the solid or the elusive hexatic phase. No detailed measurements have been reported in a quantizing field but experiments by Surewaard[28] and Stone[21] have found only small and subtle effects in the magnetoresistivity at the phase transition.

The many-electron theory[10] should also apply to the 2-D solid as short range order becomes long range order. Saitoh[29] has also calculated σ_{xx} for the solid, with qualitatively similar results. Consequently it seems probable that only second order effects will be seen on solidification or melting though complications may arise because of the rapid increase in the inertial effective mass at low frequencies due to the formation of electron dimples from the electron-ripplon coupling. This would be equivalent to the addition of an inductive term $j\omega m^*/ne^2$ to the resistivity tensor.

CYCLOTRON RESONANCE

The cyclotron resonance linewidth is closely related to the conductivity and has been measured by Edel'man[30] and Wilen and Gianneta[31] in the fluid and solid phases. A general feature of their results is that the linewidth W_{CR} is much greater than predicted on the independent electron theory[32]. The many-electron theory of Dykman[33] gives a good account of many aspects of these experiments. Wilen and Giannetta also found an extra density dependent linewidth which may have the same origin as the excess scattering in the zero field resistivity[4] attributed to viscoelastic modes by Mehrotra[7]. Wilen and Giannetta found no change in W_{CR} at the crystallisation temperature. This rather surprising result seems to confirm the importance of short range order in the fluid phase. The temperature dependence of W_{CR} changed at $\hbar\omega_c/kT > 1$at the onset of the quantum limit. Thus, in general terms, the cyclotron resonance and magnetotransport measurements are consistent.

SUMMARY AND FUTURE WORK

The system of 2-D electrons on helium is relatively simple in principle but exhibits a wide range of phenomena. It is an excellent 'model' system to investigate the properties of a 2DEG, particularly the effects of electron correlations. The plasma parameter $\Gamma = e^2(\pi n)^{1/2}/4\pi\varepsilon_0 kT$ can be varied over a wide range from classical gas to correlated fluid to electron solid at $\Gamma_m = 127 \pm 3$. The application of a magnetic field yields new information; the extrme quantum limit is particularly interesting as the conduction process is quite different from that in zero field, theoretical approximations can be made and it is accessible experimentally. Three specific areas of interest are the independent electron regime, the effects of electron correlations and the effects of collective modes on the electron transport in a magnetic field.

The Independent Electron Regime

Before looking at the effects of correlations it would be preferable to understand and confirm the properties of independent electrons, which are theoretically the easiest to calculate. Ideally this requires $\Gamma > 10$ and $\omega_0\tau < 1$ where ω_0 is the plasma frequency at a wavevector $(\pi n)^{1/2}$. These conditions require low density and high temperatures. In practice, as shown in Figure 7, many electron effects are likely to be small above 1.2K. The theory of Ando, Fowler and Stern[8] has been confirmed by van der Heijden et al.[14,16] for $\hbar\omega_c/kT < 4$. It would be interesting to extend this work into the region where $\hbar\omega_c > W > kT$ (all the experiment and discussion above relate to $\hbar\omega_c > kT > W$). This requires strong scattering and high fields but could be achieved at 2K and 14T. From eq.(8) in this region $\sigma_{xx} = (ne/B)(2W/\pi kT)^{1/2}$, lower than eq.(9) because the electrons are confined to the tail of the lowest Landau level in a region of low density of states. Since now $\sigma_{xx} > \sigma_{yx} \approx ne/B$, the Hall effect $\rho_{xy} = (B/ne)\pi kT/2W$ deviates strongly from the classical result, as pointed out by van der Heijden et al[14]. There may also be a contribution to the electron diffusion from the energy dependence of the density of states $dg(E)/dE$. Results in this region would be very sensitive to the shape of the Landau level, and also localisation effects. At lower temperatures, measurements at very low electron densities would be most interesting, to study the independent electron regime over as wide a range as possible.It is also noteworthy that weak localisation has been observed by van der Heijden et al.[34] in low fields, < 1 T. Although the helium atoms are moving, they remain stationary long enough to produce quantum interference between the different routes for elastic scattering from several atoms.

Electron correlations

For $n > 10^{12}$ m^{-2} the electron-electron correlation time is much less than the single electron scattering time, for almost all the experimental range of electrons on superfluid helium. In the gas and fluid phases the electrons will be in the "complete control" regime[5,6] and will, for instance, drift coherently with a displaced Maxwellian distribution. Buntar' et al.[5] have shown how this can modify the effective mobility in zero field due to electron-ripplon scattering. Spatial correlations will increase with increasing Γ giving short range order in the fluid and long range order in the solid.

Experimentally in a magnetic field the magneto-resistance is well below the independent electron results and, at high fields, approaches the theory by Dykman and Khazan[10], based on short range order. This gives a powerful experimental probe of many-electron effects in the fluid and a systematic study is required over a wide range of n, T, B and F_z. It is interesting that effects due to short-range order seem to be continuous through the fluid-solid phase transition, as demonstrated by the cyclotron resonance results.

Collective modes

In zero field, excess scattering (as compared to the independent electron theory) is observed as a function of density in the electron fluid and also in the solid with a loss peak at the transition. This can be attributed to ripplon scattering from collective electron modes such as viscoelastic waves in the fluid and phonon modes in the solid, or from dislocations or other defects. In a magnetic field Dykman[33] has shown that in the high density limit, for $\omega_0 > \omega_c$ (but still $\hbar\omega_c >> kT$) there should be a contribution to the scattering from the transverse phonon modes and hence measurements of σ_{xx} should be sensitive to long-range order. In the low density limit, $\omega_0 < \omega_c$, the short-range order effects should dominate in both the fluid and solid. Further experimental and theoretical work is required in this area.

REFERENCES

1.Y.Iye, J. Low Temp.Phys. 40:441 (1980).

2.M.Saitoh, J.Phys.Soc.Japan 42:201 (1977).
3.Yu.P.Monarkha, Fiz.Nizk.Temp.2:1232 (1976). [Sov.J. Low Temp.Phys.2:600 (1976)]; G.Beni and P.M.Platzman, Phys.Rev.Lett. 36:626 (1976), 36:1350(E) (1976).
4.R.Mehrotra, C.J.Guo, Y.Z.Ruan, D.B.Mast and A.J.Dahm, Phys.Rev. B29:5239 (1984).
5.V.A.Buntar',Yu.Z.Kovdyra,V.N.Grigor'ev, Yu.P.Monarkha and S.S.Sokolov, Fiz.Nizk.Temp.13:789 (1987). [Sov.J. Low Temp.Phys.13:451 (1987)].
6.V.A.Buntar',V.N.Grigoriev, O.I.Kirichek, Yu.Z.Kovdyra,Yu.P.Monarkha and S.S.Sokolov, J.Low Temp.Phys. 79:323 (1990).
7.R.Mehrotra, J.Low Temp.Phys. 79:311 (1990).
8.T.Ando, A.B.Fowler and F.Stern, Rev.Mod.Phys. 54:437 (1982).
9.M.Saitoh, Solid State Commun. 52:63 (1984).
10.M.I.Dykman and L.S.Khazan, Sov.Phys.JETP 50:747 (1979).
11.W.T.Sommer and D.J.Tanner, Phys.Rev.Lett. 27:1345 (1971).
12.R.Mehrotra and A.J.Dahm,J.Low Temp.Phys. 67:641 (1987).
13.L.Wilen and R.Giannetta, J.Low Temp.Phys. 72:353 (1988).
14.R.W. van der Heijden, H.M.Gijsman and F.M.Peeters, J.Phys.C:Solid State Phys. 21:L1165 (1988).
15.M.J.Lea, A.O.Stone and P.Fozooni, Europhys.Letts 7:641 (1988).
16.R.W. van der Heijden, M.C.M. van de Sanden, J.H.G.Surewaard, A.Th.A.M. de Waele, H.M.Gijsman and F.M.Peeters, Europhys.Letts 6:75 (1988).
17.M.J.Lea, in press.
18.D.B.Mast,A.J.Dahm and A.L.Fetter, Phys.Rev.Lett. 54:1706 (1985); D.C.Glattli, E.Y.Andrei, G.Deville, G.Poitrenaud and F.I.B.Williams, Phys.Rev.Lett. 54:1710 (1985).
19.A.O.Stone, P.Fozoni, M.J.Lea and M.Abdul-Gader, J.Phys.Condens.Matter 1:2743(1989).
20.O.I.Kirichek, B.A.Buntar', V.N.Grigor'ev and Yu.Z.Kovdyra, Kharkov Report 33-88 (1988); V.N.Grigor'ev, O.I.Kirichek, Yu.Z.Kovdyra and Yu.P.Momarkha, Fiz.Nizk.Temp.16:394 (1990).
21.A.O.Stone, Ph.D. Thesis, University of London (1990).
22.M.I.Dykman, J.Phys.C Solid State Phys. 15:7397(1982); private communication.
23.P.Fozooni, M.J.Lea, A.O.Stone and J.Frost, 19th. Intern.Conf.Low Temp.Phys. (1990).
24.M.J.Lea, J.Frost, A.O.Stone and P.Fozooni,19th. Intern.Conf.Low Temp.Phys. (1990).
25.A.Isihara, Solid State Physics, 42:271(1989).
26.A.O.Stone, M.J.Lea, P.Fozooni and J.Frost, J.Phys.CM 2:485(1990).
27.M.A.Stan and A.J.Dahm, Phys.Rev B40:8995(1990).
28.J.Surewaard, Unpublished Report, Technishe Universiteit, Eindhoven (1988).
29. M.Saitoh,J.Phys.Soc.Japan 56:706(1987).
30.S.Edel'man, Zh.Eksp.Teor.Fiz. 77:673(1979) [Sov.Phys.JETP 50:338(1979)].
31.L.Wilen and R.Giannetta,Phys.Rev.Letts. 60:231 (1988).
32.M.Saitoh, J.Phys.C 16:6983(1983).
33.M.I.Dykman, Sov.J. Low Temp.Phys. 6:268 (1980).
34.A.M.L.Janssen, R.W. van der Heijden, A.Th.A.M. de Waele and H.M.Gijsman and F.M.Peeters, Surf.Sci. 229:365(1990).

VORTEX CREATION IN SUPERFLUID HELIUM-4

P. V. E. McClintock[1] and R. M. Bowley[2]

[1] Physics Department
Lancaster University
Lancaster
LA1 4YB

[2] Physics Department
The University
Nottingham
NG7 2RD

INTRODUCTION

The hydrodynamics and the transport properties of bulk superfluid ^{4}He are dominated by quantized vortices[1–3]. Many aspects of their behaviour, for example the physics both of individual vortex lines[3] and of random tangles of vorticity (quantum turbulence)[3–6], can now be regarded as being reasonably well understood. A notable gap in our physical understanding, however, relates to the long-standing question of how the vortex lines appear in the liquid in the first place. The present paper attempts to address this question in the light, particularly, of recent experimental results.

The discovery[7] that all samples of He II seem invariably to contain at least a few vortex lines means that critical velocities measured in flow experiments will usually refer to the expansion of pre-existing vorticity rather than to the *creation* of vortices; there are possible exceptions in the work on flow through very tiny orifices[8–10] although, even there, the situation must be regarded as ambiguous[11,12]. A completely different approach to the vortex nucleation problem, which is not subject to this ambiguity, involves the use of negative ions in probes. The ions[13] are sufficiently small (radii $\sim$ 1 nm) that they are most unlikely to be affected by the remanent vorticity.

Although ions[13] have been used since the early 1960s to study diverse aspects of He II, it was not originally appreciated that measurements[3,14,15] of vortex nucleation rates could be markedly influenced[16] by even the tiny proportion ($\sim 10^{-7}$) of ^{3}He in naturally occurring well-gas helium. More recent investigations, to be described below, have employed samples

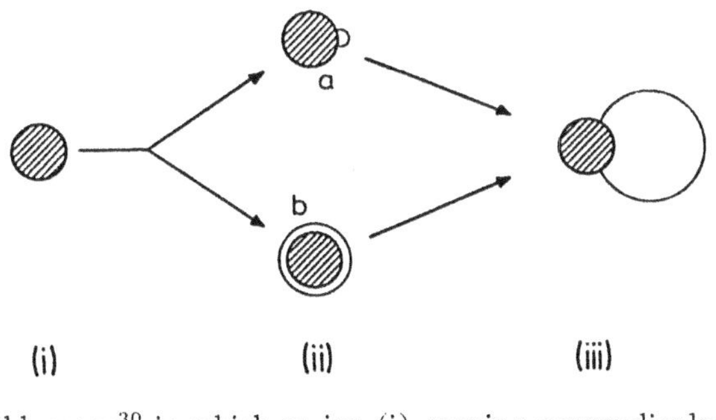

Fig. 1 Two possible ways[30] in which an ion (i), moving perpendicularly to the plane of the paper, might create (ii) a charged vortex ring (iii).

of isotopically purified[17] ^4He. Several theories of vortex creation have been proposed: an excellent discussion and overview is given in chapter 8 of the recent book[3] by R J Donnelly, to which reference may be made for further information. Here, we note the salient features of the theories relevant to the physical circumstances and experiments to be discussed below.

The basic experimental observation, which any theory must be able to account for, is that sketched in Fig 1. Thus a bare ion (i) makes a transition (ii) to a charged vortex ring[18] (iii), but it is not immediately obvious how the process occurs. Until recently, models of (ii) could be grouped into (a) those for which a vortex loop is envisaged to grow continuously[19,20] from the ion and (b) those for which there is a quantum transition[18,21,22] from a state with no vortex to a state with an encircling vortex ring. The best elements of the earlier theories were subsequently combined by Muirhead, Vinen and Donnelly (MVD)[23] to create a model in which the initial vortex is a loop of finite size: it is supposed to appear spontaneously as the result of a macroscopic quantum tunnelling event that enables the system to overcome an opposing energy barrier[24]. In what follows we will discuss the experimental results in the context, particularly, of the MVD model.

In common with earlier theories[18,21,22], MVD start from an assumption that the vortex creation process involves a quantum transition in which impulse must be locally conserved; they then calculate the corresponding change ΔE in the energy of the system as a function of the radius R of the created vortex loop or ring. The energetically most favourable configuration turns out to be for the formation of a loop (rather than an encircling ring) on the "equator" of the ion, with its axis orientated in a "north-south" direction, for an ion moving in a "north-south" direction. Their calculated ΔE as a function of the radius of the loop is plotted for different velocities in Fig 2. The shaded region is then regarded as the energy barrier which the system must overcome in order for a vortex to be created.

To gain an overview of the present state of the subject, it is essential to understand in a fairly detailed way the nature of the electric induction technique currently being used to measure vortex nucleation rates. This is described in the next section. The available experimental evidence is then summarised, analysed and discussed. Finally an attempt is made to draw conclusions and to identify the important questions that still remain to be resolved.

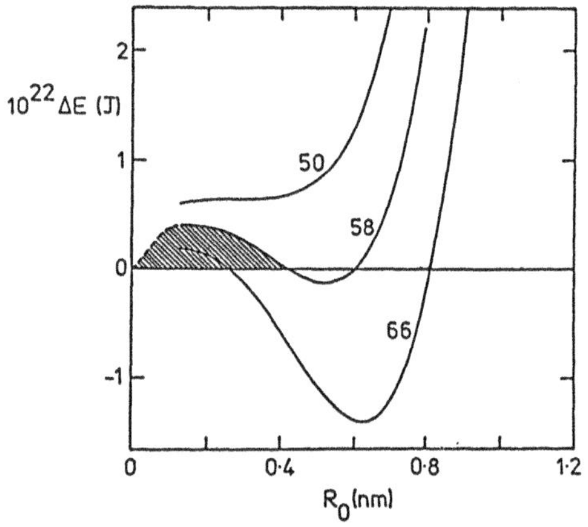

Fig. 2 Change in energy ΔE at constant impulse when a vortex loop of radius R_0 is formed in the equatorial plane by a negative ion for three different ionic velocities (ms^{-1}), after MVD[23]. For the middle curve, where the ion slightly exceeds the critical velocity, the energy barrier is represented by the hatched area.

THE ELECTRIC INDUCTION TECHNIQUE

All results to be presented and discussed below were acquired on the basis of an electric induction technique[16,25] in which thin disks of ions could be propagated within an electrode geometry as sketched in Fig 3(a). The measurements were made with relatively strong electric fields (typically $10^3 < E < 10^6$ Vm^{-1}) between G_3 and the collector, the average velocity of the ions being limited by roton emission[26–29] just above the Landau critical velocity which, in turn, required[26] that the experiments be carried out at elevated pressures $P >\sim 11$ bar.

In the absence of vortex nucleation events, the movement of a charge disk at a constant average velocity in the region between G_3 and the collector would induce a constant current in the latter (assuming that it is maintained at a constant potential). Those ions which "decay" to charged vortex rings[18], on the other hand, suffer an almost discontinuous decrease in velocity from $\sim$ 60 ms^{-1} to a few mms^{-1}. Their contribution to the collector current correspondingly vanishes; and, if vortex nucleation is a probabilistic process like radioactive decay, an exponentially decaying current is to be expected. This is exactly what is observed in the experiments, as shown in Fig 3(b) and (c).

In what follows, it is important to bear in mind that, although the roton-emission-limited average velocities of the bare ions are constant for a given pressure and electric field, the instantaneous velocity of each ion as a function of time is a saw-tooth[28] with an amplitude of a few ms^{-1}. We will tacitly assume, to start with, first that roton creation and vortex creation are quite separate processes, and do not interfere with each other and, secondly, that vortex creation can usefully be discussed in terms of the average velocity (and barrier height) at which it occurs. Each of these assumptions is obviously open to question.

SUMMARY OF THE EXPERIMENTAL EVIDENCE

The experiments to be discussed have extended over more than a decade. The more easily interpreted data are, paradoxically, those obtained quite recently at relatively low pressures: these will be presented first; followed by the earlier results obtained closer to the solidification pressure. Finally, a brief account is given of studies of the influence of ^{3}He impurities.

Low Pressure

Measurements of the vortex nucleation rate ν have been obtained for pressure P down to 12 bar. The dependence of ν on temperature T is shown in Fig 4. It exhibits a temperature independent regime below about 0.2 K and an exponential rise of ν with T

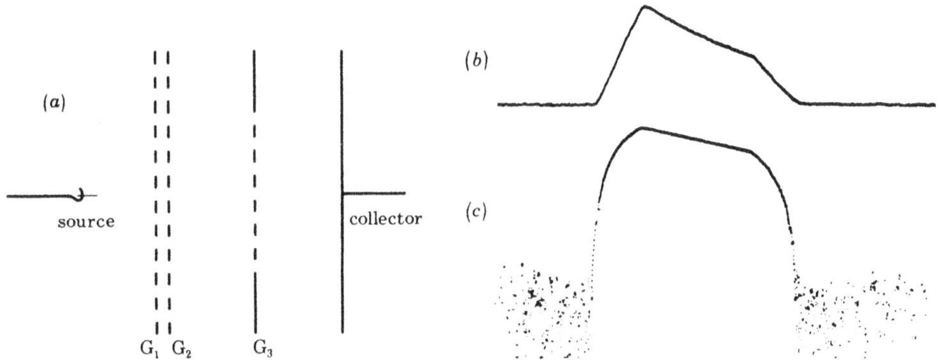

Fig. 3 (a) Electrode structure used[25] for measurements of the vortex nucleation rate by the electric induction technique. (b) Typical induction signal in the collector due to the approaching disk of charge. (c) Logarithm of the signal in (b), demonstrating the exponential shape of its falling top.

at higher temperatures. These results are quantitatively exactly what might be expected on the basis of the MVD model: quantum tunnelling through the barrier (Fig 2) at low temperatures, and thermal activation over it at higher temperatures. The data can be fitted accurately by a relation of the form

$$\nu = \nu_0(E, 0) + AT\exp(-E_b/k_B T) \qquad (1)$$

where $\nu_0(E,0)$ is the temperature-independent value of ν for $T < 0.2$ K, and A and E_b are constants. The fitted value of E_b/k_B is almost independent of E, and equal to 2.4 K. This is a very small energy. It implies that the initial vortex must itself be very small and, at most, a loop of radius a few Å. It is also in good agreement with the barrier heights predicted by MVD, and therefore lends strong support of their model.

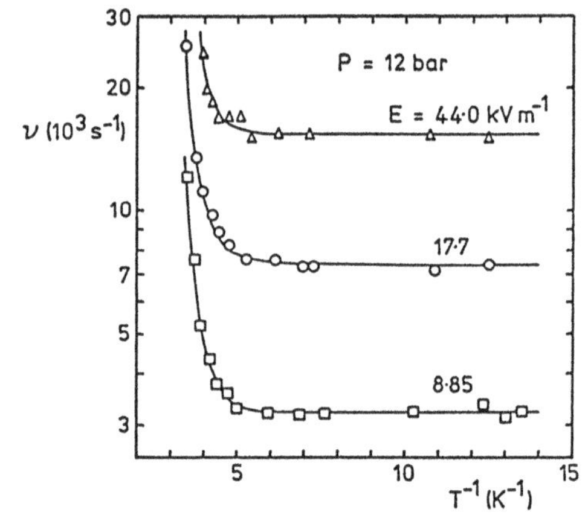

Fig. 4 Vortex nucleation rates[35] ν measured (points) for negative ions in isotopically pure ^{4}He, plotted against reciprocol temperature for three electric fields E. The curves represent fits of (1) to the data.

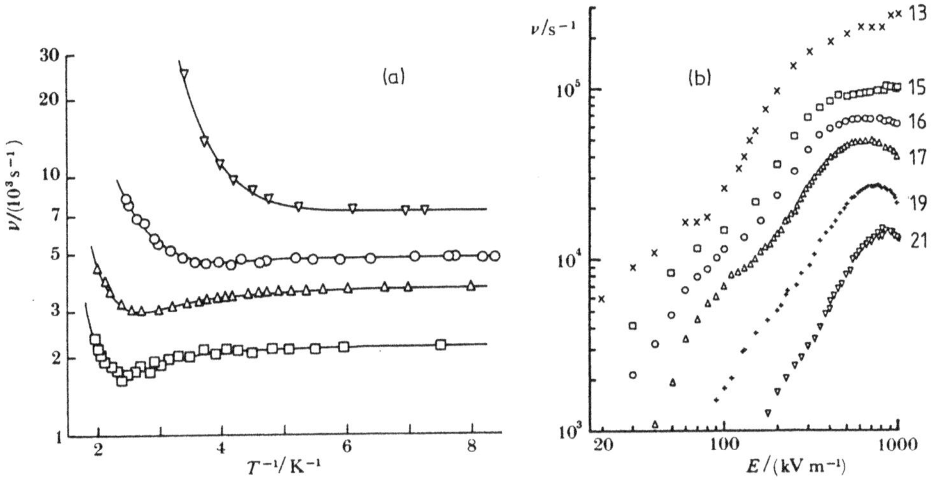

Fig. 5 (a) Vortex nucleation rates [35] ν measured (points) for a range of electric fields E and pressures P, plotted against reciprocol temperature: $\nabla P = 12$ bar, $E = 1.77 \times 10^4$ Vm^{-1}; O 15 bar, 3.10×10^4 Vm^{-1}; Δ17 bar, 6.20×10^4 Vm^{-1}; □ 19 bar, 9.70×10^4 Vm^{-1}. The curves represent fits of (5) to the data. (b) Vortex nucleation rates[35] measured in the temperature-independent regime below 140 mK for several pressures P (bar).

Higher Pressures

The appealing simplicity of the results of Fig 4 is lost as soon as P is raised above ~ 14 bar. A well-defined minimum then appears in plots of $\nu(T^{-1})$ as shown in Fig 5(a), and the minimum becomes more pronounced as the pressure increases. It is also found that (if one ignores the existence of the minimum, which was inadvertently the case with the earlier work above 0.3 K) the value of the constant E_b derived by fitting an equation of the form (1), but with a temperature independent prefactor, to the data becomes closely equal to the roton energy gap at $P \sim 25$ bar[25]. The latter observation is taken to demonstrate the existence of a roton-assisted vortex nucleation process at high pressures.

The variation of ν with E in the temperature-independent ($T < 0.2$ K) regime exhibits distinct structure at all pressures as shown in Fig 5(b); the structure becomes more marked at the lower pressures. It is also evident that $\nu(E)$ passes through a maximum at $\sim 1\mathrm{MVm}^{-1}$ and then decreases again at higher fields. The position of the maximum is dependent on pressure, falling at its minimum field of $E \sim 600\ \mathrm{kVm}^{-1}$ near 17 bar. Extension of the measurements to even larger electric fields (Fig 6) has shown[30] that ν decreases to zero above $\sim 2\mathrm{MVm}^{-1}$ and that some structure also occurs on the high field side of the maximum.

Effect of ^{3}He

As already remarked above, tiny proportions x_3 of ^{3}He exert a remarkably powerful influence on the nucleation process, and often do so in a markedly nonlinear manner as shown, for example, in Fig 7(a). It is found[31,32] that the data below 0.5 K at 23 bar are well described by a power law expansion

$$\nu(x_3) = \nu_4 + \nu' x_3 + \nu'' x_3^2 \tag{2}$$

where ν_4 represents the value in pure ^{4}He and where terms beyond the quadratic are found not to be needed. The physical significance of ν' is that it represents the ^{3}He contribution to the nucleation rate (per unit concentration) that would be measured in the limiting case of very low concentrations. Its measured variation with E is shown for three values of T in Fig 7(b).

DISCUSSION

The temperature dependence of ν at the lower pressures is readily interpreted in terms of interactions with elementary excitations. The 12 bar data (Fig 4) show evidence of thermal activation well below 0.4 K where the number of rotons in the liquid is negligible. The process in question must therefore involve phonons.

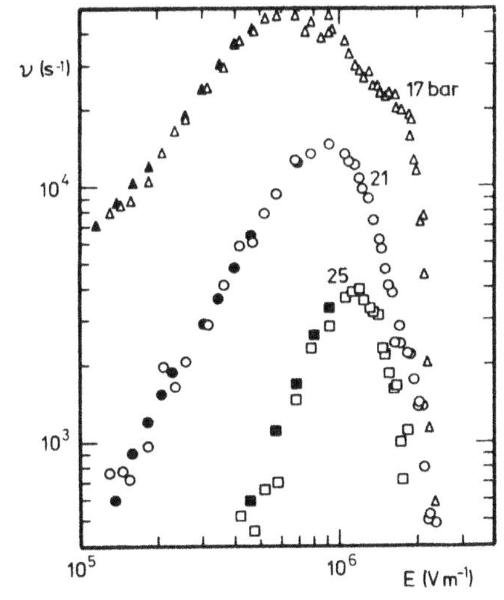

Fig. 6 Vortex nucleation rates ν measured[30] at 0.3 K as a function of electric field E, up to very strong fields.

A phonon of wave vector $\mathbf{k}$ can lead to the creation of a vortex loop on the ion if its energy exceeds $E_b + \hbar\mathbf{k}.\mathbf{v}$ where $\mathbf{v}$ is the velocity of the ion. The sound velocity $c \gg |\mathbf{v}|$ so, to order v/c, one can ignore the $\hbar\mathbf{k}.\mathbf{v}$ term. When averaged over angles, this procedure will give rise to an error or order v^2/c^2. We suppose that interactions with phonons in the requisite energy range cause vortices to form on the ions at a rate $cn_p\sigma$ where σ is the cross section for the process and the number density n_p of phonons of energy larger than E_b is

$$n_p = \frac{1}{2\pi^2}\int_{E_b/\hbar c}^{\infty} dk\, k^2 \exp(-\hbar ck/k_BT)$$

$$\approx \frac{k_BTE_b^2}{2\pi^2\hbar^3c^3}\exp(-E_b/k_BT) \tag{3}$$

If we assume that σ is a constant, independent of phonon energy, we obtain the AT prefactor of (1), with

$$A = \sigma k_B E_b^2/2\pi^2\hbar^3c^2 \tag{4}$$

By fitting (1) to the 12 bar data, and using a sound velocity[33] of $c = 312\ \mathrm{ms}^{-1}$, we find $\sigma = 2\text{Å}^2$. This result represents a plausible cross section for a process involving a vortex loop of the size envisaged.

At higher pressures, the temperature dependence of ν is more complicated (Fig. 5a) and so (1) is clearly inapplicable. To proceed further we need to account for the minimum in $\nu(T)$, whose existence appears to imply that the tunnelling term $\nu_0(E,T)$ decreases with increasing temperature. We can suggest two possible explanations. First, the increased phonon scattering with rising temperature could reduce the average velocity $\bar{v}$ of the ion, thereby increasing the effective energy barrier (Fig. 2) and so reducing the nucleation rate. This idea is easily tested[34,35] by plotting $\nu_0(E,T)$ against $\bar{v}$ in the low temperature limit, using values of $\bar{v}$ measured directly[25] in the experiments, and then using the plot to determine the change in $\bar{v}$ that would be necessary to account for the observed minimum in $\nu(T)$ at constant E. The result, for the minimum plotted with an expanded ordinate scale in Fig. 8(a), is shown by the small filled circles of Fig. 8(b). Comparison with the *actual* values of $\bar{v}$ measured from the same signals as $\nu(T)$, and shown by the open circles, demonstrates immediately that the model is wrong: the reduction in $\bar{v}$ caused by phonon scattering is far too small to account for the minimum in $\nu(T)$.

There is a second, quite different, mechanism by which phonons could affect the tunnelling rate. Phonon scattering from the vortex loop *as it tunnels* could reduce[36] the

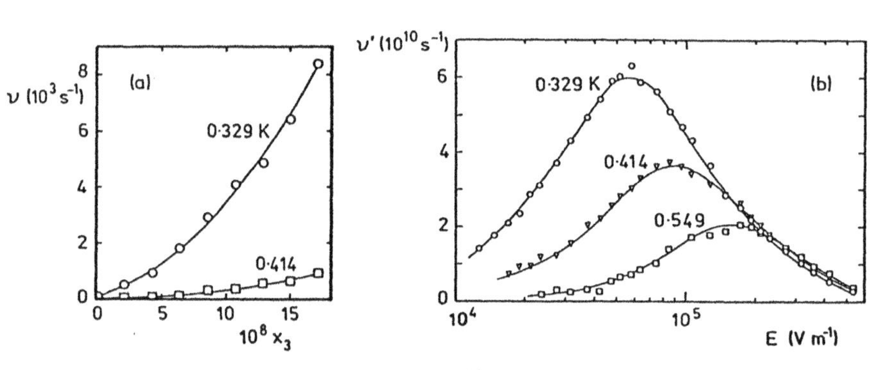

Fig. 7 (a) Vortex nucleation rates ν measured[31] as a function of ^{3}He concentration x_3 for two temperatures with an electric field $E = 1.27 \times 10^4\ \mathrm{Vm}^{-1}$. (b) Experimental values (points) of the linear coefficient ν' in (2) plotted as a function of electric field for three temperatures; the full curves represent fits of a theoretical model[32] to the data.

tunnelling rate by a factor of $\exp(-\phi\eta D^2)$ where η is a dissipation coefficient, D is the distance the vortex must tunnel through the barrier and ϕ is a numerical factor. The effect of dissipation on the tunnelling rate was considered by MVD, but in relation to roton *emission*, which is almost temperature independent. Here we are considering the influence of phonon *scattering*, which will obviously depend on the thermal phonon density and will therefore be strongly temperature dependent. For simplicity, we suppose that η varies as T^3 with the phonon number density, in which case the nucleation rate should be describable by

$$\nu(T) = \nu_0 \exp(-BT^3) + AT \exp(-E_b/k_B T) \tag{5}$$

The quantities E_b and B may be expected to be almost independent of applied electric field, but to vary smoothly with pressure.

It is found[35] that (5) provides a good fit to the $\nu(T)$ data at the higher pressures and so, in the interests of consistency, we have fitted it to all the data, over the whole range of pressures: full details of the data and the procedures used in fitting the theory are given in Ref. 35. The full curves through the experimental points in Fig. 5(a) show some typical fits of (5) to the data. The average value of E_b obtained by fitting the theory for fields below 10^5 Vm^{-1} is plotted as a function of pressure in Fig. 9(a). The existence of the shallow minimum near 15 bar depends on the values of E_b at 12 and 13 bar, both of which are believed to be reliable; the minimum therefore appears to be a real, albeit somewhat puzzling, observation. It was not predicted by MVD. The most important feature, however, is the rapid increase of E_b with P above 15 bar, leading towards a value equal to the roton energy gap Δ measured in the earlier experiments[25] near the solidification pressure.

The other important quantity derived by fitting (5) to the data is the cross section σ for the phonon-assisted vortex nucleation process. The fitted values are plotted as a function of pressure in Fig. 9(b). It is evident that σ increases slowly with increasing E, but by far the most striking feature is the very rapid decrease in σ with increasing P. The largest value of $\sim 15\text{Å}^2$, measured at 12 bar, is considerably less than the geometrical cross section of the ion ($\sim 500\text{Å}^2$).

The field dependence of ν in the temperature-independent regime below 140 mK is of considerable interest in that it enables estimates to be made of the critical velocities for vortex nucleation. According to the MVD model, there will be a critical velocity v_ν above which vortex loops can be produced; whether or not the system then creates vortices at an

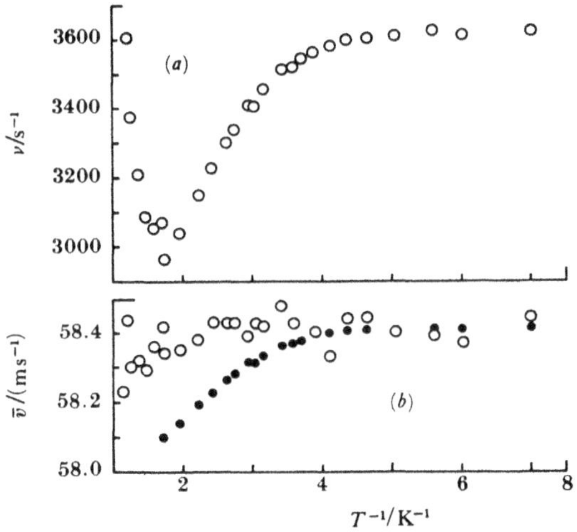

Fig. 8 (a) Vortex nucleation rate ν at 17 bar as a function of reciprocal temperature [34,35]. (b) The corresponding drift velocities (open circles) $\bar{v}$ measured[34,35] from the same induction signals. The small filled circles represent the values that would be needed to generate the dip in $\nu(T^{-1})$ shown in (a).

observable rate will depend on how rapidly it can tunnel through the potential barrier. As the velocity v of the ion increases beyond v_v, the creation rate is expected by MVD to rise very rapidly. The simplest way to treat the system theoretically is to suppose[25] that there is an effective critical velocity v_{v1} for vortex creation and that, for $v > v_{v1}$, vortices are created at the constant rate R_1. Thus the instantaneous production rate $R(v)$ will vary with velocity as

$$R(v) = R_1 \theta(v - v_{v1}) \tag{6}$$

By use of the calculated velocity distribution function[37] for the ions it can readily be shown[38] that (6) leads to

$$\nu_0 \sim \exp[-\alpha m^*(v_{v1} - v')^3/3eE] \tag{7}$$

for relatively weak electric fields. Here v' is the critical velocity for roton pair creation[28] and m^* is the effective mass of the ion. It is clear from (7) that ν_0 will increase with E and that it will decrease with P because of the corresponding increase [28] in $(v_{v1} - v')$: behaviour that is very much in accord with experimental observation. Values of v_{v1} can readily be deduced[25] by fitting (7) to the data, and turn out to be $\sim$ 60 ms^{-1} and to increase very weakly with increasing pressure; one would expect them to be slightly larger than the true critical velocities v_v.

The structure in $\nu(E)$ in stronger electric fields suggests that a second nucleation mechanism becomes operative at a higher critical velocity; the simple form (6) is clearly inadequate to describe such processes, but can readily be extended to

$$R(v) = R_1 \theta(v - v_{v1}) + R_2 \theta(v - v_{v2}) \tag{8}$$

This relation enables the $\nu(E)$ data to be fitted quite closely, an example being shown in Fig. 10. For fields approaching 1 MV m^{-1}, however, the data always fall below the fitted curve: this is a precursor of the giant fall in $\nu(E)$ seen in Fig. 6. (The effect can, to some extent, be accommodated[25] by the introduction of a third critical velocity v_{v3} at which $R(v)$ falls to zero, but there is really no physical justification for this procedure). Values of R_1, R_2, v_{v1} and v_{v2} obtained by fitting the data in this way are given in Table 1.

The complicated and potent influence of ^{3}He on the nucleation process (Fig. 7) has been analysed in detail elsewhere[31,32,39]. It can readily be accounted for in terms of the tunnelling model if it is assumed that ^{3}He atoms are attracted[40] towards the ion and bound

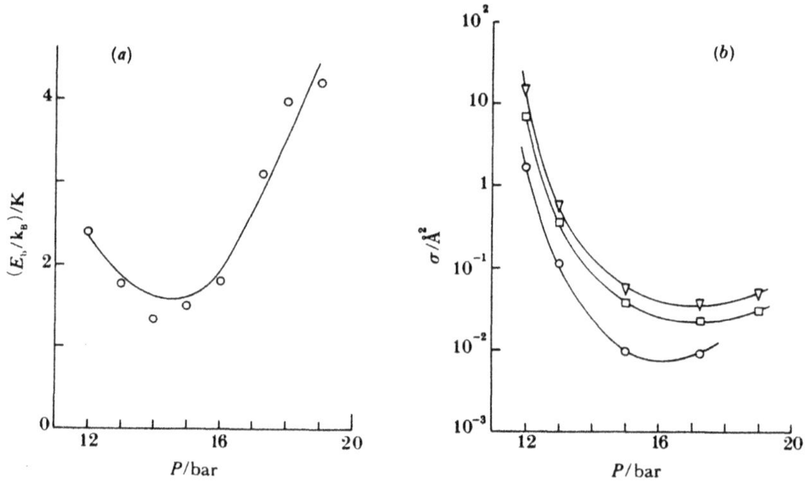

Fig. 9 (a) The energy barrier height E_b deduced[35] by fitting (5) to the experimental $\nu(E,T)$ data, plotted as a function of pressure P. (b) Cross section for the phonon assisted vortex nucleation process, deduced[35] by fitting (5) to the data, plotted against P, for electric fields E : O 0.880×10^3 Vm^{-1}; □ 4.40×10^4 Vm^{-1}; ∇8.00×10^4 Vm^{-1}. In both sections of the figure, the curves are guides to the eye.

in a Shikin level[41] on its surface. The size of the effective MVD energy barrier is modified because of the possibility that the ^{3}He atom can then transfer to the vortex loop as it is created. The highly complicated variation of ν with E and T (Fig. 7b) can be interpreted in relation to the residence time of the ^{3}He atom on the ion before being re-emitted with a pair of rotons; the nonlinear dependence on ^{3}He concentration (Fig. 7a) is related to the growing probability of two or more ^{3}He atoms being resident simultaneously. The average ^{3}He binding energy on the ion derived by fitting this model to the data is (-2.5 ± 0.2)K, only slightly larger than the limited low temperature binding energy (-2.22 ± 0.03 K) of ^{3}He on a plane He II surface under the saturated vapor pressure[42]. This very reasonable result lends further support to the MVD model.

CONCLUSION

The conclusions that may currently be drawn from ion-based research on vortex creation in He II can be summarized as follows:-

1. Vortex nucleation in He II is impeded by an energy barrier, vindicating an early suggestion by Vinen[24].
2. Below ~ 0.3 K, the dominant nucleation mechanism involves a form of macroscopic quantum tunnelling through the barrier; at higher temperatures, a thermally activated mechanism becomes prominent, kicking the system over the barrier.
3. The thermally activated mechanism involves a phonon-assisted vortex nucleation process, important at low pressures, and a roton- assisted process that becomes dominant at high pressures.
4. The experimental barrier heights are consistent with the MVD model; their small magnitudes confirm the MVD conclusion that the initial vortex is a tiny loop rather than a large encircling ring.
5. The minima in some of the $\nu(T)$ curves are not due to a reduction in $\bar{v}$ caused by phonon scattering. It is suggested, instead, that they are attributable to phonon damping of the tunnelling mechanism.
6. The structure in some of the $\nu(E)$ curves is interpreted in terms of the onset of a second nucleation mechanism. It is proposed that this corresponds to tunnelling into the first excited state of the nascent vortex loop.
7. The drastic effect of ^{3}He on the nucleation process can be understood partly in terms of a reduction in the critical velocity due to the energy released when the ^{3}He transfers from a Shikin level on the outside of the ion to the core of the nascent vortex; the increase of ~ 10^3 in the rate constant due to the presence of a trapped ^{3}He atom may be attributed to a modification of the ion/He II interface.

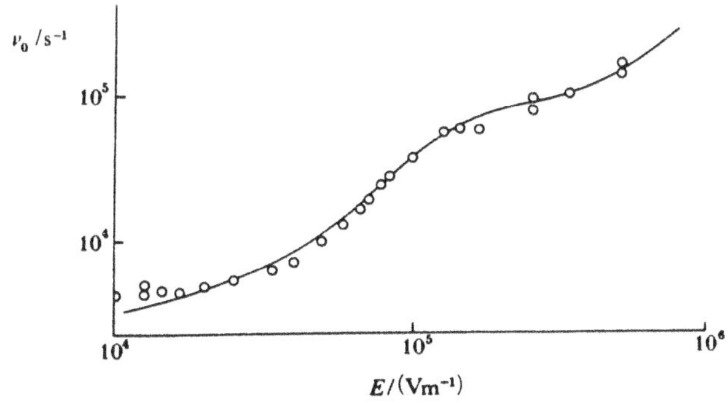

Fig. 10 Comparison of measured (points) and fitted (curve) nucleation rates[35] for $T <$ 140 mK and P = 13 bar, as a function of electric field E. The curve represents a fit of (7) and (8) to the data.

Table 1. Parameter values deduced by fitting (8) to the experimental vortex nucleation data of Fig. 5(b).

P/bar	$R_1/(10^3 s^{-1})$	$R_2/(10^3 s^{-1})$	$v_{v1}/(ms^{-1})$	$v_{v2}/(ms^{-1})$
13	29	760	61	82
15	34	280	58	76
16	28	190	58	75
17	23	220	58	76
19	20	77	60	76
21	10	48	60	76

Several problems still remain to be resolved. In particular, we would like to know answers to the following:

1. Why does the barrier height E_b vary with pressure (Fig. 9a) in such a dramatic way? There is nothing in the theory of MVD, or in the physics of negative ions, that would have led one to anticipate the rapid rise in E_b above 15 bar. (We note that an equally striking pressure variation of the roton pair-emission matrix element[29], over a similar pressure range, also remains unexplained).

2. What causes the spontaneous (tunnelling) nucleation rate R_1, corresponding to the first critical velocity, to be $\sim 2 \times 10^4$ s^{-1}, a value that is a great deal smaller than typical frequencies in He II and certainly much less than the cyclotron frequency, the natural frequency scale for the problem? Is the reduction caused by dissipation, perhaps the dissipation associated with roton emission?

3. Why does the nucleation rate drop again in high electric fields (Fig. 6)? The drop affects both the spontaneous and thermally activated nucleation rates. It could be associated either with an increase in dissipation with the velocity of the ion, or with a tendency of the ion to evade capture by the nascent vortex (or, equivalently, with an instability of the initial ion-ring complex in strong electric fields).

4. What causes the rapid pressure variation of the cross section for the phonon assisted nucleation process (Fig. 9b)? It is this effect that causes the disappearance of the minimum in $\nu(T)$ as P is reduced (Fig. 5a).

Finally, we would remark that, although the work discussed here relates to the special (very well characterised) case of vortex formation by a moving negative ion, we anticipate that the main features of the results should be of much wider relevance. In particular, we expect that our central conclusion, that the process involves macroscopic quantum tunnelling through a small energy barrier, should be quite generally applicable to vortex creation in He II.

ACKNOWLEDGEMENTS

It is a pleasure to acknowledge the assistance and encouragement of our collaborators including, particularly, T Ellis, P C Hendry, N S Lawson, F E Moss, G G Nancolas, P C E Stamp and C D H Williams. The work was supported by the Science and Engineering Research Council (UK).

REFERENCES

1. H London, "Low Temperatures: A Report on the International Conference in Cambridge", Physical Society, London (1947); and p151 of F London, "Superfluids", Wiley, New York (1954).
2. W F Vinen, The detection of a single quantum of circulation in liquid helium II, *Proc. Roy. Soc. (Lond.) A* **260**:218 (1961).

3. R J Donnelly, "Quantized vortices in He II", CUP, Cambridge (1990) and references therein.
4. K W Schwarz, Turbulence in superfluid helium: steady homogeneous counterflow, *Phys. Rev. B* **18**:245 (1978); Three-dimensional vortex dynamics in superfluid ^{4}He, *ibid* **31**:5782 (1985) and **38**:2 (1988).
5. J T Tough, Superfluid turbulence, *in* "Progress in Low Temperature Physics", vol. VIII, D F Brewer, ed., pp 133-219, North-Holland, Amsterdam (1982).
6. J T Tough, The effect of coloured quadratic noise on a turbulent transition in liquid He II, *in* "Noise in Nonlinear Dynamical Systems", vol. III, F Moss and P V E McClintock, ed., pp 1-21, CUP, Cambridge (1989).
7. D D Awschalom and K W Schwarz, Observation of a remanent vortex line density in superfluid helium, *Phys. Rev. Lett.* **52**:49 (1984).
8. G B Hess, Critical velocities in superfluid helium flow through 10μm diameter pinholes, *Phys. Rev. Lett.* **27**:977 (1971).
9. B P Beeken and W Zimmermann, Variation of the critical order parameter phase difference with temperature from 0.4 K to 1.9 K in the flow of superfluid ^{4}He through a tiny orifice, *Phys. Rev. B* **35**:1630 (1987).
10. O Avenel and E Varoquaux, Observation of singly quantized dissipation events obeying the Josephson frequency relation in the critical flow of superfluid ^{4}He through an aperture, *Phys. Rev. Lett.* **55**:2704 (1985); Varoquaux, M W Meisel and O Avenel, Onset of the critical velocity regime in superfluid ^{4}He at low temperature, *Phys. Rev. Lett.* **57**:2291 (1986).
11. K W Schwarz, Dissipative flow of liquid ^{4}He in the limit of absolute zero, *Phys. Rev. Lett.* **57**:1448 (1986); and Temperature dependence of discrete dissipative events in superfluid ^{4} He, *Phys. Rev. Lett.* **59**:1167 (1987).
12. O Avenel and E Varoquaux, Avenel and Varoquaux reply, *Phys. Rev. Lett.* **59**: 1168 (1987).
13. A L Fetter, Ions and vortices in liquid helium, *in* "The Physics of Liquid and Solid Helium", K H Bennemann and J B Ketterson, ed., pp 207-305, Wiley, New York (1976).
14. R Zoll and K W Schwarz, New features of the vortex-ring transition, *Phys. Rev. Lett.* **31**:1440 (1973).
15. R Zoll, Study of the vortex-ring transition in superfluid ^{4}He, *Phys. Rev. B* **14**: 2913 (1976).
16. R M Bowley, P V E McClintock, F E Moss and P C E Stamp, Vortex nucleation in isotopically pure superfluid ^{4}He, *Phys. Rev. Lett.* **44**:161 (1980).
17. P C Hendry and P V E McClintock, Continuous flow apparatus for preparing isotopically pure ^{4}He, *Cryogenics* **27**:131 (1987).
18. G W Rayfield and F Reif, Quantized vortex rings in superfluid helium, *Phys. Rev.* **136**:A1194 (1964).
19. G W Rayfield, Evidence for a peeling model of vortex ring formation by ions in liquid helium, *Phys. Rev. Lett.* **19**:1371 (1967).
20. R J Donnelly and P H Roberts, Stochastic theory for the nucleation of quantized vortices in superfluid helium, *Phil. Trans. Roy. Soc. (Lond.) A* **271**:41 (1970).
21. K W Schwarz and P S Jang, Creation of quantized vortex rings by charge carriers in superfluid helium, *Phys. Rev. A* **8**:3199 (1973).
22. R M Bowley, Nucleation of vortex rings by negative ions in liquid helium at low temperatures, *J. Phys. C* **17**:595 (1984).
23. C M Muirhead, W F Vinen and R J Donnelly, The nucleation of vorticity by ions in superfluid ^{4}He I. Basic theory, *Phil. Trans. Roy. Soc. (Lond.) A* **311**:433 (1984).
24. W F Vinen, "Liquid Helium: Proceedings of the International School of Physics Enrico Fermi", course XXI, pp 336-355, Acadmic Press, New York (1963).
25. R M Bowley, P V E McClintock, F E Moss, G G Nancolas and P C E Stamp, The breakdown of superfluidity in liquid ^{4}He III: nucleation of quantized vor- *Phil. Trans. Roy. Soc. (Lond.) A* **307**:201 (1982).
26. G W Rayfield, Roton emission from negative ions in helium II, *Phys. Rev. Lett.* **16**:934 (1966).

27. A Phillips and P V E McClintock, Breaking the roton barrier: an experimental study of motion faster than the Landau critical velocity for roton creation in He II, *Phys. Rev. Lett.* **33**:1468 (1974).
28. D R Allum, P V E McClintock, A Phillips and R M Bowley, The breakdown of superfluidity in liquid ^{4}He: an experimental test of Landau's theory, *Phil. Trans. Roy. Soc. (Lond.) A* **284**:179 (1977).
29. T Ellis and P V E McClintock, The breakdown of superfluidity in liquid ^{4}He V: measurement of the Landau critical velocity for roton creation, *Phil. Trans. Roy. Soc. (Lond.)* A **315**:259 (1985).
30. G G Nancolas and P V E McClintock, Quenching of the ion/ vortex-ring transition in He II by intense electric fields, *Phys. Rev. Lett.* **48**:1190 (1982).
31. R M Bowley, G G Nancolas and P V E McClintock, Vortex nucleation in ultradilute superfluid ^{3}He/^{4}He solutions, *Phys. Rev. Lett.* **52**:659 (1984).
32. G G Nancolas, R M Bowley and P V E McClintock, The breakdown of superfluidity in liquid ^{4}He IV: influence of ^{3}He isotopic impurities on the nucleation of quantized vortex rings, *Phil. Trans. Roy. Soc. (Lond.) A* **313**:537 (1985).
33. J S Brooks and R J Donnelly, The calculated thermodynamic properties of superfluid helium-4, *J. Phys. Chem. Ref. Data* **6**:51 (1977).
34. P C Hendry, N S Lawson, C D H Williams, P V E McClintock and R M Bowley, Inhibition of vortex nucleation by phonons in He II, *in* "Elementary Excitations in Quantum Fluids", K Ohbayashi and M Watabe, ed., Springer-Verlag, Berlin (1989).
35. P C Hendry, N S Lawson, P V E McClintock, C D H Williams and R M Bowley, The breakdown of superfluidity in liquid ^{4}He VI: Macroscopic quantum tunnelling by vortices in isotopically pure He II, *Phil. Trans. Roy. Soc. (Lond.) A* (in press).
36. A O Caldeira and A J Leggett, Influence of dissipation on quantum tunnelling in macroscopic systems, *Phys. Rev. Lett.* **46**:211 (1981).
37. R M Bowley and F W Sheard, Motion of negative ions at supercritical drift velocities in liquid ^{4}He at low temperatures, *Phys. Rev. B* **16**:244 (1977).
38. R M Bowley, Vortex nucleation by negative ions in liquid ^{4}He, *J. Phys. C* **9**:L367 (1976).
39. C M Muirhead, W F Vinen and R J Donnelly, The nucleation of vorticity by ions in ^{4}He II. Theory of the effect of dissolved ^{3}He, *Proc. Roy. Soc. (Lond.) A* **402**:225 (1985).
40. A J Dahm, Evidence for condensation of ^{3}He atoms on the surface of bubbles in liquid ^{4}He, *Phys. Rev.* **180**:259 (1969).
41. V B Shikin, Interaction between impurity excitation and negative ions in liquid helium *Sov. Phys. JETP* **37**:718 (1973).
42. D O Edwards and W F Saam, The free surface of liquid helium, *in* "Progress in Low Temperature Physics", vol VII A, D F Brewer, ed., pp 283-369, North-Holland, Amsterdam (1978).

IONS TRAPPED BELOW THE SURFACE OF SUPERFLUID HELIUM

W F Vinen

School of Physics and Space Research
University of Birmingham
Birmingham B15 2TT, UK

INTRODUCTION

The aim of this paper is to review recent progress in the study of pools of ions trapped below the surface of superfluid ^4He. A brief description will be given of the way in which the ions are trapped, and it will be explained why the study of the behaviour of the pools is of interest. Two recent experiments will be described in more detail: a study of the ripplon-limited mobility of the ions at very low temperatures; and the detection and study of the Wigner crystallization of the pools. Both these experiments are interesting in their own right, but they relate also to the behaviour of capillary waves or ripplons on the surface of the helium and are therefore relevant to one particular type of excitation associated with the helium.

ION POOLS BELOW THE SURFACE OF SUPERFLUID HELIUM

When a charged particle is placed below the surface of a liquid it is repelled from the surface by an image charge, the magnitude of which depends on the difference of dielectric constant between the liquid and the vapour above it. Addition of a uniform external electric field tending to push the ion towards the surface leads then to a net potential acting on the charged particle that has a minimum just below the surface; with an external field of 5kV m^{-1} the minimum is at a depth of about 44 nm. The well-known positive or negative 'ions' that can be created in liquid helium can then be trapped in the resulting potential well, trapping in the horizontal direction being achieved with a suitable fringing component of the external electric field.

At temperatures below about 1K pools of ions with densities of typically 10^{11} m^{-1} can be created without difficulty[1,2], and they form two-dimensional fluids or solids in which the particles interact through Coulomb forces. The pools are not strictly two-dimensional in the sense that the ions have some vertical thermal motion, unless the temperature is below about 3 mK, but the extent of the vertical motion is much less than the spacing between the ions.

Experimental study of these pools is of interest for two reasons. First, they offer an opportunity to study in detail the properties of a two-dimensional Coulomb fluid or solid under very clean and well-defined conditions; studies of the mechanism of melting and of viscosity and visco-elastic effects in the fluid phase are examples of work that is potentially very valuable. Secondly, the ions can be used as probes in the study of the superfluid helium itself. Ions have of course played an important role in studies of this latter type for many years, but the experiments have almost always involved the free motion of ions across a volume of liquid under the influence of steady accelerating

Excitations in Two-Dimensional and Three-Dimensional Quantum Fluids
Edited by A.G.F. Wyatt and H.J. Lauter, Plenum Press, New York, 1991

electric fields; in the surface trap the ions can be moved in a more closely controlled way and they can also be used to study the surface of the helium.

Care should be taken to distinguish between experiments on ions trapped below the surface of the helium, with which this paper is concerned, and those on electrons trapped above the surface. Many interesting experiments have been reported on this latter system (see other contributions at this Workshop and the review by Dahm and Vinen[3]), but it differs in a number of important ways from the ion system (for example, the ripplon interactions are very different and have very different effects); studies of the two systems therefore complement one another in a useful way.

Among studies of the trapped ion system that have been completed is the first detailed investigation over a wide temperature range of the interaction between the ions and the phonons in the helium[4]. The experimental results are in very good agreement with an updated version of the theory of Baym *et al*[5]. Interaction between the ions and the helium determines the ionic effective masses; detailed studies of these effective masses have also been completed, and the discovery of a curious temperature dependence of the mass of the positive ion[1,6] has provoked a vigorous theoretical debate[7,8,9,10]. More recently, an experimental study of the interaction between the ions and the surface ripplons has been completed[4], and this will be described briefly in the next section.

RIPPLON-LIMITED IONIC MOBILITY

At temperatures above about 100mk the mobility of the trapped ions, which can be measured by observing the damping of plasma resonances in the pools, is limited by roton or phonon scattering. Below about 100 mK, however, the observed mobility changes its behaviour and becomes strongly dependent on the depth at which the ions are trapped. At these low temperatures the density of bulk excitations has become very small, and the mobility is determined by scattering of the quantized capillary waves, or ripplons, on the surface of the liquid.

The detailed results of a study of this ripplon-limited mobility will be reported at the 19th International Conference on Low Temperatures Physics in Brighton, LT19 (see also reference 4). In summary, the results are as follows. Within the experimental error the inverse mobility is proportional to the temperature, down to the lowest temperature studied, which was about 10 mK. The mobility increases strongly with increasing depth (z_0) at which the ions are trapped, the mobility being closely proportional to $z_0{}^{7/2}$ over a range of depths from 20 nm to 100 nm. It has still to be clearly established whether or not the mobility is dependent of the ion number density (n_0). According to Hannahs & Williams[11] there is quite a strong dependence, but the more recent work of Barenghi *et al*[4] suggests that the dependence is quite weak if it exists at all, except perhaps when n_0 is less than about 10^{-11} m^{-2}. The absolute mobilities measured by Hannahs & Williams are significantly lower than those measured by Barenghi *et al*, which suggests that the former values were affected by some spurious broadening of the plasma resonant linewidths, which may have been density dependent. Further experiments are required.

The theory of the ripplon-limited mobility is not yet in a satisfactory state. The ion-helium system can be modelled as a charged sphere submerged below the surface of a classical ideal incompressible inviscid dielectric fluid, and the scattering process responsible for the ripplon-limited mobility can be modelled by the classical scattering of capillary waves from an array of the submerged spheres (Vinen, to be published). Interaction between the capillary wave and a sphere arises from electrostatic forces between the surface of the fluid and the charge on the sphere and from hydrodynamic forces between the moving fluid and the sphere. Calculations carried out so far relate to the mobility of an isolated ion, although the existence of the other ions must be taken into account in order to remove a divergence in the scattering cross-section at very low ripplon wavenumbers. Most of the cross-section arises from resonant scattering at the frequency of vertical oscillation of the ion in the electric potential well that traps it. The result of these calculations is a mobility that is inversely proportional to the temperature and proportional to $z_0^{7/2}$, as is observed, but is about one order of magnitude too large. It is possible that this discrepancy arises because the scattering is really from a strongly interacting fluid or solid of ions, rather than from an isolated ion. However, the way in which the mobility depends on the trapping depth does seem to be characteristic of the resonant scattering that we have described, and one would not expect this type of scattering to be affected by the ion-ion correlations; furthermore, as will be mentioned in the next section, the mobility seems to be unaffected by a crystallization of the pool, so that, if ion-ion correlations are indeed important, it can be only those of short range that are relevant, since it is only these that could be relatively unaffected by the crystallization. Further work is required, both to establish experimentally whether the mobility is really density dependent and to refine the theory.

CRYSTALLIZATION OF THE ION POOLS

Wigner crystallization of the ion pools is expected to take place when a parameter Γ $(=n_0^{1/2}e^2/4\pi^{1/2}\epsilon_0 k_B T)$, which is a measure of the ratio of the electrostatic potential energy of interaction of the ions to their thermal kinetic energy, exceeds a critical value of about 130 (Morf, reference 12). A crystallization of this type is easily seen in the electron system above the helium surface, where it leads *via* electron-ripplon interactions to large changes in both the plasma dispersion relation and the electron mobility[13,14]. However, these effects are either absent or unobservably small in the ion system, and detection of the crystallization of the ion system, which should occur in typical ion pools at about 100 mK, has proved to be quite difficult.

The crystallization has, however, been detected in a recent experiment. The details have been described elsewhere[15], and only the principle of the method will be mentioned here. A radio frequency electric field is applied to the ions, which causes

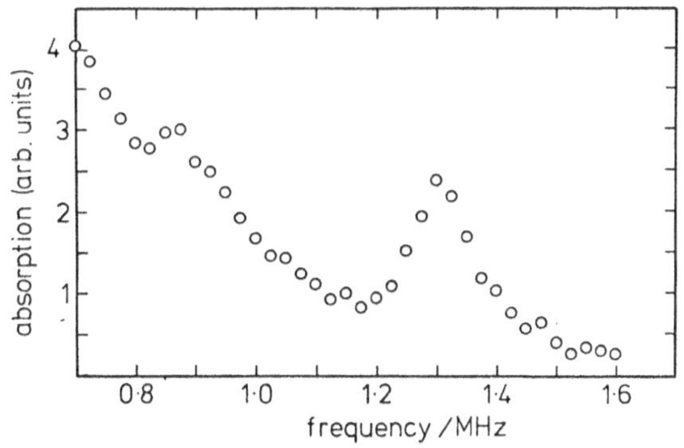

Fig. 1 Absorption of rf power by the ion system, plotted as a function of frequency, in the capillary wave interference experiment.

each ion to oscillate in position and hence to radiate capillary waves. The extent to which power is absorbed from the rf field depends on the extent to which the capillary waves radiated from different ions interfere constructively, and this is turn depends on the arrangement of the ions and on the wavelength of the capillary waves relative to the ionic spacing, this wavelength being determined through the ripplon dispersion relation by the frequency at which the ions are being driven. The problem is essentially similar to Bragg diffraction. If the ions form a crystal and if the wavenumber of the capillary waves is equal to a reciprocal lattice vector of the crystal there will be constructive interference and strong absorption of the rf power. "Bragg peaks" were indeed seen at the appropriate frequencies as the frequency of the driving field was varied, as shown in figure 1, and the peaks disappeared at the predicted melting temperature. The technique is a potentially valuable not only for detecting crystallization but also, if it can be sufficiently refined, for obtaining information about the form of the dynamic structure factor, in both the crystal and the fluid phases of the ion system[16]. Detailed studies with this new technique have yet to be carried out, but some interesting observations have already been made; for example, the apparent suppression of crystallization if too large a spatial density gradient is imposed on the fluid phase[15]. It is also interesting that crystallization is not accompanied by any observable anomaly in the mobility of the ions.

THE ATTENUATION OF CAPILLARY WAVES ON THE SURFACE OF SUPERFLUID HELIUM (RIPPLON DAMPING)

The capillary wave interference technique described in the preceding section depends for its success on a sufficiently small damping of the capillary waves; the distance over which a capillary wave is damped out must be large compared with the inter-ion spacing. Indeed the damping of the capillary waves must make a contribution to the width of the "Bragg peaks" shown in figure 1, so that the width of these peaks must contain some information about the magnitude of this damping.

The theory of ripplon damping has been considered by Saam[17] and by Gould & Wong[18]. Saam's calculations relate to the lifetime of an isolated ripplon (wavenumber q) at T = 0, this lifetime being determined by a process in which the ripplon decays into two ripplons. Saam's calculations, which yield the lifetime

$$\tau = 6.16\rho/\hbar q^5 ,$$

will presumably apply to ripplons at a finite temperature as long as the ripplon frequency satisfies the condition $\hbar\omega >> k_BT$ (ρ is the density of the helium). The calculations of Gould & Wong relate to the lifetime of low-frequency ($\hbar\omega << k_BT$) ripplons at a finite temperature, when the absorption and emission of thermal ripplons become important, and they yield the result

$$\tau = 2.66(\rho/\hbar)(\hbar\sigma^{1/2}/\rho^{1/2}k_BT)^{10/3} ,$$

where σ is the surface tension of the helium. The two lifetimes can be joined to one another smoothly, as shown in figure 2. The validity of the calculations has recently been checked by Williams & Jones (private communication; Williams[19]), who find that there is an error of a factor of two in the lifetime calculated by Saam, and that care is needed in carrying through the Gould-Wong calculation to avoid an infinite damping from the emission or absorption of very long wavelength ripplons. This divergence can be eliminated if the capillary-wave dispersion relation includes the effect of gravity, which is important at low wavenumbers:

$$\omega^2 = (\sigma/\rho)q^3 + gq .$$

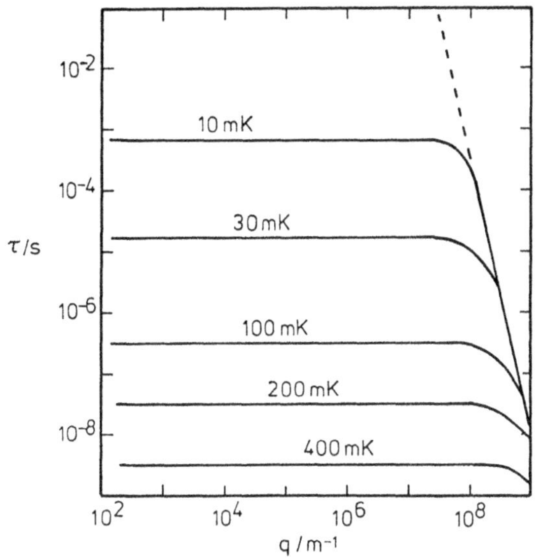

Fig. 2 The lifetime of ripplons according to the calculations of Saam[17] and of Gould and Wong[18] plotted against ripplon wavenumber for various temperatures. The predictions of the two calculations have been joined by smooth curves.

Measurements of the width of the "Bragg peaks" observed in the capillary wave interference experiment show that at a temperature of 130 mK the effective lifetime of ripplons of wavenumber 5×10^6 m^{-1} cannot be less than approximately 1.5×10^{-5} s. This can be compared with the lifetime of 1.31×10^{-7} s given by the calculations of Gould & Wong, so that there is a serious discrepancy. It is possible that the Gould–Wong lifetime relates predominantly to low angle scattering of a ripplon, which would be relatively ineffective in broadening the Bragg peak. Alternatively, the Gould–Wong calculation is probably valid only in the case where the mean free path of the thermal ripplons is large compared with the wavelength of the ripplon of which the lifetime is being calculated (cf the electron–phonon interaction, which requires a different treatment according as the value of ql is less than or greater than unity, where q is the phonon wavevector and l is the electron mean free path); this condition maybe violated in the capillary wave interference experiments, and an appropriately modified calculation, in which one must think of the attenuation of a low-frequency capillary wave as being caused by the viscosity of the thermal ripplons, has not yet been carried out.

In the experiments that have been carried out so far the width of the "Bragg peaks" has been found to be independent of temperature. It is possible that this width is related to the damping of the capillary waves, but it is possible also that it is related to the quality of the Wigner crystal; for example, to a non-uniformity in the lattice spacing caused by a non-uniformity in the holding electric field. It should be emphasized that there can be no strict long-range order in two dimensions, the order being destroyed by long wave-length phonons (see, e.g., reference 20). The Bragg peaks must therefore be broadened on this account, but by an extent that is much less than that observed.

SUMMARY

A brief account has been given of recent experimental and theoretical work on the behaviour of two-dimensional arrays of ions trapped below the surface of superfluid helium, particular attention having been drawn to the effects of ripplons on this system and on the information about the ripplons that can be obtained from the experiments.

REFERENCES

1. M.L. Ott-Rowland, V. Kotsubo, J. Theobald and G.A. Williams, Phys.Rev.Letters, 49:1708 (1982).
2. C.F. Barenghi, C.J. Mellor, C.M. Muirhead and W.F. Vinen, J.Phys.C: Solid State Phys., 19:1135 (1986).
3. A.J. Dahm and W.F. Vinen, Physics Today, 40:43 (1987).
4. C.F. Barenghi, C.J. Mellor, J. Meredith, C.M. Muirhead, P.K.H. Sommerfeld and W.F. Vinen, Phil.Trans. in the press.
5. G. Baym, C.J. Barrera and C.J. Pethick, Phys.Rev.Letters, 22:20 (1969).
6. C.J. Mellor, C.M. Muirhead, J. Traverse & W.F. Vinen, J.Phys.C: Solid State Phys., 21:325 (1988).
7. V. Elser and P.M. Platzman, Phys.Rev.Letters, 61:177 (1988).
8. V. Elser and P.M. Platzman, Phys.Rev.Letters, 64:103 (1990).
9. Tao Pang, Phys.Rev.Letters, 61:849 (1988).
10. Tao Pang,Phys.Rev.Letters, 64:104 (1990).
11. A. Hannahs and G.A. Williams, Phys.Rev., in the press.
12. R.H. Morf, Phys.Rev.Letters, 43:931 (1979).
13. C.C. Grimes and G. Adams, Phys.Rev.Letters, 42:795 (1979).
14. R. Mehrotra, B.M. Guenin and A.J. Dahm, Phys.Rev.Letters, 48:641 (1982).
15. C.J. Mellow and W.F. Vinen, Surface Science, 229:368 (1990).
16. W.F. Vinen and C.J. Mellor, Proceedings of the EPS Solid State Physics Conference, Lisbon (1990).
17. W.F. Saam, Phys.Rev., A8:1918 (1973).
18. H. Gould and V.K. Wong, Phys.Rev., B18:2124 (1978).
19. A.J.E. Williams, M.Sc. Thesis, University of Birmingham (1990).
20. N.D. Mermin, Phys.Rev., 176:250 (1968).

DISCUSSION ON IONS/ELECTRONS

Chairmen: A. J. Dahm, W. F. Vinen

Putterman to Vinen : Have you considered the Kelvin drag-induced-mass of a particle moving near the surface? This excess mass should vary as $\upsilon^2/\nu_{ripplon}(q = Z^{-1}) \sim T$ by dimensional analysis. Here υ is the velocity of the ion and Z is the distance from the surface.

Vinen : Experimentally one finds, for the positive ion only, a contribution to the effective mass that is proportional to T and therefore probably to the square of the thermal velocity. However, the magnitude of this contribution is independent of trapping depth and cannot therefore be due to interaction with the surface. It has been suggested by Platzmann that the effect is associated with the fact that the positive ion may not be spherical.

Maris to Vinen : What was the temperature at which you measured the lifetime of the ripplons?

Vinen : We measured a lower limit on the lifetime at 100mK.

Maris : Presumably from that measurement you could estimate the average lifetime of thermal ripplons because presumably the discrepancy between experiment and theory was $\omega\tau$.

Vinen : Yes, in principle. I have in fact tried to calculate the damping of a long-wavelength capillary wave by the thermal ripplons by treating these ripplons as a viscous gas. This naive approach did not work and one needs to do a better calculation based on a solution of the Boltzmann equation for the ripplons.

Maris : Presumably you must go beyond a viscous approximation because you have a second sound type system.

Vinen : I do not think that this is important.

Leiderer to Vinen : Would it also be possible to investigate the liquid-solid interface at low temperatures with ions?

Vinen questioned whether ions could be trapped at the solid-liquid interface, but it was agreed that they could be trapped. There is an energy barrier opposing the movement of the ion into the solid so that an external electric field pressing the electron against the solid would lead to trapping.

Vinen : I have not given serious thought to this problem.

Stringari to Vinen : Have you proposed to study the effect of ^{3}He on the surface?

Vinen : Yes, we intend to try.

Dahm : In reference to the mode observed by Vinen - one can see from the theoretical work on Vilk and Monarkha that it is a coupled phonon-ripplon mode. They expand the lattice potential in the harmonic approximation for deviations of the electrons from their lattice sites for motion both in the plane and perpendicular to the plane with the surface moving with the electrons. This gives phonons, ripplons and coupling between them. One can observe that these can be driven by either a parallel or perpendicular excitation field.

McClintock : The study of ions trapped at the surface can only be done at STP whereas our work can only be done under pressure, so that the two methods will complement each other.

Vinen : We believe that we have seen vortex nucleation at high ion velocities, because we find that the ions suddenly disappear from the system when the ions are driven at a critical velocity near the velocity thought to be the vortex nucleation velocity. But we must understand the non-linear effects associated with the high amplitude plasma oscillations before we can study vortex nucleation by this method.

Donnelly : One might trap the ions in a Penning trap to study vortex nucleation. This could be used at any pressure.

Lauter to McClintock : Can you give us a picture of the transition from rotons to vortex rings.

McClintock : We treat the rotons and vortex rings as completely separate excitations.

Varoquaux to McClintock : When the vortex creation velocity is close to the Landau critical velocity, wouldn't you observe a smooth transition from one to the other.

McClintock : You cannot easily separate the two processes because they are occurring simultaneously.

Balibar to Putterman : Is the evaporation of electrons from the surface driven by mechanical vibrations?

Platzman : Goodkind added vibrations and there was no change in the evaporation rate. There should not be any nice trend of the evaporation rate with electric field if it was affected by vibrations.

Godfrin to Platzman : Can you measure the temperature of the electron gas? Is is sufficiently coupled to the ripplons to ensure that it is at the same temperature? The fact that the time constant is short does not necessarily mean that ΔT is small, if a continuous heat leak couples to electrons, as is seen in semiconductors in the same temperature range.

Platzman : The estimates of equilibrium times are short. The results are independent of the time between cooling and taking the measurements.

Lea : There is a paper in which a rapid change in evaporation rate is reported at $\Gamma = 47$. The authors attribute this to a change in electron correlations.

Platzman : There is an instability which we have studied. They may have observed this. It has nothing to do with electron correlations, but rather with a chain reaction in the electron escape rate. The instability boundary does not depend only on Γ, but on the pressing field and temperature.

Dahm to Platzman : How low in temperature must one go to observe tunnelling of electrons from the surface?

Platzman : At a temperature given by $T < 0.06 n_e^{3/4}$ where n_e is the electron density in units of 10^8 cm^{-2}. It is about 10 - 30 mK for a density of 10^9 cm^{-2}.

Ferrell to Platzman : Is it possible that the discrepancy between the single-electron low temperature evaporation rate and experiment could be attributed to a very small temperature-dependent prefactor? One is reminded of the very small prefactor that is involved in the problem of the resistance of a one-dimensional superconductor.

Platzman : No, because the experimentally observed temperature independent evaporation rate is of the same order of magnitude as the predicted temperature dependent single-electron rate.

Halley : One needs an enhancement, not a depression, of the evaporation rate.

Platzman : Could I make a comment? Strictly speaking the Wigner transition is a quantum transition at $T = 0$. What has been observed for electrons on helium surface is classical freezing.

Vinen : Has anyone seen any evidence for the hexatic phase in the electron system?

Dahm : The French-Rutgers group has measured the shear modulus and specific heat through the melting transition Their results suggest a Kosterlitz-Thouless transition. The hexatic phase should accompany this transition. The only other experimental results on properties in the region where the hexatic phase should exist are the variation of the effective mass and a resistivity anomaly by our group. Theoretical work is needed to relate our results to an identification of the phase of the electrons.

Lauter to Leiderer : Can you say something about the nature of the surface during layering? What is the last layer? Is it a dilute layer becoming dense? Is it a liquid-gas coexistence? Should the surface tension oscillate with layering?

Leiderer : The analysis of the electron mobility has only been done in the submonolayer regime where it is interpreted as due to density fluctuations in the two-dimensional gas. Oscillations show that some remnant remains at highest coverages. The wiggles we observed at the ninth layer do not mean a sharp layer. It just means that the structure is reminiscent of the solid wall of the hydrogen substrate. Also, we see no evidence from our mobility measurements for the formation of a solid helium layer on hydrogen.

Platzman to Leiderer : Have you ever tried to grow thin (~ 100Å) films of hydrogen and then annealing?

Leiderer : we are trying to do this. Our first results suggest that electrons punch through thin quench annealed films.

Platzman : Perhaps you should try different substrates.

Glyde : What is the state of the calculations of the phase diagram for two-dimensional electrons?

Platzman : Ceperley calculated the one high density point at $T = 0$. We know the phase diagram in the Kosterlitz-Thouless classical regime. Some Japanese have attempted to calculate the first deviations from the Kosterlitz-Thouless regime. The rest of the phase diagram is more or less an artist's conception.

FINAL DISCUSSION SESSION

Chairmen: H. J. Lauter, A. F. G. Wyatt

In this session delegates were asked to pose question and direct comments to problems that need to be tackled in the future.

The discussion was started by the chairman (AW) asking how important is it to know the size of the superfluid fraction, n_o. Is it a key question or is it a non question as none of the physical properties of liquid ^{4}He seem to depend on it? Glyde thought it was a fundamental question and coupled to whether we have Bose-Einstein condensation. Vinen gave a clear problem for the future; to find the necessary and sufficient conditions for superfluidity. Platzman did not see why a condensate was necessary as ^{4}He is not an ideal Bose gas. (Earlier in the conference most people had agreed that a condensate was necessary for a macroscopic wave function, but the size of n_o did not matter as long as there was macroscopic occupation of the zero momentum state).

Donnelly wondered if anyone could devise a method of measuring n_o in clear and direct method as the present methods were very indirect. The method should be free of the final-states effect. Platzman thought that this could be achieved by doing neutron scattering on a thin ^{4}He film. The idea is that the scattered atom would go out into the vacuum without scattering with any other ^{4}He atoms, so getting rid of the final states effect. The conditions should be that the atom mean free path is long compared with the thickness of the film and that the film is thick enough to represent bulk ^{4}He. Glyde did not think that this idea would work at all, as the atom mean free path is very short. Platzman retorted that the mean free path is of order of 100Å but Glyde thought that this was physically impossible. Griffin did not think that any new physics would come out of a better knowledge of n_o. He said that Sokol's value of $n_o \sim 10\%$ agrees with Monte-Carlo calculations of it. What was needed was new experiments that show that $S(q,\omega)$ is related to the macroscopic wave function. (See articles by Glyde and Griffin in these proceedings).

Returning to the initial question, Sokol commented that n_o. is the microscopic order parameter and is only indirectly related to superfluid effects. This is exemplified in ^3He-^4He mixtures. As ^3He is added to ^4He, both the superfluid fraction and the transition temperature go down but calculations show that n_o increases. Wyatt suggested that a microscopic calculation of the superfluid density as a function of temperature might be a good candidate for future work.

The discussion then turned to the nature of the roton. It was pointed out by the chairman that we had heard three descriptions of a roton, i, the correlated basis function calculation as described in Campbell's article in these proceedings, ii, the field theoretic description of Glyde and Griffin which is reviewed in their articles in these proceedings and iii, the small vortex ring picture favoured by Putterman and Williams.

Platzman pragmatically commented that if we know the wave function of a roton from a variational calculation then we could just plot it out and so have a picture of it. It was said that this had been done and the result is very close to Feynman's picture. But it was felt that the variational wave function may not tell one about the long range nature of the backflow,as that depended on how large the variational wave state is chosen to be.

Griffin said that a roton exists above and below T_λ and it was basically a ^4He atom with various complications. The weight of the roton in $S(q,\omega)$ depends on n_o. In fact n_o allows the roton to be studied by neutron scattering. Wyatt then asked about the conflict that had surfaced on the first day between the neutron and light scattering measurements of the roton at $T > T_\lambda$. Ohbayashi reported a clear but broad roton signal above T_λ whereas the neutron scattering signal was zero above T_λ. Griffin and Glyde both said that there was not any conflict and that when the analysis of both groups was done consistently there would not be any difference. Griffin hoped that there would be a way of studying rotons above T_λ.

Donnelly said that no one had discussed path-integral methods for studying rotons but Campbell replied that although people have tried it is hard to do and it had not been successful yet.

The subject was then changed to a discussion of roton mode changing, e.g. the scattering of a R^+ roton to a R^- roton. Wyatt suggested that finding the probability for this process is a good problem for the future. Mezei commented that one does not only need to know about the excitations but also the interactions between them as one goes up in temperature. Only in this way will a full picture be obtained. This approach has not been represented here. He emphasised that we need to be able to make this description with only a few parameters compared the amount of experimental information. He thought we were not making much progress if theories had to have many adjustable parameters, even if they did give good agreement with experimental data. There was clear sympathy with this view.

Edwards explained that the reason that roton mode changing essentially did not occur when rotons scattered from kinks was because the kinks had a very small mass. (See the article by Edwards in this proceedings). Conservation of energy and momentum parallel to the kink prevented mode changing in all but ` 2% of the processes. Edwards considered that the question of mode changing when a roton hits a wall or the free surface was very interesting. Particularly the free surface as it is such a well defined system. He wondered if it could be demonstrated whether rotons contribute to the Kapitza conductance.

Donnelly asked Edwards if he knew any reason why mode changing could not occur in roton collisions. This had been used in the calculation of viscosity which agreed well with the measurements of Brewer and Edwards. Edwards replied that he did not think anyone knew the answer.

Someone then asked that if a roton was a small vortex ring, then could there be a whole spectrum of rotons of different size? Griffin thought that neutron scattering results ruled this out. Donnelly commented that if the Jones-Roberts vortex ring was shrunk so that it was essentially all core then it would lie near the Pitaevskii shoulder of the roton dispersion curve. It would decay into two rotons at 50° to each other. Platzman wondered if a charge could be put on a roton so that it could be made to grow. But it was pointed out that ions have dimensions ~ 20Å so the scale is wrong.

Glyde said that in the Landau-Khalatnikov theory of roton-roton scattering, the roton was treated as a point particle. This quasiparticle picture was the one that Griffin and he was emphasising. (See their articles in this proceedings).

Mezhov-Deglin thought that shock waves in liquid ^{4}He would be an interesting study, Donnelly said that experiments have been done on second sound shock waves which have been reported in the Annual Review of Fluid Mechanics by Liebmann, however this work does not reveal anything microscopic.

The chairman (HL) suggested we turn to ^{3}He and Scherm said that neutron scattering showed broad features, spin fluctuations at low energies and zero sound at higher energies. They are both folded into the 2 particle-hole excitations. To make progress we need to model the spin fluctuation paramagnons and the two particle-hole excitations. The various contributions could be separated if we know the line shapes for these and for zero sound at higher energies. He said that he plans to measure the line shape of the spin fluctuations at low energies and wave numbers.

It was agreed that present theories of ^{3}He are not good enough, but it is a very difficult problem. Krotscheck explained that methods that normally work at low energies breakdown completely when one looks at fine effects at the fermi surface. He said that he got somewhere in calculating the magnetic susceptibility from a microscopic approach which is published, but a totally consistent picture of the Fermi - liquid theory for ^{3}He is not available. He commented that the phenomenological calculations were just an exercise in fitting one set of numbers with another set.

Asked to comment on mixtures, Bowley posed the question why it is so difficult to cool ^{3}He-^{4}He mixtures below 120μk. Harrison said this was an incredible technical challenge due to the large thermal boundary resistance between the mixture and the container. Bowley agreed but wondered if there was also some heat leak that could not be got rid of. Harrison replied that the intrinsic heat leak is of the order of 10^{-14} watts.

On the problem of the interaction potential between ^3He atoms in mixtures, Bowley considered that this would be solved fairly soon.

Griffin returned to an earlier topic. He pointed out that measurements of $S(q,\omega)$ are probing density fluctuations. He considered this a great deficiency in the subject as we are not probing the elementary excitations and suggested we need to measure the single particle Green function or any other correlation function other than the density-density one. Putterman pointed out that if one makes a change from Eulerian to Lagrangian coordinates, the density fluctuations are the atom displacements and those are very fundamental quantities. He went on to say that the conference had heard that if one starts with the 2 particle interaction, the hard core and the Van der Waal's tail, n_o and the phonon-roton spectrum can be calculated. On the other hand starting from classical field theory and Landau quantised hydrodynamics, out comes the two fluid model, including the singled valueness gives the 2D KT transition. Also in Williams' talk we saw how even the λ transition might be interpreted as fluctuations of the excitations of this classical fluid. He wondered what else there was to do if we could start with two particle interactions and end up at the λ point with a critical exponent of $^2/3$. Glyde replied that the important test is the temperature dependence. Putterman suggested that explicit links be developed between classical hydrodynamics, phenomenological models and microscopic theories.

Bowley thought that spin polarised ^3He will be a rewarding field to study; will it be a superfluid and what will the phase diagram look like? Walraven thought the same was true for spin polarised H which had not been discussed at the workshop. He wondered how the different dynamics of the particles in this system will enter into the theory of the condensate. Edwards pointed out that we do not even know the energy of the ground state of spin polarised ^3He. Campbell commented that it would be nice to have a quantum monte-carlo calculation of this but pure unpolarised ^3He was giving enough trouble. Platzman thought that the issue was interesting and worthy of future study.

Griffin remembered that some five years ago it had been argued by Reppy and co-workers that ^4He in vycor glass behaved as 3D Bose gas and wondered what the position was now. Reppy replied that the data showed that the critical region contracted as $\sqrt{T_c}$ in this system and so it was difficult to get into the weakly interacting gas regime. Below 100mK there was a crossover to a different behaviour. There was no conflict with the vortex picture in the gas, it would still show the same critical behaviour. Griffin asked if there were any other measurements besides ρ_s and Reppy replied that they were working on the specific heat. However, there is a problem that the substrate is not neutral i.e. not mirror-like. It is clear that there are localisation effects going on as when you add He to a substrate the superfluid mass does not grow linearly, with added He, from the initial onset of superfluidity, but here is a power law development at very low temperatures. Griffin thought work on Bose condensed gases should be pushed. Maris mentioned that there were other proposed systems, positronium, moleculer H_z and the ongoing controversy about excitations.

Williams commenting on Bose gas system, said that with proper calibration systems follow the KT critical line but with finite size broadening which is based on the size of the vortex core. In thin films it is the increase in core size which broadens the transition so in that sense the system is becoming more dilute; but it is a dilution of an interacting gas not an ideal gas.

In a change of subject Bozler pointed out that the mechanism for the Kapitza conductance in the millikelvin range is not understood. Obsolete structures of the interface were being used. As so much more is known now we should go back and incorporate them into the models. At present we have the extremes of liquid in pores as studied by Pickett and semi infinite planes. He also thought that 2D magnetism of ^{3}He is in its infancy and that better models of indirect exchange are needed.

Golub commented that not much has been said about excitation lifetimes in ^{4}HE except for the paper by Mezei and some extent his work. If Glyde and Griffin are correct that the nature of the excitations is fundamentally different in the roton-maxon and the phonon regions we might expect some indication in the behaviour of the excitation widths in neutron scattering. The results of Halley concerning the unexpected complex behaviour of g_3 are also relevant.

Godfrin commented on the outstanding problems of nuclear magnetism of ^{3}He films. A much larger range of densities can be explored with ≤ monolayer films than in 3D. The theories used in 2D, Landau theory and effective mass theory, are still very crude and show logarithmic divergences. For multilayer films there is the question of how does one incorporate in Guyer's model, i) the layering, which arises from strong interactions, as is shown in Krotscheck's calculation and ii) the external potential from the substrate. In solid films there is the fundamental question of identifying the exchange processes, e.g. multiple spin exchange, indirect exchange and other mechanism induced by commensurability. He wondered whether there can be a 2D antiferromagnetic phase?

Ferrell kindly proposed a vote to thanks to the organisers and the meeting closed.

INDEX

GPSR Compliance

The European Union's (EU) General Product Safety Regulation (GPSR) is a set of rules that requires consumer products to be safe and our obligations to ensure this.

If you have any concerns about our products, you can contact us on ProductSafety@springernature.com

In case Publisher is established outside the EU, the EU authorized representative is:

Springer Nature Customer Service Center GmbH
Europaplatz 3
69115 Heidelberg, Germany

Batch number: 10838292

Printed by Printforce, the Netherlands